ANNUAL REVIEW OF MICROBIOLOGY

ANNUAL REVIEW
OF MICROBIOLOGY

VOLUME 53, 1999

L. NICHOLAS ORNSTON, *Editor*
Yale University

ALBERT BALOWS, *Associate Editor*
Centers for Disease Control, Atlanta

E. PETER GREENBERG, *Associate Editor*
University of Iowa, Iowa City

www.AnnualReviews.org   science@annurev.org   650-493-4400

ANNUAL REVIEWS
4139 El Camino Way • P.O. BOX 10139 • Palo Alto, California 94303-0139

ANNUAL REVIEWS
Palo Alto, California, USA

International Standard Serial Number: 0066-4227
International Standard Book Number: 0-8243-1153-1
Library of Congress Catalog Card Number: 49-432

Annual Review and publication titles are registered trademarks of Annual Reviews.
⊗ The paper used in this publication meets the minimum requirements of American National
Standards for Information Sciences—Permanence of Paper for Printed Library Materials,
ANSI Z39.48-1992.

TYPESET BY TECHBOOKS, FAIRFAX, VA
PRINTED AND BOUND IN THE UNITED STATES OF AMERICA

PREFACE

Annual Reviews may not be the first place to look for breaking news, but it is a valuable resource for news that has stood the test of time. Immediacy and validity achieve an optimal blend in the Annual Reviews homepage www:AnnualReviews. org where comprehensive reviews are keyed to the present literature. Try it once, and you will be hooked forever. Our authors are challenged to present scientific novelty that will endure. Their success is uncanny in microbiology where we constantly are reminded that the illusion of constancy is achieved at the cost of constant change. The constant theme in these diverse articles is how creativity emerges from the shifting interplay between resources and opportunities. This theme is played in the selection of authors by the editorial committee and, here too, change is our constant companion. We shall miss Barry Marrs and Janet Butel who completed their terms on the editorial committee with the design of this volume, and we welcome Judy Wall and Lynn Enquist to our deliberations. The Annual Reviews staff must be given collective credit for maintaining quality during shifts in personnel, but particular appreciation is due to NancyLee Donham for her grace and style while bringing this volume to completion.

Nick Ornston
Editor

CONTENTS

Annu. Rev. Microbiol. 1999. 53:1–42

TRANSFORMATION OF LEUKOCYTES BY *THEILERIA PARVA* AND *T. ANNULATA*

Dirk Dobbelaere and Volker Heussler

*Department of Molecular Pathology, Institute of Animal Pathology, University of Berne,
CH-3012 Berne, Switzerland; e-mail: dirk.dobbelaere@itpa.unibe.ch*

Key Words parasite, signal transduction, apoptosis, nuclear factor-κB

■ **Abstract** *Theileria parva* and *T. annulata* provide intriguing models for the study of parasite-host interactions. Both parasites possess the unique property of being able to transform the cells they infect; *T. parva* transforms T and B cells, whereas *T. annulata* affects B cells and monocytes/macrophages. Parasitized cells do not require antigenic stimulation or exogenous growth factors and acquire the ability to proliferate continuously. In vivo, parasitized cells undergo clonal expansion and infiltrate both lymphoid and non-lymphoid tissues of the infected host. *Theileria*-induced transformation is entirely reversible and is accompanied by the expression of a wide range of different lymphokines and cytokines, some of which may contribute to proliferation or may enhance spread and survival of the parasitized cell in the host. The presence of the parasite in the host-cell cytoplasm modulates the state of activation of a number of signal transduction pathways. This, in turn, leads to the activation of transcription factors, including nuclear factor-κB, which appear to be essential for the survival of *Theileria*-transformed T cells.

CONTENTS

0066-4227/99/1001-0001$08.00

1

INTRODUCTION

Theileria parasites affect a wide range of mammals, mainly ruminants, and are particularly important as pathogens of domestic cattle, sheep, and goats in the tropical and subtropical regions of the Old World. They are transmitted by ixodid ticks, and the diseases they cause are among the most serious constraints to livestock production.

In East and Southern Africa, *Theileria parva* causes East Coast fever that results in the death of up to 1 million cattle per year, with economic losses amounting to hundreds of millions of dollars. In regions where the disease is endemic, losses, predominantly among calves, can reach 50%. In such areas, growth and milk yield can be severely impaired and fertility reduced (125). In regions where the disease is newly introduced, mortality can be as high as 80–100%, with exotic cattle breeds, imported to upgrade cattle production, showing higher susceptibility than indigenous cattle.

In Northern Africa, Southern Europe, the Middle East, and large parts of Asia, *T. annulata* causes tropical theileriosis (also called Mediterranean theileriosis), which also can cause severe losses to the cattle industry. In addition to the direct and indirect losses attributed to theileriosis, considerable costs are incurred through control measures that rely predominantly on combating ticks through the use of acaricides.

BIOLOGY OF THE PARASITE

Theileria parasites are obligate intracellular protozoa belonging to the class of Sporozoea, which is part of the phyllum Apicomplexa (94, 181). A detailed account of several aspects of the parasite life cycle and distribution has been given elsewhere (115). *Theileria* parasites are transmitted trans-stadially by two- or three-host ticks. *Rhipicephallus appendiculatus* (the brown ear tick) is the main tick vector of *T. parva*. The natural vectors for *T. annulata* are ticks of the genus *Hyalomma*. Larval or nymphal ticks feeding on an animal carrying piroplasms in its blood acquire the infection. Continuation of the life cycle in the tick involves the production of gametes, which fuse to generate a zygote. Zygotes enter the intestinal epithelium, where they undergo reduction divisions to form mobile kinetes that can be found in the hemolymph approximately at the time ticks molt to the next instar. Kinetes migrate to the salivary glands, where they undergo nuclear division without accompanying cytokinesis, resulting in a complex syncytium called the sporoblast. When the tick takes the next blood meal on a new host, sporogony is completed with the production of large numbers of infective sporozoites that completely fill the infected salivary glands cells and may number up to 50,000 per infected acinus cell. Sporozoites are injected into the next host when the tick takes a blood meal in its subsequent instar (nymph or adult, respectively).

Entry of *Theileria* Sporozoites into the Target Cell

The attachment and entry of *Theileria* sporozoites have been topics of a number of elegant electron microscopic studies (66, 68, 153, 174). Entry of *T. parva* sporozoites involves a defined series of sequential, but separable steps that differ in important details from invasion by other apicomplexans such as *Plasmodium* and *Toxoplasma* species (153, 174). Sporozoites attach to the target cell by binding to one or more sites on the lymphocyte plasma membrane. This initial sporozoite-lymphocyte interaction is a chance event that can occur at 0–20°C. The exact nature of the lymphocyte molecules that are recognized by the sporozoite are not yet known, but it is likely that putative receptors consist of a complex of different surface proteins, possibly involving major histocompatibility complex (MHC) class I molecules. Monoclonal antibodies that react with common determinants on MHC class I molecules and with β-2 microglobulin have been shown to inhibit sporozoite entry by specifically preventing this initial binding event. Furthermore, in mutant bovine cell lines from which one or both MHC class I haplotypes had been lost, sporozoite binding and entry clearly correlated with the level of class I surface expression (156). Enzymatic modifications of the lymphocyte surface demonstrated that trypsin-insensitive glycoproteins containing O- and N-linked carbohydrates as well as phospholipase-sensitive molecules on the host cell surface were critical to sporozoite entry (151). Once the sporozoite is attached, invasion takes place within minutes. Shaw et al have proposed that the initial sporozoite binding event triggers the mobilization of intrasporozoite Ca^{2+}, and they also established that no de novo protein synthesis appears to be required (150, 151). Inhibitor studies have demonstrated that entry was dependent on G-protein signaling and that protein kinase activities in both the sporozoite and the host cell were essential to sporozoite invasion (150, 152, 153). Fawcett and associates (67, 68) concluded that the sporozoite invasion process occurs by a process of "passive endocytosis" that requires little energy, can occur at low temperatures, and does not require viable sporozoites. These findings contradict later observations by Shaw et al (155), who showed that all stages of the invasion process subsequent to attachment are temperature dependent and require the participation of live intact sporozoites and host cells. During entry, parasite and target-cell membranes come into apposition, and entry occurs by the progressive circumferential "zippering" of sporozoite ligands with receptors on the lymphocyte surface (67, 151). This is accompanied by the loss of elements of the sporozoite surface coat (185), resulting in the sporozoite becoming fully internalized.

Sporozoite surface proteins that appear to play an essential role in the invasion process have been identified. The sporozoite surface protein p67 (48) is synthesized during sporogony (49, 51); monoclonal antibodies directed against this molecule, although allowing sporozoites to bind to the target lymphocytes, potently block entry (50). Shaw et al (156) have shown that parasite entry can be competitively inhibited by soluble p67. Studies with recombinant p67 to isolate the relevant host cell molecule(s) with which the *T. parva* sporozoite interacts have not yet been successful. Clearly a crucial component in the sporozoite entry process, p67 is presently also being tested as a vaccine candidate against East Coast fever (123).

A sporozoite surface antigen of *T. annulata*, SPAG-1, shows a remarkable degree of molecular mimicry of the extracellular matrix protein elastin; it contains both repetitive motifs PGVGV and VGVAPG. The presence of VGVAPG repeats is interesting because this is the ligand for elastin receptors that are expressed on a range of cell types, including macrophages and monocytes, the major class of host target cells (85). Although the elastin receptor is a potential ligand for sporozoite entry, it was found that *T. annulata* sporozoites infected cells independently of whether or not they expressed the receptor (27).

The state of activation of the target cell also appears to affect susceptibility to sporozoite invasion. Lalor demonstrated that mitogenic stimulation of the target-cell populations increased the frequency with which transformed cell lines became established on infection with *T. parva* (P Lalor, unpublished observation). For *Leishmania* infection, it has been shown that sandfly-derived components dramatically enhance parasite infectivity (175). Tick-derived factors also appear to increase the establishment of *Theileria* infection. Shaw and colleagues (154) showed that short-term preincubation of lymphocytes with tick salivary gland extract enhanced host-cell susceptibility. Likewise, compounds that induce lymphocyte proliferation and the growth factor interleukin-2 (IL-2) resulted in increased susceptibility.

All intracellular parasites face the problem of survival in the often hostile intracellular environment. A number of parasites—such as *Toxoplasma*, *Leishmania*, and *Plasmodium* species—remain in a parasitophorous vacuole, which is modified in different ways to prevent phagolysosomal destruction. *Theileria* species join the ranks of *Trypanosoma cruzi* and *Listeria* species by "escaping" into the cytoplasm (8). Immediately after entry, the sporozoite is tightly surrounded by the host cell membrane (66, 67). The parasite discharges the contents of two secretory granules, the rhoptries and microspheres (also called micronemes), resulting in the separation of the enclosing lymphocyte membrane from the parasite plasma membrane. The host cell membrane is dissolved, and, on release of the parasite into the cytoplasm, an orderly array of microtubules is formed surrounding the organism. The entire process of attachment, entry, and release into the cytoplasm is completed within 30 min. During the next few days, the parasite develops into a multinucleated schizont that causes the host cell to undergo blastogenesis.

In vivo, macroschizont-infected cells undergo rapid proliferation, resulting in their clonal expansion. The phase of rapid proliferation is followed by extensive lysis of parasite-infected and uninfected cells; it is the latter process that constitutes the main pathological feature of East Coast fever and related theilerioses. Some macroschizonts differentiate further into microschizonts, and merozoites are formed; these are released when the infected cell ruptures. Merozoites, in turn, invade red blood cells where they develop into piroplasms. Piroplasms of *T. parva* undergo little if any division; therefore, anemia is not a feature of East Coast fever. Piroplasm division can be more pronounced in other *Theileria* species, such as *T. annulata* and the closely related *T. mutans*; in these cases, anemia usually accompanies the disease. Infected erythrocytes are taken up by feeding ticks, allowing completion of the life cycle.

Target Cells for *Theileria parva* and *Theileria annulata* Infection

Although *T. parva* and *T. annulata* both transform mammalian cells, they differ significantly in the range of cells they infect. A clear picture of the cells that become transformed by both parasites was not possible before the development of specific reagents that allowed bovine cells of lymphoid and myeloid origin to be analyzed. In addition, phenotypic alterations, which are most pronounced in *T. annulata*-infected cells, further complicated the issue.

Early studies of the target-cell types for infection were restricted to infected cell lines and parasitized cells isolated from infected cattle. The absence of reactivity with reagents specific for bovine immunoglobulins (52, 62) and the fact that a number of parasitized cell lines reacted with a cell surface determinant that was restricted to a subpopulation of T lymphocytes (136) led to the initial assumption that *T. parva* only infects and transforms a restricted T-cell subpopulation. The development of lymphocyte-sorting techniques and in vitro infection with sporozoites allowed a more thorough analysis of the range of target cells. Thus, it was shown that *T. parva* is capable of infecting and transforming purified B and T cells with similar frequencies in vitro and that neither monocytes nor granulocytes give rise to transformed cell lines (12, 46). However, infection of mixed-cell populations consistently resulted in the establishment of transformed T-cell lines, suggesting a selective bias. A phenotypic analysis of cultures of peripheral blood mononuclear cells (PBMC) infected in vitro with saturating concentrations of sporozoites revealed that parasitized B cells were abundant in cultures after 1 week, but they were subsequently overgrown by T cells (121).

Shaw and coworkers (154) also quantitatively examined the susceptibility of cell types known to be present at the tick attachment site. Apart from lymphocytes, sporozoites were also found to bind and enter macrophages and afferent lymph-veiled cells (dendritic cells) that are present in significant numbers at the tick attachment site in cattle. These cells are believed to migrate via the afferent lymphatics to the draining lymph node, where they differentiate into interdigitating dendritic cells. Entry of macrophages and dendritic cells was not by phagocytosis; instead, sporozoites entered these cells as they do lymphocytes. Interestingly, schizont development and accumulation occurred in dendritic cells, but this was not accompanied by concomitant transformation (186). Binding to fibroblasts, granulocytes, or erythrocytes could not be observed. Based on the above observations, it can be concluded that *T. parva*-induced transformation is restricted to cells of lymphocyte origin.

Syfrig and colleagues (168) investigated *T. parva* sporozoite binding to goat lymphocytes. Under normal conditions, sporozoites of *T. parva* did not attach to caprine cells. Interestingly, however, they found that proteolytic cleavage of surface proteins enhanced the susceptibility of lymphocytes to invasion by sporozoites, suggesting that receptor(s) for *T. parva* sporozoites might be present on these cells but are not easily accessible. Electron microscopy revealed that the process of sporozoite binding and entry into protease-treated goat lymphocytes was very similar to that observed in bovine cells. Schizonts did not develop, however, and lymphocyte proliferation was not induced, confirming that cell entry by sporozoites and cellular transformation are separate processes. In the same work, it was found that parasite interactions with bovine host cells could be partially inhibited by one of two monoclonal antibodies to BoCD45R. Proteolysis of the lymphocyte surface removed CD45R (but not the MHC class I determinants) and enhanced sporozoite binding. Although further work is clearly required to elucidate the exact nature of sporozoite-host cell interactions, the authors ventured the hypothesis that CD45R and CD45R antibodies may nonspecifically prevent the apposition of sporozoite and lymphocyte membranes. Perhaps another explanation is that activation of CD45, which is a phosphatase, counters the kinase activity that appears to be required for sporozoite entry (152, 153).

Despite the many similarities between *T. parva* and *T. annulata*, they infect different cells of the immune system. Initial studies revealed that all *T. annulata*-

infected lines were positive for MHC class II, although the amount of class II antigen expressed varied between lines. In addition, all lines were negative for a macrophage/monocyte marker, for surface immunoglobulin M (IgM), and the bovine T-cell markers boCD4 or boCD8 (163). In comparative infection experiments with cell populations purified by fluorescent-activated cell sorting, it was found that monocytes were infected very efficiently by *T. annulata* but could not be infected by *T. parva*. B cells were infected much more efficiently by *T. annulata* than *T. parva*. Interestingly, once infected and transformed, virtually all reactivity was lost for anti-surface IgM and the anti-monocyte monoclonal antibodies (27, 80, 162). An analysis of cells isolated from cattle with tropical theileriosis, by using anti-CD11b monoclonal antibodies, confirmed that cells of myeloid origin are targets for infection and transformation by *T. annulata* (72). In contrast to *T. parva*, *T. annulata* also infects and transforms peripheral blood cells from sheep and goats (causing mild infections in these species).

CHARACTERISTICS OF *THEILERIA*-TRANSFORMED CELLS

Phenotype of *Theileria*-Transformed Cells

Cell lines generated by infection of bone marrow cells, bone marrow-derived macrophages, and monocyte-derived macrophages with *T. annulata* were assessed for surface marker expression and function by Sager and coworkers (146). Transformed lines expressed MHC class I and II, CD44, CD45, and the myeloid marker DH598. However, the surface markers CD14, CD11b, M-M7, TH57A, and, to a lesser extent, CD11a/CD18, CD11c, and ACT(B) were down-regulated. In the same work, Sager and coworkers provided the first strong evidence that macrophages transformed by *T. annulata* undergo a process of parasite-induced dedifferentiation. Transformed cells failed to express macrophage functions such as Fc-receptor–mediated phagocytosis, inducible oxidative burst, lipopolysaccharide-induced TNF-α and nitric oxide generation, and up-regulation of procoagulant activity. Upon elimination of the parasite by BW720c treatment (see below), however, cells reacquired monocyte lineage properties, such as the up-regulation of CD14 and the capacity to ingest opsonized sheep red blood cells and bacteria. In a detailed analysis of a set of clones, the existence of three phenotypes of *T. annulata*-transformed cell lines could be discerned (144). The first was characterized by surface expression of Ig-M, CD21, and the B-cell epitopes B-B2 and B-B8; the Ig heavy-chain gene was rearranged and Ig mRNA expressed. The second phenotype could be distinguished from the first by the absence of Ig heavy-chain expression and reduced surface expression of B-cell markers (CD21, B-B2, B-B8). The third phenotype showed an absence of all of the above B-cell markers, including surface IgM, and a lack of Ig heavy-chain gene rearrangement. The latter clones could be maintained for several weeks after elimination of *T. annulata* by BW720c treatment, and they reacquired a macrophage-like phenotype. These findings imply that parasite-induced dedifferentiation is restricted to monocyte/ macrophages.

Dedifferentiation in *T. parva*-infected cells appears to be less pronounced, but there are indications that it also occurs, albeit at a slower pace. Cell lines obtained from B cells were found to express Ig, but only a variable proportion of cells within each cloned cell line expressed surface Ig (46, 119). After 6 months in culture, however, only a small percentage of the cell lines were still positive (119). Furthermore, many CD4$^+$CD8$^-$-derived cell lines showed CD8 expression with prolonged time in culture. *T. parva* infection and transformation need not result in a direct loss of effector function, however. When alloreactive cytotoxic T cell clones were transformed by *T. parva*, they retained their antigen-specific cytotoxicity for up to 4 months (14).

By using specific antibodies against lymphocytes and parasite epitopes, it became possible to type *T. parva*-infected cells during the course of an infection. In cells collected from peripheral blood and lymph nodes, >99% of the infected cells in cattle with patent parasitosis expressed T-cell markers. Small numbers of infected cells positive for sIg could be detected only in the advanced stages of East Coast fever. Interestingly, during the second week of infection, a population of CD4$^+$CD8$^+$ cells appeared that could represent at least 15% of the efferent lymph lymphocytes. Although T cells expressing α/β or γ/δ T-cell receptors (TCR) can be infected and transformed in vitro, the majority of parasitized cells in the tissues of infected cattle are α/β T cells. Morrison and coworkers (121) attempted to determine in more detail whether the cell type infected with *T. parva* influenced the pathogenicity of the parasite. In an initial approach, autologous cloned cell lines of different phenotypes were first generated by infection in vitro and then inoculated into cattle. This failed to resolve the issue because the prolonged period of culture required to clone and characterize the cell lines resulted in attenuation of the parasitized cells. On the other hand, when cattle were inoculated with purified populations of CD2$^+$, CD4$^+$, or CD8$^+$ autologous T cells that had been pre-incubated in vitro with *T. parva* sporozoites, severe clinical reactions with high levels of parasitosis and similar severity were produced. Infected B cells, however, gave rise to mild self-limiting infections even when administered at a 10-fold–higher dose. Thus, although parasite-dependent induction of continuous proliferation of the lymphocytes is an essential step in the pathogenesis of Theileriosis, the cell type infected by *T. parva* also influences the pathogenicity.

Continuous In Vitro Proliferation and Tumorigenicity

Theileria-infected cells show many characteristics of tumor cells. In vitro, they proliferate in an uncontrolled manner, can be cloned by limiting dilution, and cease to require the addition of exogenous growth factors. In vivo, they show an increased ability to migrate, infiltrate, and proliferate in non-lymphoid as well as lymphoid tissues, generating lesions that resemble multicentric lymphosarcomas.

Infiltration is most apparent in the lungs and the gastrointestinal tract. Infiltration of the lungs is commonly associated with a severe pulmonary edema (97). As the disease progresses, extensive lymphocytolysis occurs, affecting uninfected as well as infected cells, resulting in a marked depletion of lymphoid tissues.

Theileria-infected cells form tumors when injected into immunocompromised SCID (70), irradiated (95), or nude mice (96). In athymic nude mice, parasitized bovine cells became widely disseminated in the host's tissues and organs, in some cases causing death. When compared with *T. parva*-transformed T cells, tumors of *T. annulata*-infected cells showed more vigorous growth in SCID mice; and when injected intraperitoneally, infected cells colonized abdominal organs, in particular mesenteric tissue (69).

Induction by *Theileria*-Infected Cells of Proliferative Responses in Uninfected Cells

One of the striking features of *Theileria* infection in cattle is the marked proliferative response of uninfected lymphocytes that can be observed around the time the first parasitized cells are detected in the lymph nodes (118). In the case of East Coast fever, the proliferating cells are predominantly T cells (63). In vivo studies have shown that parasite-transformed cells induce potent proliferative responses in autologous PBMC (82, 132); this phenomenon is called the autologous *Theileria* mixed-lymphocyte reaction. PBMC from naive and immune animals respond with equal magnitude, suggesting the involvement of a nonspecific, mitogenic component (82).

Research on *T. annulata* revealed a similar picture. Infection of macrophages with *T. annulata* leads to an augmentation of their antigen-presenting capability in vitro, but infected cells can induce proliferation of autologous resting T cells from naive animals, irrespective of their antigen specificity or memory status (23, 81, 140). Interestingly, "nonspecific" proliferation was not seen with ovalbumin-specific T-cell lines (81), suggesting that a certain TCR specificity may exist. Although it was initially proposed that this may be caused by increased MHC class II expression (81), in later work no correlation was found between the level of MHC class II expression and levels of induced T-cell proliferation (23). Instead, it could be shown by semiquantitative polymerase chain reaction (PCR) technology that the degree of T-cell proliferation induced by infected cells directly correlated with the amount of mRNA for the T-cell stimulatory cytokines IL-1 α and IL-6. Campbell and coworkers (29) demonstrated that both CD4$^+$ and CD8$^+$ T cells from naive PBMC were induced to proliferate and express the activation markers IL-2R and MHC class II when cultured with *T. annulata*-infected cells. CD4$^-$ T cells also contributed to the response, but they were incapable of being directly activated by *T. annulata*-infected cells. In addition, antibodies directed against MHC class II and CD4 inhibited the induction of T-cell proliferation by schizont-infected cells. Antibodies against MHC class I and CD8, on the other hand, did not block activation (33). Preliminary observations by Campbell et al (33) suggest that lymphocytes expressing two or three TCR Vβ families (found on ~10% of peripheral T cells) appear to be the targets for *T. annulata*-induced activation. Other work (40) indicates that the proliferating cells are mainly WC1$^+$ γ/δ T cells. Collins and colleagues (40) found that the majority of WC1$^+$ γ/δ T cells in freshly isolated PBMC expressed low levels of CD25 that increased when stimulated with *T. annulata*-infected cells. Purified WC1$^+$ γ/δ T cells failed to proliferate when cultured with irradiated *T. annulata*-infected cells, and they produced only a small proliferative response to IL-2. They proliferated strongly, however, in response to irradiated or lightly fixed *Theileria*-infected cells in combination with IL-2. These findings show that WC1$^+$ γ/δ T cells recognize a surface determinant on *T. annulata*-infected cells that, together with a second signal (which can be provided by exogenous IL-2), stimulates their proliferation.

From in vitro studies, it is clear that T cells are not stimulated in a conventional manner. It is likely that this intrinsic ability of *Theileria*-infected cells to induce naive autologous T cells to proliferate is also responsible for the proliferative responses observed in vivo. It has furthermore been proposed that both the "inappropriate" activation of T cells (29) and the augmented antigen-presenting cell activity (81) may play an important role in the pathogenesis of theileriosis and contribute to the failure of the host to mount an effective immune response in vivo (33).

Theileria thus shares the capacity to induce unspecific T-cell proliferation with a number of viruses, among them Epstein-Barr virus (166), that are capable of inducing cellular transformation (92), and also to cause pronounced autologous mixed-lymphocyte reaction. In viral models, the involvement of superantigens has been demonstrated. Although candidate superantigens have yet to be identified in *Theileria*, superantigen-like activity in *T. annulata* has been proposed by Campbell and coworkers (32).

Establishment of *Theileria*-Infected Cell Lines

The most outstanding feature of *Theileria*-infected cells is that they acquire the ability to proliferate in an uncontrolled manner. In fact, *Theileria* is the only eukaryote known to transform another eukaryote. Considering the lack of other similar examples among parasites, it is perhaps not surprising that several decades elapsed between the early work on East Coast fever (172) and the elucidation of the mode of replication of the parasite in bovine lymphocytes. Early workers believed that the parasite, after infecting lymphocytes, multiplied by schizogony, with the particles of the schizonts breaking up into merozoites, which then infected other lymphocytes. It was not until prolonged cultivation of *T. parva* and *T. annulata* (178) could be achieved that Hulliger et al (93) demonstrated that the distribution of *Theileria* into both daughter cells occurred during mitotic division; during this process, the parasite became associated with fibers of the mitotic spindle. It was also demonstrated that the rate of replication of the parasites appeared to be in balance with that of their host cells.

After a number of initial attempts at cultivating *T. parva* in spleen explants in plasma clots (179), Malmquist et al (110) succeeded in establishing permanent cell lines from cells isolated by biopsy from lymphoid organs of infected animals (109). This enabled a search to be launched for theilericidal drugs, using large-scale in vitro screening. Another important milestone in *Theileria* research, with important consequences for our understanding of immune responses and cell biology, was the discovery that lymphocytes could also be infected in vitro with sporozoites isolated from infected ticks (22). Combined with fluorescent-activated cell sorting, this technology allowed researchers to examine the effect of infection and transformation on well-defined purified cell populations (reviewed in 119).

How Does the Parasite Persist in Continuously Dividing Cells?

Studies on cultured *T. parva*-infected cells have shown that parasite and host cell DNA synthesis was asynchronous (98). Pulse-labeling with ^{3}H-thymidine revealed that the major component of parasite DNA synthesis occurred while the host cell was in early M-phase, with a minor component detectable during interphase. These findings were supported by electron microscopic observations indicating that important reorganizational changes occur in the parasite while the host cell goes through metaphase (183). A distinct G2-phase appears to be missing from the parasite cell cycle because division of the schizonts immediately follows DNA synthesis. The lack of synchrony of host and parasite S-phase suggests that the parasite regulates its DNA synthesis independently of the cell. This concept is supported by the fact that compounds that block host cell metabolism at different levels do not affect parasite multiplication. Treatment with puromycin, actinomycin D, or cycloheximide all result in "super-infection," with infected cells containing several-fold more schizont nuclei than their untreated counterparts. Clearly, parasites can continue to divide in the absence of host cell DNA synthesis and division. The opposite does not apply, however, because host cells stop proliferating when parasite replication is blocked.

Although the parasite and host cell S-phases are clearly not synchronized, parasite nuclear division appears to be synchronized to host cell metaphase (98). This finding is in contradiction with earlier work by Hulliger et al (93), who found no proof for synchronization of parasite division with any stage of the host cell cycle. The actual segregation of parasite nuclei over the two daughter cells was described as a random event (93, 183), and according to this model, an equal distribution of parasites over the separating daughter cells was not necessarily guaranteed. Daughter cells endowed with the parasite continue to proliferate, whereas cells that fail to acquire parasites rapidly lose their lymphoblastoid appearance and finally apoptose. Also, when parasite distribution over the two daughter cells is blocked by treatment with cytochalasin, cells that fail to receive a parasite stop proliferating. Division over the daughter cells seems to be an efficient process, however, because in normal cultures up to 95% of the cells usually harbor parasites.

Work by Hulliger et al (93), Stagg et al (164), and Vickerman & Irvin (183) has shown that the schizont becomes closely associated with the spindle microtubules and that separate components are drawn passively into each daughter cell as the spindle contracts (Figure 1. See color insert). The nature of the interaction between the parasite surface and the host cell cytoskeleton is still unknown. Parasite surface proteins, such as PIM (polymorphic immunodominant molecule) or QP protein (15, 177), which are expressed on the surface of schizonts, often contain repetitive elements; such proteins may constitute likely candidates for interactions with host cell cytoskeleton components.

Elimination of the Parasite Blocks Transformation

Theileria-transformed cells differ in two important aspects from other transformed cells. First, *Theileria*-induced transformation does not depend on permanent alterations of the host cell genome such as mutations, translocations, or integration of foreign sequences as observed with certain viruses. This in all probability also explains the second important difference: transformation induced by *Theileria* is entirely reversible. When theilericidal drugs are added to *Theileria*-transformed cultures, the cured cells stop proliferating and will either revert to the resting state or undergo apoptosis, even after several decades in culture (45, 71). The development of a drug for the treatment of East Coast fever spanned more than a decade. By in vitro screening, a range of naphthoquinone-derived theilericidal compounds were developed that specifically inhibit the parasite respiratory system. The most efficacious drug is BW720C (also called buparvaquone), which cures cattle infected with *T. parva* and *T. annulata* (113). With the development of theilericidal drugs, another important tool became available that now also allowed an "on/off" analysis of the molecular changes that accompany the transformation of the infected host cell.

Life After (Parasite) Death?

The fate of cells after elimination of the parasite can differ significantly and appears to depend on the cell type and parasite with which they were infected. An additional important factor is the culture conditions under which the cured cells are maintained. For instance, when BW720c was added to a culture of *T. parva*-infected T cells seeded at optimal density, proliferation continued with the same kinetics as that measured for infected cells for up to 4–5 days. Upon passage, however, growth rapidly ceased, and cured cells apoptosed (45, 71). Apoptosis was accelerated if contact between the cells was disrupted by centrifugation and medium changes (45). Enhanced apoptosis was also observed when cells were resuspended in their original medium or in conditioned medium; this suggests that cellular membranes provide an important stimulus for proliferation and survival of the cured cells. This was confirmed in experiments showing that TpM(803) T cells require cell-cell contact for optimal growth. Contact stimulation could be provided by either infected cells or cured cells, but the ability to respond to surface stimulation was critically dependent on the presence of the parasite (47). Time-lapse video microscopy revealed that TpM(803) cells lost their ability to re-establish cell-cell interactions within 48 h of parasite elimination (D Dobbelaere, unpublished observation). The addition of recombinant IL-2 to the culture medium of cured cells, however, allowed them to proliferate for several additional days, and it also enhanced cellular survival, consistent with the anti-apoptotic function of this lymphokine (6). Such IL-2-treated cells would finally return to a resting phenotype and then apoptose. Proliferation beyond 7–10 days after elimination of the parasite required regular phorbol ester stimulation, the continuous presence of IL-2, and cell densities exceeding 10^6 cells/ml. Under these conditions, cured TpM(803) T cells have been maintained in culture for up to 3 months.

Cells cured of *T. annulata* infection by treatment with BW720c appear to undergo apoptosis more readily. Sager et al (146) demonstrated that macrophages cease to divide upon elimination of the parasite and demonstrated the reappearance of a differentiated phenotype in the viable cell population remaining in culture. By using conditioned medium from bovine lung fibroblasts—probably containing the mononuclear phagocyte growth and survival factor M-CSF—the survival of cured cells could be enhanced considerably (144). It was also found that the capacity to redifferentiate was restricted to cells that had been passaged for relatively few times (<10). With increasing numbers of passages, the efficiency of redifferentiation and survival of BW720c-treated cells decreased significantly. Interestingly, clones of transformed cells that initially expressed B-cell markers invariably failed to redifferentiate, and the cells died within 10–15 days. Perhaps also in this case it will be possible for this hurdle to be overcome once the appropriate culture medium conditions have been defined.

Theileria-Infected Cells Become Independent of Exogenous Growth Factors: Interleukin-2 or No Interleukin-2?

In a crucial early experiment, Brown & Logan (26) demonstrated that T cells that became transformed by infection with *T. parva* ceased to require antigenic stimulation and become independent of exogenous growth factors. Instead, the *T. parva*-transformed cells themselves were shown to produce a factor with IL-2-like activity. Dilution experiments indicated that the growth of *T. parva*-infected TpM(803) T cells was dependent on the secretion of a growth factor with IL-2-like activity, which was secreted into the culture medium as long as the cells harbored the parasite in their cytoplasm (45, 47). It was also demonstrated that TpM(803) T cells expressed high-affinity IL-2 receptors (IL-2R) on their surface in a parasite-dependent manner (41, 46). Furthermore, it was shown that upon elimination of the parasite, T cells again required the addition of IL-2 to the culture medium, in order to survive. These data provided strong support for the hypothesis that TpM(803) T cells might generate their own growth factors and proliferate via an IL2/IL-2R-mediated autocrine mechanism. Confirming this hypothesis was not straightforward, however. Heussler et al (90) demonstrated that

IL-2 mRNA was present only in low levels in five cell lines, and no transcripts could be detected by Northern blot analysis in four other cell lines. RT-PCR (reverse transcription-polymerase chain reaction) performed on polyadenylated RNA, however, showed that IL-2 transcripts could be detected in all *T. parva*-infected cells lines, independently of whether they were of B- or T-cell origin, or expressing α/β or γ/δ TCRs (90). Both IL-2 mRNA expression (90) and release of IL-2 activity into the culture medium (26, 47) were shown to be parasite-dependent. Interestingly, no IL-2 transcripts could be detected in a B-cell line transformed by bovine leukosis virus, indicating that the IL-2-inducing effect is specific for *T. parva*. In contrast to the low levels of IL-2 mRNA, IL-2Rα-chain mRNA could readily be demonstrated by Northern blot analysis in *T. parva*-infected—but not in cured—cells (41, 46). Flow cytometric analysis with anti-IL-2R antibodies showed that all *T. parva*-infected cell lines expressed the IL-2Rα-chain. A constant feature of IL-2R expression is that, at any one time, only a subset of the cells are positive. The percentage of positive cells differs between cell lines, but often also fluctuates between experiments; this also applies to a number of other surface markers (46). Contrary to what had been proposed, combined flow cytometric analysis of surface expression and DNA content clearly showed that expression was not cell-cycle dependent: the distribution of IL-2R-positive and -negative cells containing 2n or 4n amounts of DNA was identical (D Dobbelaere, & T Prospero, unpublished observation). These data suggest that other mechanisms, for instance continuous stimulation followed by internalization and re-expression, regulate surface expression.

In contrast to IL-2Rα expression in *T. parva*-transformed cells (41, 46), expression in *T. annulata*-infected cells appears not to be dependent on the presence of the parasite because IL-2Rα mRNA expression persisted, and even increased, when cells were treated with BW720c (4). Based on a characterization of the IL-2R in *T. annulata*-infected cells (89), it was proposed that proliferation in response to IL-2 is mediated predominantly via an intermediate-affinity IL-2R (4).

The finding that IL-2 and IL-2R mRNA could be detected in a wide range of *T. parva*-transformed T- and B-cell lines strongly pointed towards the involvement of an IL-2/IL-2R–mediated autocrine loop. Experiments with antisense IL-2Rα-chain RNA expression provided evidence that this might indeed be the case for TpM(803) T cells (56). Inhibition experiments with anti-IL-2 antibodies (47, 90) confirmed that an autocrine loop participates, at least in part, in the continuous proliferation of TpM(803) T cells and a number of other cell lines. Importantly, inhibition could not be demonstrated in all cell lines, and complete growth arrest was also never obtained. At least one of the T-cell lines and a B-cell line were not affected at all by anti-IL-2 antibodies, even though they expressed IL-2 mRNA in a parasite-dependent manner. It is conceivable that in these cell lines, IL-2 associates with the IL-2R before reaching the surface, as has been shown in another system for the growth factor IL-3 (21, 53). Alternatively, the lack of inhibition may reflect the fact that these cell lines proliferate via another autocrine mechanism, involving other growth factor(s) and receptor(s). Finally, it is also possible that a number of cell lines have entirely lost their requirement for growth factors.

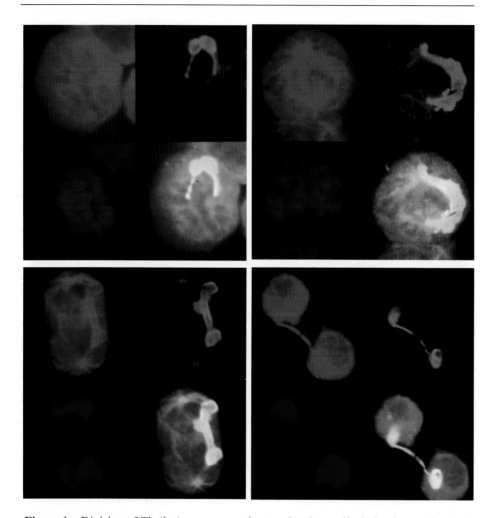

Figure 1 Division of *Theileria parva* over the two daughter cells during host cell mitosis. As mitosis starts, the *T. parva* schizont becomes closely associated with the spindle microtubules, and the parasite is drawn passively into each daughter cell as the spindle contracts. Host cell tubulin is detected by phycoerythrin-labeled antibodies (red); the parasite is stained by using FITC-labeled antibodies (green); and chromosomes are stained blue (Hoechst d). An overlay of the three pictures is also shown. Division is shown starting in the top left panel and continues through to cytokinesis in the bottom right panel. (Courtesy of Mark Carrington and Charles Nichols, Cambridge, UK.)

In earlier work on *T. annulata*, it was proposed that *T. annulata*-infected cells could produce a factor that possessed the biological activities of IL-2, because the supernatants could enhance the proliferation of concanavalin A-stimulated peripheral blood leukocytes (4). The factor was never identified, however, and it is possible that components other than IL-2 were responsible for this stimulation. In a more recent manuscript, Shayan et al (158) proposed that IL-2 might not be required for the growth of *T. annulata*- and *T. parva*-infected cells. This was based on the fact that antibodies directed against IL-2 did not inhibit proliferation and, contrary to the findings by others (23, 90, 114), the authors failed to demonstrate the presence of IL-2 mRNA by RT-PCR in all cell lines. The reason for this discrepancy is not yet clear. Taking into account the very low level of IL-2 mRNA expression, the answer for these differences might be technical.

Other Lymphokines and Cytokines Secreted by *Theileria*-Infected Cells

IL-2 is certainly not the only lymphokine expressed by *T. parva*-infected cells. McKeever and colleagues (114), using a multiplex PCR system, investigated the patterns of cytokine mRNA expressed by 19 uncloned and cloned parasite-infected lymphoblast cell lines. Considerable variation was observed in the cytokine profiles of these lines. Apart from IL-10, which was universally expressed, a lineage-specific pattern of cytokine mRNA expression that could be associated with infection could not be found. In a comparative analysis (23), *T. parva*- and *T. annulata*-infected cell lines and clones were assayed for cytokine mRNA expression by RT–PCR. Cells infected with *T. parva* were shown to produce mRNA for IL-1α, IL-2, IL-4, IL-10, and IFNα, whereas *T. annulata*-infected cells produced mRNA specific for IL-1α IL-1β, IL-6, IL-10, TNFα, and IL-12 (33), but not for IL-2 or IL-4. This was confirmed by Collins et al (40), who found expression of the same cytokines in *T. annulata*-infected cells but failed to demonstrate IFNγ, IL-2, IL-4, and IL-7. These findings suggest that IL-2 may not be directly involved in the continuous proliferation of *T. annulata*-infected cells, as was also proposed by Shayan and colleagues (158).

The constitutive expression of biologically functional IL-2Rs is likely to convey a selective growth advantage to the parasitized cell. It can be expected that IL-2, secreted by polyclonal activation of autologous uninfected cells (82, 132, 135), will support the proliferation of the parasitized cells in a paracrine manner, contributing to their spread and proliferation. IL-2Rs are also expressed in several *T. annulata*-infected cell lines (3, 4, 46, 89), and enhanced response to rIL-2 has been demonstrated that varied from cell line to cell line, however (4), and was most pronounced when cells were cultured at low density (4, 46).

Biologically active IFN has been detected in the tissue culture supernatants of a number *T. annulata*- and *T. parva*-transformed cell lines (5, 44, 64). Neutralization and inactivation experiments, combined with Northern blot analysis, demonstrated that IFNγ expression was restricted to the *T. parva*-infected cell lines. IFN expression was examined in more detail by Sager et al (145). *T. parva*-transformed cells produced exclusively IFNγ, whereas *T. annulata*-infected cells expressed only the type-I interferon IFNβ, regardless of the origin of the transformed cells. Upon exposure to double-stranded RNA, which induces IFN production in other systems, a 10–5,000-fold increase in IFN activity could be noted. The amount of IFNβ mRNA was increased, but mRNA coding for IFNα, IFNω, or IFNγ was not detected. In contrast primary macrophages, from which many of the tested lines were derived, expressed IFNα, IFNβ, and IFNω mRNA to similar degrees when stimulated by lipopolysaccharides or poly(I:C). Thus, *T. annulata* appears to turn on constitutive IFNβ gene transcription while silencing the genes coding for IFNα and IFNω.

The Effect of Cytokines on *Theileria parva-* and *Theileria annulata*-Infected Cells

Cytokines and lymphokines have been shown to affect the viability of a range of intracellular pathogens (7, 84, 188). The effect of cytokines on the establishment and proliferation of parasitized cell lines has also been studied. DeMartini et al (44) found that neither rIFNγ nor rIL-2 altered the proportion of cells initially developing schizonts, but both enhanced the establishment of *T. parva*–infected cell lines by about twofold. In contrast, rTNFα resulted in a 33% decrease in the proportion of schizont-infected cells. Interestingly, these effects on the establishment of parasitized cell lines were no longer apparent 12 days following infection. Preston et al (138), on the other hand, found that rTNFα, rIFNγ, IFNα, rIL-1, and rIL-2 significantly inhibited the establishment of transformed cells upon infection with three different stocks of *T. annulata* and one *T. parva* (Muguga) stock in vitro. The findings by Preston et al suggest that cytokines could help in resistance to infections by preventing the further development of trophozoite-infected cells (trophozoites are the transient precursors of the schizont). Perhaps more importantly, both groups found that none of these cytokines inhibited the proliferation of *T. annulata-* or *T. parva*-transformed cell lines, once they had become established. Instead, Preston et al demonstrated that rTNFα and rIL-2 consistently enhanced the proliferation of macroschizont-infected cell lines, whereas DeMartini et al found that bovine rIL-2 increased the proliferation of infected B-cell and α/β T-cell clones, but not that of γ/δ T cell clones.

These in vitro findings indicate that none of the above cytokines will directly help to resolve primary infections by inhibiting the growth of macroschizont-infected cells. On the contrary, they could play a role in the pathogenesis of *Theileria* infections by promoting the proliferation of macroschizont-infected cells and the associated lymphoid hyperplasia. This is in agreement with findings by Campbell et al (28, 31), who found greatly elevated levels of IFNγ in efferent lymph from infected lymph nodes. IFNγ production did not correlate with protection against the parasite, however, because infected cells flourished during peak IFNγ production and only very small amounts of IFNγ were produced during the effective immune response in immune animals. In subsequent work (30), it was shown that the production of IFNγ in vivo appears to be tightly controlled by the parasite, in that the induction of cytokine production by T cells was not initiated until the parasite had developed beyond the IFNγ-sensitive trophozoite stage.

Nitric Oxide

Recently, Sager et al (146) demonstrated that transformation of macrophages by *T. annulata* was accompanied by a reduced ability to express inducible nitric oxide (NO) synthase mRNA and to produce NO in response to lipopolysaccharide stimulation. This is not entirely unexpected, because sustained production of NO has been shown to endow macrophages with cytostatic or cytotoxic activity against viruses, bacteria, fungi, protozoa, and also tumor cells (reviewed in (108)). Indeed, Visser et al (183a) have demonstrated that the establishment of macroschizont-infected cell lines in vitro was suppressed either by incubating sporozoites with S-nitroso-N-acetyl- DL-penicillamine (SNAP), an NO-releasing molecule, prior to invasion of PBMC, or by pulsing developing cultures of trophozoite-infected cells with SNAP. In this work, the proliferation of established macroschizont-infected cell lines was not affected by low concentrations of NO. Higher levels of NO, on the other hand, significantly inhibited the proliferation of *T. annulata*-infected cell lines in vitro (139), causing the macroschizonts to disappear and host cells to become apoptotic. In this regard, it is worth noting that uninfected PBMC of cattle with tropical theileriosis or East Coast fever synthesized NO spontaneously in vitro (183a). NO was also induced when PBMC of immune, but not of naive, cattle were cultured with *T. annulata*-infected cell lines. These results point to NO, produced by macrophages, as a possible mediator of anti-*T. annulata* activity that might contribute the protective immune mechanisms. NO has also been implicated in vasodilation and, if produced in excess, may induce cell and tissue damage. Therefore, in addition to contributing to protection of cattle against *T. annulata* or *T. parva*, NO may play a prominent role in the pathogenesis of tropical theileriosis and East Coast fever.

PARASITE INTERFERENCE WITH HOST CELL SIGNAL TRANSDUCTION PATHWAYS

Under physiological conditions, cells are subject to the tight control of a complex set of regulatory signaling pathways that not only control cellular activation, differentiation, and proliferation, but also control down-regulation of activation, exit from the cell cycle, and apoptosis. In *Theileria*-transformed cells, activation and proliferation are constitutively induced, but cells escape the down-regulatory mechanisms and cell death. Different biochemical pathways regulate these events, and those that have been studied in *Theileria*-transformed cells will be discussed below.

Src-Related Kinases

Src-family protein tyrosine kinases are activated following engagement of different classes of cellular receptors and participate in signaling pathways that control a diverse spectrum of receptor-induced biological activities (reviewed extensively by Thomas & Brugge, 173). They mediate early signaling events of important T-cell, B-cell, and macrophage/monocyte membrane receptors. These include T- and B-cell antigen receptors as well as Fc receptors, co-receptors such as CD4 or CD8, and a wide range of additional surface receptors, independently of whether they are transmembrane or GPI anchored. In addition, Src kinases have been implicated in adhesion events regulated by integrins, cadherins, selectins, and cellular adhesion molecules and their respective co-receptors. Importantly, Src-related kinases also contribute to signaling through a wide range of cytokine receptors, including many of the interleukin receptors. The ability of viral and constitutively activated forms of Src family members to induce DNA synthesis and cell proliferation implies that activation of the wild-type kinases can stimulate pathways leading to cell proliferation. Under physiological conditions, however, activation of Src family kinases by cellular receptors is only transient, and Src kinase activation by stimulation of cellular receptors alone is insufficient to stimulate cell proliferation.

Initial experiments in several laboratories [Dobbelaere (Berne), Langsley (Paris), Bröker (Hamburg) (71)] have shown that the proliferation of *Theileria*-transformed cells is blocked by tyrosine kinase inhibitors such as herbimycin A and genistein, indicating indirectly that *Theileria*-transformed cells rely on tyrosine phosphorylation for their continuous proliferation. In preliminary work on *T. parva*-infected TpM(803) T cells (57), the Src-related kinases Fyn and Lck showed reduced electrophoretic mobility in polyacrylamide gel electrophoresis, reflecting differences in phosphorylation that can be observed upon activation. This was reversed upon elimination of the parasite, but in vitro kinase assays to determine the state of activation were not performed at that time. Later work revealed increased activity of both Lck and Fyn, compared with uninfected lymph node T cells (D Dobbelaere, Y Galley, & G Hagens, unpublished observation), suggesting that the presence of the parasite could, directly or indirectly, alter the state of activation of two kinases with important functions in T-cell activation and proliferation. Fich et al (71) monitored the modulation of Src-family kinases in a different set of *T. parva*-transformed T-cell lines in more detail. Fyn had high activity in all cell lines tested; in addition, two novel bands that were weakly phosphorylated coprecipitated with Fyn. In contrast to Fyn, enzymatic activity of Lck was low compared with primary concanavalin A-activated bovine T cells, Jurkat T cells, or human T cells transformed by *Herpesvirus saimiri*. Furthermore, Lck could not be demonstrated in one of the *T. parva*-transformed cell lines, indicating that the enzyme is not essential for the *Theileria*-dependent growth of the T cells. In addition to Fyn and Lck, weak enzymatic activity of a splice variant of Lyn was observed after infection of T cells with *T. parva*, but none of the other Src-family kinases showed detectable autophosphorylation. Another group, however, found

Hck to be constitutively active in *T. parva*-infected B cells (M Baumgartner & G Langsley, in preparation). Surprisingly, all three groups found that Src-related kinase activity increased upon elimination of the parasite. This was found to be the case for Lck (71), for Hck, and also for Fyn. Although Lck and Fyn are activated in parasitized cells, the disappearance of the high molecular weight forms of both enzymes upon elimination of the parasite, described by Eichhorn & Dobbelaere (57), clearly does reflect their loss of activity.

Thus, although no constant pattern can be detected in the different cell lines, infection with *Theileria* clearly results in a modulation of Src-family kinase activity, with higher activity in infected cells than in uninfected lymph node cells. The observations by Fich et al (71) suggest that Lck does not play an essential role in the transformation mechanism. It is worth noting, however, that Lck need not absolutely be linked to a particular pathway. There is considerable redundancy in Src kinase activation, both with respect to the ability of any one Src kinase to be activated by multiple receptors and the ability of one receptor to activate multiple Src family kinases (173). It is therefore likely that Src kinases other than Lck could compensate for the lack of Lck.

As mentioned earlier, Src-related kinases can associate with a wide range of receptors, and it is not yet known whether they are associated with the same or different molecules in infected and cured cells. The fact that Src-related kinase activity is not reduced upon elimination of the parasite might reflect its participation in down-regulatory or apoptotic pathways.

Casein Kinase II

Protein phosphorylation plays an important role in the regulation of cellular growth and proliferation and is thus also thought to participate in tumorigenesis. It has been proposed that casein kinase II (CK2) contributes significantly to these changes (122). CK2 is highly pleiotropic and phosphorylates more than 160 proteins at sites specified by multiple acidic residues (137). CK2 is required for viability and for cell cycle progression, and it is especially elevated in proliferating tissues, either normal or transformed.

Substantial alterations in phosphoprotein and protein kinase activity profiles have been described in *T. annulata*- and *T. parva*-infected leukocytes (55, 128). OleMoiYoi and colleagues (128) carried out a detailed analysis of CK2 in *T. parva*-transformed T cells. In initial studies, a comparison of in vitro protein kinase activities between uninfected, IL-2–dependent T lymphoblasts and *T. parva*-infected lymphocytes revealed a 5–12-fold increase in total phosphorylation and also the induction of a group of *Theileria* infection-specific phosphoproteins. A serine/threonine kinase with substrate and effector specificities of CK2 was shown to contribute to the phosphorylation of these substrates. Further analysis of *T. parva*-infected lymphocytes revealed marked increases of bovine CK2 at the transcriptional, translational, and functional levels, indicating that bovine CK2 is constitutively activated; and it was proposed that this kinase might be an important element in the signal-transducing pathways activated by *T. parva* (126). Subsequent studies by Ahmed and coworkers showed that CK2 mRNA expression was down-regulated when the parasite was eliminated by BW720c treatment (157). In addition, it was shown that thymidine incorporation of *T. annulata*-infected cells could be inhibited by 50% in the presence of antisense CK2 oligonucleotides, suggesting that proliferation was dependent on functional CK2 (157).

A *T. parva* CK2 catalytic α-subunit has been cloned (129). The predicted amino acid contains a 99-amino acid sequence at the N-terminus, which accommodates a sequence motif with features characteristic of signal peptides. The exact location of parasite CK2α still remains to be demonstrated, but it was proposed that the protein could be inserted into the parasite plasma membrane or be secreted into the host cell cytoplasm (127). This way, parasite CK2α would be in an ideal position to phosphorylate lymphocyte substrates without being subjected to the normal controls.

CK2 received further attention when it was shown that CK2α could serve as an oncogene when expressed in lymphocytes of transgenic mice (148). It was found that 6% of transgenic mice expressing dysregulated CK2α developed T-cell lymphomas after a relatively long latent period. This was accompanied by a diffuse spread of the tumor cells and infiltration of diverse non-lymphoid tissues, a pattern that is reminiscent of the lesions observed in East Coast fever. Co-expression of a c-*myc* transgene in addition to CK2 resulted in neonatal leukemia. At first sight, these similarities make it tempting to speculate that dysregulated expression of CK2 is one of the major events leading to lymphocyte transformation induced by *T. parva*, but there are two important discrepancies that should not be overlooked and that make a link between these two systems tenuous. First, lymphomas in CK2α-transgenic mice started to appear only after 6 months, implying that CK2α did not act independently, but that malignant transformation depended on secondary events such as the activation of oncogenes or the deletion of tumor suppressors. This long latency is in stark contrast to the fast *T. parva*-induced lymphocyte transformation (which can be observed in vitro and in vivo) and the tissue infiltration (which starts within a week after infection). Second, there seemed to be no correlation between the levels of CK2α expression and the transformed state of the T cells. CK2α was barely up-regulated in the transgenic mice, whereas it was clearly up-regulated at all levels in *T. parva*-infected cells (128).

Considering the broad range of functions of CK2, it is almost inevitable that it should play a role in the proliferation of *Theileria*-infected cells. CK2 is regulated by upstream events that can be triggered by a wide range of stimuli, such as growth factors, hormones, or mitogenic stimulation. In lymphocytes, for instance, exposure of T cells to mitogenic stimulation (76) and induction of cell-cycle progression in B cells by phorbol esters and ionomycin (43) are accompanied by increases in CK2 enzyme activity. The pathways that emanate from these stimuli involve a wide spectrum of different kinases [(including Lck, Fyn, Hck, Raf, c-JUN-NH$_2$-terminal kinases (JNK), etc)] that also show altered patterns of activation in *Theileria*-transformed cells. For all these kinases, including CK2, the actual role in *Theileria*-induced transformation still remains to be defined. It is obvious, however, that CK2 should join the ranks of the other kinases that are up-regulated in activated or proliferating cells and that have been shown to be associated with malignant transformation in other systems.

Mitogen-Activated Protein Kinases and the Activation of Transcription Factors

Nearly all cell surface receptors for polypeptide hormones, cytokines, or growth factors use one or more of the mitogen-activated protein kinase (MAPK) or extracellular signal regulated protein kinase (ERK) cascades for transducing signals to the nucleus of the cell. By stimulating transcriptional activators, they induce alterations in gene expression, allowing appropriate responses to environmental stimuli. In addition, two types of MAPKs have also been identified that mediate responses to cellular stress such as heat shock, osmotic shock, cytokines, protein synthesis inhibitors, antioxidants, UV irradiation, and DNA-damaging agents. They are the JNKs (also called stress-activated protein kinases) and members of the p38 family. Altogether, well over a dozen different members of the MAPK family have been discovered [reviewed in (141)]. Their pleiotropic properties have made it difficult to dissect the individual actions of each of these kinases, and it has become clear that their activation is also subject to the activation of numerous other pathways in the cell.

Among the substrates of MAPKs are members of the AP-1 family of transcription factors, which consists of homo- and heterodimers of Jun, Fos, or ATF (activating transcription factor) bZIP (basic region leucine zipper) proteins (102). AP-1 components are essential for the regulation of cellular proliferation and differentiation, but evidence is accumulating that AP-1 is also involved in apoptosis. Increased levels of activated AP-1 were first shown in *T. annulata*-infected leukocytes (16), and this was associated with increased levels of c-*fos* mRNA. A more recent analysis revealed AP-1 complexes in *T. annulata*-infected B cells and macrophages, as well as in *T. parva*-infected B cells (35). Judged from their different mobilities in electrophoretic mobility shift assays, it would appear that these complexes differ in composition. Increased levels of AP-1 and ATF-2 DNA-binding activity could also be demonstrated in *T. parva*-transformed T cells compared to cells cured of the parasite (19), and complexes were shown by supershift analysis to contain c-Jun, JunD, c-Fos, and ATF-2.

To determine which MAPK pathways contribute to the activation of these transcription factors, the state of activation of members of the MAPK family was studied in different *Theileria*-transformed cell lines. In lymphocytes, ERK activation is usually associated with antigen receptor stimulation (34, 100, 176) and the induction of proliferation. Contrary to expectations, ERK activity was only detectable at background levels in *T. parva*-transformed T or B cells, and the same applied to different *T. annulata*-transformed cell lines (36, 75). MEK, the upstream ERK kinase (MAPK kinase), was also inactive (35). The ERK activation pathways were shown to be functional, however, because either treatment with phorbol esters or antibody-mediated CD3/CD4 stimulation could induce a strong burst of ERK activity (35, 75). ERK has been implicated in the TCR-mediated signaling that participates in the control of IL-2 expression (187). The lack of ERK activity in *T. parva*-transformed T cells could thus help to explain why only low levels of IL-2 mRNA can be detected in *T. parva*-transformed cells (90).

In contrast to ERK, JNKs were shown to be constitutively activated in a parasite-dependent manner in *T. parva*-transformed TpM(803) T cells (75). Chaussepied et al (35) extended these observations to other cell lines and demonstrated constitutive JNK activity in a number of *T. parva*- and *T. annulata*-transformed B cell- and macrophage-derived cell lines. Immunoblot analysis with antibodies specific for phosphorylated c-Jun confirmed that activated c-Jun translocated to the nucleus of parasitized, but not of cured cells (19). Interestingly, in one cell line, TBL3—a *Theileria*-infected derivative of a bovine leukosis virus-transformed cell line—JNK activity could not be detected. Unlike c-Jun and c-Fos, which are induced at the transcriptional level, ATF-2 expression is constitutive (102). Its activation is regulated posttranslationally by phosphorylation and involves nuclear translocation. In TpM(803) T cells, phosphorylation of ATF-2 is also parasite-dependent and is downregulated within 24 h of parasite elimination (19). JNKs phosphorylate c-Jun, but can also phosphorylate and stimulate the transcriptional activity of ATF-2. The same positive regulatory sites can be phosphorylated by p38. However, the sites in the c-Jun activation domain are phosphorylated by the JNKs only. In several *T. parva*- and *T. annulata*-transformed cell lines, p38 was found not to be activated (19, 35), although the pathway was shown to be inducible by osmotic shock (19). Moreover, treatment with the p38 inhibitor SB203580 did not affect ATF-2 phosphorylation, suggesting that JNK and not p38 is in all likelihood responsible for constitutive ATF-2 activation in TpM(803) T cells (19). In addition, JNK appears to be responsible for AP-1 activation. Using a *T. annulata*-transformed macrophage cell line, Chaussepied and colleagues (35) showed that overexpression of Pyst1 (a phosphatase that specifically inactivates ERK), did not inhibit AP-1 transcriptional activity, whereas CL100 (which inactivates ERK, JNK, and p38) resulted in reduced AP-1–dependent reporter gene expression. Taking into account the absence of p38 activity, these observations firmly point towards permanent JNK activation being responsible for constitutive AP-1 and ATF-2 activation.

Theileria can thus be added to the list of JNK-inducing agents, but what are the biological consequences of permanent JNK activation for the infected cell? JNK can phosphorylate TCR/Elk1—usually considered to be a substrate of ERK—which results in the induction of c-*fos* expression (102). This way, JNK may compensate for the lack of ERK activation. JNK also activates ATF-2, which associates with c-Jun to form a heterodimer that regulates c-*jun* expression. Taken together, persistently activated JNK may thus explain the increased levels of c-*fos* and c-*jun* mRNA in *Theileria*-transformed leukocytes (35). Striking similarities can be found with human T-cell leukemia virus type 1(HTLV-1)-induced transformation.

Transformation, either by HTLV-1 in human T lymphocytes, or by HTLV-1–derived *tax* in transgenic mice fibroblasts, is accompanied by constitutive activation of ATF-2. Furthermore, HTLV-1–mediated tumorigenesis is associated with constitutively activated JNK (191). As mentioned above, Jun forms heterodimers with one of several members of the Fos and ATF-2 subfamilies. In this regard, it has recently been shown that ectopic expression of a combination of c-Jun/ATF-2 is capable of inducing growth factor independence (182), a general characteristic of *Theileria*-transformation (26, 45).

As mentioned above, AP-1 regulates a large array of genes that control cellular differentiation and proliferation. One of the consequences of constitutive AP-1 (and NF-κB) activation could be the induction of metalloprotease-9 gene expression (194). Upon infection with *T. annulata*, leukocytes express different metalloproteinase activities (15a), one of them MMP9 (16). Metalloprotease-9 is frequently localized in both tumor stroma and tumor cells, particularly at the tumor invasion front. Metalloprotease-9 could therefore also contribute to the metastatic properties of *T. annulata*-infected leukocytes (2, 160).

Theileria Infection Induces Constitutive NF-κB Activation

The transcription factor NF-κB contributes to the regulation of inducible expression of numerous cytokines and adhesion molecules, as well as that of genes that regulate proliferation or apoptosis. In this capacity, it has been shown to act as an evolutionarily conserved, coordinating element in the response to stress, injury, and infection. General aspects of NF-κB relating to its function in the regulation of immune responses have been reviewed extensively elsewhere (10, 13, 78). NF-κB is a dimer, which consists of members of the Rel protein family. Each family member contains an N-terminal 300 amino acid conserved region known as the rel homology domain. Not only is this region responsible for dimerization and DNA-binding, it also contains a nuclear localization sequence that is masked by interaction of NF-κB with its cytoplasmic inhibitors (IκBs), of which IκBα and IκBβ are the most prominent. IκBs sequester NF-κB in the cytoplasm, and activation of NF-κB requires dissociation from IκB. This involves IκB phosphorylation by IκB kinases (IKK) (summarized in 165), IκB ubiquitination, and proteasomal degradation. Released NF-κB then translocates to the nucleus, where it binds directly to specific decameric sequences (called κB motifs) that are located in the promotor and enhancer regions of a select number of genes.

In T cells, maximal activation and sustained NF-κB depends on at least two signals, the binding of an antigen to the TCR and CD28 co-stimulation. A single stimulus, such as that provided by phorbol ester, induces only IκBα phosphorylation, resulting in rapid but transient NF-κB activation (25). The down-regulation of NF-κB activation is based on the fact that NF-κB induces transcription of the *IκBα* gene with de novo synthesized IκBα, replenishing the IκBα pool that was depleted by degradation during the activation process. In the presence of dual signals, NF-κB activation is not only enhanced, it also results in the degradation of IκBβ. IκBβ degradation is associated with persistent nuclear expression of NF-κB (86), a process that can also be induced by viruses or bacteria (83, 87).

Early studies by Ivanov et al (99) have shown that NF-κB is constitutively activated in *T. parva*-transformed cells. Activation was dependent on the continuous presence of the parasite in the host cell cytoplasm, because elimination of the organism resulted in the disappearance of the transcription factor from the host cell nucleus. Recent studies (130) demonstrated that NF-κB activation in *T. parva*-transformed T cells involves the constitutive degradation of both IκBα and IκBβ. When de novo IκB protein synthesis was inhibited by treatment of the cells with cycloheximide, both IκBα and IκBβ were observed to be degraded as long as the cell harbored the parasite. In the absence of the parasite, however, no more degradation could be observed. The constitutive activation of NF-κB observed in *T. parva*-infected cells strongly resembles that of WEHI 231 early mature B cells (134). In both cases, NF-κB DNA binding complexes are found in the nucleus, despite the abundance of cytosolic IκBα. It has been shown in other systems (167) that degradation of IκBβ is followed by its subsequent resynthesis as a hypophosphorylated protein. This protein was shown to facilitate transport of a portion of NF-κB to the nucleus in a manner that protects it from cytosolic IκBα. In WEHI 231 cells, this hypophosphorylated IκBβ is found in a stable complex with NF-κB in the cytosol and is also detected in NF-κB DNA binding complexes in the nucleus. Several forms of IκBβ can be found in the nucleus of *T. parva*-transformed cells, and it is possible that hypophosphorylated IκBβ also protects NF-κB from IκBα, thus allowing the continuous nuclear import of this transcription factor.

Bandshift analyses revealed a complex pattern of distinct NF-κB complexes that could differ markedly according to nucleotide sequence of the κB recognition sequence that was used. In supershift and cross-linking experiments, p50, p65, and RelB have been demonstrated up until now, but this does not exclude that other members of the NF-κB family of proteins are also part of the complex (J Machado, manuscript in preparation). Indeed, Western blot analysis of cytoplasmic and nuclear extracts prepared from infected and cured cells demonstrated nuclear translocation of all known members of the NF-κB family.

NF-κB activation in *T. parva*-transformed T cells appears to differ from classic activation pathways in that it is not inhibited by the antioxidant N-acetyl-cysteine (NAC) (130) or a range of other antioxidants (J Machado & D Dobbelaere, unpublished information). On the other hand, the alkylating agent α-tosyl-phenylalanyl-chloromethyl-ketone, which is widely used as a serine protease inhibitor and which inhibits IκBα phosphorylation, completely inhibited NF-κB translocation and transcriptional activity (91). The lack of inhibition by NAC suggests that the IκB degradation pathway is not dependent on reactive oxygen intermediates. Although a detailed comparison has not been carried out, *T. parva*-transformed T cells may differ in this respect from *T. annulata*-transformed cells. Chaussepied & Langsley (36) reported that NF-κB in *T. annulata*-infected cells is inhibited by nordihydroguaiarectic acid, a nonspecific inhibitor of lipo-oxygenase. Low doses of nordihydroguaiarectic acid also inhibited proliferation of *T. annulata*-transformed B cells and monocytes, as well as *T. parva*-transformed B cells.

The Unknown Route to Constitutive NF-κB Activation

The way in which NF-κB activation is induced in *Theileria*-transformed cells is not yet known; at this stage, one can only speculate about possible mechanisms. In normal cells, NF-κB can be activated by a wealth of different signaling pathways involved in immune function and development (149). These encompass pathways involved in innate immune responses, including the cytokines TNFα, IL-1α, the chemotactic peptide fMet-Leu-Phe, and the recently identified homologue of *Drosophila* Toll. Pathways triggered in adaptive immune responses include those emanating from T- or B-cell antigen receptors, as well as signals delivered through CD2, CD28 (in the case of T cells), or CD40 (in the case of B cells). These pathways are normally blocked by antioxidants, and the lack of inhibition by NAC indicates that the NF-κB activation pathway in *T. parva*-transformed T cells occurs via an unorthodox mechanism (130). A tempting hypothesis would be that *Theileria* bypasses the physiological pathways that converge and regulate the activation of the multi-protein IKK complex, inducing constitutive IκB phosphorylation and degradation. This could occur by direct activation of the IKK by parasite-derived molecules, potentially involving an interaction with one of the recently identified IKK complex components (147). IKK-γ (143), a homologue of NEMO (NF-κB essential modifier) (192), is part of the IKK holoenzyme complex and is required for stimulation of IKK activity by NF-κB–inducing kinase (NIK) or the MAPK/ERK kinase (MEKK1), both of which are involved in different NF-κB activation pathways (101, 107, 165). The carboxyterminal domain of NEMO/IKK-γ contains a leucine zipper that could provide the interaction site for upstream signal-transmitting factors that activate IKK-α or IKK-β. Such upstream activators could conceivably be of parasite origin. A second target candidate for interaction with a parasite protein could be IKK-complex–associated protein (39), a scaffolding protein that complexes directly with IKK and also binds the upstream kinase NIK. Of course, it cannot be excluded that constitutive NF-κB activation occurs in a more indirect manner, by activating upstream components of the IKK activation pathway. MEKK-1, originally known as a component of the JNK pathway, has also been shown to regulate NF-κB activation (106, 107), thus explaining how many inducers that activate NF-κB also activate JNK. The Rho family of small GTPases are critical components in the regulation of signal transduction cascades from extracellular stimuli to the cell nucleus, including the JNK signaling pathway. Rho, CDC42, and Rac-1 are upstream activators of MEKK-1 and have also been shown to activate NF-κB (133). Parasite-mediated activation of small GTPases, possibly through interference with the corresponding exchange factors (116), could thus account for both JNK and NF-κB activation. Finally, the presence of the parasite in the host cell cytoplasm could induce the expression of one or more host cell surface receptors and their ligands, which, through continuous triggering, initiate constitutive NF-κB activation. To persist, such a pathway would have to be resistant to down-regulatory mechanisms and not be inhibited by NAC.

A number of interesting parallel examples of persistent NF-κB activation have been described in virus-infected cells, and this is often associated with a transforming function (65, 79). For example, the Tax oncoprotein of HTLV-1 chronically activates NF-κB, and it has been shown that Tax associates with IKK complexes containing the catalytic subunits IKK-α and IKK-β (37, 77, 180). Yin et al (193) reported that Tax binds to and stimulates the kinase activity of MEKK1 (106). The Epstein-Barr virus oncoprotein latent infection membrane protein 1 (LMP1) functions as a constitutively aggregated pseudo-tumor necrosis factor receptor that constitutively activates NF-κB through two sites in its C-terminal cytoplasmic domain. NF-κB activation by LMP1 is mediated by a pathway that includes NIK, IKK-α, and IKK-β (169).

What Are the Biological Consequences of Constitutive NF-κB Activation in *Theileria*-Infected Cells?

The role of Rel proteins in oncogenic transformation in other systems is well established (see 79 for a comprehensive review) and involves transforming forms of Rel proteins such as v-Rel, lyt-10, or truncated forms of c-Rel as well as viral activators of NF-κB (61). Viral products such as HTLV-1–derived Tax can induce transformation, and a direct role for constitutive IκBα and IκBβ degradation resulting in permanent NF-κB activation in this process has been demonstrated in mice transgenic for *tax* (42). Whether a transforming Rel protein of parasite origin is involved in *Theileria*-induced transformation has not yet been addressed and can therefore not be excluded. Whatever the origin or mechanism of activation, constitutively activated NF-κB can be expected to contribute to the proliferation of *T. parva*-transformed T cells in different ways. NF-κB regulates a restricted group of inducible genes that are directly involved in the activation and proliferation of leukocytes (10, 78, 159). NF-κB participates in inducing the transcription of the IL-2 and IL-2Rα–chain genes, both of which are constitutively expressed in *T. parva*-infected cells and could contribute to the proliferation of *T. parva*-transformed lymphocytes. The proto-oncogene c-*myc* is an immediate early gene encoding a protein that functions as a transcription factor. Because transcriptional regulation of c-*myc* is mediated through two κB sites (54), constitutive NF-κB activity may also contribute to c-*myc* expression in *T. parva*-transformed T cells (59). Expression of c-*myc*, which is exquisitely sensitive to growth factors, is required for quiescent cells to enter the cell cycle (17, 60). In addition, the G1- to S-phase transition also requires c-Myc (88). The c-Myc protein activates trancription of Cdc25, a phosphatase that activates cyclin-dependent kinases and that appears to be necessary for c-*myc*-induced cell cycle activation (74).

Several pro-inflammatory cytokines are expressed by *Theileria*-transformed cells (23, 114). TNFα, TNFβ, and IL-1 not only activate NF-κB, but they themselves are induced by NF-κB. Other genes induced by NF-κB, some of which are expressed in *Theileria*-transformed cells, include acute phase response proteins, IL-6, IFNβ, and various chemokines. *Theileria*-infected cells have been shown to migrate and proliferate in non-lymphoid as well as lymphoid tissues (96, 97, 118, 120). Migration involves the expression of cell surface adhesion proteins, such as ICAM-1, VCAM-1, MAdCAM-1, and E-selectin. From this point of view, it can be expected that NF-κB activation could also contribute to the migratory activity of *Theileira*-infected cells. Parasitized cells display a typical "clumping" pattern, and it is likely that the loss of this phenotype upon elimination of the parasite results from the down-regulation of NF-κB activation that can be observed in cured cells (99, 130).

McKeever et al have shown that in vitro infection with *T. parva* is associated with IL-10 mRNA expression in all bovine lymphocyte lineages (114); this also appears to be the case for *T. annulata*-transformed cell lines (23). It is interesting to note that NF-κB–dependent IL-10 expression can also be detected in patients with adult T-cell leukemia caused by HTLV-I infection (117), and NF-κB is in all likelihood also involved in the up-regulation of IL-10 expression in *Theileria*-transformed cells. IL-10 is a pleiotropic cytokine that has potent inhibitory effects on macrophages and T cells, and it contributes to the regulation of proliferation and differentiation of B cells. Whereas IL-10 expression is itself regulated by NF-κB, it is also capable of inhibiting NF-κB activation. This way, IL-10 markedly reduces NF-κB activity induced in PBMC by stimulation of the TCR/CD3 complex (142). The inhibitory activity on T lymphocytes is mediated by monocytes and might be ascribed to a lack of cooperation between accessory cells and T lymphocytes, resulting from down-regulation of a co-stimulatory molecule, such as CD80. IL-10 has also been shown to inhibit IFNγ-induced transcription of the ICAM-1 gene (which is also regulated by NF-κB), and this may result in impaired leukocyte recruitment (161). Finally, and perhaps most importantly, IL-10 has also been shown to protect IL-2–starved T cells from apoptosis (170). This raises the possibility that IL-10 contributes to the survival of *T. parva*-transformed T cells that migrated into IL-2–poor environments, as may be the case for non-lymphoid tissues. These observations raise the possibility that IL-10, derived from parasitized cells, may influence the immune responses of naive cattle to infection and that IL-10 expression by *Theileria*-transformed cells could also contribute to the pathology of theileriosis.

NF-κB Prevents Apoptosis of *Theileria*-Transformed T Cells

Tissue homeostasis requires a balance between cell proliferation and death. Apoptosis is an important mechanism for maintaining homeostasis and also for the control of cancer cells (reviewed in 104). Apoptosis and proliferation are linked by cell cycle regulators, and many stimuli delivered to a cell can induce either cell proliferation or cell death, depending on the context in which they were delivered. Although NF-κB plays a central role in the regulation of cellular differentiation and proliferation, it has also been implicated in the process of apoptosis. NF-κB may play a dual, apparently opposite, role in apoptosis discussed in (9, 11, 101). Compelling evidence has recently been presented for a protective role of NF-κB against apoptosis induced by TNF or chemotherapeutic agents (reviewed in 78). This involves the induction of genes that are required for cell survival (38, 189). On the other hand, it has also been proposed that NF-κB activation is required for apoptosis; this appears to apply in particular to apoptotic pathways that depend on de novo protein synthesis (103).

The potential role for constitutive NF-κB activity in protection against apoptosis in *Theileria*-transformed cells has been examined. When treated with potent inhibitors of NF-κB (such as α-tosyl-phenylalanyl-chloromethyl-ketone) or BAY 11-7082 which block IκBα degradation and subsequent nuclear translocation of NF-κB, *T. parva*-transformed T cells undergo rapid apoptosis (91, 91a). Furthermore, expression of dominant negative mutants of IκBα (which cannot be degraded because they lack key phosphorylation sites) or overexpression of mutant forms of p65, lacking the transactivation domain, both induced apoptosis in TpM(803) T cells (V Heussler & D Dobbelaere, in press). This critical dependence on persistent NF-κB activity is very reminiscent of Ras-transformed cells, which require NF-κB to avoid spontaneous apoptosis (112). Similar findings were made for a B lymphoma cell line (190). *T. parva*-transformed T cells have been shown to be resistant against TNF-induced apoptosis (44), and it is reasonable to assume that constitutive NF-κB is responsible for this protection.

Is There an Important Role for PI3-K in Proliferation and Survival of *Theileria*-Transformed Cells?

Phosphoinositide kinases phosphorylate phosphatidylinositol, a component of eukaryotic cell membranes, resulting in the generation of phosphoinositides that are involved in the regulation of diverse cellular processes. These include proliferation, survival, cytoskeletal organization, vesicle trafficking, and glucose transport (reviewed in 111). One of these phosphoinositide kinases, PI3-K, is activated upon stimulation of a wide range of growth factors. Activation of PI3-K is sufficient to promote entry into S-phase and promotes cellular changes characteristic of oncogenic transformation (105). PI3-K activation alone, however, is not sufficient to provide for progression through the entire cell cycle. PI3-K–dependent signals have been implicated in the promotion of cell survival by cytokines. PI3-K thus not only contributes to cellular proliferation, but, through its downstream effector PKB/AKT, also participates in maintaining cell survival (73). PKB/AKt promotes the expression of c-*myc* and the anti-apoptotic genes *Bcl*-2 and *Bag*-2 (6).

The most prominent candidates for inducing PI3-K activation in *T. parva*-transformed T cells are the IL-2R and CD28. The IL-2R is the most important T-cell growth factor receptor, and its stimulation by IL-2 triggers cell-cycle transit through the G1/S-phase cell-cycle checkpoint. This involves up-regulation of different cyclins and the down-regulation of the cyclin-dependent kinase inhibitor p27^{KIP1}. One of the key events during G1 is the activation of the cell-cycle regulator E2F, which is a transcription factor. The activation of E2F occurs through

PKB/AKT, which is regulated by IL-2 via a PI3-K-dependent pathway (20). CD28 stimulation also activates PI3-K (131, 184). Apart from enhancing IL-2 production, CD28 regulates T-cell survival by augmenting the expression of the intrinsic survival factor Bcl-xL (18), which is capable of overriding Fas-induced cell death (124). Bcl-2 and Bcl-xL expression can be detected in TpM(803) T cells (V Heussler & D Dobbelaere, unpublished information) and may participate in the resistance against different apoptotic signals.

Although a firm link between surface receptors and PI3-K activation has not yet been established in *Theileria*-transformed cells, indications are that PI3-K activity occupies an essential function in *Theileria*-induced transformation. Consistent with the important role of PI3-K in cellular proliferation and maintaining cell survival, the PI3-K inhibitors wortmannin and LY294002 both potently blocked proliferation of TpM(803) cells (V Heussler & D Dobbelaere, unpublished data). Whether this is also accompanied by the induction of apoptosis, however, has not yet been established. The same inhibitors also block proliferation of *T. parva*-transformed B cells and the constitutive AP-1 activity present in these cells (M Baumgartner & G Langsley, unpublished result). These observations strongly suggest that signals potentially emanating from a PI3-K–linked surface receptor are required for continuous growth.

The Immunosuppressants FK506 and Rapamycin

The macrolide antibiotics FK506 and rapamycin are potent immunosuppressants that find application in the field of organ transplantation. Extensive use of these compounds has been made in studies on signal transduction pathways that control lymphocyte activation and growth. Despite their similarities in structure and affinity for the same intracellular receptor, FK506-binding protein 12 (FKBP12), these two compounds interfere with entirely different steps in the T-cell activation and proliferation pathways. FK506 disrupts an early signaling pathway that emanates from the TCR and involves the Ca^{2+}/calmodulin-dependent activation of the serine-threonine phosphatase, calcineurin, a step that is essential for the transcriptional activation of the IL-2 and other cytokine genes. Rapamycin, on the other hand, inhibits the late events required for the progression of T cells from G1- to S-phase of the cell cycle, induced by IL-2 engagement of the IL-2R. In agreement with the findings that *T. parva* bypasses TCR-dependent signaling to induce continuous T-cell proliferation, the FK506-related compound ascomycin did not inhibit the proliferation of *T. parva*-transformed T cells (75). Likewise, cyclosporin A, which targets the same pathway, interfered only marginally with proliferation of TpM(803) T cells (58). *T. annulata*-infected cells, however, may differ in this respect from *T. parva*-transformed T cells because cyclosoporin A was shown to inhibit the proliferation of at least one cell line (3).

In contrast to ascomycin or cyclosporin A, rapamycin potently inhibited the proliferation of the *T. parva*-transformed T cell line TpM(803) (75). The rapamycin-FKBP12 complex targets a protein named mammalian target of rapamycin, mTOR (also called FRAP, RAFT1, or RAPT1) (1) and references therein). A downstream component of one of the pathways activated by PI3-K, mTOR is a member of a newly defined family of phosphoinositide-3-kinase (PI3-K)–related kinases that phosphorylate proteins rather than lipids. One of the targets of mTOR is the kinase p70 S6 kinase (p70^{S6K}), which phosphorylates the 40S ribosomal protein S6. A second PI3-K-dependent stimulus provided by PDK1, an additional kinase downstream of PI3-K, is also required for full activation of p70^{S6K}. Phosphorylation of S6 by activated p70^{S6K} enhances translation of mRNAs bearing 5'-terminal oligopolypyrimidine tracts, and among these, several transcripts encoding components of the translation machinery itself can be found. Rapamycin thus blocks a pathway coupling cytokine receptors to selective increases in translation. Because treatment with rapamycin also abrogates the activation of p70^{S6K} in lymphoid cells (24), it can be expected that this also applies to TpM(803) T cells. Differences may exist between T and B cells, however. Whereas blocking PI3-K with wortmannin or LY294002 results in complete inhibition of proliferation of *T. parva*-transformed TPM(409) B cells, rapamycin only partly inhibited proliferation, suggesting that the PKB/AKT component of the PI3-K–activated signaling pathways is predominantly important for proliferation in these cells (M Baumgartner & G Langsley, unpublished observation).

How Does *Theileria* Talk to the Host Cell? Speculations

In the case of *T. parva*-transformed T cells, it would be reasonable to assume that one of the possible mechanisms by which *T. parva* induces transformation is by constitutively activating antigen receptor–mediated signaling pathways. In TpM(803) T cells, however, no tyrosine phosphorylation of the TCRζ and CD3ε-chains, typical of TCR stimulation, could be detected (75). Proliferation was also not inhibited by compounds that block early TCR-dependent signaling pathways. These findings show that not only the requirement for antigen-specific triggering, but also the downstream components of the TCR-dependent signal transduction pathways are bypassed by the parasite. The lack of CD3ε and TCRζ phosphorylation would also argue against the involvement of a superantigen because superantigen-mediated T-cell activation also uses TCR-dependent signaling pathways. Additional observations indirectly support these findings. Stimulation of T cells is normally accompanied by the internalization of the TCR/CD3 complex (171). In TpM(803) T cells, expression of the TCR/CD3 complex was found to be stable, whereas the IL-2R displayed the typical cycling pattern of a constitutively activated receptor (41, 46). Furthermore, TpM(803) T-cells were inhibited by compounds that block the progression, but not the initiation, of T-cell proliferation.

Taking into consideration the fact that *T. parva* and *T. annulata* are very closely related organisms, and considering the many similarities that can be observed associated with *T. parva*- and *T. annulata*-induced transformation, it could be argued that both parasites use an identical mechanism to induce transformation. If this were the case, then the target for transformation should be a pathway that is conserved in T and B lymphocytes as well as monocytes/macrophages. It is therefore

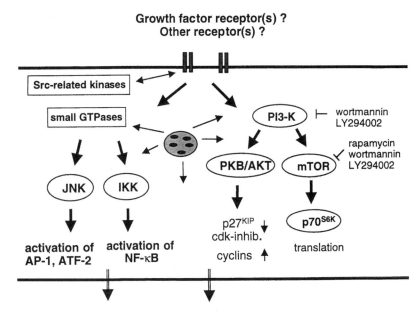

Figure 2   Different host cell signal transduction pathways are constitutively activated in *Theileria*-transformed cells. Several members of the Src-related kinases family are active in *Theileria*-transformed cells. JNK is the only mitogen-activated protein kinase found to be constitutively activated and is responsible for the activation of AP-1 and ATF-2. NF-κB is activated via an unorthodox pathway in *Theileria parva*-transformed T cells. These transcription factors regulate the expression of a range of genes encoding proteins that participate in proliferation and cell survival. Some of these could be growth factors or receptors and their ligands, which trigger signaling pathways that activate the cell cycle. The NF-κB pathway is critical and prevents spontaneous apoptosis of *T. parva*-transformed T-cells. Experiments with inhibitors of the PI3-K pathway suggest that PI3-K–dependent signaling is crucial for continuous proliferation.

intriguing that dendritic cells, which are closely related to macrophages/monocytes, can be infected by *T. parva* and allow the development of schizonts but do not become transformed (186). It is conceivable that a *Theileria*-derived molecule interferes with one or more components that are common to the signaling pathways that govern the proliferation of lymphocytes and monocytes/macrophages (see Figure 2). Such molecules could be released by the parasite into the host cell cytoplasm or be expressed on the parasite surface where they are exposed to the host cell cytoplasm, allowing their interaction with regulators of host cell signaling pathways. The nature of these molecules is still subject to speculation. Proteins with kinase or phosphatase activity could be plausible candidates, because phosphorylation is one of the most predominant regulators of signal transduction cascades. The permanent activation of positive regulatory elements or persistent inactivation of negative regulatory components by either phosphorylation or dephosphorylation are possible scenarios. The transforming moiety need not necessarily be a protein or peptide. Lipid-derived products have also been shown to exert agonistic or antagonistic effects on signal transduction pathways. A considerable number of parasite surface proteins are anchored in the parasite surface membrane by glycosylphosphotidylinositol lipids. It cannot be excluded that parasite-derived glycolipids, or breakdown products generated upon processing of glycosylphosphatidylinositol-anchored proteins, are released into the host cell cytoplasm and interfere with host cell signal transduction pathways.

To elucidate the mechanism by which *Theileria* transforms the host cell, it will be necessary to carry out a systematic analysis to pinpoint the molecular target(s) in the transformed cell. The use of specific inhibitors and the expression of dominant negative mutants that block individual components of signaling pathways might help to define the level at which *Theileria* interferes with a particular pathway. Considering the size of the parasite genome and the large number of parasite genes that are potentially expressed in the infected cells, retroviral transfer and expression of a parasite cDNA library in leukocytes might be the only way by which a "transforming" gene of parasite origin can be identified.

ACKNOWLEDGMENTS

We thank Isabel Roditi and Paula Fernandez for critical discussions and corrections. Part of this work was supported by several grants of the Swiss National Science Foundation, the Federal Office for Education and Science, and the Swiss Cancer League to Dirk Dobbelaere.

Visit the Annual Reviews home page at http://www.AnnualReviews.org

LITERATURE CITED

1. Abraham RT. 1998. Mammalian target of rapamycin: immunosuppressive drugs uncover a novel pathway of cytokine receptor signaling. *Curr. Opin. Immunol.* 10:330–36

2. Adamson RE, Hall FR. 1996. Matrix metalloproteinases mediate the metastatic phenotype of *Theileria annulata*-transformed cells. *Parasitology* 113:449–55

3. Ahmed JS, Rintelen M, Hartwig H, Schein E. 1991. Effect of cyclosporin A on the proliferation of bovine lymphocytes to concanavalin A and on the growth of *Theileria annulata*-infected bovine cells. *Trop. Med. Parasitol.* 42:375–80

4. Ahmed JS, Rintelen M, Schein E, Williams RO, Dobbelaere D. 1992. Effect of buparvaquone on the expression of interleukin

2 receptors in *Theileria annulata*-infected cells. *Parasitol. Res.* 78:285–90

5. Ahmed JS, Wiegers P, Steuber S, Schein E, Williams RO, Dobbelaere D. 1993. Production of interferon by *Theileria annulata*- and *T. parva*-infected bovine lymphoid cell lines. *Parasitol. Res.* 79:178–82

6. Ahmed NN, Grimes HL, Bellacosa A, Chan TO, Tsichlis PN. 1997. Transduction of interleukin-2 antiapoptotic and proliferative signals via Akt protein kinase. *Proc. Natl. Acad. Sci. USA* 94:3627–32

7. Alexander J, Scharton-Kersten TM, Yap G, Roberts CW, Liew FY, Sher A. 1997. Mechanisms of innate resistance to *Toxoplasma gondii* infection. *Philos. Trans. R. Soc. London Ser. B* 352:1355–59

8. Andrews NW, Webster P. 1991. Phagolysosomal escape by intracellular pathogens. *Parasitol. Today* 8:28–32

9. Baeuerle PA, Baltimore D. 1996. NF-κB: ten years after. *Cell* 87:13–20

10. Baeuerle PA, Henkel T. 1994. Function and activation of NF-κB in the immune system. *Annu. Rev. Immunol.* 12:141–79

11. Baichwal VR, Baeuerle PA. 1997. Activate NF-κB or die? *Curr. Biol.* 7:R94–96

12. Baldwin CL, Black SJ, Brown WC, Conrad PA, Goddeeris BM, et al. 1988. Bovine T cells, B cells, and null cells are transformed by the protozoan parasite *Theileria parva*. *Infect. Immun.* 56:462–67

13. Baldwin CL, Malu MN, Kinuthia SW, Conrad PA, Grootenhuis JG. 1986. Comparative analysis of infection and transformation of lymphocytes from African buffalo and Boran cattle with *Theileria parva* subsp. *parva* and *T. parva* subsp. *lawrencei*. *Infect. Immun.* 53:186–91

14. Baldwin CL, Teale AJ. 1987. Alloreactive T cell clones transformed by *Theileria parva* retain cytolytic activity and antigen specificity. *Eur. J. Immunol.* 17:1859–62

15. Baylis HA, Allsopp BA, Hall R, Carrington M. 1993. Characterisation of a glutamine- and proline-rich protein (QP protein) from *Theileria parva*. *Mol. Biochem. Parasitol.* 61:171–78

15a. Baylis HA, Megson A, Brown CG, Wilkie GF, Hall R. 1992. *Theileria annulata*-infected cells produce abundant proteases whose activity is reduced by long-term cell culture. *Parasitology* 105:417–23

16. Baylis HA, Megson A, Hall R. 1995. Infection with *Theileria annulata* induces expression of matrix metalloproteinase 9 and transcription factor AP-1 in bovine leucocytes. *Mol. Biochem. Parasitol.* 69:211–22

17. Bishop JM, Eilers M, Katzen AL, Kornberg T, Ramsay G, Schirm S. 1991. MYB and MYC in the cell cycle. *Cold Spring Harb. Symp. Quant. Biol.* 56:99–107

18. Boise LH, Minn AJ, Noel PJ, June CH, Accavitti MA, et al. 1995. CD28 costimulation can promote T cell survival by enhancing the expression of Bcl-X$_L$. *Immunity* 3:87–98

19. Botteron C, Dobbelaere D. 1998. AP-1 and ATF-2 are constitutively activated via the JNK pathway in *Theileria parva*-transformed T-cells. *Biochem. Biophys. Res. Commun.* 246:418–21

20. Brennan P, Babbage JW, Burgering BM, Groner B, Reif K, Cantrell DA. 1997. Phosphatidylinositol 3-kinase couples the interleukin-2 receptor to the cell cycle regulator E2F. *Immunity* 7:679–89

21. Browder TM, Abrams JS, Wong PM, Nienhuis AW. 1989. Mechanism of autocrine stimulation in hematopoietic cells producing interleukin-3 after retrovirus-mediated gene transfer. *Mol. Cell. Biol.* 9:204–13

22. Brown CG, Stagg DA, Purnell RE, Kanhai GK, Payne RC. 1973. Letter: Infection and transformation of bovine lymphoid cells *in vitro* by infective particles of *Theileria parva*. *Nature* 245:101–3

23. Brown DJ, Campbell JD, Russell GC, Hopkins J, Glass EJ. 1995. T cell activation by *Theileria annulata*-infected macrophages correlates with cytokine production. *Clin. Exp. Immunol.* 102:507–14

24. Brown EJ, Beal PA, Keith CT, Chen J, Shin

TB, Schreiber SL. 1995. Control of p70 S6 kinase by kinase activity of FRAP in vivo. *Nature* 377:441–46

25. Brown K, Park S, Kanno T, Franzoso G, Siebenlist U. 1993. Mutual regulation of the transcriptional activator NF-κB and its inhibitor, IκB-α. *Proc. Natl. Acad. Sci. USA* 90:2532–36

26. Brown WC, Logan KS. 1986. Bovine T-cell clones infected with *Theileria parva* produce a factor with IL 2-like activity. *Parasite Immunol.* 8:189–92

27. Campbell JD, Brown DJ, Glass EJ, Hall FR, Spooner RL. 1994. *Theileria annulata* sporozoite targets. *Parasite Immunol.* 16:501–5

28. Campbell JDM, Brown DJ, Nichani AK, Howie SEM, Spooner RL, Glass EJ. 1997. A non-protective T-helper-1 response against the intra-macrophage protozoan *Theileria annulata*. *Clin. Exp. Immunol.* 108:463–70

29. Campbell JDM, Howie SEM, Odling KA, Glass EJ. 1995. *Theileria annulata* induces abberrant T cell activation *in vitro* and *in vivo*. *Clin. Exp. Immunol.* 99:203–10

30. Campbell JDM, Nichani AK, Brown DJ, Glass EJ, Spooner RL. 1998. A stage-specific, parasite-induced, "window" of *in vivo* interferon-γ production is associated with pathogenesis in *Theileria annulata* infection. *Ann. NY Acad. Sci.* 849:152–54

31. Campbell JDM, Nichani AK, Brown DJ, Howie SEM, Spooner RL, Glass EJ. 1997. Parasite-mediated steps in immune response failure during primary *Theileria annulata* infection. *Trop. Anim. Health Prod.* 29:S133–35

32. Campbell JDM, Russell GC, Nelson RE, Spooner RL, Glass EJ. 1997. *Theileria annulata* "superantigen" activity–TCR usage by responding bovine CD4+ cells from uninfected donors. *Biochem. Soc. Trans.* 25:S277

33. Campbell JDM, Spooner RL. 1999. Macrophages behaving badly: infected cells and subversion of immune responses to

Theileria annulata. *Parasitol. Today.* 15:10–16

34. Cantrell D. 1996. T cell antigen receptor signal transduction pathways. *Annu. Rev. Immunol.* 14:259–74

35. Chaussepied M, Lallemand D, Moreau M-F, Adamson R, Hall R, Langsley G. 1999. Upregulation of Jun and Fos family members and permanent JNK activity lead to constitutive AP-1 activation in *Theileria*-transformed leukocytes. *Mol. Biochem. Parasitol.* 94:215–26

36. Chaussepied M, Langsley G. 1996. *Theileria* transformation of bovine leukocytes: a parasite model for the study of lymphoproliferation. *Res. Immunol.* 147:127–38

37. Chu ZL, DiDonato JA, Hawiger J, Ballard DW. 1998. The tax oncoprotein of human T-cell leukemia virus type 1 associates with and persistently activates IκB kinases containing IKKα and IKKβ. *J. Biol. Chem.* 273:15891–94

38. Chu ZL, McKinsey TA, Liu L, Gentry JJ, Malim MH, Ballard DW. 1997. Suppression of tumor necrosis factor-induced cell death by inhibitor of apoptosis c-IAP2 is under NF-κB control. *Proc. Natl. Acad. Sci. USA* 94:10057–62

39. Cohen L, Henzel WJ, Baeuerle PA. 1998. IKAP is a scaffold protein of the IκB kinase complex. *Nature* 395:292–96

40. Collins RA, Sopp P, Gelder KI, Morrison WI, Howard CJ. 1996. Bovine γ/δ TcR+ T lymphocytes are stimulated to proliferate by autologous *Theileria annulata*-infected cells in the presence of interleukin-2. *Scand. J. Immunol.* 44:444–52

41. Coquerelle TM, Eichhorn M, Magnuson NS, Reeves R, Williams RO, Dobbelaere DA. 1989. Expression and characterization of the interleukin 2 receptor in *Theileria parva*-infected bovine lymphocytes. *Eur. J. Immunol.* 19:655–59

42. Coscoy L, Gonzalez Dunia D, Tangy F, Syan S, Brahic M, Ozden S. 1998. Molecular mechanism of tumorigenesis in mice transgenic for the human T cell leukemia

virus *tax* gene. *Virology* 248:332–41

43. DeBenedette M, Snow EC. 1991. Induction and regulation of casein kinase II during B lymphocyte activation. *J. Immunol.* 147:2839–45

44. DeMartini JC, Baldwin CL. 1991. Effects of γ interferon, tumor necrosis factor α, and interleukin-2 on infection and proliferation of *Theileria parva*-infected bovine lymphoblasts and production of interferon by parasitized cells. *Infect. Immun.* 59:4540–46

45. Dobbelaere DAE, Coquerelle TM, Roditi IJ, Eichhorn M, Williams RO. 1988. *Theileria parva* infection induces autocrine growth of bovine lymphocytes. *Proc. Natl. Acad. Sci. USA* 85:4730–34

46. Dobbelaere DAE, Prospero TD, Roditi IJ, Kelke C, Baumann I, et al. 1990. Expression of Tac antigen component of bovine interleukin-2 receptor in different leukocyte populations infected with *Theileria parva* or *Theileria annulata. Infect. Immun.* 58:3847–55

47. Dobbelaere DAE, Roditi IJ, Coquerelle TM, Kelke C, Eichhorn M, Williams RO. 1991. Lymphocytes infected with *Theileria parva* require both cell-cell contact and growth factor to proliferate. *Eur. J. Immunol.* 21:89–95

48. Dobbelaere DAE, Shapiro SZ, Webster P. 1985. Identification of a surface antigen on *Theileria parva* sporozoites by monoclonal antibody. *Proc. Natl. Acad. Sci. USA* 82:1771–75

49. Dobbelaere DAE, Spooner PR. 1985. Production in ascites fluid of biosynthetically labelled monoclonal antibody to *Theileria parva* sporozoites. *J. Immunol. Methods* 82:209–14

50. Dobbelaere DAE, Spooner PR, Barry WC, Irvin AD. 1984. Monoclonal antibody neutralizes the sporozoite stage of different *Theileria parva* stocks. *Parasite Immunol.* 6:361–70

51. Dobbelaere DAE, Webster P, Leitch BL, Voigt WP, Irvin AD. 1985. *Theileria parva*:

expression of a sporozoite surface coat antigen. *Exp. Parasitol.* 60:90–100

52. Duffus WP, Wagner GG, Preston JM. 1978. Initial studies on the properties of a bovine lymphoid cell culture line infected with *Theileria parva. Clin. Exp. Immunol.* 34:347–53

53. Dunbar CE, Browder TM, Abrams JS, Nienhuis AW. 1989. COOH-terminal-modified interleukin-3 is retained intracellularly and stimulates autocrine growth. *Science* 245:1493–96

54. Duyao MP, Buckler AJ, Sonenshein GE. 1990. Interaction of an NF-κB-like factor with a site upstream of the c-myc promoter. *Proc. Natl. Acad. Sci. USA* 87:4727–31

55. Dyer M, Hall R, Shiels B, Tait A. 1992. *Theileria annulata*: alterations in phosphoprotein and protein kinase activity profiles of infected leukocytes of the bovine host, Bos taurus. *Exp. Parasitol.* 74:216–27

56. Eichhorn M, Dobbelaere D. 1995. Partial inhibition of *Theileria parva*-infected T-cell proliferation by antisense IL2R α-chain RNA expression. *Res. Immunol.* 146:89–99

57. Eichhorn M, Dobbelaere DAE. 1994. Induction of signal transduction pathways in lymphocytes infected by *Theileria parva. Parasit. Today* 10:469–72

58. Eichhorn M, Magnuson NS, Reeves R, Williams RO, Dobbelaere DAE. 1990. IL-2 can enhance the cyclosporin A-mediated inhibition of *Theileria parva*-infected T cell proliferation. *J. Immunol.* 144:691–98

59. Eichhorn M, Prospero TD, Heussler VT, Dobbelaere DA. 1993. Antibodies against major histocompatibility complex class II antigens directly inhibit the growth of T cells infected with *Theileria parva* without affecting their state of activation. *J. Exp. Med.* 178:769–76

60. Eilers M, Schirm S, Bishop JM. 1991. The MYC protein activates transcription of the α-prothymosin gene. *EMBO J.* 10:133–41

61. Eliopoulos AG, Rickinson AB. 1998.

Epstein-Barr virus: LMP1 masquerades as an active receptor. *Curr. Biol.* 8:R196–98

62. Emery DL. 1981. Kinetics of infection with *Theileria parva* (East Coast fever) in the central lymph of cattle. *Vet. Parasitol.* 9:1–16

63. Emery DL, MacHugh ND, Morrison WI. 1988. *Theileria parva* (Muguga) infects bovine T-lymphocytes *in vivo* and induces coexpression of BoT4 and BoT8. *Parasite Immunol.* 10:379–91

64. Entrican G, McInnes CJ, Logan M, Preston PM, Martinod S, Brown CG. 1991. Production of interferons by bovine and ovine cell lines infected with *Theileria annulata* or *Theileria parva*. *Parasite Immunol.* 13:339–43

65. Farrell PJ. 1998. Signal transduction from the Epstein-Barr virus LMP-1 transforming protein. *Trends Microbiol.* 6:175–77

66. Fawcett D, Musoke A, Voigt W. 1984. Interaction of sporozoites of *Theileria parva* with bovine lymphocytes *in vitro*. I. Early events after invasion. *Tissue Cell* 16:873–84

67. Fawcett DW, Doxsey S, Stagg DA, Young AS. 1982. The entry of sporozoites of *Theileria parva* into bovine lymphocytes *in vitro*. Electron microscopic observations. *Eur. J. Cell. Biol.* 27:10–21

68. Fawcett DW, Stagg DA. 1986. Passive endocytosis of sporozoites of *Theileria parva* in macrophages at 1–2 degrees C. *J. Submicrosc. Cytol.* 18:11–19

69. Fell AH, Preston PM. 1993. Proliferation of *Theileria annulata* and *Theileria parva* macroschizont-infected bovine cells in SCID mice. *Int. J. Parasitol.* 23:77–87

70. Fell AH, Preston PM, Ansell JD. 1990. Establishment of *Theileria*-infected bovine cell lines in SCID mice. *Parasite Immunol.* 12:335–39

71. Fich C, Klauenberg U, Fleischer B, Broker BM. 1998. Modulation of enzymatic activity of Src-family kinases in bovine T cells

transformed by *Theileria parva*. *Parasitology* 117:107–15

72. Forsyth LM, Jackson LA, Wilkie G, Sanderson A, Brown CG, Preston PM. 1997. Bovine cells infected *in vivo* with *Theileria annulata* express CD11b, the C3bi complement receptor. *Vet. Res. Commun.* 21:249–63

73. Franke TF, Kaplan DR, Cantley LC. 1997. PI3K: downstream AKTion blocks apoptosis. *Cell* 88:435–37

74. Galaktionov K, Chen XC, Beach D. 1996. Cdc25 cell-cycle phosphatase as a target of c-myc. *Nature* 382:511–17

75. Galley Y, Hagens G, Glaser I, Davis W, Eichhorn M, Dobbelaere D. 1997. Jun NH2-terminal kinase is constitutively activated in T cells transformed by the intracellular parasite *Theileria parva*. *Proc. Natl. Acad. Sci. USA* 94:5119–24

76. Geahlen RL, Harrison ML. 1984. Induction of a substrate for casein kinase II during lymphocyte mitogenesis. *Biochim. Biophys. Acta* 804:169–75

77. Geleziunas R, Ferrell S, Lin X, Mu Y, Cunningham ETJ, et al. 1998. Human T-cell leukemia virus type 1 Tax induction of NF-κB involves activation of the IκB kinase α (IKKα) and IKKβ cellular kinases. *Mol. Cell Biol.* 18:5157–65

78. Ghosh S, May MJ, Kopp EB. 1998. NF-κB and Rel proteins: evolutionarily conserved mediators of immune responses. *Annu. Rev. Immunol.* 16:225–60

79. Gilmore TD, Koedood M, Piffat KA, White DW. 1996. Rel/NF-κB/IκB proteins and cancer. *Oncogene* 13:1367–78

80. Glass EJ, Innes EA, Spooner RL, Brown CG. 1989. Infection of bovine monocyte/macrophage populations with *Theileria annulata* and *Theileria parva*. *Vet. Immunol. Immunopathol.* 22:355–68

81. Glass EJ, Spooner RL. 1990. Parasite-accessory cell interactions in theileriosis. Antigen presentation by *Theileria annulata*-infected macrophages and production of continuously growing antigen-

presenting cell lines. *Eur. J. Immunol.* 20:2491–97

82. Goddeeris BM, Morrison WI. 1987. The bovine autologous *Theileria* mixed leucocyte reaction: influence of monocytes and phenotype of the parasitized stimulator cell on proliferation and parasite specificity. *Immunology* 60:63–69

83. Good L, Sun SC. 1996. Persistent activation of NF-κB/Rel by human T-cell leukemia virus type 1 tax involves degradation of IκBβ. *J. Virol.* 70:2730–35

84. Green SJ, Scheller LF, Marletta MA, Seguin MC, Klotz FW, et al. 1994. Nitric oxide: cytokine-regulation of nitric oxide in host resistance to intracellular pathogens. *Immunol. Lett.* 43:87–94

85. Hall R, Hunt PD, Carrington M, Simmons D, Williamson S, et al. 1992. Mimicry of elastin repetitive motifs by *Theileria annulata* sporozoite surface antigen. *Mol. Biochem. Parasitol.* 53:105–12

86. Harhaj EW, Maggirwar SB, Good L, Sun SC. 1996. CD28 mediates a potent costimulatory signal for rapid degradation of IκBβ which is associated with accelerated activation of various NF-κB/Rel heterodimers. *Mol. Cell Biol.* 16:6736–43

87. Hauf N, Goebel W, Serfling E, Kuhn M. 1994. *Listeria monocytogenes* infection enhances transcription factor NF-κB in P388D1 macrophage-like cells. *Infect. Immun.* 62:2740–47

88. Heikkila R, Schwab G, Wickstrom E, Loke SL, Pluznik DH, et al. 1987. A c-myc antisense oligodeoxynucleotide inhibits entry into S phase but not progress from G0 to G1. *Nature* 328:445–49

89. Herrmann T, Ahmed JS, Diamantstein T. 1989. The intermediate-affinity interleukin (IL)2 receptor expressed on *Theileria annulata*-infected cells comprises a single IL 2-binding protein. Partial characterization of bovine IL2 receptors. *Eur. J. Immunol.* 19:1339–42

90. Heussler VT, Eichhorn M, Reeves R, Magnuson NS, Williams RO, Dobbelaere DA. 1992. Constitutive IL-2 mRNA expression in lymphocytes, infected with the intracellular parasite *Theileria parva*. *J. Immunol.* 149:562–67

91. Heussler VT, Fernandez PC, Machado J Jr, Botteron C, Dobbelaere DAE. 1999. N-acetylcysteine blocks apoptosis induced by N-α-tosyl-L-phenylalanine chloromethyl ketone in transformed T-cells. *Cell Death Differ.* In press

91a. Heussler VT, Machado J Jr, Fernandez PC, Botteron C, Chen C-G, Pearse MJ, Dobbelaere DAE. 1999. The intracellular parasite *Theileria parva* protects T-cells from apoptosis. *Proc. Natl. Acad. Sci. USA.* In press

92. Huber BT, Hsu PN, Sutkowski N. 1996. Virus-encoded superantigens. *Microbiol. Rev.* 60:473–82

93. Hulliger L, Wilde JKH, Brown CGD, Turner L. 1964. Mode of multiplication of *Theileria* in cultures of bovine lymphocytic cells. *Nature, London* 203:728–30

94. Irvin AD. 1987. Characterization of species and strains of *Theileria*. *Adv. Parasitol.* 26:145–97

95. Irvin AD, Brown CG, Kanhai GK, Kimber CD, Crawford JG. 1974. Response of whole-body irradiated mice to inoculation of *Theileria parva*-infected bovine lymphoid cells. *J. Comp. Pathol.* 84:291–300

96. Irvin AD, Brown CGD, Kanhai GK, Stagg DA. 1975. Comparative growth of bovine lymphosarcoma cells and lymphoid cells infected with *Theileria parva* in athymic (nude) mice. *Nature* 255:713–14

97. Irvin AD, Morrison WI. 1987. Immunopathology, immunology and immunoprophylaxis of Theileria infection. In *Immune responsis in parasitic infections: Immunology, Immunopathology and Immunoprophylaxis*, ed. EJL Soulsby, III:223–74. Boca Raton FL: CRC

98. Irvin AD, Ocama JG, Spooner PR. 1982. Cycle of bovine lymphoblastoid cells parasitised by *Theileria parva*. *Res. Vet. Sci.* 33:298–304

99. Ivanov V, Stein B, Baumann I, Dobbelaere

DA, Herrlich P, Williams RO. 1989. Infection with the intracellular protozoan parasite *Theileria parva* induces constitutively high levels of NF-κB in bovine T lymphocytes. *Mol. Cell. Biol.* 9:4677–86

100. Izquierdo M, Bowden S, Cantrell D. 1994. The role of Raf-1 in the regulation of extracellular signal-regulated kinase 2 by the T cell antigen receptor. *J. Exp. Med.* 180:401–6

101. Karin M. 1998. The NF-κB activation pathway: its regulation and role in inflammation and cell survival. *Cancer J. Sci. Am.* 4:(Suppl. 1)S92–99

102. Karin M, Liu Z, Zandi E. 1997. AP-1 function and regulation. *Curr. Opin. Cell Biol.* 9:240–46

103. Kasibhatla S, Brunner T, Genestier L, Echeverri F, Mahboubi A, Green DR. 1998. DNA damaging agents induce expression of Fas ligand and subsequent apoptosis in T lymphocytes via the activation of NF-κB and AP-1. *Mol. Cell.* 1:543–51

104. King KL, Cidlowski JA. 1998. Cell cycle regulation and apoptosis. *Annu. Rev. Physiol.* 60:601–17

105. Klippel A, Escobedo MA, Wachowicz MS, Apell G, Brown TW, et al. 1998. Activation of phosphatidylinositol 3-kinase is sufficient for cell cycle entry and promotes cellular changes characteristic of oncogenic transformation. *Mol. Cell. Biol.* 18:5699–711

106. Lee FS, Hagler J, Chen ZJ, Maniatis T. 1997. Activation of the IκBα kinase complex by MEKK1, a kinase of the JNK pathway. *Cell* 88:213–22

107. Lee FS, Peters RT, Dang LC, Maniatis T. 1998. MEKK1 activates both IκB kinase α and IκB kinase β. *Proc. Natl. Acad. Sci. USA* 95:9319–24

108. MacMicking J, Xie QW, Nathan C. 1997. Nitric oxide and macrophage function. *Annu. Rev. Immunol.* 15:323–50

109. Malmquist WA, Brown CGD. 1974. Establishment of *Theileria parva*-infected

lymphoblastoid cell lines using homologous feeder layers. *Res. Vet. Sci.* 16:134–35

110. Malmquist WA, Nyindo MBA, Brown CGD. 1970. East Coast fever: cultivation *in vitro* of bovine spleen cell lines infected and transformed by *Theileria parva*. *Trop. Anim. Health Prod.* 2:139–45

111. Martin TFJ. 1998. Phosphoinositide lipids as signaling molecules: common themes for signal transduction, cytoskeletal regulation, and membrane trafficking. *Annu. Rev. Cell Dev. Biol.* 14:231–64

112. Mayo MW, Wang CY, Cogswell PC, Rogers Graham KS, Lowe SW, et al. 1997. Requirement of NF-κB activation to suppress p53-independent apoptosis induced by oncogenic Ras. *Science* 278:1812–15

113. McHardy N, Wekesa LS, Hudson AT, Randall AW. 1985. Anti-theilerial activity of BW720C (buparvaquone): a comparison with parvaquone. *Res. Vet. Sci.* 39:29–33

114. McKeever DJ, Nyanjui JK, Ballingall KT. 1997. *In vitro* infection with *Theileria parva* is associated with IL10 expression in all bovine lymphocyte lineages. *Parasite Immunol.* 19:319–24

115. Mehlhorn H, Shein E. 1984. The piroplasms: life cycle and sexual stages. *Adv. Parasitol.* 23:37–103

116. Montaner S, Perona R, Saniger L, Lacal JC. 1998. Multiple signalling pathways lead to the activation of the nuclear factor-κB by the Rho family of GTPases. *J. Biol. Chem.* 273:12779–85

117. Mori N, Gill PS, Mougdil T, Murakami S, Eto S, Prager D. 1996. Interleukin-10 gene expression in adult T-cell leukemia. *Blood* 88:1035–45

118. Morrison WI, Buscher G, Murray M, Emery DL, Masake RA, et al. 1981. *Theileria parva*: kinetics of infection in the lymphoid system of cattle. *Exp. Parasitol.* 52:248–60

119. Morrison WI, Goddeeris BM, Brown

WC, Baldwin CL, Teale AJ. 1989. *Theileria parva* in cattle: characterization of infected lymphocytes and the immune responses they provoke. *Vet. Immunol. Immunopathol.* 20:213–37

120. Morrison WI, Lalor PA, Goddeeris BM, Teale AJ. 1986. Theileriosis: antigens and host-parasite interactions. In *Parasite Antigens, Towards New Strategies for Vaccines*, ed. TW Pearson, pp. 167–213. New York: Marcel Dekker

121. Morrison WI, MacHugh ND, Lalor PA. 1996. Pathogenicity of *Theileria parva* is influenced by the host cell type infected by the parasite. *Infect. Immun.* 64:557–62

122. Muganda PM, Fischer A, Bernal RA. 1995. Identification of a casein kinase activity found elevated in human cytomegalovirus transformed cells. *Biochem. Biophys. Res. Commun.* 207: 740–46

123. Musoke A, Morzaria S, Nkonge C, Jones E, Nene V. 1992. A recombinant sporozoite surface antigen of *Theileria parva* induces protection in cattle. *Proc. Natl. Acad. Sci. USA* 89:514–18

124. Noel PJ, Boise LH, Thompson CB. 1996. Regulation of T cell activation by CD28 and CTLA4. *Adv. Exp. Med. Biol.* 406:209–17

125. Norval RAI, Perry BD, Young AS. 1992. *The Epidemiology of Theileriosis in Africa*. New York: Academic

126. ole MoiYoi OK. 1989. *Theileria parva*: an intracellular protozoan parasite that induces reversible lymphocyte transformation. *Exp. Parasitol.* 69:204–10

127. ole MoiYoi OK. 1995. Casein kinase II in theileriosis. *Science* 267:834–36

128. ole MoiYoi OK, Brown WC, Iams KP, Nayar A, Tsukamoto T, Macklin MD. 1993. Evidence for the induction of casein kinase II in bovine lymphocytes transformed by the intracellular protozoan parasite *Theileria parva*. *EMBO J.* 12:1621–31

129. ole MoiYoi OK, Sugimoto C, Conrad PA,

Macklin MD. 1992. Cloning and characterization of the casein kinase II α subunit gene from the lymphocyte-transforming intracellular protozoan parasite *Theileria parva*. *Biochemistry* 31:6193–202

130. Palmer GH, Machado JJ, Fernandez P, Heussler V, Perinat T, Dobbelaere DA. 1997. Parasite-mediated nuclear factor-κB regulation in lymphoproliferation caused by *Theileria parva* infection. *Proc. Natl. Acad. Sci. USA* 94:12527–32

131. Parry RV, Reif K, Smith G, Sansom DM, Hemmings BA, Ward SG. 1997. Ligation of the T cell co-stimulatory receptor CD28 activates the serine-threonine protein kinase protein kinase B. *Eur. J. Immunol.* 27:2495–501

132. Pearson TW, Lundin LB, Dolan TT, Stagg DA. 1979. Cell-mediated immunity to *Theileria*-transformed cell lines. *Nature* 281:678–80

133. Perona R, Montaner S, Saniger L, Sanchez Perez I, Bravo R, Lacal JC. 1997. Activation of the nuclear factor-κB by Rho, CDC42, and Rac-1 proteins. *Genes Dev.* 11:463–75

134. Phillips RJ, Ghosh S. 1997. Regulation of IκBβ in WEHI 231 mature B cells. *Mol. Cell Biol.* 17:4390–96

135. Pinder M, Kar S, Withey KS, Lundin LB, Roelants GE. 1981. Proliferation and lymphocyte stimulatory capacity of *Theileria*-infected lymphoblastoid cells before and after the elimination of intracellular parasites. *Immunology* 44:51–60

136. Pinder M, Withey KS, Roelants GE. 1981. *Theileria parva* parasites transform a subpopulation of T lymphocytes. *J. Immunol.* 127:389–90

137. Pinna LA, Meggio F. 1997. Protein kinase CK2 ("casein kinase-2") and its implication in cell division and proliferation. *Prog. Cell Cycle Res.* 3:77–97

138. Preston PM, Brown CGD, Richardson W. 1992. Cytokines inhibit the development of trophozoite-infected cells of *Theileria*

annulata and *Theileria parva* but enhance the proliferation of macroschizont-infected cell lines. *Parasite Immunol.* 14:125–41

139. Richardson JO, Forsyth LM, Brown CGD, Preston PM. 1998. Nitric oxide causes the macroschizonts of *Theileria annulata* to disappear and host cells to become apoptotic. *Vet. Res. Commun.* 22:31–45

140. Rintelen M, Schein E, Ahmed JS. 1990. Buparvaquone but not cyclosporin A prevents *Theileria annulata*-infected bovine lymphoblastoid cells from stimulating uninfected lymphocytes. *Trop. Med. Parasitol.* 41:203–07

141. Robinson MJ, Cobb MH. 1997. Mitogen-activated protein kinase pathways. *Curr. Opin. Cell Biol.* 9:180–86

142. Romano MF, Lamberti A, Petrella A, Bisogni R, Tassone PF, et al. 1996. IL-10 inhibits nuclear factor-κB/Rel nuclear activity in CD3-stimulated human peripheral T lymphocytes. *J. Immunol.* 156:2119–23

143. Rothwarf DM, Zandi E, Natoli G, Karin M. 1998. IKK-γ is an essential regulatory subunit of the IκB kinase complex. *Nature* 395:297–300

144. Sager H, Bertoni G, Jungi TW. 1998. Differences between B cell and macrophage transformation by the bovine parasite, *Theileria annulata*: a clonal approach. *J. Immunol.* 161:335–41

145. Sager H, Brunschwiler C, Jungi TW. 1998. Interferon production by *Theileria annulata*-transformed cell lines is restricted to the β family. *Parasite Immunol.* 20:175–82

146. Sager H, Davis WC, Dobbelaere DA, Jungi TW. 1997. Macrophage-parasite relationship in theileriosis. Reversible phenotypic and functional differentiation of macrophages infected with *Theileria annulata*. *J. Leukocyte Biol.* 61:459–68

147. Scheidereit C. 1998. Signal transduction. Docking IκB kinases. *Nature* 395:225–26

148. Seldin DC, Leder P. 1995. Casein kinase II α transgene-induced murine lymphoma: relation to theileriosis in cattle. *Science* 267:894–97

149. Sha WC. 1998. Regulation of immune responses by NF-κB/Rel transcription factors. *J. Exp. Med.* 187:143–46

150. Shaw MK. 1995. Mobilization of intrasporozoite Ca2+ is essential for *Theileria parva* sporozoite invasion of bovine lymphocytes. *Eur. J. Cell Biol.* 68:78–87

151. Shaw MK. 1996. Characterization of the parasite-host cell interactions involved in *Theileria parva* sporozoite invasion of bovine lymphocytes. *Parasitology* 113:267–77

152. Shaw MK. 1996. *Theileria parva* sporozoite entry into bovine lymphocytes involves both parasite and host cell signal transduction processes. *Exp. Parasitol.* 84:344–54

153. Shaw MK. 1997. The same but different: the biology of *Theileria* sporozoite entry into bovine cells. *Int. J. Parasitol.* 27:457–74

154. Shaw MK, Tilney LG, McKeever DJ. 1993. Tick salivary gland extract and interleukin-2 stimulation enhance susceptibility of lymphocytes to infection by *Theileria parva* sporozoites. *Infect. Immun.* 61:1486–95

155. Shaw MK, Tilney LG, Musoke AJ. 1991. The entry of *Theileria parva* sporozoites into bovine lymphocytes: evidence for MHC class I involvement. *J. Cell Biol.* 113:87–101

156. Shaw MK, Tilney LG, Musoke AJ, Teale AJ. 1995. MHC class I molecules are an essential cell surface component involved in *Theileria parva* sporozoite binding to bovine lymphocytes. *J. Cell Sci.* 108:1587–96

157. Shayan P, Ahmed JS. 1997. *Theileria*-mediated constitutive expression of the casein kinase II-α subunit in bovine lymphoblastoid cells. *Parasitol. Res.* 83:526–32

158. Shayan P, Schop B, Conze G, Schein E, Ahmed JS. 1999. Is interleukin-2 necessary for the autocrine proliferation of *Theileria*-infected bovine cells? *Parasitol. Res.* 85:409–12

159. Siebenlist U. 1997. NF-κB/IκB proteins. Their role in cell growth, differentiation and development. Madrid, Spain, July 7–10, 1996. *Biochim. Biophys. Acta* 1332:R7–13

160. Somerville RP, Adamson RE, Brown CG, Hall FR. 1998. Metastasis of *Theileria annulata* macroschizont-infected cells in SCID mice is mediated by matrix metalloproteinases. *Parasitology* 116:223–28

161. Song S, Ling Hu H, Roebuck KA, Rabbi MF, Donnelly RP, Finnegan A. 1997. Interleukin-10 inhibits interferon-γ-induced intercellular adhesion molecule-1 gene transcription in human monocytes. *Blood* 89:4461–69

162. Spooner RL, Innes EA, Glass EJ, Brown CGD. 1989. *Theileria annulata* and *T. parva* infect and transform different bovine mononuclear cells. *Immunology* 66:284–88

163. Spooner RL, Innes EA, Glass EJ, Millar P, Brown CGD. 1988. Bovine mononuclear cell lines transformed by *Theileria parva* or *Theileria annulata* express different subpopulation markers. *Parasite Immunol.* 10:619–29

164. Stagg DA, Chasey D, Young AS, Morzaria SP, Dolan TT. 1980. Synchronization of the division of *Theileria* macroschizonts and their mammalian host cells. *Ann. Trop. Med. Parasitol.* 74:263–65

165. Stancovski I, Baltimore D. 1997. NF-κB activation: the IκB kinase revealed? *Cell* 91:299–302

166. Sutkowski N, Palkama T, Ciurli C, Sekaly RP, Thorley-Lawson DA, Huber BT. 1996. An Epstein-Barr virus-associated superantigen. *J. Exp. Med.* 184:971–80

167. Suyang H, Phillips R, Douglas I, Ghosh S. 1996. Role of unphosphorylated, newly synthesized IκBβ in persistent activation of NF-κB. *Mol. Cell. Biol.* 16:5444–49

168. Syfrig J, Wells C, Daubenberger C, Musoke AJ, Naessens J. 1998. Proteolytic cleavage of surface proteins enhances susceptibility of lymphocytes to invasion by *Theileria parva* sporozoites. *Eur. J. Cell Biol.* 76:125–32

169. Sylla BS, Hung SC, Davidson DM, Hatzivassiliou E, Malinin NL, et al. 1998. Epstein-Barr virus-transforming protein latent infection membrane protein 1 activates transcription factor NF-κB through a pathway that includes the NF-κB-inducing kinase and the IκB kinases IKKα and IKKβ. *Proc. Natl. Acad. Sci. USA* 95:10106–11

170. Taga K, Cherney B, Tosato G. 1993. IL-10 inhibits apoptotic cell death in human T cells starved of IL-2. *Int. Immunol.* 5:1599–608

171. Telerman A, Amson RB, Romasco F, Wybran J, Galand P, Mosselmans R. 1987. Internalization of human T lymphocyte receptors. *Eur. J. Immunol.* 17:991–97

172. Theiler A. 1904. Rhodesian Tick fever. *Transvaal Agric. J.* 2:421–28

173. Thomas SM, Brugge JS. 1997. Cellular functions regulated by Src family kinases. *Annu. Rev. Cell. Dev. Biol.* 13:513–609

174. Tilney LG, Tilney MS, Shaw MK. 1994. *Theileria*: strategy of infection and survival. In *Baillières Clinical Infectious Diseases*. London: Bailière Tindall

175. Titus RG, Ribeiro JM. 1988. Salivary gland lysates from the sand fly *Lutzomyia longipalpis* enhance *Leishmania* infectivity. *Science* 239:1306–8

176. Tordai A, Franklin RA, Patel H, Gardner AM, Johnson GL, Gelfand EW. 1994. Cross-linking of surface IgM stimulates the Ras/Raf-1/MEK/MAPK cascade in human B lymphocytes. *J. Biol. Chem.* 269:7538–43

177. Toye PG, Goddeeris BM, Iams K,

Musoke AJ, Morrison WI. 1991. Characterization of a polymorphic immunodominant molecule in sporozoites and schizonts of *Theileria parva*. *Parasite Immunol.* 13:49–62

178. Tsur I, Adler S. 1963. Growth of *Theileria annulata* schizonts in monolayer tissue cultures. *J. Protozool. Suppl.* 10:36

179. Tsur I, Neitz WO, Pols JW. 1957. The development of Koch bodies of *Theileria parva* in tissue cultures. *Refuah Vet.* 14:53

180. Uhlik M, Good L, Xiao G, Harhaj EW, Zandi E, et al. 1998. NF-κB-inducing kinase and IκB kinase participate in human T-cell leukemia virus I Tax-mediated NF-κB activation. *J. Biol. Chem.* 273: 21132–36

181. Uilenberg G. 1997. General review of tick-borne diseases of sheep and goats world-wide. *Parassitologia* 39:161–65

182. van Dam H, Huguier S, Kooistra K, Baguet J, Vial E, et al. 1998. Autocrine growth and anchorage independence: two complementing Jun-controlled genetic programs of cellular transformation. *Genes Dev.* 12:1227–39

183. Vickerman K, Irvin AD. 1981. Association of the parasite of East Coast fever (*Theileria parva*) with spindle microtubules of the bovine lymphoblast. *Trans. R. Soc. Trop. Med. Hyg.* 75:329

183a. Visser AE, Abraham A, Sakyi LJ, Brown CG, Preston PM. 1995. Nitric oxide inhibits establishment of macroschizont-infected cell lines and is produced by macrophages of calves undergoing bovine tropical theileriosis or East Coast fever. *Parasite Immunol.* 17:91–102

184. Ward SG, June CH, Olive D. 1996. PI3-kinase: a pivotal pathway in T-cell activation? *Immunol. Today* 17:187–97

185. Webster P, Dobbelaere DA, Fawcett DW. 1985. The entry of sporozoites of *Theileria parva* into bovine lymphocytes *in vitro*. Immunoelectron microscopic observations. *Eur. J. Cell. Biol.* 36:157–62

186. Wells CW, McKeever DJ. 1998. The entry and development of *Theileria parva* sporozoites within bovine dendritic cells from the afferent lymph. *Proc. Conf. Electron Microsc., Dublin, 1996, Comm. Eur. Soc. Microsc.*, Vol. 3, *Biology*, pp. 597–98. Brussels: Comm. Eur. Soc. Microsc.

187. Whitehurst CE, Geppert TD. 1996. MEK1 and the extracellular signal-regulated kinases are required for the stimulation of IL-2 gene transcription in T cells. *J. Immunol.* 156:1020–29

188. Wolf SF, Sieburth D, Sypek J. 1994. Interleukin 12: a key modulator of immune function. *Stem Cells (Dayt)* 12: 154–68

189. Wu M, Arsura M, Bellas RE, FitzGerald MJ, Lee HY, et al. 1996. Inhibition of c-myc expression induces apoptosis of WEHI 231 murine B cells. *Mol. Cell. Biol.* 16:5015–25

190. Wu M, Lee HY, Bellas RE, Schauer SL, Arsura M, et al. 1996. Inhibition of NF-κB/Rel induces apoptosis of murine B cells. *EMBO J.* 15:4682–90

191. Xu X, Heidenreich O, Kitajima I, McGuire K, Li QH, et al. 1996. Constitutively activated JNK is associated with HTLV-1 mediated tumorigenesis. *Oncogene* 13:135–42

192. Yamaoka S, Courtois G, Bessia C, Whiteside ST, Weil R, et al. 1998. Complementation cloning of NEMO, a component of the IκB kinase complex essential for NF-κB activation. *Cell* 93:1231–40

193. Yin MJ, Christerson LB, Yamamoto Y, Kwak YT, Xu S, et al. 1998. HTLV-I Tax protein binds to MEKK1 to stimulate IκB kinase activity and NF-κB activation. *Cell* 93:875–84

194. Yokoo T, Kitamura M. 1996. Dual regulation of IL-1β-mediated matrix metalloproteinase-9 expression in mesangial cells by NF-κB and AP-1. *Am. J. Physiol.* 270:F123–30

Annu. Rev. Microbiol. 1999. 53:43–70

ADDICTION MODULES AND PROGRAMMED CELL DEATH AND ANTIDEATH IN BACTERIAL CULTURES

Hanna Engelberg-Kulka[1] and Gad Glaser[2]

Departments of [1]Molecular Biology and [2]Cellular Biochemistry, The Hebrew University Hadassah-Medical School, Jerusalem 91010, Israel; e-mail: hanita@cc.huji.ac.il

Key Words apoptosis, plasmids, postsegregational killing, protein degradation, bacteriophage λ

For we are merely the leaf and the husk
The great death, contained in each of us,
That is the fruit around which everything revolves.

R. M. Rilke
The Book of Hours (1902)

■ **Abstract** In bacteria, programmed cell death is mediated through "addiction modules" consisting of two genes. The product of the second gene is a stable toxin, whereas the product of the first is a labile antitoxin. Here we extensively review what is known about those modules that are borne by one of a number of *Escherichia coli* extrachromosomal elements and are responsible for the postsegregational killing effect. We focus on a recently discovered chromosomally borne regulatable addiction module in *E. coli* that responds to nutritional stress and also on an antideath gene of the *E. coli* bacteriophage λ. We consider the relation of these two to programmed cell death and antideath in bacterial cultures. Finally, we discuss the similarities between basic features of programmed cell death and antideath in both prokaryotes and eukaryotes and the possibility that they share a common evolutionary origin.

CONTENTS

0066-4227/99/1001-0043$08.00

43

INTRODUCTION

Programmed cell death, defined as an active process that results in cell suicide, is recognized as an essential mechanism in multicellular organisms. Generally, programmed cell death is required for the elimination of superfluous or potentially harmful cells (for reviews, see 49, 79). In eukaryotes, programmed cell death is classically known as apoptosis (55), a term that originally defined the morphological changes that characterize cell death. Today, the phrase "programmed cell death" has evolved to refer to any form of cell death mediated by an intracellular death program, no matter what triggers it and whether or not it displays all of the characteristic features of apoptosis (for review, see 49).

In bacteria, programmed cell death is mediated through a unique genetic system. It consists of a pair of genes that specify for two components, a stable toxin and an unstable antitoxin that prevents the lethal action of the toxin. Until recently, such genetic systems for bacterial programmed cell death have been found mainly in *Escherichia coli* on low–copy-number plasmids and are responsible for what is called the postsegregational killing effect; that is, they are responsible for the death of plasmid-free cells. When bacteria lose the plasmid(s) (or other extrachromosomal elements), the cured cells are selectively killed because the unstable antitoxin is degraded faster than the more stable toxin. Yarmolinsky and colleagues have called such plasmid-borne pairs of genes "addiction modules," because they cause the bacterial host to be addicted to the continued presence of the "dispensable" genetic element (60, 121). Addiction modules are responsible for the lethal consequences of plasmid withdrawal. Along with other very precise mechanisms for preventing plasmid loss (replication control, plasmid partition, and resolution of multimers) (for reviews, see 48, 80, and 117), the stability of low–copy-number plasmids in the host bacterial cells is maintained by the mechanism for killing plasmid-free bacteria provided by the addiction modules.

Two different classes of addiction modules have been identified in bacteria: (*a*) systems in which both products, the stable toxins and the unstable antidotes, are proteins [named by Jensen & Gerdes proteic killer gene systems (51)] and (*b*) systems in which, again, the stable toxin is a protein synthesized from a stable mRNA but the antidotes are small unstable antisense RNA molecules (37). In cells harboring plasmids bearing such addiction modules, the antisense RNAs prevent the translation of the stable toxin-encoding mRNAs. However, in plasmid-free cells, the unstable antisense RNA is degraded, allowing the translation of the toxins and the subsequent death of the plasmid-free cell (105).

Here we focus on proteic addiction modules. We do not review the addiction modules specifying for antitoxins that are antisense RNAs (belonging to the *hok/sok* family) because they have recently been extensively reviewed elsewhere (36). Among the proteic addiction modules we discuss are the best character-ized systems of extrachromosomal elements including *ccdAB* of plasmid F, *kis/kid* of plasmid R1, *pemI/K* of plasmid R100, *parDE* of RK2/RP4, and *phd-doc* of prophage P1 (Table 1). This topic has been partially reviewed by Jensen & Gerdes (51) and by Couturier and colleagues (20). Here we particularly focus on a recently discovered regulatable addiction module located on the *E. coli* chro-mosome, on an antideath gene of bacteriophage λ, and on their relation to pro-grammed cell death and antideath in bacterial cultures. Finally, we discuss the similarities of programmed cell death and antideath in prokaryotes and eukaryotes and the possibility that they share a common evolutionary origin.

PROTEIC ADDICTION MODULES: Definition and General Properties

The best characterized proteic addiction modules have striking organizational and functional parallels. These include (Figure 1, Table 1): (*a*) a proteic addiction module harbors two adjacent genes; (*b*) the product of one is a long-lived and toxic protein, whereas the product of the second is a short-lived protein that antagonizes the toxic effect of the first; (*c*) the antitoxic protein is encoded by the upstream gene in the module; (*d*) the toxic and antitoxic proteins are coexpressed; (*e*) the antitoxic protein is synthesized in excess; (*f*) the toxic and antitoxic proteins are small (toxic proteins are in the range of 100–130 amino acids, and antitoxic proteins are in the range of 70–85 amino acids); (*g*) the toxic and antitoxic proteins interact; (*h*) the antitoxic protein is degraded by a specific bacterial protease; and (*i*) the addiction module is autoregulated at the level of transcription either by a complex formed between the toxin and antitoxin or by the antitoxin alone. Located on various extrachromosomal elements or on the *E. coli* chromosome, the proteic addiction modules are quite similar in genetic structure and function; however, they rarely share sequence homology. In addition, these addiction modules also differ in the natures of their toxic and antitoxic proteins, in the bacterial protease that degrades the antitoxic protein, and in the cellular targets of the toxic proteins.

The *ccdAB* Addiction Module of Plasmid F

The first system in which a genetic element was found to be responsible for killing plasmid-free segregants was the *ccd* locus of plasmid F (50, 81). The 95-kb con-jugated plasmid F has a very low copy number and is found in the cell at about one copy per chromosome. Originally, *ccd* stood for couples cell division, that is, a system coupling cell division to plasmid proliferation, thereby acting as a plasmid rescue system (73, 75, 81). Today, *ccd* stands for control cell death (50). Of all the proteic addiction modules, the *ccd* locus has been the one most studied, and it can be considered as a paradigm. The *ccd* locus consists of two genes, *ccdA* and *ccdB* (also known as *H* and *G* or *letA* and *letB*), which encode the 72-amino-acid-

TABLE 1 The most studied proteic *Escherichia coli* "addiction modules"[a]

Name of addiction module (borne on...)	Other nomenclatures used	Name of genetic locus	Antitoxin (no. of amino acids)	Toxin (no. of amino acids)	Cellular target of toxin	Protease involved in degradation of antitoxins	Autoregulation: proteins involved *in vivo*
ccdAB (of F plasmid)	*H/G; letAB* (of F)	*ccd*(F)	CcdA/LetA (72aa)	CcdB/LetB (101aa)	Gyrase	Lon	CcdA + CcdB
pemIK (of plasmid R100)	*kis/kid* (of R1)	*pem* (R100) or *parD* (R1)	PemI/Kis (84aa)	PemK/Kid (110aa)	DnaB	Lon	PemI + PemK
parDE (of plasmid RK2/RP4)	—	*par* (RK2)	ParD (83aa)	ParE (103aa)	Unknown	Unknown	ParD
phd-doc (of P1 plasmid)	—	—	Phd (73aa)	Doc (126aa)	Unknown (*)	ClpPX	Phd + Doc
mazEF (of *E. coli*)	*chpAIK*	*maz* or *chpA* (on *E. coli* chromosome)	MazE (82aa)	MazF (111aa)	Unknown	ClpPA	MazE + MazF

*See text

[a] —, None.

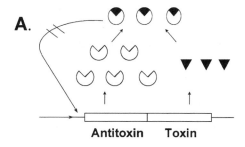

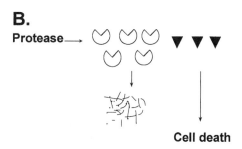

Figure 1 Schematic illustration of the general characteristics of proteic "addiction modules" and the fate of their products. Addiction modules consist of two adjacent genes that are coexpressed. These two genes specify for a stable toxic protein (▼) and for a labile antitoxic protein (♡). In each genetic module, the upstream gene specifies for the antitoxin and the downstream gene specifies for the toxin. (*A*) Under conditions of continuous expression of the addiction module. Both products are synthesized. The antitoxins form complexes with the toxins, thereby neutralizing them to prevent cell killing. In all known cases, the addiction module is negatively autoregulated by the toxin-antitoxin complex at the level of transcription. (*B*) Under conditions in which expression of the addiction module is prevented. For a plasmid-borne module, this can occur by the loss of the plasmid itself and hence of the module; for a chromosomal addiction system, this can occur by the action of the regulatory element affecting its expression. The toxin and antitoxin molecules, synthesized before their de novo synthesis was prevented, have a different fate: the antitoxins are degraded by specific proteases, leaving the toxins free to cause cell death.

long 8.7-kDa protein CcdA and the 101-amino-acid-long 11.7-kDa protein CcdB (14, 75). These two proteins are involved in the toxic-antitoxic mechanism that enables plasmid F to be maintained stably in the cell; CcdB is toxic to the cell, and CcdA is CcdB's unstable antidote (54, 73, 75). The 41 carboxy-terminal residues of CcdA are sufficient for its antitoxic activity (11). The F plasmid contains two additional operons, *srnB* (<u>s</u>tabile <u>RN</u>A degradation) (3) and *flm* (<u>F</u> leading <u>m</u>aintenance) (62), which function independently as postsegregational killing systems. These two killing systems are addiction modules in which the antidotes are antisense RNAs (36). When present on an intact F plasmid, *ccdAB* plays a relatively minor role in postsegregational killing; however, when present on a mini-F plasmid or when cloned with a heterologous replicon, the presence of *ccdAB* results in the killing of >90% of the plasmid-free segregants (50, 73, 75). In the absence of CcdA, the production of the CcdB protein causes cell filamentation, induction of the SOS pathway, and, ultimately, cell death (50, 54, 75). Similar responses were observed when the synchronous loss of a *ccdAB*-bearing plasmid was induced (77, 100). The *ccd*-induced SOS response, but not cell killing, requires the presence of the host enzymes RecA and RecBC (6). CcdA probably prevents the lethal action of CcdB by binding to it, thus forming a tight complex (104).

Studies on the *ccd* operon revealed the finding that it is negatively autoregulated at the level of transcription by a complex of the antitoxic and the toxic proteins. This has become one of the characteristics of the proteic addiction module (Table 1). Both CcdA and CcdB proteins are required for repression and binding to the operator(s) in the *ccd* promoter (24, 103, 104). CcdA and CcdB bind at several sites spaced over 113 bp overlapping the *ccd* promoter (24, 93, 103, 104).

In bacteria that have lost the F plasmid, it has been proposed that cell death is brought about by the differential loss of activity of the CcdA and CcdB proteins. Because the active half-life of CcdA is shorter than that of CcdB, in newborn plasmid-free bacteria the persistence of the toxic CcdB protein would lead to cell death (50). Later it was confirmed that cell killing in bacteria which have lost the F plasmid is indeed based on the relative instability of the CcdA protein, which has been shown both in vivo and in vitro to be degraded by the *E. coli* Lon protease and has a shorter lifetime than does CcdB (115, 116). Lon is a multimer of identical subunits that represents a major class of ATP-dependent proteases in which the ATPase domain and proteolytic domain are encoded within a single polypeptide (40). CcdB, the toxic partner, prevents degradation of CcdA by Lon. Since CcdB also inhibits the ability of CcdA to enhance the ATPase activity of Lon, it may be that Lon recognizes protein-bonding domains that become exposed when their partner is absent (116). Lon-dependent degradation of CcdA is relatively slow, like the degradation of the antitoxic proteins of other addiction modules by their respective proteases (115).

The best characterized cellular target of the toxic component of an addiction module is that of the *ccd* system of plasmid F; this target consists of the A subunits of the *E. coli* DNA gyrase (GyrA). This was first shown by genetic analysis of *E. coli* mutants resistant to the killing effect of CcdB (12, 74). Bernard & Couturier (12) found that in seven independent isolates the mutation mapped in the *gyrA* gene. Sequencing one of these GyrA mutants revealed an amino acid substitution of Arg462 to Cys. The fact that all of the independently isolated CcdB-resistant mutants all map to *gyrA* is strong evidence that *gyrA* is the target of the CcdB protein. In that same study, in a merodiploid strain, the CcdB-sensitive phenotype was found to be dominant over the resistant phenotype. This dominance of sensitivity over resistance has also been observed with quinolone drugs (45, 70) and indicated that rather than being a simple inhibitor of gyrase function (12), CcdB may poison wild-type gyrase. In a separate study, Miki and colleagues (74) isolated nine CcdB-resistant mutants and showed that three of them map in *gyrA* and the other six map to the *groE* genes. This suggested that the GroES chaperone may be involved in the interaction between CcdB and DNA gyrase or in CcdB folding. More recently, other *E. coli* mutants have been isolated that can survive low concentrations of CcdB (78). The relation of the mutated genes to DNA gyrase, CcdB toxicity, or both is not yet clear.

That the GyrA subunit of the *E. coli* DNA gyrase is the target of the CcdB protein has also been revealed by biochemical studies (12, 13, 66, 67). The *E. coli* DNA gyrase is a tetramer formed by the association of two GyrA and two GyrB subunits (A_2B_2). This tetramer catalyzes negative supercoiling at the expense of ATP hydrolysis (35). The GyrB subunits are responsible for ATP binding and hydrolysis. The GyrA subunits form the catalytic core of the enzyme that enables the DNA breaking-rejoining reaction, that is, the introduction of a transient double-strand break in the DNA, the passage of another piece of DNA through the break, and the annealing of the double strands (10, 88). The intermediates of this breaking-rejoining reaction are called "cleaved complexes." Maki and colleagues (67) studied the supercoiling activity of DNA gyrase and have shown that in cells overproducing CcdB both the free form of the GyrA subunit and the tetrameric form A_2B_2 of DNA gyrase are inactivated. This inactivation seems to be caused by CcdB protein binding to GyrA. Furthermore, protein CcdA is able to fully reactivate the inactivated gyrase or the GyrA subunit. Since protein CcdB and quinolone antibiotics seemed to poison DNA gyrase similarly, Bernard & Couturier (12) measured DNA gyrase cleavage of plasmid DNA in CcdB-overproducing cells. They found that in such cells, plasmid DNA was only partially cleaved. As in the case of quinolone drugs, the cleavage by DNA gyrase in CcdB-overproducing cells was observed only when the cells were treated with the strong protein denaturant SDS. Furthermore, overproduction of the CcdB protein in a *gyrA462* strain that tolerates the CcdB killing effect did not lead to CcdB-induced DNA cleavage. Moreover, in in vitro studies, they observed that purified CcdB, like quinolone antibiotics, induces DNA cleavage by DNA gyrase and furthermore that CcdA reverses this effect (13). While all these findings support the notion that GyrA protein is the target of CcdB, they do not clarify the mechanism(s) of CcdB action on gyrase. In particular, one question remains: under biological physiological conditions, what is the primary cause of cell killing by CcdB?

As described above, until recently, CcdB had been reported to act on DNA gyrase in two distinct modes. According to one, CcdB inactivates DNA gyrase by forming a CcdB-DNA gyrase complex, leading to the relaxation of supercoiled DNA (66, 67). According to the other, like quinolone drugs, CcdB poisons gyrase by freezing an intermediate step in the breaking-rejoining reaction, which results in double-stranded DNA cleavage in the presence of SDS (12, 13). Recently, in more detailed studies of CcdB-induced DNA cleavage, purified CcdB was shown not to affect supercoiling (95). In fact, most CcdB-induced cleavage occurred after many cycles of ATP-driven breakage and reunion when the DNA had become highly supercoiled (95). Furthermore, CcdB was found to stabilize a cleaved complex of DNA gyrase and DNA but not in the same manner as do the quinolone drugs (21, 95). For example, it was shown (21) that, although quinolone drugs can induce cleavage in relatively short DNA molecules, DNA cleavage by CcdB requires a DNA molecule at least ∼160 bp long. They also found that when linear DNA is the substrate, CcdB cleavage of DNA requires ATP hydrolysis. This requirement for ATP hydrolysis suggested the involvement of a strand passage event, so they proposed that CcdB and quinolones affect different intermediate steps of the cleaved complexes; while quinolone drugs can trap a cleaved complex without the involvement of strand passage, CcdB traps a post–strand-passage intermediate. On the other hand, like the gyrase-quinolone-DNA complex (118), the CcdB-gyrase-DNA complex can also inhibit the passage of RNA polymerases (21). This was shown by using an in vitro transcription assay in which the CcdB-gyrase-DNA complex was found to block the transcription of the T7 polymerase. An important feature of this process is the finding that it also requires ATP. The fact that CcdA, the antidote of CcdB, prevented CcdB-induced blocking of RNA polymerase further suggests that these in vitro results are a correct reflection of at least part of CcdB action in vivo.

Based on the results of earlier genetic and biochemical studies, it appears that a crucial role in the CcdB-GyrA interaction is played by GyrA Arg462 and by the last three amino acids of the CcdB C-terminus. The crystal structure of a large fragment of GyrA revealed that Arg462 points into the central hole of the GyrA dimer (76), suggesting that CcdB binds into this hole. Recently, the crystal structure of CcdB has been determined, confirming that CcdB also exists in dimer form (63). However, based on the crystal structure of GyrA (76), the diameter of the central hole of the GyrA dimer is a little too small; to accommodate the CcdB dimer, the GyrA dimer must open up to some degree. To solve this problem, Couturier and colleagues (20) proposed two possible mechanisms: either (*a*) CcdB interacts with GyrA before the GyrA dimer is formed or (*b*) CcdB interacts with GyrA when gyrase is cycling on the DNA. When the crystal structure of the CcdB-GyrA complex is elucidated, we shall have a better understanding of the mode of action of CcdB.

It is not yet clear which of CcdB's biological effects is the primary cause of its cytotoxic effect under physiological conditions in vivo. As discussed above, three distinct phenomena have been described as related to the action of CcdB on DNA gyrase: the relaxation of negatively supercoiled DNA, DNA cleavage, and interference with the passage of RNA polymerases along the DNA. Based on their observation that plasmid DNA in CcdB-overproducing cells is extensively relaxed, Maki and colleagues (66, 67) suggested that in vivo CcdB modulates the supercoiling activity of DNA gyrase. Because CcdB was overexpressed in these experiments, it represented 20% of the total cell protein. Because, under normal in vivo conditions, such high concentrations of CcdB are unlikely, it is possible that normal physiological levels of CcdB have little effect on the cellular levels of supercoiling. In addition, it is well known that bacterial cells are also able to control cellular supercoiling levels by altering the expression of gyrase and topoisomerase I (71, 109). Thus, even if CcdB inhibits in vivo supercoiling by binding GyrA, bacteria may be able to compensate for small changes in the level of supercoiling by increasing the expression of the *gyrA* and *gyrB* genes.

Recall that DNA cleavage by CcdB was observed only when the strong protein-denaturing agent SDS was added to the medium. Thus it is also questionable whether DNA cleavage, the second phenomenon reported to be related to the mode of action of CcdB and GyrA, is responsible for CcdB-mediated cell killing in vivo (12, 13). However, it does seem that the additional ability of the CcdB-gyrase complex on DNA to form a barrier for the passage of RNA polymerase and possibly of DNA polymerase (21) may have implications for the bactericidal action of CcdB protein. It has been shown in vivo that CcdB can indeed inhibit DNA replication (50) and can also induce cell filamentation (75) and the formation of anucleate cells (45). Thus, CcdB may kill bacteria that have lost the F plasmid by trapping the DNA gyrase that is bound to the DNA, thus blocking the passage of polymerases.

In summary, several processes have been identified for the involvement of each of the toxic and antitoxic proteins of the *ccd* system. CcdB is involved in three processes: (*a*) it poisons the DNA gyrase complex, (*b*) it interacts with CcdA, and (*c*) it represses its own synthesis and that of CcdA by forming a CcdA-CcdB repressor complex that binds to the *ccd* promoter-operator. On the other hand, through their interaction, CcdA inactivates the toxic CcdB. In addition, CcdA is a substrate for the *E. coli* protease Lon. The domains involved in each process and the structure-function analysis of CcdA and CcdB have yet to be clarified. Based on mutational analysis, it appears that the last three amino acids of CcdB play a key role in the poisoning process but are not involved in its autoregulation (5). It has also been shown that a truncated CcdA protein retaining only its 41 C-terminal residues loses its autoregulatory activity but retains its antitoxic activity (11, 93). Thus, it seems that the autoregulatory activities of CcdB and CcdA reside in their N-terminal regions, while the toxic and antitoxic activities, respectively, reside in their C-terminal regions (Table 2).

The *kis/kid* of Plasmid R1 and pemI/K of Plasmid R100

The stable maintenance of the two closely related *inc*FII low–copy-number plasmids, plasmid R1 and plasmid R100, also involves addiction module-mediated mechanisms. Their addiction modules are located near the origin of replication of each of the plasmids and are called *parD* for plasmid R1 and *pem* for plasmid R100. According to our present knowledge, *parD* and *pem* are identical, but they were discovered independently. Diaz-Orejas and colleagues (15) discovered the *parD* locus that consists of the genes *kis/kid*, which stands for <u>ki</u>lling <u>s</u>uppression and <u>ki</u>lling <u>d</u>eterminant; Ohtsubo and colleagues (113) discovered the *pem* locus that

TABLE 2 The effect on the activities of the toxins and antitoxins of *Escherichia coli* addiction modules as a result of alterations at their C or N termini

	Toxicity	Antitoxicity	Autoregulation	References
Toxin CcdB wt	+		+	5
Toxin Ccd B Δ3 aa at C terminus	−		+	5
Antitoxin CcdA wt		+	−	11
Antitoxin CcdA Δ31 aa at N terminus (leaving 41 aa at C terminus)		+	−	11, 93
PemI wt		+	+	92
PemI (modified N terminus)		+	−	92

consists of *pemI/K*, which stands for plasmid emergency maintenance inhibitor or killing. These addiction systems specify for the 84-amino-acid, 9.3-kDa antitoxic protein called Kid or PemI and for the 110-amino-acid, 12-kDa toxic protein Kis or PemK (16, 113). For simplicity, we shall consistently use the name *pemI/K* for this system.

The module *pemI/K* maintains plasmid stabilization by killing plasmid-free segregants: The toxic PemK protein kills, and the antitoxic PemI protein neutralizes the lethal activity of PemK protein. A mutation in *pemI* was shown to cause cell death by the action of the *pemK* gene product (16). Cell death was also explained in the *pemI/K* system by the differential stabilities of PemK and PemI. Direct evidence for the preferential turnover of the PemI protein has been shown to be due to Lon-dependent proteolysis (110). This was the first reported case of an unstable proteic antitoxic partner in an addiction system.

pemI/K transcription and autoregulation were studied by Diaz Orejas and colleagues (90) and by Tsuchimoto & Ohtsubo (112). The *pemI/K* is transcribed as a small operon from a single promoter. In addition to the single polycistronic message coding for the two proteins of the system (PemI and PemK), there is a short transcript, coding only for the antitoxic protein PemI. *pemI/K* is negatively autoregulated at the level of transcription by the concerted action of both PemI and PemK proteins that bind cooperatively to the *pem* promoter region, probably as a complex, thus repressing their own synthesis. In addition, protein PemI by itself mediates a weak autoregulation of *pemI/K*.

Diaz Orejas and colleagues (92) have shown that DnaB is probably the cellular target of PemK. They have purified PemK and PemI as a C-LYT-PemI fusion protein. They have shown both in vivo and in vitro that PemK is a potent inhibitor of DNA replication that probably acts on protein DnaB at the level of initiation. They also showed that PemI protein can neutralize the inhibition of DNA replication mediated by PemK, which is parallel to the PemI neutralization of the killing activity of PemK. These results suggest that protein PemK is toxic because it inhibits DnaB-dependent DNA replication. This study also revealed that PemI and PemK interact by forming a tight complex. Such a physical interaction could explain the killing activity of PemK, how it is neutralized by PemI, and the coordinated action of the two proteins in the regulation of the *pem* promoter (see above). The C-LYT-PemI fusion protein, in which the N terminus of PemI is altered, is still able to interact with PemK and to conserve its antagonistic activity on the inhibition of DNA replication mediated by PemK. However, C-LYT-PemI does not act as a corepressor of the *pem* promoter. These findings indicate that the amino terminus of PemI is important for regulation but not for antitoxic activity (92; Table 2). A similar situation was reported for the antitoxic protein CcdA of the *ccd* system (11, 93). CcdA is significantly homologous in both its C-terminal and N-terminal domains to the corresponding regions of PemI (91). As in the case of PemI, the CcdA amino terminus region is involved in the autoregulation process, while it is not required for its antitoxic activity. Instead, the C-terminal domain of CcdA is required for its antitoxic activity (11, 93). Based on the homologies between CcdA and PemI, as well as on the functional domains required for the antitoxic and autoregulation activity of CcdA, it has been suggested that the carboxy-terminal domain of PemI is involved in neutralizing the killer protein PemK (92).

Studies on the effect of *pemI/K* on the stable maintenance of plasmid R100 in various *E. coli* strains revealed that the primary action of *pemK* gene product is to inhibit cell division of plasmid-free segregants but not to kill the segregants directly. On a secondary level, the inhibition of cell division leads to the death of the plasmid-free segregants in some strains but not in others (52, 111). According to this view, the primary effect of the *pem* system is inhibition of cell division rather than killing, and the inhibition of cell division confers a selective disadvantage to the cured cells (51).

The *parDE* of Plasmid RK2/RP4

The broad-host-range plasmid RP4 is indistinguishable from RK2 (17). RP4, which we will refer to as RK2, is a large 60-kb plasmid, which, depending on the bacterial host species, is estimated to be found in the cell at four to seven copies per chromosome. In spite of its relatively low copy number, this plasmid is stably inherited in almost all gram-negative bacteria (for review, see 106). RK2 has been shown to carry a region formed by the *par* locus (not to be confused with *parD* of plasmid R1; see Table 2). The *par* locus of RK2 can lead to plasmid stabilization in a replicon-independent manner in a variety of gram-negative bacteria (38, 84, 94). The RK2 *par* region encodes five genes organized in two divergently transcribed operons, *parCBA* and *parDE* (85). The two promoters, arranged back-to-back and located within the intergenic region between *parC* and *parD*, mediate transcription of these operons. Both *parCBA* and *parDE* are negatively autoregulated at the level of transcription by the gene products ParA and ParD, respectively (23, 28). Furthermore, both *parCBA* and *parDE* are involved in the extremely efficient plasmid stabilization mechanisms. The *parCBA* operon encodes a system for the resolution of plasmid multimers through site-specific recombination (29). Nevertheless, it seems that multimer resolution alone is unlikely to play a significant role in the stable maintenance by the *parCBA* operon. Instead, it has been proposed that *parCBA* encodes an additional stability mechanism, a partition system, which ensures that each daughter cell receives a plasmid copy at cell division. Further evidence (99) supports the involvement of such a mechanism. On the other hand, the *parDE* operon has been shown to contribute to plasmid stabilization via the mechanism of postsegregational killing of plasmid-free daughter cells (85).

parDE is an addiction module in which the 83-amino-acid 9-kDa ParD protein is the antitoxic component and the 103-amino-acid 12-kDa ParE protein is the toxic component; both of these proteins are required for plasmid stabilization (85). Cross-linking studies and a protein-binding assay indicate a physical interaction between ParE and ParD; in solution, ParE exists as a dimer, and as a dimer it binds to the dimeric form of ParD to form a tetrameric complex (53). The formation of this ParD-ParE complex presumably leads to ParD neutralizing the toxic activity of ParE (86). In addition, the dimeric form of ParD also has an autoregulatory role; it represses transcription of *parDE* by binding to a discrete sequence of 48 bp within the parDE promoter region (23, 28, 86). On the other hand, it has been found that the toxic protein ParE was not required for the binding of ParD to the promoter (86). As was more recently reported (53), ParE also binds to the *parDE* promoter but only in the presence of the autoregulatory protein ParD. Based on these results, it has been suggested that the dimeric protein ParD may have two separate functional domains, one for binding DNA and the other for binding ParE. However, because in vivo ParE is not required for full autorepression of *parDE* by ParD (23), it is still not clear why ParE binds to the promoter together with ParD.

In *E. coli*, plasmid stabilization by the *parDE* operon has been shown to be accompanied by growth inhibition and the filamentation of plasmid-free segregants (87, 99). It appears that in certain host strains and in certain media this growth arrest results from the death of plasmid-free cells (87). By using a different host-vector system, it was possible to observe plasmid-free cell killing that was less host and media dependent (52). These results support the conclusion that *parDE* mediates plasmid stabilization by killing of plasmid-free segregants. Furthermore, properties of *parDE* mutants indicate that protein ParE is responsible for killing of plasmid-free cells and that protein ParD neutralizes the toxic activity of the protein ParE (87). As yet, the cellular target of ParE is not known. Initial genetic studies suggest that the cellular target of ParE differs from that of the *ccdAB* system; the *gyrA462* mutation that affects the postsegregational killing mediated by *ccdAB* (see above; 12) did not affect that of *parDE* (87). In addition, the differential stability of the killer protein and its antidote, which is a key characteristic of an addiction module, has not yet been reported for the ParE and ParD proteins. As described above, the Lon protease participates in the degradation of CcdA and PemI. A mutation in *lon* that affects postsegregational killing by *pemI/K* and *ccdAB* (see above; 110, 115, 116) did not affect the *parDE*-mediated activities (87). Therefore, it seems that Lon does not participate in ParD degradation.

Finally, the relative separate and combined contributions of the *parCBA* and *parDE* regions for stable maintenance of the broad-host-range plasmid RK2 have been studied (27, 99). The results indicate that together *parCBA* and *parDE* are highly effective in various strains of gram-negative bacteria grown under various conditions. Moreover, it is clear that relative contributions of the *parCBA* and *parDE* to RK2 stabilization very much depend on the organism in question, on growth temperature, and on plasmid copy number. In addition, it has been shown that the action of *parCBA* itself can lead to plasmid stabilization, not via cell killing and resolution of plasmid multimers, but probably by a partition mechanism. On the other hand, *parDE* functions to stabilize plasmid by the post-segregational killing mechanism. Furthermore, changes in plasmid copy number can affect the relative contribution of postsegregational killing and partitioning to plasmid stabilization. At a relatively low plasmid copy number, the *parCBA*-encoded partitioning mechanism appears to be less efficient, thereby resulting in an increased number of plasmid-free cells produced upon cell division. Under these conditions, postsegregational killing provides an effective backup mechanism of plasmid maintenance in a cell-growing population. Thus, Helinsky and colleagues suggested that the killing mechanism encoded by *parDE* functions as a secondary or backup system when plasmid copy number becomes so low that the stability function encoded by *parCBA* is no longer effective (99). On the other hand, at least in certain *E. coli* strains under condition of normal growth and plasmid copy number, the *parCBA*-mediated partition mechanism acts as the predominant plasmid stabilization system. Thus, the combined presence of the gene module *parCBA* and *parDE* of plasmid RK2 has evolved into a highly effective and broad-host-range stabilization system.

phd-doc of Plasmid Prophage P1

Bacteriophage P1 lysogenizes *E. coli* as a low-copy-number plasmid that is maintained with a loss of frequency of $\sim 10^{-5}$/cell per generation (89). As for other low–copy-number plasmids, this remarkable stability can be attributed to the combined effects of a partition system that ensures segregation of at least one plasmid to each daughter cell (reviewed in 48, 80, 117), and a plasmid-encoded addiction module that selectively kills plasmid-free segregants (60). In fact, as a result of work on the P1 module, called *phd-doc*, it has been suggested that all plasmids with the ability to reduce postsegregational killing can be thought of as addictive agents (60).

The *phd-doc* addiction module of plasmid prophage P1 was detected by Yarmolinsky and colleagues (60). This module forms an operon that is organized similarly to systems described above; the antitoxic gene *phd* (prevents host death) precedes the toxic gene *doc* (death on curing). *phd-doc* encodes a stable 126-amino-acid-long 13.5-kDa toxic protein, Doc, and an unstable 73-amino-acid-long 8.1-kDa antitoxin, Phd (60). The cellular target of Doc is not yet known. However, based on recent studies, it is assumed to be a step in protein synthesis (M Yarmolinsky, personal communication). Phd is unstable in vivo; Phd is slowly degraded by the serine protease ClpPX with a half-life of ~2 h (61). ClpP proteases form a family in which a proteolytic subunit, ClpP, can associate with at least one or other specific subunits bearing the ATPase activity, ClpX or ClpA (for review, see 42). It has been found that *E. coli* mutants defective in either subunit of ClpPX protease survive the loss of a plasmid that harbors the P1 addiction module *phd-doc* (61). Defects in *clpA* or *lon* lead to cell death upon the loss of the same plasmid. These results suggest that (*a*) Phd is specifically degraded by the ClpPX protease, and (*b*) the instability of the antitoxic protein is essential for postsegregational killing. On the loss of the P1 plasmid, the unstable antitoxin Phd continues to be degraded, but is not replenished. This permits the unlimited toxic activity of Doc, which, in turn, leads to cell death.

Like that of other previously described addiction modules, the expression of *phd-doc* is also subjected to an autoregulatory circuit operating on the transcriptional level (64). In vivo, in the absence of the expression of *doc*, the expression of *phd* is sufficient to repress the transcription of a *lacZ* reporter almost 10-fold. Consistent with this observation, DNAase I footprinting showed that Phd binds a perfect 19-bp palindromic DNA sequence and, at higher concentrations, an adjacent imperfect palindrome. These palindromic sites are located between the -10 region of the putative promoter and the start codon of *phd*. The palindromic nature of the DNA sites protected from DNase I suggests that, like many DNA-binding proteins, Phd might bind as a dimer. In addition, the electrophoretic mobility of DNA containing the promoter region is retarded in the presence of Phd and further retarded in the presence of both Phd and Doc. Also, when *doc* is coexpressed with *phd*, repression of the *lacZ* reporter is enhanced more than 100-fold. Thus, both products of the *phd-doc* operon participate in its autoregulation. However, as recently shown, in its own right, Doc is not a repressor but rather enhances repression by binding to Phd, thus mediating cooperative interactions between adjacent Phd binding sites (65). When the copy number of the operon is increased, expression of a *lacZ* reporter fused to the promoter of the operon decreases, suggesting a role of autoregulation for plasmid stabilization (64).

The *Escherichia coli* Chromosomal Addiction Module *mazEF*, Which Is Regulated by ppGpp

In *E. coli*, the synthesis of the signal molecule guanosine-3′,5′-bispyrophosphate (ppGpp) is governed by at least two pathways. In the first, the synthesis of ppGpp is activated by the stringent response to amino acid starvation. The enzyme responsible for this pathway, RelA, is encoded by the *relA* gene; RelA is activated by uncharged tRNA and thereby by limitation to amino acid availability, or by an inhibition of amino acylation (for review, see 18). The second pathway for ppGpp synthesis is activated by carbon source limitation and is mutant *spoT* dependent (for review, see 18).

We have shown that the *E. coli relA* gene is part of an operon in which a pair of genes called *mazE* and *mazF* is located downstream from the *relA* gene

(72). Sequence analysis revealed that the predicted protein products of *mazE* and *mazF* are partially homologous to those of the addiction module *pemI/K*, which is borne by plasmid R100; MazE and PemI share 34% identical and 69% conserved residues (68). Masuda et al, who described this homology, changed the name of the genes in the pair, *mazE* and *mazF* (located in *relA* operon at 60 min of the *E. coli* chromosome), to *chpAI* and *chpAK*, respectively, and called the locus *chpA* (68). In addition, they found another chromosomal homologue of the *pem* pair of genes, called *chpB*, which is located at 95.7 min in the *E. coli* chromosome (69); it is composed of *chpBI* and *chpBK* (68). Results of experiments carried out with plasmids bearing the genes *chpA* and *chpB* indicate that *chpAK* and *chpBK* encode growth inhibitors, whereas *chpAI* and *chpBI* encode suppressors for the inhibitory functions of *chpAK* and *chpBK*, respectively (68).

The chromosomal *mazEF* gene pair encodes for the 82-amino-acid 9.4-kDa protein MazE and for 111-amino-acid, 12.1-kDa protein MazF (68, 72). We have shown that the *mazEF* system has all the properties required for an addiction module (most of these properties are described in 2). (*a*) MazF is toxic, and MazE is antitoxic. MazF affects not only cell growth, as previously described (68), but also affects cell viability. MazE protects the bacterial cells from the toxic effect of MazF. The cellular target of MazF is not yet known. (*b*) MazF is long-lived, while MazE is a labile protein degraded in vivo by the ClpPA serine protease. In fact, it is the first cellular specific substrate of ClpPA described so far. Like the in vivo degradation of the antitoxic proteins of other described addiction modules, MazE is also degraded slowly with a half-life of ~30 min. The specific degradation of MazE by ClpPA was shown by the use of *E. coli* mutants with a defect in genes coding for either Lon or one of the subunits of ClpPX or ClpPA protease. MazE is not degraded in either a *clpP⁻* or a *clpA⁻* mutant, but it is degraded in both *clpX⁻* and *lon⁻* strains. Thus, in spite of the partial homology between PemI and MazE (68), PemI is an in vivo substrate for Lon (110), whereas MazE is a substrate for ClpPA (2). (*c*) MazE and MazF interact. This was shown by their migration in native gels (2) and, recently, by the use of the yeast two hybrid technique (T Fisher, H. Engelberg-Kulka, & G. Glaser, unpublished results). (*d*) MazE and MazF are coexpressed. (*e*) the antitoxic protein is synthesized in excess (I Marianovsky & G Glaser, unpublished results). (*f*) *mazEF* is negatively autoregulated at the level of transcription by the combined action of both MazE and MazF proteins. They bind cooperatively to the promoter region of *mazEF*, probably by forming a complex, thus repressing their own synthesis. However, MazE by itself can mediate a weak autoregulation of *mazEF* (I Marianovsky & G Glaser, unpublished results). In addition, the *mazEF* system has a unique property. Its expression is inhibited by high concentrations of ppGpp (42).

The properties of the "*mazEF* module" described here suggest a model for programmed cell death in *E. coli* (Figure 2A) (2). Under conditions of nutritional starvation, the level of ppGpp increases. During amino acid starvation, this is achieved by the involvement of the product of *relA*, and, during carbon limitation, by an alternative pathway (for review, see 18). The ppGpp inhibits the coexpression

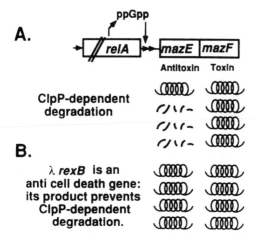

Figure 2 A model for the *E. coli rel maszEF*-mediated cell death (*A*) and the anti-death effect of λRexB (*B*). (*A*) Under conditions of nutritional starvation, the level of ppGpp increases. During amino acid starvation, this is achieved by the interaction of the tRNA with the product of *relA* (18). ppGpp inhibits the coexpression of *mazE* and *mazF*. MazF is a long-lived toxic protein, whereas MazE is an antitoxic labile protein that is degraded by the ClpPA protease. Therefore, when the cellular level of ppGpp is increased, the concentration of MazE is decreased more rapidly than that of MazF, and thereby MazF can exert its toxic effect and cause cell death (2). (*B*) λRexB antagonizes the ClpP family of proteases. As a result, it inhibits the degradation of the antitoxic protein MazE and thereby prevents cell death (30).

of *mazE* and *mazF*. Because MazE is a labile protein, its cellular concentration is decreased more rapidly than that of MazF, leaving MazF to exert its toxic effect and cause cell death. Thus, *mazEF* is a regulatable addiction module.

This model is further supported by the results of experiments suggesting that *mazEF*-mediated cell death is induced by ppGpp. This induction was carried out in two different ways: (*a*) by using a truncated *relA* gene under the control of an IPTG inducible promoter (2) and (*b*) by creating conditions of amino acid starvation by adding the serine analogue serine hydroxamate to the growth medium of a *relA*$^+$ *E. coli* strain (30). Similar results were obtained in these two sets of experiments: only ∼10% of the cells survived after abrupt induction by ppGpp. This effect is *mazEF*-mediated and *clpP*$^-$-dependent. Cell survival increases to 60% when the strain is deleted for *mazEF* or mutated to *clpP*$^-$.

It is interesting to note that another chromosomal proteic addiction module has been described recently that may be related to programmed cell death, specifically during amino acid starvation (39). We refer to the *E. coli* K12 *relBE* system that is located at 34.4 min on the bacterial chromosome; *relE* encodes the toxin, and *relB* encodes the antitoxin. Moreover, *relBE* is autoregulated by RelB, and RelE acts as a corepressor. It has not yet been shown that RelB is less stable than RelE, nor has the cellular target of RelE been identified. However, data base researching has revealed *relBE* homologues in an additional locus, *dinJ-yafQ* of the *E. coli* K12 chromosome, and also on the chromosomes of *Haemophilus influenzae* and *Vibrio cholerae*. Previously, it was shown that the *relB* operon, in which *relBE* of *E. coli* K12 is located, exhibits a delayed stringent response to amino acid starvation (for review, see 18). Thus, *relBE* together with *mazEF* (and probably other systems as well) may form a network of addiction modules responsible for programmed cell death under conditions of nutritional starvation.

As generally viewed for extrachromosomal elements, the addiction module rends the bacterial host addicted to the continued presence of this genetic element. The new concept offered by *mazEF* is that the continued expression of the addiction system is required in order to prevent cell death. The cells are addicted to the presence of the short-lived polypeptide because its de novo synthesis is essential for cell survival (Figures 1 and 2).

rexB OPERON OF BACTERIOPHAGE λ IS AN ANTI–CELL DEATH GENE

The *rex* operon of bacteriophage λ is responsible for the exclusion of the development of several unrelated phages (for reviews, see 19 and 98). The first described Rex function was the exclusion of the development of phage T4*r*II mutants by the λ prophage. In fact, the name *rex* comes from <u>r</u>II <u>ex</u>clusion. The *rex* operon is composed of two genes, *rexA* and *rexB*, which, together with the *c*I repressor gene, are located in the immunity region of λ (for review, see 19). The genes *rexA* and *rexB* can be expressed coordinately with the *c*I repressor gene from promoters p_{RM} and p_{RE} (9, 47, 59). In addition, there is a third promoter, p_{LIT}, that overlaps the region encoding the carboxyl terminus of *rexA*. Transcription from p_{LIT} results in the synthesis of the *lit* mRNA that permits the expression of *rexB* without that of *rexA* (46, 47, 59). This shift from coordinate to discoordinate expression of *rexB* over *rexA* implies that λ*rexB* has another function, independent of that of *rexA* (59, 82). In an earlier publication, we reported on another function for the product of *rexB*. It prevents the in vivo degradation of the short-lived protein λO (96), known to be involved in λ DNA replication (25, 34, 120). We have suggested that RexB may act as an inhibitor of the *E. coli* protease involved in λO degradation. The protease responsible for λO degradation has since been characterized as the ATP-dependent serine protease ClpPX (8, 41, 119).

We have recently shown that the product of λ*rexB* also inhibits the degradation of two additional substrates of the ClpP family of protease (30). These are the antitoxic labile proteins of two addiction modules: (*a*) the short-lived Phd protein of plasmid prophage P1, which is a substrate for ClpPX, and (*b*) the short-lived MazE protein of the *E. coli rel* operon, which is a substrate for ClpPA. In contrast, the degradation of the short-lived protein CII of bacteriophage λ is not prevented by the product of λ*rexB*. λCII is degraded by the ATP-dependent protease FtsH

(HflB) (56, 97, 108). Therefore, our results suggest that λ*rexB* specifically acts against the ClpP family of proteases and not as a general antagonist of the ATP-dependent proteases. Moreover, the stabilization of P1 Phd and *E. coli* MazE by λ*rexB* product also has functional consequences. This stabilization prevents the killing mediated by the two addiction modules P1 *phd-doc* and *E. coli mazEF*. In the *phd-doc* system, the product of λ*rexB* prevents postsegregational killing, and in the *E. coli mazEF* system, it prevents ppGpp-mediated killing. In both systems, the anti–cell death effect of λ*rexB* product is accomplished either when the gene is located on a plasmid or when the *E. coli* strain is lysogenized by phage λ carrying an active *rexB* gene. Moreover, a nonsense mutation in a plasmid-borne λ*rexB* gene abolishes the anti-cell-death effect of its product, and a similar effect is observed when the insertion of an omega interposon inactivates the *rexB* gene of a λ lysogen (30).

As described above, the *rel mazEF* addiction module seems to be responsible for programmed cell death in starved *E. coli* cells. The *rel mazEF* addiction module is regulated by the signal molecule ppGpp and thereby by amino acid starvation (Figure 2*A*) (2, 30) or by carbon source starvation. The stabilization by λ*rexB* of "addiction" proteins, which are substrates of the ClpP proteases (like MazE or Phd), suggests that λRexB is an anti–cell death protein. Based on our model, λRexB antagonizes degradation by ClpP, thus preventing programmed cell death mediated by the *E. coli* chromosomal addiction module *rel mazEF* (Figure 2*B*) (30).

When the host cells die, the virus is eliminated too. However, as a "selfish entity," it might have evolved ways to inhibit the death of cells it infects. Such a strategy would be particularly useful for a bacterial virus like λ that can also exist in a lysogenic state in which a prophage is incorporated into the *E. coli* chromosome. In fact, the anti–cell death gene λ*rexB* is one of the few genes expressed in the lysogenic state of the phage (46, 47, 59). We have shown that cells lysogenized by λ that includes an active λ*rexB* gene survive under conditions of nutrient stress, like amino acid starvation, when the *mazEF* addiction module is present on the chromosome (30). We therefore proposed that in addition to its well-known function of phage exclusion, under conditions of nutrient starvation the λ*rex* operon also prevents the *mazEF*-dependent death of lysogenized cells. Thus, the *rex* operon can be considered as the "survival operon" of phage λ (30).

SIMILARITY OF PROGRAMMED CELL DEATH AND ANTI–CELL DEATH IN *ESCHERICHIA COLI* AND IN EUKARYOTES

In some ways, the proteic addiction module-mediated cell death in *E. coli* and the antideath process by phage λRexB are similar to programmed cell death in eukaryotes and antideath processes by their viruses (for reviews on the eukaryotic systems, see 1, 44, 49). In eukaryotic cell death program(s), there is a crucial role for specific proteases. These are ICE-like proteases, called caspases, that cleave their substrates at specific loci following aspartic acid residues (for recent reviews on caspases, see 22, 107). As we have discussed above, proteases have also been shown to have a central role in cell death mediated by the *mazEF* addiction module of the *E. coli* chromosome and also those of the *E. coli* extrachromosomal elements like *phd-doc* of phage P1, *ccdA-ccdB* of F factor, and *pemI-pemK* of plasmid R100 (Table 1). In addition, these addiction modules bear striking functional resemblance to the basic principle of the genetic module that allows in eukaryotic cells, the subtle regulation of programmed cell death; the executioner protein of the suicidal process on one hand and the antideath protein belonging to the Bcl-2 family (for a recent review, see 1) on the other hand. Moreover, the cowpox virus and the baculovirus have an anti-host-death strategy similar to that which we describe here for *λrexB* (Figure 2*B*). The 38-kDa CrmA protein of the cowpox virus (83) and the p35 protein of the baculovirus (7) prevent caspases from carrying out their programmed cell death and are currently the best-known caspase inhibitors. An additional class of viral inhibitors is the IAP (inhibitors of apoptosis) family of proteins, whose precise caspase targets remain elusive (107, 114).

CONCLUSIONS AND FUTURE DIRECTIONS

"To be or not to be, that is the question!" The principal decision of each cell is to choose life or to choose death. An addiction module is a genetically well-designed simple "biological bomb" (121) that enables the cell to make such a decision at definite times and physiological conditions. In the addiction module system, the straightforward choice is death, which is facilitated by a stable intracellular toxin that causes cell suicide, that is, death from within. Choosing life requires a dynamic antagonistic process. The cell can survive only in the presence of an active process, requiring either the continued production of an unstable antitoxin or some process that would prevent its degradation. Thus, the principle of toxicity-antitoxicity of the proteic addiction module is attractive on several levels. This notion of toxicity-antitoxicity suggests fundamental questions for future research of which we shall mention several.

1. The toxins of addiction modules are biotechnologically important because they represent an enormous reservoir of potentially useful antibiotics. In view of the widespread development of multidrug resistance among bacterial pathogens, the intracellular induction of such antibiotics could be used as the basis for alternative therapies and thus help to solve what has become a major public health problem.

2. Both the toxins and antitoxins specified by the various addiction modules are interesting candidates for structural-functional analyses. All the toxic and the antitoxic proteins of the extrachromosomal and chromosomal

addiction modules studied so far are surprisingly small. The toxins are in the range of 100–130 amino acids, and the antitoxins are in the range of 70–85 amino acids (Table 1). Until now, structural analysis has been partially accomplished only for the toxic protein CcdB, which was recently crystallized (63) and whose cellular target is also known (for review, see 20). Similar analyses of CcdA and of toxins and antitoxins of other addiction modules should help us to understand their multiple analogous functions and also the relation of their structural similarities or dissimilarities to their functions. For example, predictions of secondary structure suggest that there is a short β-sheet region in the N terminus of all tested antitoxins, including Phd (64), CcdA (93), ParD (29), PemI (91), and (G Prag, G Glaser, & H Engelberg-Kulka, unpublished data). All these antitoxins participate in the negative transcription autoregulation of their respective addiction modules (Table 1), as is characteristic for such modules. It has therefore been suggested that the antitoxic partners of addiction modules may bind to the DNA specifically via β-sheet structures (64).

3. The antitoxic partners of addiction modules are also interesting because they are substrates for specific proteases. For example, MazE is the only chromosomally encoded cellular substrate described so far that is specifically degraded by the ClpPA protease (2). MazE is therefore a promising candidate for studies on the specificity of the *E. coli* mini-proteosome ClpP family of proteases, which have recently undergone structural analysis and have attracted a lot of attention (for review, see 42).

4. Addiction modules have been found primarily on a number of *E. coli* extrachromosomal elements (Table 1) and have been described as being responsible for the postsegregational killing effect. However, in the last 5 years, several different addiction modules have been found to be located on bacterial chromosomes, primarily in *E. coli*. Among these are proteic addiction modules (2, 39, 68, 69) and addiction modules belonging to the *hok/sok* family of genes (reviewed in 36). The presence of addiction modules on the bacterial chromosome raises obvious questions about their origin and whether there is a role for programmed cell death in bacterial cultures. The finding that the expression of the chromosomal addiction module *mazEF* is regulated by ppGpp and thereby by amino acid starvation provides some insights into these problems (2, 30). Our results suggest that *mazEF* has a role in programmed cell death under nutritional stress (Figure 2A). Our model is further supported by our more recent findings (30) showing that λ*rexB*, one of the few genes expressed in the lysogenic state of bacteriophage λ, prevents the nutritional stress-induced cell death mediated by *mazEF* (Figure 2B). The results of future work should reveal whether other chromosomal addiction modules might also have a role in programmed cell death under similar or other stressful conditions both in

E. coli and in other organisms. Moreover, it will be interesting to discover whether other lysogenous phages in *E. coli* or other bacteria carry an antideath gene(s) like the *rexB* gene of the *E. coli* bacteriophage λ.

In the orthodox view of single-celled organisms like bacteria, programmed cell death does not make adaptive sense. For multicellular organisms, however, programmed cell death makes a lot of sense. In contrast to the orthodox view, growing experimental evidence suggests that bacteria seldom behave as isolated organisms. Despite the apparent simplicity of the bacterial cells, as bacterial cultures, they manifest multicellular behavior (26). A well-known example for such behavior is the ability of bacteria to communicate with one another via small signal molecules, including some derivatives of homoserine lactone and various peptides (reviewed in 33, 43, 57, 102). These molecules control several kinds of phenotypic expression in a manner dependent on the cell density of the bacterial culture. Because they provide an index of population densities, these signal molecules have been called quorum sensors (32). Recently, we have found that *E. coli* cell death mediated by *mazEF* is dependent on the cell density of the bacterial culture. Moreover, in the supernatant of dense *E. coli* cultures, an as yet uncharacterized factor has been found that when added to diluted cultures results in cell death (R Hazan & H Engelberg-Kulka, unpublished data). This finding supports our view that programmed cell death in bacteria is one of the fundamental characteristics of the multicellular behavior of bacteria. Accordingly, a given bacterial population is expected to be differentiated; under stress conditions, a part of the population may have a specific function and, having served that function, will die. For example, suppose that only a subpopulation of a given *E. coli* culture responds to nutrient starvation by producing ppGpp. In that case, only that particular subpopulation would die through action of the *mazEF* addiction module. Death of a subpopulation may provide food for the surviving cells (2) or fulfill other useful functions for the benefit of the multicellular-like organisms, such as the elimination of unwanted cells.

The results of future studies should help us to understand how programmed cell death mediated through chromosomal addiction modules is related to the multicellular characteristics of bacterial cultures. We have suggested that several basic features of programmed cell death mediated by the bacterial proteic addiction modules are similar to the process of programmed cell death in eukaryotes. Thus, future studies on chromosomal and extrachromosomal bacterial cell death addiction modules should contribute to our general understanding of the evolution of the process of programmed cell death. The results of recent studies suggest that mitochondria have a central role in the control of programmed cell death in mammalian cells (apoptosis) (for reviews, see 44, 101). This implication that mitochondria may be centrally involved in apoptosis provides the basis for several versions of a theory on the evolution of apoptosis (4, 31, 58). For example, apoptosis may have evolved along with the endosymbiotic incorporation of aerobic bacteria (the precursors of mitochondria) into ancestral unicellular eukaryotes. When invading the early eukaryotic cells, the pro-mitochondria may have used (4) or developed (58) one or more addiction module(s) for stabilizing the host-parasite microsystem. Thus, it is possible that historically addiction modules have enabled the evolution of the eukaryotic cell.

ACKNOWLEDGMENTS

We thank Martine Couturier and Michael Yarmolinsky for providing us manuscripts before publication and for personal communication of unpublished results. We thank Ronen Hazan for fruitful discussions. We are grateful to F. R. Warshaw-Dadon for her critical reading of the manuscript, to Nathalie Pekerman for her excellent secretarial assistance, and to Myriam Reches for her help with the proofreading. This work was supported by a grant from the German-Israel Foundation for Scientific Research and Development (GIF) awarded to Hanna Engelberg-Kulka.

Visit the Annual Reviews home page at http://www.AnnualReviews.org

LITERATURE CITED

1. Adams JM, Cory S. 1998. The Bcl-2 protein family: arbiters of cell survival. *Science* 281:1322–26
2. Aizenman E, Engelberg-Kulka H, Glaser G. 1996. An *Escherichia coli* chromosomal "addiction module" regulated by guanosine-3′5′-bispyrophosphate: a model for programmed bacterial cell death. *Proc. Natl. Acad. Sci. USA* 93:6059–63
3. Akimoto S, Ono K, Ono T, Ohnishi Y. 1986. Nucleotide sequence of the F plasmid gene *srnB* that promotes degradation of stable RNA in *Escherichia coli*. *FEMS Microbiol. Lett.* 33:241–45
4. Ameisen JC. 1998. The evolutionary origin and role of programmed cell death in single celled organisms: a new view of executioners, mitochondria, host-pathogen interactions, and the role of death in natural selection. In *When Cells Die*, ed. RA Lockshin, Z Zakeri, JL Tilly, pp. 3–56. New York: Wiley-Liss
5. Bahassi EM, Salmon MA, Van-Melderen L, Bernard P, Couturier M. 1995. F plasmid CcdB killer protein: *ccdB* gene mutants coding for non-cytotoxic proteins which retain

their regulatory functions. *Mol. Microbiol.* 15:1031–37
6. Bailone A, Brandenburger A, Levine A, Pierre M, Dutreix M, et al. 1984. Indirect SOS induction is promoted by ultraviolet light-damaged miniF and requires the miniF *lynA* locus. *J. Mol. Biol.* 179:367–90
7. Beidler DR, Tewari M, Friesen PD, Poirer G, Dixit VM. 1995. The Baculovirus p35 protein inhibits FAS and Tumor Necrosis Factor-induced apoptosis. *J. Biol. Chem.* 270:16526–28
8. Bejarano I, Klemes Y, Schoulaker-Schwarz R, Engelberg-Kulka H. 1993. Energy dependent degradation of λO protein in *Escherichia coli*. *J. Bacteriol.* 175:7720–23
9. Belfort M. 1978. Anomalous behavior of bacteriophage λ polypeptides in polyacrylamide gels: resolution, identification, and control of the λ*rex* gene product. *J. Virol.* 28:270–78
10. Berger JM, Gamblin SJ, Harisson SC, Wang JC. 1996. Structure and mechanism of DNA topoisomerase II. *Nature* 379:225–32
11. Bernard P, Couturier M. 1991. The 41 carboxy-terminal residues of the miniF plas-

mid CcdA protein are sufficient to antagonize the killer activity of the CcdB protein. *Mol. Gen. Genet.* 226:297–304

12. Bernard P, Couturier M. 1992. Cell killing by the F plasmid CcdB protein involves poisoning of the DNA-topoisomerase II complex. *J. Mol. Biol.* 226:735–45

13. Bernard P, Kézdy KE, Van Melderen L, Steyaert J, Wyns L, et al. 1993. The F plasmid CcdB protein induces efficient ATP-dependent DNA cleavage by gyrase. *J. Mol. Biol.* 234:534–41

14. Bex F, Karoui H, Rokeach L, Drèze P, Garcia L, et al. 1983. Mini-F encoded proteins: identification of a new 10.5 kilodalton species. *EMBO J.* 2:1853–61

15. Bravo A, de Torrontegui G, Diaz R. 1987. Identification of components of a new stability system of plasmid R1, ParD, that is close to the origin of replication of this plasmid. *Mol. Gen. Genet.* 210:101–10

16. Bravo A, Ortega S, de Torrontegui G, Diaz R. 1988. Killing of *Escherichia coli* cells modulated by components of the stability system *parD* of plasmid R1. *Mol. Gen. Genet.* 215:146–51

17. Burkardt HJ, Riess G, Puhler A. 1979. Relationship of group P1 plasmids revealed by heteroduplex experiments: RP1, RP4, R68 and RK2 are identical. *J. Gen. Microbiol.* 114:341–48

18. Cashel M, Gentry DR, Hernandez VZ, Vinella D. 1996. The stringent response. In *Escherichia coli and Salmonella: Cellular and Molecular Biology*, ed. FC Neidhardt, R Curtiss III, JL Ingraham, ECC Ling, KB Low, et al, pp. 1458–96. Washington, DC: ASM Press

19. Court D, Oppenheim AB. 1983. Phage lambda's accessory genes. In *The Bacteriophage lambda II*, ed. RW Hendrix, JW Roberts, FW Stahl, RA Weisberg, pp. 251–76. Cold Spring Harbor, NY: Cold Spring Harbor Laboratory Press

20. Couturier M, Bahassi EM, Van Melderen L. 1998. Bacterial death by DNA gyrase poisoning. *Trends Microbiol.* 6:269–75

21. Critchlow SE, O'Dea MH, Howells AJ, Couturier M, Gellert M, et al. 1997. The interaction of the F plasmid killer protein, CcdB, with DNA gyrase: induction of DNA cleavage and blocking of transcription. *J. Mol. Biol.* 273:826–39

22. Cryns V, Yuan J. 1998. Proteases to die. *Genes Dev.* 12:1551–70

23. Davis TL, Helinski DR, Roberts RC. 1992. Transcription and autoregulation of the stabilizing functions of broad-host-range plasmid RK2 in *Escherichia coli, Agrobacterium tumefaciens* and *Pseudomonas aeruginosa. Mol. Microbiol.* 6:1981–94

24. de Feyter R, Wallace C, Lane D. 1989. Autoregulation of the *ccd* operon in the F plasmid. *Mol. Gen. Genet.* 218:481–86

25. Dodson M, McMacken R, Echols H. 1989. Specialized nucleoprotein structures at the origin of replication of bacteriophage lambda. *J. Biol. Chem.* 264:10719–25

26. Dworkin M, Shapiro J., eds. 1997. *Bacteria as Multicellular Organisms.* New York: Oxford Univ. Press

27. Easter CL, Sobecky PA, Helinski DR. 1997. Contribution of different segments of the *par* region to stable maintenance of the broad-host-range plasmid RK2. *J. Bacteriol.* 179:6472–79

28. Eberl L, Givskov M, Schwab H. 1992. The divergent promoters mediating transcription of the *par* locus of plasmid RP4 are subject to autoregulation. *Mol. Microbiol.* 6:1969–79

29. Eberl L, Kristensen CS, Givskov M, Grohmann E, Gerlitz M, et al. 1994. Analysis of the multimer resolution system encoded by the *parCBA* operon of broad-host-range plasmid RP4. *Mol. Microbiol.* 12:131–41

30. Engelberg-Kulka H, Reches M, Narasimhan S, Schoulaker-Schwarz R, Klemes Y, et al. 1999. *rexB* bacteriophage λ as an anti cell death gene. *Proc. Natl. Acad. Sci. USA* 95:15481–86

31. Frade JM, Michaelidis TM. 1997. Origin of eukaryotic programmed cell death: a

consequence of aerobic metabolism? *BioEssays* 19:827–32

32. Fuqua WC, Winans SC, Greenberg EP. 1994. Quorum sensing in bacteria: the LuxR/luxI family of cell density responsive transcriptional regulators. *J. Bacteriol.* 176:269–75

33. Fuqua WC, Winans SC, Greenberg EP. 1996. Census and concensus in bacterial ecosystems: the LuxR-LuxI family of quorum-sensing transcriptional regulator. *Annu. Rev. Microbiol.* 50:727–51

34. Furth ME, Wickner SH. 1983. Lambda DNA replication. In *The Bacteriophage Lambda II*, ed. RW Hendrix, JW Roberts, FW Stahl, RA Weisberg, pp. 145–73. Cold Spring Harbor, NY: Cold Spring Harbor Lab. Press

35. Gellert M, Mizuuchi K, O'Dea MH, Nash HA. 1976. DNA gyrase: an enzyme that introduces superhelical turns into DNA. *Proc. Natl. Acad. Sci. USA* 73:3872–76

36. Gerdes K, Gultyaev AP, Franch T, Pedersen K, Mikkelsen ND. 1997. Antisense RNA-regulated programmed cell death. *Annu. Rev. Genet.* 19:49–61

37. Gerdes K, Helin K, Christensen OW, Lobner-Olesen A. 1988. Translational control and differential RNA decay are key elements regulating postsegregational expression of the killer protein encoded by the *parB* locus of plasmid R1. *J. Mol. Biol.* 203:119–29

38. Gerlitz M, Hrabak O, Schwab H. 1990. Partitioning of broad-host-range plasmid RP4 is a complex system involving site-specific recombination. *J. Bacteriol.* 172:6194–203

39. Gotfredsen M, Gerdes K. 1998. The *Escherichia coli relBE* genes belong to a new toxin-antitoxin gene family. *Mol. Microbiol.* 29:1065–76

40. Gottesman S. 1996. Proteases and their targets in *Escherichia coli*. *Annu. Rev. Genet.* 30:465–506

41. Gottesman S, Clark WP, deCrecy-Layard V, Maurizi MR. 1993. ClpX, an alternative subunit for the ATP dependent Clp protease of *Escherichia coli*. *J. Biol. Chem.* 268:22618–26

42. Gottesman S, Maurizi MR, Wickner S. 1997. Regulatory subunits of energy dependent proteases. *Cell* 91:435–38

43. Gray KM. 1997. Intercellular communication and group behavior in bacteria. *Trends Microbiol.* 5:184–88

44. Green DR, Reed JC. 1998. Mitochondria and apoptosis. *Science* 281:1309–12

45. Hane MW, Wood TH. 1969. *Escherichia coli* K-12 mutants resistant to nalidixic acid: genetic mapping and dominance studies. *J. Bacteriol.* 99:238–41

46. Hayes S, Bull HJ, Tulloch J. 1997. The Rex phenotype of altruistic cell death following infection of a λ lysogen by T4*rII* mutants is suppressed by plasmids expressing OOP RNA. *Gene* 189:35–42

47. Hayes S, Szybalski W. 1973. Control of short leftward transcription from the immunity and ori regions in induced coliphage lambda. *Mol. Gen. Genet.* 126:257–90

48. Hiraga S. 1992. Chromosome and plasmid partition in *Escherichia coli*. *Annu. Rev. Biochem.* 61:283–306

49. Jacobson MD, Weil M, Raff MC. 1997. Programmed cell death in animal development. *Cell* 88:347–54

50. Jaffé A, Ogura T, Hiraga S. 1985. Effects of the *ccd* function of the F plasmid on bacterial growth. *J. Bacteriol.* 163:841–49

51. Jensen RB, Gerdes K. 1995. Programmed cell death in bacteria: proteic plasmid stabilization systems. *Mol. Microbiol.* 17:205–10

52. Jensen RB, Grohmann E, Schwab H, Diaz-Orejas R, Gerdes K. 1995. Comparison of *ccd* of F, *parDE* of RP4, and *parD* of R1 using a novel conditional replication control system of plasmid R1. *Mol. Microbiol.* 17:211–20

53. Johnson EP, Ström AR, Helinski DR. 1996. Plasmid RK2 toxin protein ParE: purification and interaction with the ParD antitoxin protein. *J. Bacteriol.* 178:1420–29

54. Karoui H, Bex F, Drèze P, Couturier M. 1983. *Ham22*, a miniF mutation which is lethal to host cell and promotes *recA*-dependent induction of lambdoid prophage. *EMBO J.* 2:1863–68

55. Kerr JFR, Wyllie AH, Curie AR. 1972. Apoptosis: a basic biological phenomenon with wide-ranging implication in tissue kinetics. *Br. J. Cancer.* 26:239–57

56. Kihara A, Akiyama Y, Ito K. 1997. Host regulation of lysogenic decision in bacteriophage λ: transmembrane modulation of FtsH (HflB), the cII degrading protease, by HflKC (HflA). *Proc. Natl. Acad. Sci. USA* 94:5544–49

57. Kleerebezem M, Quadri LEN, Kuipers OP, de Vos WM. 1997. Quorum sensing by peptides pherhormones and two-component signal transduction system, in gram-positive bacteria. *Mol. Microbiol.* 24:895–904

58. Kroemer G. 1997. Mitochondrial implication in apoptosis: towards an endosymbiont hypothesis of apoptosis evolution. *Cell Death Differ.* 4:443–45

59. Landsman J, Kroger M, Hobom G. 1982. The *rex* region of bacteriophage lambda: two genes under three way control. *Gene* 20:11–24

60. Lehnherr H, Maguin E, Jafri S, Yarmolinsky MB. 1993. Plasmid addiction genes of bacteriophage P1: *doc*, which causes cell death on curing of prophage, and *phd*, which prevents host death when prophage is retained. *J. Mol. Biol.* 233:414–28

61. Lehnherr H, Yarmolinsky MB. 1995. Addiction protein Phd of plasmid prophage P1 is a substrate of the ClpXP serine protease of *Escherichia coli. Proc. Natl. Acad. Sci. USA* 92:3274–77

62. Loh SM, Cram DS, Skurray RA. 1988. Nucleotide sequence and transcriptional analysis of a third function (Flm) involved in F plasmid maintenance. *Gene* 66:259–68

63. Loris R, Dao-Thi M-H, Bahassi EM, Van Melderen L, Poortmans F, et al. 1999. Crystal structure of Ccdb, a topoisomerase poison from *E. coli. J. Mol. Biol.* 285:1667

64. Magnuson R, Lehnherr H, Mukhopadhyay G, Yarmolinsky MB. 1996. Autoregulation of the plasmid addiction operon of bacteriophage P1. *J. Biol. Chem.* 271:18705–10

65. Magnuson R, Yarmolinsky MB. 1999. Corepression of the P1 addiction operon by Phd and Doc. *J. Bacteriol.* 180:6342

66. Maki S, Takiguchi S, Horiuchi T, Sekimizu J, Miki T. 1996. Partner switching mechanisms in activation and rejuvenation of *Escherichia coli* DNA gyrase by F plasmid proteins, LetD (CcdB) and LetA (CcdA). *J. Mol. Biol.* 256:473–82

67. Maki S, Takiguchi S, Miki T, Horiuchi T. 1992. Modulation of DNA supercoiling activity of *Escherichia coli* DNA gyrase by F plasmid proteins. *J. Biol. Chem.* 267:12244–51

68. Masuda Y, Miyakawa K, Nishimura Y, Ohtsubo E. 1993. *chpA* and *chpB, Escherichia coli* chromosomal homologs of the *pem* locus responsible for stable maintenance of plasmid R100. *J. Bacteriol.* 175:6850–56

69. Masuda Y, Ohtsubo E. 1994. Mapping and disruption of the *chpB* locus in *Escherichia coli. J. Bacteriol.* 176:5861–63

70. Maxwell A. 1992. The molecular basis of quinolone action. *J. Antimicrob. Chemother.* 30:409–16

71. Menzel R, Gellert M. 1983. Regulation of the genes for *E. coli* DNA gyrase: homeostatic control of DNA supercoiling. *Cell* 34:105–13

72. Metzger S, Ben-Dror I, Aizenman E, Schreiber G, Toone M, et al. 1988. The nucleotide sequence and characterization of the *relA* gene of *Escherichia coli. J. Biol. Chem.* 263:15699–704

73. Miki T, Chang ZT, Horiuchi T. 1984. Control of cell division by sex factor F in *Escherichia coli.* II. Identification of genes for inhibitor protein and trigger protein on the 42.84-43.6 F segment. *J. Mol. Biol.* 174:627–46

74. Miki T, Park JA, Nagao K, Murayama N,

Horiuchi T. 1992. Control of segregation of chromosomal DNA by sex factor F in *Escherichia coli. J. Mol. Biol.* 225:39–52

75. Miki T, Yoshioka K, Horiuchi T. 1984. Control of cell division by sex factor F in *Escherichia coli.* I. The 42.84-43.6 F segment couples cell division of the host bacteria with replication of plasmid DNA. *J. Mol. Biol.* 174:605–25

76. Morais-Cabral JH, Jackson AP, Smith CV, Shikotra N, Maxwell A, et al. 1997. Crystal structure of the breakage-reunion domain of DNA gyrase. *Nature* 388:903–6

77. Mori H, Ogura T, Hiraga S. 1984. Prophage λ induction caused by mini-F plasmid genes. *Mol. Gen. Genet.* 196:185–93

78. Murayama N, Shimuzu H, Takiguchi S, Baba Y, Amino H, et al. 1996. Evidence for involvement of *Escherichia coli* genes *pmBA, csrA* and a previously unrecognized gene *tldD*, in the control of DNA gyrase by *letD* (*ccdB*) of sex factor F. *Mol. Biol.* 256:483–502

79. Nagata A. 1997. Apoptosis by death factor. *Cell* 88:355–65

80. Nordström K, Austin SJ. 1989. Mechanisms that contribute to the stable segragation of plasmids. *Annu. Rev. Genet.* 23:37–69

81. Ogura T, Hiraga S. 1983. Mini-F plasmid genes that couple host cell division to plasmid proliferation. *Proc. Natl. Acad. Sci. USA* 80:4784–88

82. Parma DH, Snyder M, Sobolevski S, Nawroz M, Brody E, et al. 1992. The *rex* system of bacteriophage λ: tolerance and altruistic cell death. *Genes Dev.* 6:497–510

83. Ray CA, Black RA, Kronheim SR, Greenstreet TA, Sleath PR, et al. 1992. Viral inhibition of inflammation: cowpox virus encodes an inhibitor of the interleukin-1 beta converting enzyme. *Cell* 69:597–604

84. Roberts RC, Burioni R, Helinski DR. 1990. Genetic characterization of the stabilizing functions of a region of broad-host-range plasmid RK2. *J. Bacteriol.* 172:6204–16

85. Roberts RC, Helinski DR. 1992. Definition of a minimal plasmid stabilization system from the broad-host-range plasmid RK2. *J. Bacteriol.* 174:8119–32

86. Roberts RC, Spangler C, Helinski DR. 1993. Characteristics and significance of DNA binding activity of plasmid stabilization protein ParD from the broad-host-range plasmid RK2. *J. Biol. Chem.* 268:27109–17

87. Roberts RC, Ström AR, Helinski DR. 1994. The *parDE* operon of the broad-host-range plasmid RK2 specifies growth inhibition associated with plasmid loss. *J. Mol. Biol.* 237:35–51

88. Roca J. 1995. The mechanisms of DNA topoisomerases. *Trends Biochem. Sci.* 20:156–60

89. Rosner JL. 1972. Formation, induction, and curing of bacteriophage P1 lysogens. *Virology* 48:679–80

90. Ruiz-Echevarria MJ, Berzal-Harranz A, Gerdes K, Diaz-Orejas R. 1991. The *kis* and *kid* genes of the *parD* maintenance system from plasmid R1 form an operon that is autoregulated at the level of transcription by the coordinated action of the Kis and Kid proteins. *Mol. Microbiol.* 5:2685–93

91. Ruiz-Echevarria MJ, de Torrontegui G, Giménez-Gallego G, Diaz-Orejas R. 1991. Structural and functional comparison between stability systems *parD* of plasmid R1 and Ccd of plasmid F. *Mol. Gen. Genet.* 225:335–62

92. Ruiz-Echevarria MJ, Giménez-Gallego G, Sabariegos-Jareño R, Diaz-Orejas R. 1995. Kid, a small protein of the *parD* stability system of plasmid R1, is an inhibitor of DNA replication acting at the initiation of DNA synthesis. *J. Mol. Biol.* 247:568–77

93. Salmon MA, Van Melderen L, Bernard P, Couturier M. 1994. The antidote and autoregulatory functions of the F plasmid CcdA protein: a genetic and biochemical survey. *Mol. Gen. Genet.* 244:530–38

94. Saurugger PN, Hrabak O, Schwab H, Lafferty RM. 1986. Mapping and cloning of

the *par*-region of broad-host-range plasmid RP4. *J. Biotechnol.* 4:333–43

95. Scheirer KE, Higgins NP. 1997. The DNA cleavage reaction of DNA gyrase. *J. Biol. Chem.* 24:27202–9

96. Schoulaker-Schwarz R, Dekel-Gorodetsky L, Engelberg-Kulka H. 1991. An additional function for bacteriophage lambda *rex*: the *rexB* product prevents degradation of the λO protein. *Proc. Natl. Acad. Sci. USA* 88:4996–5000

97. Shotland Y, Koby S, Teff D, Mansur N, Oren DA, et al. 1997. Proteolysis of the phage λ CII regulatory protein by FtsH (HflB) of *Escherichia coli*. *Mol. Microbiol.* 24:1303–10

98. Snyder L, Kaufman G. 1994. T4 phage exclusion mechanisms. In *Molecular Biology of Bacteriophage T4*, ed. JW Drake, KN Kreuzer, G Mosig, DH Hall, FA Eiserling, et al, pp. 391–96. Washington, DC: ASM Press

99. Sobecky PA, Easter CL, Bear PD, Helinski DR. 1996. Characterization of the stable maintenance properties of the *par* region of broad-host-range plasmid RK2. *J. Bacteriol.* 178:2086–93

100. Sommer S, Bailone A, Devoret R. 1985. SOS induction by thermosensitive replication mutants of miniF plasmid. *Mol. Gen. Genet.* 198:456–64

101. Susin SA, Zamzami N, Kroemer G. 1998. Mitochondria as regulators of apoptosis: doubt no more. *Biochem. Biophys. Acta* 1366:151–65

102. Swift S, Throup JP, Williams P, Salmond GPC, Steart GSAB. 1996. Quorum sensing: a population density component in the determination of bacterial phenotype. *Trends Biochem. Sci.* 21:214–19

103. Tam JE, Kline BC. 1989. Control of the *ccd* operon in plasmid F. *J. Bacteriol.* 171:2353–60

104. Tam JE, Kline BC. 1989. The F plasmid *ccd* autorepressor is a complex of CcdA and CcdB proteins. *Mol. Gen. Genet.* 219:26–32

105. Thisted T, Nielsen AK, Gerdes K. 1994. Mechanism of post-segregational killing: translation of Hok, SrnB and Pnd mRNAs of plasmid R1, F and R483 is activated by 3′-end processing. *EMBO J.* 13:1950–59

106. Thomas CM, Helinski DR. 1989. Vegetative replication and stable inheritance of IncP plasmids. In *Promiscuous Plasmids of Gram Negative Bacteria*, ed. CM Thomas, pp. 1–25, London: Academic

107. Thornberry NA, Lazebnik Y. 1998. Caspases: enemies within. *Science* 281: 1312–16

108. Tomoyasu T, Gamer J, Bakau B, Kanemori M, Mori M, et al. 1995. *Escherichia coli* FtsH is a membrane-bound, ATP-dependent protease which degrades the heat-shock transcription factor σ32. *EMBO J.* 14:2551–60

109. Tse-Dinh Y-C. 1985. Regulation of the *Escherichia coli* DNA topoisomerase I gene by DNA supercoiling. *Nucleic Acids Res.* 13:4751–63

110. Tsuchimoto S, Nishimura Y, Ohtsubo E. 1992. The stable maintenance system *pem* of plasmid R100: degradation of PemI protein may allow PemK protein to inhibit cell growth. *J. Bacteriol.* 174:4205–11

111. Tsuchimoto S, Ohtsubo E. 1989. Effect of the *pem* system on stable maintenance of plasmid R100 in various *Escherichia coli* hosts. *Mol. Gen. Genet.* 215:463–68

112. Tsuchimoto S, Ohtsubo E. 1993. Autoregulation by cooperative binding of the *pemI* and *pemK* proteins to the promoter region of the *pem* operon. *Mol. Gen. Genet.* 237:81–88

113. Tsuchimoto S, Ohtsubo H, Ohtsubo E. 1988. Two genes, *pemK* and *pemI*, responsible for stable maintenance of resistance plasmid R100. *J. Bacteriol.* 170:1461–66

114. Uren AG, Coulson EJ, Vaux DL. 1998. Conservation of baculovirus inhibitor of apoptosis repeat proteins (BIRPs) in viruses, nematodes, vertebrates and yeasts. *Trends Biochem. Sci.* 23:159–62

115. Van Melderen L, Bernard P, Couturier M. 1994. Lon-dependent proteolysis of CcdA is the key control for activation of CcdB in plasmid-free segregant bacteria. *Mol. Microbiol.* 11:1151–57

116. Van Melderen L, Thi MHD, Lecchi P, Gottesman S, Couturier M, et al. 1996. ATP-dependent degradation of CcdA by Lon protease. *J. Biol. Chem.* 271:27730–38

117. Williams DR, Thomas CM. 1992. Active partitioning of bacterial plasmids. *J. Gen. Microbiol.* 138:1–16

118. Willmott CJR, Critchlow SE, Eperon IC, Maxwell A. 1994. The complex of DNA gyrase and quinolone drugs with DNA forms a barrier to transcription by RNA polymerase. *J. Mol. Biol.* 242:351–63

119. Wojtkowiak D, Georgopoulos C, Zylicz M. 1993. Isolation and characterization of ClpX, a new ATP-dependent specificity component of the Clp protease of *Escherichia coli. J. Biol. Chem.* 268:22609–17

120. Wold MS, Mallory JB, Roberts JD, LeBowitz JH, McMacken R. 1982. Initiation of bacteriophage lambda DNA replication *in vitro* with purified lambda replication proteins. *Proc. Natl. Acad. Sci. USA* 79:6176–80

121. Yarmolinsky MB. 1995. Programmed cell death in bacterial population. *Science* 267:836–37

Annu. Rev. Microbiol. 1999. 53:71–102

Wolbachia Pipientis: Microbial Manipulator of Arthropod Reproduction

R. Stouthamer[1], J. A. J. Breeuwer[2], and G. D. D. Hurst[3]

[1]*Laboratory of Entomology, Wageningen Agricultural University, 6700 EH Wageningen, Netherlands;* [2]*Department of Fundamental and Applied Ecology, University of Amsterdam, Kruislaan 320, 1098 SM Amsterdam, Netherlands;* [3]*Department of Biology, University College London, Wolfson House, London NW1 2HE, United Kingdom*

Key Words   cytoplasmic incompatibility, parthenogenesis, male killing, feminization, α-proteobacteria

■ **Abstract**   The α-proteobacterium *Wolbachia pipientis* is a very common cytoplasmic symbiont of insects, crustaceans, mites, and filarial nematodes. To enhance its transmission, *W. pipientis* has evolved a large scale of host manipulations: parthenogenesis induction, feminization, and male killing. *W. pipientis*'s most common effect is a crossing incompatibility between infected males and uninfected females. Little is known about the genetics and biochemistry of these symbionts because of their fastidious requirements. The affinity of *W. pipientis* for the microtubules associated with the early divisions in eggs may explain some of their effects. Such inherited microorganisms are thought to have been major factors in the evolution of sex determination, eusociality, and speciation. *W. pipientis* isolates are also of interest as vectors for the modification of wild insect populations, in the improvement of parasitoid wasps in biological pest control, and as a new method for interfering with diseases caused by filarial nematodes.

INTRODUCTION

Bacteria belonging to the genus *Wolbachia* have recently been recognized to infect a high proportion of insects, mites, isopods, and filarial nematodes. These intracellular α-proteobacteria were reported for the first time in 1924, by Hertig & Wolbach (45), as the unnamed rickettsia in the ovaries of the mosquito *Culex pipiens*, and they were formally named in 1936 by Hertig (44) as *Wolbachia pipientis* in honor of his collaborator Wolbach. Until 1970, hardly any work on these bacteria was reported. In 1971, Yen & Barr (162) discovered that *W. pipientis* in mosquitoes caused a crossing incompatibility between infected males and uninfected females. Uninfected eggs fertilized by sperm from infected males died. Interest in this group increased when it was found that the infection and its effect were not limited to mosquitoes but were also present in several other insect species

0066-4227/99/1001-0071$08.00   **71**

(5, 71, 90, 102). Most importantly perhaps is the fact that such incompatibilities occurred in *Drosophila* species (48).

With the availability of molecular techniques such as the polymerase chain reaction (PCR), the work on these bacteria has rapidly accelerated. Just in the last 5 years, it has become evident that these bacteria are very common and have important effects on their hosts. At least 16% of neotropic insects are infected with *W. pipientis* (157); in some insect groups they are very common. For instance, 50% of the Indonesian ant species are infected with *W. pipientis* (149). Surveys have shown that, in addition to the insects, *W. pipientis* is common in mites, terrestrial isopods, and filarial nematodes. In spider mites and predatory mites (Acari), 6 of the 16 species and 4 of the 7 species were found to be infected, respectively (14). In terrestrial isopods, 35% of the species (7) are infected, and 9 of 10 species of filarial nematodes are infected (2). The common occurrence of *W. pipientis* in these groups also led to a survey in molluscs, but none of the species tested were infected (119).

Not only are these bacteria widespread, but the unusual effects they impart on their hosts have also been a reason for the extensive attention *W. pipientis* has received over the last decade. *W. pipientis* manipulates the host biology in many sometimes unexpected ways, such as parthenogenesis (135), in which infected virgin females produce daughters; feminization (110), in which infected genetic males reproduce as females; and male killing (56, 61), in which infected male embryos die while female embryos develop into infected females. The best studied, and perhaps the most common, effect of *W. pipientis* is cytoplasmic incompatibility (CI). In its simplest form, a cross between an infected male and an uninfected female results in the mortality of the embryos. More complicated cases of CI involve bidirectional incompatibility, when two forms of the same species are incompatible because they are infected with different *Wolbachia* strains. Finally, two other effects are enhancing the fecundity or fertility of their hosts (39) and pathenogenicity (87). The manipulation of the host's biology and the expected response of the host's genes to such manipulations cause these infections to have important implications for the evolution of sex determination (104), speciation (16, 58), and eusociality (53). From the applied perspective, *W. pipientis* is of interest as a tool to genetically transform insects (4) for the modification of their disease-transmitting abilities. Filarial worms may be controlled by interfering with their *Wolbachia* symbionts (3). Parasitoids used in biological control of insects may be more effective when infected with parthenogenesis *W. pipientis* (130).

Proof of *Wolbachia* Involvement in the Host's Phenotype

Koch's postulates in the classic sense have not been fulfilled for any effect attributed to *W. pipientis*. It has not been possible to culture *W. pipientis* in a cell-free medium, and only in a single case has *W. pipientis* been grown in a tissue culture (95). Reinfection experiments have been successful in some species by taking infected egg cytoplasm and injecting that cytoplasm into uninfected eggs or embryos (10, 11, 24, 38, 40, 94). In long-lived species, transfers have been possible by injecting infected cytoplasm into adults or larval stages (105). Molecular techniques that show if a particular microorganism is present can replace some of the steps of the traditional Koch's postulates. The following may be seen as increased levels of certainty that a particular bacterium causes a particular host phenotype:

1. Hosts expressing the phenotype are infected with a particular bacterium, whereas hosts not expressing the phenotype are free of it.

2. Feeding antibiotics or exposure to elevated temperatures leads to the disappearance of the host phenotype and infection with the bacterium either immediately or in the subsequent generations. The antibiotics rifampicin, tetracycline, and sulfamethoxazole have been successfully used to kill *W. pipientis* in infected wasps, whereas gentamycin, penicillin G, and erythromycin did not cause the effects of the infection to disappear (135).

3. When uninfected hosts are infected with inoculum collected from an infected host, the bacterium is present, and the phenotype will be expressed in the same or in subsequent generations.

This method will show the absolute correlation between the presence of a particular bacterium and its effect on the host's phenotype. However, in those cases in which more than one symbiont is present, it again becomes more difficult to show an effect of a particular symbiont. Many insects have obligatory symbioses with prokaryotes; without the presence of these symbionts, such insects fail to reproduce or even to grow. In such cases, it will only be possible to show that a particular effect is related to a *Wolbachia* infection if *W. pipientis* can be removed without removing the obligatory symbionts.

Morphological Description of *Wolbachia* pipientis

Hertig (44) gives a detailed morphological description of the bacteria. They have the general characteristics of rickettsiae. The bacteria are dimorphic, with very small irregularly formed rodlike (0.5–1.3 μm in length) and coccoid forms (0.25–0.5 μm in diameter) that exist next to very large forms (1–1.8 μm in diameter) containing one to several of the smaller forms. The level of pleiomorphy appears to increase with the age of the host cell (159). *W. pipientis* is present in a vacuole enveloped by three layers of membranes. The outer layer is of host origin, followed by the outer cell wall of the bacteria; the innermost layer consists of the plasma membrane of the bacteria (79). Intracellular bacteria are commonly surrounded by multiple membranes; these membranes are thought to play a role in the host's control over the prokaryote (159). In the mosquito, Hertig (44) observed *W. pipientis* mainly in the cytoplasm of cells in the reproductive organs. Occasionally, these bacteria were found in the Malpighian tubules and once in muscle tissues next to the body cavity. Subsequent studies have shown that *W. pipientis* is found in high numbers in the ovaries and testes. In several species—the fly *Drosophila simulans*

(79), the parasitoid wasp *Dahlbominus fuscipennis* (20), and the woodlouse *Armadillidium vulgare* (108)—*W. pipientis* is also found in the nervous tissue. Such close association with the nervous tissue may allow for a very direct influencing by *W. pipientis* of the host's behavior. In the woodlouse *A. vulgare, W. pipientis* is also commonly found in the hemocytes (108). In the ovaries, *W. pipientis* appears to be present in the highest abundance in the nurse cells, where multiplication takes place (79, 166). The contents of the nurse cells enter the developing egg through cytoplasmic bridges. Inside the eggs, *W. pipientis* associates with microtubules. This association is thought to be important in causing the effects that *W. pipientis* has on its hosts in the case of CI and parthenogenesis (72).

The number of *W. pipientis* per host may vary substantially. An infected female of the crustacean *A. vulgare* harbors between 66,000 and 164,000 bacteria (106). In the minute wasp (of the genus *Trichogramma*), 250–670 bacteria per egg have been counted (136). A single egg of individual *Drosophila simulans* from Riverside, California, may contain as many as 500,000 *Wolbachia*, and male flies are estimated to harbor 36.5×10^6 bacteria (9).

PHYLOGENY OF *WOLBACHIA PIPIENTIS*

W. pipientis belongs to the group of α-proteobacteria, and its closest known relatives are all rickettsialike bacteria that cause arthropod-borne diseases of mammals, such as *Cowdria* and *Anaplasma* species (92, 132). The clade containing *W. pipientis* has been divided into four groups (A–D) (2, 158). The groups A and B contain the insect, mite, and crustacean *Wolbachia*, whereas groups C and D harbor the filarial nematode *Wolbachia*. Groups A and B have been estimated to have diverged 60 MYA (158), and they separated from group C and D ~100 MYA (2). Because the nematodes and the arthropods diverged >600 MYA, Bandi et al (2) suggest that one of the following must have happened around ~100 MYA: a horizontal transmission event of *Wolbachia* between arthropods and nematodes or the independent acquisition of *Wolbachia* from a third organism. It is unlikely that the *Wolbachia* bacteria were acquired at that time from a free-living form because the group of bacteria to which *W. pipientis* belongs is thought to have acquired an intracellular lifestyle >100 MYA (148).

There have been a large number of publications on the phylogeny of this group, using various genes for estimating the relationships. The first generally used *Wolbachia* gene was the *16S* rDNA(15, 92, 114, 132), followed by the *ftsZ* gene (cell division gene) (37, 120, 158), again followed by the *groEl* gene (bacterial heat shock protein) (84) and the *wsp* gene (cell surface protein) (145, 167). The latter gene is now used for the classification of *W. pipientis*. The extensive interest in the phylogeny of the group is caused by several unexpected features of *Wolbachia* phylogeny: the lack of correlation between *Wolbachia* phylogeny and that of its hosts and the fact that closely related *Wolbachia* bacteria can cause quite different effects on their hosts.

Horizontal Transfer of *Wolbachia pipientis* over Time

The lack of congruence between the phylogenies of the arthropod hosts and their *Wolbachia* symbionts became clear with the first few phylogenies published (15, 92), indicating that horizontal transmissions between hosts must happen rather frequently. The manner in which the horizontal transmission between species takes place is unknown for most species. In woodlice, however, blood-to-blood contact between individuals is sufficient to allow for horizontal transfer (105).

Phylogenies have also been used to find possible proof for recent horizontal transmission of *Wolbachia* species. The idea is to find cases where two quite different arthropod species share the same *Wolbachia* strain and are also associated in their occurrence or ecology. The first example of potential horizontal transfer was reported by Werren et al (158), who found that the insect parasitoid *Nasonia* and its fly host *Sarcophaga* each contain *Wolbachia* that are very similar. A comparable case was reported for *Wolbachia* from the parasitoid wasp *Trichogramma bourarache* and its moth host *Ephestia kuehniella* (145). Additional studies have been done to search for evidence of recent horizontal transmission. In a group of parasitoid wasps all sharing the same infected host species, no evidence was found for horizontal transmission among the parasitoids and their host (121).

The group of parthenogenesis-inducing *Wolbachia* bacteria found in many species of *Trichogramma* wasps are all closely related and form a monophyletic group (121). By comparing the phylogeny of the hosts (wasps) and the symbionts, it became evident that closely related *Trichogramma* wasps can harbor less-related *Wolbachia* bacteria. Thus, on an evolutionary timescale, horizontal transmission must occur quite frequently among the different *Trichogramma* species. Such transfers may occur when hosts are shared by different *Trichogramma* species (121).

Multiple Origins of the *Wolbachia* Phenotypes?

None of the various genes used for the phylogeny have been able to make the *Wolbachia* strains associated with a particular host-effect monophyletic (110, 132, 145, 158). The various effects are spread over the clades A and B. Often practically identical *Wolbachia* strains (based on the DNA sequence) can have quite different effects in different host species. The lack of association of the host effect with the phylogeny has resulted in a number of hypotheses on how this came about. The first hypothesis is that there is an effect of the host on the expression of *Wolbachia*. For instance, a *Wolbachia* strain causing incompatibility may result in parthenogenesis in another host. Little evidence exists for this hypothesis, although host effects of *Wolbachia* are evident from many studies. Generally, however, the effect that the *Wolbachia* cause in their new host is either a modulation of the effect already caused in the original host (98), or they cause lethality in their new host (65). The second possibility is that the transition from one effect to the other is rather easily attained. It is generally assumed that CI was the ancestral effect of *Wolbachia*. Possibly, the mutation from CI to some sex-ratio distortion may occur frequently. Thus far, there is no direct evidence for this hypothesis either. Finally, there is a possibility that the genes responsible for the effects caused by the *Wolbachia* are not located on the bacterial chromosome but on a plasmid or bacteriophage. Horizontal transmission of such mobile DNA from one *Wolbachia* strain to the next may have resulted in the observed pattern. However, there is also little evidence for this hypothesis. Phage-like particles have been observed several times in electron microscopic pictures of the *Wolbachia* strains of the mosquito *C. pipiens* (161), several other *Culex* species (89), and the western corn root worm *Diabotrica virgifera* (32). In other mosquito species (*Aedes*), no such phage-like particles have been detected (160). No molecular evidence exists as yet for the presence of phage or plasmids in the genus *Wolbachia*.

Naming of *Wolbachia* Species

Most authors have refrained from formally naming the different *Wolbachia* forms now known. In a few cases, additional *Wolbachia* strains have been named *W. postica* (52), *W. trichogrammae* (80), and *W. popcorn* (87), but none of these names is officially recognized. Only the original description of *W. pipientis* stands, as does the name *W. persica*; however, this latter species clearly does not belong to the same group as *W. pipientis*. Weisburg et al (148) showed that *W. persica* is closely related to *Francisella* spp., which are γ-proteobacteria. The only obvious differences among *Wolbachia* groups A–D are in the DNA sequence of the various genes studied.

Several systems have been adopted for naming the *Wolbachia* strains; for instance, Rousset & Stordeur (113) suggest naming the various *Wolbachia* strains causing CI-related effects in *Drosophila* spp. as *w*-followed by the name of the host from which the bacteria were collected. For instance, *w*RI stands for *Wolbachia* isolate of *D. simulans* collected in Riverside, California. Although this system works well for the intensively studied *Wolbachia* bacteria in *Drosophila* species, a more general system was necessary to classify the many *Wolbachia* strains found in other species. Recently, a system based on the level of similarity in the *wsp* gene sequence has been proposed (167). In this system, *Wolbachia* are grouped by reference strains; all members belonging to the group should not differ >2.5% in their *wsp* sequence from the reference strain. The group name generally consists of the first three letters of name of the reference species. This system offers a method of dividing the *Wolbachia* clades A and B, now referred to as supergroup A and B, into many groups. Until now, two publications (145, 167) have contributed groups. Within supergroup A, 10 groups have been designated; in supergroup B, 9 groups have been recognized. The grouping criterion of 97.5% similarity will result in some conflicts between groups when more *wsp* sequences are determined, as was already shown (145). The increased number of groups may also result in a more homogeneous host phenotype associated with a group. For instance, for *Drosophila* species, *Wolbachia* strains belonging to the same group have a similar CI type. The predictive value for other phenotypes appears to be less (145, 167).

PHENOTYPIC EFFECTS OF *WOLBACHIA PIPIENTIS* ON THEIR HOSTS

Cytoplasmic Incompatibility

Introduction   The most common effect that *W. pipientis* can have on arthropod host reproduction is CI. The CI phenotype results in aberrant offspring production between strains carrying different cytoplasmic factors because of disruption of the normal kinetics of sperm chromosomes shortly after fertilization. Typically, the paternal chromosomes are eliminated, which renders the developing embryo haploid. These embryos eventually die in diploid species and some haplodiploid mite species, whereas they develop into normal (haploid) males in other haplodiploid species such as wasps (13, 37, 63; see also 93). Hence, the two phenotypic effects of *Wolbachia*-induced CI: mortality or male-biased sex ratios among offspring.

CI is widespread in insects and has been reported in different insect orders, including Coleoptera, Diptera, Homoptera, Hymenoptera, Orthoptera, and Lepidoptera (37; see also 93). Recently, *Wolbachia*-induced CI has been described outside the insects, in several mite species (Arachnidae: Acari) (13, 62), and in an isopod (Crustacea) species (107, 110). This bias toward insects is most likely because other arthropod groups are less well studied.

The effect of CI on crossability is typically unidirectional: The incompatible cross is between infected males and uninfected females, whereas the reciprocal cross between uninfected males and infected females is compatible and produces normal progeny. In addition, bidirectional incompatibilities have been reported between infected strains in the mosquito *C. pipiens* (75, 82), mosquitoes in the *Aedes scutellaris* group (33), *D. simulans* (94), species of wasps in the genus *Nasonia* (16), and likely between species of crickets in the genus *Gryllus* (37). Microorganism-mediated incompatibility, especially bidirectional incompatibility, is of special interest because it may play a role in speciation by facilitating reproductive isolation (58, 151). Several factors have been identified that influence incompatibility or crossing type and expression of the CI phenotype, including *Wolbachia* strains, double versus single infections, bacterial density, host genotype and age, and environmental factors. These factors can interact and generate complex incompatibility relationships between geographic host strains (33, 75, 82, 85, 88, 112, 126).

Cytological Mechanism of Cytoplasmic Incompatibility   Little is known about the mechanism of microorganism-induced CI. In insects, several excellent cytological studies on the early events after fertilization revealed a sequence of aberrant events during early embryo development (22, 64, 72, 73, 101, 115). Normally, after sperm has entered the egg, sperm chromatin decondenses to form the paternal pronucleus. Presumably, sperm-specific histonelike proteins are removed and replaced by maternal histones (101). Replication follows, chromosomes condense for mitosis, and spindle attachment occurs. Next, paternal and maternal pronuclei fuse to form the diploid nucleus of the zygote. In eggs from incompatible crosses, however, only the female pronucleus forms individual chromosomes and undergoes the first cleavage division. The paternal pronucleus does not condense in individual chromosomes but reappears as a diffuse tangled chromatin mass and tends to get fragmented during the first mitotic division. Using genetic markers in crossing experiments it was confirmed that the paternal chromosomes were eliminated in fertilized eggs from incompatible crosses (102). The outcome is that, despite fertilization, embryos remain effectively haploid. In diploid organisms, such embryos show irregular development and eventually die (22, 64, 73, 94); in haplodiploid organisms such as *Nasonia* species, they develop into males (16, 101, 115). In haplodiploid species, females normally develop from fertilized eggs and are diploid, whereas males develop from unfertilized eggs and are haploid. Thus, the failure of syngamy and deviant behavior of paternal chromatin does not necessarily interfere with mitotic division of maternal chromosomes (101, 115). Subsequent fate of the paternal chromatin mass has not been determined. However, occasionally, fragments are incorporated into the daughter nuclei and may be stably transmitted. Evidence comes from the observation of extra chromosomal pieces in spermatogonia of some male progeny resulting from incompatible crosses and associated aberrant segregation of phenotypic markers (116, 117).

Paternal chromosome destruction in incompatible crosses is consistent with both the aberrant development and eventual death in diploids and production of all male progeny in haplodiploids. The CI phenotype in haplodiploid spider mites of the genus *Tetranychus* seems inconsistent with the cytological observations in insects (13). This may be due to the unusual, holokinetic chromosome structure in spider mites. Such chromosomes do not have a centromere to which the microtubules attach during meiosis/mitosis; instead, microtubules can attach along the entire chromosome. In contrast to fragments of centromeric chromosomes, which are likely to lack the centromere, fragments of holokinetic chromosomes may remain capable of connecting to microtubules and being incorporated into the daughter nuclei. However, proper segregation of paternal chromosomes is likely to be disturbed, and aneuploid nuclei are generated. Depending on the degree of aneuploidy, several CI phenotypes may be expected, ranging from early embryo mortality (i.e. nonhatching) to adult female offspring, which are sterile or have highly reduced fecundity (13). This interpretation is consistent with the occasional chromosome fragments observed in *Nasonia* species, which appear to contain a centromere (116, 117).

Thus, cytological observations in a number of species reveal that eggs produced in incompatible crosses are normally fertilized, but syngamy of maternal and paternal pronuclei is aborted. Depending on chromosome structure, the paternal chromosomes are probably lost, resulting in haploid embryos.

The fact that CI is widespread in arthropods and that it causes the same phenotype both at the chromosomal and organismal levels in a wide variety of arthropods suggests that *Wolbachia* bacteria interfere with fundamental, but conserved, molecular and developmental processes.

Molecular Interactions between Cytoplasmic Incompatibility*–Wolbachia *Strains and Host Despite the above logic, little is known about the molecular mechanism of CI. One of the first mechanisms put forward postulated that CI consists of two components analogous to the restriction-modification defense system in bacteria: modification of sperm chromosomes in males and rescuing of these chromosomes in infected *Wolbachia* eggs. *Wolbachia* bacteria are absent in mature sperm of infected males (5, 18), yet uninfected eggs differentiate between sperm from infected and uninfected males. This is inferred from both crossing experiments and cytological observations on uninfected males and infected males: sperm of the latter are incompatible. Thus, *Wolbachia* bacteria somehow modify sperm; in the male, they either produce a product that disrupts normal processing of sperm chromosomes in the egg (unless rescued) or they act as a sink by binding a host product necessary for normal processing of sperm chromosomes in the fertilized egg (72, 73, 101). This "imprinting" difference between sperm from infected and uninfected males does not play a role if the egg is infected with the same microorganism; both types of sperm are compatible with infected eggs. Apparently, *Wolbachia* bacteria only "rescue" sperm chromosomes that have been modified by the same *Wolbachia* strain.

Werren (152) described the modification (mod)–rescue (res) mechanism of CI-*Wolbachia* bacteria in genetic terms: mod$^+$ res$^+$ *Wolbachia* bacteria, which can induce CI by modifying sperm chromosomes but can rescue these when in the egg, and mod$^-$ res$^-$, which cannot induce CI. Assuming that *Wolbachia* strains vary in modification and rescue components (in other words, multiple alleles exist for each locus), the modification-rescuing mechanism can also explain other CI relationships: (*a*) bidirectional incompatibility between infected strains if each is infected with a different *Wolbachia* variant, and (*b*) unidirectional incompatibility between infected strains if one strain is doubly infected and the other only harbors one of two *Wolbachia* variants (6, 96, 112). In addition, some *Wolbachia* variants do not seem to cause CI and apparently have lost the capability of modification and imprinting of host chromosomes. Theoretically, there is a third kind of *Wolbachia* strain, mod$^-$ res$^+$, which can rescue imprinted sperm chromosomes in the egg but are incapable of modifying the sperm chromosomes (60, 100, 141). Indeed, not much later, the third kind of *Wolbachia* was found in *Drosophila* species (8, 86). Sperm of males from certain infected strains were compatible with uninfected eggs, consistent with mod$^-$. However, eggs of females from these strains were compatible with sperm from CI-inducing infected strains, consistent with the res$^+$ *Wolbachia* phenotype. This is strong evidence for the modification-rescuing mechanism of CI. The existence of mod$^-$ res$^+$ *Wolbachia* bacteria also points out that *Wolbachia* strains classified as having no effect or being neutral may in fact be mod$^-$ res$^+$ and not mod$^-$ res$^-$. Standard crosses to uninfected tester strains cannot distinguish between these two kinds of *Wolbachia* strains, and previously reported neutral or no-effect *Wolbachia* strains may in fact have the mod$^-$ res$^+$ genotype. A few other conclusions can be drawn from crossing experiments. Apparently, the uninfected host egg has a default response/reaction to *Wolbachia*-modified sperm. Such a system may normally be used in the recognition of conspecific sperm or may prevent fertilization by foreign DNA.

Recently, Braig et al (12) and Sasaki et al (118) have started to examine protein synthesis by *Wolbachia* bacteria in a *Drosophila* host in vivo by selective labeling of prokaryotic proteins and subsequent gel electrophoresis. In this way, they have already identified a 26- to 28-kDa protein, named wsp, which has some homology to outer surface proteins. At this point, it is unclear whether this protein is involved in *Wolbachia*-host interactions. Interestingly, however, *Wolbachia* bacteria obtained from different *Drosophila* strains that vary in the expression of CI also vary in the length of the *wsp* genes: *Wolbachia* bacteria from infected strains that do not show CI all shared a common deletion. The elucidation of the complete genomic map of the intracellular *Rickettsia prowazeki* (1) may provide exciting starting points for further research on the molecular mechanism of CI-*Wolbachia* and interactions between *Wolbachia* bacteria and host.

Factors Influencing Incompatibility and Expression of Cytoplasmic Incompatibility   So far, most of the work that has been done to unravel the complexity of CI relationships and understand the evolutionary dynamics of CI-*Wolbachia* comes from studies on *Drosophila* and *Nasonia* species. The studies involve crossing experiments combined with molecular identification of *Wolbachia* strains and/or transfection experiments. The interaction between host genotype and *Wolbachia* strain is of great interest because it may help explain the taxonomic distribution of CI-*Wolbachia*; for example, it may provide insight as to why some species are infected and other closely related species are not, as well as insight into the dynamics and evolutionary consequences for *Wolbachia*-host systems.

It is clear that *Wolbachia* strains play an important role on the phenotype expressed by the host and that the action of infections may be independent of the host genome (25). Host strains belonging to the same species but infected with different *Wolbachia* variants are almost always bidirectionally incompatible (46, 88, 96, 112). Introduction of *Wolbachia* strains in a novel host genetic background by introgression experiments or artificial transfer of *Wolbachia* strains between host species also showed that some *Wolbachia* strains act independently of the host genome (11, 17, 26, 37, 113). Apparently, many modification and corresponding rescue alleles exist in *Wolbachia* bacteria. Furthermore, crossing studies have demonstrated both CI-inducing and non–CI-inducing strains (38, 46, 114).

In addition, host genotype may influence the incompatibility relationships. To distinguish host genotypic effects from *Wolbachia* strain effects, *Wolbachia* strains were exchanged between the host species via microinjection in *Drosophila* or introgression in *Nasonia* species. CI in *D. simulans* is typically very strong, whereas it is weak in *Drosophila melanogaster* (46). *Wolbachia* strains from *D. simulans* in a *D. melanogaster* nuclear background resulted in low levels of incompatibility, similar to that found in crosses between naturally infected *D. melanogaster* strains, rather than strong CI as seen in its original host (10). The reciprocal transfer of *Wolbachia* strains from *D. melanogaster* into *D. simulans* induced high levels of CI in the recipient host *D. simulans* (98). Similarly, in *Nasonia* strains, a *Wolbachia* variant expresses partial incompatibility in its original host *Nasonia vitripennis* but expresses complete CI after introgression into the nuclear background of *Nasonia giraulti* (6). Weak expression of CI is proposed as evidence for existence of repressing host genotypes (6, 25, 141).

Molecular identification of *Wolbachia* bacteria revealed that some hosts harbor more than one *Wolbachia* strain (15, 37, 85, 88, 112, 122, 157, 158). Double infections can both result in bidirectional and unidirectional incompatibility (96, 112). For example, in *Nasonia* species, males of double-infected hosts can be unidirectionally incompatible with females of lines with one of the *Wolbachia* variants (96). Double infections represent an interesting problem; that is, how are double infections maintained in the host and faithfully transmitted, without rapidly losing one or the other *Wolbachia* variant owing to stochastic loss or differences in replication rates?

Several studies have demonstrated that expression of CI is correlated with bacterial numbers in eggs (9, 10, 17, 26, 123) and proportion of infected cysts in testes of *Drosophila* species (18, 98, 126). Males of a strain showing higher infection densities are incompatible with females from strains with lower bacterial densities, but the reciprocal cross is compatible. A dosage effect may explain these results such that *Wolbachia* bacteria at low density in the embryo are unable to rescue sperm from males with high densities (17). However, any relationship with density breaks down when different *Wolbachia* strains or double infections are considered (38, 46, 122). It is unclear how bacterial densities are regulated; probably both *Wolbachia* strain and host genotypes play a role, as is suggested by the *Wolbachia* transfection experiments. Typically, CI is much stronger in laboratory strains than among field strains (27, 47, 143). In addition, the almost perfect maternal transmission of *Wolbachia* bacteria in the laboratory is lowered in field populations, resulting in the production of uninfected ova (143). This clearly indicates the importance of environmental factors on the evolution of *Wolbachia*-host interactions. Several environmental factors may influence CI levels and transmission rates of *Wolbachia* strains: temperature (49, 50, 128, 140), naturally occurring antibiotics (128, but see 143), food quality (128), larval density (27, 122), host age (18), and larval diapause (96). Most environmental factors reduce bacterial density in eggs or transmission efficiency to sperm cysts and consequently lower the strength of incompatibility to uninfected strains. In double-infected *Nasonia* species, extension of diapause stage resulted in the singly infected adults (96). More environmental factors are likely to be found, depending on the particulars of the ecology of the host.

Localization of Wolbachia *Bacteria Inside Host Tissue*   *Wolbachia* bacteria are associated with the syncytial nuclei and concentrate around the pole of mitotic spindles (22, 72, 73, 94). In the parasitoid wasp of the genus *Nasonia*, *Wolbachia* bacteria are localized at the posterior end of the egg and become incorporated into the pole cells, which bud off from the rest of the cytoplasm. The pole cells typically develop into germ-line tissue. *Wolbachia* bacteria appear to be absent in other parts of the egg and early syncytial embryo (16, 101). In newly laid *Drosophila* eggs, however, *Wolbachia* bacteria are initially evenly distributed in the thin cortical layer and scattered in the inner yolk region (21, 72, 94). During first cleavage divisions in the early embryo, the bacteria redistribute around the syncytial nuclei and concentrate around the poles of the mitotic spindles, suggesting that the microtubules and centrosomes play a role in localizing *Wolbachia* bacteria (21, 72). This may be an important evolutionary feature, in particular during oogenesis or spermiogenesis, to ensure that each daughter cell receives *Wolbachia* bacteria. Little is known about the regulation of *Wolbachia* cell division during development of their host. Limited growth seems to occur during early *D. melanogaster* embryogenesis (73, 87). *Wolbachia* distribution in tissues other than the reproductive system has not been extensively studied. An exception is "popcorn." While screening for brain gene mutations in *D. melanogaster*, Min & Benzer (87) found a virulent *Wolbachia* variant that greatly reduces adult life span. It is quiescent during the fly's development but starts to multiply rapidly in adult tissue, causing degeneration of a variety of tissues, resulting in premature death. Apparently, this *Wolbachia* variant does not cause CI when crossed to uninfected females.

Parthenogenesis-Inducing *Wolbachia* Strains

The induction of parthenogenesis by parthenogenesis-inducing (PI) *Wolbachia* bacteria seems almost a perfect manipulation of the host's reproduction in favor of the cytoplasmically inherited symbiont. Because males are not transmitters of such symbionts, they are a "waste" from the perspective of the symbiont; making them superfluous can be seen as the ultimate manipulation. PI *Wolbachia* strains are restricted to the insect order Hymenoptera (wasps). The method in which the PI *Wolbachia* bacteria allow infected females to produce female offspring from unfertilized eggs is through a modification of the first mitotic division (133). In infected eggs, the first mitotic division is aborted in the anaphase, leading to a diploid nucleus in an unfertilized egg. Hymenoptera and a number of other insect groups have a particular sex determination system (arrhenotoky), in which males arise from haploid eggs and females arise from diploid eggs. Uninfected females generally determine the sex of their offspring by either fertilizing their egg (diploid, female) or by leaving it unfertilized (haploid, male). In some species, both infected and uninfected individuals coexist, and mating still takes place. The infected females are then still able to fertilize their (infected) eggs. In these eggs, made diploid by fertilization, the *Wolbachia* bacteria do not interfere with the mitotic events (133). In Hymenoptera, we have evidence of *Wolbachia* involvement in parthenogenesis in at least 40 species (131); however, many more cases remain to be studied.

Little is known about the biochemical aspects of the PI *Wolbachia* infection. Detailed cytogenetic work on the possible mechanical interference of the bacteria with the spindles in the first mitosis remains to be done. Besides influencing the mitosis, the *Wolbachia* bacteria also influence the offspring production in the laboratory. Infected females generally produce fewer offspring than uninfected conspecifics (134). This effect seems to be associated with those species where the infected females co-occur with uninfected individuals in populations. The negative influence of the infection appears to be less in those species where all individuals are infected. The transmission of *Wolbachia* bacteria in many species decreases with the age of females and/or the number of eggs she has laid and results in the production of male offspring by older females. In addition, the *Wolbachia* expression is influenced by the rearing temperature of the mothers. Females reared at high temperatures start to produce more male offspring (80, 135). If infected females of many species are reared at 28°C, they start to produce some intersexes, i.e. offspring that are partly male and partly female (131). Intersexes are formed when some tissues become diploid during embryogenesis and some remain haploid. It is assumed that rearing temperature leads to a reduction in the *Wolbachia* titer. Such reduced *Wolbachia* titer may lead to an abortion of the mitotic anaphase of one of the nuclei in the second or later mitotic division (131).

Feminizing *Wolbachia* Strains

Feminizing symbionts, bacteria, and protists that alter their host's normal pattern of sex determination, such that individuals that would have developed into males develop as females, have been found in three groups: marine amphipod crustaceans, terrestrial isopod crustaceans, and one lepidopteran insect (*Ostrinia furnacalis*, the Asian corn borer). The feminizing symbionts identified in the former group are protists (19, 36, 139). In contrast, the feminizing traits in *A. vulgare* and *O. furnacalis* are curable with antibiotics (70, 108), and they are associated with the presence of *Wolbachia* bacteria (114; S Hoshizaki, personal communication). Feminizing *Wolbachia* bacteria were then found in two further species of isopod (65). Using *Wolbachia*-specific PCR tests across a wide range of species, Bouchon and coworkers (7) have revealed *Wolbachia* bacteria to be common in isopod Crustacea, with 35% of the species tested being infected. In their survey, they tested more than one individual of a species, recording the sex of the individual and the location from which the individual was collected. Many of the *Wolbachia* infections were found to be either solely found in females or at least more prevalent in females than males, indicating that these *Wolbachia* strains are commonly associated with feminization. This conclusion was corroborated by the ability of many of the *Wolbachia* strains to induce intersexuality (partial feminization) after artificial inoculation into an uninfected male host. Prevalence varied between species, but, with the exception of one species that was fixed for *Wolbachia* infection, prevalence was between 10% and 50% in the majority of cases. There was also evidence that, within a species, prevalence varied over space.

The *Wolbachia* strains in isopod crustaceans all fall within the B group. All but one of the *Wolbachia* strains from oniscoid isopods appear to form a monophyletic clade, but some of the strains, especially from other isopod groups, are more distantly related and suggest that the *Wolbachia* strains of isopods are polyphyletic (7). These conclusions are based on the sequence of 16S rDNA, and the sequence of *ftsZ* and *wsp* genes will prove helpful in fully resolving this issue.

Basic details of the mode of action of *Wolbachia* bacteria in *A. vulgare* are known. In *A. vulgare*, individuals develop as females unless they are masculinized by the action of the androgenic gland, which produces an androgenic hormone that induces male differentiation (83). When inherited from the female parent, *Wolbachia* bacteria in some way prevent the formation of the androgenic gland and thus ensure female development (78). Further, when *Wolbachia* bacteria are injected into adult male *A. vulgare* with differentiated gonads, a feminization response is evident. The males acquire an intersex phenotype, differentiating female sexual characteristics. The androgenic hormone is still active in these intersex individuals (66), indicating that *Wolbachia* bacteria in the adult do not affect the hormone-producing capabilities of the androgenic gland. Rather, they affect the response of the host to the hormone.

Maternally inherited elements are also selected to produce a bias in the sex ratio at fertilization. In *N. vitripennis*, for instance, a maternally inherited agent termed *msr* biases the primary sex ratio toward the production of female offspring (124). The nature of the agent is not known in this case. In another haplodiploid group, chiggers of the genus *Leptotrobidium*, the bacterium *Orientia tsutsugamushi*, is reported to be associated with a sex-ratio bias, although the nature of this bias is not known (109, 137). By extension from these cases, it is possible for maternally inherited agents to produce female-biased primary sex ratios in haplodiploid species. Although not yet recorded for a *Wolbachia* strain, it is a potential phenotype of which workers should be aware.

Male-Killing *Wolbachia* Strain

Maternally inherited factors that kill male progeny during embryogenesis were the first of the "unusual" cytoplasmic effects recorded in animals (81) and have since been recorded in over 20 species of insects (54). Where the antibiotic sensitivity of these traits has been tested, they have been found to be curable with antibiotics; in only one case (still a subject of debate) has a bacterium not been considered to underlie the trait. Using the sequence of PCR-amplified 16S rDNA, six different bacteria have so far been identified as being associated with male-killing traits. These derive from a wide range of the eubacteria: two mollicutes from the genus *Spiroplasma*, a member of the *Flavobacteria-Bacteroides* group, a member of the

γ group of proteobacteria, and two from the α group of proteobacteria, members of the genera *Rickettsia* and *Wolbachia* (41, 54, 55, 153, 156).

Male killing is thus unusual among the reproductive manipulations exhibited by *Wolbachia* bacteria in being a commonly found phenotype within eubacteria in general. This contrasts with PI and CI, which have been uniquely associated with *Wolbachia* bacteria. The systematic diversity of male killers has given rise to speculation that this trait is, for some reason, more easily evolved than other manipulations of host reproduction. However, details of the method by which males are killed are still lacking, and the study of male-killing *Wolbachia* strains is still in its infancy.

To date, male-killing *Wolbachia* have been found in two taxa. The first of these is *Adalia bipunctata*, the two-spot ladybird beetle. This species is known to be infected with two other species of male killer, and the *Wolbachia* male killer has so far been recorded only in Russian populations of this species. The male-killing *Wolbachia* bacteria infect $\sim$20–30% of *A. bipunctata* females from Moscow. The other species infected with male-killing *Wolbachia* bacteria is *Acraea encedon*, an African butterfly. Upward of 80% of females of this species may be infected (61), and to date *Wolbachia* bacteria are the only male-killing agents found.

It is not possible from present data to conclude whether these represent one or two transitions to male-killing behavior within the B clade of *Wolbachia* bacteria. What is certain is that the two host species differ in their system of sex determination; *A. bipunctata* is male heterogametic, whereas *A. encedon* is female heterogametic. This indicates that *Wolbachia* bacteria are relatively unconstrained with respect to the range of hosts in which they can effect the male-killing phenotype. In turn, this suggests that male-killing *Wolbachia* bacteria will turn out to be common, at least within insects.

Fecundity and Fertility-Modifying *Wolbachia* Strains

Another method to enhance the transmission of symbionts is to increase the offspring production of infected individuals. In the parasitoid wasp *T. bourarachae*, a *Wolbachia* infection causes enhanced offspring production (39). The infected line produces approximately twice the number of offspring as a "cured" line. No other effects of this infection have been detected. In two other cases, CI *Wolbachia* bacteria also appear to enhance the offspring production of the infected females. In *D. simulans*, transitory reduced offspring production was reported after the flies had been cured (99). However, three generations after the antibiotic treatment the cured flies produced equal numbers of offspring as the infected flies. The explanation for this phenomenon is unclear. A second case of enhanced offspring production was found in the CI *Wolbachia* bacteria of the wasp *N. vitripennis*. The wasps in the generation after the antibiotic treatment produced significantly more offspring than the wasps that were still infected. This effect was found in a line infected with two different *Wolbachia* strains (129). In an experiment carried out simultaneously with a single-infected line, no difference in offspring production was found. *Wolbachia* infection in a stalk-eyed fly (*Sphyracephala beccarii*) does not cause any detectable incompatibility or fecundity effect; however, males cured from the infection had a substantially reduced fertility (42). In the flour beetle *Tribolium confusum*, an effect of the CI *Wolbachia* bacteria on male fertility has also been reported. In females mated with both an infected and an uninfected male, the sperm of infected males fertilized the majority of the eggs (146).

POPULATION BIOLOGY OF *WOLBACHIA* BACTERIA

Being maternally inherited, the population biology of *Wolbachia* bacteria is relatively straightforward. Two factors impede the spread of *Wolbachia* bacteria. First, maternal inheritance may be imperfect, such that while all the daughters of uninfected females are uninfected, only a proportion of the daughters of infected females are infected. This loss may be induced in the environment by exposure to high temperatures (127) or to naturally occurring antibiotics (128). Also, as discussed below, inefficient transmission may be caused by host genetic factors. Second, there may be a direct physiological cost to infection, such that the lifetime fecundity of an infected female is less than that of an uninfected one.

Without either manipulation of host reproduction or a positive contribution to host physiology, *Wolbachia* infections would be lost from current populations. The manipulations of host reproduction produced by *Wolbachia* bacteria lead to a relative increase in the number of surviving daughters produced by infected individuals. This is transparent for feminizing and PI *Wolbachia* strains. In a similar vein, male killing may increase the number of surviving daughters produced by an infected female. Where siblings compete for food, the death of males is accompanied by an increase in the survival of their sisters (59, 125). Alternatively, if there is cannibalism of unhatched eggs by siblings, as in *A. bipunctata*, then the death of males provides an initial meal to their hosts' sisters (57). In addition, death of males may lower the rate of inbreeding suffered by infected females, and this too may increase the survivorship and fecundity of infected females over that of uninfected ones (150).

In the case of feminizing and male-killing *Wolbachia* strains, the infection spreads to equilibrium prevalence if transmission is inefficient but may cause host extinction through lack of males if both symbiont transmission and host manipulation occur with near-perfect efficiency. The spread of feminizing *Wolbachia* infections in female heterogametic species is also accompanied by the increase to fixation of the male-determining sex chromosome (138), as seen in *A. vulgare* (67, 68). In the case of PI *Wolbachia* strains, high-transmission efficiency coupled with high efficiency of host conversion to parthenogenesis leads to the transition of the host to asexuality. This is known to occur in several species of Hymenoptera (97, 135, 165) and is probably a common cause of asexuality in this group.

The dynamics of CI are less intuitive. When CI-inducing *Wolbachia* bacteria are at low prevalence, there are few infected males in the population. At this point, uninfected females only have a low probability of losing progeny because of incompatibility. When *Wolbachia* bacteria are at higher prevalence, uninfected females are more likely to mate with males carrying a *Wolbachia* infection and thus are more likely to have progeny dying through incompatibility. Clearly, the amount of death of uninfected females that occurs through incompatibility is proportional to the prevalence of the infection. This positive frequency dependence accounts for the rapid spread of the trait through populations once prevalence reaches 10%–20%, as has been witnessed in the case of the Riverside strain of *Wolbachia* in *D. simulans* (142) and in the delphacid bug, *Laodelphax striatellus* (51).

When prevalence is very low, the uninfected condition may be favored. This is because losses of uninfected individuals from incompatibility are outweighed by the generation of uninfecteds following inefficiency in *Wolbachia* transmission. There is thus a threshold prevalence above which CI-inducing *Wolbachia* bacteria increase in prevalence and below which they decrease.

This leads to the question, how does the *Wolbachia* infection frequency reach this threshold level? There are two possible answers to this question. First, stochastic increases in frequency (drift) may take the *Wolbachia* infection frequency above the threshold for deterministic spread. This will occur most commonly in small populations. Alternatively, the population may be subdivided. To exemplify this, consider the case in which the threshold for invasion is 0.75%. If a population size was 1000 females, then the initial infected female would be at prevalence 0.1%, and its loss would be likely. However, if the population of 1000 were subdivided into 10 populations of 100, which exchanged just a small proportion of individuals each generation, then the initial infected female would be in a population of 100, i.e. at 1%. Because this is above the threshold for invasion, it would be likely to spread within this subpopulation. Infected individuals would migrate to other subpopulations and the infection spread generally across the range of the species.

The above describes the population biology of a single infection in an otherwise uninfected population. What will happen when either new incompatible types, or a double-infected type, arises? Clearly, a new singly infected type that is not compatible with previous types will not spread through an infected population. However, it may spread through uninfected populations, leading to different populations of the species bearing different CI *Wolbachia*. Bidirectional incompatibility will exist between these populations, and several examples of populations bearing different incompatibility types of *Wolbachia* bacteria have been described [see (25) for details of infections in the *D. simulans* system]. Bidirectional incompatibility is not stable if the two differently infected populations come perfectly into mixis (23, 111). However, if a zone of contact exists, then under certain ecological circumstances, the two populations bearing different infections can be maintained stably, in a similar manner to hybrid zones between races of a species (141).

The spread of a new dually infected type through a previously singly infected population is straightforward. If this type is compatible with all other cytoplasms (which it will be if it bears the previous single infection) and it produces incompatibility with the previous type, then it can spread (subject to it reaching a threshold level in the population). Interestingly, the presence of a dually infected class where dually infected females are compatible to all types, and dually infected males incompatible with all but the dually infected class, allows polymorphism in *Wolbachia* cytotype to exist (35). The dually infected class is most fit, but singly infected lineages are continuously generated through inefficient transmission.

If *Wolbachia* strains do not distort the sex ratio and do not cause any CI (i.e. are "no effect" *Wolbachia* strains), then their maintenance in the population may depend on the bacteria either gaining horizontal transmission, as has been found for members of the related genus *Rickettsia* (31), or providing a physiological benefit to their host, as has been suggested in nematodes (3). Such physiological benefits may exist even in the presence of sex-ratio distortion. However, they are somewhat less likely in such circumstances because (with the exception of certain strains inducing parthenogenesis) such infections are polymorphic and the host is therefore unable to depend on the presence of the bacterium, impeding coevolution between host and bacterium, therefore reducing the potential contribution of the symbiont to the host.

EVOLUTIONARY DYNAMICS OF *WOLBACHIA* INFECTIONS

Evolutionary Dynamics of Feminizing, Parthenogenesis-Inducing and Male-Killing *Wolbachia* Strains

The above describes the population biology of *Wolbachia* strains and is a scenario in which the host is depicted as a passive background upon which *Wolbachia* strains increase or decrease in prevalence. However, for *Wolbachia* bacteria that distort host sex ratio or sexuality, their presence produces selection on the host. *Wolbachia* bacteria that distort the sex ratio or sexuality produce populations that are female biased. In such populations, males have higher per capita reproductive success than females. Thus, when feminizing *Wolbachia* bacteria have spread into a population, there is selection on the host for genes that prevent their action and transmission, because these promote the production of males.

Empirical studies bear out this prediction. Host genes preventing the transmission of feminizing *Wolbachia* bacteria are known in *A. vulgare* (69). Although host genes preventing the feminizing action of *Wolbachia* bacteria have not been proven to occur, there is some evidence for their presence in *Porcellionides pruinosis*. Rigaud (103) notes the presence of functional infected males in this species (65), which strongly suggests the presence of genes preventing the action of *Wolbachia* bacteria. Further, selection may promote the production of a male-biased primary sex ratio directly (43, 150), although there is as yet to our knowledge no direct evidence of this in *Wolbachia*-isopod associations.

In a similar fashion, PI strains, when polymorphic, are also parasitic, preventing the production of males in a population in which males are rare, and there is evidence that selection has promoted repressor elements in such species (131). Male-killing *Wolbachia* bacteria are the most clearly parasitic, with infected females producing only a fraction of the total progeny produced by an uninfected female, and all of these are female. Resistance genes are predicted in these systems, although they are yet to be discovered in either of the *Wolbachia*-host interactions documented to date.

Feminizing *Wolbachia* bacteria may also be important in the evolution of host sex determination systems (103). In female-heterogametic species, the spread of *Wolbachia* infection tends to produce loss of the female-determining chromosome, leaving the species with a sex determination system based on the presence of the bacteria and host genes affecting their expression and transmission (138). This is seen in various populations of *A. vulgare* (67, 68).

Evolutionary Dynamics of Cytoplasmic Incompatibility-Inducing *Wolbachia* Strains

The strength of incompatibility produced by *Wolbachia* bacteria in a cross between infected male and uninfected female is subject to selection. It has been shown that, if weaker incompatibility in crosses is associated with a reduced cost to females of possessing *Wolbachia* infection and there is no association between the strength of incompatibility and resistance to it, then *Wolbachia* strains producing weaker incompatibility may spread (141). This would result in the production of *Wolbachia* bacteria resistant to CI, but not causing it, as are found in *D. simulans* (8, 86, 155).

Frank (34) has shown that the result of selection on the strength of CI produced by *Wolbachia* bacteria depends on three factors. First, there is the strength of the correlation between incompatibility and cost. If strong incompatibility is associated with a high cost, this will favor weakened incompatibility. Second, there is the intensity of kin-kin interactions in the host. If populations are dense, such that *Wolbachia* bacteria within an area are closely related, then the incompatibility of *Wolbachia* bacteria will benefit itself, rather than less-related strains that may have lowered incompatibility. Thus, strong kin-kin interactions select for strong incompatibility. Third, there is the transmission efficiency of the *Wolbachia* bacteria. Inefficient transmission leads to the continual production of uninfecteds, which selects for high incompatibility.

Incompatibility may therefore be strengthened toward perfect penetrance (if there is no correlation between CI strength and cost), maintained at significant levels of penetrance (if there is some correlation but transmission is inefficient or kin-kin interactions strong), or lost (if there is a correlation and transmission is perfectly efficient or if transmission is good but kin-kin interactions weak). Clearly, if incompatibility is weakened or lost and there are no beneficial effects of infection, then the *Wolbachia* infection may disappear from the population.

The commonness of CI-producing *Wolbachia* strains among species will depend on two factors. First, it will depend on the number of host species-*Wolbachia* interactions in which high incompatibility levels are selected for and can thus maintain CI-causing *Wolbachia* over significant periods. Some species (those with dense populations) can be permanent reservoirs of high CI-causing *Wolbachia* strains, and some *Wolbachia*-host interactions, by virtue of the low transmission efficiency produced, select for high CI. Second, it will depend on the rate of horizontal transmission of infections. CI *Wolbachia* bacteria may be maintained in a range of species by virtue of horizontal transmission of CI-competent *Wolbachia* bacteria between species. Horizontal transmission is well known for *Wolbachia* bacteria (92, 154, 158), and the rate of transfer of CI-causing *Wolbachia* bacteria between species will be an important determinant of their commonness.

Wolbachia Bacteria, Cytoplasmic Incompatibility, and Speciation

As has been previously mentioned, different populations of a species may become infected with different *Wolbachia* strains, each of which causes CI but some of which are mutually incompatible, in that crosses between individuals bearing different strains fail. Bidirectional incompatibility produced by the possession of different *Wolbachia* strains by individuals of different populations makes the individuals from the different populations reproductively isolated. This reproductive isolation can be near complete, as is witnessed in crosses between the parasitoid wasps *N. vitripennis* and *N. giraulti* (16). *Wolbachia* bacteria causing CI thus have the potential to act as agents causing speciation and have been dubbed agents of "infectious speciation" (28).

As is the nature with studies of speciation, there is as yet no direct evidence linking the presence of *Wolbachia* bacteria to a speciation event. The evidence required would be the presence of infected sibling species. Crosses between these sibling species are inviable only by virtue of the possession of *Wolbachia* bacteria. In the case of *N. vitripennis-N. giraulti* crosses, for instance, although F1 progeny are viable in the absence of *Wolbachia* bacteria, the action of nuclear genes causes hybrid breakdown by the F2 (17). Thus, in this case, it is possible that either the *Wolbachia* strains produced the initial speciation (with the populations later diverging at nuclear loci creating hybrid breakdown) or divergence of nuclear genes occurred first, followed by the spread of different *Wolbachia* strains through the two already-isolated populations. Alternatively, a combination of *Wolbachia* bacteria and nuclear genes may have been important in producing isolation. Simply speaking, the problem is that usually found in speciation biology; speciation occurs owing to some nuclear genetic/cytoplasmic divergence in the past, and this is followed by the buildup of other nuclear genes/cytoplasmic factors causing incompatibilities, such that it is impossible to tell which of the currently present factors was originally important in producing isolation.

We cannot therefore delineate the importance of *Wolbachia* bacteria in speciation empirically. Its importance is a matter of current debate (58, 151). One of the major issues is whether *Wolbachia* bacteria alone can produce complete reproductive isolation. Incomplete penetrance of the CI phenotype and incomplete transmission of the bacterium allow gene flow to occur between differently infected populations. Thus, *Wolbachia* strains producing complete reproductive isolation may be the exception rather than the rule. However, it may produce sufficient reproductive isolation to select for assortative mating of the host by incompatibility type (host population), in a process termed reinforcement.

In addition to the incomplete nature of *Wolbachia*-induced reproductive isolation, the reproductive isolation produced by *Wolbachia* bacteria may be transient. CI may wane in intensity over time owing to selection on the bacterium (see above), and horizontal transmission of *Wolbachia* strains between differently infected strains could create a dually infected individual, compatible with all. This cytotype would spread, restoring compatibility and removing reproductive isolation. In this case, *Wolbachia* bacteria would be important in speciation only if incompatibility remained long enough to allow the divergence of the two populations at nuclear genes, producing nuclear incompatibility between them.

APPLICATIONS

Cytoplasmic Incompatibility as a Method for Modifying Pest Populations

Even before the causative agent of CI was known, experiments were already done to use the CI caused by *Wolbachia* bacteria as a method for mosquito control (76). The basic idea here is to release vast quantities of males that will render the females with which they mate sterile because the incompatible matings result in no offspring. Although the experiments both in the laboratory and the field were promising, the vast amounts of work in separating the males from the females made these techniques inapplicable on a large scale. No recent work has been done to apply *Wolbachia* bacteria in these sterile-insect techniques.

Other ideas that have been tried are to use bidirectionally incompatible *Wolbachia* strains for population replacement (77). The goal of this technique is to replace the existing population with another, less-harmful population of the same species. This method has also been tested on a small scale and proved to be successful (29, 30). The problem with this method is that it requires an absolute incompatibility between the two lines. If through the production of compatible sperm by older males the genotype of the existing population enters the released population, the replacement fails.

The method receiving the most attention recently is to use CI *Wolbachia* bacteria as a driving factor to bring new traits into existing populations. The invasion of *Wolbachia* bacteria in *D. simulans* populations in California showed that other cytoplasmic factors hitchhike along with the spreading *Wolbachia* infection (142, 144). In this case, the factor was a mitochondrial variant in which the first infection with *Wolbachia* bacteria must have taken place. Wild populations could be transformed with desirable traits by coinfecting a genetically modified cytoplasmically inherited factor such as a virus or a symbiotic bacterium with a CI *Wolbachia* bacterium (4). The CI *Wolbachia* bacteria would be the driving force, whereas the other cytoplasmic factor would express the desirable gene. For this to be successful, the cytoplasmic factor containing the desirable gene should remain linked to the *Wolbachia* infection. If the cytoplasmic factor becomes unlinked, the *Wolbachia* bacteria spread without the desired genetic transformation taking place. For the spreading of the trait, it would be better to transform the *Wolbachia* bacteria. Initially, this was thought not to be feasible because (*a*) transforming *Wolbachia* bacteria is difficult because they cannot be cultured easily, and (*b*) *Wolbachia* bacteria were thought to be abundant only in the host's reproductive tissues. Many of the desirable traits such as interference with arthropod-borne diseases require its expression in other tissues like the gut or the hemolymph. Recent studies indicate that *Wolbachia* bacteria may not be limited only to the reproductive tissues but may occur throughout the host.

Improvement of Wasps Through Introduction of Parthenogenesis *Wolbachia* Bacteria

In biological control using parasitoid wasps, pest insects are controlled because the parasitoid larvae develop by eating the pest insect. PI *Wolbachia*-infected wasps may be better at controlling the pest than the uninfected populations of the same parasitoid species, because all the offspring will consist of females. Three potential advantages exist for the PI *Wolbachia* bacteria-infected wasps: (*a*) the production costs in mass rearing per female is less, (*b*) infected wasps may have a higher population growth rate, and (*c*) infected wasps may be able to depress the pest insect population to a lower level (130). Efforts to transfect the PI *Wolbachia* bacteria from an infected species to species without the infection have met with little success. Only in the parasitoid wasp *Trichogramma dendrolimi* has the PI *Wolbachia* bacteria from another *Trichogramma* species been introduced successfully (40). However, the newly acquired infection only leads to a very low penetrance of the parthenogenesis phenotype. Fewer than 0.5% of the offspring of infected virgin females were daughters.

CONCLUDING REMARKS

There is a strong impression that *Wolbachia* bacteria are unique among inherited bacteria of insects in the range of host manipulations that have evolved. While there is taxonomic diversity in the inherited bacteria present in insects [proteobacteria, flavobacteria, and mollicutes have all been found (155)], these currently fall into three main camps: they have epidemiologically significant levels of horizontal transmission, are beneficial to their host, or kill male hosts during embryogenesis. Other than *Wolbachia* bacteria, no bacteria have been observed to produce feminization, CI, and PI, and only for *Wolbachia* bacteria have so many phenotypes been observed.

Is the genus *Wolbachia* really unique? It could be argued that, in this age of PCR testing for *Wolbachia* presence, our impression is an artifact of looking for *Wolbachia* bacteria and then for phenotypes, rather than looking for phenotypes and then identifying the agent responsible. It is notable that in the areas where phenotypes have always preceded symbiont identification (male killing, beneficial effects), many different agents have been observed, only one of which is *Wolbachia* bacteria.

However, this criticism is not entirely fair. Many records of CI were obtained before the advent of *Wolbachia*-specific PCR tests (50, 52, 71, 75, 91, 115, 147, 162), and these have all been subsequently found to be associated with *Wolbachia* bacteria. Similarly, symbiont-induced parthenogenesis was identified three times through phenotype (135, 164, 165), and all three cases were later found to be associated with *Wolbachia* strains (132, 158, 163). In the case of feminization, there are fewer data. However, the recent case of feminization in *O. furnacalis*, the Asian corn borer, was identified first from phenotype and antibiotic treatment (70) and was later found to be associated with *Wolbachia* presence (S. Hoshizaki, personal communication). Thus, although it can never be certain that only *Wolbachia* strains cause PI, CI, and feminization, the evidence does suggest that the majority of these cases in insects will turn out to be associated with *Wolbachia* presence. In addition, we can firmly state that *Wolbachia* bacteria do show unusual plasticity in the manipulations they achieve and the range of hosts in which they achieve them.

The plasticity of *Wolbachia* bacteria raises many questions, as yet answered only partially at best. What is it that gives *Wolbachia* bacteria such amazing plasticity? Is there one major innovation that has been modified several times or many different innovations? What is the molecular basis of their interaction with host chromosomes? What is the genetic basis of differences in phenotype?

There is a great temptation to believe that there is one innovation in *Wolbachia* bacteria that has been modified to produce different reproductive manipulations of the host. Two of the mechanisms by which *Wolbachia* bacteria produce their responses are at least superficially similar. CI is created through condensation of the paternal chromosome set. PI is produced by a modification of chromosome behavior during the first mitotic divisions of the host. Indeed, it could also be the means by which male killing is effected, although no studies of this have been carried out to date. The exception here is feminization in isopods, in which chromosome manipulation has not been implicated, although its role in preventing the formation of the androgenic gland has not to our knowledge been investigated. It will clearly be instructive to look at the mechanistic basis of male death in the case of early male killing (is it caused by widespread chromosome condensation?) and also to examine the root causes of feminization.

When the basic cause of *Wolbachia* manipulations comes to light, it will then be time to dissect the molecular details of interaction with the host. These questions have started to be addressed for CI-inducing *Wolbachia* bacteria in *Drosophila* spp.

(118). What chemicals are being produced by *Wolbachia* bacteria and what are their targets?

The greatest impediment to the study of this bacterium and its interaction with its host is its current refractoriness to cell-free in vitro cultures. This complicates analysis of *Wolbachia* genetics, because we cannot easily perform experiments investigating the effect of defined mutant *Wolbachia* strains on *Wolbachia*-host interaction. At present, we are uncertain even as to many of the fundamental biological features of this bacterium, such as whether it possesses plasmids or phages. The uncovering of these basic biological features is necessary before we can evaluate the potential role of plasmid and phage in producing transfer of phenotypic effects between *Wolbachia* strains.

One hope is that a full genome sequence will be obtained. With this, potentially important genes can be identified, expressed, and characterized in vitro, and their pattern of expression can be observed in vivo. From this, we may gain insights without the presence of defined mutant strains.

Visit the Annual Reviews home page at http://www.AnnualReviews.org

LITERATURE CITED

1. Andersson SGE, Zomorodipur A, Andersson JO, Sicheritz-Ponten T, Alsmark UCM, et al. 1998. The genome sequence of *Rickettsia prowazekii* and the origin of mitochondria. *Nature* 396:133–40.
2. Bandi C, Anderson TJC, Genchi C, Blaxter ML. 1998. Phylogeny of *Wolbachia*-like bacteria in filarial nematodes. *Proc. R. Soc. London Ser. B* 265:2407–13
3. Bandi C, McCall JW, Genchi C, Corona S, Venco L, et al. 1999. Effects of tetracycline on the filarial worms *Brugia pahangi* and *Dirofilaria immitis* and their bacterial endosymbionts *Wolbachia*. *Int. J. Parasitol.* In press
4. Beard CB, O'Neill SL, Tesh RB, Richards FF, Aksoy S. 1993. Modification of arthropod vector competence via symbiotic bacteria. *Parasitol. Today* 9:179–83
5. Binnington KL, Hoffmann AA. 1989. *Wolbachia*-like organisms and cytoplasmic incompatibility in *Drosophila simulans*. *J. Invert. Pathol.* 54:344–52
6. Bordenstein SR, Werren JH. 1998. Effects of A and B *Wolbachia* and host genotype on interspecies cytoplasmic incompatibility in *Nasonia*. *Genetics* 148:1833–44
7. Bouchon D, Rigaud T, Juchault P. 1998. Evidence for widespread *Wolbachia* infection in isopod crustaceans: molecular identification and host feminization. *Proc. R. Soc. London Ser. B* 265:1081–90
8. Bourtzis K, Dobson SL, Braig HR, O'Neill SL. 1998. Rescuing *Wolbachia* have been overlooked... *Nature* 391:852–53
9. Bourtzis K, Nirgianaki A, Markakis G, Savakis C. 1996. *Wolbachia* infection and cytoplasmic incompatibility in *Drosophila* species. *Genetics* 144:1063–73
10. Boyle L, O'Neill SL, Robertson HM, Karr TL. 1993. Interspecific and intraspecific horizontal transfer of *Wolbachia* in *Drosophila*. *Science* 260:1796–99
11. Braig HR, Guzman H, Tesh RB, O'Neill SL. 1994. Replacement of the natural *Wolbachia* symbiont of *Drosophila simulans* with a mosquito counterpart. *Nature* 367:453–55
12. Braig HR, Zhou WG, Dobson SL, O'Neill SL. 1998. Cloning and characterization of a gene encoding the major surface protein of the bacterial endosymbiont *Wolbachia pipientis*. *J. Bacteriol.* 180:2373–78

13. Breeuwer JAJ. 1997. *Wolbachia* and cytoplasmic incompatibility in the spider mites *Tetranychus urticae* and *T. turkestani*. *Heredity* 79:41–47

14. Breeuwer JAJ, Jacobs G. 1996. *Wolbachia*: intracellular manipulators of mite reproduction. *Exp. Appl. Acarol.* 20:421–34

15. Breeuwer JAJ, Stouthamer R, Barns SM, Pelletier DA, Weisburg WG, Werren JH. 1992. Phylogeny of cytoplasmic incompatibility microorganisms in the parasitoid wasp genus *Nasonia* based on 16S ribosomal DNA sequences. *Insect Mol. Biol.* 1:25–36

16. Breeuwer JAJ, Werren JH. 1990. Microorganisms associated with chromosome destruction and reproductive isolation between two insect species. *Nature* 346:558–60

17. Breeuwer JAJ, Werren JH. 1993. Effect of genotype on cytoplasmic incompatibility between two species of *Nasonia*. *Heredity* 70:428–36

18. Bressac C, Rousset F. 1993. The reproductive incompatibility system in *Drosophila simulans*: Dapi-staining analysis of the *Wolbachia* symbionts in sperm cysts. *J. Invert. Pathol.* 61:226–30

19. Bulnheim H-P, Vavra J. 1968. Infection by the microsporidian *Octaspora effeminans* sp.n., and its sex determining influence in the amphipod *Gammarus duebeni*. *J. Parasitol.* 54:241–48

20. Byers JR, Wilkes A. 1970. A rickettsialike microorganism in *Dahlbominus fuscipennis*: observations on its occurrence and ultrastructure. *Can. J. Zool.* 48:959–64

21. Callaini G, Dallai R, Riparbelli MG. 1997. *Wolbachia*-induced delay of paternal chromatin condensation does not prevent maternal chromosomes from entering anaphase in incompatible crosses of *Drosophila simulans*. *J. Cell. Sci.* 110:271–80

22. Callaini G, Riparbelli MG, Dallai R. 1994. The distribution of cytoplasmic bacteria in the early *Drosophila* embryo is mediated by astral microtubules. *J. Cell. Sci.* 107:673–82

23. Caspari E, Watson GS. 1959. On the evolutionary importance of cytoplasmic sterility in mosquitoes. *Evolution* 13:568–70

24. Chang NW, Wade MJ. 1994. The transfer of *Wolbachia pipientis* and reproductive incompatibility between infected and uninfected strains of the flour beetle, *Tribolium confusum*, by microinjection. *Can. J. Microbiol.* 40:978–81

25. Clancy DJ, Hoffmann AA. 1996. Cytoplasmic incompatibility in *Drosophila simulans*: evolving complexity. *TREE* 11:145–46

26. Clancy DJ, Hoffmann AA. 1997. Behavior of *Wolbachia* endosymbionts from *Drosophila simulans* in *Drosophila serrata*, a novel host. *Am. Nat.* 149:975–88

27. Clancy DJ, Hoffmann AA. 1998. Environmental effects on cytoplasmic incompatibility and bacterial load in *Wolbachia*-infected *Drosophila simulans*. *Entomol. Exp. Appl.* 86:13–24

28. Coyne JA. 1992. Genetics of speciation. *Nature* 355:511–15

29. Curtis CF. 1976. Population replacement in *Culex fatigans* by means of cytoplasmic incompatibility. 2. Field cage experiments with overlapping generations. *Bull. WHO* 53:107–19

30. Curtis CF, Adak T. 1974. Population replacement in Culex fatigans by means of cytoplasmic incompatibility. 1. Laboratory experiments with non-overlapping generations. *Bull. WHO* 51:249–55

31. Davis MJ, Ying Z, Brunner BR, Pantoja A, Ferwerda FH. 1998. Rickettsial relative associated with papaya bunchy top disease. *Curr. Microbiol.* 36:80–84

32. Degrugillier ME, Degrugillier SS. 1991. Nonoccluded, cytoplasmic virus particles and rickettsia-like organisms in testes and spermathecae of *Diabrotica virgifera*. *J. Invert. Pathol.* 57:50–58

33. Dev V. 1986. Non-reciprocal fertility among species of the *Aedes (Stegomyia)*

scutellaris group. *Experientia* 42:803–6

34. Frank SA. 1997. Cytoplasmic incompatibility and population structure. *J. Theor. Biol.* 184:327–30

35. Frank SA. 1998. Dynamics of cytoplasmic incompatibility with multiple *Wolbachia* infections. *J. Theor. Biol.* 192:213–18

36. Ginsburg-Vogel T, Carre-Lecuyer MC, Fried-Montaufier MC. 1980. Transmission expérimentale del la thélygenie liee a l'intersexualité chez *Orchestia gammarellus* (Pallas); analyse des génotypes sexuels dans la descendance des femelles thélygenes. *Arch. Zool. Exp. Gen.* 122:261–70

37. Giordano R, Jackson JJ, Robertson HM. 1997. The role of *Wolbachia* bacteria in reproductive incompatibilities and hybrid zones of *Diabrotica* beetles and *Gryllus* crickets. *Proc. Natl. Acad. Sci. USA* 94:11439–44

38. Giordano R, O'Neill SL, Robertson HM. 1995. *Wolbachia* infections and the expression of cytoplasmic incompatibility in *Drosophila sechellia* and *D. mauritiana*. *Genetics* 140:1307–17

39. Girin C, Bouletreau M. 1995. Microorganism-associated variation in host infestation efficiency in a parasitoid wasp *Trichogramma bourarachae*. *Experientia* 52:398–402

40. Grenier S, Pintureau B, Heddi A, Lassabliere F, Jager C, et al. 1998. Successful horizontal transfer of *Wolbachia* symbionts between *Trichogramma* wasps. *Proc. R. Soc. London. Ser. B* 265:1441–45

41. Hackett KJ, Lynn DE, Williamson DL, Ginsberg AS, Whitcomb RF. 1986. Cultivation of the *Drosophila* sex-ratio spiroplasm. *Science* 232:1253–55

42. Hariri AR, Werren JH, Wilkinson GS. 1998. Distribution and reproductive effects of *Wolbachia* in stalk-eyed flies. *Heredity* 81:254–60

43. Hatcher MJ, Dunn AM. 1995. Evolutionary consequences of cytoplasmically inherited feminizing factors. *Philos. Trans. R. Soc.* 348:445–56

44. Hertig M. 1936. The rickettsia, *Wolbachia pipiens* (gen. et sp.n.) and associated inclusions of the mosquito, *Culex pipiens*. *Parasitology* 28:453–86

45. Hertig M, Wolbach SB. 1924. Studies on rickettsia-like microorganisms in insects. *J. Med. Res.* 44:329–74

46. Hoffmann AA, Clancy D, Duncan J. 1996. Naturally-occurring *Wolbachia* infection in *Drosophila simulans* that does not cause cytoplasmic incompatibility. *Heredity* 76:1–8

47. Hoffmann AA, Hercus M, Dagher H. 1998. Population dynamics of the *Wolbachia* infection causing cytoplasmic incompatibility in *Drosophila melanogaster*. *Genetics* 148:221–31

48. Hoffmann AA, Turelli M. 1988. Unidirectional incompatibility in *Drosophila simulans*: inheritance, geographic variation and fitness effects. *Genetics* 119:435–44

49. Hoffmann AA, Turelli M, Harshman LG. 1990. Factors affecting the distribution of cytoplasmic incompatibility in *Drosophila simulans*. *Genetics* 126:933–48

50. Hoffmann AA, Turelli M, Simmons GM. 1986. Unidirectional incompatibility between populations of *Drosophila simulans*. *Evolution* 40:692–701

51. Hoshizaki S, Shimada T. 1995. PCR-based detection of *Wolbachia*, cytoplasmic incompatibility microorganisms, infected in natural populations of *Laodelphax striatellus* in central Japan: has the distribution of *Wolbachia* spread recently? *Insect Mol. Biol.* 4:237–43

52. Hsiao C, Hsiao TH. 1985. *Rickettsia* as the cause of cytoplasmic incompatibility in the alfalfa weevil, *Hypera postica*. *J. Invert. Pathol.* 45:244–46

53. Hurst GDD. 1997. *Wolbachia*, cytoplasmic incompatibility, and the evolution of eusociality. *J. Theor. Biol.* 184:99–100

54. Hurst GDD, Hammarton TC, Bandi C, Majerus TMO, Bertrand D, et al. 1997. The

diversity of inherited parasites of insects: the male-killing agent of the ladybird beetle *Coleomegilla maculata* is a member of the Flavobacteria. *Genet. Res.* 70:1–6

55. Hurst GDD, Hammarton TC, Obrycki JJ, Tamsin MO, Majerus LE, et al. 1996. Male-killing bacterium in a fifth ladybird beetle, *Coleomegilla maculata* (Coleoptera:Coccinellidae). *Heredity* 77:177–85

56. Hurst GDD, Jiggins FM, Van der Schulenburg JHG, Bertrand D, West SA, et al. 1999. Male-killing *Wolbachia* in two species of insect. *Proc. R. Soc. London Ser. B.* In press

57. Hurst GDD, Majerus MEN, Walker LE. 1992. Cytoplasmic male killing elements in *Adalia bipunctata*. *Heredity* 69:84–91

58. Hurst GDD, Schilthuizen M. 1998. Selfish genetic elements and speciation. *Heredity* 80:2–8

59. Hurst LD. 1991. The incidences and evolution of cytoplasmic male killers. *Proc. R. Soc. London Ser. B* 244:91–99

60. Hurst LD, McVean GT. 1996. Clade selection, reversible evolution and the persistence of selfish elements: the evolutionary dynamics of cytoplasmic incompatibility. *Proc. R. Soc. London Ser. B* 263:97–104

61. Jiggins FM, Hurst GDD, Majerus MEN. 1998. Sex ratio distortion in *Acraea encedon* is caused by a male-killing bacterium. *Heredity* 81:87–91

62. Johanowicz DL, Hoy MA. 1995. Molecular evidence for a *Wolbachia* endocytobiont in the predatory mite *Metaseiulus occidentalis*. *J. Cell. Biochem.* Suppl. 21A:198

63. Johanowicz DL, Hoy MA. 1998. Experimental induction and termination of nonreciprocal reproductive incompatibilities in a parahaploid mite. *Entomol. Exp. Appl.* 87:51–58

64. Jost E. 1970. Untersuchungen zur Inkompatibilitat im *Culex-pipiens*-Komplex. *Wilhelm Roux' Arch.* 166:173–88

65. Juchault P, Frelon M, Bouchon D, Rigaud

T. 1994. New evidence for feminizing bacteria in terrestrial isopods: evolutionary implications. *C. R. Acad. Sci. III Paris* 317:225–30

66. Juchault P, Legrand JJ. 1985. Contribution à l'étude du mechanisme de l'état réfractaire à l'hormone androgène chez les *Armadillidium vulgare* hérbergeant une bactérie féminisante. *Gen. Comp. Endocrinol.* 463–67

67. Juchault P, Legrand JJ, Mocquard JP. 1980. Contribution à l'étude qualitative et quantitative des facteurs contrôlant le sexe dans les populations du Crustacé Isopode terrestre *Armadillidium vulgare*. 1-La population de Niort (Deux-Sèvres). *Arch. Zool. Exp. Gen.* 121:3–127

68. Juchault P, Mocquard JP. 1993. Transfer of a parasitic sex factor to the nuclear genome of the host: a hypothesis on the evolution of sex-determining mechanisms in the terrestrial isopod *Armadillidium vulgare* Latr. *J. Evol. Biol.* 6:511–28

69. Juchault P, Rigaud T, Mocquard JP. 1992. Evolution of sex-determining mechanisms in a wild population of *Armadillidium vulgare*: competition between two feminizing parasitic sex factors. *Heredity* 69:382–90

70. Kageyama D, Hoshizaki S, Ishikawa Y. 1998. Female biased sex ratio in the Asian corn borer, *Ostrinia furnacalis*: evidence for the occurrence of feminizing bacteria in an insect. *Heredity* 81:311–16

71. Kellen WR, Hoffmann DF. 1981. *Wolbachia* sp. a symbiont of the almond moth, *Ephestia cautella*: ultrastructure and influence on host fertility. *J. Invert. Pathol.* 37:273–83

72. Kose H, Karr TL. 1995. Organization of *Wolbachia pipientis* in the *Drosophila* fertilized egg and embryo revealed by anti-*Wolbachia* monoclonal antibody. *Mech. Dev.* 51:275–88

73. Lassy CW, Karr TL. 1996. Cytological analysis of fertilization and early embryonic development in incompatible crosses

of *Drosophila simulans. Mech. Dev.* 57:47–58

74. Laven H. 1951. Crossing experiments with *Culex* strains. *Evolution* 5:370–75

75. Laven H. 1957. Vererbung durch Kerngene und das Problem der ausserkaryotischen Vererbung bei *Culex pipiens*. II. Ausserkaryotische Vererbung. *Z. Indukt. Abstamm. Vererbungsl.* 88:478–516

76. Laven H. 1967. Eradication of *Culex pipiens fatigans* through cytoplasmic incompatibility. *Nature* 261:383–84

77. Laven H, Aslamkhan M. 1970. Control of *Culex pipiens pipiens* and *C. p. fatigans* with integrated genetical systems. *Pak. J. Sci.* 22:303–12

78. LeGrand J-J, LeGrand-Hamelin E, Juchault P. 1987. Sex determination in Crustacea. *Biol. Rev.* 62:439–70

79. Louis C, Nigro L. 1989. Ultrastructural evidence of *Wolbachia* Rickettsiales in *Drosophila simulans* and their relationships with unidirectional cross-incompatibility. *J. Invert. Pathol.* 54:39–44

80. Louis C, Pintureau B, Chapelle L. 1993. Research on the origin of unisexuality: thermotherapy cures both rickettsia and thelytokous parthenogenesis in a *Trichogramma* species. *C. R. Acad. Sci. III Paris* 316:27–33

81. Lus YY. 1947. Some rules of reproduction in populations of *Adalia bipunctata*. II. Non-male strains in populations. *Dokl. Akad. Nauk. SSSR* 57:951–54

82. Magnin M, Pasteur N, Raymond M. 1987. Multiple incompatibilities within populations of *Culex pipiens* in southern France. *Genetica* 74:125–30

83. Martin G, Juchault P, Sorokine O, Van Dorsselaer A. 1990. Purification and characterization of androgenic hormone from the terrestrial isopod *Armadillidium vulgare. Gen. Comp. Endocrinol.* 80:349–54

84. Masui S, Sasaki T, Ishikawa H. 1997. GroE-homologous operon of *Wolbachia*, an intracellular symbiont of arthropods: a new approach for their phylogeny. *Zool. Sci.* 14:701–6

85. Merçot H, Llorente B, Jacques M, Atlan A, Montchamp-Moreau C. 1995. Variability within the Seychelles cytoplasmic incompatibility system in *Drosophila simulans. Genetics* 141:1015–23

86. Merçot H, Poinsot D. 1998. . . . and discovered on mount Kilimanjaro. *Nature* 391:853–53

87. Min KT, Benzer S. 1997. Wolbachia, normally a symbiont of drosophila, can be virulent, causing degeneration and early death. *Proc. Natl. Acad. Sci. USA* 94:10792–96

88. Montchamp-Moreau C, Ferveur J-F, Jacques M. 1991. Geographic distribution and inheritance of three cytoplasmic incompatibility types in *Drosophila simulans. Genetics* 129:399–407

89. Ndiaye M, Mattei X, Thiaw OT. 1995. Extracellular and intracellular rickettsia-like microorganisms in gonads of mosquitoes. *J. Submicrosc. Cytol. Pathol.* 27:557–63

90. Noda H. 1984. Cytoplasmic incompatibility in a plant ricehopper. *J. Hered.* 75:345–48

91. Noda H. 1984. Cytoplasmic incompatibility in allopatric field populations of the small brown planthopper, *Laodelphax striatellus*, in Japan. *Entomol. Exp. Appl.* 35:263–67

92. O'Neill SL, Giordano R, Colbert AME, Karr TL, Robertson HM. 1992. 16S rRNA phylogenetic analysis of the bacterial endosymbionts associated with cytoplasmic incompatibility in insects. *Proc. Natl. Acad. Sci. USA* 89:2699–702

93. O'Neill SL, Hoffman AA, Werren JH, eds. 1997. *Influential Passenger: Inherited Microorganisms and Arthropod Reproduction*, pp. 1–214. New York: Oxford Univ. Press

94. O'Neill SL, Karr TL. 1990. Bidirectional incompatibility between conspecific populations of *Drosophila simulans. Nature* 348:178–80

95. O'Neill SL, Pettigrew MM, Sinkins SP, Braig HR, Andreadis TG, et al. 1997. In vitro cultivation of *Wolbachia pipientis* in an *Aedes albopictus* cell line. *Insect Mol. Biol.* 6:33–39

96. Perrot-Minnot M-J, Guo LR, Werren JH. 1996. Single and double infections with *Wolbachia* in the parasitic wasp *Nasonia vitripennis*: effects on compatibility. *Genetics* 143:961–72

97. Pijls JWAM, van Steenbergen HJ, van Alphen JJM. 1996. Asexuality cured: the relations and differences between sexual and asexual *Apoanagyrus diversicornis*. *Heredity* 76:506–13

98. Poinsot D, Bourtzis K, Markakis G, Savakis C, Merçot H. 1998. *Wolbachia* transfer from *Drosophila melanogaster* into *D. simulans*: host effect and cytoplasmic incompatibility relationships. *Genetics* 150:227–37

99. Poinsot D, Merçot H. 1997. *Wolbachia* infection in *Drosophila simulans*: does the female host bear a physiological cost. *Evolution* 51:180–86

100. Prout T. 1994. Some evolutionary possibilities for a microbe that causes incompatibility in its host. *Evolution* 48:909–11

101. Reed KM, Werren JH. 1995. Induction of paternal genome loss by the paternal-sex-ratio chromosome and cytoplasmic incompatibility bacteria (*Wolbachia*): a comparative study of early embryonic events. *Mol. Reprod. Dev.* 40:408–18

102. Richardson PM, Holmes WP, Saul GB II. 1987. The effect of tetracycline on reciprocal cross incompatibility in *Mormoniella* [=*Nasonia*] *vitripennis*. *J. Invert. Pathol.* 50:176–83

103. Rigaud T. 1997. Inherited microorganisms and sex determination of arthropod hosts. See Ref. 93, pp. 81–102.

104. Rigaud T, Juchault P. 1993. Conflict between feminizing sex ratio distorters and an autosomal masculinizing gene in the terrestrial isopod *Armadillidium vulgare*. *Genetics* 133(2):247–52

105. Rigaud T, Juchault P. 1995. Success and failure of horizontal transfer of feminizing *Wolbachia* endosymbionts in woodlice. *J. Evol. Biol.* 8:249–55

106. Rigaud T, Juchault P, Mocquard J. 1991. Experimental study of temperature effects on the sex ratio of broods in terrestrial Crustacea *Armadillidium vulgare* Latr. Possible implications in natural populations. *J. Evol. Biol.* 4:603–17

107. Rigaud T, Rousset F. 1996. What generates the diversity of *Wolbachia*-arthropod interactions? *Biodivers. Conserv.* 5:999–1013

108. Rigaud T, Souty-Grosset C, Raimond R, Mocquard J, Juchault P. 1991. Feminizing endocytobiosis in the terrestrial crustacean *Armadillidium vulgare* LATR. (Isopoda): recent acquisitions. *Endocytobiosis Cell. Res.* 7:259–73

109. Roberts LW, Rapmund G, Cadigan FC. 1977. Sex ratios in *Rickettsia tsutsugamushi*-infected and noninfected colonies of *Leptotrombidium*. *J. Med. Entomol.* 14:89–92

110. Rousset F, Bouchon D, Pintureau B, Juchault P, Solignac M. 1992. *Wolbachia* endosymbionts responsible for various alterations of sexuality in arthropods. *Proc. R. Soc. London Ser. B* 250:91–98

111. Rousset F, Raymond M, Kjellberg F. 1991. Cytoplasmic incompatibilities in the mosquito *Culex pipiens*: how to explain a cytotype polymorphism? *J. Evol. Biol.* 4:69–81

112. Rousset F, Solignac M. 1995. Evolution of single and double *Wolbachia* symbioses during speciation in the *Drosophila simulans* complex. *Proc. Natl. Acad. Sci. USA* 92:6389–93

113. Rousset F, Stordeur E. 1994. Properties of *Drosophila simulans* strains experimentally infected by different clones of the bacterium *Wolbachia*. *Heredity* 72:325–31

114. Rousset F, Vauhin D, Solignac M. 1992. Molecular identification of *Wolbachia*,

the agent of cytoplasmic incompatibility in *Drosophila simulans*, and variability in relation to host mitochondrial type. *Proc. R. Soc. London Ser. B* 247:163–68

115. Ryan SL, Saul GB II. 1968. Post-fertilization effect of incompatbility factors in *Mormoniella*. *Mol. Gen. Genet.* 103:29–36

116. Ryan SL, Saul GB II, Conner GW. 1985. Aberrant segregation of R-locus genes in male progeny from incompatible crosses in *Mormoniella*. *J. Hered.* 76:21–26

117. Ryan SL, Saul GB II, Conner GW. 1987. Separation of factors containing R-locus genes in *Mormoniella* stocks derived from aberrant segregation following incompatible crosses. *J. Hered.* 78:273–75

118. Sasaki T, Braig HR, O'Neill SL. 1998. Analysis of *Wolbachia* protein synthesis in *Drosophila* in vivo. *Insect Mol. Biol.* 7:101–5

119. Schilthuizen M, Gittenberger E. 1998. Screening mollusks for *Wolbachia* infection. *J. Invert. Pathol.* 71:268–70

120. Schilthuizen M, Honda J, Stouthamer R. 1998. Parthenogenesis-inducing *Wolbachia* in *Trichogramma kaykai* (Hymenoptera: Trichogrammatidae) originates from a single infection. *Ann. Entomol. Soc. Am.* 91:410–14

121. Schilthuizen M, Stouthamer R. 1997. Horizontal transmission of parthenogenesis inducing microbes in *Trichogramma* wasps. *Proc. R. Soc. London Ser. B* 264:361–66

122. Sinkins SP, Braig HR, O'Neill SL. 1995. *Wolbachia* superinfections and the expression of cytoplasmic incompatibility. *Proc. R. Soc. London Ser. B* 261:325–30

123. Sinkins SP, Braig HR, O'Neill SL. 1995. *Wolbachia pipientis*: bacterial density and unidirectional cytoplasmic incompatibility between infected populations of *Aedes albopictus*. *Exp. Parasitol.* 81:284–91

124. Skinner SW. 1982. Maternally-inherited sex ratio in the parasitoid wasp *Nasonia vitripennis*. *Science* 215:1133–34

125. Skinner SW. 1985. Son-killer: a third extrachromosomal factor affecting the sex ratio in the parasitoid wasp, *Nasonia* (= *Mormoniella*) *vitripennis*. *Genetics* 109:745–59

126. Solignac M, Vautrin D, Rousset F. 1994. Widespread occurrence of the proteobacteria *Wolbachia* and partial cytoplasmic incompatibility in *Drosophila melanogaster*. *C. R. Acad. Sci. III Paris* 317:461–70

127. Stevens L. 1989. Environmental factors affecting reproductive incompatibility in flour beetles, genus *Tribolium*. *J. Invert. Pathol.* 53:78–84

128. Stevens L, Wicklow DT. 1992. Multispecies interactions affect cytoplasmic incompatibility in *Tribolium* flour beetles. *Am. Nat.* 140:642–53

129. Stolk C, Stouthamer R. 1995. Influence of a cytoplasmic-incompatibility-inducing *Wolbachia* on the fitness of the parasitoid wasp *Nasonia vitripennis*. *Proc. Sect. Exp. Appl. Entomol. Neth. Entomol. Soc.* 7:33–37

130. Stouthamer R. 1993. The use of sexual versus asexual wasps in biological control. *Entomophaga* 38:3–6

131. Stouthamer R. 1997. *Wolbachia*-induced parthenogenesis. See Ref. 93, pp. 102–24

132. Stouthamer R, Breeuwer JAJ, Luck RF, Werren JH. 1993. Molecular identification of microorganisms associated with parthenogenesis. *Nature* 361:66–68

133. Stouthamer R, Kazmer DJ. 1994. Cytogenetics of microbe-associated parthenogenesis and its consequences for gene flow in *Trichogramma* wasps. *Heredity* 73:317–27

134. Stouthamer R, Luck RF. 1993. Influence of microbe-associated parthenogenesis on the fecundity of *Trichogramma deion* and *T. pretiosum*. *Entomol. Exp. Appl.* 67:183–92

135. Stouthamer R, Luck RF, Hamilton WD.

1990. Antibiotics cause parthenogenetic *Trichogramma* to revert to sex. *Proc. Natl. Acad. Sci. USA* 87:2424–27

136. Stouthamer R, Werren JH. 1993. Microbes associated with parthenogenesis in wasps of the genus *Trichogramma*. *J. Invert. Pathol.* 61:6–9

137. Takahashi M, Urakami H, Yoshida Y, Furuya Y, Misumi H, et al. 1997. Occurrence of high ratio of males after introduction of minocycline in a colony of *Leptotrombidium fletcheri* infected with *Orientia tsutsugamushi*. *Eur. J. Epidemiol.* 13:79–86

138. Taylor DR. 1990. Evolutionary consequences of cytoplasmic sex ratio distorters. *Evol. Ecol.* 4:235–48

139. Terry RS, Dunn AM, Smith JE. 1997. Cellular distribution of a feminizing microsporidian parasite: a strategy for transovarial transmission. *Parasitology* 115:157–63

140. Trpis M, Perrone JB, Reissig M, Parker KL. 1981. Control of cytoplasmic incompatibility in the *Aedes scuttelaris* complex. *J. Hered.* 72:313–17

141. Turelli M. 1994. Evolution of incompatibility-inducing microbes and their hosts. *Evolution* 48:1500–13

142. Turelli M, Hoffmann AA. 1991. Rapid spread of an inherited incompatibility factor in California *Drosophila*. *Nature* 353:440–42

143. Turelli M, Hoffmann AA. 1995. Cytoplasmic incompatibility in *Drosophila simulans*: dynamics and parameter estimates from natural populations. *Genetics* 140:1319–38

144. Turelli M, Hoffmann AA, McKechnie S-W. 1992. Dynamics of cytoplasmic incompatibility and mtDNA variation in natural *Drosophila simulans* populations. *Genetics* 132:713–23

145. van Meer MMM, Witteveldt J, Stouthamer R. 1999. Detailed phylogeny of *Wolbachia* clones involved in the alteration of its host's phenotype based on the wsp gene. *Insect Mol. Biol.* In press

146. Wade MJ, Chang NW. 1995. Increased male fertility in *Tribolium confusum* beetles after infection with the intracellular parasite *Wolbachia*. *Nature* 373:72–64

147. Wade MJ, Stevens L. 1985. Microorganism mediated reproductive isolation in flour beetles. *Science* 227:527–28

148. Weisburg WG, Dobson ME, Samuel JE, Dasch GA, Mallavia LP, et al. 1989. Phylogenetic diversity of the Rickettsiae. *J. Bacteriol.* 4202–6

149. Wenseleers T, Ito F, van Borm S, Huybrechts R, Volckaert F, et al. 1998. Widespread occurrence of the microorganism *Wolbachia* in ants. *Proc. R. Soc. London Ser. B* 265:1447–52

150. Werren JH. 1987. The coevolution of autosomal and cytoplasmic sex ratio factors. *J. Theor. Biol.* 124:317–34

151. Werren JH. 1997. *Wolbachia* and speciation. In *Endless Forms: Species and Speciation*, ed. D Howard, S Berlocher. New York: Oxford Univ. Press

152. Werren JH. 1997. Biology of *Wolbachia*. *Annu. Rev. Entomol.* 42:587–609

153. Werren JH, Hurst GDD, Zhang W, Breeuwer JAJ, Stouthamer R, Majerus MEN. 1994. Rickettsial relative associated with male-killing in the ladybird beetle (*Adalia bipunctata*). *J. Bacteriol.* 176:388–94

154. Werren JH, Jaenike J. 1995. *Wolbachia* and cytoplasmic incompatibility in mycophagous *Drosophila* and their relatives. *Heredity* 75:320–26

155. Werren JH, O'Neill SL. 1997. The evolution of heritable symbionts. See Ref. 93, pp. 1–41.

156. Werren JH, Skinner SW, Huger AM. 1986. Male-killing bacteria in a parasitic wasp. *Science* 231:990–92

157. Werren JH, Windsor D, Guo L. 1995. Distribution of *Wolbachia* among neotropical arthropods. *Proc. R. Soc. London Ser. B* 262:197–204

158. Werren JH, Zhang W, Guo LR. 1995. Evolution and phylogeny of *Wolbachia*:

reproductive parasites of arthropods. *Proc. R. Soc. London Ser. B* 261:55–63

159. Wright JD. 1979. *The etiology and biology of cytoplasmic incompatibility in the Aedes scutellaris group.* PhD thesis. Univ. Calif., Los Angeles. 199 pp.

160. Wright JD, Barr AR. 1980. The ultrastructure and symbiotic relationships of *Wolbachia* of mosquitos of the *Aedes scutellaris* group. *J. Ultrastruct. Res.* 72:52–64

161. Wright JD, Sjostrand FS, Portaro JK, Barr AR. 1978. The ultrastructure of the Rickettsia-like microorganism *Wolbachia pipientis* and associated virus-like bodies in the mosquito *Culex pipiens. J. Ultrastruct. Res.* 63:79–85

162. Yen JH, Barr AR. 1971. New hypothesis of the cause of cytoplasmic incompatibility in *Culex pipiens. Nature* 232:657–58

163. Zchori-Fein E, Faktor O, Zeidan M,

Gottlieb Y, Czosnek H, Rosen D. 1995. Parthenogenesis-inducing microorganisms in *Aphytis. Insect Mol. Biol.* 4:173–78

164. Zchori-Fein E, Rosen D, Roush RT. 1994. Microorganisms associated with thelytoky in *Aphytis liganensis. Int. J. Insect Morphol. Embryol.* 23:169–72

165. Zchori-Fein E, Roush RT, Hunter MS. 1992. Male production by antibiotic treatment in *Encarsia formosa*, in asexual species. *Experientia* 48:102–5

166. Zchori FE, Roush RT, Rosen D. 1998. Distribution of parthenogenesis-inducing symbionts in ovaries and eggs of *Aphytis. Curr. Microbiol.* 36:1–8

167. Zhou W, Rousset F, O'Neill SL. 1998. Phylogeny and PCR-based classification of *Wolbachia* strain using *wsp* gene sequences. *Proc. R. Soc. London Ser. B* 265:509–15

Annu. Rev. Microbiol. 1999. 53:103–28

AEROTAXIS AND OTHER ENERGY-SENSING BEHAVIOR IN BACTERIA

Barry L. Taylor[1], Igor B. Zhulin[2], and Mark S. Johnson[3]

Department of Microbiology and Molecular Genetics, School of Medicine, Loma Linda University, Loma Linda, California 92350; e-mail: [1]blTaylor@som.llu.edu, [2]iZhulin@som.llu.edu, [3]mJohnson@som.llu.edu

Key Words   oxygen sensing, energy taxis, phototaxis, chemotaxis, PAS domains

■ **Abstract**   Energy taxis is widespread in motile bacteria and in some species is the only known behavioral response. The bacteria monitor their cellular energy levels and respond to a decrease in energy by swimming to a microenvironment that reenergizes the cells. This is in contrast to classical *Escherichia coli* chemotaxis in which sensing of stimuli is independent of cellular metabolism. Energy taxis encompasses aerotaxis (taxis to oxygen), phototaxis, redox taxis, taxis to alternative electron acceptors, and chemotaxis to a carbon source. All of these responses share a common signal transduction pathway. An environmental stimulus, such as oxygen concentration or light intensity, modulates the flow of reducing equivalents through the electron transport system. A transducer senses the change in electron transport, or possibly a related parameter such as proton motive force, and initiates a signal that alters the direction of swimming. The Aer and Tsr proteins in *E. coli* are newly recognized transducers for energy taxis. Aer is homologous to *E. coli* chemoreceptors but unique in having a PAS domain and a flavin-adenine dinucleotide cofactor that is postulated to interact with a component of the electron transport system. PAS domains are energy-sensing modules that are found in proteins from archaea to humans. Tsr, the serine chemoreceptor, is an independent transducer for energy taxis, but its sensory mechanism is unknown. Energy taxis has a significant ecological role in vertical stratification of microorganisms in microbial mats and water columns. It plays a central role in the behavior of magnetotactic bacteria and also appears to be important in plant-microbe interactions.

CONTENTS

0066-4227/99/1001-0103$08.00

103

INTRODUCTION

The cellular energy level is a critical determinant of growth and survival of the cell. Bacteria, like other cells, use a complex array of regulatory strategies to maintain optimal energy levels. It is not surprising to discover bacterial behavior that is also directed to maintaining energy. In this chapter, we review the current knowledge of energy-sensing behavior in bacteria with an emphasis on responses to oxygen (aerotaxis), for which the signal-transducing proteins were recently discovered in *Escherichia coli* (19, 127), and *Halobacterium salinarum* (29). This topic was previously reviewed in this series in 1983 (153), and we have minimized duplication of information presented previously. However, an important paradigm shift will be obvious from even a casual comparison of the two reviews. The earlier review focused on the sensing of proton motive force in bacterial behavior. Since then it has become clear that bacteria use a variety of energy-related parameters to guide them to environments that support optimal energy levels.

HISTORICAL PERSPECTIVE ON BEHAVIORAL STUDIES

Anton van Leeuwenhok observed bacteria through his single-lens microscope in 1676 and was fascinated by the movement of these "leaving creatures" (45). The first systematic studies of bacterial motility responses to oxygen, light, and nutrients appeared in 1881–1884 (for a history of behavioral studies in bacteria, see 5, 15, 46, 57, 153). The following is a brief overview of historical findings that are relevant to this review of energy-sensing behavior.

Aerotaxis

Aerotaxis was first reported by Engelmann (48), who also demonstrated that microaerophilic *Spirillum tenue* were attracted by low oxygen concentrations and repelled by high oxygen concentrations (49). Later, Beijerinck observed that each motile species of bacteria formed a band ("Atmungsfiguren") in an oxygen gradient; the position of the band was species dependent (3, 14). Beijerinck was the first to suggest that bacteria seek an optimal concentration of oxygen. Jennings & Crosby (76) observed the process of band formation. *Spirillum* species around illuminated algae became trapped in a zone that had an optimal concentration of oxygen. Whenever the bacteria reached the edge of the zone, the direction of their movement was reversed and they returned to the zone, forming a band. Band formation in capillary tubes was used to survey aerotaxis in various bacterial species (8). An aerotactic response has now been demonstrated in representatives of all major taxonomic groups of bacteria as well as in archaea.

Phototaxis

Engelmann was the first to describe phototaxis in bacteria. When an illuminated suspension of *Bacterium photometricum* (*Chromatium*) was suddenly darkened, the swimming bacteria backed up, stopped for a few seconds, and then resumed motion (50). Engelmann called this response a shock reaction ("Schreckbewegung"). The response was dependent on the time rate of change of light intensity, not the direction of illumination or spatial differences in intensity (50). A basic similarity between the photosynthetic and phototactic action spectra in the purple nonsulfur bacterium *Rhodospirillum rubrum* led Manten (102) to propose that the photophobic reaction resulted from the abrupt decrease in the rate of photosynthesis when the cells swam into the dark. Clayton (34) compared phototaxis, aerotaxis, and chemotaxis, concluding that the tactic response in *R. rubrum* is mediated by a step down in the rate of metabolism (see also 153). No photophobic response was observed if there was a source of metabolic energy in the dark.

Chemotaxis

Bacterial responses to chemicals were first described by Pfeffer, who identified attractants and repellents for various microbial species and showed that chemicals that attract one species may repel others (3, 31, 119). Adler (1) determined that chemotaxis in *E. coli* is independent of uptake or metabolism of the chemical stimulus. The binding of a chemoeffector molecule to a chemotaxis receptor is sufficient to initiate the chemotaxis response (2, 88). This metabolism-independent sensory transduction mechanism for chemotaxis in *E. coli* is in contrast to the energy- and metabolism-dependent phototaxis and aerotaxis studied by TW Engelmann (48–50), MW Beijerinck (14), and RK Clayton (34). Most modern investigations of bacterial behavior have focused on the signaling pathways for metabolism-independent behaviors, and these have been extensively reviewed (5, 25, 47, 99, 101, 117, 118, 152). The energy-sensing behaviors covered in this review are dependent on metabolism of the stimulus. The bacteria sense a change in metabolism (energy) and not the stimulus per se.

ENERGY SENSING IN CELLS

Energy Sensing as a Survival Strategy

The advantage for survival of sensing energy is illustrated in the following scenario. In a water column, there is a source of amino acids, which is a strong chemoattractant for bacteria, for example, decaying detritus. As the amino acids diffuse out from the source, they establish a chemotactic gradient. Bacteria within the sphere of diffusing amino acids respond to the chemotactic stimuli, swim up the gradient, and cluster around the source. Here they rapidly oxidize the amino acids that provide both carbon and nitrogen. In the cluster, the density of the bacteria may increase to 10^9 cells/ml and be visible to the naked eye (153). At this density, bacteria quickly consume the oxygen, and unless the bacteria are within 1 mm of an air-water interface, the diffusion of oxygen will be inadequate to maintain aerobic metabolism. If they did not have an aerotaxis response, the bacteria clustered around the source of amino acids would be trapped in an anaerobic, growth-limiting environment. However, the bacteria sense a decrease in cell energy when oxygen becomes rate limiting for respiration. Energy taxis then overrides chemotaxis toward amino acids and stimulates the cells to follow the oxygen gradient and leave the hypoxic environment, even though they are moving away from the highest concentrations of carbon and nitrogen sources. Energy taxis may be the *E. coli* equivalent of the human adrenalin-mediated flight response. It gives the bacteria a chance to escape to a microenvironment that supports optimal energy levels.

True oxygen taxis (where diatomic oxygen per se is the attractant) could guide bacteria to an oxygen-rich environment that may not have a metabolizable carbon source, where the bacteria would eventually starve. Energy taxis will not attract bacteria to an oxic environment that lacks a carbon source and cannot maintain cellular energy. However, bacteria that were solely dependent on energy taxis could be attracted to sugars and organic acids that support energy production and be deprived of an adequate source of nitrogen. Fully energized *E. coli* preferentially navigate by chemotaxis to amino acids and other constituents for cell building. It is only when cell energy levels are threatened that the bacteria give priority to energy taxis.

How Cells Sense Energy

By common consent, the term "transcriptional regulation by oxygen" is used to describe physiological phenomena and seldom implies a molecular mechanism of oxygen sensing. This can be confusing to the uninitiated. The term usually refers to differential gene expression in bacteria grown anaerobically with respect to the same strain grown aerobically. Only a few regulatory pathways are known to sense diatomic oxygen. For example, the FixL protein in *Sinorhizobium meliloti* senses oxygen directly via a heme cofactor and represses expression of nitrogen-fixation proteins that are inactivated by oxygen (58). A preponderance of evidence indicates that the SoxR protein senses superoxide and nitric oxide radicals via an iron sulfur cofactor (73). Most "oxygen-sensing pathways" for transcriptional regulation and behavioral responses are likely to sense a metabolic or energy change resulting from oxygen decrease or increase. The electron transport system in *E. coli* is essential for the initial response to an oxygen or energy decrease in aerotaxis and in ArcB regulation of oxidative metabolism (reviewed in 157, 158).

The precise mechanisms that bacteria and other cells use to sense cellular energy or electron transport remain enigmatic. ATP is apparently not the primary energy signal. When faced with a loss of oxygen or other conditions that threaten energy production, cells initiate regulatory changes that will conserve energy long before there is a significant reduction in ATP concentration. It takes several minutes even in the presence of respiratory inhibitors to significantly deplete ATP levels (79, 140).

The passage of electrons through the bacterial electron transport system is closely coupled to the translocation of protons across the cytoplasmic membrane and formation of the proton motive force (for a review, see 69). The latter is the source of energy for active transport of molecules into and out of the cell, pH homeostasis, secretion of membrane proteins, motility, and the synthesis of ATP. As such the proton motive force is more critical to the viability of the cell than ATP. There is a close coupling between the proton motive force, electron transport, redox states of respiratory complexes, and the ATP/(ADP + Pi) ratio. A change in one of these components is reflected in changes in the other components. As a result, a bacterium could sense energy changes by monitoring any of the components. The parameters of the energy profile that are perceived by a cell will vary with the energy component that is sensed. A proton motive force of 200 mV can be generated in 5–10 ms after electron transport is initiated (69). Compared with that, changes in ATP concentration are slow. The proton motive force consists of $\Delta\psi$, the membrane potential and ΔpH, the osmotic component. Changes in ΔpH are slower than changes in membrane potential owing to the buffering capacity of the cell (69).

There are a variety of potential energy sensors being investigated. PAS domains are a recently identified superfamily of sensory modules in proteins that are putative or known energy sensors (122, 158, 176). In *E. coli* the aerotaxis transducer Aer has a flavin-adenine dinucleotide (FAD) cofactor associated with a PAS domain that senses electron transport (19, 127). The Tsr chemotaxis receptor in *E. coli* also senses energy changes and has been suggested as a possible proton motive force sensor (127). The FNR protein regulates synthesis of fumarate and nitrate reductases via an iron sulfur center (98). An FNR-like protein (FLP) in

Lactobacillus casei uses oxidation and reductions of thiol residues as a transcriptional switch (61). There is no agreement as to the energy component that is sensed in vivo by FNR and FLP. For a review of regulation of transcription by oxygen/redox/energy, see 98.

ENERGY TAXIS

Sensing the redox status of the electron transport system provides a more versatile energy-sensing mechanism for *E. coli* than sensing of oxygen per se. Optimal electron transport requires an adequate source of electron donor (NADH or FADH$_2$) and oxygen or an alternative electron acceptor as a sink for electrons (56) (Figure 1). Environmental factors that limit the electron source or sink or interrupt electron flow between them will cause a decrease in electron transport and cellular energy.

Aerotaxis is a response to changes in respiratory electron flow that result from an increase or decrease in oxygen concentration (92, 142, 143, 153). In the absence of oxygen, alternative electron acceptors such as fumarate and nitrate support electron flow (Figure 1). This is the basis for electron acceptor taxis (55, 153, 156). In the

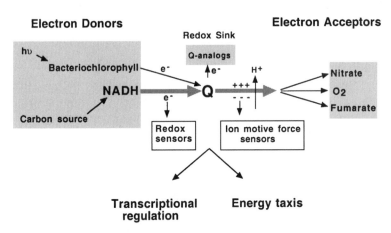

Figure 1 Scheme showing energy sensing through the electron transport system. The rate of electron transport may be limited by the availability of electron donors or acceptors or by diversion of electrons from the electron transport system to a redox sink. Electron transport in bacteria is coupled to the translocation of protons across the cytoplasmic membrane, forming membrane potential ($\Delta \psi$) and ΔpH components of the proton motive force. Exchange of protons creates a secondary ion motive force. Energy sensors could monitor changes in redox potential of the electron carriers or changes in one of the components of the ion motive force. The sensors then transduce the changes in electron transport into signals that regulate transcriptional and behavioral responses. Abbreviation: Q, ubiquinone or menaquinone.

presence of air, the thermodynamically preferred acceptor (oxygen) is used and exposure to fumarate or nitrate does not elicit a behavioral response, confirming a prediction of the energy taxis hypothesis (92). Phototaxis in photosynthetic bacteria, such as *R. sphaeroides* and *Rhodospirillum centenum* is linked to changes in the electron transport system that controls both aerotaxis and phototaxis (55, 66, 132).

When *E. coli* electron transport is depleted by the lack of a carbon source, proline (33) and glycerol (173) restore electron flow and membrane potential, and elicit behavioral responses (Figure 1). Chemotaxis to carbon and energy sources is dependent on their metabolism and does not occur in the presence of an alternative source of energy (173). It serves as an emergency response in *E. coli*, but Armitage and coworkers have shown chemotaxis to a carbon source to be the major behavioral response in *R. sphaeroides* (77).

In a spatial redox gradient, *E. coli* and *Azospirillum brasilense* cells migrate to form a sharply defined band at a preferred redox potential (18, 65). In temporal assays, redox molecules such as ubiquinone and menaquinone analogs, which elicit redox taxis, interrupt electron flow by diverting electrons from the electron transport system (Figure 1). This results in decreased oxygen uptake and proton motive force. The magnitude of the redox response is dependent on the reduction potential of the stimulus molecule that is accepting electrons from the electron transport system (18).

Studies of energy taxis in *E. coli* have come full circle. After evidence for a central role of electron transport in aerotaxis, other electron transport-dependent behavioral responses were predicted and then demonstrated. Identification of *aer* and *tsr* as structural genes for the aerotaxis transducers provided definitive support for the energy taxis hypothesis when all energy taxis responses were abolished in the *aer tsr* double mutant (127).

The preferred oxygen concentration in aerotaxis has been determined for four bacterial species. These species seek oxygen concentrations in the following ranges: *Bacillus subtilis*, an obligate aerobe, 200 μM (164); *E. coli*, a facultative anaerobe, 50 μM (VA Bespalov, IB Zhulin & BL Taylor, unpublished data); *Azospirillum brasilense*, a microaerophile, 4 μM (170); and *Desulfovibrio vulgaris*, an aerotolerant anaerobe, 0.4 μM (80). For comparison, the oxygen concentration of air-saturated water at 760 torr and 25°C is 258 μM (130). Bacteria swim into an aerotactic band at the preferred oxygen concentration for the species, but when they cross either boundary to exit the aerotactic band, they experience oxygen concentrations that are too high or too low and return to the band. It has been uncertain whether band formation requires two signals for separate responses to high and low oxygen or can be achieved with a single signal. A surprising relationship between oxygen concentration and membrane potential in *A. brasilense* (170) and *E. coli* (VA Bespalov, IB Zhulin & BL Taylor, unpublished data) suggests that energy sensing is the only signal needed for band formation. As oxygen concentration is increased, the membrane potential in the bacteria increases until it reaches a maximum at the preferred oxygen concentration and then declines at higher oxygen concentrations. Aerotaxis apparently guides bacteria to a concentration of oxygen that is optimal for energy production. By this hypothesis, bacteria leaving the aerotactic band through the high-oxygen or low-oxygen boundary will experience a loss of energy and return to the band. An earlier hypothesis that bacteria are repelled in high-oxygen concentrations by free radicals derived from oxygen (142) now appears less tenable (171).

TRANSDUCTION MECHANISMS FOR AEROTAXIS AND ENERGY TAXIS

Chemotaxis

The pathways for energy taxis feed into the central signaling pathway for chemotaxis. We present here a brief overview of signaling in chemotaxis of enteric bacteria. For additional information, see Eisenbach (47), Manson (101), and Stock & Surette (152).

Salmonella typhimurium and *E. coli* swim by rotating their flagella at 18,000 rpm, using "motors" that are fueled by the proton motive force (16, 42, 90). Unlike kinesin and myosin, the flagellar motors are reversible, and bacteria randomly change direction by briefly reversing the direction of rotation (91). Chemotaxis controls reversal of the motors in response to external stimuli. Bacteria that are moving in a favorable direction suppress direction changes, and those moving in an unfavorable direction change the direction of swimming to increase the probability of moving in a favorable direction. This biases the random walk motility so that the bacterial population moves in a favorable direction (17).

Four homologous chemotaxis receptors are encoded by the *tsr*, *tar*, *trg*, and *tap* genes in *E. coli* (28, 152). Two transmembrane sequences divide each receptor into a periplasmic sensing domain and a cytoplasmic signaling domain (Figure 2). The dimeric, active form of the periplasmic domain of the Tar receptor is a four-helix bundle that binds aspartate asymmetrically (103). Linear displacement of the α-4 helix by aspartate induces a conformational change in the highly conserved signaling domain in the cytoplasm. Coiled-coil K1 and R1 methylation regions flank the highly conserved domain (85, 152). Adaptation to an attractant increase results from methylation within the K1 and R1 regions by the CheR methyltransferase (87, 149). The signal transduction mechanism is believed to be similar in the Tsr, Trg, and Tap receptors.

A phosphorelay pathway from the receptors to the switch on the flagellar motors is similar to other two-component regulatory systems in bacteria (27, 72, 117, 152). The CheA histidine kinase is autophosphorylated by ATP, followed by transfer of the phosphoryl moiety to the cognate response regulator, CheY (Figure 2). Binding of phospho-CheY to the switch on the flagellar motor is the signal for reversal of the motor (35, 126). A CheA-CheW complex docks with the chemotaxis receptor, enabling the receptor to regulate the autophosphorylation of CheA and thereby the phosphorylation of CheY (26). Adaptation to a decrease in attractant concentration

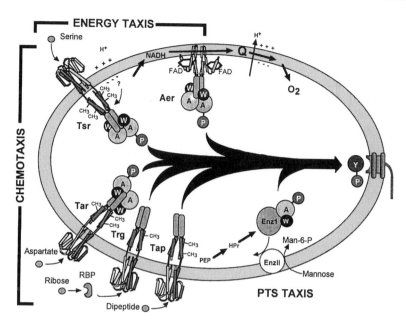

Figure 2   Scheme showing the convergence of the signal transduction pathways in *E. coli* for energy taxis, chemotaxis, and phosphotransferase (PTS) taxis. The Aer energy transducer interacts with the electron transport system through a flavin-adenine dinucleotide (FAD)-containing PAS domain. The Tsr protein transduces an energy-related signal by an unknown mechanism. Tsr is also a receptor for chemotaxis, as are the Tar, Trg, and Tap proteins. Chemoreceptors mostly bind molecules that are in the periplasmic space and induce a conformational change in a highly conserved signaling domain in the cytoplasm. Adaptation to external stimuli is accomplished by methylation or demethylation of K1 and R1 domains of the chemoreceptors, but Aer is not methylated. PTS taxis requires transport and phosphorylation of a sugar by enzyme II (EnzII) with enzyme I (EnzI) and HPr as phosphoryl carriers. The pathways converge when signaling domains of enzyme I, Aer, and the chemoreceptors associate with a CheA-CheW (AW) complex and regulate phosphorylation of CheA, which modulates the phosphorylation state of the CheY (Y) response regulator. Phospho-CheY binds to the switch of the flagellar motor and is the signal for reversal of the motors. Interaction of Trg and Tap with AW is not shown. Abbreviation: RBP, ribose-binding protein.

occurs when CheA phosphorylates the CheB methylesterase, stimulating demethylation of the receptor (151, 152).

Chemotaxis to sugars transported by the phosphoenolpyruvate-phosphotransferase system has a mechanism that is intermediate between chemotaxis and metabolism-dependent energy taxis. Transport, but not metabolism, of the sugar attractant is required for a behavioral response (124, 155). During transport by a sugar-specific enzyme II, the sugar is phosphorylated by a phosphate group donated by phosphoenolpyruvate and relayed by HPr and enzyme I. The phosphorylation state of enzyme I regulates autophosphorylation of CheA (97). This is apparently an allosteric mechanism and there is no phosphotransfer from enzyme I to CheA (78).

The Aerotaxis Transducer Aer in *Escherichia coli*

Aerotaxis is observed in *E. coli* in the absence of the four known chemotaxis receptors, Tsr, Tar, Trg, and Tap (114), but the *cheA*, *cheW*, and *cheY* genes are essential, suggesting that the aerotaxis transducer, like chemotaxis receptors, regulates the CheA histidine kinase (134). The Aer protein, recently identified in *E. coli* as a transducer for aerotaxis and other energy taxis responses, has a carboxyl domain that is homologous to those of the chemotaxis receptors (Figure 2) (19, 127, 157). In a strain lacking the chemotaxis receptors, overexpression of Aer restored some reversal of the flagellar motors (19). This suggests that the signaling domain of Aer interacts with the CheA-CheW complex, regulating the chemotaxis phospho-relay (Figure 2).

The FAD-containing, sensing domain of Aer is a PAS domain that is similar to the redox-sensing domain of the NifL protein of *Azotobacter vinelandii* and some other oxygen and light (redox) sensors (19, 127). The predicted topology of the Aer transducer has two cytoplasmic domains anchored to the membrane by one central hydrophobic sequence (Figure 2). The sequence is flanked by three positively charged residues at the amino end and by an amphipathic sequence at the carboxyl end (127) that typically anchor hydrophobic sequences to the cytoplasmic surface of the membrane (138). This prediction is supported by the finding that all known PAS domains are located in the cytoplasm (158). Cross-linking of Aer in vivo indicates that the aerotaxis transducer is present as a dimer (MS Johnson, JV Mello, BL Taylor, unpublished observation).

A model for signal transduction by Aer is shown in Figure 2. Oxidation and reduction of FAD generate the on and off signals for aerotaxis. The PAS domain is postulated to interact with a component of the electron transport system. Redox changes in FAD reflect redox potential changes in the electron transport system that result from an increase or decrease in oxygen concentration. The PAS domain signals the redox status of FAD to the highly conserved signaling domain in the C-terminus. In the presence of CheW, the highly conserved domain is an input domain for regulating the histidine kinase activity of CheA.

The Aer protein lacks a site for methyltransferase binding, and putative methylation sites in Aer are not conserved (127). A methylation-independent adaptation is observed in temporal gradient assays of aerotaxis where there is a relatively large change in oxygen concentration (114), but this may not occur in the shallow oxygen gradients that bacteria encounter in their natural habitat. Mathematical modeling of bacterial behavior in a spatial oxygen gradient indicates that band formation does not require adaptation (A Mogilner & IB Zhulin, unpublished data).

PAS Domains and Sensing Mechanisms

PAS domains similar to the N-terminal domain of Aer meet the criteria of a long-sought sensor that measures proton motive force or a similar parameter that reads the energy status of a cell (11, 60, 153). A superfamily of PAS domains has recently been identified in proteins from bacteria to humans, including histidine and serine/threonine kinases, chemoreceptors and photoreceptors for taxis and tropism, circadian clock proteins, voltage-activated ion channels, cyclic nucleotide phosphodiesterases, and regulators of responses to hypoxia (122, 157, 158, 176).

The role of the PAS domain in signal transduction in aerotaxis has been explored by using cysteine replacement mutagenesis of Aer. Serial replacement of 40 residues in the PAS domain, including the highly conserved amino acids, produced mutants with various defective phenotypes (AV Repik, A Rebbapragada, JÖ Haznedar, MS Johnson, BL Taylor, unpublished data). In addition to mutants with no aerotactic responses, the signaling in some mutants was locked in the signal-on (continuous reversal) mode. One mutant had inverted responses to oxygen and redox stimuli. That is, it reacted to attractants as repellants and vice versa. Some mutant proteins did not bind FAD. These results clearly demonstrate the importance of the PAS domain in signal transduction in Aer, but structural studies are needed in order to draw conclusions about the signal transduction mechanism.

Tsr: a Multifunctional Protein

The serine receptor Tsr in *E. coli* is also a transducer for aerotaxis (127). Aerotaxis is restored in the *aer tsr* double mutant by expression of either Aer or Tsr from a plasmid, demonstrating that Aer and Tsr are independent transducers for aerotaxis. The *aer tsr* double mutant is also deficient in all forms of energy taxis that have been identified in *E. coli* (127). This confirms that the same proteins transduce the signals for aerotaxis and other energy taxis responses.

Tsr is a multifunctional protein with intriguing properties. It is the dominant chemotaxis receptor in *E. coli* because there are ~1600 copies/cell, compared with 900 Tar and 150 Trg and Tap proteins (71). In addition to detecting serine and energy (redox), Tsr senses temperature and the repellents leucine, indole, and weak acids (100, 148). The periplasmic domain (Figure 1) senses external pH and cytoplasmic residues sense internal pH and indole (86, 144).

The Tsr transduction mechanism in aerotaxis and energy taxis is unknown. Unlike Aer, Tsr has no known cofactor and is not a likely candidate for sensing redox or oxygen per se. Evidence that the proton motive force can mediate aerotaxis responses (141) is the basis for a proposal that Tsr senses changes in proton motive force (127).

A mutation in the Tsr receptor can invert the response of *E. coli* to weak acids so that they become attractants rather than repellents (86, 110). Deletion of the *cheB* gene (protein methylesterase) results in overmethylation of Tsr, which also inverts the response to weak acids (86) and to oxygen (39). A model that explains this inversion has been presented (154). Similar inverted responses for thermotaxis in *E. coli* (107, 111) and phototaxis in *Halobacterium salinarum* (84) suggest that the signal transduction pathway for behavioral responses in these bacteria requires only a simple modification to generate inverted responses. For additional discussion of the mechanism of inverted responses, see Jung & Spudich (83) and Olson et al (115).

The dominant role of Tsr in *E. coli* may explain the preferential detection of some responses by Tsr. Thermosensing, originally thought to be unique to Tsr (100), is observed in the Tar receptor when cells are deleted for Tsr (107) and in the Trg and Tap receptors when they are overexpressed (112); Trg and Tap can also function as intracellular pH sensors (165). It remains to be determined whether Tar, Trg, or Tap can sense redox (energy) changes when they are overexpressed.

Transduction Mechanisms in Other Bacterial Species

In *R. sphaeroides*, energy taxis is the predominant type of behavior and electron transport-sensing mediates aerotaxis (6), various photoresponses (66, 132), electron acceptor taxis (55), and possibly taxis toward metabolized substrates (77). These unique responses are an interplay between true taxis and kinesis, variation of the speed of flagellar rotation (123), modulated by membrane-associated and cytoplasmic transducers expressed under environmental control (70). At least three redundant taxis operons consisting of homologs of *E. coli* chemotaxis genes comprise the most complex signal transduction circuits discovered in a bacterium (68, 132, 162).

Sinorhizobium meliloti has a swimming behavior similar to that of *R. sphaeroides* (62, 121). Because all naturally occurring amino acids induce positive chemotaxis in rhizobia, it is likely that metabolism (energy)-dependent signals are dominant also in these bacteria. Several putative transducers in the cytoplasm of *S. meliloti* (63) may respond to intracellular signals. *S. meliloti* cells sense and respond to changes in oxygen concentration by changing their swimming speed proportionally to changes in ion motive force (172). Kinesis is the main component of their overall behavior (121). In a spatial gradient of oxygen, *S. meliloti* form pseudobands of nonmotile cells; motile cells are evenly distributed between the source of oxygen and the pseudoband (172). A new signal transduction component was found in this bacterium: in the absence of the CheZ phosphatase, one of the two CheY response regulators (CheY1) assumes the role of a "phosphatase" (146, 147).

Azospirillum brasilense is the first bacterial species in which redox taxis was reported (65). Aerotaxis, the major response in *A. brasilense* (10, 65, 128, 170), overrides other behaviors and attracts the bacteria to an oxygen concentration that supports a maximal energy level in cells (170). Chemotaxis genes similar to those of other bacteria were recently identified in *A. brasilense* (D Hauwaerts, SK Das, J Vanderleyden, IB Zhulin, unpublished data).

B. subtilis is a representative of gram-positive (low G+C) bacteria. The basic mechanism of motility and chemotaxis in *B. subtilis* is similar to that of *E. coli*, although there is a unique adaptation system (133). In a spatial gradient of oxygen, *B. subtilis* cells form an aerotactic band at 21% dissolved oxygen (164). The aerotactic response is sensitive to respiratory inhibitors. In contrast to *E. coli*, *B. subtilis* cells are not repelled by high oxygen concentrations and methylation of a transducer is required for aerotaxis (142, 164). None of the PAS domains identified in a completely sequenced genome of *B. subtilis* is in a chemotaxis transducer (175). It is possible that the transducer in *B. subtilis* for aerotaxis and energy taxis is a methylated chemotaxis receptor similar to Tsr.

The archeon *Halobacterium salinarum* has sensory rhodopsins as specialized receptors for phototaxis. This metabolism-independent photobehavior was reviewed recently (150). A strain of *H. salinarum* lacking sensory rhodopsin photoreceptors demonstrates phototaxis that is mediated by proton-pumping bacteriorhodopsin (20, 166). The phototaxis signal was a light-dependent change in membrane potential (64). Membrane potential appears to be a common signal for both aerotaxis and bacteriorhodopsin-mediated phototaxis in *H. salinarum* (21, 94).

Aerotaxis in *H. salinarum* is dependent on a methyl-accepting chemotaxis receptor (94). A large family of 13 putative chemoreceptors has been identified in *H. salinarum* (135, 167), including HtrVIII, which functions as a sensor for aerotaxis (29). The carboxyl domain of the HtrVIII transducer is typical of bacterial chemotaxis transducers (93), whereas the N-terminal sensing domain is similar to the heme-binding domains of the mitochondrial cytochrome *c* oxidase (29). A strain deleted for the *htrVIII* gene lacks aerotaxis, and a strain overproducing the protein shows enhanced aerotaxis. During adaptation to oxygen, the HtrVIII transducer is demethylated. We have identified a PAS domain in another putative transducer of *H. salinarum*, the HtrIII (HtA) protein (158). This soluble transducer may be a second aerotaxis (energy) sensor in *H. salinarum*.

In *Pseudomonas putida*, a protein similar to the Aer aerotaxis transducer (GenBank accession no. AF079997) from *E. coli* was identified recently (N Nichols & C Harwood, personal communication). We identified PAS domains in two putative chemotaxis transducers, AF1034 and AF1045, from the *Archaeoglobus fulgidis* genome (175), and in the putative chemotaxis transducer McpA from *Agrobacterium tumefaciens* (158). The DcrA protein from *D. vulgaris*, proposed as an aerotaxis transducer (53), was subsequently shown to have a PAS domain (122).

ECOLOGICAL ROLE OF ENERGY TAXIS

Vertical Stratification in Microbial Mats and Water Columns

Microbial mats are stratified layers of bacteria found in hot springs, hydrothermal vent systems, hypersaline and other extreme habitats, shallow marine environments, and fresh water. An upper layer, several millimeters thick, is usually made up of cyanobacteria. Next are layers of anoxygenic phototrophic bacteria, and nonphotosynthetic sulfide-oxidizing bacteria form bands distant from the surface

(36, 82, 104, 108). These compact ecosystems are characterized by very steep vertical gradients of oxygen and various ions (sulfate, sulfide, hydrogen), in addition to chemicals that fluctuate in the diurnal and seasonal rhythms (41, 129, 161).

Most bacterial species within the mats are motile, and active vertical movements of various bacteria are observed in response to changing gradients of oxygen and other chemicals. Cyanobacteria, such as *Oscillatoria* and *Spirulina* spp., are capable of gliding motility and migrate downward in the hypersaline mat in the early morning and remain there until dusk, then migrate upward (54). Light is the major stimulus that regulates migration of cyanobacteria; however, this migration cannot be explained only in terms of light-regulated behavior. A variety of tactic responses such as sulfide taxis, aerotaxis, and phototaxis are proposed to contribute to the overall behavior of cyanobacteria in the mat (54). These are energy-dependent responses and probably signal through the electron transport system (see also 109).

Aerotaxis appears to be the dominant response guiding *Beggiatoa* spp. to optimal habitats. The sulfide-oxidizing *Beggiatoa* spp. migrate vertically in microbial mats as a band, following a preferred low oxygen concentration (54, 108).

Sulfate-reducing bacteria are found predominantly in sediments and microbial mats. They dominate the anaerobic mineralization of organic matter in marine sediments (81). In fresh water, sulfate concentrations are lower. Because electron donors reach the sediment surface by sedimentation, many sulfate-reducing bacteria must live near the oxic layer. Sass and collaborators (137) found two vertical distributions of sulfate reducing bacteria: one at the oxic-anoxic interface and one deeper in the reduced sediment layer of a freshwater lake. Cells at the oxic surface exhibited higher oxygen tolerance as well as higher oxygen respiration than those isolated from the anoxic sediment. Once considered as strict anaerobes, many sulfate reducers are found in an oxic surface layer of cyanobacterial mats (52, 159) and in oxic zones of water columns (160). Some sulfate-reducing bacteria (notably, *Desulfovibrio* spp.) not only survive oxygen stress (38) but can also respire by using oxygen (40, 44, 136), generate a proton motive force from oxygen reduction (44), and grow by using oxygen as a sole electron acceptor (80). *D. vulgaris* Hildenborough forms an aerotactic band at 0.04% dissolved oxygen, a concentration, that also supports its growth (80). It is unlikely that *D. vulgaris* would be attracted to this concentration of oxygen if it did not provide a growth advantage. Other *Desulfovibrio* isolates appear to be attracted by a higher oxygen concentration (89). Active respiration in aerotactic bands can remove oxygen and protect the integrity of the anoxic environment for species that can reduce sulfate only under anoxic conditions (80, 89). In so doing, aerotaxis in sulfate-reducing bacteria may contribute not only to the ecology of a particular species but also to maintaining the integrity of the microbial consortium.

Taxis in the Open-Ocean

The open ocean is an environment that is dominated by large-scale mixing driven by wind and thermohaline processes (120). However, stratified marine microbial ecosystems are formed in microzones surrounding phytoplankton, and taxis toward nutrient gradients around phytoplankton is likely to be critical in maintaining such ecosystems (105). Speed and acceleration of marine bacteria from natural open-ocean assemblages are ≤10-fold higher than those of cultured *E. coli* strains (105, 106). However, chemokinetic effects cannot account for the distribution of bacteria around the source of nutrients. Mathematical modeling predicts that chemokinesis in the absence of taxis leads to a dispersal of a bacterial population (30, 123).

Surprisingly, the open-ocean isolates of cyanobacteria do not display any kind of photobehavior. They are, however, attracted to simple nitrogenous substrates (163). The mechanism of this behavior is unknown, but we postulate that it is likely to be a metabolism-dependent response that allows motile open-ocean cyanobacteria to reach and maintain themselves in microzones and patches. As mentioned above, photoresponses are dominant in shallow-water cyanobacteria, where they assemble the bacteria into horizontal layers that are not destroyed by wind and multiscale mixing processes.

Rhizobacteria: Energy Taxis in Plant-Microbe Interactions

Chemotaxis has been demonstrated in various plant root-associated bacteria (for a review, see 174) and is proposed to play a major role in establishing symbiotic and associative relationships between plants and microorganisms, as well as in pathogenesis (4, 32, 43, 116, 139, 168, 174). Rhizobacteria are able to travel significant distances in soil between the plant roots but only if the strain is motile (12).

Several lines of evidence suggest that taxis in plant-associated bacteria is metabolism dependent (7, 169). In plant-associated *Azospirillum* spp., aerotaxis and taxis toward rapidly used substrates (malate, succinate) are the dominant responses (9, 10, 65, 169, 170). In symbiotic rhizobia in which various metabolizable substrates cause positive chemotaxis (4, 131), putative chemoreceptors are cytoplasmic proteins that respond to an intracellular signal (7, 63). When metabolism-dependent chemotaxis was measured, various species and strains of azospirilla and rhizobia responded to different spectra of attractants correlated with chemicals typical of the rhizosphere of the host plant (174). The carbon sources that support the fastest growth are also the best chemoattractants, which is typical of energy-dependent taxis.

In addition to energy taxis toward metabolizable substrates in plant root exudates, redox taxis may play a major role in plant-microbe interactions. The redox state of the rhizosphere is one of the most important parameters for maintaining this ecological system. Plant roots release various redox chemicals that have a crucial role in the biology of the plant. For example, roots of sorghum exude sorgoleone, a complex derivative of p-hydroquinone, and several substituted *p*-benzoquinones (113). These chemicals may suppress weed growth by inhibiting the mitochondrial electron transport system at the bc1-complex (125) or by creating an unfavorable redox environment for the weeds (145). We have recently demonstrated that bacteria are repelled by substituted benzo- and naphtho-quinones in negative redox taxis (18). *Azospirillum* spp., the first bacteria in which redox taxis has been described (65), are present in the rhizosphere of many plants, including sorghum. Some *Azospirillum* isolates possess laccase, a polyphenol oxidase not found previously in bacteria (59). We have recently demonstrated that a laccase-positive strain of *Azospirillum* spp. is three- to fivefold more tolerant of the inhibitory action of oxidized quinones than laccase-negative isolates, owing to its ability to bypass the bc1-containing branch of the electron transport system (G Alexandre, R Bally, BL Taylor, IB Zhulin, submitted for publication).

Magneto-aerotaxis

Magnetotactic bacteria orient and navigate by following the Earth's geomagnetic lines (22, 23). In the Northern Hemisphere, they are predominantly north seeking, but in the Southern Hemisphere, they are predominantly south seeking (24). This aligns the bacteria to swim downward along the inclined geomagnetic field in both hemispheres. Thus, a three-dimensional random walk is reduced to one dimension, yielding a theoretical energy advantage over nonmagnetic organisms. Magnetotactic bacteria migrate toward microaerobic or anaerobic environments, away from the highly oxygenated water surface. When the preferred oxygen concentration is reached, migration stops and a band or veil is formed, often in the oxic-anoxic transition zone (13). The magnetic field orients the bacteria, but aerotaxis determines where the downward migration stops. Consequently, the term magnetotaxis has been replaced by magneto-aerotaxis (51).

Two mechanisms of aerotactic signaling were demonstrated in magnetotactic bacteria by using glass capillaries (51). Signaling in *Magnetospirillum magnetotacticum* appears typical of other aerotactic organisms. The signal elicits a reversal in swimming direction if the bacterium swims into a region where the oxygen concentration is above or below the ideal oxygen level. This is known as axial magneto-aerotaxis. However, cells of the bilophotrichous marine coccus, strain MC-1, are locked into persistent swimming parallel to the magnetic field at higher than preferred oxygen concentrations and persistent swimming antiparallel to the magnetic field at lower than preferred oxygen concentrations. Because the oxygen gradient is oriented antiparallel to the geomagnetic field in the Northern Hemisphere, this two-state mechanism, known as polar magneto-aerotaxis, keeps cells of strain MC-1 at the preferred oxygen concentration.

The difference between axial and polar magnetotaxis is apparent when the local magnetic field is switched 180° from antiparallel to parallel to the oxygen gradient. Cells of *M. magnetotacticum* continue to form an aerotactic band (51). Although the bacteria rotate when the field is rotated, the aerotaxis signal induces a change in direction when a cell swims away from the optimum oxygen concentration with the result that the cell returns to the band. In contrast, cells of strain MC-1 in an aerotactic band rapidly disperse when the magnetic field is rotated 180°. This is a result of the fixed response to oxygen concentrations and because the oxygen gradient is now parallel to the magnetic field. Cells in the oxic side of the band are locked into persistent swimming parallel to the magnetic field, and those in the anoxic side are locked into persistent swimming antiparallel to the magnetic field and migrate away from the optimum concentration after rotation of the field.

Rhodobacter Species: Versatile Bacteria

In terms of their physiology and ecology, *Rhodobacter* spp. are probably the most versatile bacteria (75). They are photoheterotrophs under anaerobic conditions in the light, and they grow equally well under aerobic dark conditions. Photoautotrophic growth is possible with a sulfur compound or molecular hydrogen as an electron donor. Fermentation of sugars occurs under anaerobic dark conditions, and sugars and organic acids also support anaerobic respiration with a variety of electron acceptors. The ability to fix N_2 and CO_2 complements the physiological repertoire. *Rhodobacter* spp. are found at the top and bottom of lakes, in soil, and even in some brackish environments because they are salt and pH tolerant. Energy taxis provides a way for *Rhodobacter* cells to integrate the various metabolic signals that they receive from the environment.

AEROTAXIS AS A MODEL FOR OXYGEN SENSING IN HIGHER ORGANISMS

PAS domains are present in various proteins that sense hypoxia in higher organisms, including humans. This raises the interesting question of what we may learn about oxygen sensing mechanism in mammals from studies of energy sensing in bacteria. Cellular responses to prolonged hypoxia in humans include a shift to glycolytic metabolism of carbohydrates, permanent restructuring of the cells' blood supply, and stimulation of red blood cell and hemoglobin production (67). The hypoxia response is under control of a class of bHLH-PAS transcription factors, the best known of which is hypoxia-inducible factor 1. Hypoxia stabilizes hypoxia-inducible factor 1 against ubiquitin-dependent proteolysis but the PAS domains are not essential for hypoxia sensing (74). They are, however, involved in protein-protein interactions.

Oxygen-sensing ion channels participate in fast cellular responses to hypoxia, including oxygen sensing by arterial chemoreceptors in the carotid bodies and in other cardiocirculatory and respiratory responses to low oxygen (37, 96). They are associated with the pathophysiology of hypertension, cardiac arrhythmias, and ischemic neuronal damage (95, 158). The mechanism of oxygen sensing by K^+ channels remains elusive. Recently, a PAS domain, a possible oxygen sensor, was identified in the N terminus of voltage-activated K^+ channels of the ether a go-go (*eag*) family (122, 176). As these and other hypoxia-sensing systems are investigated in higher organisms, it is likely that similarities to prokaryotic systems will be identified, but it is too early to predict those similarities.

ACKNOWLEDGMENTS

We are grateful to Judith Armitage, Nancy Nichols and Caroline Harwood for providing information before publication and to Sean Bulloch, Joshua Haznedar, and Anu Rebbapragada for help in preparation of the manuscript. Work in the authors' laboratories is supported by grants from the National Institutes of Health (BLT) and the US Department of Agriculture (IBZ).

Visit the Annual Reviews home page at http://www.AnnualReviews.org

LITERATURE CITED

1. Adler J. 1969. Chemoreceptors in bacteria. *Science* 166:1588–97
2. Adler J. 1975. Chemotaxis in bacteria. *Annu. Rev. Biochem.* 44:341–56
3. Adler J. 1988. Chemotaxis: old and new. *Bot. Acta* 101:93–100
4. Ames P, Schluederberg SA, Bergman K. 1980. Behavioral mutants of *Rhizobium meliloti. J. Bacteriol.* 141:722–27
5. Armitage JP. 1997. Three hundred years of bacterial motility. In *Further Milestones in Biochemistry*, ed. MG Ord, LA Stocken, 3:107–64. Greenwich, CT: JAI
6. Armitage JP, Ingham C, Evans MCW. 1985. Role of proton motive force in phototactic and aerotactic responses of *Rhodopseudomonas sphaeroides. J. Bacteriol.* 161:967–72
7. Armitage JP, Schmitt R. 1997. Bacterial chemotaxis: *Rhodobacter sphaeroides* and *Sinorhozobium meliloti*—variations on a theme? *Microbiology* 143:3671–82
8. Baracchini O, Sherris JC. 1959. The chemotactic effect of oxygen on bacteria. *J. Pathol. Bacteriol.* 77:565–74
9. Barak R, Nur I, Okon Y. 1983. Detection of chemotaxis in *Azospirillum brasilense. J. Appl. Bacteriol.* 53:399–403
10. Barak R, Nur I, Okon Y, Henis Y. 1982. Aerotactic response of *Azospirillum brasilense. J. Bacteriol.* 152:643–49
11. Baryshev VA, Glagolev AN, Skulachev VP. 1981. Sensing of $\Delta\mu H^+$ in phototaxis of *Halobacterium halobium. Nature* 292:338–40
12. Bashan Y, Holguin G. 1994. Root-to-root travel of the beneficial bacterium *Azospirillum brasilense. Appl. Environ. Microbiol.* 60:2120–31
13. Bazylinski DA. 1995. Structure and function of the bacterial magnetosome. *ASM News* 61:337–43
14. Beijerinck MW. 1893. Ueber Atmungsfiguren beweglicher Bakterien. *Zentrabl. Bakteriol. Parasitenkd.* 14:827–45
15. Berg HC. 1975. Chemotaxis in bacteria. *Annu. Rev. Biophys. Bioeng.* 4:119–36
16. Berg HC, Anderson RA. 1973. Bacteria swim by rotating their flagellar filaments. *Nature* 245:380–82
17. Berg HC, Brown DA. 1972. Chemotaxis in *Escherichia coli* analyzed by three-dimensional tracking. *Nature* 239:500–4
18. Bespalov VA, Zhulin IB, Taylor BL. 1996. Behavioral responses of *Escherichia coli* to changes in redox potential. *Proc. Natl. Acad. Sci. USA* 93:10084–89
19. Bibikov SI, Biran R, Rudd KE, Parkinson JS. 1997. A signal transducer for aerotaxis in *Escherichia coli. J. Bacteriol.* 179:4075–79
20. Bibikov SI, Grishanin RN, Kaulen AD, Marwan W, Oesterhelt D, Skulachev VP. 1993. Bacteriorhodopsin is involved in halobacterial photoreception. *Proc. Natl. Acad. Sci.* 90:9446–50
21. Bibikov SI, Skulachev VP. 1989. Mechanisms of phototaxis and aerotaxis in *Halobacterium halobium. FEBS Lett.* 243:303–6

22. Blakemore RP. 1975. Magnetotactic bacteria. *Science* 190:377–79
23. Blakemore RP. 1982. Magnetotactic bacteria. *Annu. Rev. Microbiol.* 36:217–38
24. Blakemore RP, Frankel RB, Kalmijn AJ. 1980. South-seeking bacteria in the southern hemisphere. *Nature* 286:384–85
25. Bourret RB, Borkovich KA, Simon MI. 1991. Signal transduction pathways involving protein phosphorylation in prokaryotes. *Annu. Rev. Biochem.* 60:401–41
26. Bourret RB, Davagnino J, Simon MI. 1993. The carboxy-terminal portion of the CheA kinase mediates regulation of autophosphorylation by transducer and CheW. *J. Bacteriol.* 175:2097–101
27. Bourret RB, Hess JF, Borkovich KA, Pakula AA, Simon MI. 1989. Protein phosphorylation in chemotaxis and two-component regulatory systems of bacteria. *J. Biol. Chem.* 264:7085–88
28. Boyd A, Krikos A, Simon MI. 1981. Sensory transducers of *E. coli* are encoded by homologous genes. *Cell* 26:333–43
29. Brooun A, Bell J, Freitas T, Larsen RW, Alam M. 1998. An archaeal aerotaxis transducer combines subunit I core structures of eukaryotic cytochrome *c* oxidase and eubacterial methyl-accepting chemotaxis proteins. *J. Bacteriol.* 180:1642–46
30. Brown S, Poole PS, Jeziorska W, Armitage JP. 1993. Chemokinesis in *Rhodobacter sphaeroides* is the result of a long term increase in the rate of flagellar rotation. *Biochim. Biophys. Acta* 1141:309–12
31. Bünning E. 1989. Ahead of his time: Wilhelm Pfeffer. In *Early Advances in Plant Biology*. Ottawa, Canada: Carlton Univ. Press. pp. 49–53
32. Caetano-Anolles G, Wrobel-Boerner E, Bauer WD. 1992. Growth and movement of spot inoculated *Rhizobium meliloti* on the root surface of alfalfa. *Plant Physiol.* 98:1181–89
33. Clancy M, Madill KA, Wood JM. 1981. Genetic and biochemical requirements for chemotaxis to L-proline in *Escherichia coli. J. Bacteriol.* 146:902–6
34. Clayton RK. 1958. On the interplay of environmental factors affecting taxis and motility in *Rhodospirillum rubrum. Arch. Microbiol.* 29:189–212
35. Clegg DO, Koshland DEJ. 1984. The role of a signaling protein in bacterial sensing: behavioral effects of increased gene expression. *Proc. Natl. Acad. Sci. USA* 81:5056–60
36. Cohen Y, Rosenberg E. 1989. *Microbial Mats. Physiological Ecology of Benthic Microbial Communities.* Washington, DC: Am. Soc. Microbiol.
37. Cornfield DN, Reeve HL, Tolarova S, Weir EK, Archer S. 1996. Oxygen causes fetal pulmonary vasodilation through activation of a calcium-dependent potassium channel. *Proc. Natl. Acad. Sci. USA* 93:8089–94
38. Cypionka H, Widdel F, Pfenning N. 1985. Survival of sulfate-reducing bacteria after oxygen stress, and growth in sulfate-free oxygen-sulfide gradients. *FEMS Microbiol. Ecol.* 31:39–45
39. Dang CV, Niwano M, Ryu J-I, Taylor BL. 1986. Inversion of aerotactic response in *Escherichia coli* deficient in cheB protein methylesterase. *J. Bacteriol.* 166:275–80
40. Dannenberg S, Kroder M, Dilling W, Cypionka H. 1992. Oxidation of H_2, organic compounds and inorganic sulfur compounds coupled to reduction of O_2 or nitrate by sulfate-reducing bacteria. *Arch. Microbiol.* 158:93–99
41. De Wit R, Jonkers HM, Vand den Ende FP, Van Gemerden H. 1989. In situ fluctuations of oxygen and sulphide in marine microbial sediment ecosystems. *Neth. J. Sea Res.* 23:271–81
42. DeRosier DJ. 1998. The turn of the screw: the bacterial flagellar motor. *Cell* 93:17–20
43. Dharmatilake AJ, Bauer WD. 1992. Chemotaxis of *Rhizobium meliloti* towards nodulation gene-inducing compounds from alfalfa roots. *Appl. Environ. Microbiol.* 58:1153–58

44. Dilling W, Cypionka H. 1990. Aerobic respiration in sulfate-reducing bacteria. *FEMS Microbiol. Lett.* 71:123–28

45. Dobell C. 1932. *Antony van Leeuwenhoek and his "Little Animals."* London: John Bale, Danielsson

46. Doetsch RN. 1971. Functional aspects of bacterial flagellar motility. *Crit. Rev. Microbiol.* 1:73–103

47. Eisenbach M. 1996. Control of bacterial chemotaxis. *Mol. Microbiol.* 20:903–10

48. Engelmann TW. 1881a. Neue Methode zur Untersuchung der Sauerstoffausscheidung pflanzlicher und tierischer Organismen. *Pflugers Arch. Gesammte Physiol.* 25: 285–92

49. Engelmann TW. 1881b. Zur Biologie der Schizomyceten. *Pflugers Arch. Gesammte Physiol.* 26:537–45

50. Engelmann TW. 1883. Bacterium photometricum: ein beitrag zur vergleichenden physiologie des licht und fabensinnes. *Pfluegers Arch. Gesamte Physiol. Menschen Tiere* 42:183–86

51. Frankel RB, Bazylinski DA, Johnson MS, Taylor BL. 1997. Magneto-aerotaxis in marine coccoid bacteria. *Biophys. J.* 73:994–1000

52. Fründ C, Cohen Y. 1992. Diurnal cycles of sulfate reduction under oxic conditions in cyanobacterial mats. *Appl. Environ. Microbiol.* 58:70–77

53. Fu R, Wall JD, Voordouw G. 1994. DcrA, a c-type heme-containing methyl-accepting protein from *Desulfovibrio vulgaris* Hildenborough, senses the oxygen concentration or redox potential of the environment. *J. Bacteriol.* 176:344–50

54. Garcia-Pichel F, Mechling M, Castenholz RW. 1994. Diel migrations of microorganisms within a benthic, hypersaline mat community. *Appl. Environ. Microbiol.* 60:1500–11

55. Gauden DE, Armitage JP. 1995. Electron transport-dependent taxis in *Rhodobacter spaeroides*. *J. Bacteriol.* 177:5853–59

56. Gennis RB, Stewart V. 1996. Respiration. In *Escherichia coli and Salmonella: Cellular and Molecular Biology*, ed. FC Neidhardt, R Curtiss III, JL Ingraham, ECC Lin, KB Low, et al, Vol. 1, pp. 217–61. Washington, DC: Am. Soc. Microbiol. 2nd ed.

57. Gest H. 1995. Phototaxis and other sensory phenomena in purple photosynthetic bacteria. *FEMS Microbiol. Rev.* 16:287–94

58. Gilles-Gonzalez MA, Ditta GS, Helinski DR. 1991. A haemoprotein with kinase activity encoded by the oxygen sensor of *Rhizobium meliloti*. *Nature* 350:170–72

59. Givaudan A, Effosse A, Faure D, Potier P, Bouillant M-L, Bally R. 1994. Polyphenol oxidase in *Azospirillum lipoferum* isolated from rice rhizosphere: evidence for laccase activity in non-motile strains of *Azospirillum lipoferum*. *FEMS Microbiol. Lett.* 108:205–10

60. Glagolev AN. 1980. Reception of the energy level in bacterial taxis. *J. Theor. Biol.* 82:171–85

61. Gostick DO, Green J, Irvine AS, Gasson MJ, Guest JR. 1998. A novel regulatory switch mediated by the FNR-like protein of *Lactobacillus casei*. *Microbiology* 144:705–17

62. Götz R, Schmitt R. 1987. *Rhizobium meliloti* swims by unidirectional, intermittent rotation of right-handed flagellar helices. *J. Bacteriol.* 169:3146–50

63. Greck M, Platzer J, Sourjik V, Schmitt R. 1995. Analysis of a chemotaxis operon in *Rhizobium meliloti*. *Mol. Microbiol.* 15: 989–1000

64. Grishanin RN, Bibikov SI, Altschuler IM, Kaulen AD, Kazimirchuk SB, et al. 1996. Dy-mediated signalling in the bacteriorhodopsin-dependent photoresponse. *J. Bacteriol.* 178:3008–14

65. Grishanin RN, Chalmina II, Zhulin IB. 1991. Behaviour of *Azospirillum brasilense* in a spatial gradient of oxygen and in a 'redox' gradient of an artificial electron acceptor. *J. Gen. Microbiol.* 137:2781–85

66. Grishanin RN, Gauden DE, Armitage

JP. 1997. Photoresponses in *Rhodobacter sphaeroides*: role of photosynthetic electron transport. *J. Bacteriol.* 179:24–30

67. Guillemin K, Krasnow MA. 1997. The hypoxic response: huffing and HIFing. *Cell* 89:9–12

68. Hamblin PA, Maguire BA, Grishanin RN, Armitage JP. 1997. Evidence for two chemosensory pathways in *Rhodobacter sphaeroides*. *Mol. Microbiol.* 26:1083–96

69. Harold FM, Maloney PC. 1996. Energy transduction by ion currents. See Gennis & Stewart 1996, pp. 283–306

70. Harrison DM, Skidmore J, Armitage JP, Maddock JR. 1999. Localisation and environmental regulation of MCP-like proteins in *Rhodobacter sphaeroides*. *Mol. Microbiol.* 31:885–92

71. Hazelbauer GL, Harayama S. 1983. Sensory transduction in bacterial chemotaxis. *Int. Rev. Cytol.* 81:33–70

72. Hess JF, Bourret RB, Simon MI. 1988. Histidine phosphorylation and phosphoryl group transfer in bacterial chemotaxis. *Nature* 336:139–43

73. Hidalgo E, Ding H, Demple B. 1997. Redox signal transduction via iron-sulfur clusters in the SoxR transcription activator. *Trends Biochem. Sci.* 22:207–10

74. Huang LE, Gu J, Schau M, Bunn HF. 1998. Regulation of hypoxia-inducible factor 1alpha is mediated by an O2-dependent degradation domain via the ubiquitin-proteasome pathway. *Proc. Natl. Acad. Sci. USA* 95:7987–92

75. Imhoff JF, Truper HG, Pfenning N. 1984. Rearrangement of the species and genera of the phototrophic "purple non-sulfur bacteria." *Int. J. Syst. Bacteriol.* 34:340–43

76. Jennings MS, Crosby JH. 1901. Studies on reactions to stimuli in unicellular organisms. VII. The manner in which bacteria react to stimuli, especially to chemical stimuli. *Am. J. Physiol.* 6:31–37

77. Jeziore-Sassoon Y, Hamblin PA, Bootle-Wilbraham CA, Poole PS, Armitage JP. 1998. Metabolism is required for chemo-

taxis to sugars in *Rhodobacter sphaeroides*. *Microbiology* 144:229–39

78. Johnson MS, Rowsell EH, Taylor BL. 1995. Investigation of transphosphorylation between chemotaxis proteins and the phosphoenolpyruvate:sugar phosphotransferase system. *FEBS Lett.* 374:161–64

79. Johnson MS, Taylor BL. 1993. Comparison of methods for specific depletion of ATP in *Salmonella typhimurium*. *Appl. Environ. Microbiol.* 59:3509–12

80. Johnson MS, Zhulin IB, Gapuzan MR, Taylor BL. 1997. Oxygen-dependant growth of the obligate anaerobe *Desulfovibrio vulgaris Hildenborough*. *J. Bacteriol.* 179: 5598–601

81. Jørgensen BB. 1977. The sulfur cycle of a coastal marine sediment. *Limnol. Oceanogr.* 22:814–32

82. Jørgensen BB, Revsbech NP. 1983. Colorless sulfur bacteria, *Beggiatoa* spp. and *Thiovulum* spp. in O_2 and H_2S microgradients. *Appl. Environ. Microbiol.* 45:1261–70

83. Jung K-H, Spudich JL. 1996. Protonatable residues at the cytoplasmic end of transmembrane helix-2 in the signal transducer HtrI control photochemistry and function of sensory rhodopsin I. *Proc. Natl. Acad. Sci. USA* 93:6557–61

84. Jung K-H, Spudich JL. 1998. Suppressor mutation analysis of the sensory rhodopsin-I transducer complex: insights into the color-sensing mechanism. *J. Bacteriol.* 180:2033–42

85. Kehry MR, Dahlquist FW. 1982. The methyl-accepting chemotaxis proteins of *Escherichia coli*. Identification of the multiple methylation sites on methyl- accepting chemotaxis protein I. *J. Biol. Chem.* 257:10378–86

86. Kihara M, Macnab RM. 1981. Cytoplasmic pH mediates pH taxis and weak-acid repellent taxis of bacteria. *J. Bacteriol.* 145: 1209–21

87. Kort EN, Goy MF, Larsen SH, Adler J. 1975. Methylation of a membrane protein

involved in bacterial chemotaxis. *Proc. Natl. Acad. Sci. USA* 72:3939–43

88. Koshland DEJ. 1977. Sensory response in bacteria. *Adv. Neurochem.* 2:277–341

89. Krekeler D, Teske A, Cypionka H. 1998. Strategies of sulfate-reducing bacteria to escape oxygen stress in a cyanobacterial mat. *FEMS Microbiol. Ecol.* 25:89–96

90. Larsen SH, Adler J, Gargus JJ, Hogg RW. 1974. Chemomechanical coupling without ATP: the source of energy for motility and chemotaxis in bacteria. *Proc. Natl. Acad. Sci. USA* 71:1239–43

91. Larsen SH, Reader RW, Kort EN, Tso W-W, Adler J. 1974. Change in direction of flagellar rotation is the basis of the chemotactic response in *Escherichia coli*. *Nature* 249:74–77

92. Laszlo DJ, Taylor BL. 1981. Aerotaxis in *Salmonella typhimurium*: role of electron transport. *J. Bacteriol.* 145:990–1001

93. Le Moual H, Koshland DEJ. 1996. Molecular evolution of the C-terminal cytoplasmic domain of a superfamily of bacterial receptors involved in taxis. *J. Mol. Biol.* 261:568–85

94. Lindbeck JC, Goulbourne EA Jr, Johnson MS, Taylor BL. 1995. Aerotaxis in *Halobacterium salinarium* is methylation-dependent. *Microbiology* 141:2945–53

95. Lopez-Barneo J. 1996. Oxygen-sensing by ion channels and the regulation of cellular functions. *Trends Neurosci.* 19:435–40

96. Lopez-Barneo J, Lopez-Lopez JR, Urena J, Gonzalez C. 1988. Chemotransduction in the carotid body: K^+ current modulated by PO2 in type I chemoreceptor cells. *Science* 241:580–82

97. Lux R, Jahreis K, Bettenbrock K, Parkinson JS, Lengeler JW. 1995. Coupling the phosphotransferase system and the methyl-accepting chemotaxis protein-dependent chemotaxis signaling pathways of *Escherichia coli*. *Proc. Natl. Acad. Sci. USA* 92:11583–87

98. Lynch AS, Lin ECC. 1996. Responses to molecular oxygen. In *Escherichia coli and Salmonella: Cellular and Molecular Biology*, eds. FC Neidhardt, R Curtiss III, ECC Lin, KB Low, B Magasanik, et al, Vol. 1, pp. 1526–38. Washington, DC: Am. Soc. Microbiol. 2nd ed.

99. Macnab RM. 1987. Motility and chemotaxis. In *Escherichia coli and Salmonella typhimurium*, eds. FC Neidhardt, JL Ingraham, KB Low, B Magasanik, M Schaechter, HE Umbarger, Washington, DC: Am. Soc. Microbiol. pp. 732–59

100. Maeda K, Imae Y. 1979. Thermosensory transduction in *Escherichia coli*: inhibition of the thermoresponse by L-serine. *Proc. Natl. Acad. Sci. USA* 76:91–95

101. Manson MD. 1992. Bacterial motility and chemotaxis. *Adv. Microbiol. Physiol.* 33:277–346

102. Manten A. 1948. Phototaxis in the purple bacterium *Rhodospirillum rubrum* and the relationship between phototaxis and photosynthesis. *Antonie van Leeuwenhoek* 14:65–86

103. Milburn MV, Prive GG, Scott WG, Yeh J, Jancarik J, et al. 1991. Three-dimensional structures of the ligand-binding domain of the bacterial aspartate receptor with and without a ligand. *Science* 254:1342–47

104. Mir J, Martinez-Alonso M, Esteve I, Guerrero R. 1991. Vertical stratification and microbial assemblage of a microbial mat in Ebro Delta (Spain). *FEMS Microbiol. Ecol.* 86:59–68

105. Mitchell JG, Okubo A, Fuhrman JA. 1985. Microzones surrounding phytoplankton form the basis for a stratified marine microbial ecosystem. *Nature* 316:58–59

106. Mitchell JG, Pearson L, Dillon S, Kantalis K. 1995. Natural assemblages of marine bacteria exhibiting high-speed motility and large accelerations. *Appl. Environ. Microbiol.* 61:4436–40

107. Mizuno T, Imae Y. 1984. Conditional inversion of the thermoresponse in *Escherichia coli*. *J. Bacteriol.* 159:360–67

108. Møller MM, Nielsen LP, Jørgensen BB. 1985. Oxygen responses and mat formation by *Beggiatoa* sp. *Appl. Environ. Microbiol.* 50:373–82

109. Murvanidze GV, Glagolev AN. 1982. Electrical nature of the taxis signal in cyanobacteria. *J. Bacteriol.* 150:239–44

110. Muskavitch MA, Kort EN, Springer MS, Goy MF, Adler J. 1978. Attraction by repellents: an error in sensory information processing by bacterial mutants. *Science* 201:63–65

111. Nara T, Kawagishi I, Nishiyama S, Homma M, Imae Y. 1996. Modulation of the thermosensing profile of the *Escherichia coli* aspartate receptor tar by covalent modification of its methyl-accepting sites. *J. Biol. Chem.* 271:17932–36

112. Nara T, Lee L, Imae Y. 1991. Thermosensing ability of Trg and Tap chemoreceptors in *Escherichia coli. J. Bacteriol.* 173:1120–24

113. Netzly DH, Riopel JL, Ejeta G, Butler LG. 1988. Germination stimulants of witchweed (*Striga asiatica*) from hydrophobic root exudate of sorghum (*Sorghum bicolor*). *Weed Sci.* 36:441–46

114. Niwano M, Taylor BL. 1982. Novel sensory adaptation mechanism in bacterial chemotaxis to oxygen and phosphotransferase substrates. *Proc. Natl. Acad. Sci. USA* 79:11–15

115. Olson KD, Zhang X-N, Spudich JL. 1995. Residue replacements of buried aspartyl and related residues in sensory rhodopsin I: D201N produces inverted phototaxis signals. *Proc. Natl. Acad. Sci. USA* 92:3185–89

116. Parke D, Ornston LN, Nester EW. 1987. Chemotaxis to plant phenolic inducers of virulence genes is constitutively expressed in the absence of the Ti plasmid in *Agrobacterium tumefaciens. J. Bacteriol.* 169:5336–38

117. Parkinson JS. 1993. Signal transduction schemes of bacteria. *Cell* 73:857–71

118. Parkinson JS, Kofoid EC. 1992. Communication modules in bacterial signaling proteins. *Annu. Rev. Genet.* 26:71–112

119. Pfeffer W. 1883. Locomotorische richtungsbewegungen durch chemische rieze. *Ber. Dtsch Bot. Ges.* 1:524–33

120. Pickard GL. 1963. Descriptive physical oceanography. New York: Pergamon

121. Platzer J, Sterr W, Hausmann M, Schmitt R. 1997. Three genes of a motility operon and their role in flagellar rotary speed variation in *Rhizobium meliloti. J. Bacteriol.* 179:6391–99

122. Ponting CP, Aravind L. 1997. PAS: a multifunctional domain family comes to light. *Curr. Biol.* 7:R674–R677

123. Poole PS, Brown S, Armitage JP. 1991. Chemotaxis and chemokinesis in *Rhodobacter sphaeroides*: modelling of the two effects. *Binary* 3:183–90

124. Postma PW, Lengeler JW, Jacobson GR. 1996. Phosphoenolpyruvate: carbohydrate phosphotransferase systems. See Gennis & Stewart 1996, pp. 1149–74

125. Rasmussen JA, Hejl AM, Einhellig FA, Thomas JA. 1998. Sorgoleone from root exudate inhibits mitochondrial functions. *J. Chem. Ecol.* 18:197–207

126. Ravid S, Matsumura P, Eisenbach M. 1986. Restoration of flagellar clockwise rotation in bacterial envelopes by insertion of the chemotaxis protein CheY. *Proc. Natl. Acad. Sci. USA* 83:7157–61

127. Rebbapragada A, Johnson MS, Harding GP, Zuccharelli AJ, Fletcher HM, et al. 1997. The Aer protein and the serine chemoreceptor Tsr independently sense intracellular energy levels and transduce oxygen, redox, and energy signals for *Escherichia coli* behavior. *Proc. Natl. Acad. Sci. USA* 94:10541–46

128. Reiner O, Okon Y. 1986. Oxygen recognition in aerotactic behaviour of *Azospirillum brasilense* Cd. *Can. J. Microbiol.* 32:829–34

129. Revsbech NP, Jorgensen BB, Blackburn HT, Cohen Y. 1983. Microelectrode

studies of the photosynthesis and O_2, H_2S and pH profiles of a microbial mat. *Limnol. Oceanogr.* 28:1062–74

130. Reynafarje B, Costa LE, Lehninger AL. 1985. O_2 solubility in aqueous media determined by a kinetic method. *Anal. Biochem.* 145:406–18

131. Robinson JB, Bauer WD. 1993. Relationships between C_4 dicarboxylic acid transport and chemotaxis in *Rhizobium meliloti*. *J. Bacteriol.* 175:2284–91

132. Romagnoli S, Armitage JP. 1999. The role of the chemosensory pathways in transient changes in swimming speed of *Rhodobacter sphaeroides* induced by changes in photosynthetic electron transport. *J. Bacteriol.* 181:34–9

133. Rosario ML, Ordal GW. 1996. CheC and CheD interact to regulate methylation of *Bacillus subtilis* methyl-accepting chemotaxis proteins. *Mol. Microbiol.* 21: 511–18

134. Rowsell EH, Smith JM, Wolfe A, Taylor BL. 1995. CheA, CheW, and CheY are required for chemotaxis to oxygen and sugars of the phosphotransferase system in *Escherichia coli*. *J. Bacteriol.* 177:6011–14

135. Rudolph J, Nordmann B, Storch K-F, Gruenberg H, Rodewald K, Oesterhelt D. 1996. A family of halobacterial transducer proteins. *FEMS Microbiol. Lett.* 139:161–68

136. Santos H, Fareleira P, Xavier AV, Chen L, Liu MY, Legall J. 1993. Aerobic metabolism of carbon reserves by the obligate anaerobe *Desulfovibrio gigas*. *Biochem. Biophys. Res. Commun.* 195: 551–57

137. Sass H, Cypionka H, Babenzien H-D. 1997. Vertical distribution of sulfate-reducing bacteria atg the oxic-anoxic interface in sediments of the oligotrophic Lake Stechlin. *FEMS Microbiol. Ecol.* 22: 245–55

138. Seligman L, Bailey J, Manoil C. 1995. Sequences determining the cytoplasmic lo-

calization of a chemoreceptor domain. *J. Bacteriol.* 177:2315–20

139. Shaw CH, Ashby AM, Brown A, Royal C, Loake GJ. 1988. *virA* and *virG* are the Ti-plasmid functions required for chemotaxis of *Argobacterium tumefaciens* towards acetosyringone. *Mol. Microbiol.* 2: 413–17

140. Shioi J-I, Galloway RJ, Niwano M, Chinnock RE, Taylor BL. 1982. Requirement of ATP in bacterial chemotaxis. *J. Biol. Chem.* 257:7969–75

141. Shioi J-I, Taylor BL. 1984. Oxygen taxis and proton motive force in *Salmonella typhimurium*. *J. Biol. Chem.* 259:10983–88

142. Shioi J, Dang CV, Taylor BL. 1987. Oxygen as attractant and repellent in bacterial chemotaxis. *J. Bacteriol.* 169:3118–23

143. Shioi J, Tribhuwan RC, Berg ST, Taylor BL. 1988. Signal transduction in chemotaxis to oxygen in *Escherichia coli* and *Salmonella typhimurium*. *J. Bacteriol.* 170:5507–11

144. Slonczewski JL, Macnab RM, Alger JR, Castle AM. 1982. Effects of pH and repellent tactic stimuli on protein methylation levels in *Escherichia coli*. *J. Bacteriol.* 152:384–99

145. Smith CE, Ruttledge T, Zeng Z, O'Malley RC, Lynn DG. 1996. A mechanism for inducing plant development: the genesis of a specific inhibitor. *Proc. Natl. Acad. Sci. USA* 93:6986–91

146. Sourjik V, Schmitt R. 1996. Different roles of CheY1 and CheY2 in the chemotaxis of *Rhizobium meliloti*. *Mol. Microbiol.* 22:427–36

147. Sourjik V, Schmitt R. 1998. Phosphotransfer between CheA, CheY1, and CheY2 in the chemotaxis siganl transduction chain of *Rhizobium meliloti*. *Biochemistry* 37:2327–2335

148. Springer MS, Goy MF, Adler J. 1977. Sensory transduction in *Escherichia coli*: two complementary pathways of information processing that involve methy-

lated proteins. *Proc. Natl. Acad. Sci. USA* 74:3312–16

149. Springer WR, Koshland DEJ. 1977. Identification of a protein methyltransferase as the cheR gene product in the bacterial sensing system. *Proc. Natl. Acad. Sci. USA* 74:533–37

150. Spudich JL. 1998. Variations on a molecular switch: transport and sensory signalling by archaeal rhodopsins. *Mol. Microbiol.* 28:1051–58

151. Stock JB, Koshland DEJ. 1978. A protein methylesterase involved in bacterial sensing. *Proc. Natl. Acad. Sci. USA* 75:3659–63

152. Stock JB, Surette MG. 1996. Chemotaxis. See Gennis & Stewart 1996, pp. 1103–29

153. Taylor BL. 1983. Role of proton motive force in sensory transduction in bacteria. *Annu. Rev. Microbiol.* 37:551–73

154. Taylor BL, Johnson MS. 1998. Rewiring a receptor: negative output from positive input. *FEBS Lett.* 425:377–81

155. Taylor BL, Lengeler JW. 1990. Transductive coupling by methylated transducing proteins and permeases of the phosphotransferase system in bacterial chemotaxis. In *Membrane Transport and Information Storage*, eds. RC Aloia, CC Curtain, LM Gordon. New York: Wiley-Liss. pp. 69–90

156. Taylor BL, Miller JB, Warrick HM, Koshland DEJ. 1979. Electron acceptor taxis and blue light effect on bacterial chemotaxis. *J. Bacteriol.* 140:567–73

157. Taylor BL, Zhulin IB. 1998. In search of higher energy: metabolism-dependent behaviour in bacteria. *Mol. Microbiol.* 28:683–90

158. Taylor BL, Zhulin IB. 1999. PAS domains: internal sensors of oxygen, redox and light. *Microbiol. Mol. Biol. Rev.* In press

159. Teske A, Ramsing NB, Habicht K, Fukui M, Kuver J, et al. 1998. Sulfate-reducing bacteria and their activities in cyanobacterial mats of solar lake (Sinai, Egypt). *Appl. Environ. Microbiol.* 64:2943–51

160. Teske A, Wawer C, Muyzer G, Ramsing NB. 1996. Distribution of sulfate-reducing bacteria in a stratified fjord (Mariager Fjord, Denmark) as evaluated by most-probable-number counts and denaturing gradient gel electrophoresis of PCR-amplified ribosomal DNA fragments. *Appl. Environ. Microbiol.* 62:1405–15

161. Van Gemerden H. 1993. Microbial mats: a joint venture. *Mar. Geol.* 113:3–25

162. Ward MJ, Bell AW, Hamblin PA, Packer HL, Armitage JP. 1995. Identification of a chemotaxis operon with two cheY genes in *Rhodobacter sphaeroides*. *Mol. Microbiol.* 17:357–66

163. Willey JM, Waterbury JB. 1989. Chemotaxis toward nitrogenous compounds by swimming strains of marine *Synechococcus* spp. *Appl. Environ. Microbiol.* 55:1888–94

164. Wong LS, Johnson MS, Zhulin IB, Taylor BL. 1995. Role of methylation in aerotaxis in *Bacillus subtilis*. *J. Bacteriol.* 177:3985–91

165. Yamamoto K, Macnab RM, Imae Y. 1990. Repellent response functions of the Trg and Tap chemoreceptors of *Escherichia coli*. *J. Bacteriol.* 172:383–88

166. Yan B, Cline SW, Doolittle WF, Spudich JL. 1992. Transformation of a Bop⁻Hop⁻ Sop-I⁻Sop-II⁻ *Halobacterium halobium* mutant to Bop⁺: effects of bacteriorhodopsin photoactivation on cellular proton fluxes and swimming behaviour. *Photochem. Photobiol.* 56:553–61

167. Zhang W, Brooun A, McCandless J, Banda P, Alam M. 1996. Signal transduction in the archaeon *Halobacterium salinarium* is processed through three subfamilies of 13 soluble and membrane-bound transducer proteins. *Proc. Natl. Acad. Sci. USA* 93:4649–54

168. Zhulin IB, Armitage JP. 1992. The role

of taxis in the ecology of *Azospirillum.*
Symbiosis 13:199–206

169. Zhulin IB, Armitage JP. 1993. Motility, chemokinesis and methylation-independent chemotaxis in *Azospirillum brasilense. J. Bacteriol.* 175:952–58

170. Zhulin IB, Bespalov VA, Johnson MS, Taylor BL. 1996. Oxygen taxis and proton motive force in *Azospirillum brasilense. J. Bacteriol.* 178:5199–204

171. Zhulin IB, Johnson MS, Taylor BL. 1997. How do bacteria avoid high oxygen concentrations? *Biosci. Rep.* 17:335–42

172. Zhulin IB, Lois AF, Taylor BL. 1995. Behavior of *Rhizobium meliloti* in oxygen gradients. *FEBS Lett.* 367:180–82

173. Zhulin IB, Rowsell EH, Johnson MS, Taylor BL. 1997. Glycerol elicits energy taxis

of *Escherichia coli* and *Salmonella typhimurium. J. Bacteriol.* 179:3196–201

174. Zhulin IB, Taylor BL. 1995. Chemotaxis in plant-associated bacteria: the search for the ecological niche. In *Azospirillum VI and Related Microorganisms,* ed. I Fendrik, M Del Gallo, J Vanderleyden, M de Zamaroczy, Vol. G37:451–59. Berlin: Springer-Verlag

175. Zhulin IB, Taylor BL. 1998. Correlation of PAS domains with electron transport-associated proteins in completely sequenced microbial genomes. *Mol. Microbiol.* 29:1522–23

176. Zhulin IB, Taylor BL, Dixon R. 1997. PAS domain S-boxes in archaea, bacteria and sensors for oxygen and redox. *Trends Biochem. Sci.* 22:331–33

Annu. Rev. Microbiol. 1999. 53:129–54

In Vivo Genetic Analysis of Bacterial Virulence

Su L. Chiang,[1] John J. Mekalanos,[1] and David W. Holden[2]

[1]Department of Microbiology and Molecular Genetics and Shipley Institute of Medicine, Harvard Medical School, Boston, Massachusetts 02115; e-mail: jmekalan@warren.med.harvard.edu, su_chiang@hms.harvard.edu
[2]Department of Infectious Diseases, Imperial College of Science, Technology and Medicine, Hammersmith Hospital, London W12 ONN, United Kingdom; e-mail: dholden@ic.ac.uk

Key Words in vivo expression technology, signature-tagged mutagenesis, differential fluorescence induction, pathogenesis, GAMBIT

■ **Abstract** In vitro assays contribute greatly to our understanding of bacterial pathogenesis, but they frequently cannot replicate the complex environment encountered by pathogens during infection. The information gained from such studies is therefore limited. In vivo models, on the other hand, can be difficult to use, and this has to some extent diminished the incentive to perform studies in living animals. However, several recently developed techniques permit in vivo examination of many genes simultaneously. Most of these methods fall into two broad classes: in vivo expression technology and signature-tagged mutagenesis. In vivo expression technology is a promoter-trap strategy designed to identify genes whose expression is induced in a specific environment, typically that encountered in a host. Signature-tagged mutagenesis uses comparative hybridization to isolate mutants unable to survive specified environmental conditions and has been used to identify genes critical for survival in the host. Both approaches have so far been used exclusively for investigating pathogen-host interactions, but they should be easily adaptable to the study of other processes.

CONTENTS

0066-4227/99/1001-0129$08.00

129

INTRODUCTION

The virulence of bacterial pathogens (their ability to produce morbidity and mortality in a host) is a complex, multifactorial process requiring the coordinated activity of many bacterial gene products. Infections may be described generally as proceeding in a sequence that begins with attachment to and colonization of the host, followed in the case of some pathogens by invasion of host tissues or cells. To multiply and persist within the host, a pathogen must then be able to circumvent the host's immune system and obtain nutrients for itself. Exit from the host and transmission to new hosts are subsequent stages in the infectious cycle, and a pathogen may at any point during infection produce factors that cause damage to the host (25, 38).

A variety of in vitro systems have been developed that simulate certain aspects of the infectious process, enabling the development of screens to study bacterial gene expression and the behavior of mutant strains in physiological conditions that reflect the situation in vivo (24). These include the use of specific culture conditions to mimic the host environment and tissue culture assays for adhesion, invasion, or cytotoxicity. For example, studies of bacterial responses to changes in pH (55), temperature (44, 49), and iron levels (10, 29, 39), and analysis of host cell invasion (27, 51) and survival in macrophages (7, 12, 13, 23, 26) have all been used to identify and characterize bacterial virulence determinants.

In vitro assays have been enormously useful and continue to provide much information on the mechanisms of bacterial pathogenesis, but it is obvious that they cannot accurately reproduce all aspects of the host-pathogen interaction. A pathogen may encounter several radically different environments in the host, and it may therefore have very different requirements at various points during infection, particularly in the context of a developing immune response. Consequently, a gene that seems important in in vitro studies may not be important in vivo, and genes that appear unimportant in an in vitro assay may play a critical role during a natural infection.

For these reasons, in vivo experimental models are highly desirable. They permit direct assessment of a pathogen's ability to colonize and survive in a living host and to cause disease or damage. Animal models nevertheless have their own limitations, being generally labor intensive, expensive, and otherwise unwieldy, and these issues present a considerable barrier to undertaking large-scale in vivo experiments. This is perhaps one reason that many searches for novel virulence determinants have focused on identifying factors that are coregulated with known virulence determinants, rather than attempting to conduct generalized screens in animals. Nevertheless, some screening of individual mutant strains for altered virulence has been carried out on a limited scale with animal infection models, using either randomly chosen transposon mutants (11) or strains affected in cell surface or extracellular proteins (48).

Recently, however, several methods have been developed that greatly simplify in vivo analysis of large numbers of strains. A number of these can be classified as in vivo expression technology (IVET) methods. These are promoter-trap strategies designed to identify promoters that are specifically activated in the host, and many IVET procedures permit positive selection for such promoters. Another method that has been used to perform in vivo analysis is signature-tagged mutagenesis (STM), which relies on comparative hybridization to identify mutants unable to survive in the host. In this review, we summarize adaptations of these techniques that have been used in different bacteria, compare the genes identified by IVET and STM in *Salmonella typhimurium*, *Staphylococcus aureus*, and *Vibrio cholerae*, and discuss the advantages and disadvantages of the two approaches. Possible future development and applications of these and several recently developed methods are also considered.

EARLY SCREENS FOR IN VIVO-INDUCED GENES

Upon entering the host, many pathogenic organisms find themselves in a situation that must differ significantly from any encountered in the environmental reservoir. Bacteria respond to this change in circumstances by modulating their patterns of gene expression accordingly, downregulating the expression of genes that are no longer necessary, and upregulating those that are specifically required for survival in the host (e.g. nutrient acquisition or evasion of host defenses). It therefore seemed probable that at least some in vivo-induced genes would play a critical role in pathogenesis, and several promoter-trap strategies were used to identify genetic loci whose expression is induced in host environments.

Conceptually, finding in vivo-induced genes is potentially a simple process if it is possible to generate appropriately selectable or screenable gene fusions and obtain a host organism that is amenable to brute-force screening. For example, one of the first screens for host-induced genes was carried out in the plant pathogen *Xanthomonas campestris* (56). A library of *X. campestris* DNA fusions to a promoterless chloramphenicol resistance gene was generated and introduced into *X. campestris* on plasmids. Eleven hundred of the resultant strains were individually tested for ability to produce disease symptoms in chloramphenicol-treated turnip seedlings. Of the 19 strains found to be virulent in treated seedlings, 14 were also highly sensitive to chloramphenicol in vitro, indicating that these 14 strains harbored plasmids carrying host-inducible fusions. Mutations were subsequently created in and around two of the host-inducible genes, and the mutant strains were tested for effects on pathogenicity. Mutations in one of the genes led to delayed symptom expression in plants (57), whereas mutations in the other gene had no discernible phenotype (58).

Another example of brute-force screens for in vivo-induced genes involved individual scoring of 2550 *Listeria monocytogenes* Tn*917-lac* mutants for genes expressed at higher levels in macrophage-like cells than in laboratory medium (37). This resulted in the identification of five genes that had as much as 100-fold induction within macrophages. Three of these genes were nucleotide biosynthetic genes, one (*arpJ*) was involved in arginine uptake, and the fifth was *plcA*, the gene encoding the previously identified virulence factor phosphatidylinositol-phospholipase C (16). Mutations in the nucleotide biosynthetic genes did not result in increased LD_{50} values in mice, although the purine mutation tested did reduce the number of bacteria recoverable from the liver. The *arpJ* mutation caused a twofold increase in LD_{50} as well as a decrease in bacterial load in the liver, whereas the *plcA* mutant showed a 25-fold increase in LD_{50} (46) and decreased bacterial load in both liver and spleen.

These results provide excellent examples of how promoter-trap screens can identify in vivo-induced genes in host-pathogen models amenable to selection or biochemical assay. However, since each strain carrying a given gene fusion was tested individually, the procedures described above were relatively laborious. This was especially true in the *Xanthomonas* study, in which each strain was inoculated by hand into antibiotic-treated seedlings. Therefore, although the goal of identifying host-inducible genes was achieved in both cases, it was clear that significant reductions in labor would represent a critical advance in the development of similar methods for studying bacterial gene expression in the host.

IN VIVO EXPRESSION TECHNOLOGY

In the last five years, a variety of additional methods have been formulated to isolate genes whose expression is induced in the host, all of which have increased efficiency compared with the examples discussed above. Such techniques were generally termed "in vivo expression technology" (IVET) methods, and although initial interest was understandably focused on host-induced genes, IVET could presumably be adapted to study the induction of microbial genes in response to any condition.

The first IVET methods used promoterless reporter genes whose products confer a phenotype that can be positively selected in the host (42). Both auxotrophic and antibiotic selections were used to this end. The recently developed differential fluorescence induction (DFI) method (67) also permits positive selection for host-induced promoters, although the selection is carried out by fluorescence-activated cell sorting (FACS) of organisms recovered from host cells or animals. Resolvase IVET uses genetic recombination as a reporter activity and requires screening for host-induced promoters after bacteria are recovered from host tissues. Its advantage is that it can in theory detect promoters that are only weakly or transiently induced during infection (14, 15).

Auxotrophic Selection

The original IVET selection (42) was performed by creating transcriptional fusions of random fragments of the *S. typhimurium* chromosome with a promoterless *purA* gene and introducing this library onto the chromosome of an *S. typhimurium* Δ*purA* strain via homologous recombination at the chromosomal fragment (Figure 1). Because purines are limiting for growth of *S. typhimurium* in the mouse, only those strains expressing *purA* from fused promoters would survive. It should be noted that the integration event resulting from a single crossover does not lead to disruption of the wild-type locus on the chromosome, thereby permitting analysis of genes essential for growth in vivo. Bacteria representing the pool of chromosomal fusions were then injected intraperitoneally into BALB/c mice, and the surviving pools were recovered 3 days later and screened on laboratory medium for clones with low promoter activity. Several strains carrying promoters meeting the criteria of in vivo expression and in vitro inactivity were, on subsequent analysis, found to have severe virulence defects as assayed by oral LD_{50}, thereby validating the ability of IVET to isolate virulence genes.

Similar IVET selections were carried out in *Pseudomonas aeruginosa*, both in vivo in BALB/c mice (70) and in vitro to find genes induced by respiratory mucus collected from cystic fibrosis (CF) patients (69). The latter study was undertaken on the premise that *P. aeruginosa* isolates from CF patients are phenotypically different from isolates from natural environments and that CF respiratory mucus might contain substances that induce expression of CF patient-specific virulence factors. Both selections successfully identified novel loci specifically induced under their respective conditions, including genes with no known homologs. Two specific loci were identified independently by both studies. One of these genes encoded the proposed virulence determinant FptA, a protein involved in iron acquisition (5), and the gene product of the other, *np20*, was similar to ferric uptake regulatory proteins. Insertional disruption of *np20* in a wild-type genetic background caused an ~100-fold increase in LD_{50}. The other mouse-induced *P. aeruginosa* loci displaying homology to known proteins appeared to fall into the two general categories of gene regulation and amino acid biosynthesis. Further information regarding the virulence contributions of these genes and the contributions of the mouse *ivi* genes without known homologs is not yet available.

Antibiotic Selection

The IVET method described above obviously requires the existence of an attenuating and complementable auxotrophy, which unfortunately may not be readily available in all microbial systems. However, a variation on the basic principle was established in which expression of the reporter gene provided resistance to the antibiotic chloramphenicol, which could be administered to the host. Using this

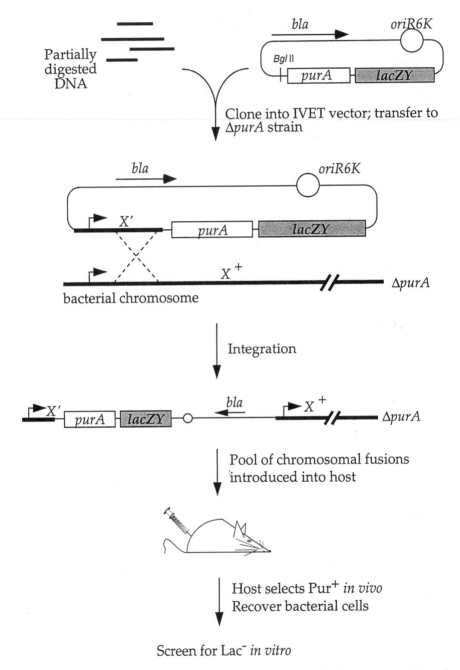

Figure 1   Schematic representation of auxotrophic IVET selection strategy. Adapted from Reference 42.

method, it should be possible to carry out selection for *ivi* genes in any tissue in which the antibiotic concentration can be made sufficiently high to select against strains not expressing the resistance gene. Adjustment of the antibiotic dosage may permit isolation of *ivi* promoters with different levels of activity, and variation of the timing of antibiotic administration might allow investigators to identify *ivi* genes that are expressed at a particular time or place during infection.

The first application of antibiotic-based *ivi* selection was also carried out in *S. typhimurium*, in both BALB/c mice and in cultured macrophages (43). Taken together, the *purA* (42) and antibiotic IVET selections identified >100 *ivi* genes in *S. typhimurium*. Several of these were known virulence determinants, but more than half had either no homologs or none with known function. For example, one of the in vivo-induced genes identified by Heithoff et al (31) was *phoP*, which is known to autoregulate its own expression as well as the expression of multiple virulence genes that are induced after invasion into macrophages (31, 50). Mutations in many of the *ivi* genes had no significant effect on LD_{50}, but some mutant strains showed reduced ability to persist in the spleen (31).

Thirty of the *S. typhimurium ivi* genes identified to date are located in regions of atypical base composition. Hybridization analysis showed that these *ivi*-containing regions are specific to the *Salmonellae* but that several are serovar specific. Although some were present in all salmonellae, others were present only in broad host-range serovars (*S. typhimurium* and *S. newport*), and a few were found in all serovars except the host-adapted serovar *S. typhi*. Two of the regions also contain mobile genetic elements or insertionlike sequences, and deletion of certain regions resulted in colonization defects as assessed by competition assays in BALB/c mice. These observations raise the possibility that these regions might have been acquired by horizontal transmission and may have contributed to the evolution of serovars with different host and tissue specificities (21).

Antibiotic-based IVET selection was also successfully used to detect *ivi* genes or host-responsive elements (*hre*) in *Yersinia enterocolitica* (71). Selection was performed in the Peyer's patches of chloramphenicol-treated mice after peroral infection, and the subset of prototrophic strains that were unable to grow on laboratory medium containing chloramphenicol was retained for further analysis. The fusions in these 404 strains were defined as *hre* fusions and were found to fall into 61 different allelic groups. Sequence analysis of 48 *hre* genes showed that about half had significant similarity to known genes, a few were similar to genes with unknown function, and 18 had no similarity to any sequence in public DNA and protein databases. Insertion mutations were constructed in four *hre* genes, and these mutants demonstrated increased LD_{50}, decreased persistence in host tissues, or both.

Genetic Recombination as a Reporter for In Vivo Activity

The paramount advantage of the preceding IVET variations (auxotrophic complementation or antibiotic selection) is the use of positive selection to isolate *ivi* gene fusions from a pool of fusion strains, thereby largely circumventing the labor-intensive nature of individually screening for such loci. However, both methods favor the identification of genes that are expressed at high levels throughout the infection, because the stringent selective pressures would tend to prevent the survival of strains with fusions to promoters that are expressed weakly or only transiently during infection. Stringent selections might also favor the isolation of promoters that had mutated to higher activity during infection.

To address this problem, an IVET system was developed in which the reporter is $\gamma\delta$ resolvase, which catalyzes irreversible recombination between specific DNA sequences, termed *res* sites. By constructing a system in which resolvase activity results in permanent excision from the chromosome of a tetracycline resistance gene flanked by *res* sites (Figure 2), this method permits detection of promoter activity even if the promoter is active only briefly during infection. Any expression of the resolvase reporter results in a heritable change (i.e. conversion from tetracycline resistance to tetracycline sensitivity) that can be detected by replica plate screening after the bacteria are recovered from the animal (14). Although the resolvase IVET method does not have the benefits of positive selection, theoretically it should be much more sensitive than the previous IVET systems. On the other hand, it may not be able to distinguish between strong and weak induction.

The application of resolvase IVET in *V. cholerae* led to the identification of 13 *ivi* fusions (15). Analysis of the sequences fused to the resolvase reporter determined that some were homologous to genes known to be involved in amino acid biosynthesis and general metabolism, whereas others either had homologs with unknown function or no homologs at all. Two *ivi* fusions appeared to be to antisense transcripts whose gene products are involved in cell motility. Insertion mutants of all 13 loci were tested in infant mouse competition assays, and three *ivi* mutants demonstrated moderate but reproducible colonization defects.

Resolvase IVET was also used in the gram-positive bacterium *Staphylococcus aureus* (40). Owing to the lack of a suitable stable integrating plasmid, the fusion library was not recombined onto the chromosome. A total of 45 *ivi* genes were identified by using the murine renal abscess model. Several were previously known staphylococcal genes, including *agrA*, which is involved in regulation of several virulence factors and is known to be autoregulated (35, 53). The remaining *ivi* genes either had similarity to nonstaphylococcal genes or had no known similarities. Eleven *ivi* genes, representing all three classes, were mutated in the parental genetic background and tested for virulence, and seven of these mutants showed reduced ability to persist in the mouse. Six of these seven attenuating mutations were in genes without homologs in public databases.

Differential Fluorescence Induction   A promising new method for identifying genes induced during infection is DFI (66, 67). Developed in the *S. typhimurium* system, DFI uses expression of green fluorescent protein as the reporter for promoter activity and relies on FACS to carry out the selection for active gene fusions. Random fragments of chromosomal *S. typhimurium* DNA were cloned upstream

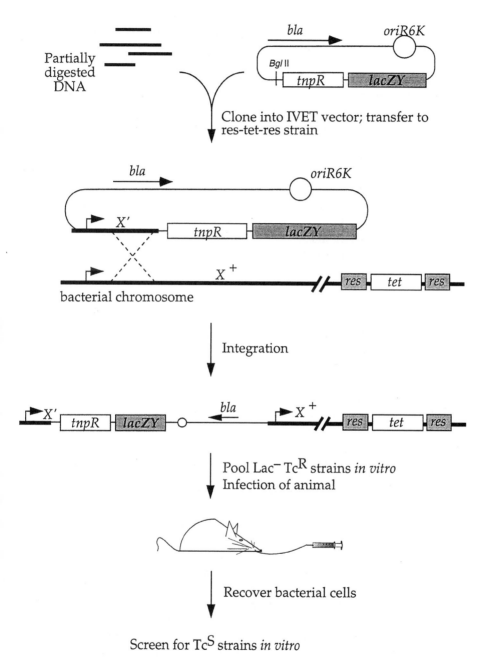

Figure 2 Schematic representation of resolvase IVET strategy. Adapted from Reference 15.

of a promoterless *gfp* gene, and the resultant library was introduced into *S. typhimurium*. To reduce technical difficulties, the library of fusions was maintained on plasmids and not recombined onto the chromosome. After the library of clones was used to infect macrophages, FACS was used to isolate macrophages containing bacteria with active *gfp* fusions. These bacteria were recovered from the macrophages, grown in tissue culture medium, and then re-sorted to obtain clones with low fluorescence (Figure 3). As many as 50% of the promoters thus isolated were confirmed to have host cell-dependent activity on subsequent analysis.

Of 14 macrophage-inducible genes identified by DFI, 8 had bacterial homologs of known function, some of which had previously described roles in virulence. The remaining six genes either had no known bacterial homologs or had homologs with no known function. At least two of these novel loci contribute to virulence, as determined in competition assays testing spleen colonization in BALB/c mice, and both of these loci were regulated by the PhoP/PhoQ two-component regulatory system, which modulates the expression of several macrophage-inducible virulence factors in *S. typhimurium* (4, 8, 30, 50).

SIGNATURE-TAGGED MUTAGENESIS

A different approach to studying bacterial pathogenesis in vivo is STM, a comparative hybridization technique that uses a collection of transposons, each one modified by the incorporation of a different DNA sequence tag. The tags are short DNA segments that contain a 40-bp variable central region flanked by invariant "arms" that facilitate the coamplification and labeling of the central portions by polymerase chain reaction (PCR). When the tagged transposons are used to mutagenize an organism, each individual mutant can in theory be distinguished from every other mutant based on the different tags carried by the transposons in its genome. The use of DNA tags to monitor the fate of different cells in a mixed population was originally used to study the distribution of neuronal clones in the cerebral cortex, by employing retroviruses marked with DNA segments of different sizes and restriction patterns (68).

In STM, mutagenized bacterial strains are stored individually in arrays (usually in the wells of microtiter dishes), and colony or dot blots are made from these arrays. Pools of mutants are then subjected to a selective process such as infection of an animal, and PCR is used to prepare labeled probes representing the tags present in the preselection (input) and postselection (output) pools. Hybridization of the tags from the input and output pools to the colony or dot blots permits the identification of mutants that are unable to survive the selective process, because the tags carried by these mutants will not be present in the output pools. These strains can then be recovered from the original arrays (Figure 4), and the nucleotide sequence of DNA flanking the transposon insertion point can be determined.

In the original method, the suitability of tags was checked before use by amplification, labeling, and hybridization to colony blots representing the tags used

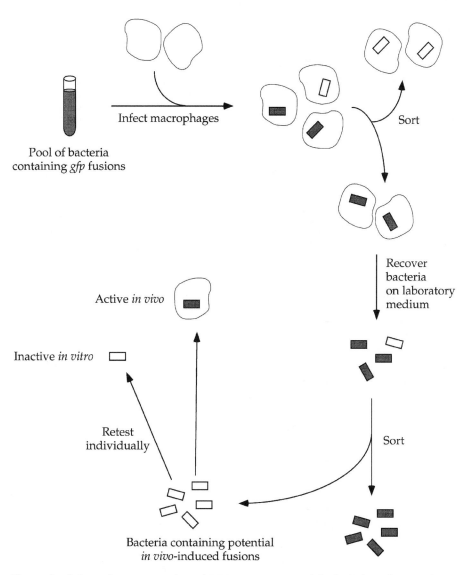

Figure 3   Schematic representation of DFI strategy, as used for isolation of macrophage-induced genes. Adapted from material kindly supplied by T McDaniels and S Falkow (Stanford University).

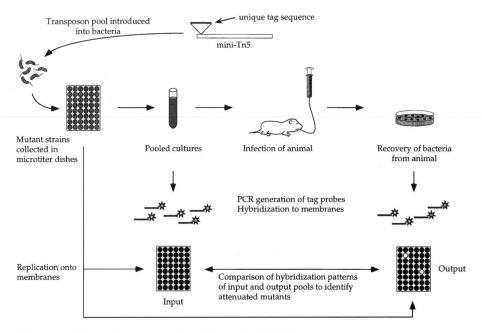

Figure 4   Schematic representation of the original STM strategy.

to make the probes. Mutants whose tags failed to yield clear signals on autoradiograms were discarded, and those that gave good signals were reassembled into new pools for animal infection studies (32). The method was subsequently modified to avoid this prescreening process (45). In this version of STM, a series of tagged transposons is selected before mutagenesis, based on efficient tag amplification and labeling and lack of cross-hybridization to other tags. These modified transposons are then used separately to generate a large number of bacterial mutants that are arrayed based on the tags they carry (Figure 5). Because the same tags can be used to generate an infinite number of mutants, the need to prescreen mutant strains for the suitability of the tags they carry is obviated. A second advantage is that, because the identity of the tag in each mutant is known, hybridization analysis can be done by plasmid or tag DNA dot blots rather than colony blots. This increases the sensitivity of the assay and allows the use of nonradioactive detection methods (45).

STM relies on the ability of the pathogen in question to replicate in vivo as a mixed population and can be expected to identify only virulence genes whose mutant phenotypes cannot be trans-complemented by other virulent strains present in the same inoculum. When STM is applied to a bacterial pathogen for the first time, a number of parameters must be considered to obtain reproducible identification of mutants attenuated in virulence from different animals inoculated with the same pool of mutants.

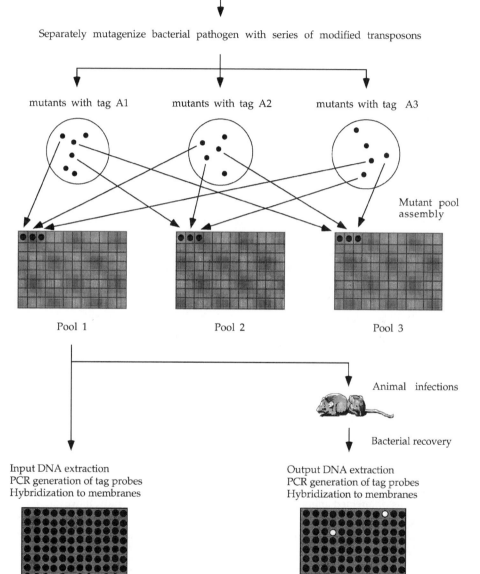

Preselect a series of tagged transposons based on efficient hybridization, labeling, and lack of cross-hybridization

Separately mutagenize bacterial pathogen with series of modified transposons

mutants with tag A1   mutants with tag A2   mutants with tag A3

Mutant pool assembly

Pool 1   Pool 2   Pool 3

Animal infections

Bacterial recovery

Input DNA extraction
PCR generation of tag probes
Hybridization to membranes

Output DNA extraction
PCR generation of tag probes
Hybridization to membranes

Figure 5   Schematic representation of the revised STM strategy.

Pool Complexity

As the complexity of the pool (the number of different mutant strains) increases, so must the probability that virulent mutants will fail to be recovered in sufficient numbers to yield hybridization signals, and this could lead to false identification of attenuated mutants. For *S. typhimurium* inoculated into mice by the intraperitoneal route, pools of 96 different mutants gave reproducible hybridization signals, whereas pools of 192 did not (32). With *V. cholerae*, even pools of 96 different strains did not give reproducible results, and it was necessary to reduce the pool complexity to 48 strains (17).

Inoculum Dose

If the inoculum dose is too low, there may be insufficient cells of any one virulent mutant to initiate a successful infection. For instance, a given input containing two differentially marked, wild-type strains can yield markedly different output ratios of the two strains after an infection cycle initiated by a small inoculum. Such events are reflective of a "bottleneck" in the infection process that selects individual cells stochastically that then grow out as the infection proceeds. On the other hand, if the dose is too high, the animal's immune defenses may be overwhelmed, resulting in the growth of mutant strains that would otherwise be attenuated. In *S. typhimurium*, it was found that, with a pool of 96 different mutants, an inoculum of 10^4 cells ($\sim$100 cells/mutant) gave variable hybridization patterns from animal to animal (DWH, unpublished observations), whereas an inoculum of 10^5 cells ($\sim$1000 cells/mutant) gave reproducible hybridization patterns and an attenuated virulence frequency of $\sim$4% (32). These results are consistent with studies of *S. typhimurium* and *S. paratyphi* in mice, which showed (more than 30 years ago) that bacterial cells cause infection by independent rather than synergistic action (47).

Route of Inoculum Administration

The route of administration of bacterial inoculum also influences the numbers of bacterial strains that reach the target organ(s) and tissues, hence the reproducibility of tag hybridization signals. For example, if inoculated by the intraperitoneal route, 10^5 *S. typhimurium* cells representing a pool of 96 mutants yield reproducible hybridization signals for the vast majority of strains recovered from the spleens of infected animals. If the same inoculum is given orally, however, only a small percentage of mutants are subsequently found in the spleens, and the identity of these varies from animal to animal (J Shea, DW Holden, unpublished observations). Evidently, the majority of cells in the inoculum fail to cross the gut epithelium, either because they are rapidly cleared from the small intestine or because the M cells of Peyer's patches, through which the majority of bacteria are thought to gain access to the deeper tissues of the host (19, 20, 62), represent an infection bottleneck, and only a relatively small number of bacteria proceed to cause systemic disease. These observations suggest that, apart from its use in studies of bacterial virulence, STM might also prove to be useful in studies of the population dynamics of virulent strains during the course of infection. These types of studies have hitherto been restricted by the small number of markers available for strain identification (47, 52).

Duration of Infection

Another important aspect of the STM screening process concerns the postinoculation time point at which bacteria are recovered to prepare tags for hybridization analysis. If this time period is short, virulent cells may have had insufficient time to outgrow the attenuated strains to a degree that is reflected in a clear difference in hybridization signal intensity of tags on the blots. On the other hand, if the period is too long, there may be a risk that some virulent strains may simply outgrow other virulent strains in a nonspecific manner.

The parameters described above are obviously interrelated and must be optimized empirically for each pathogen-host interaction, to obtain reproducible hybridization patterns with tags recovered from at least two animals infected with the same pool of mutants.

From IVET, STM, and earlier studies (47), it is clear that, if the inoculum dose is sufficiently high, systemic *S. typhimurium* infection of the mouse involves multiplication of many of the cells present in the inoculum, rather than clonal expansion from one or a small number of cells in the inoculum. By comparing the results of STM with results from virulence tests with individual mutants at lower doses (11), it is possible to determine whether trans-complementation of mutant defects by virulent mutant strains occurs to a significant degree and whether inoculation with a mutant pool at a dose several orders of magnitude higher than the wild-type LD_{50} (<10 cells by the intraperitoneal route) overwhelms the immune response and results in the growth of strains that would otherwise be attenuated. The virulence of 330 individual Mud*J* transposon mutants was tested by intraperitoneal inoculations at a dose of 10^3 bacteria, and it was found that 1.2% had LD_{50}s >1000-fold higher than that of the parental strain (11). In the initial STM screen using mTn*5* mutagenesis of the same *S. typhimurium* strain in the same mouse strain, 3.4% of 1152 mutants were identified as attenuated, and the LD_{50}s of $>70\%$ of these strains are >1000-fold higher than that of the parental strain. There is therefore no evidence from the *S. typhimurium*-mouse interaction that mixed infections of virulent and attenuated strains inoculated at high dosages lead to a lower level of attenuated-mutant recovery than would be observed with single-strain infections at a lower dose.

The original application of STM in *S. typhimurium* (32) by intraperitoneal inoculation of mice resulted in the identification of a new pathogenicity island, SPI2, containing at least 31 genes predicted to encode proteins of a type III secretion system that is specific to the salmonellae (33, 60). Genes in SPI2 were independently identified by a genome comparison approach (54) and by DFI (67), and the SPI2 type III secretion system appears to be required for replication of bacterial cells in macrophages (18, 33, 54). Two of the SPI2 mutants were inoculated by the peroral route and were shown to be severely attenuated as evidenced by significantly increased LD_{50} values (60). This result, along with the recovery of known virulence factors by STM (32), shows that although *Salmonella* infections are not acquired intraperitoneally in nature, this route of inoculation does provide information relevant to natural infection. By the same token, it is not surprising that genes important for survival in the gut and for translocation across the gut epithelium (28) were not identified by STM screening.

Virulence in *Staphylococcus aureus* has been studied by using the modified STM methodology described above (22, 45, 59). In the study by Mei et al (45), Tn*917* mutants were tested in a murine model of bacteremia. The majority of loci from 50 mutants that were identified as attenuated were predicted by sequence similarity to be involved in cell surface metabolism (e.g. peptidoglycan cross-linking and transport functions), nutrient biosynthesis, and cellular repair processes, but most of the remainder had no known function. A slightly larger signature-tagged mutant bank was constructed by using the same transposon and tested in models of bacteremia, abscess and wound formation, and endocarditis (22). This enabled the identification of various genes affecting growth and virulence in specific disease states, as well as 18 that are important in at least three of the infection models. Many of these genes appear to be involved in the same kinds of processes as those identified in the earlier study (45); indeed, seven of the genes identified by Mei et al (45) were also found by Coulter et al (22).

STM was also used to isolate colonization-defective mutants of *V. cholerae* (17), and the screen resulted in the identification of a number of genetic loci critical for colonization of the infant mouse intestine. As expected, several of these genes were previously known to be involved in biogenesis of the toxin coregulated pilus, which is absolutely required for efficient colonization in both infant mice and humans (6, 34, 36, 64, 65). Mutations in purine, biotin, and lipopolysaccharide biosynthetic genes were also found to cause severe colonization defects. Two loci identified by STM appear to encode phosphotransferases, and mutations in these genes affect coordinate regulation of virulence factors in *V. cholerae*. Other identified loci had no previously known function in pathogenesis, and one had no homology to any known genes.

A further modification of the basic STM method involves hybridization of tags to high-density arrays, in an approach termed molecular bar coding (61). Its potential feasibility was demonstrated in a pilot study with 11 auxotrophic *Saccharomyces cerevisiae* deletion strains to monitor the depletion of some of these strains in media lacking the relevant metabolite. Molecular bar coding appears to be quantitative, and it may be particularly useful for studying mutant strains with subtle phenotypic defects. It may also be capable of processing very large numbers of strains simultaneously because the tag population is monitored by hybridization to a high-density oligonucleotide array, but it should be noted that, although this could potentially permit simultaneous analysis of thousands of strains, pool complexity would still be subject to biological constraints such as in vivo bottlenecks.

ADVANTAGES AND DISADVANTAGES OF IVET AND STM

The studies described above demonstrate that IVET is quite capable of finding novel virulence genes, although the rate of success rather depends on the definition of virulence gene. Not all *ivi* mutations result in pronounced virulence defects as evidenced by vastly increased LD_{50} values or complete inability to survive in the host, but many do cause decreased ability to persist in host tissues. They could also be responsible for damage to the host, which has not been assessed in most IVET studies. It may be that many *ivi* loci make small individual contributions to virulence, and their effects may be additive or synergistic. There have been no published studies examining the effects of multiple *ivi* mutations in a single strain, presumably owing to technical considerations, but this could be a fruitful approach eventually.

The most significant disadvantage of IVET is that, in most of its current incarnations, it discriminates perhaps too strongly against genes that are expressed in vitro. These are almost invariably removed from the pool at some point, although there is no reason to expect that in vitro-expressed genes would not be important for either survival in the host or to cause damage to the host.

With IVET methods, it is necessary to bear in mind that the in vitro conditions may have a profound influence on the nature of the genes isolated. For example, if essential biosynthetic genes are induced in response to the lack of a particular nutrient, then the presence or absence of that nutrient in the in vitro situation may determine whether these biosynthetic genes are identified by IVET as host inducible. Growth on minimal media would cause such genes to be expressed in vitro, which in turn would lead to their elimination from consideration. On rich media, however, the genes might be expressed at a low level, and they would be identified as *ivi* loci if their expression were subsequently induced in the nutrient-limited host. A similar argument applies to any gene, and because many virulence genes are already known to be regulated by environmental signals, the choice of in vitro conditions becomes a major consideration when using IVET to search for virulence factors.

To date, no published IVET strategy has attempted to identify genes whose expression must be downregulated during infection, although this could be a valuable approach. In *Bordetella bronchiseptica*, for instance, it appears that flagella are not produced during infection in rat and rabbit models, and forced expression of flagella during rat infection in fact results in decreased colonization (1). Although flagella are not required for successful infection in rats and so perhaps would not commonly be described as virulence factors, the knowledge that ectopic production of flagella reduces colonization surely increases our understanding of the infectious process.

The preceding discussion makes it clear that, although most current IVET methods aim to detect increases in promoter activity, ideally IVET should be capable of studying both increases and decreases in promoter activity. It would also be desirable to be able to quantitate such changes in expression level. Studies of this nature may be possible with antibiotic IVET selections, which theoretically permit identification of *ivi* promoters with different levels of activity through variation of antibiotic levels and administration. No such studies have yet been carried out, and it has not been determined whether the levels of antibiotic can be controlled at a sufficiently fine level both in vitro and in vivo for this method to be implemented easily. However, DFI provides a simpler way to accomplish the same goals, because *gfp* fusions are not required for survival during infection and strains carrying such fusions can be efficiently and arbitrarily sorted by their green fluorescent protein activity. Even more exciting is the finding that active *gfp* fusions can be detected by FACS analysis of homogenized tissue from infected animals (67). This indicates that DFI is useful not only in cell culture models of virulence, but that it might also be used to isolate active fusions directly from animals.

Every strain identified by STM is by definition attenuated for survival under the specified conditions, regardless of the expression patterns of the gene mutated in that strain. It is therefore a more direct method than IVET for isolating genes required for survival in the host, because genes identified by IVET must be mutated subsequently to demonstrate their requirement for virulence. On the other hand, STM does not select positively for mutants bearing the desired traits. The host animal selects against the interesting mutants, but these can be identified only postinfection by hybridization screening. Therefore, although STM is generally much less laborious than traditional "one-mutant, one-animal" screens because of its ability to screen mutants in pools, STM is not as straightforward a selection method as IVET.

The majority of mutant strains identified by STM in *S. typhimurium*, *S. aureus*, and *V. cholerae* have subsequently been shown by LD_{50} tests or competition analysis to be important for growth in vivo. Very occasionally, however, strains have been isolated with weak output hybridization signals but for which no apparent virulence defect could be demonstrated (R Mundy & DW Holden, unpublished observations). The reason for this is not known.

Not surprisingly, only a subset of the genes identified by IVET as host-induced were found to have a substantial role in virulence as assessed by LD_{50} or competition assays. In *S. typhimurium*, IVET identified previously known virulence genes (such as *phoP*) and several novel genes whose inactivation did not produce a noticeable virulence defect in LD_{50} assays (31). Of 11 *S. aureus ivi* genes that were mutated, 7 of the corresponding mutant strains had a virulence defect (40). In *V. cholerae*, 3 of the 13 identified *ivi* genes had a demonstrable role in colonization (15).

It is curious that, whereas many of the genes identified by IVET and STM in *S. typhimurium* are clearly "virulence determinants" in the classical sense of the term, many of the *S. aureus* genes identified by IVET and STM appear to have more fundamental roles in bacterial metabolism. Although IVET successfully identified the *S. aureus* virulence factor *agrA*, genes for other known virulence determinants, such as toxins and extracellular matrix-binding proteins (41), have not yet been identified by these screens (22, 40, 45). There are several possible explanations for this. First, it may be that these factors are expressed at too high a level in vitro to qualify as in vivo-induced. Second, for STM, transposon muta-genesis is not fully random and may have favored mutation of certain areas of the chromosome over others. Third, mutation of genes encoding toxins may result in trans-complementable phenotypes. Fourth, no STM screen to date has examined more than ~1500 different mutants, so the screens are probably not saturating. It should also be noted that, depending on the sensitivity of the PCR/hybridization protocol for detecting such changes in tag populations, STM might not identify mutations causing small or even moderate reductions in survival. This is a partic-ularly important consideration because this category includes mutations in genes that are critical for causing disease but do not appreciably affect survival of the bacterium in the host. For example, it is the action of cholera toxin that is primarily responsible for the lethality of cholera (9), but deletion of the cholera toxin genes does not affect the bacterium's ability to colonize the host (64).

The types of genes that could in theory be identified by IVET and STM can be summarized as follows: STM should identify a subset of genes that are required for growth in vivo; IVET should identify some genes that are required for growth in vivo and others that are not, because not all genes that are expressed in vivo are required for survival in vivo, and some genes that are required for growth in vivo may also be expressed in vitro; STM would not be expected to identify genes that are essential for bacterial growth, nor would IVET unless the genes were expressed at a sufficiently low level in vitro. The results of the IVET and STM studies reported to date support these predictions. Moreover, based on the studies in *S. typhimurium*, *S. aureus*, and *V. cholerae*, there seems to be little overlap between the genes identified by IVET and STM, so these two approaches appear to be genuinely complementary.

GAMBIT

Essential genes are by definition required for growth or viability in vitro. Because one would expect such genes to be expressed under all conditions, they would not be identified by most IVET methods unless they were expressed at extremely low levels in vitro. This particular category of *ivi* genes may have been documented by Lowe et al (40) in that they identified several *S. aureus* genes that were in vivo-induced but that could not be disrupted in subsequent analyses. STM is equally unable to assess the role of essential genes in pathogenesis, because transposon insertions in these loci would be expected almost invariably to be lethal. This is a critical issue because essential genes are prime targets for antimicrobial strategies. For these reasons, a systematic and efficient means of studying essential genes is certain to contribute greatly to our understanding of pathogenic processes.

The recently developed GAMBIT method was designed to identify essential genes in naturally transformable organisms whose genomes have been sequenced (2). The name GAMBIT stands for "genomic analysis and mapping by in vitro transposition," and the procedure is outlined in Figure 6. A specific region of the chromosome is amplified by extended-length PCR, and the product is subjected to

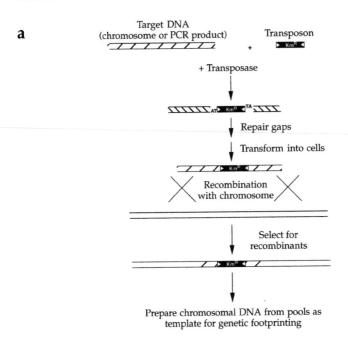

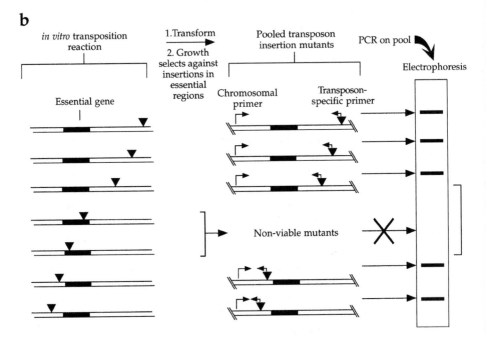

Figure 6 Schematic diagram of the two steps required for GAMBIT. (*a*) Strategy for production of chromosomal mutations by in vitro transposon mutagenesis. (*b*) Genetic footprinting for detection of essential genes. Reprinted from Reference 2.

in vitro transposon mutagenesis. The resultant pool of mutagenized DNA is then transformed into bacteria, which are then grown under selective conditions (e.g. on defined medium or in an animal). PCR is then performed on the postselection pool using a transposon-specific primer and a primer to a known location on the chromosome. Subsequent analysis of the PCR products allows determination of which genes in that region of the chromosome are required for survival under those selective conditions. This type of PCR analysis of transposon insertions has been termed "genetic footprinting" and was first tested in *Saccharomyces cerevisiae* (63).

The ability of GAMBIT to identify in vitro essential genes was confirmed in both *Haemophilus influenzae* and *Streptococcus pneumoniae* (2), and GAMBIT has already been applied to the problem of identifying genes essential for growth of *H. influenzae* in a mouse model (3). Like STM, the use of GAMBIT in animal models constitutes a negative selection in which certain mutants are eliminated by selection in the animal. These mutants are recognized by the loss of PCR products corresponding to insertions in the in vivo essential genes that are represented in the preinfection inoculum. A particularly attractive aspect of GAMBIT analysis is its ability to target specific genes or regions of the chromosome. Although it is necessary to design quite a large number of PCR primers ($\sim$130 primers per Mb of target DNA) to apply GAMBIT to entire genomes, the facility with which this method defines essential genes should make it enormously useful in both the study of microbial pathogenesis and the development of antimicrobial drugs. Finally, the development of efficient DNA transformation methods should enable the adaptation of this system for the analysis of bacteria that are not naturally competent.

CONCLUDING REMARKS

In recent years, sequences of entire bacterial genomes have been obtained with greater rapidity and ease than thought possible only a short time ago, and this is revolutionizing our understanding of and experimental approaches to the study of bacterial virulence. With good annotation, genomic sequences will constitute a powerful genetic "infrastructure" capable of providing not only the sequence of all of an organism's genes but also functional information for some of them. It is nevertheless clear that, for the majority of genes, it is not possible to determine the biochemical function of their products from their DNA sequences. Therefore, a continuing need exists for gene expression and mutational studies to provide phenotypes that can be used to characterize the functions of these genes. Such studies are also necessary for genes with known function that were not previously suspected to play a role in pathogenesis.

The value of the IVET and STM methods is that they allow these types of analysis to be performed simultaneously on a relatively large number of genes during an actual infection. STM is most useful for determining outright which loci contribute strongly to survival in the host, whereas IVET strategies are capable of providing more subtle information regarding the expression patterns of genes during infection. Comparison of the IVET and STM results in *S. typhimurium, S. aureus,* and *V. cholerae* shows that they are complementary approaches. The integrated use of these approaches is already well underway in a variety of organisms, because current work in this field includes STM analyses in *Streptococcus pneumoniae, P. aeruginosa, Y. enterocolitica,* and *Legionella pneumophila* and the application of DFI in *L. pneumophila* and *Bartonella henselae.* The information obtained from such studies will undoubtedly contribute to a more comprehensive understanding of bacterial pathogenesis.

ACKNOWLEDGMENTS

Work in the Holden laboratory was supported by the MRC (U.K.), Wellcome Trust (U.K.), Pharmacia and Upjohn, Inc., and SmithKline Beecham plc. Work in the Mekalanos laboratory was supported by National Institutes of Health grants AI18045 and AI26289 (to JJM).

Visit the Annual Reviews home page at http://www.AnnualReviews.org.

LITERATURE CITED

1. Akerley BJ, Cotter PA, Miller JF. 1995. Ectopic expression of the flagellar regulon alters development of the Bordetella-host interaction. *Cell* 80:611–20

2. Akerley BJ, Rubin EJ, Camilli A, Lampe DJ, Robertson HM, et al. 1998. Systematic identification of essential genes by in vitro mariner mutagenesis. *Proc. Natl. Acad. Sci. USA* 95:8927–32

3. Akerley BJ, Rubin EJ, Lampe DJ, Mekalanos JJ. 1998. *PCR-mediated detection of growth-attenuated mutants in large pools generated by in vitro transposon mutagenesis.* Presented at Am. Soc. Microbiol., Gen. Meet., 98th, Atlanta

4. Alpuche Aranda CM, Swanson JA, Loomis WP, Miller SI. 1992. *Salmonella typhimurium* activates virulence gene transcription within acidified macrophage phagosomes. *Proc. Natl. Acad. Sci. USA* 89: 10079–83

5. Ankenbauer RG, Quan HN. 1994. FptA, the Fe(III)-pyochelin receptor of *Pseudomonas aeruginosa*: a phenolate siderophore receptor homologous to hydroxamate siderophore receptors. *J. Bacteriol.* 176:307–19

6. Attridge SR, Voss E, Manning PA. 1993. The role of toxin-coregulated pili in the pathogenesis of *Vibrio cholerae* O1 El Tor. *Microb. Pathog.* 15:421–31

7. Baorto DM, Gao Z, Malaviya R, Dustin ML, van der Merwe A, et al. 1997. Survival of FimH-expressing enterobacteria in macrophages relies on glycolipid traffic. *Nature* 389:636–39

8. Belden WJ, Miller SI. 1994. Further characterization of the PhoP regulon: identification of new PhoP-activated virulence loci. *Infect. Immun.* 62:5095–101

9. Bennish ML. 1994. Cholera: pathophysiology, clinical features, and treatment. In *Vibrio cholerae and Cholera: Molecular to Global Perspectives,* ed. IK Wachsmuth, PA Blake, Ø Olsvik, pp. 229–255. Washington, DC: Am. Soc. Microbiol.

10. Böckmann R, Dickneite C, Middendorf B, Goebel W, Sokolovic Z. 1996. Specific binding of the *Listeria monocytogenes* transcrip-

tional regulator PrfA to target sequences requires additional factor(s) and is influenced by iron. *Mol. Microbiol.* 22:643–53

11. Bowe F, Lipps CJ, Tsolis RM, Groisman E, Heffron F, et al. 1998. At least four percent of the *Salmonella typhimurium* genome is required for fatal infection of mice. *Infect. Immun.* 66:3372–77

12. Buchmeier NA, Heffron F. 1989. Intracellular survival of wild-type *Salmonella typhimurium* and macrophage-sensitive mutants in diverse populations of macrophages. *Infect. Immun.* 57:1–7

13. Buchmeier NA, Heffron F. 1990. Induction of *Salmonella* stress proteins upon infection of macrophages. *Science* 248:730–32

14. Camilli A, Beattie D, Mekalanos J. 1994. Use of genetic recombination as a reporter of gene expression. *Proc. Natl. Acad. Sci. USA* 91:2634–38

15. Camilli A, Mekalanos JJ. 1995. Use of recombinase gene fusions to identify *Vibrio cholerae* genes induced during infection. *Mol. Microbiol.* 18:671–83

16. Camilli A, Tilney L, Portnoy D. 1993. Dual roles of *plcA* in *Listeria monocytogenes* pathogenesis. *Mol. Microbiol.* 8:143–57

17. Chiang SL, Mekalanos JJ. 1998. Use of signature-tagged transposon mutagenesis to identify *Vibrio cholerae* genes critical for colonization. *Mol. Microbiol.* 27:797–806

18. Cirillo DM, Valdivia RH, Monack DM, Falkow S. 1998. Macrophage-dependent induction of the *Salmonella* pathogenicity island 2 type III secretion system and its role in intracellular survival. *Mol. Microbiol.* 30:175–88

19. Clark MA, Hirst BH, Jepson MA. 1998. Inoculum composition and Salmonella pathogenicity island 1 regulate M-cell invasion and epithelial destruction by *Salmonella typhimurium. Infect. Immun.* 66:724–31

20. Clark MA, Jepson MA, Simmons NL, Hirst BH. 1994. Preferential interaction of *Salmonella typhimurium* with mouse Peyer's patch M cells. *Res. Microbiol.* 145:543–52

21. Conner CP, Heithoff DM, Julio SM, Sinsheimer RL, Mahan MJ. 1998. Differential patterns of acquired virulence genes distinguish *Salmonellan* strains. *Proc. Natl. Acad. Sci. USA* 95:4641–45

22. Coulter SN, Schwan WR, Ng EYW, Langhorne MH, Ritchie HD, et al. 1998. *Staphylococcus aureus* genetic loci impacting growth and survival in multiple infection environments. *Mol. Microbiol.* 30:393–404

22a. Darwin AJ, Miller VL. 1999. Identification of *Yersinia enterocolitica* genes affecting survival in an animal host using signature-tagged transposon mutagenesis. *Mol. Microbiol.* 32:51–62

23. De Groote MA, Ochsner UA, Shiloh MU, Nathan C, McCord JM, et al. 1997. Periplasmic superoxide dismutase protects *Salmonella* from products of phagocyte NADPH-oxidase and nitric oxide synthase. *Proc. Natl. Acad. Sci. USA* 94:13997–4001

24. DiRita VJ, Mekalanos JJ. 1989. Genetic regulation of bacterial virulence. *Annu. Rev. Genet.* 23:455–82

25. Falkow S. 1996. The evolution of pathogenicity in *Escherichia, Shigella,* and *Salmonella.* In *Escherichia coli and Salmonella: Cellular and Molecular Biology,* ed. FC Neidhardt, R Curtiss III, JL Ingraham, ECC Lin, KB Low, et al, pp. 2723–29. Washington, DC: Am. Soc. Microbiol.

26. Fields PI, Swanson RV, Haidaris CG, Heffron F. 1986. Mutants of *Salmonella typhimurium* that cannot survive within the macrophage are avirulent. *Proc. Natl. Acad. Sci. USA* 83:5189–93

27. Gahring LC, Heffron F, Finlay BB, Falkow S. 1990. Invasion and replication of *Salmonella typhimurium* in animal cells. *Infect. Immun.* 58:443–48

28. Galan JE. 1996. Molecular genetic bases of *Salmonella* entry into host cells. *Mol. Microbiol.* 20:263–71

29. Goldberg MB, DiRita VJ, Calderwood SB.

1990. Identification of an iron-regulated virulence determinant in *Vibrio cholerae*, using Tn*phoA* mutagenesis. *Infect. Immun.* 58:55–60

30. Groisman EA, Chiao E, Lipps CJ, Heffron F. 1989. *Salmonella typhimurium phoP* virulence gene is a transcriptional regulator. *Proc. Natl. Acad. Sci. USA* 86:7077–81

31. Heithoff DM, Conner CP, Hanna PC, Julio SM, Hentschel U, et al. 1997. Bacterial infection as assessed by *in vivo* gene expression. *Proc. Natl. Acad. Sci. USA* 94:934–39

32. Hensel M, Shea JE, Gleeson C, Jones MD, Dalton E, et al. 1995. Simultaneous identification of bacterial virulence genes by negative selection. *Science* 269:400–3

33. Hensel M, Shea JE, Waterman SR, Mundy R, Nikolaus T, et al. 1998. Genes encoding putative effector proteins of the type III secretion system of *Salmonella* pathogenicity island 2 are required for bacterial virulence and proliferation in macrophages. *Mol. Microbiol.* 30:163–74

34. Herrington DA, Hall RH, Losonsky G, Mekalanos JJ, Taylor RK, et al. 1988. Toxin, toxin-coregulated pili, and the *toxR* regulon are essential for *Vibrio cholerae* pathogenesis in humans. *J. Exp. Med.* 168:1487–92

35. Ji G, Beavis R, Novick RP. 1997. Bacterial interference caused by autoinducing peptide variants. *Science* 276:2027–30

36. Kaufman MR, Taylor RK. 1994. The toxin-coregulated pilus: biogenesis and function. In *Vibrio cholerae and Cholera: Molecular to Global Perspectives*, ed. IK Wachsmuth, PA Blake, Ø Olsvik, pp. 187–202. Washington, DC: Am. Soc. Microbiol.

37. Klarsfeld AD, Goossens PL, Cossart P. 1994. Five *Listeria monocytogenes* genes preferentially expressed in infected mammalian cells: *plcA, purH, purD, pyrE* and an arginine ABC transporter gene, *arpJ*. *Mol. Microbiol.* 13:585–97

38. Lipsitch M, Moxon ER. 1997. Virulence and transmissibility of pathogens: what is the relationship? *Trends Microbiol.* 5:31–37

39. Litwin CM, Calderwood SB. 1993. Role of iron in regulation of virulence genes. *Clin. Microbiol. Rev.* 6:137–49

40. Lowe AM, Beattie DT, Deresiewicz RL. 1998. Identification of novel staphylococcal virulence genes by *in vivo* expression technology. *Mol Microbiol* 27:967–976

41. Lowy FD. 1998. *Staphylococcus aureus* infections. *N. Engl. J. Med.* 339:520–32

42. Mahan MJ, Slauch JM, Mekalanos JJ. 1993. Selection of bacterial virulence genes that are specifically induced in host tissues. *Science* 259:686–88

43. Mahan MJ, Tobias JW, Slauch JM, Hanna PC, Collier RJ, et al. 1995. Antibiotic-based IVET selection for bacterial virulence genes that are specifically induced during infection of a host. *Proc. Natl. Acad. Sci. USA* 92:669–73

44. Maurelli AT. 1989. Temperature regulation of virulence genes in pathogenic bacteria: a general strategy for human pathogens? *Microb. Pathog.* 7:1–10

45. Mei J-M, Nourbakhsh F, Ford CW, Holden DW. 1997. Identification of *Staphylococcus aureus* virulence genes in a murine model of bacteremia using signature-tagged mutagenesis. *Mol. Microbiol.* 26:399–407

46. Mengaud J, Dramsi S, Gouin E, Vazquez-Boland JA, Milon G, et al. 1991. Pleiotropic control of *Listeria monocytogenes* virulence factors by a gene which is autoregulated. *Mol. Microbiol.* 5:2273–83

47. Meynell GG, Stocker BAD. 1957. Some hypotheses on the aetiology of fatal infections in partially resistant hosts and their application to mice challenged with *Salmonella paratyphi-B* or *Salmonella typhimurium* by intraperitoneal injection. *J. Gen. Microbiol.* 16:38–58

48. Miller I, Maskell D, Hormaeche C, Johnson K, Pickard D, et al. 1989. Isolation of orally attenuated *Salmonella typhimurium*

following *TnphoA* mutagenesis. *Infect. Immun.* 57:2758–63

49. Miller JF, Mekalanos JJ, Falkow S. 1989. Coordinate regulation and sensory transduction in the control of bacterial virulence. *Science* 243:916–22

50. Miller SI, Kukral AM, Mekalanos JJ. 1989. A two-component regulatory system (*phoP phoQ*) controls *Salmonella typhimurium* virulence. *Proc. Natl. Acad. Sci. USA* 86:5054–58

51. Miller VL. 1995. Tissue-culture invasion: fact or artefact? *Trends Microbiol.* 3:69–71

52. Moxon RE, Murphy PA. 1978. *Haemophilus influenzae* bacteremia resulting from survival of a single organism. *Proc. Natl. Acad. Sci. USA* 75:1534–36

53. Novick RP, Projan SJ, Kornblum J, Ross HF, Ji G, et al. 1995. The *agr* P2 operon: an autocatalytic sensory transduction system in *Staphylococcus aureus*. *Mol. Gen. Genet.* 248:446–58

54. Ochman H, Soncini FC, Solomon F, Groisman EA. 1996. Identification of a pathogenicity island required for *Salmonella* survival in host cells. *Proc. Natl. Acad. Sci. USA* 93:7800–4

55. Olson ER. 1993. Influence of pH on bacterial gene expression. *Mol. Microbiol.* 8:5–14

56. Osbourn AE, Barber CE, Daniels MJ. 1987. Identification of plant-induced genes of the bacterial pathogen *Xanthomonas campestris* pathovar *campestris* using a promoter-probe plasmid. *EMBO J.* 6:23–28

57. Osbourn AE, Barber CE, Daniels MJ. 1988. Identification and analysis of plant-induced genes of *Xanthomonas campestris*. In *Molecular Genetics of Plant-Microbe Interactions*, ed. R Palacios, DPS Verma, pp. 237–74. St. Paul, MN: Am. Phytopathol. Soc.

58. Osbourn AE, Clarke BR, Daniels MJ. 1990. Identification and DNA sequence of a pathogenicity gene of *Xanthomonas*

campestris. *Mol. Plant-Microbe Interact.* 3:280–85

59. Schwan WR, Coulter SN, Ng EYW, Langhorne MH, Ritchie HD, et al. 1998. Identification and characterization of the PutP proline permease that contributes to in vivo survival of *Staphylococcus aureus* in animal models. *Infect. Immun.* 66:567–72

60. Shea JE, Hensel M, Gleeson C, Holden DW. 1996. Identification of a virulence locus encoding a second type III secretion system in *Salmonella typhimurium*. *Proc. Natl. Acad. Sci. USA* 93:2593–97

61. Shoemaker DD, Lashkari DA, Morris D, Mittmann M, Davis RW. 1996. Quantitative phenotypic analysis of yeast deletion mutants using a highly parallel molecular bar-coding strategy. *Nat. Genet.* 14:450–56

62. Siebers A, Finlay BB. 1996. M cells and the pathogenesis of mucosal and systemic infections. *Trends Microbiol* 4:22–29

63. Smith V, Botstein D, Brown PO. 1995. Genetic footprinting: a genomic strategy for determining a gene's function given its sequence. *Proc. Natl. Acad. Sci. USA* 92:6479–83

64. Taylor RK, Miller VL, Furlong DB, Mekalanos JJ. 1987. Use of *phoA* gene fusions to identify a pilus colonization factor coordinately regulated with cholera toxin. *Proc. Natl. Acad. Sci. USA* 84:2833–37

65. Thelin KH, Taylor RK. 1996. Toxin-coregulated pilus, but not mannose-sensitive hemagglutinin, is required for colonization by *Vibrio cholerae* O1 El Tor biotype and O139 strains. *Infect. Immun.* 64:2853–56

66. Valdivia RH, Falkow S. 1996. Bacterial genetics by flow cytometry: rapid isolation of *Salmonella typhimurium* acid-inducible promoters by differential fluorescence induction. *Mol. Microbiol.* 22:367–78

67. Valdivia RH, Falkow S. 1997. Fluorescence-based isolation of bacterial genes expressed within host cells. *Science* 277:2007–11

68. Walsh C, Cepko CL. 1992. Widespread dispersion of neuronal clones across functional regions of the cerebral cortex. *Science* 255:434–40

69. Wang J, Lory S, Ramphal R, Jin S. 1996. Isolation and characterization of *Pseudomonas aeruginosa* genes inducible by respiratory mucus derived from cystic fibrosis patients. *Mol. Microbiol.* 22:1005–12

70. Wang J, Mushegian A, Lory S, Jin S. 1996. Large-scale isolation of candidate virulence genes of *Pseudomonas aeruginosa* by *in vivo* selection. *Proc. Natl. Acad. Sci. USA* 93:10434–39

71. Young GM, Miller VL. 1997. Identification of novel chromosomal loci affecting *Yersinia enterocolitica* pathogenesis. *Mol. Microbiol.* 25:319–28

Annu. Rev. Microbiol. 1999. 53:155–87

THE INDUCTION OF APOPTOSIS BY BACTERIAL PATHOGENS

Yvette Weinrauch and Arturo Zychlinsky

*Department of Microbiology, Skirball Institute and Kaplan Cancer Center,
New York University School of Medicine, New York, New York 10016;
e-mail: zychlinsky@saturn.med.nyu.edu*

Key Words   toxins, bacteria, infections, effectors, virulence

■ **Abstract**   Apoptosis is a highly regulated process of cell death that is required for the development and homeostasis of multicellular organisms. In contrast to necrosis, apoptosis eliminates individual cells without inducing an inflammatory response. Activation or prevention of cell death could be a critical factor in the outcome of an infection. Programmed cell death has been observed as a response to infection by a wide range of animal and plant pathogens and is mediated by an array of pathogen-encoded virulence determinants. Pathogen-induced modulation of the host cell-death pathway may serve to eliminate key immune cells or evade host defenses that can act to limit the infection. Alternatively, suppression of the death pathway may facilitate the proliferation of intracellular pathogens.

CONTENTS

0066-4227/99/1001-0155$08.00   **155**

INTRODUCTION

Cell death is critical in morphogenesis and provides specificity to the immune and nervous systems during development. Apoptosis also plays a crucial role in the turnover or homeostasis of some organs. Defects in apoptosis during embryogenesis lead to developmental abnormalities, whereas interference with homeostatic apoptosis results in pathologies such as tumors or functional deficiencies.

Apoptosis was initially defined by two different lines of research. First, morphological studies described apoptosis as a series of rapid, specific events that resulted in the systematic dismantling of cellular components. Membrane blebbing, nuclear condensation, and DNA fragmentation are characteristic features that lead to the clearance of the apoptotic debris. Second, genetic studies in the nematode *Caenorhabditis elegans* showed that apoptosis is genetically programmed. Programmed cell death occurs during the normal course of development, and the analysis of mutants defective for cell death has led to isolation and characterization of several genes that show remarkable conservation from nematodes to mammals.

We initially review the molecular mechanisms of apoptosis, some of which are relevant to bacteria-induced apoptosis. We then discuss the mechanisms and potential role of pathogens in activating programmed cell death in the host.

ACTIVATION OF APOPTOSIS

Receptor-Mediated Signaling

The death machinery of cells is strictly regulated through the interaction of pro- and anti-apoptotic molecules. The apoptotic machinery is activated when this balance is upset by the withdrawal of survival signals like growth factors, the intervention of conflicting signals during the cell cycle, or the recognition of a specific molecule at the cell surface. Of these three possibilities, the mechanism of receptor-mediated signaling is the best defined.

Death receptors belonging to the tumor necrosis factor (TNF) receptor (TNFR) super family are transmembrane proteins localized to the cytoplasmic membrane. Their extracellular domains are responsible for ligand binding (22, 176, 205, 241, 245). They also have "death domains" (DDs) in their cytoplasmic tails, which enable the death receptors to engage the apoptotic machinery. Paradoxically, in some instances, these domains mediate interactions that suppress the apoptotic stimulus.

The well-characterized death receptors TNFR and CD95 (Fas) bind the cytokine TNF-α and CD95 ligand (CD95L), respectively. Upon ligand binding, the DDs of CD95 interact with the Fas-associated DD protein, an adaptor molecule that is recruited to the receptor complex (31, 46) (Figure 1). The Fas-associated DD, which contains a "death effector domain," then binds to an analogous domain in caspase-8 (30, 174). Caspases are proteases that play a pivotal role in the activation of apoptosis, and they are described in more detail below.

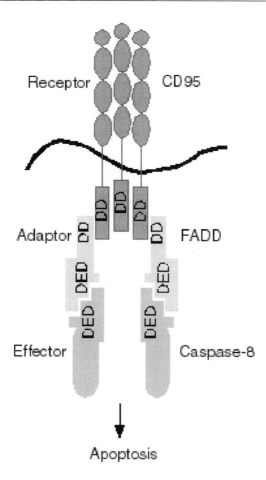

Figure 1   Signaling for apoptosis by CD95. DD, Death domain; DED, death effector domain.

Signaling through TNFR requires an additional adapter molecule to interact with the Fas-associated DD, called TNFR-associated–DD protein (97). TNFR-associated–DD protein can also bind to both TNFR-associated factor 2 (96, 204) and to a receptor interaction protein (95, 237). Both TNFR-associated factor 2 and receptor interaction protein stimulate pathways that culminate in the activation of transcription via nuclear factor κB (NF-κB) and JNK/AP-1 and act as antiapoptotic signals (60, 149, 155, 162, 179, 200, 222, 257).

Cell Cycle

Cells in multicellular organisms respond to DNA damage by triggering a plethora of cellular responses. These responses include DNA repair, growth arrest, and apoptosis. Choosing the appropriate response is critical to the survival of the organism. Mutations in the DNA repair machinery predispose the organism to malignancies (64). A critical regulator of the cellular response to DNA damage is the transcription factor encoded by the *p53* tumor suppressor gene. Studies show that p53 promotes apoptosis by the activation of specific target genes (13, 209) by undefined mechanisms.

The link between cell death and cell proliferation came from studies on the oncoprotein C-myc (65, 220) and E1A (250), the principal growth-promoting protein encoded by adenovirus. These two proteins are potent inducers of cell proliferation and also play a role in the apoptotic pathway.

Mitochondria: Key Organelles in Apoptosis

Mitochondria play a major role in the orchestration of apoptosis. Nevertheless, oxidative phosphorylation, the major metabolic contribution of these organelles, does not appear to be relevant in this phenomenon, because mitochondrial DNA is not required for apoptosis (107).

Cell-free systems have uncovered the dependence of some apoptotic pathways on mitochondria (180) or factors released by mitochondria (149). During apoptosis, cytochrome *c* and probably other factors are released from the mitochondria to the cytoplasm (127, 253). Cytochrome *c* is required for the formation of the proapoptotic complex that also includes a *C. elegans* Ced-4 homolog Apaf-1 and a pro caspase-9 (142).

Disruption of the mitochondria might also contribute to cell death, not necessarily apoptosis, by disrupting the energy metabolism of the cell (76). Loss of the integrity of the mitochondria leads to the production of reactive oxygen species. These species have been detected late in the apoptosis initiated by many stimuli (33). A direct role for reactive oxygen species in apoptosis remains controversial.

EFFECTOR MOLECULES OF APOPTOSIS

Caspases

Caspases were the first effector molecules identified in the apoptosis pathway. They form a family of cysteine proteases that cleave after an aspartic acid (6). Caspases appear to be similar in sequence and structure (181). In the nematode *C. elegans*, the gene *ced*-3 encodes a cysteine protease that is an effector of cellular suicide and is homologous to the mammalian protein interleukin-1β-converting enzyme or caspase-1 (6, 252, 255). Caspases are synthesized as zymogens that have little catalytic activity (Figure 2). The zymogen consists of a variable-length amino-terminal domain that does not form part of the mature enzyme and is probably involved in the regulation of the zymogen's activation. The carboxy-terminal portion of the zymogen is ultimately cleaved into two peptides of 20 and 10 kDa that associate to form the active enzyme. Caspase zymogens are all cleaved at

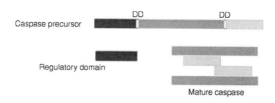

Figure 2 Caspases. The zymogen is cleaved by a caspase at the specific aspartates to cleave off the amino-terminal regulatory domain and form the mature enzyme with the two carboxy-terminal peptides.

caspase-specific sites, suggesting that their activation is dependent on autocatalysis, the mature enzyme, or other caspases.

There are three ways in which caspases participate in apoptosis. First, caspases can inactivate inhibitors. An example of this mechanism involves the caspase-activated deoxyribonuclease (CAD) and its inhibitor. CAD is the nuclease responsible for DNA fragmentation during apoptosis. CAD exists in an inactive complex with its inhibitor in normal cells; however, during apoptosis, CAD is liberated by caspase-mediated cleavage of its inhibitor (62, 150). Second, caspases can dissociate the effector and regulatory domains of other proteins. Examples of this activity include the cleavage of the proapoptotic protein BIK and proteins involved in the processes of DNA repair, mRNA splicing, and DNA replication (55, 62). Third, caspases contribute to apoptosis by destroying structural proteins like laminins (186, 233), which form the scaffolding of the nucleus, and cytoskeletal proteins (131).

Caspase zymogens accumulate in cells without causing any damage. This implies that activation of caspases is very tightly regulated. Caspase activation might be regulated by oligomerization or inhibitors (23). There is good biochemical evidence that close proximity of caspase zymogens promotes maturation of the enzyme. Whether this mechanism is relevant in vivo remains to be determined (77, 175, 254).

The Bcl-2 Family

Genetic studies of programmed cell death in *C. elegans* show that *ced-9* and its mammalian counterpart, *bcl-2*, prevent cells from undergoing apoptosis. The *bcl-2* gene was identified in a chromosomal translocation that resulted in human follicular lymphoma (18, 49, 239). It is now evident that Bcl-2 and related proteins are key regulators of apoptosis (32, 80, 83, 185, 242).

There are 15 *bcl-2*–like genes identified so far. The family of proteins encoded by these genes includes Bcl-2 and Bcl-x_L, which are inhibitory, whereas other proteins, such as Bax (185) and Bik (32, 80), promote apoptosis.

Some of the pro- and anti-apoptotic family members can heterodimerize and may act by regulating the concentration of free proteins to decide the fate of the cell.

Bcl-2 family members are regulated both transcriptionally (3, 28, 41, 132, 144) and by post-translational modifications, namely phosphorylation (58, 59, 79, 106, 148, 160, 259).

Bcl-2 localizes to the cytoplasmic face of mitochondria, the endoplasmic reticulum, and the nuclear envelope (133, 199, 256). The three-dimensional structure of some family members is similar to that of bacterial pore-forming proteins, and purified Bcl-2 family members can function as pores for small molecules (172, 214). The prosurvival members of the family might act by preventing the binding of Ced-4-like proteins, a family of caspase activators, to caspases (98, 99, 189). Bcl-2 might also work by regulating the permeability of the mitochondrial membrane (142, 256, 261) to the passage of caspase activators into the cytoplasm. Proapoptotic members of this family can kill independently of caspase activity, suggesting that the pore-forming activity of these proteins might be sufficient to induce apoptosis (103, 112, 258).

The complexity of this family of proteins indicates that there might be some redundancy in their function. Indeed, although Bcl-2 is expressed in many tissues in embryos, *bcl-2* −/− mice develop normally. However in adulthood, they show upregulation of apoptosis in several tissues (243). In contrast, *bcl-x* −/− mice die in utero after massive neuronal and erythroid apoptosis (170).

INDUCTION OF APOPTOSIS BY BACTERIA

The strategies used by pathogens to activate programmed cell death may be necessary to subvert normal host defense responses. A number of pathogens are armed with an array of virulence determinants, which interact with key components of the cell death pathway of the host or interfere with the regulation of transcription factors monitoring cell survival.

These virulence factors induce cell death by a variety of mechanisms and include (*a*) pore-forming toxins, which interact with the host cell membrane and permit the leakage of cellular components, (*b*) other toxins that express their enzymatic activity in the host cytosol, (*c*) effector proteins delivered directly into host cells by a highly specialized type-III secretory system, (*d*) superantigens that target immune cells, and (*e*) other modulators of host cell death.

The aim of the rest of this review is to summarize the numerous mechanisms that extracellular and intracellular pathogenic bacteria use to induce, and in some cases inhibit, apoptosis in host cells. In some instances, the relationship between apoptotic events and pathogenesis is also explored.

Pore-Forming Toxins

There are at least three different types of apoptotic pore formers: amphiphilic toxins, such as alpha toxin from *Staphylococcus aureus*, that insert directly into the cell membrane thereby creating an ion-permeable pore; thiol-activated toxins, such as listeriolysin O from *Listeria monocytogenes*; and the repeats-in-toxin (RTX) family of pore-forming toxins. The latter can be subdivided into two classes based on their target cell specificity: (*a*) toxins such as that from *Actinobacillus actinomycetemcomitans*, with a narrow range of target cell types, and (*b*) toxins such as the *Escherichia coli* alpha-hemolysin, with a wide range of cell types.

Staphylococcus aureus   *S. aureus* is a gram-positive coccus that causes a variety of diseases, including skin lesions, food poisoning, toxic shock syndrome, endocarditis, and osteomyelitis. A number of virulence factors, including several exotoxins, have been implicated in the pathogenesis of infections with this bacterium. One of these, alpha toxin, induces apoptosis in eukaryotic cells.

Alpha toxin forms pores in many eukaryotic cell membranes (2, 26, 102, 161, 230, 231) and triggers programmed cell death in T-lymphocytes (110). At low doses the toxin binds to specific, as yet unidentified, cell surface receptors (87) and produces small pores that selectively facilitate the release of monovalent ions, resulting in characteristic DNA fragmentation and cell death (110). At high doses, alpha toxin nonspecifically absorbs to the lipid bilayer (87), forming larger pores that are Ca^{2+} permissive, which results in massive necrosis without DNA fragmentation.

In a recent study, *S. aureus* isolated from a case of bovine mastitis was shown to invade bovine mammary epithelial cells (MAC-T), escape from the endosome, and induce apoptosis (21). Cell death depended on factors regulated by the global virulence gene regulators Agr and Sar (249). Staphylococcal virulence factors that are regulated by Agr and Sar include hemolysins and exotoxins (196), one or more of which may be involved in effecting the release of *S. aureus* from the endosome and mediating host cell death.

Listeria monocytogenes   The gram-positive rod *L. monocytogenes* is an opportunistic food-borne human and animal pathogen responsible for serious infections in immunocompromised individuals, pregnant women, neonates, and others. After contact with target cells, this invasive pathogen is internalized into a phagolysosome and escapes from this compartment into the cytosol. From there, it mediates the polymerization of actin filaments, possibly to facilitate cell-to-cell spread (194, 236).

L. monocytogenes kills infected hepatocytes by apoptosis. This effect is observed both in vitro in primary cells and cell lines and in vivo (202) and is associated with the release of neutrophil chemoattractants early in the inflammatory process.

The mechanism by which *L. monocytogenes* kills hepatocytes has not been defined. However, in vitro studies in murine-splenic dendritic cells suggest that cell death is mediated by listeriolysin O, a pore-forming toxin. Mutations in the structural gene (*hyl-2*) for listeriolysin abrogate dendritic-cell cytotoxicity, and purified toxin is sufficient to trigger cell death. In contrast to the effects of other intracellular pathogens, apoptosis has not been observed in macrophages infected with *L. monocytogenes* (20, 262).

Actinobacillus actinomycetemcomitans A. *actinomycetemcomitans* is a gram-negative facultative anaerobic bacterium that is the etiologic agent in diseases such as periodontitis, meningitis, and endocarditis. One of the potential virulence factors produced by this bacterium is a leukotoxin (Ltx) that generates membrane pores and kills several lymphoid cell types (232, 238) but spares fibroblasts, platelets, and endothelial and epithelial cells (238). Cells exposed to low concentrations of Ltx exhibit alterations consistent with apoptosis (130, 136).

Ltx sensitivity appears to depend on host cell surface expression of a ß2 integrin, lymphocyte function-associated antigen 1 (LFA-1), which is found predominantly on lymphocytes, monocytes, and neutrophils. Ltx binding to LFA-1 may facilitate insertion of the toxin into the cell membrane resulting in the elimination of immune cells. At high doses, as with staphylococcal alpha toxin, Ltx inflicts overwhelming damage by necrosis (152).

The RTX family of pore-forming bacterial toxins also includes the cytotoxins synthesized by *Actinobacillus pleuropneumoniae* (136, 143) and *Pasteurella haemolytica*, the latter causing cell death in bovine leukocytes (228).

Escherichia coli E. *coli* alpha-hemolysin (HlyA) is another RTX toxin that mediates cell death via LFA-1 in human immune cells (137). However, unlike *A. actinomycetemcomitans* leukotoxin, HlyA is toxic to a wide range of cell types, such as erythrocytes from many species, chicken embryo fibroblasts, rabbit granulocytes, and mouse fibroblasts (37). It is possible that cell surface molecules other than LFA-1 might contribute to HlyA-mediated cytotoxicity. The acylated form of this pore-forming toxin is the active species (130). It attacks the plasma membranes of target mammalian cells by inserting as a monomer into the bilayer to generate a transmembrane pore that is selective for cations over anions (26).

Protein Synthesis Inhibitors

The bacterial toxins in this group are proteins with intrinsic enzymatic activity. They have an A-B structure in which the B subunit mediates cell binding and the enzymatically active A subunit is translocated across the eukaryotic cell membrane to its cytosolic destination. Examples include the diphtheria toxin (Dtx) from *Corynebacterium diphtheria*, *Pseudomonas* exotoxin A, and the Shiga and Shiga-like toxins.

Corynebacterium diphtheriae C. *diphtheriae* is a gram-positive rod that causes diphtheria in humans. The disease is characterized by sore throat, fever, and the formation of a gray, fibrinous pseudomembrane on the posterior pharynx, composed of bacteria and inflammatory cells. Airway obstruction, recurrent laryngeal nerve palsy, and cardiac manifestations are complications of diphtheria.

As an extracellular pathogen, *C. diphtheriae* communicates with host cells across the plasma membrane. It secretes a classic bacterial A-B toxin called Dtx, which is encoded by a temperate bacteriophage and is synthesized in response to low iron. The B subunit of the A-B toxin mediates attachment to cells via a host receptor and facilitates delivery of the catalytic A subunit to the cytoplasm. The A domain in the cytoplasm catalyzes the transfer of the ADP-ribosyl group of NAD^+ to elongation factor 2 (EF-2). This inactivates EF-2, a critical component of the protein synthesis machinery, thereby killing the target cell (47, 183).

Dtx causes apoptosis in several epithelial and myeloid cell lines (38, 40, 128, 168, 210). The following evidence supports protein synthesis inhibition as a mechanism of Dtx-induced cell death: (*a*) Dtx mutants, deficient in ADP-ribosylating activity, are not cytotoxic (168), (*b*) cell lines with mutant EF-2, which are not susceptible to ADP ribosylation, are Dtx insensitive (129, 193), and (*c*) compounds that block ADP ribosylation of EF-2 by Dtx inhibit apoptosis (168).

Although necessary, inhibition of protein synthesis is not sufficient for Dtx toxicity. Cell lines have been established in which Dtx effectively blocks protein synthesis without inducing apoptosis (39, 128, 168). A recent study indicates that Dtx, as well as ExoA-mediated apoptosis involves the cellular apoptosis susceptibility (CAS) protein, a host nuclear transport factor that plays a role in proliferation (34, 35, 215). Antisense-mediated attenuation of CAS protein expression confers resistance to cells normally sensitive to Dtx's lethal effect but does not prevent the inhibition of protein synthesis (34, 35).

The CAS protein is highly expressed in tissues containing dividing cells and in cells exposed to proliferative stimuli (35). Scherf et al (215) suggest that phosphorylation of CAS may be required for nuclear transport of substrates involved in mitosis and/or apoptosis.

Dtx may possess other functions. The A subunit of the toxin is reported to have nuclease activity (39, 140, 168). This proposed nuclease cleaves at internucleosomal junctions, creating a prototypical apoptotic DNA ladder. However, its role in mediating cell death remains to be defined.

Pseudomonas aeruginosa   *P. aeruginosa*, a gram-negative rod found in soil, water, and occasionally in the human gut, causes sepsis, urinary-tract infections, and pneumonia. It is an opportunistic pathogen capable of infecting immunocompromised individuals and patients suffering from cystic fibrosis.

P. aeruginosa secretes several toxins, of which the key virulence factor, exotoxin A (Exo-A), induces apoptosis in a human monoblastoid cell line. Mutants in Exo-A are avirulent (128, 168, 197). Although they display no homology, Exo-A, like Dtx, ADP ribosylates EF-2 and inhibits protein synthesis. Cell lines expressing mutant EF-2 molecules are also resistant to Exo-A (197).

P. aeruginosa–induced cytotoxicity was also reported in cell cultures of MDCK epithelial cells. This activity was independent of Exo-A production (8) and dependent on a type-III protein secretion apparatus (116). Whether cytotoxicity is mediated by programmed cell death remains to be established.

Shigella dysenteriae and Enterohemorrhagic Escherichia coli   *S. dysenteriae* and enterohemorrhagic *E. coli* cause dysenteric syndromes with sequelae, including kidney and central nervous system manifestations. These bacteria produce similar toxins, called Shiga and Shiga-like toxins (Slt) and verotoxin. Shiga toxin is an A-B toxin composed of one A and five B subunits. The A subunit has N-glycosidase activity, which cleaves a single adenine residue from the 28S rRNA component of eukaryotic ribosomes, thereby disrupting ribosome function and protein synthesis (234). The B subunits facilitate delivery of the toxin by binding to globotriaosylceramide (Gb3) receptors on the cell surface (156, 234).

Shiga toxin and Slt induce apoptosis in several systems. Using a rabbit ileal loop model of intestinal inflammation, Keenan et al (121) have demonstrated that both toxins induce apoptosis by acting directly on differentiated, villus epithelial cells but spare crypt cells. This may be caused in large part by the greater content of Gb3 in villus cells than in crypt cells. The relative abundance of Gb3 receptors in kidney cells may also account for the renal sequelae caused by *S. dysenteriae* and enterohemorrhagic *E. coli* (234). In vitro, Shiga toxin and Slt kill kidneylike epithelial cells by apoptosis (104, 210). Although the binding of the B subunit to the Gb3 receptor is sufficient to trigger apoptosis (156), the holotoxin is even more potent, suggesting that the inhibition of protein synthesis and Gb3 receptor binding account for this synergy.

Type-III Secreted Proteins

Some gram-negative pathogens are equipped with highly adapted secretory machinery termed type-III secretion apparatus, which is conserved in both animal and plant pathogens. A variety of animal pathogens, such as *Shigella* spp., *Salmonella* spp., and *Yersinia* spp., as well as the plant pathogens from the genera *Erwinia*, *Pseudomonas, Xanthomonas*, and *Ralstonia*, use this protein secretion pathway to deliver a set of bacterial effector proteins into the host cell cytosol. These highly divergent proteins interact with host cell components to alter host cell signal transduction pathways (reviewed in 100).

Shigellae *Shigella* spp. cause a diarrheal disease with a variable clinical picture, the most severe form being dysentery, which is characterized by blood and mucus in the stools. Shigellosis is endemic in developing countries, where it is a significant cause of infant mortality.

The disease is transmitted via the fecal/oral route, and invasion of the colonic mucosa constitutes the essential first step in the pathogenesis. Because shigellae are not able to invade the polarized intestinal epithelial cells through the apical membrane of these cells, they initially gain access by entering M cells located within the epithelial layer (105, 192, 211, 246). After transcytosis through M cells, the bacteria are promptly phagocytosed by resident macrophages. They rapidly escape from the phagosomal compartment to the cytosol and subsequently trigger apoptosis (262, 263). *S. flexneri* induces apoptosis in macrophages both in vitro (262) (Figure 3) and in vivo (263).

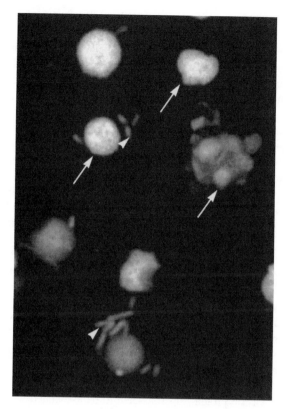

Figure 3   Induction of apoptosis by *Shigella flexneri*. Macrophages were infected in vitro with a *Shigella* isolate. After an incubation period, the cells were fixed and stained with the DNA-binding dye propidium iodide. *Arrowheads* show intracellular shigellae, and *arrows* show macrophage nuclei with the characteristic signs of apoptosis: chromatin condensation and segmentation of the nucleus.

Macrophage apoptosis results in the release of the mature form of the proinflammatory cytokine interleukin-1ß (212, 263), leading to the recruitment of polymorphonuclear leukocytes to the infected site. Transmigration of polymorphonuclear leukocytes across the intestinal epithelium compromises the integrity of the colonic epithelial barrier, allowing massive numbers of bacteria to reach the basolateral epithelial-cell membrane. Subsequent bacterial colonization of the epithelium results in the extensive mucosal damage and inflammatory infiltrates associated with shigellosis.

The mechanism of *Shigella* pathogenesis has been partially determined. Invasion and cytotoxicity genes are located on a plasmid that encodes a type-III secretion apparatus and its target invasins (135). Strains cured of this plasmid are completely noninvasive and avirulent (213). Invasion and escape from the phagosome are dependent on the expression and secretion of the invasion plasmid antigens IpaB, -C, and -D. Only IpaB is needed to initiate cell death (45).

Microinjection of purified IpaB into macrophages is sufficient to cause apoptosis (44). IpaB interacts with Interleukin-1ß–converting enzyme, or caspase-1, but not with caspase-2 or caspase-3 (86). It is interesting that macrophages isolated from *casp-1* knockout mice are not susceptible to *Shigella*-induced apoptosis but respond normally to other apoptotic stimuli. In contrast, macrophages isolated from *casp-3, casp-11*, or *p53* knockout mice, as well as those overexpressing *bcl-2* or *bcl-X$_L$*, are killed by *Shigella* (86).

Salmonella typhimurium   The genus *Salmonella* includes several species of gram-negative pathogens that cause a broad spectrum of diseases, including enteric (typhoid) fever, gastroenteritis, and bacteremia. Most of the studies to date have focused on *S. typhimurium*, because it is an important cause of food poisoning in humans.

Salmonellae also colonize the intestinal mucous membrane by passing through M cells (111). Within the mucosa, the bacteria localize to lymphoid follicles, where they are phagocytosed by resident macrophages. In contrast to *Shigella flexneri*, salmonellae do not escape from the phagolysosome. Instead, they survive and multiply within the macrophage (1, 7, 36). However, the evidence for survival in macrophages in vivo remains uncertain (43, 94).

In vitro studies show that *Salmonella* spp. can induce apoptosis in macrophage cell lines and primary cells (9, 43, 147, 165). Cytotoxicity was dependent on a type-III secretion system (165). A set of *Salmonella* virulence genes encoded by the *sip* (*Salmonella* invasion protein) operon seem to be associated with the bacteria's ability to invade and kill macrophages. The *sipEBCDA* genes exhibit extensive sequence homology to the genes of the *ipa* operon in *Shigella* spp. (84, 117, 191). There appears to be some functional similarity between the Ipa and Sip proteins. SipB is able to complement an *ipaB* mutant of *Shigella flexneri* for HeLa cell invasion and escape from the phagosome in macrophages (84). Furthermore, SipB induces apoptosis through a mechanism similar to that of IpaB (85). Purified SipB was sufficient to cause macrophage apoptosis in a caspase-1– dependent fashion. Therefore SipB is an analog of the *Shigella* invasin IpaB.

Yersiniae   *Yersinia* spp. cause a variety of diseases and are pathogenic for humans and rodents. *Y. pestis* is the causative agent of bubonic plague. *Y. pseudotuberculosis* and *Y. enterocolitica* may cause adenitis, septicemia, and gastrointestinal symptoms.

Both *Y. pseudotuberculosis* and *Y. enterocolitica* usually infect their hosts via the oral route, cross the intestinal barrier through M cells, and proliferate within lymphoid follicles (Peyer's patches) (81, 221). Subsequent colonization and growth of bacteria in the liver and spleen lead to the death of the host (89, 229).

Yersinia spp. induce apoptosis in macrophages in vitro (163, 166, 208) and in vivo (167). Unlike *Shigella* spp., which must be internalized to induce cell death, yersiniae are able to induce apoptosis from the outside of the host cells

(163, 166). Upon contact with the host cell, yersiniae deliver a number of plasmid-encoded proteins, referred to as *Yersinia* outer proteins (Yops), into the cytoplasm of eukaryotic cells via a type-III secretion system (51). These proteins have been shown to interfere with different host cellular processes, including (*a*) alteration of the cytoskeleton (203), (*b*) inhibition of phagocytosis (67), and (*c*) induction of apoptosis (163, 166, 208). Cell death was dependent on Yop translocators and the effector protein YopP (163) in *Y. enterocolitica*, or its homolog YopJ in *Y. pseudotuberculosis* (166).

The mechanism by which yersiniae trigger apoptosis is not known, but recent studies suggest that modulation of host cytokine responses may be involved. The release of cytokines is an important part of the immune response to infection, and studies with a murine model of yersiniosis provide evidence that the Yops are involved in the suppression of TNF-α and gamma interferon production in vivo (177). These cytokines are crucial in limiting the severity of *Yersinia* infection (14).

In vitro, wild-type yersiniae impair the normal TNF-α response of infected macrophages (25, 177, 178, 207). Downregulation of the TNF-α release is correlated to decreased mitogen-activated protein (MAP) kinase activity and is dependent on YopJ/P (29, 188, 207, 216). Additional studies indicate that YopJ/P is also required for *Yersinia*-mediated inhibition of NF-κB activation (206, 216). This finding is consistent with studies demonstrating that NF-κB–mediated signaling is required both for inducible cytokine expression and for antiapoptotic effects (15, 16).

Plant-pathogenic bacteria Bacteria of the genera *Erwinia*, *Pseudomonas*, *Xanthomonas*, and *Ralstonia* cause disease in many different plants. These include fire blight of rosaceous plants and soft rot caused by various *Erwinia* spp., bacterial spot diseases of peppers and tomatoes (*Xanthomonas campestris*), bacterial speck (*P. syringae*), and bacterial wilt of solanaceous plants (*Ralstonia solanacearum*).

A type-III protein secretion system, known as the Hrp system, is essential for these pathogenic bacteria to cause disease in susceptible host plants. The diseased plant exhibits typical necrotic lesions, caused by bacterial growth in the infected tissue. In nonhost plants or resistant plants, the bacteria trigger the hypersensitive response (HR), a rapid, defense-associated death of the plant cells at the site of invasion (reviewed in 5). This localized zone of tissue destruction is thought to limit the spread of the pathogen.

Whether bacterial infection leads to plant disease or to an HR-resistant phenotype is determined largely by the presence of a dominant resistant gene (R) in the host and the avirulence (*avr*) gene in the pathogen (120). It has been proposed that pathogen avirulence gene products are recognized by cognate receptors encoded by corresponding plant resistance genes (R), and the interaction between these two proteins may trigger a signaling cascade, which ultimately results in the elicitation of an HR response (17, 138, 226).

Like the Ipa, Yop, and Sip proteins, Avr proteins are delivered to host cells only upon host cell contact. Avr proteins do not appear to be secreted in culture and are largely hydrophilic proteins lacking N-terminal signal peptides. Several Avr proteins have been shown to be functional when expressed inside plant cells in an R-gene–dependent manner (reviewed in 57, 139).

The R genes from several plant species have been cloned (17, 244) and found to encode proteins that share one or more similar motifs (226). These motifs include leucine-rich repeats, found in nucleotide-binding sites, and kinase domains, both of which predict a role for the R genes as components in signal transduction pathways (24).

Very little is known about the biochemical processes by which an R-Avr interaction leads to the activation of an HR response. Some studies suggest that reactive oxygen intermediates such as H_2O_2 may be involved in triggering the HR response (141). Ca^{2+} influx has been implicated in bacterial induction of HR in tobacco (12). Protein phosphorylation is thought to be involved in the induction of an HR response via the signal transduction pathway (182).

There also appears to be an overlap between the effector proteins from plant and animal pathogens. The AvrRxv avirulence factor from *Xanthomonas campestris* pv. *vesicatoria* shares sequence homology with YopJ effector protein from *Yersinia* spp. and the AvrA virulence determinant from *Salmonella enterica* (82). The *avrA* mutant, however, exhibits the same level of virulence as the wild-type *Salmonella* sp.

Recently, evidence was provided for the presence of caspase-like protease(s) in plants, indicating that cell death in plants and animals may share common components. Caspase-specific peptide inhibitors were able to abolish tobacco HR cell death induced by *P. syringae* pv. *phaseolicola* (195).

Superantigens

Superantigens comprise a large family of microbial products that have multiple effects in vivo and cause markedly different clinical manifestations, ranging from acute toxic shock to chronic arthritis. Superantigens simultaneously interact with the major histocompatibility complex class-II molecules on antigen-presenting cells and specific Vß regions of T-cell receptors. This interaction stimulates the T cells to proliferate and produce cytokines. The expanded T-cell population is eventually depleted by apoptosis (73, reviewed in 227). The superantigens that have been characterized include exotoxins from *Staphylococcus aureus* and *Streptococcus pyogenes*.

Staphylococcus aureus   *S. aureus* expresses several exotoxins that act as superantigens and kill T-lymphocytes by apoptosis after the activation (11) and proliferation of these lymphocytes (75). *Staphylococcals* exotoxins A, B, D, and E cause clonal cells to undergo apoptosis in vitro (56, 108, 115). At this point only staphylococcal exotoxins A and B (SEB) have been shown to induce cell death in vivo (56, 118, 134, 146).

The signal by which superantigens induce cell death is still unclear. However, ATP-gated ion channels, known as purinoreceptors, may be involved in

SEB-induced apoptosis in immature thymocytes in mice (48). Reduced levels of protein kinase C activity were also observed in thymocytes from mice treated with SEB (145). This correlates with the observation that protein kinase C activation is necessary to prevent apoptosis in thymocytes induced by glucocorticoids, calcium ionophores, and anti-CD3 monoclonal antibody (109).

The role of Fas and the Fas ligand in SEB-mediated T-cell deletion remains controversial. Although some studies report that T cells from Fas-defective mice are impaired in SEB-mediated deletion (63, 217, 260), other reports have shown that Fas or the Fas ligand is not involved (75, 134).

Staphylococcal toxic shock syndrome toxin 1 is another superantigen. When added to peripheral blood mononuclear cells, toxic shock syndrome toxin 1 triggers T cells to proliferate and produce cytokines such as gamma interferon and also interacts with B cells, either directly or indirectly, to result in B-cell death (88).

Streptococcus pyogenes   Streptococci encode a number of toxins that can act as superantigens. These include the streptococcal pyrogenic exotoxins A, B, and C, mitogenic factor, and streptococcal superantigen (reviewed in 227). A direct analysis of samples from a group of patients with invasive gram-positive group-A streptococcal infections revealed characteristic selective depletion of Vβ-bearing T cells. T cells from patients with severe infections showed evidence of apoptosis. However, this pattern of Vβ depletion was not found in patients with severe gram-positive non–group-A streptococcal infections (247). It is interesting that apoptosis and the activation of caspases have also been implicated in acute meningitis caused by Streptococci (JS Braun, EI Tuomanen, personal communication).

Other Effectors

Clostridium difficile   Toxigenic strains of the gram-positive anaerobic bacterium *Clostridium difficile* are involved in the pathogenesis of antibiotic-associated diarrhea and pseudomembranous colitis.

The key virulence determinants produced by *C. difficile* are two exotoxins, toxin A and toxin B (159, 235). Both toxins are monoglucosyltransferases that catalyze the incorporation of glucose into Thr-37 of RhoA, a small GTP-binding protein of the Rho family (113, 114). The Rho family of proteins participates in various signal transduction processes and is also implicated in cell-cycle progression, cell transformation, and apoptosis (4).

Studies with animal-intestinal models suggest that toxin A is of primary importance in pathogenesis. Toxin A damages ileal and colonic epithelium and induces hemorrhagic fluid secretion (218). In vitro studies show that toxin A induces cell death in normal rat intestinal crypt cells (71), causes human intestinal epithelial cells to detach from the basement membrane (154), and results in the release of interleukin-8, a potent polymorphonuclear leukocyte chemoattractant. More recently, cell death was also observed in T cells and eosinophils after exposure to high concentrations of toxin A (153).

Fiorentini et al have recently reported that toxin B also triggers apoptosis in cultured intestinal cells (72). These authors suggest that the toxin-B–mediated inhibition of the Rho family of proteins may play a role in this process.

Bordetella pertussis   *B. pertussis*, the etiologic agent of whooping cough, produces several adhesins and toxins that are involved in pathogenesis. The adhesins include fimbriae, filamentous hemagluttinin, and pertactin. Several toxins, such as pertussis toxin and the adenylate cyclase-hemolysin (AC-Hly), interact with each other to play a role in this disease (10, 240). Only pertussis toxin and AC-Hly have been shown to be lethal in a murine respiratory model (123, 248).

Recently, in a murine respiratory model *B. pertussis* was shown to induce apoptosis of bronchopulmonary cells (78). Histological analysis of *B. pertussis*-infected lung tissue showed increased numbers of apoptotic cells in the alveolar compartments soon after infection. Furthermore, cellular accumulation in bronchoalveolar-lavage fluids and apoptosis of alveolar macrophages were both significantly attenuated in mice infected with a mutant deficient in the expression of AC-Hly.

In vitro, *B. pertussis* kills macrophage cell lines as well as murine alveolar macrophages by apoptosis (122, 124). Expression of AC-Hly is sufficient to promote cell death. AC-Hly belongs to the RTX toxin family described earlier. The toxin carries a calmodulin-sensitive AC activity and an Hly activity. After gaining entrance to the host cell by way of the pore-forming property of Hly, the AC moiety rapidly induces, after calmodulin activation, an elevation in cyclic AMP (cAMP) levels, thereby impairing the cell's chemotaxic, phagocytic, and oxidative abilities (201, 251).

Mycobacteria   *Mycobacterium tuberculosis* is the etiologic agent of the infectious disease tuberculosis, which worldwide kills more than 3 million individuals per year. The disease usually is transmitted via inhaled respiratory droplets. Once in the lung, the bacteria are taken up by alveolar macrophages, in which they reside and multiply during the course of infection. The mycobacteria are released after their host macrophages are killed, freeing them to initiate a new round of infection. The host's response to the infection culminates in the formation of a granulomatous lesion composed of macrophage-derived epithelioid giant cells and lymphocytes (reviewed in 70).

Mycobacterium avium-intracellulare is a common opportunistic pathogen that causes a disease similar to tuberculosis and primarily infects immunocompromised individuals. *M. avium-intracellulare*, like *M. tuberculosis*, lives and replicates in macrophages. In addition to persisting in these cells, these mycobacteria also induce programmed cell death.

Morphological changes consistent with apoptosis are observed in epithelioid cells in granulomata from patients with tuberculosis (54) and within caseating granulomas from clinical samples (119). There is also a significant increase in apoptotic alveolar macrophages in bronchoalveolar lavages from patients with active tuberculosis (126). These findings suggest that apoptosis plays an important role in clinical tuberculosis. It is not clear whether host or mycobacterial factors are associated with cell death in vivo.

Some studies suggest that apoptosis may be a macrophage defense mechanism to infection by mycobacteria. Apoptosis, but not necrosis of human monocytes, is shown to limit the growth of *M. bovis bacillus Calmette-Guérin* (164) and *M. avium* (74). However, other studies suggest that mycobacterial factor(s) may be involved in modulating cell death (19, 66, 119). These factors have not been defined.

Helicobacter pylori   *H. pylori* is a gram-negative bacterium that colonizes human gastric epithelium and is etiologically associated with the formation of peptic ulcers, and strongly associated with gastric lymphomas and gastric carcinomas (27, 52, 158).

Several groups have documented that *H. pylori* kills gastric cells by apoptosis (42, 69, 157, 169, 190). In addition, gastric biopsy specimens display a significant correlation between *H. pylori* infection and apoptosis (157, 169).

Both bacterial and host factors contribute to gastric disease. The diversity of the disease is thought to result from the interaction of *H. pylori* with a number of surface molecules on gastric epithelial cells (52). A secreted vacuolating cytotoxin, VacA, is believed to play a role in the pathogenesis of *H. pylori* infections. This toxin shows biological and structural similarity to the A-B family of toxins (53, 151) and manifests a cytotoxic activity via differential receptor-mediated interactions (187). Whether VacA mediates cytotoxicity by programmed cell death remains to be established.

Legionella pneumophila   *L. pneumophila*, a gram-negative facultative intracellular bacterium, is the causative agent of Legionnaire's disease. This organism survives and multiplies within monocytes and macrophages (90). Inside the cell, *L. pneumophila* resides within a specialized phagosome that does not acidify and fails to fuse with primary or secondary lysosomes (91, 92). Here the bacteria continue to grow and eventually lyse the host cell, releasing the bacteria to initiate a new infection cycle.

Both intracellular and extracellular *L. pneumophila* are known to have cytopathic effects on macrophages (101). Müller et al have demonstrated apoptosis in a human pre-monocytic leukemia cell line infected with *L. pneumophila* (173).

Recent studies indicate that *Legionella*-induced cytotoxicity after a high multiplicity of infection may be caused by osmotic lysis from pore formation in the macrophage membrane (125). Factors encoded by the *dot/icm* gene cluster (reviewed in 219) of *Legionella* spp. are essential for altering phagosome properties. They are also required for pore formation (125).

Segal et al (219) and Kirby et al (125) suggest that the interaction of *Legionella* spp. with host cells may be mediated by a type-III secretory system, analogous to those found in *Shigella*, *Salmonella*, and *Yersinia* spp. (125, 219). *L. pneumophila* requires contact with the eukaryotic cell for pore formation (125). Identification and characterization of effector molecules may shed light on the determinants required to mediate apoptosis.

Chlamydiae   Chlamydiae are intracellular bacterial pathogens that cause a spectrum of clinically important diseases such as conjunctivitis, trachoma, pneumonia, and the most common sexually transmitted bacterial infection. The four main species currently recognized, *C. trachomatis*, *C. psittaci*, *C. pneumoniae*, and *C. pecorum*, cause infections in both humans and animals.

The development of chlamydial disease is largely caused by persistent intracellular infection by the organism. A typical chlamydial infection cycle is initiated by the entry of an infectious extracellular elementary body (EB) into epithelial cells. The EBs are internalized into vacuoles that avoid fusion with host cell lysosomes. After 8–12 h, the EBs differentiate into noninfectious but metabolically active reticulate bodies, which multiply and differentiate back to EBs. Cells begin to lyse and release infectious EBs at 48–72 h postinfection (reviewed in 171). Host cytokine responses are not observed until 20–24 h after chlamydial infection and persist throughout the growth cycle of the pathogen (198).

The unique, biphasic infection cycle of chlamydiae may account for their proapoptotic and anti-apoptotic activities. Programmed cell death was observed in epithelial cells and macrophages infected with *C. psittaci* (184). Blocking bacterial entry or bacterial protein synthesis inhibited cell death. The antiapoptotic activity of *Chlamydia* spp. demonstrated by Fan et al (68) is believed to be important during the initial stages of invasion. *Chlamydia*-infected host cells are resistant to a variety of proapoptotic stimuli such as TNF-α, Fas antibody, and the kinase inhibitor staurosporine. Furthermore, resistance is due to chlamydial and not host protein synthesis. The antiapoptotic activity correlates with the blockade of mitochondrial cytochrome *c* release and downstream caspase activation (68).

It is interesting that genes encoding a type-III secretion system have recently been identified in *Chlamydia* spp. (93). Whether the bacteria use this putative secretion pathway to mediate their pro- and anti-apoptotic effects remains to be elucidated.

Rickettsia rickettsii   *R. rickettsii*, an obligate intracellular gram-negative bacterium, is the etiologic agent of Rocky Mountain spotted fever. During infection, the vascular endothelial cell is the primary target of infection, and the pathologic sequelae that follow may result from *R. rickettsii*-induced damage and/or the response of endothelial cells to intracellular infection.

Cultured human endothelial cells infected with *R. rickettsii* actively respond to infection by expressing both proinflammatory and procoagulant proteins (61, 223, 224) and by activation of members of the NF-κB/Rel family of transcription factors

(225). Just as suppression of NF-κB correlates with *Yersinia*-induced apoptosis of macrophages (206, 216), activation of NF-κB plays a critical role in protecting against apoptosis during *R. rickettsii* infection. A recent study reported that the inhibition of apoptosis in endothelial cells infected with *R. rickettsii* was dependent on the activation of NF-κB (50). In the presence of a specific inhibitor of NF-κB or a dominant negative form of this transcription factor, apoptosis was observed in infected, but not in uninfected, endothelial cells and embryonic fibroblasts. It has yet to be shown whether the NF-κB–protective mechanism of this organism functions during endothelial cell infection in vivo.

CONCLUSIONS

There is increasing evidence that pathogens use different mechanisms to impinge on the cell death pathway. These include (*a*) engagement of specific host receptors by bacterial ligands such as superantigens, (*b*) activation of second-messenger pathways by a bacterial effector (*B. pertussis*), (*c*) direct interaction of pathogen effector molecules with the host cell death machinery to trigger apoptosis (*Shigella* spp. and *Salmonella* spp.), (*d*) the alteration of host membrane permeability by pore-forming toxins, and (*e*) the inhibition of host protein synthesis by bacterial toxins.

It is interesting that there are pro- and anti-apoptotic host cell mechanisms of apoptosis that currently are not known to be targeted by bacterial pathogens. For example, the activation of the Bcl-2 family of proteins or the specific release of mitochondrial cytotoxic factors has not been implicated.

There are at least three possible reasons for pathogens to modulate the host cell death pathway. First, and most obvious, would be to eliminate key defense cells such as lymphocytes, monocytes, and neutrophils, which would otherwise participate in the destruction of the pathogen. Second, for intracellular pathogens such as *R. rickettsii*, inhibition of host apoptosis provides a safe haven for bacterial proliferation. In some cases a niche within the host cell provides the pathogen with a cloak of invisibility from the immune system. Third, for some pathogens such as mycobacteria, triggering of apoptosis may be a host response to reduce or inhibit bacterial proliferation. Finally, it appears that the induction of apoptosis can be important in circumscribing the infection. For plant pathogens, the "hypersensitive response," a defense-associated death of the plant cells at the site of invasion, is thought to limit the spread of the pathogen. Similarly, *Shigella*-induced apoptosis results in an inflammation that serves to clear and possibly localize the infection as well as provide a vehicle to promote bacterial spread in the intestinal epithelium.

In summary, much progress has been made in understanding the role of programmed cell death in response to different infections. There is little doubt that, as more of the mechanisms are defined, further insight into the complex relationship between hosts and pathogens will be gained.

ACKNOWLEDGMENTS

We thank Antonios Aliprantis, Andrea Guichon, David Hersh, and Hubert Hilbi for critically evaluating this manuscript, Jeremy Moss for helpful information, and Chi Wong for formatting. Studies presented in this review were supported by grants from the National Institutes of Health (AI 37720 and AI 42780) to AZ.

Visit the Annual Reviews home page at http://www.AnnualReviews.org

LITERATURE CITED

1. Abshire K, Neidhardt F. 1993. Growth rate paradox of *Salmonella typhimurium* within macrophages. *J. Bacteriol.* 175:3744–48
2. Ahnert-Hilger G, Bhakdi S, Gratzl M. 1985. Minimal requirements for exocytosis: a study using PC12 cell permeabilized with staphylococcal alpha-toxin. *J. Biol. Chem.* 260:12730–34
3. Akashi K, Kondo M, von Freeden-Jeffry U, Murray R, Weissman IL. 1997. Bcl-2 rescues T lymphopoiesis in interleukin-7 receptor-deficient mice. *Cell* 89:1033–41
4. Aktories K. 1997. Rho proteins: targets for bacterial toxins. *Trends Microbiol.* 7:282–88
5. Alfano JR, Collmer A. 1996. Bacterial pathogens in plants: life up against the wall. *Plant Cell* 8:1683–98
6. Alnemri ES, Livingston DJ, Nicholson DW, Salvesen G, Thornberry NA, Wong WW, Yuan J. 1996. Human ICE/CED-3 protease nomenclature. *Cell* 87:171
7. Alpuche-Aranda CM, Racoosin EL, Swanson JA, Miller SI. 1994. *Salmonella* stimulate macrophage macropinocytosis and persist within spacious phagosomes. *J. Exp. Med.* 179:601–8
8. Apodaca G, Bomsel M, Lindstedt R, Engel J, Frank D, et al. 1995. Characterization of *Pseudomonas aeruginosa*-induced MDCK cell injury: glycosylation defective host cells are resistant to bacterial killing. *Infect. Immun.* 63:1541–51
9. Arai T, Hiromatsu K, Nishimura H, Kimura Y, Kobayashi N, et al. 1995. Endogenous interleukin-10 prevents apoptosis in macrophages during *Salmonella* infection. *Bio-chem. Biophys. Res. Commun.* 213:600–7
10. Arico B, Nuti S, Scarlato V, Rappuoli R. 1993. Adhesion of *Bordetella pertussis* to eukaryotic cells requires a time-dependent export and maturation of filamentous hemagglutinin. *Proc. Natl. Acad. Sci. USA* 90:9204–8
11. Aroeira L, Moreno M, Martinez C. 1996. In vivo activation of Y cell induction into the primed phenotype and programmed cell death by staphylococcal enterotoxin B. *Scand. J. Immunol.* 43:545–50
12. Atkinson MM, Keppler LD, Oralandi EW, Baker JC, Mischki CF. 1990. Involvement of plasma membrane calcium influx in bacterial induction of the K+/H+ and hypersensitive responses in tobacco. *Plant Physiol.* 92:215–21
13. Attardi LD, Lowe SW, Brugarolas J, Jacks T. 1996. Transcriptional activation by p53, but not induction of the p21 gene, is essential for oncogene-mediated apoptosis. *EMBO J.* 15:3693–701
14. Autenrieth IB, Heesemann J. 1992. In vivo neutralization of tumor necrosis factor alpha and interferon-gamma abrogates resistance to Yersinia enterocolitica infection in mice. *Med. Microbiol. Immunol.* 181:333–38
15. Baeuerle P, Henkel T. 1994. Function and activation of NF-kB in the immune system. *Annu. Rev. Immunol.* 12:141–79
16. Baichwal V, Baeuerle P. 1997. Activate NF-kB or die? *Curr. Biol.* 7:94–96
17. Baker B, Zambryski P, Staskawicz B, Dinesh-Kumar SP. 1997. Signaling in plant-microbe interactions. *Science* 276:726–33

18. Bakhshi A, Jensen JP, Goldman P, Wright JJ, McBride OW, Epstein AL, Korsmeyer SJ. 1985. Cloning the chromosomal breakpoint of t(14;18) human lymphomas: clustering around JH on chromosome 14 and near a transcriptional unit on 18. *Cell* 41: 899–906

19. Balcewicz-Sablinska MK, Keane J, Kornfeld H, Remold HG. 1998. Pathogenic *Mycobacterium tuberculosis* evades apoptosis of host macrophages by release of TNF-R2, resulting in inactivation of TNF-a. *J. Immunol.* 161:2636–41

20. Barsig J, Kaufmann SH. 1997. The mechanism of cell death in *Listeria monocytogenes*-infected murine macrophages is distinct from apoptosis. *Infect. Immun.* 65: 4075–81

21. Bayles KW, Wesson CA, Liou LE, Fox LK, Bohach GA, Trumble WR. 1998. Intracellular S*taphylococcus aureus* escapes the endosome and induces apoptosis in epithelial cells. *Infect. Immun.* 66:336–42

22. Beg AA, Baltimore D. 1996. An essential role for NF-kB in preventing TNFa-induced cell death. *Science* 274:782–84

23. Beltrami E, Jesty J. 1995. Mathematical analysis of activation thresholds in enzyme-catalyzed positive feedbacks: application to the feedbacks of blood coagulation. *Proc. Natl. Acad. Sci. USA* 92:8744–48

24. Bent AF. 1996. Plant disease resistance genes: function meets structure. *Plant Cell* 8:1757–71

25. Beuscher HU, Rödel F, Forsberg Å, Röllinghoff M. 1995. Bacterial evasion of host immune defense: *Yersinia enterocolitica* encodes a suppressor for tumor necrosis factor alpha expression. *Infect. Immun.* 63:1270–77

26. Bhakdi S, Mackman N, Menestrina G, Gray L, Hugo F, et al. 1998. The hemolysin of Escherichia coli. *Eur. J. Epidemiol.* 4:135–43

27. Blaser MJ, Perez-Perez GI, Kleanthous H, Cover TL, Peek RM, et al. 1995. Infection with *Helicobacter pylori* strains possessing *cagA* is associated with an increased risk of developing adenocarcinoma of the stomach. *Cancer Res.* 55:2111–15

28. Boise LH, Minn AJ, Noel PJ, June CH, Accavitti MA, Lindsten T, Thompson CB. 1995. CD28 costimulation can promote T cell survival by enhancing the expression of Bcl-XL. *Immunity* 3:87–98

29. Boland A, Cornelis GR. 1998. Role of YopP in suppression of tumor necrosis factor alpha release by macrophages during *Yersinia* infection. *Infect. Immun.* 66:1878–84

30. Boldin MP, Goncharov TM, Goltsev YV, Wallach D. 1996. Involvement of MACH, a novel MORT1/FADD-interacting protease, in Fas/APO-1- and TNF receptor-induced cell death. *Cell* 85:803–15

31. Boldin MP, Mett IL, Varfolomeev EE, Chumakov I, Shemer-Avni Y, Camonis JH, Wallach D. 1995. Self-association of the "death domains" of the p55 tumor necrosis factor (TNF) receptor and Fas/APO1 prompts signaling for TNF and Fas/APO1 effects. *J. Biol. Chem.* 270:387–91

32. Boyd JM, Gallo GJ, Elangovan B, Houghton AB, Malstrom S, Avery BJ, et al. 1995. Bik, a novel death-inducing protein shares a distinct sequence motif with Bcl-2 family proteins and interacts with viral and cellular survival-promoting proteins. *Oncogene* 11:1921–28

33. Bredesen DE. 1995. Neural apoptosis. *Ann. Neurol.* 38:839–51

34. Brinkmann U, Brinkmann E, Gallo M, Pastan I. 1995. Cloning and characterization of a cellular apoptosis susceptibility gene, the human homologue to the yeast chromosome segregation gene CSE1. *Proc. Natl. Acad. Sci. USA* 92:10427–31

35. Brinkmann U, Brinkmann E, Gallo M, Scherf U, Pastan I. 1996. Role of CAS, a human homologue to the yeast chromosome segregation gene CSE1, in toxin and tumor necrosis factor mediated apoptosis. *Biochemistry* 35:6891–99

36. Buchmeier N, Heffron F. 1989. Intracellular survival of wild-type *Salmonella typhimurium* and macrophage-sensitive mutants in diverse populations of macrophages. *Infect. Immun.* 57:1–7

37. Cavalieri S, Bohach G, Snyder I. 1984. *Escherichia coli* alpha hemolysin: characteristics and probable role in pathogenicity. *Microbiol. Rev.* 48:326–43

38. Chang M, Bramhal J, Graves S, Bonavida B, Wisnieski B. 1989. Internucleosomal DNA cleavage precedes diphtheria toxin induced cytolysis. *J. Biol. Chem.* 264:15261–67

39. Chang M, Baldwin R, Bruce C, Wisnieski B. 1989. Second cytotoxic pathway of diphtheria toxin suggested by nuclease activity. *Science* 246:1165–68

40. Chang M, Wisnieski B. 1990. Comparison of the intoxication pathways on tumor necrosis factor and diphtheria toxin. *Infect. Immun.* 58:2644–50

41. Chao JR, Wang JM, Lee SF, Peng HW, Lin YH, Chou CH, et al. 1998. *mcl-1* is an immediate-early gene activated by the granulocyte-macrophage colony-stimulating factor (GM-CSF) signaling pathway and is one component of the GM-CSF viability response. *Mol. Cell. Biol.* 18:4883–98

42. Chen G, Sordillo EM, Ramey WG, Reidy J, Holt PR, et al. 1997. Apoptosis in gastric epithelial cells is induced by *Helicobacter pylori* and accompanied by increased expression of BAK. *Biochem. Biophys. Res. Commun.* 239:626–32

43. Chen LM, Kaniga K, Galan JE. 1996. *Salmonella* spp. are cytotoxic for cultured macrophages. *Mol. Microbiol.* 21:1101–15

44. Chen Y, Smith MR, Thirumalai K, Zychlinsky A. 1996. A bacterial invasin induces macrophage apoptosis by directly binding ICE. *EMBO J.* 15:3853–60

45. Chen Y, Zychlinsky A. 1994. Apoptosis induced by bacterial pathogens. *Microb. Pathog.* 17:203–12

46. Chinnaiyan AM, O'Rourke K, Tewari M, Dixit VM. 1995. FADD, a novel death domain-containing protein, interacts with the death domain of Fas and initiates apoptosis. *Cell* 81:505–12

47. Choe S, Bennett M, Fugii G, Curmi K, Kantardjieff K, et al. 1992. The crystal structure of diphtheria toxin. *Nature* 357:216–22

48. Chvatchko Y, Valera S, Aubry J, Renno T, Buell G, et al. 1996. The involvement of an ATP gated ion channel, P2X1, in thymocyte apoptosis. *Immunity* 5:275–83

49. Cleary ML, Smith SD, Sklar J. 1986. Cloning and structural analysis of cDNAs for bcl-2 and a hybrid bcl- 2/immunoglobulin transcript resulting from the t(14;18) translocation. *Cell* 47:19–28

50. Clifton DR, Goss RA, Sahni SK, van Antwerp D, Baggs RB, et al. 1998. NF-kB-dependent inhibition of apoptosis is essential for host cell survival during *Rickettsia rickettsii* infection. *Proc. Natl. Acad. Sci. USA* 95:4646–51

51. Cornelis G, Wolf-Watz H. 1997. The *Yersinia* Yop virulon: a bacterial system for subverting eukaryotic cells. *Mol. Microbiol.* 23:861–67

52. Cover TL, Blaser MJ. 1992. *Helicobacter pylori* and gastroduodenal disease. *Annu. Rev. Med.* 43:135–45

53. Cover TL, Hanson PI, Heuser JE. 1997. Acid-induced dissociation of VacA, the *Helicobacter pylori* vacuolating cytotoxin, reveals its pattern of assembly. *J. Cell. Biol.* 138:759–69

54. Cree I, Nurbhai S, Milne G, Beck J. 1987. Cell death in granulomata: the role of apoptosis. *J. Clin. Pathol.* 40:1314–19

55. Cryns V, Yuan J. 1998. Proteases to die for. *Genes Dev.* 12:1551–70

56. D'Adamio L, Awad K, Reinherz E. 1993. Thymic and peripheral apoptosis of antigen-specific T cells might cooperate in establishing self tolerance. *Euro. J. Immunol.* 23:747–53

57. Dangel JL. 1994. The enigmatic avirulence genes of phytopathogenic bacteria. *Curr. Top. Microbiol. Immunol.* 192:99–118

58. Datta SR, Dudek H, Tao X, Masters S, Fu H, et al. 1997. Akt phosphorylation of BAD couples survival signals to the cell-intrinsic death machinery. *Cell* 91:231–41

59. del Peso L, Gonzalez-Garcia M, Page C, Herrera R, Nunez G. 1997. Interleukin-3-induced phosphorylation of BAD through the protein kinase Akt. *Science* 278:687–89

60. DiDonato JA, Hayakawa M, Rothwarf DM, Zandi E, Karin M. 1997. A cytokine-responsive IkB kinase that activates the transcription factor NF-kB. *Nature* 388:548–54

61. Drancourt M, Alessi M-C, Levy P-Y, Juhan-Vague I, Raoult D. 1990. Secretion of tissue-type plasminogen activator and plasminogen activator inhibitor by *Rickettsia conorii-* and *Rickettsia rickettsii-*infected cultured endothelial cells. *Infect. Immun.* 58:2459–63

62. Enari M, Sakahira H, Yokoyama H, Okawa K, Iwamatsu A, et al. 1998. A caspase-activated DNase that degrades DNA during apoptosis, and its inhibitor ICAD. *Nature* 391:43–50

63. Ettinger R, Panka D, Wang J, Stanger B, Ju S, Rothstein A. 1995. Fas ligand mediated cytotoxicity is directly responsible for apoptosis of normal CD4+ T cells responding to a bacterial superantigen. *J. Immunol.* 154:4320–28

64. Evan G, Littlewood T. 1998. A matter of life and cell death. *Science* 281:1317–22

65. Evan GI, Wyllie AH, Gilbert CS, Littlewood TD, Land H, Brooks M, et al. 1992. Induction of apoptosis in fibroblasts by c-myc protein. *Cell* 69:119–28

66. Eviks IS, Emerson CL. 1997. Temporal effect of tumor necrossis factor-a on murine macrophages infected with *Mycobacterium avium*. *Infect. Immun.* 65:2100–6

67. Fällmann M, Andersson K, Hakansson S, Magnusson K-E, Stendahl O, Wolf-Watz H. 1995. *Yersinia pseudotuberculosis* inhibits Fc receptor-mediated phagocytosis in J774 cells. *Infect. Immun.* 63:3117–24

68. Fan T, Lu H, Hu H, Shi L, McClarty GA, et al. 1998. Inhibition of apoptosis in Chlamydia-infected cells: blockade of mitochondrial cytochrome c release and caspase activation. *J. Exp. Med.* 187:487–96

69. Fan XJ, Crowe SE, Behar S, Gunasena H, Ye G, Haeberle H, et al. 1998. The effect of Class 11 major histocompatibility complex expression on adherence of *Helicobacter pylori* and induction of apoptosis in gastric epithelial cells: a mechanism for T helper cell type 1-mediated damage. *J. Exp. Med.* 187:1659–69

70. Fenton MJ, Vermeulen MW. 1996. Immunopathology of tuberculosis: roles of macrophages and monocytes. *Infect. Immun.* 64:683–90

71. Fiorentini C, Donelli G, Nicotera P, Thelestam M. 1993. *Clostridium difficile* toxin A elicits Ca^{2+}-independent cytotoxic effects in cultured normal rat intestinal crypt cells. *Infect. Immun.* 61:3988–93

72. Fiorentini C, Fabbri A, Falzano L, Fattorossi A, Matarrese P, et al. 1998. *Clostridium difficile* toxin B induces apoptosis in intestinal cultured cells. *Infect. Immun.* 66:2660–65

73. Fleischer B. 1994. Superantigens produced by infectious pathogens: molecular mechanism of action and biological significance. *Int. J. Clin. Lab. Res.* 24:193–37

74. Fratazzi C, Arbeit R, Carini C, Remold H. 1997. Programmed cell death of Mycobacterium avium serovar 4-infected macrophages prevents the mycobacteria from spreading and induces mycobacterial growth inhibition by freshly added, uninfected macrophages. *J. Immunol.* 158:432–37

75. Gonzalo J, Baixeras E, Garcia A, Chandy A, Rooijen N, et al. 1994. Differential in vivo effects of a superantigen and an anti-

body targeted to the same T cell receptor. *J. Immunol.* 152:1597–8

76. Green DR, Reed JC. 1998. Mitochondria and apoptosis. *Science* 281:1309–12

77. Gu Y, Wu J, Faucheu C, Lalanne JL, Diu A, et al. 1995. Interleukin-1ß converting enzyme requires oligomerization for activity of processed forms in vivo. *EMBO J.* 14:1923–31

78. Gueirard P, Druilhe A, Pretolani M, Guiso N. 1998. Role of adenylate cyclase-hemolysin in alveolar macrophage apoptosis during Bordetella pertussis infection in vivo. *Infect. Immun.* 66:1718–25

79. Haldar S, Basu A, Croce CM. 1997. Bcl2 is the guardian of microtubule integrity. *Cancer Res.* 57:229–33

80. Han J, Sabbatini P, White E. 1996. Induction of apoptosis by human Nbk/Bik, a BH3-containing protein that interacts with E1B 19K. *Mol. Cell. Biol.* 16:5857–64

81. Hanski C, Kutsrchka H, Schmoranzer H, Naumann M, Stallmach A, et al. 1989. Immunohistochemical and electron microscopic study of the interaction of *Yersinia enterocolitica* serotype 08 with intestinal mucosa during experimental enteritis. *Infect. Immun.* 57:673–78

82. Hardt W, Galan J. 1997. A secreted Salmonella protein with homology to an avirulence determinant of plant pathogenic bacteria. *Proc. Natl. Acad. Sci. USA* 94:9887–92

83. Hengartner MO, Horvitz HR. 1994. *C. elegans* cell survival gene *ced-9* encodes a functional homolog of the mammalian proto-oncogene *bcl-2*. *Cell* 76:665–76

84. Hermant D, Menard R, Arricau N, Parsot C, Popoff MY. 1995. Functional conservation of the *Salmonella* and *Shigella* effectors of entry into epithelial cells. *Mol. Microbiol.* 17:781–89

85. Hersh D, Monack D, Smith M, Ghori N, Falkow S, Zychlinsky A. 1999. The *Salmonella* invasin SipB induces macrophage apoptosis by binding to caspase-1. *Proc. Natl. Acad. Sci. USA* 96:2396–401

86. Hilbi H, Moss J, Hersh D, Chen Y, Arondel J, et al. 1998. Shigella-induced apoptosis is dependent on caspase-1 which binds to IpaB. *J. Biol. Chem.* 273:32895–900

87. Hildebrand A, Roth M, Bhakdi S. 1991. Staphylococcal alpha-toxin: dual mechanism of binding to target cells. *J. Biol. Chem.* 266:17195–200

88. Hofer M, Newell K, Duke R, Schlievert P, Freed J, Leung D. 1996. Differential effects of staphylococcal toxic shock syndrome toxin-1 on B cell apoptosis. *Proc. Natl. Acad. Sci. USA* 93:5425–30

89. Holmstrom A, Rosqvist R, Wolf-Watz H, Forsberg A. 1995. Virulence plasmid-encoded YopK is essential for *Yersinia pseudotuberculosis* to cause systemic infection in mice. *Infect. Immun.* 63:2269–76

90. Horowitz MA, Silverstein SC. 1980. Legionnaires' disease bacterium (*Legionella pneumophila*) multiplies intracellularly in human monocytes. *J. Clin. Invest.* 66:441–50

91. Horwitz MA. 1983. The legionnaires' disease bacterium (*Legionella pneumophila*) inhibits lysosome-phagosome fusion in human monocytes. *J. Exp. Med.* 158:1319–31

92. Horwitz MA. 1984. Phagocytosis of the legionnaires' disease bacterium (*Legionella pneumophila*) occurs by a novel mechanism: engulfment with a pseudopod coil. *Cell* 36:27–33

93. Hsia R-C, Pannekeok Y, Ingerowski E, Bavoil PM. 1997. Type III secretion genes identify a putative virulence locus of *Chlamydia*. *Mol. Microbiol.* 25:351–59

94. Hsu H. 1989. Pathogenesis and immunity in murine salmonellosis. *Microbiol. Rev.* 53:390–409

95. Hsu H, Huang J, Shu HB, Baichwal V, Goeddel DV. 1996. TNF-dependent recruitment of the protein kinase RIP to the TNF receptor-1 signaling complex. *Immunity* 4:387–96

96. Hsu H, Shu HB, Pan MG, Goeddel DV. 1996. TRADD-TRAF2 and TRADD-FADD interactions define two distinct TNF

receptor 1 signal transduction pathways. *Cell* 84:299–308

97. Hsu H, Xiong J, Goeddel DV. 1995. The TNF receptor 1-associated protein TRADD signals cell death and NF-kB activation. *Cell* 81:495–504

98. Hu Y, Benedict MA, Wu D, Inohara N, Nunez G. 1998. Bcl-XL interacts with Apaf-1 and inhibits Apaf-1-dependent caspase-9 activation. *Proc. Natl. Acad. Sci. USA* 95:4386–91

99. Huang DC, Adams JM, Cory S. 1998. The conserved N-terminal BH4 domain of Bcl-2 homologues is essential for inhibition of apoptosis and interaction with CED-4. *EMBO J.* 17:1029–39

100. Hueck CJ. 1998. Type III protein secretion systems in bacterial pathogens of animals and plants. *Microbiol. Mol. Biol. Rev.* 62:379–433

101. Husmann LK, Johnson W. 1994. Cytotoxicity of extracellular *Legionella pneumophila. Infect. Immun.* 62:2111–14

102. Ikigae H, Nakae T. 1987. Interaction of the alpha-toxin of S. aureus with the liposome membrane. *J. Biol. Chem.* 262:2156–60

103. Ink B, Zornig M, Baum B, Hajibagheri N, James C, et al. 1997. Human Bak induces cell death in *Schizosaccharomyces pombe* with morphological changes similar to those with apoptosis in mammalian cells. *Mol. Cell. Biol.* 17:2468–74

104. Inward C, Williams J, Chant I, Crocker J, Milford D, et al. 1995. Verocytotoxin-1-Induces apoptosis in vero cells. *J. Infect.* 30:213–18

105. Islam D, Veress B, Bardhan PK, Lindberg AA, Christensson B. 1997. In situ characterization of inflammatory responses in the rectal mucosae of patients with shigellosis. *Infect. Immun.* 65:735–49

106. Ito T, Deng X, Carr B, May WS. 1997. Bcl-2 phosphorylation required for anti-apoptosis function. *J. Biol. Chem.* 272:11671–73

107. Jacobson MD, Burne JF, King MP, Miyashita T, Reed JC, Raff MC. 1993. Bcl-2 blocks apoptosis in cells lacking mitochondrial DNA. *Nature* 361:365–69

108. Jenkinson E, Kingston R, Smith C, Williams G, Owen J. 1989. Antigen induced apoptosis in developing T cells: a mechanism for negative selection of the T cell repertoire. *Eur. J. Immunol.* 19:2175–77

109. Jin LW, Inaba K, Saitoh T. 1992. The involvement of protein kinase C in activation-induced cell death in T-cell hybridoma. *Cell. Immunol.* 144:217–27

110. Jonas D, Walev I, Berger T, Liebetrau M, Palmer M, Bhakdi S. 1994. Novel path to apoptosis small transmembrane pores created by staphylococcal alpha toxin T lymphocytes evokes internucleosomal degradation. *Infect. Immun.* 62:1304–12

111. Jones B, Ghouri N, Falkow S. 1994. *Salmonella typhimurium* initiates murine infection by penetrating and destroying the specialized epithelial M cells of Peyer's patches. *J. Exp. Med.* 180:15–23

112. Jurgensmeier JM, Krajewski S, Armstrong RC, Wilson GM, Oltersdorf T, Fritz LC, et al. 1997. Bax- and Bak-induced cell death in the fission yeast *Schizosaccharomyces pombe. Mol. Biol. Cell.* 8:325–39

113. Just I, Selzer J, Rex J, von Eichel-Streiber C, Mann M, Aktories K. 1995. The enterotoxin from Clostridium difficile (ToxA) monoglucosylates the Rho protein. *J. Biol. Chem.* 270:13392–96

114. Just I, Selzer J, Wilm M, von Eichel-Streiber C, Mann M, Aktories K. 1995. Glycosylation of Rho proteins by *Clostridium difficile* toxin B. *Nature* 375:500–3

115. Kabelitz D, Wesselborg S. 1992. Life and death of a superantigen reactive human CD4+ T cell clone: staphylococcal enterotoxins induce death by apoptosis but simultaneously trigger a proliferative response in the presence of HLA-DR+

antigen presenting cells. *Int. Immunol.* 4:1381–88

116. Kang PJ, Hauser AR, Apodaca G, Fleiszig SMJ, Weiner-Kronish J, et al. 1997. Identification of Pseudomonas aeruginosa genes required for epithelial cell injury. *Mol. Microbiol.* 24:1249–62

117. Kaniga K, Tucker SC, Trollinger D, Galan JE. 1995. Homologs of the *Shigella* IpaB and IpaC invasins are required for *Salmonella typhimurium* entry into cultured epithelial cells. *J. Bacteriol.* 177:3965–71

118. Kawabe Y, Ochi A. 1991. Programmed cell death and extrathymic reduction of Vb8+ CD4+ T cells in mice tolerant to *Staphylococcus aureus* enterotoxin B. *Nature* 349:245–48

119. Keane J, Balcewisz-Sablinska MK, Remold HG, Chupp GL, Meek BB, et al. 1997. Infection by *Mycobacterium tuberculosis* promotes human alveolar macrophage apoptosis. *Infect. Immun.* 65:298–304

120. Keen NT. 1990. Gene-for-gene complementarity in plant pathogen interactions. *Annu. Rev. Genet.* 24:447–63

121. Keenan K, Dharpnack D, Formal S, O'Brien A. 1986. Morphologic evaluation of the effects of Shiga toxin and E. coli Shiga like toxin on the rabbit intestine. *Am. J. Pathol.* 125:69–80

122. Khelef N, Guiso N. 1995. Induction of macrophage apoptosis by *Bordetella pertussis* adenylate cyclase-hemolysin. *FEMS Microbiol. Lett.* 134:27–32

123. Khelef N, Sakamoto H, Guiso N. 1992. Both adenylate cyclase and hemolytic activities are required by *Bordetella pertussis* to initiate infection. *Microb. Pathog.* 12:227–35

124. Khelef N, Zychlinsky A, Guiso N. 1993. *Bordetella pertussis* induces apoptosis in macrophages: role of adenylate cyclase-hemolysin. *Infect. Immun.* 61:4064–71

125. Kirby JE, Vogel JP, Andrews HL, Isberg RR. 1998. Evidence of pore-forming ability by *Legionella pneumophila*. *Mol. Microbiol.* 27:323–36

126. Klinger K, Tchou-Wong K-M, Brändli O, Aston C, Kim R, et al. 1997. Effects of mycobacteria on regulation of apoptosis in mononuclear phagocytes. *Infect. Immun.* 65:5272–78

127. Kluck RM, Bossy-Wetzel E, Green DR, Newmeyer DD. 1997. The release of cytochrome c from mitochondria: a primary site for Bcl-2 regulation of apoptosis. *Science* 275:1132–36

128. Kochi S, Collier J. 1993. DNA fragmentation and cytolysis in U937 cells treated with diphtheria toxin or other inhibitors of protein synthesis. *Exp. Cell Res.* 208:296–302

129. Kohno K, Uchida T. 1987. Highly frequent single amino acid substitution in mammalian EF-2 results in expression of resistance to EF-2 ADP-ribosylating toxins. *J. Biol. Chem.* 262:12298–305

130. Korostroff J, Wang JF, Kieba I, Miller M, Shenker BJ, Lally ET. 1998. *Actinobacillus actinomycetemcomitans* leukotoxin induces apoptosis in HL-60 cells. *Infect. Immun.* 66:4474–83

131. Kothakota S, Azuma T, Reinhard C, Klippel A, Tang J, Chu K, et al. 1997. Caspase-3-generated fragment of gelsolin: effector of morphological change in apoptosis. *Science* 278:294–98

132. Kozopas KM, Yang T, Buchan HL, Zhou P, Craig RW. 1993. MCL1, a gene expressed in programmed myeloid cell differentiation, has sequence similarity to BCL2. *Proc. Natl. Acad. Sci. USA* 90:3516–20

133. Kroemer G. 1997. The proto-oncogene Bcl-2 and its role in regulating apoptosis. *Nat. Med.* 3:614–20

134. Kuroda K, Yagi J, Imanishi K, Yan X, Li X, et al. 1996. Implantation of IL-2 containing osmotic pump prolongs the survival of superantigen reactive T cells expanded in mice injected with bacterial superantigen. *J. Immunol.* 157:1422–31

135. LaBrec EH, Schneider H, Magnani TJ, Formal SB. 1964. Epithelial cell penetration as an essential step in the pathogenesis of bacillary dysentery. *J. Bacteriol.* 88:1503–18

136. Lally ET, Kieba I, Demuth DR, Rosenbloom J, Golub EE, Taichman NS. 1989. Identification and expression of the *Actinobacillus actinomycetemcomitans* leukotoxin gene. *Biochem. Biophy. Res. Commun.* 159:256–62

137. Lally ET, Kieba IR, Sato A, Green CL, Rosenbloom J, et al. 1997. RTX Toxins recognize a b2 integrin on the surface of human target cells. *J. Biol. Chem.* 272:30463–69

138. Lamb CJ. 1994. Plant disease resistance genes in signal perception and transduction. *Cell* 76:419–22

139. Leach JE, White FF. 1996. Bacterial avirulence genes. *Annu. Rev. Phytopathol.* 34:153–79

140. Lessnick S, Lyczak J, Bruce C, Lewis D, Kim P, et al. 1992. Localization of diphtheria toxin nuclease activity to fragment A. *J. Bacteriol.* 174:2032–38

141. Levine A, Tenhaken R, Dixon R, Lamb C. 1994. H_2O_2 from the oxidative burst orchestrates the plant hypersensitive disease resistance response. *Cell* 79:583–93

142. Li P, et al. 1997. Cytochrome c and dATP-dependent formation of Apaf-1/caspase-9 complex initiates an apoptotic protease cascade. *Cell* 91:479–89

143. Lian CJ, Rosendal S, Macinnes JI. 1989. Molecular cloning and characterization of a hemolysin gene from *Actinobacillus (Haemophilus) pleuropneumoniae*. *Infect. Immun.* 57:3377–82

144. Lin EY, Orlofsky A, Berger MS, Prystowsky MB. 1993. Characterization of A1, a novel hemopoietic-specific early-response gene with sequence similarity to *bcl-2*. *J. Immunol.* 151:1979–88

145. Lin Y, Kao S, Jan M, Cheng M, Wing L, et al. 1995. Changes of protein kinase C subspecies in staphylococcal enterotoxin B induced thymocyte apotosis. *Biochem. Biophy. Res. Commun.* 213:1132–39

146. Lin Y, Lei H, Low TLK, Shen CL, Chou LJ, Jan MS. 1992. In vivo induction of apoptosis in immature thymocytes by staphylococcal enterotoxin B. *J. Immunol.* 149:1156–63

147. Lindgren SW, Stojilkovic I, Heffron F. 1996. Macrophage killing is an essential virulence mechanism of *Salmonella typhimurium*. *Proc. Natl. Acad. Sci. USA* 93:4197–201

148. Ling YH, Tornos C, Perez-Soler R. 1998. Phosphorylation of Bcl-2 is a marker of M phase events and not a determinant of apoptosis. *J. Biol. Chem.* 273:18984–91

149. Liu X, Kim CN, Yang J, Jemmerson R, Wang X. 1996. Induction of apoptotic program in cell-free extracts: requirement for dATP and cytochrome c. *Cell* 86:147–57

150. Liu X, Zou H, Slaughter C, Wang X. 1997. DFF, a heterodimeric protein that functions downstream of caspase-3 to trigger DNA fragmentation during apoptosis. *Cell* 89:175–84

151. Lupetti P, Heuser JE, Manetti R, Massari P, Lanzavecchia S, et al. 1996. Oligomeric and subunit structure of the *Helicobacter pylori* vacuolating cytotoxin. *J. Cell. Biol.* 133:801–6

152. Magan D, Taichman N, Lally E, Wahl S. 1991. Lethal effects of *Actinobacillus actinomycetemcomitans* leukotoxin on human T lymphocytes. *Infect. Immun.* 59:3267–72

153. Mahida YR, Galvin A, Makh S, Hyde S, Sanfilippo L, et al. 1998. Effect of *Clostridium difficile* toxin A on human colonic lamina propria cells: early loss of macrophages followed by T-cell apoptosis. *Infect. Immun.* 66:5462–69

154. Mahida YR, Makh S, Hyde S, Gray T, Borriello SP. 1996. Effect of *Clostridium difficile* toxin A on human intestinal epithelial cells: induction of interleukin 8

production and apoptosis after cell detachment. *Gut* 38:337–47

155. Malinin NL, Boldin MP, Kovalenko AV, Wallach D. 1997. MAP3K-related kinase involved in NF-kB induction by TNF, CD95 and IL-1. *Nature* 385:540–44

156. Mangeney M, Lingwood C, Taga S, Caillou B, Tursz T, Wiels J. 1993. Apoptosis induced in Burkitt's lymphoma cells via Gb3/CD77, a glycolipid antigen. *Cancer Res.* 53:5314–19

157. Mannick E, Bravo L, Zarama G, Realpe J, Zhang X, et al. 1996. Inducible nitric oxide synthase, nitrotyrosine, and apoptosis in *Helicobacter pylori* gastritis: effect of antibiotics and antioxidants. *Cancer Res.* 56:3238–43

158. Marshall BJ, Warren JR. 1984. Unidentified curved bacilli in the stomach of patients with gastritis and peptic ulceration. *Lancet* 8390:1311–15

159. Mastrantonio P, Pantosti A, Cerquetti M, Fiorentini C, Donelli G. 1996. *Clostridium difficile*: an update on virulence mechanisms. *Anaerobe* 2:337–43

160. Maundrell K, Antonsson B, Magnenat E, Camps M, Muda M, Chabert C, et al. 1997. Bcl-2 undergoes phosphorylation by c-Jun N-terminal kinase/stress- activated protein kinases in the presence of the constitutively active GTP-binding protein Rac1. *J. Biol. Chem.* 272:25238–42

161. Menestrina G. 1986. Ionic channel formed by Staphylococcus aureus alphatoxin voltage dependent inhibition by divalent and trivalent cations. *J. Membr. Biol.* 19:177–90

162. Mercurio F, Zhu H, Murray BW, Shevchenko A, Bennett BL, Li J, et al. 1997. IKK-1 and IKK-2: cytokine-activated IkB kinases essential for NF-kB activation. *Science* 278:860–66

163. Mills S, Boland A, Sory MP, Van Der Smissen P, Kerbourch C, et al. 1997. *Yersinia enterocolitica* induces apoptosis in macrophages by a process requiring functional type III secretion and translocation mechanisms involving YopP, presumably acting as an effector protein. *Proc. Natl. Acad. Sci. USA* 94:12638–43

164. Molloy A, Laochumroonvorapong P, Kaplan G. 1994. Apoptosis, but not necrosis, of infected monocytes coupled with killing of intracellular Bacillus Calmette-Guerin. *J. Exp. Med.* 180:1499–509

165. Monack DM, Raupach B, Hromockyj AE, Falkow S. 1996. *Salmonella typhimurium* invasion induces apoptosis in infected macrophages. *Proc. Natl. Acad. Sci. USA* 93:9833–38

166. Monack D, Mecsas J, Ghori N, Falkow S. 1997. *Yersinia* signals macrophages to undergo apoptosis and YopJ is necessary for this cell death. *Proc. Natl. Acad. Sci. USA* 94:10385–90

167. Monack DM, Mecsas J, Bouley D, Falkow S. 1998. *Yersinia*-induced apoptosis *in vivo* aids in the establishment of a systemic infection of mice. *J. Exp. Med.* 188:2127–37

168. Morimoto H, Bonavida S. 1992. Diphtheria toxin and Pseudomonas A toxin mediated apoptosis. *J. Immunol.* 149:2089–94

169. Moss S, Calam J, Agarwal B, Wang S, Holt P. 1996. Induction of gastric epithelial apoptosis by *Helicobacter pylori*. *Gut* 38:498–501

170. Motoyama N, Wang F, Roth KA, Sawa H, Nakayama K, Nakayama K, et al. 1995. Massive cell death of immature hematopoietic cells and neurons in Bcl-x-deficient mice. *Science* 267:1506–10

171. Moulder JW. 1991. Interaction of chlamydiae and host cells in vitro. *Microbiol. Rev.* 55:143–90

172. Muchmore SW, Sattler M, Liang H, Meadows RP, Harlan JE, Yoon HS, et al. 1996. X-ray and NMR structure of human Bcl-xL, an inhibitor of programmed cell death. *Nature* 381:335–41

173. Müller A, Hacker J, Brand BC. 1996. Evidence for apoptosis of human macrophage-like HL-60 cells by *Legionella*

pneumophila infection. *Infect. Immun.* 64:4900–6

174. Muzio M, Chinnaiyan AM, Kischkel FC, O'Rourke K, Shevchenko A, Ni J, et al. 1996. FLICE, a novel FADD-homologous ICE/CED-3-like protease, is recruited to the CD95 (Fas/APO-1) death-inducing signaling complex. *Cell* 85:817–27

175. Muzio M, Stockwell BR, Stennicke HR, Salvesen GS, Dixit VM. 1998. An induced proximity model for caspase-8 activation. *J. Biol. Chem.* 273:2926–30

176. Nagata S. 1997. Apoptosis by death factor. *Cell* 88:355–65

177. Nakajima R, Brubaker RR. 1993. Association between virulence of Yersinia pestis and suppression of gamma interferon and tumor necrosis factor alpha. *Infect. Immun.* 61:23–31

178. Nakajima R, Motin VL, Brubaker RR. 1995. Suppression of cytokines in mice by A-V antigen fusion peptide and restoration of synthesis by active immunization. *Infect. Immun.* 63:3021–29

179. Natoli G, Costanzo A, Ianni A, Templeton DJ, Woodgett JR, et al. 1997. Activation of SAPK/JNK by TNF receptor 1 through a noncytotoxic TRAF2-dependent pathway. *Science* 275:200–3

180. Newmeyer DD, Farschon DM, Reed JC. 1994. Cell-free apoptosis in Xenopus egg extracts: inhibition by Bcl-2 and requirement for an organelle fraction enriched in mitochondria. *Cell* 79:353–64

181. Nicholson D, Thornberry N. 1997. Caspases: killer proteases. *Trends Biochem. Sci.* 22:299–6

182. Nürnberger T, Nennsteil D, Jabs T, Sacks WR, Hahlbrock K, Schell D. 1994. High affinity binding of a fungal oligopeptide elicitor to parsley plasma membranes triggers multiple defense responses. *Cell* 78:449–60

183. Oh K, Zhan H, Cui C, Hideg K, Collier R, Hubbel W. 1996. Organization of diphtheria toxin T domain in bilayers: a site directed spin labeling study. *Science* 273:810–12

184. Ojcius DM, Souque P, Perfettini J-L, Dautry-Varsat A. 1998. Apoptosis of epithelial cells and macrophages due to infection with the obligate intracellular pathogen *Chlamydia*. *J. Immunol.* 161:4220–26

185. Oltvai ZN, Millman CL, Korsmeyer SJ. 1993. Bcl-2 heterodimerizes in vivo with a conserved homolog, Bax, that accelerates programmed cell death. *Cell* 74:609–19

186. Orth K, Chinnaiyan AM, Garg M, Froelich CJ, Dixit VM. 1996. The CED-3/ICE-like protease Mch2 is activated during apoptosis and cleaves the death substrate lamin A. *J. Biol. Chem.* 271:16443–46

187. Pagliaccia C, Bernard M, Lupetti P, Ji X, Burroni D, et al. 1998. The m2 form of the *Helicobacter pylori* cytotoxin has cell type-specific vacuolating activity. *Proc. Natl. Acad. Sci. USA* 95:10212–17

188. Palmer LE, Hobbie S, Galan JE, Bliska J. 1998. YopJ of Yersinia pseudotuberculosis is required for the inhibition of macrophage TNF-a production and downregulation of the MAP kinases p38 and JNK. *Mol. Microbiol.* 27:953–65

189. Pan G, O'Rourke K, Dixit VM. 1998. Caspase-9, Bcl-XL, and Apaf-1 form a ternary complex. *J. Biol. Chem.* 273:5841–45

190. Peek RM Jr, Moss SF, Tham KT, Perez-Perez GI, Wang S, et al. 1997. *Helicobacter pylori* cagA+ strains and dissociation of gastrointestinal epithelial cell proliferation from apoptosis. *J. Natl. Cancer Inst.* 89:863–68

191. Pegues DA, Hantman MJ, Behlau I, Miller SI. 1995. PhoP/PhoQ transcriptional repression of *Salmonella typhimurium* invasion genes: evidence for a role in protein secretion. *Mol. Microbiol.* 17:169–81

192. Perdomo OJJ, Cavaillon JM, Huerre M,

Ohayon H, Gounon P, Sansonetti PJ. 1994. Acute inflammation causes epithelial invasion and mucosal destruction in experimental shigellosis. *J. Exp. Med.* 180:1307–19

193. Phan L, Perentesis J, Bodley J. 1993. Saccharomyces cerevisiae elongation factor 2. *J. Biol. Chem.* 268:8665–68

194. Portnoy D, Chakraborty T, Goebel W, Crossart P. 1992. Molecular determinants of Listeria monocytogenes pathogenesis. *Infect. Immun.* 60:1263–67

195. Pozo O, Lam E. 1998. Caspases and programmed cell death in the hypersensitive response of plants to pathogens. *Curr. Biol.* 8:1129–32

196. Projan SJ, Novick RP. 1997. The molecular basis of pathogenicity. In *The Staphylococci in Human Disease*, ed. KB Crossley, GL Archer, pp. 55–81. New York: Churchill Livingstone

197. Rahme LG, Stevens EJ, Wolfort SF, Shao J, Tompkins RG, Ausubel FM. 1995. Common virulence factors for bacterial pathogenicity in plants and animals. *Science* 268:1899–902

198. Rasmussen SJ, Eckmann L, Quayle AJ, Shen L, Zhang YX, et al. 1997. Secretion of proinflammatory cytokines by epithelial cells in response to *Chlamydia* infection suggests a central role for epithelial cells in chlamydial pathogenesis. *J. Clin. Invest.* 99:77

199. Reed J. 1997. Double identity for proteins of the Bcl-2 family. *Nature* 387:773–76

200. Regnier CH, Song HY, Gao X, Goeddel DV, Cao Z, Rothe M. 1997. Identification and characterization of an IkB kinase. *Cell* 90:373–83

201. Rogel A, Hanski E. 1992. Distinct steps in the penetration of adenylate cyclase toxin of *Bordetella pertussis* into sheep erythrocytes. *J. Biol. Chem.* 12:232–37

202. Rogers H, Callery MD, Deck B, Unanue E. 1996. Listeria monocytogenes induces apoptosis of infected hepatocytes. *J. Immunol.* 156:679–84

203. Rosqvist R, Forsber Å, Wolf-Watz H. 1991. Intracellular targeting of the *Yersinia* YopE cytotoxin in mammalian cells induces actin microfilament disruption. *Infect. Immun.* 59:4562–69

204. Rothe M, Pan MG, Henzel WJ, Ayres TM, Goeddel DV. 1995. The TNFR2-TRAF signaling complex contains two novel proteins related to baculoviral inhibitor of apoptosis proteins. *Cell* 83:1243–52

205. Roulston A, Reinhard C, Amiri P, Williams LT. 1998. Early activation of c-Jun N-terminal kinase and p38 kinase regulate cell survival in response to tumor necrosis factor a. *J. Biol. Chem.* 273:10232–39

206. Ruckdeschel K, Harb S, Roggenkamp A, Hornnef M, Zumbihl R, et al. 1998. Yersinia enterocolitica impairs activation of transcripton factor NF-kB: involvement in the induction of cell death and in the suppression of the macrophage tumor necrosis factor a production. *J. Exp. Med.* 187:1069–79

207. Ruckdeschel K, Machold J, Roggenkamp A, Schubert S, Pierre J, et al. 1997. *Yersinia enterocolitica* promotes deactivation of macrophage mitogen-activated protein kinases extracellular signal regulated kinase-1/2, p38, and c-Jun NH$_2$-terminal kinase. *J. Biol. Chem.* 272:15920–27

208. Ruckdeschel K, Roggenkamp A, Lafont V, Mangeat P, Heesemann J, Rouot B. 1997. Interaction of *Yersinia enterocolitica* with macrophages leads to macrophage cell death through apoptosis. *Infect. Immun.* 65:4813–21

209. Sabbatini P, Lin J, Levine AJ, White E. 1995. Essential role for p53-mediated transcription in E1A-induced apoptosis. *Genes Dev.* 9:2184–92

210. Sandvig K, Van Deurs B. 1992. Toxin induced cell lysis: protection by 3-methyladenine and cycloheximide. *Exp. Cell Res.* 200:253–62

211. Sansonetti PJ, Arondel J, Cantey JR, Prevost MC, Huerre M. 1996. Infection of rabbit Peyer's patches by *Shigella flexneri*: effect of adhesive or invasive phenotypes on follicle-associated epithelium. *Infect. Immun.* 64:2752–64

212. Sansonetti PJ, Arondel J, Cavaillon J-M, Huerre M. 1995. Role of IL-1 in the pathogenesis of experimental shigellosis. *J. Clin. Invest.* 96:884–92

213. Sansonetti PJ, Kopecko DJ, Formal SB. 1982. Involvement of a plasmid in the invasive ability of *Shigella flexneri*. *Infect. Immun.* 35:852–60

214. Sattler M, Liang H, Nettesheim D, Meadows RP, Harlan JE. 1997. Structure of Bcl-xL-Bak peptide complex: recognition between regulators of apoptosis. *Science* 275:983–86

215. Scherf U, Kalab P, Dasso M, Pastan I, Brinkman U. 1998. The hCSE1/CAS protein is phosphorylated by HeLa extracts and MEK-1: MEK-1 phosphorylation may modulate the intracellular localization of CAS. *Biochem. Biophys. Res. Commun.* 250:623–28

216. Schesser K, Splik AK, Dukuzumuremyi JM, Neurath MF, Pettersson S, Wolf-Watz H. 1998. The *yopJ* locus is required for *Yersinia*-mediated inhibition of NF-kB activation and cytokine expression: YopJ contains a eukaryotic SH2-like domain that is essential for its repressive activity. *Mol. Microbiol.* 28:1067–79

217. Scott DE, Kisch WJ, Steinberg AD. 1993. Studies of T-cell deletion and T-cell anergy following in vivo administration of SEB to normal and lupus-prone mice. *J. Immunol.* 150:664

218. Sears CL, Kaper JB. 1996. Enteric bacterial toxins: mechanisms of action and linkage to intestinal secretion. *Microbiol. Rev.* :167–215

219. Segal G, Purcell M, Shuman HA. 1998. Host cell killing and bacterial conjugation require overlapping sets of genes within a 22-kb region of the Legionella pneu-mophila genome. *Proc. Natl. Acad. Sci. USA* 95:1669–74

220. Shi Y, Glynn JM, Guilbert LJ, Cotter TG, Bissonnette RP, Green DR. 1992. Role of *c-myc* in activation-induced apoptotic cell death in T cell hybridomas. *Science* 257:212–14

221. Siebers A, Finlay BB. 1996. M cells and the pathogenesis of mucosal and systemic infections. *Trends Microbiol.* 4:22–29

222. Song HY, Regnier CH, Kirschning CJ, Goeddel DV, Rothe M. 1997. Tumor necrosis factor (TNF)-mediated kinase cascades: bifurcation of nuclear factor-kB and c-jun N-terminal kinase (JNK/SAPK) pathways at TNF receptor-associated factor 2. *Proc. Natl. Acad. Sci. USA* 94:9792–96

223. Sporn LA, Haidaris PJ, Shi R-J, Nemerson Y, Silverman DJ, Marder VJ. 1994. *Rickettsia rickettsii* infection of cultured human endothelial cells induces tissue factor expression. *Blood* 83:1527–34

224. Sporn LA, Marder VJ. 1996. Interleukin-1a production during *Rickettsia rickettsii* infection of cultured endothelial cells: potential role in autocrine cell stimulation. *Infect. Immun.* 64:1609–13

225. Sporn LA, Sahni SK, Lerner NB, Marder VJ, Silverman DJ, et al. 1997. Rickettsia rickettsii infection of cultured human endothelial cells induces NF-kB activation. *Infect. Immun.* 65:2786–91

226. Staskawicz BJ, Ausubel FM, Baker BJ, Ellis JG, Jones JDG. 1995. Molecular genetics of plant disease resistance. *Science* 268:661–67

227. Stevens DL. 1997. Superantigens: their role in infectious diseases. *Immunol. Invest.* 26:275–81

228. Stevens PK, Cruprynski CJ. 1996. *Pasteurella haemolytica* leukotoxin induces bovine leukocytes to undergo morphologic changes with apoptosis in vitro. *Infect. Immun.* 64:2687–94

229. Straley SC, Cibull ML. 1989. Differential clearance of host-pathogen interactions of

YopE⁻ and YopK⁻ and YopL⁻ *Yersinia pestis* in BALB/c mice. *Infect. Immun.* 57:1200–10

230. Suttorp N, Seeger W, Dewein E, Bhakdi S, Roka L. 1985. Staphylococcal alpha toxins stimulate synthesis of prostacyclic cultured endothelial cells from pig pulmonary arteries. *Am. J. Physiol.* 248:C127–C135

231. Suttorp N, Seeger W, Zucker-Reinmann J, Roka L, and Bhakdi S. 1987. Mechanism of leukotriene generation in polymorphonuclear leukocytes by staphylococcal alpha-toxin. *Infect. Immun.* 55:104–9

232. Taichman N, Dean R, Sanderson C. 1980. Biochemical and morphological characterization of the killing of human monocytes by a leukotoxin derived from *Actinobacillus actinomycetemcomitans. Infect. Immun.* 28:258–68

233. Takahashi A, Alnemri ES, Lazebnik YA, Fernandes-Alnemri T, Litwack G, et al. 1996. Cleavage of lamin A by Mch2 alpha but not CPP32: multiple interleukin 1ß-converting enzyme-related proteases with distinct substrate recognition properties are active in apoptosis. *Proc. Natl. Acad. Sci. USA* 93:8395–400

234. Tesh V, O'Brien A. 1991. The pathogenic mechanisms of Shiga toxin and the Shiga like toxins. *Mol. Microbiol.* 5:1817–22

235. Thelestam M, Florin I, Olarte EC. 1997. Review: *Clostridium difficile* toxins. In *Bacterial Toxins: a Laboratory Companion*, ed. K Aktories, pp. 141–58. London: Chapman & Hall

236. Tilney L, Portnoy D. 1989. Actin filaments and the growth, movement, and spread of the intracellular parasite, Listeria monocytogenes. *J. Cell Biol.* 109:1597–8

237. Ting AT, Pimentel-Muinos FX, Seed B. 1996. RIP mediates tumor necrosis factor receptor 1 activation of NF-kB but not Fas/APO-1-initiated apoptosis. *EMBO J.* 15:6189–96

238. Tsai C, McArthur W, Baehni P, Hammond B, Taichman N. 1979. Extraction and partial characterization of a leukotoxin from a plaque derived gram-negative microorganism. *Infect. Immun.* 25:427–39

239. Tsujimoto Y, Finger LR, Yunis J, Nowell PC, Croce CM. 1984. Cloning of the chromosome breakpoint of neoplastic B cells with the t(14;18) chromosome translocation. *Science* 226:1097–99

240. Tuomanen E, Weiss AA. 1985. Characterization of two adhesins of *Bordetella pertussis* for human ciliated respiratory-epithelial cells. *J. Infect. Dis.* 152:118–25

241. van Antwerp DJ, Martin SJ, Kafri T, Green DR, Verma IM. 1996. Suppression of TNFa-induced apoptosis by NF-kB. *Science* 274:787–89

242. Vaux DL, Weissman IL, Kim SK. 1992. Prevention of programmed cell death in Caenorhabditis elegans by human *bcl-2. Science* 258:1955–57

243. Veis DJ, Sorenson CM, Shutter JR, Korsmeyer SJ. 1993. Bcl-2-deficient mice demonstrate fulminant lymphoid apoptosis, polycystic kidneys, and hypopigmented hair. *Cell* 75:229–40

244. Vivian A, Gibbon MJ. 1997. Avirulence genes in plant-pathogenic bacteria: signals or weapons? *Microbiology* 143:693–4

245. Wang CY, Mayo MW, Baldwin AS Jr. 1996. TNF- and cancer therapy-induced apoptosis: potentiation by inhibition of NF-kB. *Science* 274:784–87

246. Wassef JS, Keren DF, Mailloux JL. 1989. Role of M cells in initial antigen uptake and in ulcer formation in rabbit intestinal loop model of shigellosis. *Infect. Immun.* 57:858–63

247. Watanabe-Ohnish R, Low D, McGeer A, Stevens D, Schlievert P, et al. 1995. Selective depletion of Vb-bearing T cells in patients with severe invasive group A streptococcal infection and streptococcal toxic shock syndrome. *J. Infect. Dis.* 171:78–84

248. Weiss AA, Goodwin MS. 1989. Lethal infection by *Bordetella pertussis* in the

infant mouse model. *Infect. Immun.* 57: 3757–64

249. Wesson CA, Liou LE, Todd KM, Bohach GA, Trumble WR, Bayles K. 1998. Staphylococcus aureus Agr and Sar global regulators influence internalization and induction of apoptosis. *Infect. Immun.* 66:5238–43

250. White E, Stillman B. 1987. Expression of adenovirus E1B mutant phenotypes is dependent on the host cell and on synthesis of E1A proteins. *J. Virol.* 61:426–35

251. Wolff JG, Cook H, Goldhammer AR, Berkowitz SA. 1980. Calmodulin activates prokaryotic adenylate cyclase. *Proc. Natl. Acad. Sci. USA* 77:3841–44

252. Xue D, Shaham S, Horvitz HR. 1996. The *Caenorhabditis elegans* cell-death protein CED-3 is a cysteine protease with substrate specificities similar to those of the human CPP32 protease. *Genes Dev.* 1:1073–83

253. Yang J, Liu X, Bhalla K, Kim CN, Ibrado AM, Cai J, et al. 1997. Prevention of apoptosis by Bcl-2: release of cytochrome c from mitochondria blocked. *Science* 275:1129–32

254. Yang X, Chang HY, Baltimore D. 1998. Autoproteolytic activation of procaspases by oligomerization. *Mol. Cell.* 1:319–25

255. Yuan JY, Horvitz HR. 1990. The *Caenorhabditis elegans* genes *ced-3* and *ced-4* act cell autonomously to cause programmed cell death. *Dev. Biol.* 138:33–41

256. Zamzami N, Brenner C, Marzo I, Susin SA, Kroemer G. 1998. Subcellular and submitochondrial mode of action of Bcl-2-like oncoproteins. *Oncogene* 16:2265–82

257. Zandi E, Rothwarf DM, Delhase M, Hayakawa M, Karin M. 1997. The IkB kinase complex (IKK) contains two kinase subunits, IKKalpha and IKKbeta, necessary for IkB phosphorylation and NF-kB activation. *Cell* 91:243–52

258. Zha H, Fisk HA, Yaffe MP, Mahajan N, Herman B, Reed JC. 1996. Structure-function comparisons of the proapoptotic protein Bax in yeast and mammalian cells. *Mol. Cell. Biol.* 16:6494–508

259. Zha J, Harada H, Yang E, Jockel J, Korsmeyer SJ. 1996. Serine phosphorylation of death agonist BAD in response to survival factor results in binding to 14-3-3 not BCL-X(L). *Cell* 87:619–28

260. Zhou T, Bluethmann H, Zhang J, Edwards CK III, Mountz JD. 1992. Defective maintenance of T cell tolerance to a superantigen in MRL-*lpr/lpr* mice. *J. Exp. Med.* 176:1063

261. Zou H, Henzel WJ, Liu X, Lutschg A, Wang X. 1997. Apaf-1, a human protein homologous to C. elegans CED-4, participates in cytochrome c-dependent activation of caspase-3. *Cell* 90:405–13

262. Zychlinsky A, Prevost MC, Sansonetti PJ. 1992. *Shigella flexneri* induces apoptosis in infected macrophages. *Nature* 358:167–68

263. Zychlinsky A, Thirumalai K, Arondel J, Cantey JR, Aliprantis A, Sansonetti PJ. 1996. In vivo apoptosis in *Shigella flexneri* infections. *Infect. Immun.* 64:5357–65

Annu. Rev. Microbiol. 1999. 53:189–215

POLES APART: Biodiversity and Biogeography of Sea Ice Bacteria

James T. Staley and John J. Gosink

Department of Microbiology 357242, University of Washington, Seattle, Washington 98195; e-mail: jtstaley@u.washington.edu

Key Words   polar microbiology, sea ice microbiology, psychrophilic bacteria, bacterial biodiversity

■ **Abstract**   This review introduces the subjects of bacterial biodiversity and biogeography. Studies of biogeography are important for understanding biodiversity, the occurrence of threatened species, and the ecological role of free-living and symbiotic prokaryotes. A set of postulates is proposed for biogeography as a guide to determining whether prokaryotes are "cosmopolitan" (found in more than one geographic location on Earth) or candidate endemic species. The term "geovar" is coined to define a geographical variety of prokaryote that is restricted to one area on Earth or one host species. This review discusses sea ice bacteriology as a test case for examining bacterial diversity and biogeography. Approximately 7% of Earth's surface is covered by sea ice, which is colonized principally by psychrophilic microorganisms. This extensive community of microorganisms, referred to as the sea ice microbial community (SIMCO), contains algae (mostly diatoms), protozoa, and bacteria. Recent investigations indicate that the sea ice bacteria fall into four major phylogenetic groups: the proteobacteria, the *Cytophaga-Flavobacterium-Bacteroides* (CFB) group, and the high and low mol percent gram-positive bacteria. Archaea associated with sea ice communities have also been reported. Several novel bacterial genera and species have been discovered, including *Polaromonas, Polaribacter, Psychroflexus, Gelidibacter*, and *Octadecabacter*; many others await study. Some of the gram-negative sea ice bacteria have among the lowest maximum temperatures for growth known, <10°C for some strains. The polar sea ice environment is an ideal habitat for studying microbial biogeography because of the dispersal issues involved. Dispersal between poles is problematic because of the long distances and the difficulty of transporting psychrophilic bacteria across the equator. Studies to date indicate that members of some genera occur at both poles; however, cosmopolitan species have not yet been discovered. Additional research on polar sea ice bacteria is needed to resolve this issue and extend our understanding of its microbial diversity.

0066-4227/99/1001-0189$08.00

CONTENTS

INTRODUCTION

Sea ice covers 10% of the ocean surface of Earth and harbors a unique community dominated by microorganisms. The vast scale of this extreme ecosystem for life underscores its significance; yet studies of sea ice microbiology, particularly bacteriology, are still in their infancy. One aim of this review is to introduce readers to the sea ice microbial community and what is currently known about its resident bacteria. In addition, research on bacterial biogeography is discussed as it applies to the polar sea ice environment. As a guide to future research in bacterial biogeography, a set of biogeography and coevolution postulates is proposed.

ASSESSING BACTERIAL (PROKARYOTIC) BIODIVERSITY

Biodiversity is defined as the variety and abundance of life forms that live on Earth. The basic unit of biodiversity is the species, but biodiversity is also measured as intraspecific genetic variability and by the richness of evolutionary lineages at higher taxonomic levels. The assessment of biodiversity, particularly of microbial diversity, is one of the most challenging and fascinating aspects of microbiology (2). Perhaps this aspect is best illustrated by the fact that estimates of the numbers of bacterial species that exist on Earth today range from 10,000 by some bacteriologists to more than a billion (10^9) by others. Several books have been written on biodiversity, although most of them have concentrated on plants and animals (e.g. 83). Only a brief overview of biodiversity is presented to provide an introduction to the subject as it pertains to the topic of sea ice microorganisms.

The enormous diversity of microbial life should not seem surprising given that bacteria have inhabited Earth for >3.5 Ga (10^9 years) of its 4.5 Ga existence. Prokaryotes were the first organisms, and microorganisms existed on Earth for ~3 Ga before land plants and animals evolved only 600 million years ago.

The number of bacterial species that have been described is remarkably few in comparison with plants and animals. About 6000 bacterial species have been named (37), whereas the number of named plant and animal species is more than 1 million (50, 83). Some scientists have interpreted this difference to mean that the bacteria are not particularly diverse (50). However, there are several reasons for the small number of bacterial species compared with plants and animals. One reason is that bacterial species cannot be defined by visual observation alone because most are too simple to be distinguished morphologically. Typically, strains must be cultivated in pure culture; thus an appreciable amount of laboratory work is required to characterize them and differentiate them from close relatives. In addition, little effort has been directed at bacterial taxonomy primarily because of lack of financial support for research in this field. Finally, it is not presently possible to make a fair comparison between the numbers of species of animals and plants vs bacteria given that these groups are defined differently (66, 78).

Culture-Based Approaches

Our knowledge of bacterial biodiversity stems largely from analyses of pure cultures. Until the 1970s, the bacteria were classified into various hierarchical groups by phenotypic features (37). Phenotypic methods were limited, however, because there was little basis for comparing the evolutionary relatedness of bacteria with one another or with other microorganisms and plants and animals. The shift to using molecular sequencing of conserved macromolecules has revolutionized bacterial classification and our understanding of microbial phylogeny and biodiversity.

Bacterial biodiversity has been assessed by ribosomal RNA (rRNA) sequence analyses of pure cultures of described bacteria. For the first time it became possible, due largely to the pioneering efforts of Woese (84), to compare the phylogenetic diversity of all organisms on Earth by using the same macromolecular sequence standard, the RNA from the small subunit of the ribosome (16S or 18S rRNA). A "universal tree of life" obtained from an analysis of many species clearly indicates a far greater diversity in the microbial world than in the world of plants and animals (85). Indeed, the plant and animal kingdoms appear as only two branches of the eukarya. In contrast, the eubacteria archaea, and eukaryotic microorganisms occupy tens of additional branches.

There appears to be a coarse relationship between 16S rDNA sequence and DNA-DNA reassociation values, the procedure that is used for identifying prokaryotic species. Organisms that exhibit >70% DNA-DNA reassociation are considered members of the same species (80). Stackebrandt & Goebel (64) conducted an extensive examination of the literature and reported that, at one extreme, strains can have essentially identical 16S rDNA sequences and comprise different species; whereas, at the other extreme, they can exhibit as much as 2.5% difference in 16S rDNA base composition and still be members of the same species. Thus, the correlation between DNA-DNA reassociation and 16S rDNA sequences appears to vary from genus to genus. Whereas the 16S rRNA sequence information is important from the standpoint of phylogeny, it does not have sufficient resolution for identifying bacterial species.

Community Approaches

Molecular methods have revolutionized our concept of microbial diversity on Earth (56). These methods have shown that far greater microbial diversity exists in natural ecosystems than had been previously thought from studies of pure cultures. One of the reasons for this newly discovered diversity is that microbiologists have not been able to grow the most numerous bacteria from natural habitats on artificial media in the laboratory (72). This inability to recover the most numerous organisms from natural habitats by using cultural approaches has been called the "enumeration anomaly" or the "Great Plate Count Anomaly" (70). This technical problem has resulted in an underestimation of diversity in natural habitats. For example, Hugenholz et al (39) reported the discovery of 36 major phylogenetic groups of eubacteria in natural communities, which is about threefold the number of those that have been cultivated in pure culture. Fortunately, new cultivation approaches are being developed that are helping to overcome this problem (61).

Many microbiologists have adopted molecular procedures for identification and diversity analysis. These procedures are typically accomplished by extracting DNA from a microbial community and using polymerase chain reaction (PCR) to amplify DNA sequences of interest. For bacterial biodiversity studies, universal primers for 16S rDNA are used; in theory, however, any conserved DNA region could be amplified. The PCR product is cloned, sequenced, and used in phylogenetic analysis. The result is an approximate measure of the community diversity and phylogenetic identification of its major inhabitants. Two major advantages of this procedure are that it circumvents the poor recovery found by the cultural approach and it avoids the tedious aspects of the cultivation approach (55, 56, 79, 82).

Although this procedure provides information on the phylogeny of the types of organisms present, it provides little other information. For example, even strains with identical complete 16S rRNA sequences can be separate species (21); also, information about the physiological features of the organism is not revealed. Characteristics of the organisms may be inferred from pure cultures if, and only if, they are very closely related.

A recent meeting sponsored by the American Academy of Microbiology (2) has recommended that microbiologists begin to assess the microbial diversity of natural habitats by using improved cultivation methods as well as molecular approaches.

BIOGEOGRAPHY

Biogeography is defined as the study of the global distribution of species, living or extinct. As with biodiversity, the basic unit of biogeography is the species. Species that are found in more than one geographic location on Earth are referred to as "cosmopolitan." Species that live in only one area on Earth are termed "endemic to that area."

Little emphasis has been given to the study of bacterial biogeography, which is unfortunate because bacterial biogeography is critical for several reasons. First, knowledge of biogeography will help determine how many bacterial species exist on Earth. Second, until we know the ranges of species, we cannot identify which bacteria might be threatened (66). Third, if we wish to identify the ecological role of a particular species, we need to know its distribution.

The prevailing hypothesis for bacterial biogeography is based on the axiom of the Dutch microbiologist Baas-Becking, who stated, "Alles is overal: maar het milieu selecteert" (Everything is everywhere, but the environment selects) (4). Baas-Becking attributed the first phrase of this statement to Beijerinck (6). This statement is interpreted by contemporary bacteriologists as a hypothesis that free-living bacteria are cosmopolitan in their geographic distribution. The rationale given by Baas-Becking in support of this hypothesis, which differs from evidence of plant and animal biogeography (see 11), is that bacteria are readily disseminated from one location on Earth to another by water and air currents or animal vectors such as birds that migrate between regions. Thus, through a combination of natural selection and rapid dispersal, two similar habitats on Earth would harbor the same bacterial species.

Until recently, however, it has not been possible to rigorously test the cosmopolitan distribution of bacteria because unbiased molecular biological approaches were not available. That situation has now changed. Molecular biological procedures based on analyses of the sequence of subunits of conserved macromolecules such as rRNA are having a major impact on our understanding of the phylogeny and taxonomy of microorganisms (63, 84). These same molecular approaches can now be applied to the assessment of the distribution of microbial species in natural communities.

Bacterial Biogeography Research

Much of the research on bacterial biogeography has been conducted with a photosynthetic prokaryotic group, the cyanobacteria. Recently, in a mini-review on the biodiversity and endemism of cyanobacteria from thermal environments, Castenholz (12) reported that northern hot springs in Alaska and Iceland did not harbor certain thermophilic cyanobacteria that are found in temperate-zone North American hot springs, even though they would be expected to grow in these habitats. This study, which did not include molecular analyses, concluded that Baas-Becking's cosmopolitan hypothesis was probably incorrect, at least regarding thermophilic cyanobacteria, for reasons of dissemination. First, it is difficult to disperse thermophilic species from one hot spring to another because of the great distances involved, and these prokaryotes lack special survival stages that can withstand long-distance dispersal. Second, because Iceland and Alaska are much farther north than temperate-zone hot springs, there might not have had been sufficient time since the retreat of the ice sheets for the hot springs to become uncovered and for dissemination to occur.

A recent analysis of the biogeography of cyanobacteria involved molecular procedures for analysis (26). The authors concluded that, based on 16S rRNA gene sequences and phenotypic properties, *Microcoleus chthonoplastes* is a cosmopolitan species. Strains from various locations in Europe and North America were compared. It is interesting that two European strains, MEL and EBD, had identical sequences, and two North American strains, GN5 and PCC, also had identical sequences. The European and North American sets of strains differed from one another by 0.2% base homology. However, only about one-third [550 base pairs (bp)] of the 16S rRNA gene was sequenced.

Evidence for restricted ranges was provided in a recent study of 3-chlorobenzoate–degrading bacteria isolated from soils in six regions on five continents (25). Four regions were in Mediterranean ecosystems, and two were in boreal forests. From different sites within each ecosystem, four or five gross enrichments (not extinction dilution) were prepared for 3-chlorobenzoate–degrading bacteria, from which 150 strains were isolated. PCR was used to amplify the 16S rDNA from strains that were subjected to amplified rDNA restriction analysis. In addition, the genotype of strains was assessed by PCR repetitive extragenic palindromic genomic fingerprints. Results of these analyses indicated that the majority of the genotypes were unique to the regions from which they were isolated. Some genotypes were found repeatedly from one region but not from others, which suggests that these genotypes are endemic to specific regions.

Other recent reports based on 16S rDNA sequence and phenotypic properties suggest that similar strains of archaea are found at widely disparate locations (47). However, the most similar sequences reported had 98.2% identity (48). As mentioned previously, 16S rDNA sequences do not have sufficient resolution to allow the determination of prokaryotic species (21). Therefore, this analysis is not sufficient to decide whether these species are cosmopolitan.

In the previous studies discussed, DNA-DNA hybridization experiments were not performed. Thus, the organisms may or may not be separate species based on the current accepted definition (80). However, DNA-DNA reassociation has been performed in some investigations. For example, a high degree of DNA-DNA reassociation was reported for some thermophilic archaea. Stetter et al (73) discovered that hyperthermophilic archaea isolated from Alaskan oil reservoirs showed high degree of DNA-DNA reassociation with selected *Archaeoglobus*, *Thermococcus*, and *Pyrococcus* species. By using dot blots, Stetter et al concluded that the species were the same as those from European thermal marine sources. In a separate study, DNA-DNA reassociation of a strain isolated from North Sea crude oil fields showed 100% relatedness to an *Archaeoglobus fulgidus* strain from Italian hydrothermal systems (5). These two studies comprise some of the best evidence to date supporting the cosmopolitan hypothesis of Baas-Becking.

Importance of Dispersal/Survival Stages

Two implicit assumptions of the Baas-Becking hypothesis are that all free-living bacteria can survive dispersal and that they can successfully colonize new locations. Both of these assumptions can be questioned. For example, it is doubtful that all bacteria are equally equipped to survive dispersal in that not all bacteria produce special spores and cysts that could serve as hardy dispersal stages. The best example of a bacterial survival stage is the endospore that is produced by certain gram-positive bacteria such as the genus *Bacillus*. The endospore is particularly resistant to conditions of desiccation and high temperature. In fact, the endospores of some species can withstand boiling temperatures for prolonged periods. Another example of a specialized resting/dispersal cell is the exospore produced by some of the high mole percent G + C gram-positive bacteria such as the genus *Streptomyces*. These filamentous bacteria produce these spores on an aerial mycelium, which allows for air dispersal when they are mature. Although they are not particularly heat resistant, they are resistant to desiccation and are readily dispersed by wind.

One other example of a specialized survival stage formed by some bacteria is the cyst. Cysts are produced from ordinary vegetative cells during stationary-phase growth. They are formed by certain types of gram-negative bacteria such as *Azotobacter* spp. and the myxococci. Although the cysts are less hardy than bacterial endospores, they are more resistant to desiccation and mild heat than vegetative cells are, and are therefore better suited for dispersal.

Thus, it seems reasonable that some bacteria, in particular certain gram-negative species that lack dispersal/survival stages, are more likely to have restricted ranges. In contrast, other bacterial species, especially spore-forming gram-positive bacteria, are more likely to be cosmopolitan in their distribution.

Cosmopolitan species must have the ability to colonize new locations as well as get there. Organisms with highly specialized nutrient requirements may have difficulty, especially if they arrive during periods when nutrients essential for their growth are absent. Likewise, organisms without sufficient invasive capabilities may not be able to establish themselves.

Importance of Evolution

Let us consider an alternative model for the biogeography of free-living bacteria. Assuming global dissemination occurs for bacteria and that some are better colonists, then the better a species is at colonizing the more likely it is to be cosmopolitan. In contrast, those with poor dispersal and colonization abilities would be less likely to establish new colonies. These bacteria would likely be restricted to certain locales where they would be well adapted to the habitat. Over time, these species might evolve sufficiently to become separate endemic species. For example, populations of poorly dispersible, free-living species of soil bacteria must have been separated into subgroups by continental drift during the breakup of Gondwanaland more than 100 million years ago. Over time, plate tectonic movements could have carried the subpopulations into different geographic zones. Given sufficient time for genetic drift or mutation and selection to occur, the populations on different continents could have evolved into different species even though these new locations might have similar latitudes and climates. This argument would also apply to symbiotic bacteria that live in association with plants and animals. Plate tectonic movements would have separated their hosts into subpopulations. Given sufficient time, this process could have led to the co-evolution of symbiont and host.

Another factor that influences the range of a species is the rate at which evolution occurs. Some groups, such as the planctomycetales (23), are regarded as fast-clock organisms. Other factors considered equal, fast-clock organisms should be more likely to have restricted ranges than do slow-clock phyla. Thus, microbial communities may consist of a mixture of species, some of which are endemic and some of which are cosmopolitan.

Importance of Habitat and Climate

Another basic premise of the Baas-Becking hypothesis is that at least two geographically separate habitats exist on Earth that are essentially identical. This premise seems highly unlikely, as major variations occur in physicochemical features of habitats and climate, including soil type, seasonal temperature range, rainfall, humidity, pH, and type and concentration of chemicals, in addition to numerous other factors. Also, the flora and fauna typically vary. Resident endemic plant and animal species are common globally in ecosystems. Just one of many examples is the cacti found only in desert ecosystems in the Americas, but not in Africa or Eurasia (7). Speciation of novel endemic plants and animals has occurred at disparate geographic areas, even though climatic features might be similar. Should this circumstance not also be true for microorganisms? Indeed, plant species have been reported to harbor their own unique symbiotic species of fungi associated with the leaves, bark, roots, etc (35).

Biogeography and Coevolution Postulates

If endemism occurs in bacteria, the evidence for it should be conserved in the traits of the organisms. If all free-living bacteria are cosmopolitan, then their phylogeny should not show geographic clustering. If endemism occurs, however, the phylogeny of the organisms should exhibit geographic clustering, or clades.

The individual clades may or may not be sufficiently different to warrant the designation of separate species. The occurrence of clades, however, would indicate that speciation processes are occurring that may lead to the adaptive radiation of new species.

As a means of stimulating interest in what we deem to be fundamental questions regarding bacterial biodiversity and biogeography, we propose the following. Because geographic factors may have played a major role in the speciation of free-living bacteria and perhaps symbionts as well, a term should designate such varieties. We propose the term "geovar," defined as follows: a geovar is a geographic variety of a bacterium that is endemic to a specific area or host.

We propose several postulates, the fulfillment of which would be necessary to infer whether a bacterial species is cosmopolitan. The procedure entails the use of pure cultures, although the postulates would apply equally well to simple consortia. At this time, noncultivation approaches are not adequate, although this situation may change as molecular approaches become more informative. For example, complete in situ genome sequencing would certainly be adequate! The postulates are as follows:

1. At least four bacterial strains of a purported species must be isolated from different samples taken from one ecosystem (or host). Ideally, bacteria should be isolated by using extinction dilution as well as gross enrichment methods to ensure that the greatest variety of organisms of a particular type can be selected and obtained in pure culture (20, 61).

2. These strains must be shown to be indigenous to the habitat (or host) from which they were obtained by demonstrating their growth in the habitat at some time during the annual or other periodic cycle of the environment. For some organisms, such as mat-forming cyanobacteria, observation alone may be sufficient. For most bacteria, further research may be required. This research could be accomplished by the use of procedures such as the nalidixic acid growth procedure (44), coupled with a procedure for identifying the organism, or carefully conducted seasonal distribution studies could be used such as for *Caulobacter* spp. (17). Appropriate in situ molecular approaches are also possible, such as combining tritrated-thymindine microautoradiography to demonstrate growth and fluorescent antibody or other labeling approaches to identify strains (1).

3. At least four strains of a potentially identical candidate species must be obtained from one or more other geographically separate ecosystems (or host) that are similar to the ecosystem (or host) from which the first strains were obtained, using the same procedures used to isolate the first strains. These new strains must be shown to be indigenous to these other ecosystems.

4. The two or more groups of strains from the two or more geographically separate ecosystems must be subjected to phylogenetic analyses by sequencing two or more appropriate genes. The choice of genes may differ

from one phylogenetic group to another. For example, one of the appropriate genes for enteric bacteria would appear to be for the housekeeping protein malate dehydrogenase (58). Sequences of 16S rDNA do not appear to provide sufficient resolution for most prokaryotic groups.

If the strains show no evidence of clades in either phylogenetic analysis, then they are considered cosmopolitan. In contrast, if the phylogenetic analysis indicates that geographic clustering occurs, then they are provisionally considered to be endemic to those areas (or hosts). Such strains would be called geovars.

Geovars may or may not be the same species. This distinction would need to be determined by separate analyses in postulate 5 below. Cultures of the geovars should be maintained in national culture collections for other researchers to study.

5. (optional). If one wishes to determine whether the two or more groups of strains from the two or more geographically separate ecosystems (or hosts) compose the same species, then the strains must be compared directly with one another. Tests to be conducted include but are not limited to DNA-DNA reassociation, sequencing of 16S rDNA, and phenotypic analyses that are appropriate to the phylogenetic group, including but not limited to nutrition, physiology, fatty acid composition, and cell and colony morphology.

On the one hand, if the two or more sets of strains are not geovars (i.e. do not exhibit geographic clustering) and are shown to compose the same species, then they would be designated as a single cosmopolitan species. If, on the other hand, the two or more sets of strains exhibit geographic clustering and are different species, they should be named and described as two or more separate species.

Absolute proof of endemism is difficult in plants and animals and may not be possible in bacteriology, because this would involve proving a negative; that is, not finding an organism does not mean that it is not there. However, if phylogenetic evidence from several strains supports the view that speciation events have occurred, then it can be proposed that this organism is a candidate endemic strain. In some ecosystems that have simple species compositions, it may be possible to demonstrate that endemic species exist beyond reasonable doubt.

By proposing the biogeography postulates and coining the term geovar, we hope to stimulate research in this area. Clearly, the task to assess whether bacteria are cosmopolitan is not a trivial one. Furthermore, until some guidelines are proposed, there are no parameters for comparing one investigation's results with those of another. By establishing these guidelines, we hope to stimulate rigorous assessments of bacterial distributions in natural environments.

Studies of the biodiversity and biogeography of extreme environments are in their infancy. The sea ice environments of the polar regions are prime examples of poorly known ecosystems that are beginning to be investigated.

BIODIVERSITY OF POLAR SEA ICE BACTERIA

Although sea ice is one of the most extensive habitats for microbial life on Earth, studies of its bacterial composition were not undertaken until the 1980s. Information about sea ice, its suitability as a habitat for microbial life, and the types of bacteria that reside there is discussed below.

Formation of Sea Ice

During the polar winter, vast areas in the polar oceans become sufficiently cold that the seawater freezes. Strong winds and wave action prevent the ice from forming one large sheet on the surface. Land-fast or fast ice is formed in shoreline areas, and the remaining large majority of ice consists of floating sheets termed "pack ice." All told, polar sea ice occupies >7% of Earth's surface, a figure determined by adding the largest areas occupied during the winter maxima of the north and south polar regions (49).

Sea ice is primarily seasonal. It forms in the winter months and breaks up and melts during the polar summer. Sea ice begins to form in polar regions when the surface waters reach about $-1.8°C$, the freezing temperature of seawater at a salinity of 35‰. The initial collection of randomly oriented ice crystals is called "frazil ice." The frazil ice collects in long plumes by wind and wave action to form "grease ice" (81). Eventually consolidated, circular forms called "pancake ice" or "clumped forms" are produced (54). Subsequently a surface layer of ice is produced, and long vertically oriented ice crystals form columnar congelation ice from beneath the surface.

The coldest and most variable temperatures are encountered at the air-ice interface, reaching very low temperatures when the air temperature is low. The temperature at the ice-water interface remains much more stable at about $-2.0°C$. The light intensity increases as the polar spring arrives and continues to increase until it reaches its summer peak. As the ice develops, a salting-out process occurs whereby the marine salts that are excluded from the ice are concentrated in brine pockets. As the ice rises above sea level, brine channels develop that drain the dense brine through the layer of columnar congelation ice by gravity flow to the underlying sea water. Thus, the sea ice produces microenvironments of various temperatures, salinity, nutrient concentrations, and light intensities within the column, which may be as thick as 2 m. For example, salinity in sea ice brine can range from near that of freshwater to >150‰ at the ice-sea-water interface (49). Although most ice melts out during the polar summer months, some ice persists to form multiyear ice, which can be re-colonized as a habitat (38).

The sea ice is remarkable in that it serves as a habitat for a microbial community referred to as the SIMCO. Typically, the SIMCO is found at the interface between

Figure 1 An ice core taken from McMurdo Sound, Antarctica. The dark band near the bottom (left end) of the core is the sea ice microbial community.

the ice column and the underlying seawater. The primary producers in this community are mostly pennate diatoms. Protozoa, bacteria, fungi, and invertebrates are also abundant (57).

Nature of Polar Sea Ice as a Place for Life

Sea ice is a surprisingly complex environment for microbial habitation. The community typically forms in the lower 10–20 cm of sea ice column at the ice-water interface (Figure 1). Nutrients are available from the water column, and light is available from the surface.

The sea ice contains its own dynamic and distinct microbiota. SIMCOs form in the brine inclusions of the ice that range in salinity from <10‰ to >150‰. SIMCOs are dominated by diatoms, such as *Amphiprora* and *Nitzschia* spp., which serve as the major primary producers of the sea ice microbial community (29, 38). Many bacterial epiphytes of diatoms have been reported, which suggests that heterotrophic bacteria may be deriving nutrients from diatom exudates or dying diatoms (57, 75). In addition, some flagellate species and other algal groups can also be found (28, 75). Although ice algal blooms normally occur during the summer months, autumn blooms have been reported (22). Protozoa are also indigenous to the sea ice community, but information on types and concentrations is limited (28).

Metazoa, including polychaetes, amphipods, and euphausiids (krill), are known to colonize sea ice as well (27). Krill, which serve as a major food for penguins, seals, whales, and other marine animals, feed directly on this concentrated food source by grazing at the seawater-ice interface, where the SIMCO is typically found (14). It has been estimated that from 9% to 25% of the annual productivity of the ice-covered southern oceans is caused by the SIMCO (3). Additional information on the other biota can be found in the reviews by Horner (38) and Palmisano & Garrison (57).

Polar Sea Ice Bacteriology

Bacteria are very important members of the sea ice food web community because their concentrations are enriched relative to those found in open sea water (15, 36, 74, 75, 86). Their principal role is the heterotrophic mineralization of organic matter produced in the SIMCO (43). Bacterial production rates are high, attaining levels that are 10–15% of primary production (45, 46). Assuming a growth efficiency of ∼50% (19), then 20–30% of the ice-bound primary production cycles through the heterotrophic bacteria (57).

Diversity of Sea Ice Bacteria Isolates

Only recently have studies been undertaken on the types and diversity of sea ice bacteria. What is known indicates that this community contains many unique taxa. For example, heterotrophic gas-vacuolate bacteria, not reported in other marine habitats, have been discovered in and near sea ice (31, 69, Figure 2). These comprise some of the most psychrophilic bacteria known, based on their maximum temperatures for growth (41). Very few of these bacteria grow at temperatures >20°C. Thus, *Polaromonas vacuolata*, one of the new genera and species, has an optimum growth temperature of 4°C and a maximum temperature of about 10°C (32). Other strains have temperature maxima of 7–8°C. Phylogenetic studies involving 16S rDNA sequence analyses indicate that these gas-vacuolate bacteria are members of two major eubacterial phyla, including the proteobacteria and the *Cytophaga-Flavobacterium-Bacteroides* (CFB) group (32).

Bowman et al (8) examined a large collection of sea ice bacteria isolated from polar sea ice in the Indian Ocean near the Australian base at Davis. These bacteria fell into four major phylogenetic groups—including the proteobacteria and the CFB, as reported above, and the low and high mole percent G + C grampositive bacteria, based on 16S rDNA sequence analyses. Obligate and facultative psychrophilic proteobacteria genera represented in their samples included *Colwellia, Shewanella, Marinobacter, Pseudoalteromonas, Alteromonas, Pseudomonas, Halomonas, Hypohomonas*, and *Sphingomonas*.

Three new genera of proteobacteria have been described from sea ice. *Octadecabacter* is a new genus of psychrophilic gas-vacuolate bacteria from the alphaproteobacteria (30). These rod-shaped bacteria are so named because they contain large amounts (70–80%) of octadecenoic acid (18:1) in their cell membranes.

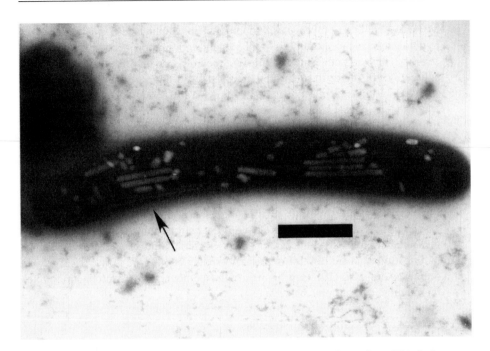

Figure 2 Transmission electron micrograph of a gas-vacuolate bacterium. This isolate was obtained from sea ice near Pt. Barrow, Alaska. Note that two gas vesicles of different-sizes occur in the cytoplasm. Most are relatively wide and short, whereas two are very thin and long. *Bar* equals 1.0 μm.

Based on 16S rDNA sequence analyses, they are most closely related to the genus *Roseobacter*, which are aerobic, bacteriochlorophyll *a*-producing bacteria (62). Unlike *Roseobacter* spp., however, *Octadecabacter* spp. are gas vacuolate and nonpigmented. They lack bacteriochlorophyll *a*, and they differ from *Roseobacter* spp. in their inability to reduce nitrate to nitrite and hydrolyze gelatin and in their carbon source utilization patterns. *Octadecabacter* spp. also grow at lower maximum temperatures, that is, <19°C, whereas *Roseobacter denitrificans* requires at least 37°C for growth.

Polaromonas vacuolata is a new genus and species of the beta-proteobacteria (40). Several strains of this species were collected in Antarctic waters off the Palmer Peninsula near the US Palmer Station on Anvers Island. 16S rDNA sequence analysis showed that these bacteria were most closely related to *Variovorax paradoxus* and *Rhodoferax fermentans*. 16S rDNA base homology demonstrated that the differences between *Polaromonas* spp. and these two strains were 5% and 6.1%, respectively. *Polaromonas* spp. have a lower G + C ratio than do *Rhodoferax* spp. (52–57 vs 60 mol%) and are non-photosynthetic and nonpigmented. Also, unlike *Rhodoferax* spp., *Polaromonas* spp. are gas vacuolate, obligate aerobes and are psychrophilic. They differ from *Variovorax* spp. in G + C ratio (52–57 vs 67–69 mol%) and have polar rather than peritrichous flagella.

Sea ice contains a variety of members of the gamma subclass of proteobacteria, several genera of which are gas vacuolate (32). The best studied members of these occur in the genus *Colwellia* (7). This genus of obligate marine bacteria is well known for its psychrophilic and barophilic members. Many species are known to degrade particulate materials such as casein, chitin, and starch. In addition, all strains studied produce the omega-3 polyunsaturated fatty acid docosahexaenoic acid (22:6ω3), ranging in concentration from 0.8 to 8.0% of the total fatty acids. This fatty acid has been previously reported only in *Moritella marinus* strains isolated from deep-sea environments. Thus, it may be associated with survival and growth in low-temperature and possibly high-pressure marine habitats (7). The sea ice environment is a habitat for four newly described species of *Colwellia*, including *C. demingii*, *C. psychrotropica*, *C. rossiensis*, and *C. horneri*. One species, *C. rossiensis*, is gas vacuolate.

The other major group of gram-negative bacteria that are known to live in sea ice is the CFB group (34). This was the first report of gas-vacuolate bacteria in the CFB group of the bacteria. Several strains were found to be similar to a previously described species, *Flectobacillus glomeratus*, isolated from the marine environment near the Davis Station in Antarctica (51). However, *F. glomeratus* cannot be a true *Flectobacillus* species because its 16S rDNA sequence differs markedly from the type strain of the species, *Flectobacillus glomeratus major* (34). *F. glomeratus* and the other strains from the sea ice grouped most closely to *Cytophaga marina* and *Flexithrix maritimus* as well as other species. Because of the major phylogenetic and phenotypic differences between the sea ice strains and *F. glomeratus* compared with known taxa in the CFB group, a new genus, *Polaribacter*, was proposed with the following new species: *P. irgensii*, *P. franzmannii*, *P. filamentus*, and *P. glomeratus* combination nov.

Polaribacter species are nonmotile, elongated rods, 0.25–1.6 μm in diameter and 2–48 μm in length. They are obligate aerobes and are facultative to obligate psychrophiles, which are pigmented orange, salmon, or pink. Gas vacuoles are produced by some strains. They use a variety of organic carbon sources but prefer to use complex carbon sources such as yeast extract and casamino acids. All strains hydrolyze starch, and some hydrolyze gelatin.

In addition, the Australian group described two new genera of CFB group from Antarctic sea ice, *Psychroflexus* (10) and *Gelidibacter* (9). *Psychroflexus* strains synthesize the unusual polyunsaturated fatty acids eicosapentaenoic acid (20:5ω3) and arachidonic acid (20:4ω6). They are members of the family *Flavobacteriaceae*. Two new species were described, including *P. torquis* and *P. gondwanense*. The latter species, *P. gondwanense*, was isolated from saline lakes near the Davis Station (18) and described initially as a *Flavobacterium* species.

Gelidibacter spp. are gliding rod-shaped–to–filamentous bacteria from the CFB phylum with DNA of low G + C content, ranging from 36 to 38 mol%. These yellow-pigmented saccharolytic bacteria degrade starch and dextran as well as DNA and various Tween compounds. One new species, *G. algens*, was described from Antarctic sea ice (9).

Various gram-positive bacteria also occur in sea ice (8, 42, 60). Members of both the low- and high-mole-percent G + C gram-positive phyla have been reported. Low-mole-percent G + C genera include *Planococcus* and *Halobacillus*; high-mole-percent G + C genera include *Arthrobacter* and *Brachybacterium*. Some of the gram-positive bacteria are true psychrophiles, which grow at <20°C, such as some members of the genus *Planococcus* (8). However, many gram-positive sea ice bacteria are facultative psychrophiles; although they grow at 0°C, they can also grow at temperatures of 30–37°C (42).

The gram-positive sea ice bacteria grow over a broad range of salinities from 0 to 250‰. The fact that they grow at very high salinities suggests that they are well adapted to live in sea ice brine from which many strains have been isolated.

Planococcus mcmeekinii is a newly described member of low-mole-percent G + C gram-positive sea ice bacteria (42). This species is most closely related to another marine bacterium originally described as "*Flavobacterium okeanokoites*" (Zobell & Upham 1944) but which has been reassigned to the genus *Planococcus* as *P. okeanokoites* (53). This motile organism has a mole percent G + C content of 35, very low in comparison with the other planococci.

The high-mole-percent G + C gram-positive sea ice bacteria are most closely related to *Arthrobacter* and *Brachybacterium* species. Although these may represent new species, DNA-DNA reassociation studies would need to be conducted to determine whether they are sufficiently different from *Arthrobacter agilis* and *Brachybacterium tyrofermentans* to justify new taxa. Both strains of high-mole-percent G + C organisms have extremely high GC ratios, 73 and 76 mol%, respectively.

It seems somewhat surprising that the gram-positive bacteria live in sea ice communities because they are normally associated with marine sediments and soils, where nutrient concentrations may be high. However, the brine habitat in sea ice would be expected to have higher concentrations of nutrients than seawater. Also, it is noteworthy that 16S rDNA sequences of gram-positive bacteria have been reported from marine water column habitats (24, 76). It is possible that these organisms could be inoculated from marine sediments by anchor ice that is formed in marine sediments but may float to the overlying ice after it has been disturbed or dislodged.

Other Sea Ice Prokaryotes

Very little is known about the incidence of autotrophic bacteria from sea ice. Ammonia-oxidizing bacteria have been reported from sea ice, suggesting that chemoautotrophs may play an important role in this community (59). Cyanobacteria or other photosynthetic prokaryotic groups that may reside in the SIMCO layer have not been reported.

DeLong's laboratory (16, 17) has reported finding archaea in Antarctic seawater and SIMCO habitats. In DeLong and coworker's study, they measured the binding of group-specific oligomers to rRNA that was extracted from the environment. On this basis, the archaea comprised as high as 34% of the total prokaryotic rRNA from some samples. The phylogenetic studies indicated that these organisms were members of two separate groups, the euryarchaeota and the crenarchaeota. None of these archaea have been cultivated as yet, but this finding is one of the most exciting discoveries in polar microbiology because it suggests that members of the crenarchaeota, which had been previously regarded as thermophiles only, are actually able to grow at freezing temperatures.

It should be noted that, until now, molecular approaches have not been used to survey bacterial diversity from polar sea ice communities. Phylogenetic information from 16S rDNA libraries from the DNA of natural sea ice communities will likely reveal additional bacterial diversity in this habitat.

Function of Gas Vacuoles in SIMCO

The ecological significance of gas vacuole formation is not yet known for sea ice bacteria. Two hypotheses can be advanced to explain their occurrence. First, sea ice is a vertically stratified habitat, similar to a thermally stratified lake in which gas-vacuolate prokaryotes abound (65, 77). Thus, it could be hypothesized that these bacteria have a selective advantage because the gas vacuoles allow the bacteria to rise to the surface of the water column. This brings the bacteria into close association with the sea ice algae, which could provide soluble and particulate organic carbon sources for bacterial growth. Indeed, in some sea ice cores, some gas-vacuolate bacteria occur above the chlorophyll a maximum in the core, which suggests that they may obtain dissolved nutrients that are not available for heterotrophs that grow at lower depths in the ice column. Consistent with this is the lower optimum salinity of many of these bacteria, which grow well at 10–15‰ (41). However, gas vesicles were not found in a cursory transmission electron microscopic examination of fixed cells taken from a sea ice sample from which a high percentage of cells were capable of gas vesicle production. This observation suggests that these bacteria may not be producing gas vesicles while growing in the SIMCO.

An alternative hypothesis is that the gas vesicles are formed after the sea ice community has been disrupted by melting. Frequently, gas vacuolation in pure cultures of heterotrophic bacteria occurs when the nutrients are depleted from the medium—thus, they are found abundantly in stationary-phase growth but not during active-cell multiplication. If this is the case in the sea ice environment, then these organisms may be producing gas vesicles as a dispersal mechanism. They would be formed after the SIMCO collapses as the ice melts. Their formation at that time would enable cells to remain high in the water column where they could colonize the newly forming sea ice as it develops during the fall. Supporting this interpretation is the fact that all of the gas-vacuolate bacteria isolated from the water column in the Palmer Peninsula waters were obtained from a 0- to 25-m depth—none were found at depths from 50 to 500 m (R Irgens & JT Staley, unpublished data). Clearly, additional studies are warranted to determine more about the reasons for gas vesicle formation in these bacteria.

BIOGEOGRAPHY OF SEA ICE BACTERIA

One of the major goals of our laboratory has been to determine whether sea ice bacteria exhibit a bipolar distribution. Finding the same species at both poles would indicate that these bacteria were cosmopolitan in distribution and would support the Baas-Becking hypothesis. We believe that the polar sea ice bacteria are particularly appropriate to study regarding biogeography because of the natural constraints imposed on dispersal.

Why Is the Polar Sea Ice Community a Test Case for Bacterial Biogeography?

We purposely selected the polar sea ice communities to test the cosmopolitan hypothesis because dispersal of psychrophilic, free-living bacteria between the north and south poles should be extremely difficult. This difficulty is not only because of the long distances between the poles but also because dispersal between the poles would require passage of these psychrophiles across the warmer latitudes near the equator. Thus, for example, it would seem improbable that the arctic tern could carry viable psychrophilic bacteria during its bi-polar migrations. One mechanism that might explain transequatorial passage is cold, deep underwater currents (13). These currents, however, take hundreds of years to carry water from one pole to the other. It is extremely doubtful that these bacteria, which have been removed from their normal habitat, could survive such a long transit. Perhaps during the Pleistocene, icebergs could have carried psychrophilic bacteria closer to the equator, although, even at its maximum extent, the ice sheets would have been separated by >8000 km. Alternatively, one could propose passage across the equator in ice crystals in the upper atmosphere, but there is no evidence to indicate that this occurs.

Biogeography of Octadecabacter, Polaribacter, and Iceobacter Species

The gas-vacuolate sea ice bacteria were selected as test organisms for polar sea ice studies. These bacteria were selected because they appear to be specialists associated with sea ice. Furthermore, they are all true psychrophiles, which cannot survive at temperatures >20°C (52) and therefore would be sensitive to transequatorial dissemination. Strains of these bacteria were isolated from both poles by using the same medium: Ordal's sea-water cytophaga medium with succinate as a supplemental carbon source (41, 67, 68). Results of that study confirmed that gas-vacuolate bacteria were also indigenous to north polar sea ice communities (33). To determine whether the same bipolar bacterial species exist, strains of gas-vacuolate sea ice bacteria from the north were compared with those from the south polar marine habitats.

Whole-cell fatty acid analysis was used initially to group the ~200 gas-vacuolate strains that had been isolated from both poles. In this manner, it was possible to readily limit the number of strains for further study. From this study, three groups were identified that contained similar bipolar representatives. One group of alpha-proteobacteria contained two strains of *Octadecabacter* spp. These two strains were then subjected to phenotypic testing and 16S rDNA sequence analysis (Figure 3).

Phenotypically, the strains were quite similar. However, some differences were noted. The arctic strain required three vitamins: thiamine, nicotinic acid, and pantothenic acid. The Antarctic strain was similar but had at least one additional unidentified requirement. The arctic strain exhibited a narrower pH range (6.5–8.5) for growth versus a pH range of 6.5–9.5 for the Antarctic strain and a somewhat higher temperature maximum. Some differences were also noted in carbon source utilization patterns.

Although their 16S rDNA sequences were similar, the *Octadecabacter* strains differed from one another at 11–13 rRNA nucleotide positions. These differences by themselves, however, are not great enough to conclude that they are members of separate species (64). To determine whether they were members of the same species, DNA-DNA reassociation experiments were conducted. The results indicated that they were 42% related to one another by this procedure, less than the >70% value that is used to describe a species (80). Thus, it was concluded that they were not the same bipolar species. The arctic strain was named *Octadecabacter arcticus*, and the Antarctic strain was designated *O. antarcticus*. These results indicate that members of the same genus can be found at both poles.

The second set of strains subjected to biogeographic analysis was a collection of four strains from the CFB genus *Polaribacter*. The two most closely related strains differed in size. One strain ranged from 0.25 to 0.5 μm in diameter, whereas the other was 0.8 to 1.6 μm. Colonies of these strains were pigmented with orange- or salmon-colored carotenoid pigments. Differences were also noted in temperature ranges for growth and carbon source utilization patterns.

The 16S rDNA sequences among all four strains differed by 18–50 nucleotide bases. Differences of 40–50 are sufficiently great to warrant separation of new species, hence giving rise to *Polaribacter irgensii* and *P. franzmanii*, but two of the strains exhibited smaller differences necessitating DNA-DNA reassociation studies. When the four strains were subjected to DNA-DNA hybridization, the highest values found indicated a 34% relatedness between the two most closely related strains. Thus, the interrelatedness of the strains was below that of the species

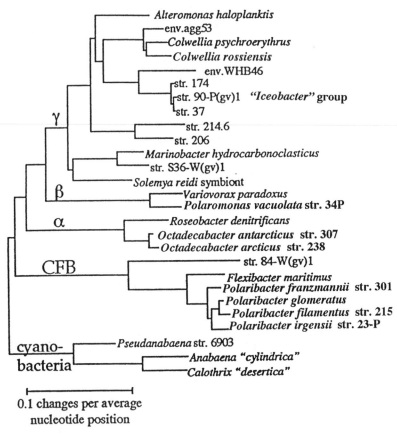

Figure 3   16S rRNA-based phylogeny of gas-vacuolate sea ice bacteria isolated from north and south polar sea ice communities. This tree was produced by using neighbor joining with pair-wise distances determined by a maximum-likelihood estimate. CFB refers to the *Cytophaga-Flavobacterium-Bacteroides* group. The scale bar indicates 0.1 changes per average nucleotide position.

based on this definition, indicating that none of these had a bipolar distribution. One of the south polar strains, *F. glomeratus*, was reclassified as *P. glomeratus*, combination nov., *Polaribacter irgensii* and *P. franzmannii* were from Antarctic sea ice, whereas *P. filamentus* was isolated from arctic sea ice. Again, representatives of the same genus were found at both poles, so this genus also exhibits a bipolar distribution.

The other group of strains closely related by fatty acid analyses was from the gamma proteobacteria. One strain of this as yet officially unnamed genus, "*Iceobacter*" strain 90-P, is from Antarctica, whereas the other two, 174 and 37, are from the Arctic. When these were subjected to 16S rDNA sequence analysis, they differed by 6–11 nucleotides. It is interesting that one of the north-south polar pairs differed by 6 nucleotides and the other by 11. The 2 north polar pairs differed from one another by 6 nucleotides.

Although these pairs have not yet been subjected to DNA-DNA reassociation, major phenotypic differences exist between the north and south polar isolates. The south polar strain has an average length of 2 μm, whereas the north polar strains range from 6 to 18 μm in length. Strain 90-P grows at temperatures as high as 20°C, whereas the north polar strains do not grow at 15°C or above. Strains 37 and 174 have much more similar carbon source utilization patterns and are non-pigmented, whereas the south polar strain is pink. These major phenotypic differences suggest that the south polar strain is likely a different species.

Our results show that many of the genera of gram-negative bacteria have a bipolar distribution. The results at the species level suggest that the sea ice bacteria we have studied are endemic to the north or south polar SIMCO environments. These results, however, are misleading. Clearly, if we apply the biogeography postulates discussed previously, we do not have a sufficient sample size. Perhaps if 1000 or 100,000 strains were examined, it could be determined whether each of these species is indeed cosmopolitan. Or it may be that there is such phenomenal diversity in the sea ice community that it will be difficult to find cosmopolitan species even though they might exist. Not finding cosmopolitan species does not mean that they do not exist.

Furthermore, if one examines the trees shown in Figure 3, it is apparent that, although the genera cluster together, bipolar grouping does not occur. For example, "*Iceobacter*" strain 37 from the Arctic is more closely related phylogenetically to the Antarctic strain 90-P(gv)1 than it is to strain 174 from the Arctic. Likewise, *Polaribacter filamentus*, an arctic strain, is found within a group of antarctic strains, including *P. glomeratus*, *P. irgensii*, and *P. franzmanii*. We interpret these results as an indication that the 16S rDNA sequences are too highly conserved to allow for the assessment of endemic polar strains and species. As suggested earlier, other less highly conserved genes should be used to construct phylogenetic trees for closely related organisms. In addition, a larger sampling of strains is necessary to determine whether bipolar clades exist.

If we are to apply the biogeographic postulates to our studies, we are led to the conclusion that sufficient data are not yet available to infer that geovars or candidate endemic species occur. These gas-vacuolate strains appear to be indigenous to the sea ice habitat based on (*a*) the high concentrations at which they occur in this community that develops each season and (*b*) the observation that they are not found by using the same cultivation procedures in temperate-zone marine habitats. However, additional closely related strains are needed to ascertain whether specific phylogenetic groups have evolved at each pole. Also, less highly conserved phylogenetic markers are necessary to assess their phylogeny.

CONCLUDING REMARKS

When people think of polar environments, they envisage polar bears, emperor penguins, Weddel seals, and other indigenous animal species. Few know of the abundant microbial life. Those that have heard of the microorganisms know them not as distinctive species but primarily because of the extreme environments in which they live. Two examples are the endolithic microbial communities that live in rocks in Antarctica and the sea ice microbial communities that occupy the frozen expanses of the oceans.

Until recently, bacterial work on sea ice has ignored the composition of the SIMCO. This tendency is surprising considering that the SIMCO occupies 7% Earth's surface on a seasonal basis and that almost all of these bacteria are likely to be novel. Now microbiologists have begun to study the types of bacteria that inhabit these environments. Studies reveal that the SIMCO environment has its own distinct species and phyla and, further, that they compose the most psychrophilic bacteria known. Members of the alpha-, beta-, and gamma-proteobacteria and CFB group have been reported. In addition, representatives of both the low- and high-mole-percent G + C gram-positive bacteria have been found. Several new bacterial genera have been reported, and many others remain to be studied. Some of the gram-negative sea ice bacteria have among the lowest maximum temperatures for growth known, $<7–10°C$ for some strains. Archaea are also known to occupy this remarkable environment.

Biogeographic studies are of paramount importance in bacteriology for understanding the diversity, the evolution, the speciation, and the niche concept as well as the existence of threatened species. Several postulates are proposed that may be helpful in guiding future research in bacterial biogeographic investigations.

North and south polar sea ice communities provide a special test case for bacterial dissemination for two reasons. First, the long distances between the polar sea ice communities mean that it is difficult to transport organisms from one pole to the next. Second, many of the obligate psychrophilic bacteria from the SIMCO do not survive at temperatures $\geq 20°C$ and would have difficulty surviving transequatorial passage. Indeed, preliminary studies of biogeography have not found bipolar species of those organisms studied. Additional work must be conducted before it is possible to determine whether these bacteria are endemic to one or the other pole.

ACKNOWLEDGMENTS

This work was supported in part by National Science Foundation grants (BSR 9006788 and DPP 8415069) to JTS and a National Research Service Award through the National Institute of General Medical Sciences (2T32 GM07270) to JJG. We appreciate the support of colleagues in this work, including Karen Junge, who studied the gram-positive sea ice bacteria, and Ann Auman, who helped characterize the "*Iceobacter*" strains. We appreciate the helpful comments of Gordon Orians and Brian Hedlund.

Visit the Annual Reviews home page at http://www.AnnualReviews.org

LITERATURE CITED

1. Amann RI, Ludwig W, Schleifer KH. 1995. Phylogenetic identification and in situ detection of individual microbial cells without cultivation. *Microbiol. Rev.* 59:143–69

2. American Academy of Microbiology. 1997. *The Microbial World: Foundation of the Biosphere.* Washington, D.C.: Am. Soc. Microbiol. 32 pp.

3. Arrigo KR, Worthen DL, Lizotte MP, Dixon P, Dieckmann G. 1997. Primary production in antarctic sea ice. *Science* 276:394–97

4. Baas-Becking LGM. 1934. *Geobiologie of Inleiding Tot de Milieukunde.* The Hague: Van Stockum & Zoon. 263 pp.

5. Beeder J, Nielsen RK, Rosnes JT, Torsvik T, Lien T. 1994. *Archaeoglobus fulgidus* isolated from hot North Sea oil field waters. *Appl. Environ. Microbiol.* 60:1227–34

6. Beijerinck MW. 1913. De infusies en de ontdekking der backteriën. In *Jaarboek van de Koninklijke Akademie v. Wetenschappen.* Amsterdam: Müller

7. Bowman JP, Gosink JJ, McCammon SA, Lewis TE, Nichols DS, et al. 1998. *Colwellia demingiae*, sp. nov., *Colwellia hornerae*, sp. nov., *Colwellia rossensis*, sp. nov., and *Colwellia psychrotropica*, sp. nov., psychrophilic Antarctic species with the ability to synthesize docosaheaenoic acid (22:6ω3). *Int. J. Syst. Bacteriol.* 48:1171–80

8. Bowman JP, McCammon SA, Brown MV, McMeekin TA. 1997. Diversity and association of psychrophilic bacteria in Antarctic sea ice. *Appl. Environ. Microbiol.* 63:3068–78

9. Bowman JP, McCammon SA, Brown MV, Nichols PD, McMeekin TA. 1997. *Psychroserpens burtonensis* gen. nov., sp. nov., and *Gelidibacter algens* gen. nov., sp. nov., psychrophilic bacteria isolated from Antarctic lacustrine and sea ice habitats. *Int. J. Syst. Bacteriol.* 47:670–77

10. Bowman JP, McCammon SA, Lewis T, Skerrat JH, Brown JL, et al. 1998. *Psychroflexus torquis* gen. nov., sp. nov., a psychrophilic species from Antarctic sea ice, and reclassification of *Flavobacterium gondwanense* (Dobson et al, 1993) as *Psychroflexus gondwanense* gen. nov. comb. nov. *Microbiology* 144:1601–9

11. Brown JH, Gibson AC. 1983. *Biogeography.* St. Louis, MO: Mosby. 643 pp.

12. Castenholz RW. 1996. Endemism and biodiversity of thermophilic cyanobacteria. *Nova Hedwigia, Beiheft* 112:33–47

13. Crowley TJ. 1986. Paleoclimatic modelling. In *Physically Based Modelling and Simulation of Climate and Climatic Change (Part 2)*, ed. ME Schlesinger, pp. 883–950. Dordrecht, The Netherlands/Norwell, MA: Kluwer Academic

14. Daly KL. 1990. Overwintering development, growth, and feeding of larval *Euphausia superba* in the antarctic marginal ice zone. *Limnol. Oceanogr.* 37:1564–76

15. Delille D, Rosiers C. 1995. Seasonal changes of Antarctic marine bacterioplankton and sea ice bacterial assemblages. *Polar Biol.* 16:27–34

16. DeLong EF. 1998. Archaeal means and extremes. *Science* 20:542–43

17. DeLong EF, Wu KY, Prezelin BB, Jovine RVM. 1994. High abundance of archaea in Antarctic marine picoplankton. *Nature* 371:695–97

18. Dobson SJ, Colwell RR, McMeekin TA, Franzmann PD. 1993. Direct sequencing of the polymerase chain reaction-amplified 16S rRNA gene of *Flavobacterium*

gondwanense sp. nov. and *Flavobacterium salgenes* sp. nov., two new species from a hypersaline antarctic lake. *Int. J. Syst. Bacteriol.* 43:77–83

19. Ducklow HW. 1983. Productivity and fate of bacteria in the oceans. *BioScience* 33:494–501

20. Ferris MJ, Ruff-Roberts AL, Kopczynski ED, Bateson MM. 1996. Enrichment culture and microscopy conceal diverse thermophilic *Synechococcus* populations in a single hot spring microbial mat habitat. *Appl. Environ. Microbiol.* 62:1045–50

21. Fox GE, Wisotzkey JD, Jurtshuk P Jr. 1992. How close is close: 16S rRNA sequence identity may not be sufficient to guarantee species identity. *Int. J. Syst. Bacteriol.* 42:166–70

22. Fritsen CH, Lytle VI, Ackley SF, Sullivan CW. 1994. Autumn bloom of Antarctic pack-ice algae. *Science* 266:782–84

23. Fuerst JA. 1995. The planctomycetes: emerging models for microbial ecology, evolution, and cell biology. *Microbiology* 141:1493–1506

24. Fuhrmann JA, McCallum K, Davis AA. 1993. Phylogenetic diversity of subsurface marine microbial communities from the Atlantic and Pacific oceans. *Appl. Environ. Microbiol.* 59:1294–1302

25. Fulthorpe RR, Rhodes AN, Tiedje JM. 1998. High levels of endemicity of 3-chlorobenzoate-degrading soil bacteria. *Appl. Environ. Microbiol.* 64:1620–27

26. Garcia-Pichel F, Prufert-Bebout L, Muyzer G. 1996. Phenotypic and phylogenetic analyses show *Microcoleus chthonoplastes* to be a cosmopolitan cyanobacterium. *Appl. Environ. Microbiol.* 62:3284–91

27. Garrison DL. 1991. Antarctic sea ice biota. *Am. Zool.* 4:17–33

28. Garrison DL, Buck KR. 1989. The biota of Antarctic pack ice in the Weddell Sea and Antarctic peninsula regions. *Polar Biol.* 10:211–19

29. Garrison DL, Sullivan CW, Ackley SF. 1986. Sea ice microbial communities in Antarctica. *BioScience* 36:243–50

30. Gosink J, Herwig RP, Staley JT. 1997. *Octadecobacter arcticus*, gen. nov., sp. nov. and *O. antarcticus,* sp. nov., nonpigmented, psychrophilic gas vacuolate bacteria from polar sea ice and water. *Syst. Appl. Microbiol.* 20:356–65

31. Gosink J, Irgens RL, Staley JT. 1993. Vertical distribution of bacteria from arctic sea ice. *FEMS Microbiol. Ecol.* 102:85–90

32. Gosink J, Staley JT. 1995. Biodiversity of gas vacuolate bacteria from Antarctic sea ice and water. *Appl. Environ. Microbiol.* 61:3486–89

33. Gosink JJ, Irgens RL, Staley JT. 1993. Vertical distribution of bacteria in arctic sea ice. *FEMS Microbiol. Ecol.* 102:85–90

34. Gosink JJ, Woese CR, Staley JT. 1998. *Polaribacter* gen. nov., with three new species, *P. irgensii*, sp. nov., *P. franzmannii* sp. nov., and *P. filamentus*, sp. nov., gas vacuolate polar marine bacteria of the Cytophaga/Flavobacterium/Bacteroides Group and reclassification of "*Flectobacillus glomeratus*" as *Polaribacter glomeratus*. *Int. J. Syst. Bacteriol.* 48:223–35

35. Hawkesworth DL. 1991. The fungal dimension of biodiversity: magnitude, significance, and conservation. *Mycol. Res.* 95:641–55

36. Helmke E, Weyland H. 1995. Bacteria in sea ice and underlaying water of the eastern Weddell Sea in midwinter. *Mar. Ecol. Prog. Ser.* 117:269–87

37. Holt JG, Krieg NR, Sneath PHA, Staley JT, Williams ST, eds. 1994. *Bergey's Manual of Determination Bactriology*, 9th Edition. Baltimore, MD: Williams & Wilkins. 787 pp.

38. Horner RA. 1985. *Sea Ice Biota*. Boca Raton, FL: CRC Press. 215 pp.

39. Hugenholz P, Goebel BM, Pace NR. 1988. Impact of culture-dependent studies on the emerging phylogenetic view of bacterial diversity. *J. Bacteriol.* 180:4765–74

40. Irgens RL, Gosink J, Staley JT. 1996. *Polaromonas vacuolata*, nov. gen. et sp., gas vacuolate bacteria from sea waters of Antarctica. *Int. J. Syst. Bacteriol.* 46:822–26

41. Irgens RL, Suzuki I, Staley JT. 1989. Gas vacuolate bacteria obtained from marine waters of Antarctica. *Curr. Microbiol.* 18:262–65

42. Junge K, Gosink J, Hoppe J, Staley JT. 1998. *Arthrobacter, Brachybacterium,* and *Planococcus* isolates identified from Antarctic sea ice brine. Description of *Planococcus mcmeekinii*, sp. nov. *Syst. Appl. Microbiol.* 21:306–14

43. Karl DM. 1993. Microbial processes in the southern oceans. In *Antarctic Microbiology*, ed. EI Friedmann. New York: Wiley-Liss. 634 pp.

44. Kogure K, Simidu U, Taga N. 1979. A tentative direct microscopic method for counting living marine bacteria. *Can. J. Microbiol.* 25:415–20

45. Kottmeier ST, Grossi SM, Sullivan CW. 1987. Sea ice microbial communities. VIII. Bacterial production in annual sea ice of McMurdo Sound, Antarctica. *Mar. Ecol. Prog. Ser.* 35:175–86

46. Kottmeier ST, Sullivan CW. 1987. Later winter primary production and bacterial production in sea ice and seawater west of the Antarctic peninsula. *Mar. Ecol. Prog. Ser.* 36:287–98

47. L'Haridon S, Reysenbach AL, Glénat Prieur D, Jeanthon C. 1995. *Nature* 377:223–24

48. Magot M. 1996. Similar bacteria in remote oil fields. *Nature* 379:681

49. Maycut GA. 1985. The ice environment. In *Sea Ice Biota*, ed. RA Horner. Boca Raton, FL: CRC Press. 215 pp.

50. Mayr E. 1998. Two empires or three? *Proc. Natl. Acad. Sci. USA* 95:9720–23

51. McGuire AJ, Franzmann PD, McMeekin TA. 1987. *Flectobacillus glomeratus*, sp. nov., a curved nonmotile, pigmented bacterium isolated from Antarctic marine environments. *Syst. Appl. Microbiol.* 9:265–72

52. Morita RY. 1975. Psychrophilic bacteria. *Bacteriol. Rev.* 29:144–67

53. Nakagawa Y, Sakane T, Yokota A. 1996. Emendation of the genus *Planococcus* and transfer of *Flavobacterium okeanokioites* (Zogell & Upham 1944) to the genus *Planococcus* as *Planococcus okeanokoites* comb. nov. *Int. J. Syst. Bacteriol.* 46:866–70

54. Nichol S, Allison I. 1997. The frozen skin of the southern ocean. *Am. Sci.* 85:426–39

55. Pace NR, Angert ER, DeLong EF, Schmidt TM, Wickham GS. 1993. New perspective on the natural microbial world. In *Industrial Microorganisms: Basic and Applied Molecular Genetics*, ed. RH Baltz, GD Hegeman, PL Skatrud, pp. 77–83. Washington, DC: Am. Soc. Microbiol.

56. Pace NR, Stahl DA, Lane DJ, Olsen GJ. 1986. The analysis of natural microbial populations by ribosomal RNA sequences. *Adv. Microb. Ecol.* 9:1–55

57. Palmisano AC, Garrison DL. 1993. Microorganisms in Antarctic sea ice. In *Antarctic Microbiology*, ed. EI Friedmann, pp. 167–219. New York: Wiley-Liss

58. Palys T, Nakamura LK, Cohan FM. Discovery and classification of ecological diversity in the bacterial world: the role of DNA sequence data. *Int. J. Syst. Bacteriol.* 47:1145–56

59. Priscu JC, Downes MT, Priscu LR, Palmisano AC, Sullivan CW. 1990. Dynamics of ammonium oxidizer activity and nitrous oxide (N_2O) within and beneath Antarctic sea ice. *Mar. Ecol. Prog. Ser.* 62:37–46

60. Rotert KR, Toste AP, Steiert JG. 1993. Membrane fatty acid analysis of Antarctic bacteria. *FEMS Microbiol.* 114:253–58

61. Schut F, de Vries EJ, Gottschal JC, Robertson BR, Harderw, et al. 1993. Isolation of typical marine bacteria by dilution culture: growth, maintenance, and characteristics of isolates under laboratory conditions. *Appl. Environ. Microbiol.* 59:2150–60

62. Shiba T. 1991. *Roseobacter litoralis*, gen. nov., sp. nov. and *Roseobacter denitrificans* sp. nov., aerobic pink-pigmented bacteria which contain bacteriochlorophyll a. *Syst. Appl. Microbiol.* 14:140–45

63. Sogin ML, Morrison HG, Hinkle G, Silberman JD. 1996. Ancestral relationships of the major eukaryotic lineages. *Microbiolog. Soc. Eur. Microbiol.* 12:17–28

64. Stackebrandt E, Goebel BM. 1994. Taxonomic note: a place for DNA-DNA reassociation and 16S rRNA sequence analysis in the present species definition in bacteriology. *Int. J. Syst. Bacteriol.* 44:846–49

65. Staley JT. 1980. The gas vacuole: an early organelle of prokaryote motility? *Orig. Life* 10:111–16

66. Staley JT. 1997. Biodiversity: Are microbial species threatened? *Curr. Opin. Biotechnol.* 8:340–45

67. Staley JT, Gosink JJ, Hedlund BP. 1996. New bacterial taxa from polar sea ice communities and culture collections. In *Culture Collections to Improve the Quality of Life*, ed. RA Samson, JA Stalpers, D van der Mei, AH Stouthamer, pp. 114–18. The Netherlands: Ponsen and Looyen, Wageningen

68. Staley JT, Gosink J, Irgens RL, Van Neerven ARW. 1994. Gas vacuolate heterotrophic bacteria. In *Trends in Microbial Ecology*, ed. R Guerrero, RC Pedros-Alio, pp. 527–30. Barcelona: Span. Soc. Microbiol.

69. Staley JT, Irgens RL, Herwig RP. 1989. Gas vacuolate bacteria found in Antarctic sea ice with ice algae. *Appl. Environ. Microbiol.* 55:1033–36

70. Staley JT, Konopka AL. 1985. Measurement of in situ activities of heterotrophic microorganisms in terrestrial habitats. *Annu. Rev. Microbiol.* 39:321–46

71. Staley JT, Konopka AL, Dalmasso JP. 1987. Spatial and temporal distribution of *Caulobacter*, spp. in two mesotrophic lakes. *FEMS Microbiol. Ecol.* 45:1–6

72. Staley JT, Lehmicke L, Palmer FE, Peet R, Wissmar RC. 1982. Impact of Mt. St. Helens' eruption on bacteriology of lakes in blast zone. *Appl. Environ. Microbiol.* 43:664–70

73. Stetter KO, Huber R, Blöchl E, Kurr M, Eden RD, et al. 1993. Hyperthermophilic archaea are thriving in deep North Sea and Alaskan oil reservoirs. *Nature* 365:743–45

74. Sullivan CW. 1985. Sea ice bacteria: reciprocal interactions of the organisms and their environment. In *Sea Ice Biota*, ed. RA Horner, pp. 22–79. Boca Raton, FL: CRC Press

75. Sullivan CW, Palmisano AC. 1984. Sea ice microbial communities: distribution, abundance, and diversity of ice bacteria in McMurdo Sound, Antarctica, in 1980. *Appl. Environ. Microbiol.* 47:788–95

76. Suzuki MT, Rapp'e MS, Haimberger ZW, Winfield H, Adair N, et al. 1997. Bacterial diversity among small-subunit rRNA gene clones and cellular isolates from the same seawater sample. *Appl. Environ. Microbiol.* 63:983–89

77. Walsby AE. 1994. Gas vesicles. *Microbiol. Rev.* 58:94–144

78. Ward DM. 1998. A natural species concept for prokaryotes. *Curr. Opin. Microbiol.* 1:271–77

79. Ward DM, Bateson MM, Weller R, Ruff-Roberts AL. 1992. Ribosomal RNA analysis of microorganisms as they occur in nature. *Adv. Microbial Ecol.* 12:219–86

80. Wayne LG, Brenner DJ, Colwell RR, Grimont PAD, Kandler O, et al. 1987. Report of the ad hoc committee on reconciliation of approaches to bacterial systematics. *Int. J. Syst. Bacteriol.* 37:463–64

81. Weeks WF, Ackley SF. 1982. The growth, structure, and photosynthesis of sea ice. *Cold Reg. Res. Eng. Lab. Monogr.* 82(l):1–131

82. Weisburg WG, Barns SM, Pelletier DA, Lane DJ. 1991. 16S ribosomal DNA am-

plification for phylogenetic study. *J. Bacteriol.* 173:697–703

83. Wilson EO. 1992. *The Diversity of Life.* Cambridge, MA: Belknap Press

84. Woese CR. 1987. Bacterial evolution. *Microbiol. Rev.* 51:221–71

85. Woese CR. 1994. There must be a prokaryote somewhere: microbiology's search for itself. *Microbiol. Rev.* 58:1–9

86. Zdanowski MK, Donachie SP. 1993. Bacteria in the sea-ice zone between Elephant Island and the South Orkneys during the Polish sea-ice zone expedition (December 1988 to 1989). *Polar Biol.* 13:245–54

Annu. Rev. Microbiol. 1999. 53:217–44

DNA UPTAKE IN BACTERIA

David Dubnau

Public Health Research Institute, New York, NY 10016;
e-mail: dubnau@phri.nyu.edu

Key Words   competence, transformation, DNA transport, type-2 secretion, pilus formation

■ **Abstract**   Natural competence is widespread among bacterial species. The mechanism of DNA uptake in both gram-positive and gram-negative bacteria is reviewed. The transformation pathways are discussed, with attention to the fate of donor DNA as it is processed by the competent cell. The proteins involved in mediating various steps in these pathways are described, and models for the transformation mechanisms are presented. Uptake of DNA across the inner membrane is probably similar in gram-positive and gram-negative bacteria, and at least some of the required proteins are orthologs. The initial transformation steps differ, as expected, from the presence of an outer membrane only in the gram-negative organisms. The similarity of certain essential competence proteins to those required for the assembly of type-4 pili and for type-2 protein secretion is discussed. Finally several hypotheses for the biological role of transformation are presented and evaluated.

CONTENTS

0066-4227/99/1001-0217$08.00

217

DEFINITIONS AND SCOPE

Natural competence is a genetically programmed physiological state permitting the efficient uptake of macromolecular DNA. It is distinct from artificial transformation involving electroporation, protoplasts, and heat shock/$CaCl_2$ treatment. This review deals only with natural competence. Recombination is not discussed. In many bacteria, competence is highly regulated, and much research has been devoted to exploring the complex control mechanisms involved. This regulatory work is not included here, and discussion is limited to the mechanism of DNA uptake and to the evolutionary significance of competence.

THE UPTAKE PATHWAY

In gram-positive bacteria, DNA must pass through the cell wall and the cytoplasmic membrane. In gram-negative bacteria, DNA must also traverse the outer membrane. Additional steps must therefore be involved in the gram-negative transformation systems, and the initial interaction of DNA with the cell surface must be different in the two types of bacteria. We begin with a description of the transformation pathway in the gram-positive organisms *Bacillus subtilis* and *Streptococcus pneumoniae* and then describe the gram-negative systems, with emphasis on *Neisseria gonorrhoeae* and *Haemophilus influenzae*. These initial descriptions are restricted to the fate of donor DNA. Discussion of the proteins involved is presented later. A summary of the transformation pathways is provided in Figure 1. Because these pathways were described in the earlier literature and have been extensively reviewed (32, 61, 76, 88, 128), they are presented briefly here.

Uptake of DNA in Gram-Positive Bacteria

Binding of DNA   The first step in transformation is the binding of double-stranded DNA to the cell surface. In *B. subtilis* it has been estimated that there are ~50 binding sites per competent cell (36, 126). Binding occurs with no detectable lag (36, 81) and without base sequence preference. Immediately after binding, intact double-stranded DNA can be recovered (38). In *S. pneumoniae* there also appears to be no base sequence specificity for transformation. It has been determined that there are 33–75 uptake sites per colony-forming unit in this organism (45). In both organisms the mass of DNA bound to the cell is proportional to the molecular size of the DNA (33, 69); binding occurs to a fixed number of sites with a probability that is independent of molecular weight.

Fragmentation   Within 30 sites of binding to *B. subtilis*, double-stranded DNA fragments are recovered that have been produced by limited cleavage of the initially bound donor molecules (7, 33, 35). These fragments are still completely accessible to added DNase. When either phage T7 DNA or higher-molecular-weight chromosomal donor DNA was used, a population of molecules with a number

Gram Positive   Gram Negative

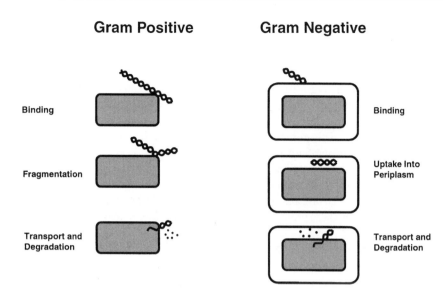

Figure 1   Transformation pathways in gram-positive (*Bacillus subtilis* and *Streptococcus pneumoniae*) and gram-negative (*Haemophilus influenzae* and *Neisseria gonorrhoeae*) bacteria.

average molecular size of ~18 kilobases (kb) was recovered from the cell surface (33, 38). In another study the estimate for the size of the surface-localized donor fragments was 13.5 kb (35). In *S. pneumoniae*, fragmentation on the cell surface has also been reported (70, 71, 98). In this organism, single-strand nicks occur first, and double-strand breaks occur subsequently. The average length between nicks is 6 kb, and that between double-strand breaks is about the same (99), somewhat less than for *B. subtilis*. It has been suggested for *B. subtilis* that the distance between binding/fragmentation sites and the stiffness of the DNA determine the distance between cuts (32). In *S. pneumoniae* the nicking step may be concomitant with binding (70). It is not known whether nicking precedes double-strand cleavage in *B. subtilis*.

It is usually assumed that uptake proceeds from a newly formed end and not from pre-existing ends. As the size of donor DNA increases, it becomes less likely that binding will occur very close to a pre-existing end. It follows that, if uptake were to proceed only from pre-existing ends, the search time needed to encounter these ends might place an unacceptable kinetic limitation on transformation. Cell-surface cleavage may therefore serve to generate DNA ends near cellular uptake sites.

Transport   In this review the term transport refers to the process of uptake across the (inner) cell membrane. Homologous single-stranded donor DNA integrates to form a heteroduplex consisting of base-paired donor and recipient strands (11, 44, 132). In both *B. subtilis* (28, 35, 106) and *S. pneumoniae* (68, 72), precursor single-stranded donor DNA can be recovered from transformed cells. Single strands are transported across the membrane, whereas the non-transported strand is degraded.

In *B. subtilis*, after a lag of 1 to 1.5 min at 37°C, bound donor chromosomal DNA becomes inaccessible to added DNase (81). At the same time, single-stranded DNA of donor origin can be recovered from lysed cells (28, 106). These apparently concomitant events indicate that the single-stranded DNA has been transported into the cytosol. The average length of these single-stranded fragments is 6–15 kb (28). This compares reasonably with the length of the double-stranded fragments recoverable from the cell surface (13.5–18 kb). These single strands have not yet formed stable base-paired complexes with the recipient chromosome. Simultaneously with the appearance of internalized single-stranded DNA and the acquisition of DNase resistance, acid-soluble degradation products appear in the medium (35). These products consist of 5'-nucleotides, nucleosides, and free bases. Because nucleotides cannot ordinarily pass freely across membranes, it is likely that the nuclease responsible for this degradative step is localized outside the membrane or within an aqueous channel. The internalized strands interact efficiently with complementary sequences in the recipient chromosome to yield heteroduplex DNA. An unligated complex in which the donor and recipient moieties are stabilized only by base pairing has been detected (6, 37). The average size of the donor moiety in this transient complex was estimated as 8.3 kb. The average size of the final integrated donor DNA was 8.5–10 kb (34, 43).

In *B. subtilis* the size correspondence between the surface-localized fragments, the internalized single strands, and the non-covalently bound and final integrated donor segments is satisfying. This correspondence is consistent with the hypothesis that these various molecular forms are related as precursors and products and that they are formed from one another in the order listed. Kinetic analysis has established that this is the case for the surface localized fragments, the internalized single strands, and the integrated donor fragments (29, 36). Without such evidence one or more of these forms could be the products of side reactions.

As noted above, single genetic markers become DNase resistant after a 1- to 1.5-min lag during the transformation of *B. subtilis*. However, marker pairs require additional time, which increases with the physical distance between the markers (130, 131). This indicates that DNA is taken up in a linear fashion. Based on this data and on newer sequence information, we estimate that DNA passes into the cytosol at the rate of ∼180 nucleotides/s at 28°C, which was the temperature used for the marker pair uptake experiment. Linear uptake fits nicely with the existence of an aqueous channel for the transport of DNA across the membrane. The nature of the lag preceding transport of a single genetic marker is unclear, although it must include the time needed for both cleavage and for transport across the membrane of the minimal segment adequate for recombination. This should take less than a minute at 28°C. However, the lag measured at this temperature is considerably longer (2.5 min) (130). It has been reported that *B. subtilis* takes up single strands with either 5'–3' or 3'–5' polarity (139). This contrasts with findings in other transformation systems in which transport occurs with 3'–5' polarity.

Transport of DNA in *S. pneumoniae* is similar to that in *B. subtilis* in most respects. Linked pairs require more time for entry than do single markers, again implying linear uptake (49). Single strands appear in the cells (68, 94), and one strand equivalent is released into the medium as low-molecular-weight products (74, 77, 98). These consist of short oligonucleotides (74). It is believed that transport initiates at a second break opposite the initial single-strand nick (70). The polarity of DNA entry in this organism is $3'-5'$ (92). In an important study carried out with radiolabeled linear and circular donor molecules, Méjean & Claverys (93) found that the degradation of the non-transported strand proceeded with $5'-3'$ polarity and at approximately the same rate as the entry of the transported strand. The uptake rate was estimated as 90–100 nucleotides/s at 31°C. It is tempting to conclude from these results that the degradation and transport processes are mechanistically coupled, although this is not certain.

Uptake of DNA in Gram-Negative Bacteria

The *H. influenzae* and *N. gonorrhoeae* uptake pathways are closely related. Efficient DNA uptake by these gram-negative bacteria requires the presence of a specific uptake sequence (25, 27, 39, 50, 127). Uptake sequences are present at frequencies far greater than those expected randomly. Uptake sequences are often found as inverted repeats between genes, acting as transcriptional terminators (129). Inspection of the *H. influenzae* genome revealed an extended 29-base-pair (bp) consensus uptake sequence. However, not all gram-negative competence systems display such specificity; *Acinetobacter calcoaceticus* is able to take up DNA from any source (87, 103). Another major difference between the gram-positive and -negative systems is the presence of an outer membrane in the latter.

Binding   A binding step, corresponding to the initial irreversible attachment of DNA to the cell surface in a form accessible to DNase or to shearing forces, has not been characterized directly in the *H. influenzae* and *N. gonorrhoeae* systems, perhaps because of the rapidity with which DNase resistance is acquired. An uptake rate of 500–1000 nucleotides/s has been measured (61). In *H. influenzae* the number of active sites for DNA uptake is only 4–8 per competent cell (30).

Fragmentation   Circular 11.5-kb plasmid DNA was used to transform *N. gonorrhoeae* (9). After the acquisition of DNase resistance, double-stranded molecules were recovered from the cells, which had been linearized at random positions. When linear plasmid DNA was used, no additional cleavage events were detected. However, when a larger plasmid DNA (42 kb) was used, more extensive double-strand cleavage was observed. This pattern is reminiscent of that observed with the gram-positive systems, and a similar mechanism may exist to introduce ends for subsequent transport across the inner membrane.

DNase Resistance   DNase resistance, used in the gram-positive systems as an index of transport across the cytoplasmic membrane, has a different interpretation in the gram-negative systems. The acquisition of DNase resistance in these systems is not concomitant with the conversion to single-stranded DNA. Instead, DNase resistance probably corresponds to entry into the periplasmic compartment or into a specialized structure. In *H. influenzae*, a particle was observed on the surface of competent cells (61). These "transformasomes" were considered to be structures in which double-stranded donor DNA is sequestered early in the transformation pathway. This important work has not been pursued in recent years, and transformasomes deserve re-evaluation.

Transport   In *H. influenzae*, transformation results in single-strand integration (101), and it is likely that one strand is degraded during transport across the inner membrane (61) just as in the gram-positive systems. However free cytoplasmic single-stranded DNA of donor origin has not been detected in this organism. Instead it is believed that only a short length of single-stranded DNA is present at a given time as the DNA crosses the inner membrane and searches out a complementary sequence in the recipient. The degradation of the nontransported strand may be concomitant with transport and integration. This would imply the existence of a degradation-transport-recombination protein complex associated with the inner membrane. Transport proceeds in a $3'-5'$ direction (8), as it does in *S. pneumoniae*. Recently a low level of single-stranded donor DNA was detected during gonococcal transformation (16). It is not certain that this material is a precursor of integrated DNA.

Acinetobacter *Transformation*   The transformation of the gram-negative *A. calcoaceticus* in some respects resembles that of the gram-positive systems. It was concluded that this organism takes up DNA from any source (103), and an initial binding step resulting in DNase-sensitive association of DNA with the cell surface was identified (107). Moreover, in *A. calcoaceticus* double-stranded molecules are converted to single strands on transport (103).

COMPETENCE PROTEINS

Several genes and proteins required for DNA uptake have been characterized. Many are conserved in the gram-positive and gram-negative systems. These proteins and their individual roles are now described. After these descriptions, an attempt is made to integrate this information with our knowledge of the transformation pathways to construct summarizing models for gram-positive and gram-negative transformation.

DNA RECEPTOR PROTEIN

In *B. subtilis*, several proteins mediate DNA binding to the cell surface. Mutants lacking these proteins are completely deficient in transformation (53); they exhibit $<10^{-7}$ of the wild-type transformation frequency. ComEA is encoded by the first open reading frame of the *comE* operon (54, 60). Non-polar null mutations in *comEA* eliminate DNA binding. An in-frame deletion in *comEA* results in the expression of a shortened protein that is detectable in western blots (60). These mutants are still able to bind DNA about half as well as the wild type, but they fail to transport DNA. Thus ComEA has more than one function in transformation. ComEA possesses a single membrane-spanning domain near its N terminus and is an integral membrane protein with its C terminus outward (60). ComEA is intimately associated with the cell wall, presumably via its C-terminal domain because it can be chemically cross-linked to wall material in vivo (YS Chung & D Dubnau, unpublished data).

A His$_6$-tag was substituted for the membrane-spanning domain of ComEA, and the resulting protein was purified (109). The His$_6$-ComEA protein bound to double-stranded DNA in gel shift assays, with an apparent K_d of ~5 × 10^{-7} M. Membrane proteins from competent cultures were tested for DNA binding in southwestern assays, and it was shown that intact ComEA was capable of binding DNA. In both assays, a marked preference for double- over single-stranded DNA was noted, and binding occurred with no apparent sequence specificity. The C terminus of ComEA shows similarity to other nucleic acid-binding proteins and contains a possible helix-turn-helix motif (60, 109). A purified protein lacking this motif had no detectable DNA-binding activity in the gel shift assay. These properties of ComEA imply that it is a DNA receptor for transformation.

A ComEA ortholog has been detected in *S. pneumoniae* and shown to be essential for transformation (15, 104). Proteins possessing similarity to the C-terminal DNA-binding domain of ComEA are widespread in nature, including those in *H. influenzae* and *N. gonorrhoeae* (109). However, there is no evidence that these proteins are involved in transformation, and, because DNA uptake in these organisms requires a specific sequence, a ComEA ortholog may not be involved. It will be interesting to see whether such a protein is needed by *A. calcoaceticus* in view of the similarity of its transformation pathway to that of the gram-positive organisms.

COMPETENCE PROTEINS RELATED TO THOSE INVOLVED IN TYPE-4 PILUS ASSEMBLY AND SECRETION: The PSTC Proteins

Another group of proteins encoded by the *comG* operon and by *comC* was shown to be required for DNA binding in *B. subtilis* (1, 53, 96). Analysis of non-polar mutants lacking each of the seven ComG proteins demonstrated that they are individually needed for binding (19). In their absence, transformation was undetectable. These proteins resemble a widespread group required in gram-negative bacteria for the assembly of type-4 pili, for the type-2 secretion pathway, and for twitching motility (58). We call them PSTC proteins (for pilus/secretion/twitching motility/competence) (Table 1).

Because the formation of a pilus is a requirement for twitching motility (57), all proteins needed for pilus formation must also be required for twitching motility,

TABLE 1 PSTC proteins required for competence, pullulanase secretion and pilus formation

Class	Competence proteins					Pul[f]	Pil[g]
	B. s.[a]	Strep.[b]	N. g.[c]	H. i.[d]	A. c.[e]		
1	ComGA (1)	CglA (107) CilD1 (16) ComYA (92)	PilT (146)	PilB (54)	N. D.	PulE (111)	PilB (104)
2	ComGB (1)	CglB (107) CilD2 (16) ComYB (92)	PilG (141)	PilC (54)	N. D.	PulF (111)	PilC (104)
3	ComGC (1) ComGD (1) ComGE (1) ComGG (1)	CgC (107) CglD (107) ComYC (92)	PilE (128)	PilA (54)	ComP (110)	PulG (119) PulH (119) PulI (119) PulJ (119)	XcpT (8, 105) XcpU (8, 105) XcpV (8, 105) XcpW (8, 105)
4	ComC (98)	CilC (16) ComYD (92)	PilD (47)	ComE (140)	N. D.	PulO (114)	PilD (104)
5	None	N. D.	PilQ (32)	ComE (140)	N. D.	PulD (58)	PilQ (93)

[a]B. subtilis.
[b]Streptococci.
[c]N. gonorrhoeae.
[d]H. influenzae.
[e]A. calcoaceticus.
[f]Pullulanase secretion representing class 2 secretion.
[g]Pilus assembly in P. aeruginosa representing type 4 pilus assembly.

which is thought to require pilus retraction (12). The reverse is not true; PilT of *N. gonorrhoeae* (143) and *Pseudomonas aeruginosa* (141) is needed for twitching motility but not for pilus formation. PSTC proteins fall into five classes, which are discussed in turn below. The ubiquity of PSTC genes and their use for DNA and protein transport and for pilus assembly in diverse species suggest an ancient mechanism, which predates the divergence of gram-positive and -negative bacteria.

Class 1, exemplified by ComGA in *B. subtilis* (1), consists of membrane-associated proteins with consensus nucleotide-binding sites. Members of this group have been implicated in transformation of both gram-negative (52, 107, 143) and gram-positive bacteria (15, 90, 104) as well as in pilus assembly (102), twitching motility (141, 143), and type-2 protein secretion (108). Class-1 proteins are also required for conjugation-related systems. One of these, TrwD, required for conjugation of the plasmid R388, has been shown to be an ATPase (116). The *B. subtilis* class-1 PSTC protein (ComGA) is located as a peripheral membrane protein on the cytosolic face of the membrane (18).

ComGB of *B. subtilis* exemplifies a second class, consisting of membrane proteins apparently with three predicted membrane-spanning segments. Class-2 proteins have been implicated in pilus assembly (102) and type-2 secretion (108), as well as in competence in gram-positive (1, 15, 90, 104) and gram-negative systems (52, 138).

The third class consists of small proteins with conserved sequences at their N-termini (2–4, 15, 104, 107, 115, 123, 124). Representatives of this class are required for competence, secretion, and pilus assembly. Four of the *B. subtilis* ComG proteins (ComGC, ComGD, ComGE, and ComGG) and two proteins from the streptococcal systems possess such conserved sequences (15, 90, 104), resembling the cleavage sites of type-4 prepilin proteins. The major pilin protein of *N. gonorrhoeae* is a member of this class and is required for transformation (124), as is an *Acinetobacter* protein (107). In nearly all of the systems that have been studied, multiple class-3 proteins are needed. An exception may be *H. influenzae*, in which a single class-3 protein was detected in a search of the genome (D Dubnau, unpublished data). Only for type-4 pilus assembly are the functions of any of these proteins understood; one of them is the precursor of the major structural protein of the pilus. In general the conservation of sequence among the members of this class is restricted to the hydrophobic N-terminal domain. However members of one subgroup required for competence, consisting of the *B. subtilis* ComGC protein, ComYC from *S. gordonii* (90), and CglC from *S. pneumoniae* (15, 104), are similar over their entire lengths.

The four *B. subtilis* class-3 pre-proteins (ComGC, ComGD, ComGE, and ComGG) are processed by the peptidase, ComC (18, 20). They are integral membrane proteins, anchored by their single predicted N-terminal membrane-spanning segments (13, 18, 20). In the absence of ComC, the pre-proteins are organized with their C-termini outside the membrane, except for ComGG, which is accessible to proteinase K only in inside-out vesicles (18). On processing, some molecules of

ComGC, ComGD, ComGG, and possibly ComGE, undergo translocation and are no longer located as integral membrane proteins (18, 20). This requires the peptidase (ComC) and is probably dependent on processing. Consequently about one fourth of each mature class-3 protein is peripherally associated with the outer face of the membrane, and another fourth of the total is released from the protoplast on removal of the cell wall. We have recently found that these proteins can be efficiently cross-linked in vivo to cell wall material (YS Chung & D Dubnau, unpublished data). Additional cross-linking experiments have shown that portions of both the mature and unprocessed ComGC are probably present as homodimers. The single cysteine residue of ComGC is apparently involved in vivo in an intramolecular disulfide bond. A portion of ComGG, on the other hand, is in a homodimer that is stabilized by a disulfide bond (18).

The fourth class of proteins in this group consists of membrane-localized peptidase/transmethylases. These cleave the class-3 proteins at a site within the N-terminal conserved sequence and, in at least some cases, transfer a methyl group from S-adenosyl methionine to the newly formed primary amino group at the N-terminus (133). Processing of the class-3 *B. subtilis* competence proteins is dependent on the peptidase ComC (18, 20).

The class-5 PSTC proteins are the secretins, which exhibit sequence similarities in their C-terminal domains (121). Secretins are required for competence (31, 136), pilus assembly (31, 91), and type-2 secretion (24). Secretins are also involved in filamentous phage maturation (122) and in the so-called type-3 secretion systems (62). These proteins are located in the outer membrane, where they form large multisubunit complexes (17, 56, 65, 100, 125). Protein pIV, the best-studied secretin, is required for the maturation of the filamentous phages. pIV forms a cylindrical complex, composed of ~14 identical subunits, with a central pore (82). Other secretins form similar structures (10, 23, 66). Secretins are almost certainly involved in permitting passage of phage particles, DNA, protein, or pili across the outer membrane. It is not surprising that they have been implicated in gram-negative (31, 136) but not in gram-positive transformation systems. A search of the completed *B. subtilis* genome failed to indicate the presence of a secretin (D Dubnau, unpublished data).

An additional protein, PilC, is required for competence in *N. gonorrhoeae* (119). Although PilC is pilus-associated, it is not absolutely required for pilus production (120) and does not resemble any of the PSTC proteins described above. Interestingly, competence can be partially restored to a *pilC* null mutant by the addition of purified PilC protein or, even better, by the addition of crude preparations of pili-containing PilC (119). These results strongly suggest that PilC functions on the cell surface. In *A calcoaceticus*, a PilC ortholog has also been identified as a competence protein (83). In both organisms this protein is required for DNA uptake to a DNase-resistant state (83, 119). In its absence, DNA binding to the *A. calcoaceticus* cell surface was also reduced, but no effect of the *pilC* null mutation on twitching motility or pilus formation was observed. A search of the

B. subtilis genome failed to reveal the presence of a PilC ortholog (D Dubnau, unpublished data).

ROLE OF THE PSTC PROTEINS

In *B. subtilis*, each of the PSTC proteins is required for the initial binding of DNA to the cell surface (13, 19). ComGF is also required for binding in *B. subtilis*, but has no known ortholog. Southwestern blotting experiments with competent cell membrane proteins, and gel shift assays with purified ComGC, ComGD, ComGE, and ComGG, failed to detect any evidence that these are DNA-binding proteins (R Provvedi & D Dubnau, unpublished data). Right side-out membrane vesicles prepared from competent cells are able to bind double-stranded DNA in a ComEA-dependent reaction (109). Although binding of DNA to intact cells requires each of the ComG proteins, binding in the vesicle system is unaffected by their absence. These experiments indicate a role for the PSTC proteins in providing access to the ComEA receptor. The cell wall in *B. subtilis*, as in most gram-positive organisms, is complex and quite thick (5). In addition to peptidoglycan, the cell wall contains an abundance of teichoic acid polyanions. DNA must therefore breach both physical and electrostatic barriers. The ComG proteins may modify the wall to increase porosity and locally shield or remove negative charges. Alternatively the pilin-like proteins may form a tunnel that traverses the wall and offers access to incoming DNA. These two modes of action are not mutually exclusive because the construction of such a channel may require cell wall modification.

In *A. calcoaceticus*, the PSTC class 3 protein so far identified is also required for binding to the cell surface (107). Knockout of the streptococcal PSTC loci confers severe transformation deficiencies (15, 90, 105). It is likely that the PSTC proteins play analogous roles in these systems.

A COMPETENCE NUCLEASE

In *S. pneumoniae*, the EndA protein is required for transport and for degradation of the nontransported strand but not for binding to the cell surface (73). It may also be needed for cleavage opposite the initial nick on the cell surface, but it is not needed to introduce the initial nick (70). EndA is an endonuclease that acts on both RNA and DNA (117). EndA is localized in the membrane and possesses an uncleaved signal sequence that probably serves to anchor the protein in the membrane, with its bulk facing outward (75, 111, 118). It was suggested that EndA is fixed asymmetrically near a pore and that degradation of one strand may occur by successive endonucleolytic cleavages as DNA moves past the active site of the nuclease (118). This mechanism is consistent with the evidence described above suggesting that entry and degradation are coupled (93). In fact EndA was shown to be present in a large membrane-localized complex, in association with uncharacterized proteins (118).

No protein with nuclease activity required for transport has been identified in other competence systems. A search of the complete *B. subtilis* genome sequence failed to reveal an EndA ortholog (D Dubnau, unpublished data). It is worth pointing out that, unlike all other known competence proteins in the gram-positive systems, EndA is constitutively expressed and represents the major endonuclease of *S. pneumoniae* (74). It is possible that each competence system has recruited its own non-dedicated nuclease.

COMPETENCE PROTEINS REQUIRED FOR TRANSPORT

It was noted above that ComEA plays a part in transport in addition to its role as the DNA receptor protein. Another protein encoded by the *comE* operon, ComEC, is required for transport but is dispensable for binding (53, 60). ComEC is predicted to be a polytopic membrane protein with >6 membrane-spanning segments. This complex topology and its role in DNA transport suggest that ComEC may form all or part of an aqueous channel. ComEC orthologs have been shown to play essential roles in transformation of *S. pneumoniae* (15, 104), *H. influenzae* (22), and *N. gonorrhoeae* (41). In *N. gonorrhoeae* a mutation in the *comEC* ortholog did not interfere with DNA binding or with the acquisition of DNase resistance, in contrast to PSTC mutations, which prevented DNA binding (40). It is likely that this mutation permitted entry into the periplasm but not into the cytosolic compartment, consistent with a common role for the *B. subtilis* and *N. gonorrhoeae* proteins.

The first ORF of the *B. subtilis comF* operon is also required for transport but not for binding (85). ComFA resembles members of the DEAD family of helicases and has a strong similarity to PriA (84), an *Escherichia coli* helicase that translocates along single-stranded DNA in the $3'-5'$ direction, by using the energy of ATP hydrolysis (80). ComFA is located in the membrane fraction of competent cells, and its solubility properties suggest that it is an integral membrane protein (85). ComFA is accessible to proteinase K only from the cytoplasmic face of the membrane. ComFA possesses a consensus nucleotide-binding sequence. Point mutations in this sequence confer the same phenotype as the loss-of-function mutation, namely an absence of DNA transport and no effect on binding (86). Interestingly some of these mutations confer dominant negative phenotypes, implying that ComFA exists in a homo- or heteromultimer. Overexpression of the wild-type ComFA protein also decreases DNA transport markedly (85). An excess of ComFA over other competence proteins may result in the formation of aberrant non-functional complexes. ComFA may be a membrane-localized DNA transporter protein that hydrolyzes ATP and translocates DNA. In the cell ComFA may be associated with ComEC and ComEA, other components of the transport apparatus. Transformation is decreased about 1000-fold in null mutants of *comFA*

(53, 84). In contrast, knockouts of other *B. subtilis* competence genes decrease transformation at least 10^7-fold. In the absence of ComFA, the other competence proteins may be able to position the end of an incoming strand near the aqueous channel. Entry might then occur inefficiently via diffusion, perhaps directionally biased by interaction with single-strand binding protein (SSB) in the cytosol. An SSB ortholog (CilA) is required for transformation in *S. pneumoniae* (15). A *comF*-like operon preceded by a competence regulatory signal [a *cin* box (15)] was also detected in *S. pneumoniae* (21). A PriA ortholog was identified in *H. influenzae* (42), and proteins with similarity to ComFA were also detected in a search of the incomplete *N. gonorrhoeae* and *N. meningitidis* databases (D Dubnau, unpublished data). No data exist concerning the possible roles of any of these proteins in transformation, and it is uncertain whether a ComFA equivalent is an essential competence protein in the gram-negative systems.

The *comFA* and *comEC* null mutations and the in-frame deletion of *comEA*, which prevents transport without a major effect on binding, also prevent the release of acid-soluble products from radiolabeled donor DNA (R Provvedi & D Dubnau, unpublished data). These results suggest that transport may be necessary for degradation, although the nuclease is most likely located outside the membrane, consistent with the appearance of the acid soluble products in the medium (35). Transport may serve to drag the incoming DNA past the nuclease active site.

ADDITIONAL COMPETENCE PROTEINS

Por, a periplasmic protein disulfide oxidoreductase, is required for the transformation of *H. influenzae* (137). In its absence, DNA binding does not occur. Por is located in the periplasm and is required for competence-associated changes in the protein composition of the membrane. It is likely that Por is needed for the correct folding of one or more competence proteins. DprA is another protein required for transformation in *H. influenzae* (63, 64), as well as in *S. pneumoniae* (15) and *B. subtilis* (P Tortosa & D Dubnau, unpublished data). In *Haemophilus*, the *dprA* mutant took up DNA into a DNase-resistant form as readily as the wild type, but transformation was severely depressed. DprA may be needed to transport DNA across the inner membrane.

As noted above, SSB (CilA) has been shown to be a competence protein in *S. pneumoniae* (15). A deletion mutant lacking the 33 C-terminal amino acid residues exhibited 10% of the wild-type transformability. It is possible that SSB plays a supporting role in uptake. For instance, as the 3′ donor DNA terminus enters the cytosol, SSB may bind, facilitating transport and formation of a DNA-RecA filament before integration. In the absence of SSB, transport and filament formation might still occur with reduced efficiency.

Another gene, *comFC* (84), has been described in *B. subtilis*. The product of this gene resembles *com-101* (78, 79) [also identified as *orfF* (136)] in *H. influenzae*.

Inactivation of ComFC in *B. subtilis* decreases transformation only 5- to 10-fold (84), does not affect DNA binding, and decreases transport only slightly (85). This protein may affect a step that follows transport. In *H. influenzae*, a knockout mutant of *com-101* bound DNA as well as the wild type, but it released much less donor DNA in acid-soluble form (78). It is difficult to reconcile this phenotype with the one observed in *B. subtilis*.

Two competence genes in *N. gonorrhoeae* may be involved in murein metabolism. *comL* encodes a protein that is associated with and perhaps covalently attached to peptidoglycan (46). Complete loss of *comL* is lethal, but a pleiotropic mutation was characterized that decreased both competence and cell volume. It appears that the N-terminal half of ComL, a lipoprotein, is sufficient to support viability but not competence. No ComL ortholog is encoded by *B. subtilis* (D Dubnau, unpublished data), although a similar protein (HI0177) is encoded by *H. influenzae* (42). Inactivation of *tpc*, another *N. gonorrhoeae* gene, is also pleiotropic, resulting in transformation deficiency and a cell separation defect (47). Extracts of a *tpc* mutant exhibited slightly decreased murein hydrolase activity. No obvious ortholog of Tpc exists in the database. It was suggested that ComL and Tpc might be murein hydrolases, responsible for providing access for DNA through the wall (46, 47).

ENERGETICS OF DNA UPTAKE

In *B. subtilis*, transformation is inhibited by a number of uncoupling reagents, and the extent of inhibition paralleled the effect of these reagents on membrane potential (51). Inhibition occurred at or before the transport step because the acquisition of DNase resistance was inhibited. The addition of arsenate to the transformation medium halved the ATP pool but had no effect on transformation. It was concluded that the "... initial stages of genetic transformation proceeded despite the drastically lowered value of the intracellular phosphorylation potential" (51). However, this conclusion must be questioned, because measurement of intracellular ATP was carried out with the bulk culture, whereas only $\sim$10% of the cells in a *B. subtilis* culture become competent for transformation. It was also concluded that both the pH gradient and the membrane potential are required for DNA uptake (51). Grinius therefore proposed that DNA binds to the cell surface, forming a nucleoprotein complex that binds protons and acquires a positive charge. The complex then electrophoreses across the membrane, releasing the protons. However, another study with inhibitors (140) has concluded that ΔpH alone is required for DNA transport in *B. subtilis*, suggesting a proton symport mechanism for DNA uptake. In *S. pneumoniae*, it was proposed that degradation of one strand provides the driving force for introduction of the intact strand (118). However, this model does not readily accommodate the observation that single-strand transformation occurs efficiently in many systems. Single-strand transformation does not use a totally different transport pathway because in *B. subtilis* it requires the competent state (D Dubnau, unpublished data). In *H. influenzae*, it was concluded that both components of the proton motive force could drive DNA uptake (14).

Although transformation requires the input of metabolic energy, no clear picture exists as to the proximal source of this energy. Because all the relevant experiments so far involve the action of inhibitors on whole cells, the available information is indirect and subject to various interpretations. For instance, is proton motive force required to drive DNA transport directly or to maintain the active state of an essential competence protein?

MODELS FOR DNA UPTAKE

The information presented above can be summarized in the form of working models that ascribe particular roles to some of the competence proteins. Before presenting these models, it is useful to compare transformation to the type-2 secretion systems in gram-negative bacteria (110). In type-2 secretion, an unfolded substrate protein is transported by the Sec system across the inner membrane. These substrate protein molecules are delivered to the periplasm, where they are folded and then transported across the outer membrane. During transformation of gram-negative bacteria, double-stranded DNA must be transported across the outer membrane in the reverse direction. Conversion to single-stranded DNA and transport across the inner membrane ensues by using proteins that are functionally analogous to the Sec proteins. Transport across the inner membrane, probably through water-filled channels, requires that the substrate macromolecules be in particular flexible states—unfolded for proteins and single-stranded for DNA. In the transformation and secretion systems, PSTC proteins appear to be required to provide access through the wall and outer membrane and therefore to fulfill similar roles in all of these systems.

The Gram-Positive Model

Figure 2 presents a working model for the transformation process in gram-positive bacteria. It draws on evidence taken from both *S. pneumoniae* and *B. subtilis*, although most of the information concerning the roles of individual proteins relies on the latter. Double-stranded DNA binds to the C-terminal domain of membrane-anchored ComEA. A nuclease then cleaves the bound DNA near the point of contact with ComEA. The newly formed DNA terminus is then delivered to the membrane-associated transport apparatus, containing ComEC and ComFA. A nuclease that degrades the non-transported strand (EndA in *S. pneumoniae*) is presumably also part of this apparatus. ComEA also plays a role in transport, possibly by contributing structurally to the transport apparatus or perhaps by playing an integral role in the delivery process. In the figure we depict the class-3 PSTC proteins encoded by *comGC, comGD, comGE,* and *comGG* as composing a structure

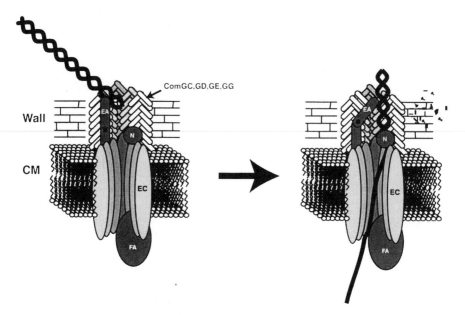

Figure 2 Transformation in gram-positive bacteria. The protein nomenclature is from *Bacillus subtilis*, but it is likely that a similar mechanism operates in *Streptococcus pneumoniae*. Double-stranded DNA binds to the receptor protein ComEA. A double-strand cleavage event is followed by delivery to a nuclease (N) that degrades one strand, releasing acid-soluble products into the medium. The nuclease has not been identified in *B. subtilis*, whereas in *S. pneumoniae* it is encoded by *endA*. The ComG proteins are depicted, forming a structure that provides access of DNA to ComEA through the wall. One possible model for the delivery step is shown in which ComEA conformation is altered so that the bound DNA can contact the nuclease. The single-strand product of nuclease action then contacts the transporter protein ComFA, and the energy of ATP hydrolysis is used to drive the DNA into the cell. We postulate that ComEC forms an aqueous pore in the membrane for this transport step. References and further details are provided in the text.

that traverses the wall, permitting access of DNA to the ComEA receptor. It is possible that these proteins serve instead to increase the porosity of the wall or that the individual ComG proteins play different roles, some increasing porosity and some composing a wall-traversing structure.

The delivery process deserves comment. One possibility is that the DNA-binding domain of ComEA is close to the transport machinery. Another is that some movement is required to establish this proximity. Two suggestive features are consistent with this idea. First, ComEA is similar to the C-terminal domain of Kid, a human kinesin-like DNA-binding protein (135). Kid also possesses a myosin-like N-terminal domain with a nucleotide-binding site that is lacking in ComEA. However, ComGA and ComFA, which reside on the inner face of the membrane (18, 85), contain nucleotide-binding sites. The binding site of ComFA is known to be required for transport (86). Perhaps one of these proteins drives a conformational change in ComEA that delivers DNA to the transport pore. It is interesting that just upstream from the DNA-binding domain of ComEA is a flexible sequence (QQGGGG). A similar sequence is conserved in the *S. pneumoniae* ortholog. When this was deleted, the mutant cells were able to bind but not to transport DNA (60). Perhaps bending of ComEA delivers DNA to the uptake apparatus. A second suggestive phenomenon is that of twitching motility, associated with type-4 pili. It is believed that this form of bacterial motion requires the assembly and disassembly of pili, and an ortholog of ComGA has been implicated in the disassembly process (141). It is possible that the proposed wall-traversing structure disassembles by an analogous process, and that this permits ComEA to bend and thereby deliver DNA to the transport machine. Because disassembly of this structure and delivery of the bound DNA to the uptake machinery would be time dependent, this type of mechanism may explain the lag observed before the acquisition of DNase resistance in *B. subtilis*.

In all of the well-studied transformation systems, a ComEC ortholog is required for DNA transport. These are polytopic membrane proteins and may form uptake pores (Figures 2, 3). In *B. subtilis* and probably in *S. pneumoniae* as well (21), ComFA is a component of this machinery, perhaps mobilizing the energy of ATP hydrolysis for this purpose. As DNA enters the cytosol it probably associates with SSB and RecA, and these binding energies may facilitate uptake.

The Gram-Negative Model

Figure 3 presents a model for transformation in *H. influenzae* and *N. gonorrhoeae*. As noted above, transformation in *A. calcoaceticus* may be different, resembling that of the gram-positive bacteria. In *N. gonorrhoeae* the phenotypes associated with loss-of-function mutations in several competence genes have been analyzed (40, 48). PilC and the secretin protein PilQ are needed for outer membrane transport, the latter to form a pore. The proposed murein hydrolases Tpc and ComL may aid access through the wall. By analogy with the *B. subtilis* results described above, the PSTC proteins allow DNA passage across the wall and periplasm. Because these proteins are needed for binding in *N. gonorrhoeae* (40), access to a DNA receptor analogous to ComEA may occur within the proposed PSTC protein complex. Double-strand cleavage by an unknown nuclease also occurs (9). ComA [in *N. gonorrhoeae* (41)] or Rec2 [in *H. influenzae* (22)] orthologs of the *B. subtilis* ComEC protein mediate transport across the inner membrane. Conversion to single-stranded DNA and degradation of one strand equivalent are probably closely coupled in time, if not mechanistically, to inner membrane transport. Because large amounts of free single-stranded DNA are not detected, the integration step may occur rapidly, preventing the accumulation of single strands. The previous discussion regarding the possible roles of SSB and of PTSC structure disassembly applies also to gram-negative transformation.

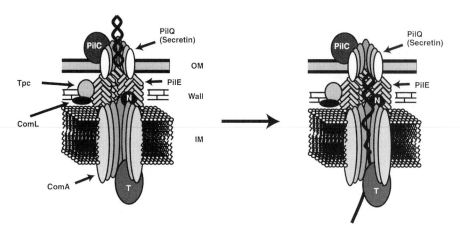

Figure 3 Transformation in gram-negative bacteria. The protein nomenclature is taken from the *Neisseria gonorrhoeae* system, but the mechanism is likely to be similar in *Haemophilus influenzae*. Double-stranded DNA enters the periplasm through an outer membrane (OM) pore consisting of the secretin protein PilQ. PilC, a pilus-associated protein, is required for this step. It is believed that the pilin protein (PilE) forms a structure that traverses the murein layer and that remodeling of the wall for transformation requires the proteins ComL and Tpc. The DNA receptor proteins and the nuclease that degrades one strand have not been identified. The intact single strand is then transported across the inner membrane (IM) through an aqueous pore, possibly formed by ComA, a ComEC ortholog. The transporter protein (T) has not been identified. References and further details are provided in the text.

WHAT USE IS COMPETENCE?

Competence is widespread. A review published five years ago listed >40 naturally transformable bacterial species (88). Transformation requires more than a dozen proteins and is often exquisitely regulated, as if the expression of this capability must be finely tuned to the needs of each organism. What is the selective force that has shaped and maintained these elaborate systems in so many species? Several hypotheses have been advanced, and as is usual with evolutionary arguments they are not easily testable and may not be mutually exclusive. These hypotheses can be characterized as DNA as food, DNA for repair, and DNA for genetic diversity.

It was proposed that competence evolved to permit the uptake of DNA as a food supply (113, 114), but this reviewer believes that this is not a major factor among the selective pressures that maintain the competence mechanism. For instance *B. subtilis* possesses a powerful nonspecific nuclease that is secreted into the medium, as well as uptake systems for the nucleolytic products. This would seem to provide an efficient route for the consumption of environmental nucleic acids.

Why would the elaborate transformation machinery evolve to meet this need, which is met by a simpler and more generally useful pathway? It should be pointed out that, because one strand equivalent is released into the medium, the competence machinery discards half of the potential foodstuff, a wasteful mechanism. The *H. influenzae* and *N. gonorrhoeae* systems exhibit uptake specificity. This does not suggest a food-gathering mechanism. It cannot be ruled out that competence has served such a function under some conditions, in some organisms, and perhaps at some point in evolutionary history, but this does not seem plausible as a general selective force.

A second proposal is that transformation serves a function in DNA repair (59, 95, 142). Lysed cells provide DNA that is taken up and used for the repair of otherwise lethal lesions. This suggestion is supported by the finding that the DNA repair machinery of *B. subtilis* is induced as part of the competence regulon (55, 89). The repair hypothesis has been criticized because DNA-damaging agents do not induce competence (112). However, such induction is not a strong prediction of the repair hypothesis. Induction of a DNA repair system before the appearance of DNA damage could provide a selective advantage. Such an induction mechanism may respond to conditions in which DNA damage is likely to occur, not to the damage itself. For instance, as *B. subtilis* approaches stationary phase, when competence is induced, its metabolism becomes more aerobic, and oxidative damage to DNA may be more likely. Such preventive induction might repair damaged DNA before serious harm is done by the secondary introduction of double-strand breaks or the synthesis of toxic proteins. In *B. subtilis*, transformation was found to increase survival in populations UV-irradiated before, but not after, the addition of DNA (59, 95, 142). This was interpreted as favoring the DNA repair hypothesis. Similar data were reported for *H. influenzae*, but transformation with a cloned fragment had the same effect on survival as total chromosomal DNA, indicating that repair was not targeted to the site of integration as predicted by the repair hypothesis (97). Foreign DNA had no effect. Somehow a recombination event increased the survival rate, raising important doubts about the validity of the earlier experiments with *B. subtilis*.

The third popular hypothesis proposes that transformation is a mechanism for exploring the fitness landscape. All genetic diversity ultimately derives from mutation, but recombination can generate new allelic combinations. Many examples of horizontal gene transfer exist, and it is likely that transformation has played a role. Two instructive examples have been reported recently. In *N. meningitidis*, sequencing of the *sodC* gene revealed the presence of two *H. influenzae* transformation uptake sequences, which form the transcriptional terminator of this virulence determinant (67). The sequence of *sodC* resembled that of the *Haemophilus* ortholog more closely than the *sod* gene of *Escherichia coli*, whereas other *Neisseria* genes were more similar to those of *E. coli* than of *H. influenzae*. This was interpreted as evidence for the horizontal acquisition of *sodC* from *Haemophilus* spp. A search of the *Neisseria* genome revealed two additional examples of the *Haemophilus* uptake sequence, both associated with *Haemophilus*-like genes. Tellingly, one of the corresponding *Haemophilus* sequences contained a *Neisseria* uptake sequence, providing a plausible mechanism for the transfer of the associated DNA from *Haemophilus* to *N. meningitidis.*

Another example of evolution by transformation was discovered in *Helicobacter pylori*, a cause of human gastritis (134). The sequences of three DNA fragments from clinical strains of this organism were determined, and it was demonstrated that polymorphisms in these loci are at linkage equilibrium. Because transformation is the only known means of genetic exchange in this organism, it is likely that competence is responsible for its panmictic population structure. It was pointed out (134) that recombination can contribute to genetic diversity in two ways, perhaps illustrated by the two examples given here. In *H pylori*, recombination may have served to prevent a reduction in diversity caused by selective sweeps (whereas selection for a mutation that increases fitness causes the loss of competing genotypes with their associated diversity) and founder effects associated with the initial spread of a small number of organisms to a new niche. Frequent recombination will rapidly disperse an advantageous mutation to many genetic backgrounds, preventing these bottleneck effects from reducing diversity. In the *Neisseria* example, on the other hand, transformation has apparently served to introduce new genes from another species.

Because transformation has actually enabled genetic exchange in natural populations, it is tempting to conclude that this explains the selection for competence. But this remains speculative, and the debate concerning the evolutionary role of transformation will continue. In fact the signals that induce competence are species-specific. This may be telling us that competence can serve a variety of needs and that each species has learned to use the ability to transport DNA to meet its particular requirements.

ACKNOWLEDGMENTS

I thank all the members of our lab for helpful comments on the manuscript and especially for countless stimulating discussions. NIH Grants GM57720 and GM43756 supported the work from our group.

Visit the Annual Reviews home page at http://www.AnnualReviews.org

LITERATURE CITED

1. Albano M, Breitling R, Dubnau D. 1989. Nucleotide sequence and genetic organization of the *Bacillus subtilis comG* operon. *J. Bacteriol.* 171:5386–404

2. Alm RA, Hallinan JP, Watson AA, Mattick JS. 1996. Fimbrial biogenesis genes of *Pseudomonas aeruginosa*: *pilW* and *pilX* increase the similarity of type 4 fimbriae to the GSP protein secretion systems and *pilY1* encodes a gonococcal PilC homolog. *Mol. Microbiol.* 22:161–73

3. Alm RA, Mattick JS. 1995. Identification of a gene, *pilV*, required for type 4 fimbrial biogenesis in *Pseudomonas aeruginosa*, whose

product possesses a pre-pilin-like sequence. *Mol. Microbiol.* 16:485–96

4. Alm RA, Mattick JS. 1996. Identification of two genes with prepilin-like leader sequences involved in type 4 fimbrial biogenesis in *Pseudomonas aeruginosa. J. Bacteriol.* 178:3809–17

5. Archibald AR, Hancock IC, Harwood CR. 1993. Cell wall structure, synthesis and turnover. In *Bacillus subtilis and Other Gram-Positive Bacteria: Physiology, Biochemistry, and Molecular Biology,* ed. AL Sonenshein, R Losick, JA Hoch, pp. 381–410. Washington DC: Am. Soc. Microbiol.

6. Arwert F, Venema G. 1973. Evidence for a non-covalently bonded intermediate in recombining during transformation in *Bacillus subtilis.* In *Bacterial Transformation,* ed. LJ Archer, pp. 203–14. New York: Academic

7. Arwert F, Venema G. 1973. Transformation in *Bacillus subtilis.* Fate of newly introduced transforming DNA. *Mol. Gen. Genet.* 123:185–98

8. Barany F, Kahn ME, Smith HO. 1983. Directional transport and integration of donor DNA in *Haemophilus influenzae* transformation. *Proc. Natl. Acad. Sci. USA* 80:7274–78

9. Biswas GD, Burnstein KL, Sparling PF. 1986. Linearization of donor DNA during plasmid transformation in *Neisseria gonorrhoeae. J. Bacteriol.* 168:756–61

10. Bitter W, Koster M, Latijnhouwers M, de Cock H, Tommassen J. 1998. Formation of oligomeric rings by XcpQ and PilQ, which are involved in protein transport across the outer membrane of *Pseudomonas aeruginosa. Mol. Microbiol.* 27:209–19

11. Bodmer W, Ganesan AT. 1964. Biochemical and genetic studies of integration and recombination in *Bacillus subtilis* transformation. *Genetics* 50:717–38

12. Bradley DE. 1980. A function of *Pseudomonas aeruginosa* PAO polar pili: twitching motility. *Can. J. Microbiol.* 26:146–54

13. Breitling R, Dubnau D. 1990. A pilin-like membrane protein is essential for DNA binding by competent *Bacillus subtilis. J. Bacteriol.* 172:1499–508

14. Bremer W, Kooistra J, Hellingwerf KJ, Konings WN. 1984. Role of the electrochemical proton gradient in genetic transformation of *Haemophilus influenzae. J. Bacteriol.* 157:868–73

15. Campbell EA, Choi S, Masure HR. 1998. A competence regulon in *Streptococcus pneumoniae* revealed by genomic analysis. *Mol. Microbiol.* 27:929–39

16. Chaussee MS, Hill SA. 1998. Formation of single-stranded DNA during DNA transformation of *Neisseria gonorrhoeae. J. Bacteriol.* 180:5117–22

17. Chen LY, Chen DY, Miaw J, Hu NT. 1996. XpsD, an outer membrane protein required for protein secretion by *Xanthomonas campestris* pv. *campestris,* forms a multimer. *J. Biol. Chem.* 271:2703–8

18. Chung YS, Breidt F, Dubnau D. 1998. Cell surface localization and processing of the ComG proteins, required for DNA binding during transformation of Bacillus subtilis. *Mol. Microbiol.* 29:905–13

19. Chung YS, Dubnau D. 1998. All seven *comG* open reading frames are required for DNA binding during the transformation of competent *Bacillus subtilis. J. Bacteriol.* 180:41–45

20. Chung YS, Dubnau D. 1994. ComC is required for the processing and translocation of ComGC, a pilin-like competence protein of *Bacillus subtilis. Mol. Microbiol.* 15:543–51

21. Claverys JP, Martin B. 1998. Competence regulons, genomics and streptococci. *Mol. Microbiol.* 29:1126–27

22. Clifton SW, McCarthy D, Roe BA. 1994. Sequence of the *rec-2* locus of *Hemophilus influenzae*: homologies to *comE*-ORF3 of *Bacillus subtilis* and *msbA* of *Escherichia coli. Gene* 146:95–100

23. Crago AM, Koronakis V. 1998. *Salmonella* InvG forms a ring-like multimer that

requires the InvH lipoprotein for outer membrane localization. *Mol. Microbiol.* 30:47–56

24. d'Enfert C, Reyss I, Wandersman C, Pugsley AP. 1989. Protein secretion by gram negative bacteria. Characterization of two membrane proteins required for pullulanase secretion by *Escherichia coli* K-12. *J. Biol. Chem.* 264:17462–68

25. Danner DB, Deich RA, Sisco KL, Smith HO. 1980. An eleven-base-pair sequence determines the specificity of DNA uptake in *Haemophilus* transformation. *Gene* 11:311–18

26. Deleted in proof

27. Danner DB, Smith HO, Narang SA. 1982. Construction of DNA recognition sites active in *Haemophilus* transformation. *Proc. Natl. Acad. Sci. USA* 79:2393–97

28. Davidoff-Abelson R, Dubnau D. 1973. Conditions affecting the isolation from transformed cells of *Bacillus subtilis* of high molecular weight single-stranded deoxyribonucleic acid of donor origin. *J. Bacteriol.* 116:146–53

29. Davidoff-Abelson R, Dubnau D. 1973. Kinetic analysis of the products of donor deoxyribonucleate in transformed cells of *Bacillus subtilis*. *J. Bacteriol.* 116:154–62

30. Deich RA, Smith HO. 1980. Mechanism of homospecific DNA uptake in *Haemophilus influenzae* transformation. *Mol. Gen. Genet.* 177:369–74

31. Drake SL, Koomey M. 1995. The product of the *pilQ* gene is essential for the biogenesis of type IV pili in *Neisseria gonorrhoeae*. *Mol. Microbiol.* 18:975–86

32. Dubnau D. 1993. Genetic exchange and homologous recombination. In *Bacillus subtilis and Other Gram-Positive Bacteria: Biochemistry, Physiology, and Molecular Genetics*, ed. AL Sonenshein, JA Hoch, R Losick, pp. 555–84. Washington, DC.: Am. Soc. Microbiol.

33. Dubnau D. 1976. Genetic transformation of *Bacillus subtilis*: a review with emphasis on the recombination mechanism. In *Microbiology*, ed. D Schlessinger, pp. 14–27. Washington, DC: Am. Soc. Microbiol.

34. Dubnau D, Cirigliano C. 1972. Fate of transforming deoxyribonucleic acid after uptake by competent *Bacillus subtilis*: size and distribution of the integrated donor sequences. *J. Bacteriol.* 111:488–94

35. Dubnau D, Cirigliano C. 1972. Fate of transforming DNA following uptake by competent *Bacillus subtilis*. III. Formation and properties of products isolated from transformed cells which are derived entirely from donor DNA. *J. Mol. Biol.* 64:9–29

36. Dubnau D, Cirigliano C. 1972. Fate of transforming DNA following uptake by competent *Bacillus subtilis*. IV. The endwise attachment and uptake of transforming DNA. *J. Mol. Biol.* 64:31–46

37. Dubnau D, Cirigliano C. 1973. Fate of transforming DNA following uptake by competent *Bacillus subtilis*. VI. Noncovalent association of donor and recipient DNA. *Mol. Gen. Genet.* 120:101–6

38. Dubnau D, Cirigliano C. 1974. Uptake and integration of transforming DNA in *Bacillus subtilis*. In *Mechanisms in Recombination*, ed. RF Grell. London/New York: Plenum

39. Elkins C, Thomas CE, Seifert HS, Sparling PF. 1991. Species-specific uptake of DNA by gonococci is mediated by a 10-base-pair sequence. *J. Bacteriol.* 173:3911–13

40. Facius D, Fussenegger M, Meyer TF. 1996. Sequential action of factors involved in natural competence for transformation of *Neisseria gonorrhoeae*. *FEMS Microbiol. Lett.* 137:159–64

41. Facius D, Meyer TF. 1993. A novel determinant (*comA*) essential for natural transformation competence in *Neisseria gonorrhoeae*. *Mol. Microbiol.* 10:699–712

42. Fleischmann RD, Adams MD, White O, Clayton RA, Kirkness EF, et al. 1995. Whole-genome random sequencing and

assembly of *Haemophilus influenzae* Rd. *Science* 269:496–512

43. Fornilli SL, Fox MS. 1977. Electron microscope visualization of the products of *Bacillus subtilis* transformation. *J. Mol. Biol.* 113:181–91

44. Fox MS, Allen MK. 1964. On the mechanism of deoxyribonucleate incorporation in pneumococcal transformation. *Proc. Natl. Acad. Sci. USA* 52:412–19

45. Fox MS, Hotchkiss RD. 1957. Initiation of bacterial transformation. *Nature* 179: 1322–25

46. Fussenegger M, Facius D, Meier J, Meyer TF. 1996. A novel peptidoglycan-linked lipoprotein (ComL) that functions in natural transformation competence of *Neisseria gonorrhoeae*. *Mol. Microbiol.* 19: 1095–105

47. Fussenegger M, Kahrs AF, Facius D, Meyer TF. 1996. Tetrapac (tpc), a novel genotype of *Neisseria gonorrhoeae* affecting epithelial cell invasion, natural transformation competence and cell separation. *Mol. Microbiol.* 19:1357–72

48. Fussenegger M, Rudel T, Barten R, Ryll R, Meyer TF. 1997. Transformation competence and type-4 pilus biogenesis in *Neisseria gonorrhoeae*—a review. *Gene* 192: 125–34

49. Gabor M, Hotchkiss R. 1966. Manifestation of linear organization in molecules of pneumococcal transforming DNA. *Proc. Natl. Acad. Sci. USA* 56:1441–48

50. Goodman SD, Scocca JJ. 1988. Identification and arrangement of the DNA sequence recognized in specific transformation of *Neisseria gonorrhoeae*. *Proc. Natl. Acad. Sci. USA* 85:6982–86

51. Grinius L. 1982. Energetics of gene transfer into bacteria. *Sov. Sci. Rev. Sect. D* 3: 115–65

52. Gwinn ML, Ramanathan R, Smith HO, Tomb JF. 1998. A new transformation-deficient mutant of *Haemophilus influenzae* Rd with normal DNA uptake. *J. Bacteriol.* 180:746–48

53. Hahn J, Albano M, Dubnau D. 1987. Isolation and characterization of Tn*917lac*-generated competence mutants of *Bacillus subtilis*. *J. Bacteriol.* 169:3104–9

54. Hahn J, Inamine G, Kozlov Y, Dubnau D. 1993. Characterization of *comE*, a late competence operon of *Bacillus subtilis* required for the binding and uptake of transforming DNA. *Mol. Microbiol.* 10:99–111

55. Haijema BJ, van Sinderen D, Winterling K, Kooistra J, Venema G, et al. 1996. Regulated expression of the dinR and recA genes during competence development and SOS induction in *Bacillus subtilis*. *Mol. Microbiol.* 22:75–85

56. Hardie KR, Lory S, Pugsley AP. 1996. Insertion of an outer membrane protein in *Escherichia coli* requires a chaperone-like protein. *EMBO J.* 15:978–88

57. Henrichsen J. 1983. Twitching motility. *Annu. Rev. Microbiol.* 37:81–93

58. Hobbs M, Mattick JS. 1993. Common components in the assembly of type-4 fimbriae, DNA transfer systems, filamentous phage and protein secretion apparatus: a general system for the formation of surface-associated protein complexes. *Mol. Microbiol.* 10:233–43

59. Hoelzer MA, Michod RE. 1999. DNA repair and the evolution of transformation in *Bacillus subtilis*. III. Sex with damaged DNA. *Genetics.* In press

60. Inamine GS, Dubnau D. 1995. ComEA, a *Bacillus subtilis* integral membrane protein required for genetic transformation, is needed for both DNA binding and transport. *J. Bacteriol.* 177:3045–51

61. Kahn ME, Smith HO. 1984. Transformation in Haemophilus: a problem in membrane biology. *J. Membr. Biol.* 81:89–103

62. Kaniga K, Bossio JC, Galan JE. 1994. The *Salmonella typhimurium* invasion genes *invF* and *invG* encode homologues of the AraC and PulD family of proteins. *Mol. Microbiol.* 13:555–68

63. Karudapuram S, Barcak GJ. 1997. The *Haemophilus influenzae dprABC* genes

constitute a competence-inducible operon that requires the product of the *tfoX* (*sxy*) gene for transcriptional activation. *J. Bacteriol.* 179:4815–20

64. Karudapuram S, Zhao X, Barcak GJ. 1995. DNA sequence and characterization of *Haemophilus influenzae dprA⁺*, a gene required for chromosomal but not plasmid DNA transformation. *J. Bacteriol.* 177: 3235–40

65. Kazmierczak BI, Mielke DL, Russel M, Model P. 1994. pIV, a filamentous phage protein that mediates phage export across the bacterial envelope, forms a multimer. *J. Mol. Biol.* 238:187–98

66. Koster M, Bitter W, de Cock H, Allaoui A, Cornelis GR, et al. 1997. The outer membrane component, YscC, of the Yop secretion machinery of *Yersinia enterocolitica* forms a ring-shaped multimeric complex. *Mol. Microbiol.* 26:789–97

67. Kroll JS, Wilks KE, Farrant JL, Langford PR. 1998. Natural genetic exchange between *Haemophilus* and *Neisseria*: intergeneric transfer of chromosomal genes between major human pathogens. *Proc. Natl. Acad. Sci. USA* 95:12381–85

68. Lacks S. 1962. Molecular fate of DNA in genetic transformation of *Pneumococcus*. *J. Mol. Biol.* 5:119–31

69. Lacks S. 1978. Steps in the process of DNA binding and entry in transformation. In *Transformation*, ed. SW Glover, LO Butter. Oxford, UK: Cotswold

70. Lacks S. 1979. Uptake of circular deoxyribonucleic acid and mechanism of deoxyribonucleic acid transport in genetic transformation of *Streptococcus pneumoniae*. *J. Bacteriol.* 138:404–9

71. Lacks S, Greenberg B. 1976. Single-strand breakage on binding of DNA to cells in the genetic transformation of *Diplococcus pneumoniae*. *J. Mol. Biol.* 101:255–75

72. Lacks S, Greenberg B, Carlson K. 1967. Fate of donor DNA in pneumococcal transformation. *J. Mol. Biol.* 29:327–47

73. Lacks S, Greenberg B, Neuberger M. 1975. Identification of a deoxyribonuclease implicated in genetic transformation of *Diplococcus pneumoniae*. *J. Bacteriol.* 123:222–32

74. Lacks S, Greenberg B, Neuberger M. 1974. Role of a deoxyribonuclease in the genetic transformation of *Diplococcus pneumoniae*. *Proc. Natl. Acad. Sci. USA* 71:2305–9

75. Lacks S, Neuberger M. 1975. Membrane location of a deoxyribonuclease implicated in the genetic transformation of *Diplococcus pneumoniae*. *J. Bacteriol.* 124:1321–29

76. Lacks SA. 1977. Binding and entry of DNA in bacterial transformation. In *Microbial Interactions; Receptors and Recognition*, ed. JL Reissig. London: Chapman & Hall

77. Lacks SA, Greenberg B. 1973. Competence for deoxyribonucleic acid uptake and deoxyribonuclease action external to cells in the genetic transformation of *Diplococcus pneumoniae*. *J. Bacteriol.* 114:152–63

78. Larson TG, Goodgal SH. 1992. Donor DNA processing is blocked by a mutation in the *com101A* locus of *Haemophilus influenzae*. *J. Bacteriol.* 174:3392–94

79. Larson TG, Goodgal SH. 1991. Sequence and transcriptional regulation of *com101A*, a locus required for genetic transformation in *Haemophilus influenzae*. *J. Bacteriol.* 173:4683–91

80. Lee MS, Marians KJ. 1990. Differential ATP requirements distinguish the DNA translocation and DNA unwinding activities of the *Escherichia coli* PriA protein. *J. Biol. Chem.* 265:17078–83

81. Levine JS, Strauss N. 1965. Lag period characterizing the entry of transforming deoxyribonucleic acid into *Bacillus subtilis*. *J. Bacteriol.* 89:281–87

82. Linderoth NA, Simon MN, Russel M. 1997. The filamentous phage pIV multimer visualized by scanning transmission electron microscopy. *Science* 278:1635–38

83. Link C, Eickernjager S, Porstendorfer D, Averhoff B. 1998. Identification and characterization of a novel competence gene, *comC*, required for DNA binding and uptake in *Acinetobacter* sp. strain BD413. *J. Bacteriol.* 180:1592–95

84. Londoño-Vallejo JA, Dubnau D. 1993. *comF*, a *Bacillus subtilis* late competence locus, encodes a protein similar to ATP-dependent RNA/DNA helicases. *Mol. Microbiol.* 9:119–31

85. Londoño-Vallejo JA, Dubnau D. 1994. Membrane association and role in DNA uptake of the *Bacillus subtilis* PriA analog ComF1. *Mol. Microbiol.* 13:197–205

86. Londoño-Vallejo JA, Dubnau D. 1994. Mutation of the putative nucleotide binding site of the *Bacillus subtilis* membrane protein ComFA abolishes the uptake of DNA during transformation. *J. Bacteriol.* 176:4642–45

87. Lorenz MG, Reipschlager K, Wackernagel W. 1992. Plasmid transformation of naturally competent *Acinetobacter calcoaceticus* in non-sterile soil extract and groundwater. *Arch. Microbiol.* 157:355–60

88. Lorenz MG, Wackernagel W. 1994. Bacterial gene transfer by natural genetic transformation in the environment. *Microbiol. Rev.* 58:563–602

89. Love PE, Lyle MJ, Yasbin RE. 1985. DNA-damage-inducible (*din*) loci are transcriptionally activated in competent *Bacillus subtilis*. *Proc. Natl. Acad. Sci. USA* 82:6201–5

90. Lunsford RD, Roble AG. 1997. *comYA*, a gene similar to *comGA* of *Bacillus subtilis*, is essential for competence-factor-dependent DNA transformation in *Streptococcus gordonii*. *J. Bacteriol.* 179:3122–26

91. Martin PR, Hobbs M, Free PD, Jeske Y, Mattick JS. 1993. Characterization of *pilQ*, a new gene required for the biogenesis of type 4 fimbriae in *Pseudomonas aeruginosa*. *Mol. Microbiol.* 9:857–68

92. Méjean V, Claverys J. 1988. Polarity of DNA entry in transformation of *Streptococcus pneumoniae*. *Mol. Gen. Genet.* 213:444–48

93. Méjean V, Claverys JP. 1993. DNA processing during entry in transformation of *Streptococcus pneumoniae*. *J. Biol. Chem.* 268:5594–99

94. Méjean V, Claverys JP. 1984. Use of a cloned DNA fragment to analyze the fate of donor DNA in transformation of *Streptococcus pneumoniae*. *J. Bacteriol.* 158:1175–78

95. Michod RE, Wojciechowski MF, Hoelzer MA. 1988. DNA repair and the evolution of transformation in the bacterium *Bacillus subtilis*. *Genetics* 118:31–39

96. Mohan S, Aghion J, Guillen N, Dubnau D. 1989. Molecular cloning and characterization of *comC*, a late competence gene of *Bacillus subtilis*. *J. Bacteriol.* 171:6043–51

97. Mongold JA. 1992. DNA repair and the evolution of transformation in *Haemophilus influenzae*. *Genetics* 132:893–98

98. Morrison DA, Guild WR. 1973. Breakage prior to entry of donor DNA in pneumococcus transformation. *Biochim. Biophys. Acta* 299:545–56

99. Morrison DA, Guild WR. 1973. Structure of deoxyribonucleic acid on the cell surface during uptake by *Pneumococcus*. *J. Bacteriol.* 115:1055–62

100. Newhall WJ, Wilde CED, Sawyer WD, Haak RA. 1980. High-molecular-weight antigenic protein complex in the outer membrane of *Neisseria gonorrhoeae*. *Infect. Immun.* 27:475–82

101. Notani N, Goodgal SH. 1966. On the nature of recombinants formed during transformation in *Hemophilus influenzae*. *J. Gen. Physiol.* 49(2):197–209

102. Nunn D, Bergman S, Lory S. 1990. Products of three accessory genes, *pilB*, *pilC*, and *pilD*, are required for biogenesis of *Pseudomonase aeruginosa* pili. *J. Bacteriol.* 172:2911–19

103. Palmen R, Vosman B, Buijsman P, Breek CK, Hellingwerf KJ. 1993. Physiological

characterization of natural transformation in *Acinetobacter calcoaceticus. J. Gen. Microbiol.* 139:295–305

104. Pestova EV, Morrison DA. 1998. Isolation and characterization of three *Streptococcus pneumoniae* transformation-specific loci by use of a *lacZ* reporter insertion vector. *J. Bacteriol.* 180:2701–10

105. Pestova EV, Morrison DA. 1997. *Streptococcus pneumoniae* chromosomal loci specifically induced during competence for genetic transformation. *Abstr. Gen. Meet. Am. Soc. Microbiol.* 97th, Miami, p. 293. Washington, DC: Am. Soc. Microbiol.

106. Piechowska M, Fox MS. 1971. Fate of transforming deoxyribonuclease in *Bacillus subtilis. J. Bacteriol.* 108:680–89

107. Porstendorfer D, Drotschmann U, Averhoff B. 1997. A novel competence gene, *comP*, is essential for natural transformation of *Acinetobacter* sp. strain BD413. *Appl. Environ. Microbiol.* 63:4150–57

108. Possot O, d'Enfert C, Reyss I, Pugsley AP. 1992. Pullulanase secretion in *Escherichia coli* K-12 requires a cytoplasmic protein and a putative polytopic cytoplasmic membrane protein. *Mol. Microbiol.* 6:95–105

109. Provvedi R, Dubnau D. 1998. ComEA is a DNA receptor for transformation of competent *Bacillus subtilis. Mol. Microbiol.* In press

110. Pugsley AP. 1993. The complete general secretory pathway in gram-negative bacteria. *Microbiol. Rev.* 57:50–108

111. Puyet A, Greenberg B, Lacks SA. 1990. Genetic and structural characterization of EndA. A membrane-bound nuclease required for transformation of *Streptococcus pneumoniae. J. Mol. Biol.* 213:727–38

112. Redfield RJ. 1993. Evolution of natural transformation: testing the DNA repair hypothesis in *Bacillus subtilis* and *Haemophilus influenzae. Genetics* 133:755–61

113. Redfield RJ. 1993. Genes for breakfast: the have-your-cake-and-eat-it-too of bacterial transformation. *J. Hered.* 84:400–4

114. Redfield RJ, Schrag MR, Dean AM. 1997. The evolution of bacterial transformation: sex with poor relations. *Genetics* 146:27–38

115. Reyss I, Pugsley AP. 1990. Five additional genes in the *pulC-O* operon of the gram-negative bacterium *Klebsiella oxytoca* UNF5023 which are required for pullulanase secretion. *Mol. Gen. Genet.* 222:176–84

116. Rivas S, Bolland S, Cabezon E, Goni FM, de la Cruz F. 1997. TrwD, a protein encoded by the IncW plasmid R388, displays an ATP hydrolase activity essential for bacterial conjugation. *J. Biol. Chem.* 272:25583–90

117. Rosenthal AL, Lacks SA. 1977. Nuclease detection in SDS-polyacrylamide gel electrophoresis. *Anal. Biochem.* 80:76–90

118. Rosenthal AL, Lacks SD. 1980. Complex structure of the membrane nuclease of *Streptococcus pneumoniae* revealed by two-dimensional electrophoresis. *J. Mol. Biol.* 141:133–46

119. Rudel T, Facius D, Barten R, Scheuerpflug I, Nonnenmacher E, et al. 1995. Role of pili and the phase-variable PilC protein in natural competence for transformation of *Neisseria gonorrhoeae. Proc. Natl. Acad. Sci. USA* 92:7986–90

120. Rudel T, van Putten JP, Gibbs CP, Haas R, Meyer TF. 1992. Interaction of two variable proteins (PilE and PilC) required for pilus-mediated adherence of *Neisseria gonorrhoeae* to human epithelial cells. *Mol. Microbiol.* 6:3439–50

121. Russel M. 1998. Macromolecular assembly and secretion across the bacterial cell envelope: type II protein secretion systems. *J. Mol. Biol.* 279:485–99

122. Russel M. 1994. Phage assembly: a paradigm for bacterial virulence factor export? *Science* 265:612–14

123. Russell MA, Darzins A. 1994. The *pilE*

gene product of *Pseudomonas aeruginosa*, required for pilus biogenesis, shares amino acid sequence identity with the N-terminus of type-4 prepilin proteins. *Mol. Microbiol.* 13:973–85

124. Seifert HS, Ajioka RS, Paruchuri D, Heffron F, So M. 1990. Shuttle mutagenesis of *Neisseria gonorrhoeae*: pilin null mutations lower DNA transformation competence. *J. Bacteriol.* 172:40–46

125. Shevchik VE, Robert-Baudouy J, Condemine G. 1997. Specific interaction between OutD, an *Erwinia chrysanthemi* outer membrane protein of the general secretory pathway, and secreted proteins. *EMBO J.* 16:3007–16

126. Singh RM. 1972. Number of deoxyribonucleic acid uptake sites in competent cells of *Bacillus subtilis*. *J. Bacteriol.* 110:266–72

127. Sisco KL, Smith HO. 1979. Sequence-specific DNA uptake in *Haemophilus* transformation. *Proc. Natl. Acad. Sci. USA* 76:972–76

128. Smith H, Danner DB, Deich RA. 1981. Genetic transformation. *Annu. Rev. Biochem.* 50:41–68

129. Smith HO, Tomb JF, Dougherty BA, Fleischmann RD, Venter JC. 1995. Frequency and distribution of DNA uptake signal sequences in the *Haemophilus influenzae* Rd genome. *Science* 269:538–40

130. Strauss N. 1965. Configuration of transforming deoxyribonucleic acid during entry into *Bacillus subtilis*. *J. Bacteriol.* 89:288–93

131. Strauss N. 1966. Further evidence concerning the configuration of transforming deoxyribonucleic acid during entry into *Bacillus subtilis*. *J. Bacteriol.* 91:702–8

132. Strauss N. 1970. Transformation of *Bacillus subtilis* using hybrid DNA molecules constructed by annealing resolved complementary strands. *Genetics* 66:583–93

133. Strom MS, Nunn DN, Lory S. 1993. A single bifunctional enzyme, PilD, catalyzes cleavage and N-methylation of proteins belonging to the type IV pilin family. *Proc. Natl. Acad. Sci. USA* 90:2404–8

134. Suerbaum S, Smith JM, Bapumia K, Morelli G, Smith NH, et al. 1998. Free recombination within *Helicobacter pylori*. *Proc. Natl. Acad. Sci. USA* 95:12619–24

135. Tokai N, Fujimoto-Nishiyama A, Toyoshima Y, Yonemura S, Tsukita S, et al. 1996. Kid, a novel kinesin-like DNA binding protein, is localized to chromosomes and the mitotic spindle. *EMBO J.* 15:457–67

136. Tomb JF, El-Hajj H, Smith HO. 1991. Nucleotide sequence of a cluster of genes involved in the transformation of *Haemophilus influenzae* RD. *Gene* 104:1–10

137. Tomb JF. 1992. A periplasmic protein disulfide oxidoreductase is required for transformation of *Haemophilus influenzae* Rd. *Proc. Natl. Acad. Sci. USA* 89:10252–56

138. Tønjum T, Freitag NE, Namork E, Koomey M. 1995. Identification and characterization of *pilG*, a highly conserved pilus-assembly gene in pathogenic *Neisseria*. *Mol. Microbiol.* 16:451–64

139. Vagner V, Claverys JP, Ehrlich SD, Méjean V. 1990. Direction of DNA entry in competent cells of *Bacillus subtilis*. *Mol. Microbiol.* 4:1785–88

140. van Nieuwenhoven MH, Hellingwerf KJ, Venema G, Konings WN. 1982. Role of proton motive force in genetic transformation of *Bacillus subtilis*. *J. Bacteriol.* 151:771–76

141. Whitchurch CB, Hobbs M, Livingston SP, Krishnapillai V, Mattick JS. 1991. Characterization of a *Pseudomonas aeruginosa* twitching motility gene and evidence for a specialized protein export system widespread in eubacteria. *Gene* 101:33–44

142. Wojciechowski MF, Hoelzer MA,

Michod RE. 1989. DNA repair and the evolution of transformation in *Bacillus subtilis*. II. Role of inducible repair. *Genetics* 121:411–22

143. Wolfgang M, Lauer P, Park HS, Brossay L, Hebert J, et al. 1998. PilT mutations lead to simultaneous defects in competence for natural transformation and twitching motility in piliated *Neisseria gonorrhoeae*. *Mol. Microbiol.* 29:321–30

Annu. Rev. Microbiol. 1999. 53:245–81

INTEGRATING DNA: Transposases and Retroviral Integrases

L. Haren, B. Ton-Hoang, and M. Chandler

Laboratoire de Microbiologie et Génétique Moléculaire, CNRS (UPR 9007) and Université Paul Sabatier, 31062 Toulouse, France; e-mail: mike@ibcg.biotoul.fr

Key Words transposons, insertion sequences, structure, transposition mechanism

■ **Abstract** Transposable elements appear quite disparate in their organization and in the types of genetic rearrangements they promote. In spite of this diversity, retroviruses and many transposons of both prokaryotes and eukaryotes show clear similarities in the chemical reactions involved in their transposition. This is reflected in the enzymes, integrases and transposases, that catalyze these reactions and that are essential for the mobility of the elements. In this chapter, we examine the structure-function relationships between these enzymes and the different ways in which the individual steps are assembled to produce a complete transposition cycle.

CONTENTS

0066-4227/99/1001-0245$08.00

INTRODUCTION

Transposable elements are discrete segments of DNA capable of moving from one locus to another in their host genome or between different genomes. They are distributed across the living world and play a fundamental role as motors of genome plasticity in all three biological kingdoms. They induce various types of genome rearrangement and are a major source of mutation. In the case of prokaryotic organisms, they are implicated in the acquisition of "accessory" functions such as resistance to antibacterial agents, catabolism of "unusual" substances, virulence, and in the control of expression of neighboring host genes (see 16, 134).

One of the major distinguishing features of transposable elements is whether their transposition relies exclusively on DNA intermediates or includes an RNA stage. DNA elements [transposons and insertion sequences (ISs)] can be found in both prokaryotes and eukaryotes, whereas those with RNA intermediates (retroviruses and retrotransposons) are restricted to eukaryotic organisms. RNA elements can be divided into those that carry long-terminal repeats (LTR) (retroviruses and LTR retrotransposons) and those that do not (LINE-type retrotransposons). Retroviruses are distinguished from retrotransposons by the presence of an envelope gene located at the 3′ end of a multigenic polyprotein reading frame; this gene is necessary for the extracellular phase of the viral life cycle (see 20). DNA elements range from the genetically complex (such as bacteriophage Mu, which, like the retroviruses, encodes an extensive set of proteins necessary for its viral life-style) to the genetically elementary ISs composed of two short terminal inverted repeats flanking a single, or two, short open reading frames. Many hundreds of RNA and DNA elements have been identified. For bacterial ISs alone, > 17 families containing >500 members are currently known (106). Although at first sight RNA and DNA elements might appear to be fundamentally different, passage of the retroviruses and LTR retrotransposons through a double-strand DNA stage provides a DNA substrate for integration that is analogous to that of the DNA transposons and is processed in a similar manner. This review focuses on the properties and activities of the enzymes that catalyze the DNA transactions necessary for the mobility of these elements. These are called transposases (Tpases) for transposons and ISs; for the LTR RNA elements, they are called integrases (IN). Discussion will be limited to a particular group that uses a conserved triad of acidic amino acids (DDE) for catalysis. Members of this enzyme family are extremely widespread and appear to represent the majority of known Tpases and IN. However, where appropriate, information concerning other well-characterized or promising transposition systems whose Tpases do not exhibit an obvious DDE motif (such as the *Drosophila melanogaster* P element and the plant transposons Ac, En/Spm) is included.

THE TRANSPOSITION CYCLE

Transposition proceeds by endonucleolytic cleavage of the phosphodiester bonds at the ends of the transposable element and transfer of these ends into a target DNA molecule. This recombination reaction requires assembly of a highly organized nucleoprotein ternary complex (the synaptic complex or transpososome), which includes Tpase or IN, the transposon ends, and target DNA. Historically, transposition mechanisms have been separated into two principal modes, conservative and replicative, based on whether the element is copied in the course of its displacement. This is dictated by the nature and order of the cleavages at the ends (Figure 1): whether the transposon is liberated from its donor backbone by double-strand cleavages or whether it remains attached following cleavage of only a single strand.

Conservative Transposition

In conservative or cut-and-paste transposition (Figures 1*A* and 1*B*), the element is excised from the donor site and reinserted into a target site without replication. This implies cleavage of both DNA strands at the ends of the element and their rejoining to target DNA to generate a simple insertion. Examples of elements shown to have adopted this strategy include Tn*7* (34), IS*10* and IS*50* (60, 83), the Tc/*mariner* family (123), and the P element (50). Transposon excision may introduce a potentially lethal double-strand break in the donor molecule that can be rescued by double-strand break repair or by gene conversion using a sister chromosome as a template (32, 51, 62). For IS*10* and probably other ISs of the IS*4*/IS*5* group (181; see 106), passive replication of the element by the host replicon activates transposition. This ensures that duplication of the IS occurs before transposition (see 83).

Replicative Transposition

Replicative transposition (Figure 1*C*) involves cleavage of only one strand at each transposon end (the transposon is not excised from the donor molecule) and transfer into a target site in such a way as to create a replication fork. Elements that use this strategy include bacteriophage Mu (94, 109), members of the Tn*3* family of bacterial transposons (144), and the IS*6* family of ISs (see 106). If transposition is intermolecular, replication from the nascent fork(s) generates cointegrates in which donor and target replicons are joined but separated by a directly repeated copy of the element at each junction. Intramolecular transposition of this type generates inversions and deletions. Resolution of the cointegrate structures to regenerate the donor and target molecules, each carrying a single copy of the element, is accomplished by recombination between the two elements. This proceeds for some transposons by site-specific recombination promoted by a specialized transposon-specific enzyme distinct from the Tpase, the "resolvase" (e.g. Tn*3* family; 144), or

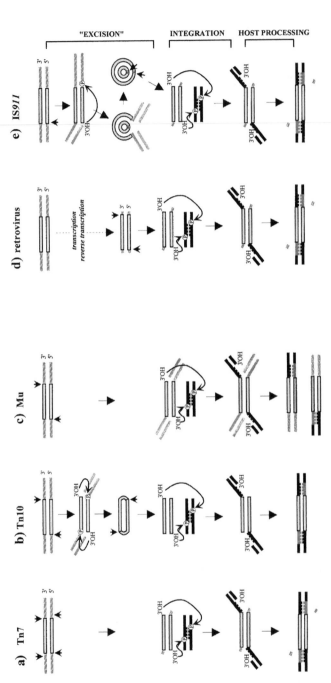

Figure 1 Types of strand cleavage and transfer. Transposon DNA is indicated by light gray boxes, and target DNA by filled boxes. Positions of strand cleavage are indicated by short vertical arrows. Nucleophilic attack of the phosphodiester bond (represented within a circle) by the active 3′ hydroxyl resulting in strand transfer is also indicated by short arrows. The "toothed" region shown in the target DNA represents the small target duplications associated with insertion. The newly synthesized DNA is indicated by dark gray. DNA polarity is shown at the top of each panel. (*A*) elements that undergo double-strand cleavage, e.g. Tn7; (*B*) elements that undergo double-strand cleavage by way of a hairpin intermediate, e.g. IS*10*; (*C*) elements that pass through a cointegrate intermediate, e.g. phage Mu (note that the newly replicated transposon strand is indicated in dark gray); (*D*) amplification by transcription and reverse transcription before insertion, e.g. retroviruses; (*E*) amplification before insertion by way of a circular DNA intermediate, e.g. IS*911*. This figure was inspired by Reference (78).

is taken in charge by the host homologous recombination system (e.g. IS*6* family; see 106).

Alternative Routes

Transcription   The LTR elements (Figure 1*D*) undergo amplification by transcription from an integrated DNA copy into RNA copies. A free double-stranded cDNA copy, designated for integration, is then synthesized by reverse transcription (20). This generally carries 2-base-pair (-bp) extensions at each end, compared with the integrated copy. Single-strand cleavage occurs at both ends to remove two terminal bases, and the recessed ends are then transferred into a target site to give simple insertions. Thus, in spite of the fact that cleavages are limited to a single strand at each end, integration is a conservative event. This is made possible because the element is separated from the donor backbone by transcription before initiation of the transposition reactions.

Circle Formation   Another alternative transposition strategy to circumvent the requirement for double-strand end cleavages has recently been described (Figure 1*E*). In the case of IS*911* IS*2* and IS*3*, Tpase promotes single-strand cleavage at one end of the transposon and its site-specific transfer to the same strand of the opposite end. This circularizes a single transposon strand, leaving the complementary strand attached to the donor backbone. This second transposon strand is then resolved to generate a circular transposon copy in which the transposon ends are abutted. The resolution mechanism is at present unclear but could involve simple cleavage and repair or replication promoted by host proteins. The covalently attached ends can then undergo simultaneous single-strand cleavage and transfer to a suitable target (154 and references therein). Elements that use this type of strategy involving site-specific strand transfer may include other members of the IS*3* family and probably those of the IS*30* and IS*21* families (see 106). Although site-specific strand transfer from one end of the element to the other generates transposon circles, it also may occur between two elements carried by the same molecule. Transfer of ends between the two IS copies in a plasmid dimer, for example, would be expected to generate head-to-tail IS dimers. This type of structure has been observed in the case of IS*21* (127), IS*2* (150), IS*30* (120), and IS*911* (C. Turlan B.T.H. and M.C., unpublished) and is extremely active in transposition.

It is clear from the pathways described above that replicative transposition per se, or more precisely replicative integration, is limited to cases such as Mu and the Tn*3* family of transposons, in which replication of the element is intimately associated with transfer of the cleaved ends into the target. In the other pathways, the preintegration intermediate is separated from the donor backbone; as a consequence, the integration event itself is conservative. The major difference between these pathways is the way in which separation of the preintegration intermediate occurs. This may simply involve double-strand cleavage at each transposon end or entail more elaborate steps such as reverse transcription or circularisation of the element, as shown in Figure 1.

MECHANISM

The development of defined in vitro systems has been essential in understanding the steps involved in transposition. Such systems are now available for several retroviruses [human immunodeficiency virus (HIV), avian sarcoma virus (ASV), and murine leukemia virus], the P element, members of the Tc/*mariner* group of eukaryotic ISs, the prokaryotic elements Mu, Tn7, and bacterial ISs of the IS4 (IS10, IS50) and IS3 (IS911) families (13, 20, 34, 60, 83, 109, 110, 126, 154). In all these cases, the cleavage and strand-joining reactions necessary for transposition are remarkably similar. They consist of a series of consecutive Tpase- or IN-catalyzed hydrolysis and transesterification reactions that require no external energy source (Figures 1 and 2). The overall reaction can be divided into three stages: detachment of the transposon from its donor site by single- or double-strand cleavage at the ends, transfer of the cleaved transposon ends to a target site, and processing of the products by host-encoded enzymes.

First-Strand Donor Cleavage

This important first chemical step initiates separation of transposon and donor DNA. For elements such as bacteriophage Mu, the retroviruses, and members of the IS3 family (Figures 1C, 1D, and 1E, respectively), the Tpases catalyze single-strand cleavage at the ends liberating a 3'OH (see 109). For the retroviruses, the equivalent reaction is called processing and results in the liberation of two bases from the 3' end of the double-stranded cDNA, leaving a two-base 5' overhang. This is a hydrolysis reaction with H_2O as an attacking nucleophile and it generates a free 3'OH (Figure 2A) (see 109, 111).

Second-Strand Donor Cleavage

For elements in which transposition occurs by excision from the donor molecule, cleavage of the complementary strand must also take place. Whereas cleavage at the 3' end appears to be common to all transposable elements analyzed to date, this is not true for second-strand cleavage. For IS10, this occurs opposite the cleaved 3' strand to generate flush transposon ends (Figure 1B); for Tn7, second-strand cleavage takes place three nucleotides to the 5' side, resulting in a three-base 5' overhang (Figure 1A). For elements of the Tc/*mariner* family and for the P element, second-strand cleavage occurs several nucleotides within the transposon (2 for Tc1/3 [162]; 3 for the *mariner* element, Himar1 [93]; and 17 for P [13]), generating a molecule with a 3' overhang.

In the case of IS10, the two breaks are not analogous: 5' cleavage occurs subsequent to 3' cleavage (17), and the free 3'OH generated by 3' cleavage is itself used

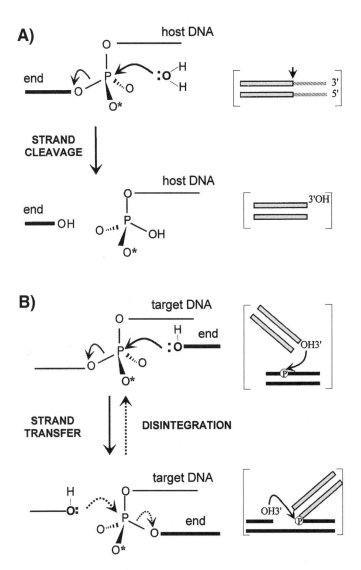

Figure 2 The chemistry of strand cleavage and transfer: (*A*) strand cleavage, (*B*) strand transfer and disintegration. Only one of the relevant DNA strands is shown in each case. The transposon DNA is represented as a bold line. Nucleophilic attack either by H_2O (Figure 2*A*) or a 3′ hydroxyl from the exposed transposon end (end) for strand transfer or from the exposed 3′ hydroxyl in the target DNA for disintegration (Figure 2*B*) is shown from the right and indicated by curved arrows. The phosphodiester bond that undergoes cleavage in these reactions is shown as a chiral form. Although the chirality of the phosphate is not normally fixed, introduction of a sulphur atom to replace a nonbridging oxygen (O*) fixes one or the other chiral forms. The hydrolysis (Figure 2*A*) and transesterification (Figure 2*B*) reactions shown here result in an inversion of chirality. The corresponding steps involving double-strand donor and target DNA together with their polarity are presented on the right of this figure for clarity.

as the nucleophile in attacking the second strand (78). This generates a hairpin structure at the transposon ends that is subsequently hydrolyzed to regenerate the 3'OH and 5' phosphate ends (Figure 1*B*). This mechanism is reminiscent of V(D)J recombination, which generates the immunoglobin repertoire, although the V(D)J hairpin is generated on the equivalent of the donor backbone ends (159). Interestingly, the V(D)J recombination system can promote inter- and intramolecular transposition and may indeed have derived from an ancestral transposon (2, 69). A similar mechanism has been proposed for the excision/transposition of a variety of other eukaryotic transposons, such as Ac, Tam*3*, hobo, and Ascot-*1* (8, 32, 33).

Few details are available concerning second-strand cleavage of the other elements. Transposon Tn*7* (Figure 1*A*) uses a different strategy. Its Tpase is composed of two distinct polypeptides, TnsA and TnsB, each dedicated to cleavage of a particular strand (137). For members of the Tc family, early studies with purified Tpase were unable to detect 3' strand cleavage but revealed specific 5' cleavage (167), showing that cleavage of the two strands can be uncoupled. As described above, for IS*911* and probably a relatively large group of bacterial ISs, the necessity for a 5' cleavage step is circumvented by the use of asymmetric strand cleavage and transfer. In the resulting circular intermediate, both ends are abutted and only two single-strand 3' cleavages are required to liberate them for strand transfer into the target (Figure 1*E*) (154).

Strand Transfer

Once cleaved, the 3' ends of the transposon are transferred into the target DNA molecule. Similar to strand cleavage, strand transfer proceeds by a Tpase- or IN-catalyzed nucleophilic attack. Here the attacking nucleophile is the 3'OH group at the free transposon end (Figure 2*B*) (see 109, 111). Like strand cleavage, the reaction does not require an external energy source, suggesting that the energy in the target phosphodiester bond is used directly in the formation of the new transposon-target joint. Experiments with Mu and HIV-1, in which chirality was imposed on the phosphate of the scissile target phosphodiester bond by substitution of a nonbridging oxygen for a sulfur atom to generate a phosphorothioate group (49, 112), imply that this reaction does not involve the formation of a covalent protein-DNA intermediate. The target phosphate underwent stereochemical inversion in the course of the reaction, implying a direct single-step in-line nucleophilic attack (Figure 2). A two-step mechanism using a protein-DNA intermediate would have regenerated the original stereochemical configuration as a consequence of the second step, as shown for recombination reactions promoted by phage λ Int (112). For HIV-1, both cleavage and strand transfer reactions behaved in the same way (49). It is the direct attack by the 3' transposon end(s) that couples transposon to target DNA without prior target cleavage. This creates a new phophodiester bond between transposon and target and leaves a 3'OH group in the target DNA at the point of insertion. Concerted insertion of each 3' transposon end into the target generally occurs in a staggered way (Figure 1). Although the transfer reaction, like the initial cleavage reaction, is isoenergetic, several in vitro transposition systems show a requirement for high-energy cofactors [ATP for Mu and Tn7 (9, 107); GTP for P (77)]. These appear to have a regulatory role, however, and do not intervene in catalysis. Furthermore, the fact that precleaved substrates can undergo relatively efficient integration (9, 14, 35, 76, 98, 135, 152) shows that the strand transfer reaction is not energetically coupled to donor strand cleavage.

IN and Tpases also catalyze a reaction called disintegration that can be considered as a reversal of integration (14, 30, 126, 167). Joining of the 3' transposon end to the target generates a branched molecule and uncovers a 3'OH on the target DNA. In the disintegraton reaction, it is this 3'OH that attacks the newly formed transposon target junction to liberate the inserted end (Figure 2B). Although this reaction has been useful in the analysis of IN and Tpase function, it is unclear whether it is biologically relevant.

Post-Transfer Processing

The staggered insertion of the transposon ends into the target site results in the creation of short, complementary single-strand regions flanking the inserted element. Those regions in which the transposon has been separated from the donor DNA (Figures 1A, 1B, 1D, and 1E) are thought to be repaired by the host repair/replication machinery to generate the characteristic short direct-target repeats, a hallmark of many types of insertion. Where the transposon remains connected to the donor backbone by its 5' ends (Figure 1C), a potential replication fork is created at each end in which the flanking 3'OH of the vector can act as a primer. Replication across the element would then result in a cointegrate molecule.

Although the transposable elements considered here appear quite diverse in their transposition cycles, their Tpases or IN catalyze similar types of reaction: hydrolysis for strand cleavage and transesterification for strand transfer. The distinctiveness of these elements is not derived from the chemistry of the reactions but rather in the way these reactions are coordinated during transposition. The fact that these enzymes carry out similar chemical reactions implies a degree of functional and structural similarity. This is examined below.

ENZYMES

Tpases and IN are multifunctional enzymes that must accomplish a series of tasks. To ensure the reactions described above, the enzymes must be able to locate the ends of the element, bring them together into a synaptic complex, recognize the phosphodiester bonds to be cleaved, ensure the incorporation of a target DNA molecule at an appropriate stage in the pathway, direct strand transfer, and relinquish their place to host enzymes recruited for post-transfer processing. These functions involve specific and nonspecific DNA binding, catalysis, the capacity to multimerize and, in some cases, the capacity to interact with accessory proteins.

Overall Organization

Initial dissection of IN and of various Tpases has often used partial proteolysis in vitro and deletion analysis in vitro and in vivo to define topologically independent functional domains. The Mu Tpase (MuA) has been divided in this way into three major domains: I, II, and III (see 94). These have been further sectioned into subdomains I(α,β,γ), II(α,β), and III(α,β), each with specific activities. Similar types of study have been undertaken with IN, in which partial proteolysis yields N-terminal and C-terminal fragments together with a central core region (see 5) and with the Tpases of IS*10* (91) and IS*50* (19) (Figure 3).

Although it is convenient to analyze Tpase and IN functions in terms of domain structure, it should be kept in mind that individual functions are not necessarily accomplished by a single isolated domain but may be assumed by several different regions of a single polypeptide or by more than one polypeptide in a multimeric complex. However, a general pattern for the functional organization of Tpases is emerging: sequence-specific DNA-binding activities are generally located in the N-terminal region, whereas the catalytic domain is often localized toward the C-terminal end (Figure 3). This arrangement has been observed for IS*1* (104, 183), IS*30* (146), Mu (see 94), Tn*3* (105), IS*50* (175), IS*903* (151), IS*911* (66, 126), Tc*1/3* (see 123), P (14), and Ac (15).

DNA Recognition

DNA recognition operates at several levels. It intervenes in discriminating the ends of the element from nontransposon DNA, in assembling the transpososome, in fitting the transposon substrate into the catalytic pocket, and in sequestering target DNA.

Transposon Ends   A key feature of Tpases is their capacity to specifically recognize and bind the transposon ends. In discussing how these enzymes may accomplish this, it is important to consider the organization of the ends themselves. With several notable exceptions (see 106), the majority of ISs exhibit short terminal inverted repeat sequences (IR) of between 10 and 40 bp, whereas other types of transposon may carry an array of such sequences. Many transposon ends are composed of two functional domains. One, located internally, ensures correct sequence-specific positioning of Tpase on the ends. This corresponds to a short nucleotide stretch within the single terminal repeats of ISs (or Tn*3* family elements) or the terminal Tpase-binding sites of certain of the more complex elements, such as Mu and Tn*7*. The other domain corresponds generally to the 2–4 terminal base pairs necessary for cleavage and strand transfer. This extreme terminal domain is identical at both ends of the element and tends to be conserved between related

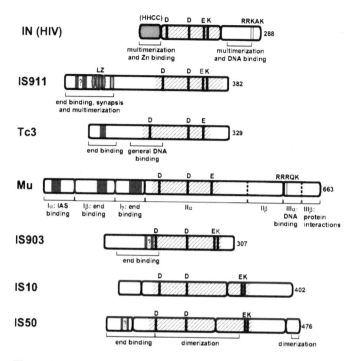

Figure 3   Transposase (Tpase) organization. The positions of protease sensitive sites are delimited by the open boxes. Potential or real helix-turn-helix (HTH) motifs are shown as dark grey boxes. Potential HTH motifs are indicated by "?." The catalytic core is indicated by pale grey and carries the DDE motif. These residues, together with a number of others referred to in the text, are indicated in uppercase letters above each Tpase molecule. LZ indicates the leucine zipper motif with the four repeating heptads observed in the IS911 Tpase. A second region involved in multimerization is also shown slightly downstream. In those cases investigated, the catalytic core is also capable of promoting multimerization. Known functions of the different Tpase regions are indicated below. Tpase alignments are centered on the second aspartate residue. The length of each protein in amino acids is indicated at the right. The function of each region, where known, is indicated under the respective proteins.

transposons. One of the most common is the terminal CA-3′ dinucleotide found at the ends of Mu, Tn7, IS30, Tn552, and the IS3 family elements, as well as the retroviral genomes (see 125). Binding sites for host proteins are also sometimes found within or close to the ends (see 106). Such proteins play a role in modulating Tpase expression or in transposition activity by influencing transpososome assembly and progression through the cycle (see "Assembling the Pieces" below).

Simple single terminal Tpase-binding sites observed for most IS elements and members of the Tn3 family of transposons are to be contrasted with multiple and asymmetric protein-binding sites carried by Mu (36), Tn7 (34) and Tn552 (133), or the eukaryote transposons Ac (89), En/Spm (57), and P (14). These are often arranged in a different way in the left and right ends and can provide a functional distinction between the ends either in the assembly or in the activity of the synaptic complex. The subterminal sites presumably ensure the correct synapsis of two ends or positioning of Tpase on the terminal site.

Tpase End Binding   The bifunctional organization of the terminal Tpase-binding sites may be reflected in the overall organization of the cognate Tpases, where the N-terminal sequence-specific DNA recognition domain is frequently topologically independent of the catalytic domain. The arrangement may allow for flexibility in the protein, enabling simultaneous contact of the N-terminal region with the internal domain and the catalytic domain with the external end domain. A common helix-turn-helix (HTH) motif is often found at the heart of the sequence-specific DNA-binding domain both in prokaryotic and eukaryotic elements (Figure 3). This is potentially able to provide binding specificity. The domain may be simple, as appears to be the case for Tpases of IS elements (see 106), or bipartite and able to recognize different DNA sequences, as found in the Mu, Ac, and Tc Tpases (15, 27, 123). In MuA, the end-binding domain corresponds to the adjacent subdomains Iβ and Iγ. Mutations in this region eliminate or reduce Mu end binding (81). Both subdomains contain an HTH motif, and each recognizes one half of the binding site (31, 140). Note that domain Iα is also a DNA-binding domain that recognizes another transposon site, the enhancer (or internal activation sequence; Figure 3) involved in transpososome assembly (see "Assembling the Pieces" below). For the Tc*1/3* Tpases, an N-terminal subdomain recognizes an internal binding site, whereas a second subdomain contacts nucleotides proximal to the terminal nucleotides. The domain resembles the paired domain found in transcription factors and, like them, also carries HTH and HTH-like motifs (122, 163, 167). For Ac, binding to the terminal sites also requires two subdomains, (one in the N-terminal and the other in the C-terminal regions of the protein) whereas binding to the internal sites requires only the second, C-terminal, subdomain (15).

Direct DNA binding of the N-terminal Tpase domain to the ends of its cognate transposon has also been observed for IS*1* (183), IS*30* (146), IS*903* (151), IS*50* (19), IS*911* (66), Tn*3* (105), the P element (96), and Ac (15). Although the protein domain responsible has not been determined, Tpase binding has also been demonstrated for Tn*1000* (174), Tn*552* (133), and one of the two Tn*7* enzymes (TnsB) but not the other (TnsA) (6).

Nonspecific Tpase Binding   Sequence "independent" DNA binding is thought to be necessary for target sequestration and to contribute to positioning of the transposon ends in the catalytic pocket. The catalytic core domain of several Tpases is involved in interaction with the terminal nucleotides of the ends (see "Catalytic Domain" below) and/or target DNA. The isolated core domain of MuA and the Tpases of IS*10* and Tc*1/3* show nonspecific DNA-binding properties (91, 117, 168; see 123). Like IN (see below), the isolated core domain of the IS*911* Tpase is able to carry out efficient disintegration, but not strand transfer, indicating that it must be capable of efficiently recognizing the branched disintegration substrate (Figure 2; 126). This type of discrimination is also likely to be true for Tc*1* (167) and P (14).

In addition, other domains of these proteins can contribute to sequence-independent DNA binding. For example, domain IIIα of the MuA Tpase (Figure 3) exhibits a nonspecific DNA-binding activity associated with a string of basic amino acids, RRRQK (177). These types of motif can be found in many Tpases, although their role has not been determined. Domain IIIα collaborates with the core domain (II) for catalysis and might stabilize interactions between the catalytic domain and its substrate (see "Assembling the Pieces" below).

Integrase-DNA Interaction   IN-DNA interactions have received considerable attention. In contrast to the Tpases, IN shows no conspicuous specific DNA-binding activity, as judged by standard gel retardation and footprinting procedures (92, 157), although cross-linking and competition experiments indicate that it is able to form stable complexes with viral DNA in vitro (44, 164). Moreover, in spite of the fact that the N-terminal domain carries an HTH motif, this appears to be involved in protein multimerization (26). The absence of pronounced binding specificity in the case of IN may reflect the difference in the biology of the LTR-RNA elements and DNA transposons. The DNA substrates of IN are contained within a large nucleoprotein particle (preintegration complex) in close contact with the enzyme (172; see 20), whereas Tpases must recognize and bind the ends of their cognate DNA elements while these are embedded in the host DNA. An important component in recognition of viral DNA ends may simply be the fact that they are the physical ends of the molecule.

Close DNA contacts have been mapped in all three proteolytically defined IN domains. Photo-cross-linking experiments have shown that the N-terminal domain is in close proximity to both viral DNA and target DNA 5′ to the site of integration (67). As for Tpases, the catalytic core domain also participates in viral and target DNA binding (48, 52, 68, 71). The use of chimeric IN proteins derived from HIV and feline immunodeficiency virus demonstrated that this domain determines target site selection (145). The observation that the isolated core domain, IN_{HIV} [50–212], is capable of catalyzing disintegration but not the forward processing or strand-transfer reactions suggests that it might have enhanced affinity for the branched substrate (25). Photo-cross-linking has also revealed an interaction with the terminal nucleotides of the viral LTRs (52, 67). This specificity is provided by several residues observed to be exposed in the crystal structure (see "Catalytic Domain" below). The C-terminal domain is also involved in viral DNA binding (48, 80, 103) and recognizes DNA features proximal to the terminal nucleotides (52, 67, 68). This domain is organized in an SH3-like fold, which produces a large cleft proposed to accept DNA (42, 100). Analysis of IN_{HIV} by mutagenesis revealed that a stretch of basic residues, R262, R263, and K264 (RRKAK, similar to the RRRQK stretch in MuA), is important for DNA binding (103). The C- and N-terminal domains appear to be involved in stabilization of IN interactions with the LTRs by allowing, for example, tighter binding of the core domain to the terminal sequences (44, 52, 68, 164).

Catalytic Domain

Some of the earliest comparisons between retroviral IN proteins and Tpases revealed a highly conserved triad of acidic amino acids with a characteristic spacing known as the DD(35)E motif and several additional conserved residues, in particular, a K or R residue seven amino acids downstream (54, 71, 80, 87, 125, 132). Subsequent studies generalized this signature to many other bacterial ISs and revealed three relatively well conserved regions (called N2, N3, and C1) centered on the D, D, and E residues, respectively (129) (Figures 3 and 4) . The triad was also detected in bacteriophage Mu [MuA (10)] and Tn7 [TnsA and B (137)] Tpases, in eukaryotic ISs of the Tc/*mariner* group (38), and in Tpases of the Tn*3* family (182). A recent compilation includes a majority of known ISs (106) (Figure 4). Although both the spacing and conservation is somewhat variable from group to group and not all groups have been analyzed in sufficient detail to confirm the biological significance of these residues, the conservation is nevertheless remarkable. One essential role of this triad is to coordinate the divalent cations necessary for catalysis (see "Catalysis").

The importance of the DDE residues has been demonstrated by site-directed mutagenesis of enzymes of several members of the family. These include different IN proteins and the Tpases of bacteriophage Mu (10), Tn7 (137), IS*10* (74), Tc*1/3* (167), and IS*911* (65).

Integrases   In the case of IN, mutation of the conserved amino acids was found to abolish processing, joining, and disintegration, indicating that all three reactions are catalyzed by a single active site (40, 47, 87, 158), although residual disintegration activity was observed in IN_{HIV} mutants in which the second D (D116)

$\longrightarrow$

Figure 4   DDE motifs in transposases (Tpases) and integrases (INs). In addition to those enzymes described in detail in the text, the figure includes representative Tpases from various insertion sequence (IS) families. The source or relevant IS family is indicated to the left. Where the enzymes shown are not those of the founding member of the family, this is included between brackets where appropriate. The DDE triad is shown in large bold letters together with the associated downstream basic lysine or arginine. The number of residues between the first and second D residue and between the second D and the E residue is shown between brackets. The regions N2, N3, and C1 are those defined by (129). Note the presence of additional relatively well conserved residues. These include a W or other hydrophobic amino acid three residues upstream from the first D, a G two residues downstream from the second D, a basic residue four residues upstream, and four residues downstream from the final E. Uppercase bold letters indicate conserved residues shown in the case of IN to cross-link to nucleotides in the viral ends.

and E (E152) residues were exchanged (47). In addition, the observation that the core fragment $IN_{HIV}[50–212]$, which carries the DD(35)E motif, is capable of catalyzing disintegration but not the processing and joining reactions indicated that it contained the active site of the enzyme (25, 88, 165). Mutation of other amino acids in the vicinity of the DD(35)E residues can change the specificity both for the type of divalent cation and for the nucleophile used in the processing reaction (47, 156). These results therefore provide strong evidence that the DD(35)E triad lies at the heart of the catalytic site.

	N2	**N3**	**C1**
HIV-1 (IN)	64 wql **D** cth (51)	116 vht **D** ngsnf (35)	152 ynpqsQgvi **E** smNKel **K**
Mu (MuA)	269 ing **D** gyl (66)	336 iti **D** ntrga (55)	392 kgwgqaKpv **E** rafgvg
Tn7 (TnsA)	28 (hgk **D** yip) (85)	114 mst **D** flvdc (34)	149 ertleKlel **E** rrywqq **K**
Tn7 (TnsB)	273 yei **D** ati (87)	361 lla **D** rgelm (34)	396 rrfdaKgiv **E** stfRtl
Tn552	166 wqa **D** htl (73)	240 fyt **D** hgsdf (35)	276 gvprgRgki **E** rffQtv
Tn3	689 asa **D** gmr (75)	765 imt **D** tagas (129)	895 riltqlNrg **E** srHava **R**
IS911 (IS3)	207 wcg **D** vty (59)	287 fhs **D** qgshy (35)	323 gncwdNspm **E** rffRsl **K**
IS10 (IS4)	97 vlv **D** wsd (63)	161 ivs **D** agfkv (130)	292 niyskRmqi **E** etfRdl **K**
IS50 (IS4)	119 siq **D** ksr (67)	188 avc **D** readi (136)	326 diythRwri **E** efHKaw **K**
IS903 (IS5)	121 lvi **D** stg (71)	193 asa **D** gaydt (65)	259 tdynrRsia **E** tamyrv **K**
IS26 (IS6)	78 whm **D** ety (59)	138 int **D** kapay (36)	173 qikyrNNvi **E** cdHgkl **K**
IS30	237 weg **D** lvs (55)	293 ltw **D** rgmel (33)	327 qspwqRgtn **E** ntNgli **R**
IS21 (IstA)	122 lqh **D** wge (61)	184 vlv **D** nqkaa (46)	230 rrartKgkv **E** rmvKyl **K**
IS630	181 fye **D** evd (80)	261 liv **D** nyiih (35)	297 vyspwvNhv **E** rlwQal **H**
IS982	112 sii **D** sfp (79)	192 vlg **D** mgylg (45)	237 nfskrRKvi **E** rvfsfl
IS256	167 lmt **D** vly (65)	233 vis **D** ahkgl (107)	341 nrlkstNli **E** rlNQev **R**
Tc1	86 iws **D** esk (90)	177 fqq **D** ndpkh (108)	286 sqspdlNpi **E** hmweele **R**

Other studies have suggested that D116 of IN_{HIV} also contributes to stable binding of DNA substrates (39, 164), whereas E152 may provide the specificity for the terminal viral A/T base pair (Figure 5A) (56). Furthermore, cross-linking experiments and directed mutagenesis have suggested that the neighboring and highly conserved K159 and the less well conserved K156 are involved in positioning the terminal viral–CA3′ bases (Figure 5A) (71), whereas another nearby residue, Q148 (and, less strongly Y143), may be involved in interactions with the 5′ nucleotide at the processed viral end (52, 56). Finally, Q62 was found to cross-link to the penultimate 5′ C nucleotide (Figure 5A) (52). Mutation at these and other neighboring residues also increases the salt sensitivity of the enzyme, suggesting that they may disrupt interactions with DNA (56).

Many of these results can now be understood from the known structure of the IN_{HIV} and IN_{ASV} catalytic domains, and mutagenic studies were indeed often directed by a structure-based approach. Although the crystal structure of intact IN has yet to be determined, that of the core domains (IN_{HIV}[50–212] and IN_{ASV} [52–207]) has been resolved (see Figure 5B). For IN_{HIV}[50–212], the mutation F185K was introduced to increase solubility. Initial studies (41) showed that the domain crystallizes as a dimer in which each monomer consists of a five-stranded β-sheet surrounded by six α-helices with the two Asp residues, D64 and D116, located close together on β1 and β4, respectively. The crystal structure of an F185H mutant subsequently revealed that E152 lies close to the Asp residues in a region (residues 141–153) that was disordered in the initial study (21). More recently, two additional mutant forms of IN_{HIV-1} were used to generate crystals in which one of the α-helices (α4) was observed to be extended to include E152. This structure showed that the side chains of all three members of the triad face toward each other (Figure 5B) (58). The crystal structure of wild-type IN_{ASV} [52–207] showed similar overall topology to that of HIV (23, 24) with some differences (including an ordered active site). The flexible nature of the IN_{HIV} region containing E152 and Q148 and other amino acids that have been shown to influence the choice of nucleophile for the processing reaction (160) suggests that it plays a role in positioning both the nucleophile and viral DNA. In the IN_{HIV} structure, the DDE triad is clustered at the surface close to the important K156 and K159 residues and other residues shown to undergo cross-linking with viral DNA ends (Figure 5B).

Tpases   Although mutational studies of Tpases have not been as extensive as for IN, they have provided complementary evidence concerning the importance of the DD(35)E motif in catalysis. For MuA, directed mutagenesis based on primary sequence alignments demonstrated that D269 (equivalent to the first of the two Asp residues) and E392 were essential for donor cleavage and strand transfer (Figure 4) (10, 82). The second Asp residue, D336, which could not be predicted from sequence alignments or localized by mutagenesis, was identified from the crystal structure of the catalytic domain (IIα and β; 131). Remarkably, one of the two subdomains (IIα; Figure 3) can be superimposed on the IN_{HIV} structure with the two active-site Asp residues located in identical positions. The third member of the

a)

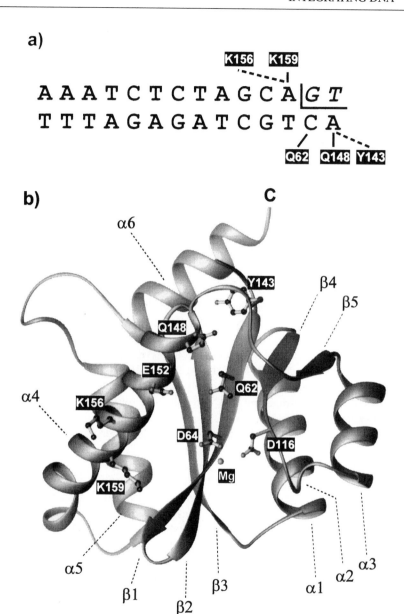

b)

Figure 5 (*A*) The HIV end. The sequence of HIV-1 ends showing the dinucleotide, which is removed during processing (in italics). The amino acids within the catalytic core that have been shown to cross-link to specific bases are indicated. (*B*) The structure of the catalytic domain of IN$_{HIV}$. The structure shown is taken from (58). Relevant amino acids are indicated as is a single Mg^{2+} ion. We thank F Dyda for providing us with this figure.

triad, E392, was found to extend away from the active site and was proposed to require activation by a conformational change induced by protein-protein or protein-DNA interactions, consistent with the tightly regulated Mu transposition cycle (see "Assembling the Pieces") (131). The second subdomain (IIβ) forms a six-stranded β barrel, and, because of a large area of positive electrostatic potential located on its surface, it has been proposed to be involved in nonspecific DNA binding.

The crystal structure of the Inh protein of IS50, a regulatory derivative of the Tpase lacking the first 55 amino acids (128) has recently been determined (37a). Here, too, the DDE triad (D119, D188, E326) forms a distinct catalytic pocket with a similar fold to that found in IN and MuA. In this case, the presence of R322, a residue that is conserved in several Tpases (Figure 4), appears to occupy the pocket. This raises the possibility that the protein may undergo a conformational change to displace R322 before catalytic activation. In the case of the related IS10 Tpase, the equivalent residue (R288) is essential for catalytic activity (18).

Although the crystal structure of the IS10 Tpase is not available, sequence alignments coupled with directed mutagenesis of 18 conserved positions identified D97, D161, and E292 as the carboxylate triad. A variety of elegant screening procedures also distinguished several additional pertinent residues. Mutation of this DDE triad (Figure 4) gave an absolute defect in all catalytic reactions, whereas mutation of the highly conserved K299 downstream of E292 was found to compromise target capture and strand transfer (18). Mutants of K299 (K299A) together with those of the partially conserved R102 (R102H) and P167 (P167S) retained their ability to induce the host SOS response, presumably as a result of their wild-type capacity to catalyze strand cleavage, but they were defective for transposition (79). Three other residues, A162, M289, and E263, were also shown to influence catalysis. Mutations A162V(T) and M289I (Figure 4) and E263K generated a hypernicking phenotype (83).

The situation for Tn7 is more complex than for other transposons because two proteins, TnsA and TnsB, are involved in the double-strand cleavages at each end and are both physically required for all cleavage and strand-transfer reactions (see "Mechanism" above). TnsB binds to the transposon ends, whereas TnsA is directed to the ends by interaction with TnsB. TnsB exhibits a well-defined DD(35)E motif (D237, D361, and E396), whereas in TnsA, the first Asp residue is less obvious (D114, E149) (Figure 4). Mutation of D114 or E149 of TnsA reduced or abolished cleavage at the 5' end while leaving 3' cleavage and 3' strand transfer unaffected. This resulted in conversion of the normal cut-and-paste Tn7 pathway into a cointegrate mode (108). Similarly, mutations of triad members in TnsB resulted in a block in 3' cleavage (and strand transfer), leaving 5' end cleavage unaffected (137).

The limited information available for other systems serves to generalize the importance of the DDE triad. As a member of the IS5 family, IS903 also carries the DDE core domain. Site-directed mutants at these residues (D121, D193, and E259) were found to exhibit severely reduced transposition in vivo (151), as were mutants of the partially conserved Y196, Y252, and R255 (Figure 4). No mechanistic analysis has yet been possible. For the Tc3 element, mutants in the potential DDE motif (D144, D231, and E266) were also severely compromised for catalysis (162).

One important observation that stems from the structural studies of the catalytic core is that all exhibit similar topologies not only within the Tpase/IN group but also with the catalytic domains of other enzymes that promote phosphoryltransfer reactions, notably RNaseH and the RuvC resolvase. This has led to the notion that such enzymes belong to a superfamily of proteins known as polynucleotidyltransferases (61, 130).

Comparison of the ensemble of structural and mutagenic data is beginning to provide a general picture of the detailed functions of the catalytic core and its interaction with the transposon ends, especially in the C1 region (Figure 4). In IN_{HIV}, this region forms $\alpha4$ (Figure 5B). K159 or K156, which cross-link to the terminal CA3′ dinucleotide, are located on the same side of the α-helix and are strongly conserved in those Tpases shown in Figure 4. Q148 also lies on the same side and appears to interact with the 5′ A on the nonprocessed strand (Figure 5A). An amide or basic amino acid is highly conserved at this or the neighboring position (Figure 4), and mutation results in severe impairment of catalysis [IN (56); IS10 (18); IS50 (WS Reznikoff, personal communication); IS903 (151)]. Additional convergences in the structure-function relationships of the catalytic domains and their role in target sequestration and positioning will certainly be forthcoming.

Multimerization

Multimerization is another fundamental property of many Tpases of both prokaryotic and eukaryotic origin (see 5, 19, 27, 66, 90, 101, 173). It is important in the assembly of the ternary synaptic complex, in ensuring collaboration between protomers for catalysis, and in regulating Tpase activity (see "Assembling the Pieces" below).

The self-association properties of Tpases are complex and still poorly understood. This is presumably because the proteins undergo a series of conformational changes during the course of the transposition reaction, and these involve changes in protein-protein interactions. Multimerization has been demonstrated physically in the case of IN (see 5) and for the Mu (95, 110), IS50 (19, 173), and IS911 (66) Tpases. It is also implied from complementation experiments in vivo and in vitro (see "Assembling the pieces" below) and from experiments using the yeast two-hybrid system (55, 75, 89, 101). In the case of bacteriophage Mu, MuA is a monomer in solution and oligomerizes only in the presence of its cognate DNA-binding sites (95, 114), but the nature of the interactions and the regions of the protein involved have not been probed. This type of information is available for a very limited number of enzymes.

For IN (Figure 3), which has been shown to form both dimers and tetramers, all three isolated domains (the N-terminal, core, and C-terminal) carry determinants for self-association and dimerize independently. It has been argued (26) that although the C-terminal and core dimers are compatible (i.e. can both bridge the same pair of full-length protein molecules), the N-terminal dimers are not and must therefore bridge two dimers to generate a tetramer. Dimerization of the N-terminal domain involves an HTH motif similar to that used for DNA recognition by DNA-binding proteins. It is stabilized by coordination of Zn^{2+} with a Cys_2-His_2 motif also located in this domain (26, 43), and Zn^{2+} has been shown to enhance tetramerization of the entire protein (184). Moreover, IN derivatives deleted for the N-terminal domain form both dimers and tetramers, suggesting that a C-terminal domain dimer may bridge a core dimer (4, 23, 41, 42, 70, 100). Although the type of active multimer of IN is still unclear, circumstantial evidence indicates that the ability to form tetramers is important for activity. Zn^{2+} not only stimulates tetramerization but also processing and strand transfer (97, 184), and mutation L241A in the C-terminal domain inhibits tetramerization, processing, and strand-transfer activities (102).

For the IS*911* Tpase, OrfAB, three distinct self-interaction regions have been identified (Figure 3). In addition to the catalytic core (homologous to IN and also able to dimerize), a leucine zipper motif also promotes multimerization. This permits binding to the IRs, whereas an adjacent region appears involved in multimerization leading to synapsis (65, 66). This coiled-coil motif is common to a majority of the members of this IS family (66).

In the IS*50* Tpase, Tnp, partial proteolysis uncovered two distinct regions capable of interacting with full-length Tnp (19). One of these regions, 114–314, overlaps the catalytic domain. The other, 441–476, lies at the extreme C-terminus, and mutations within this region (Figure 3) have significant effects on transposition activity.

ROLE OF DIVALENT METAL IONS

Catalysis

Based on a reaction mechanism proposed for several transesterification reactions (see 180), it was suggested that the acidic DDE triad plays a central role in the catalytic activity of Tpases and IN by coordinating the divalent metal ions essential for activity (80). In this two-metal ion model, one metal ion is proposed to function as a general base. It increases the partial negative charge of the incoming nucleophile by deprotonation. The second is proposed to act as a general acid that assists the leaving 3′ oxyanion and stabilizes a pentacovalent transition state by coordination of the nonbridging oxygen atoms at the cleavage site (Figure 6).

Several lines of evidence support the idea that the DDE triad is responsible for divalent cation coordination. The structure of crystals of IN_{ASV}[52–207] includes a single bound Mg^{2+} or Mn^{2+} ion (24) complexed by the two Asp residues, D64 and D121 (equivalent to IN_{HIV} D64 and D116), and four molecules of water with octahedral coordination. Recent studies with new crystal forms of the IN_{HIV} core have shown that a single Mg^{2+} ion can also be coordinated by D64 and D116

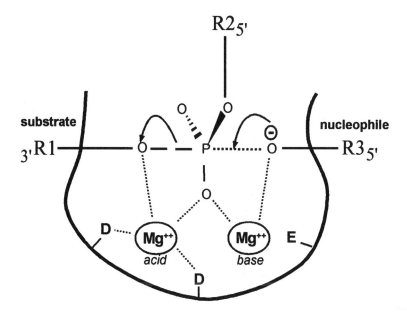

Figure 6 Role of divalent metal ions in catalysis. This figure represents the catalytic pocket containing the pentacoordinated phosphate in the (hypothetical) transition state. One Mg^{2+} ions proposed to act as a Lewis acid is shown to be coordinated with two aspartate residues, the oxyanion of the leaving group (R1 and a nonbridging oxygen). The second Mg^{2+} is also shown to complex with the nonbridging oxygen and to act as a base for deprotonization of the incoming nucleophile. The substrate is $3'R1$-$R25'$, and the product is $3'R3$-$R25'$.

(Figure 5*B*), together with two water molecules in a similar configuration (58). Other metal ions such as Ca^{2+}, Zn^{2+}, and Cd^{2+} also bind the active site of the ASV core (22). The involvement of the DDE triad in coordinating metal ions has also received support from studies on Tn7. These have exploited the fact that, compared to Mg^{2+}, Mn^{2+} shows a preference for coordination with sulfur-substituted components. Substitution of an oxygen ligand by sulfur can change the metal-ion specificity of reactions. This approach was first used in demonstrating interaction between metal ion and substrate in RNA self-splicing (37, 121), and in RNaseH activity (155). A D114C mutation in TnsA, which abolished 5′ strand cleavage in the presence of Mg^{2+}, was found to be rescued if the reaction was carried out in Mn^{2+} (137).

Although the biologically relevant cation is generally thought to be Mg^{2+}, other cations can be used in the in vitro reactions. Indeed, in the case of IN from several sources, Mn^{2+} was initially found to be the preferred cation (143, 158). However, Mn^{2+} is known to reduce the specificity of some reactions. Under optimized assay conditions, Mg^{2+} also supports a robust activity (52). In other systems, Mg^{2+} or Mn^{2+} are essential for cleavage and strand transfer and cannot be substituted by Ca^{2+} [IS*10* (73, 135), Tn7 (137), IS*911* (126)]. Although Zn^{2+} can support the

IN_{ASV} processing reaction, which uses an external nucleophile, it is inefficient for strand transfer, which uses the viral DNA $3'OH$ end (22). In addition to Mg^{2+} and Mn^{2+}, MuA can also use Zn^{2+} or Co^{2+} for donor cleavage (170). Moreover, although Ca^{2+} does not support cleavage, it is able to promote strand transfer in this case (139). That certain cations can support cleavage but not strand transfer whereas others support strand transfer but not cleavage in these systems is consistent with a two-metal ion mechanism for catalysis.

Structural Role

Besides their fundamental role as cofactors in the catalytic activities of Tpases and IN, divalent metal ions also appear to intervene in the assembly of the transpososome. For retroviruses, Mg^{2+}, Mn^{2+}, or Ca^{2+} is necessary for the formation of stable interactions between IN_{HIV} and the viral LTRs (44, 164). Mn^{2+}-induced conformational changes correlated with an enhancement of catalytic activity have been revealed by a variety of methods. These changes affect both the self-assembly of IN (45, 176) and the organization of its structural domains within the multimer, as judged by the use of monoclonal antibodies directed against each of the three domains and by changes in proteolysis patterns of the full-length protein (7).

For Mu, divalent cations are also required for assembly of an active Tpase tetramer on the ends of the element (see "Assembling the Pieces" below; 11, 114). Ca^{2+} stabilizes the Mu transpososome but does not promote first-strand cleavage. Conversely, although Zn^{2+} and Co^{2+} permit catalysis of first-strand cleavage, they are inefficient in assembly of the synaptic complex (170). Furthermore, a mutant derivative of MuA has been obtained that specifically uses Ca^{2+} for assembly of the synaptic complex and Mn^{2+} for first-strand cleavage (82). This mutation, G348D, lies within the catalytic core. These results reflect a differential action of divalent metal ions in transpososome assembly and in catalysis. For IS*10*, Ca^{2+}, which does not support catalysis, stabilizes both the synaptic complex (135) and interaction with the target (136). As for IN, the stabilizing effect of these cations may result from changes in protein conformation.

ASSEMBLING THE PIECES

We have so far discussed the different steps in the transposition process and described the different functions of IN and the various Tpases associated with these transposons. In the following text, we attempt to show how these different elements are integrated into the overall transposition process. This is understood in some detail from studies on Mu and IS*10*, in which transposition is initiated by the formation of a highly organized nucleoprotein synaptic complex (83, 114) and proceeds through a series of ordered steps showing consecutive increases in complex stability to temperature and protein-denaturing agents (135, 148). Several other systems are being actively investigated, and although certain steps may be well defined, the overall transposition pathway is generally less well understood.

End Synapsis

Tpase-mediated synapsis has been formally demonstrated for Mu (11, 139) and for IS*10* (135). In both cases, synapsis also involves accessory factors and implies a two-step mechanism leading to activation of the Tpase multimer. In a first step, a transitory constrained complex is formed. This then undergoes conformational changes during which the accessory factors are ejected and the active sites of the enzyme monomers are presumably brought closer to the reactive phosphodiester bonds (28, 110, 171).

Bacteriophage Mu has three Tpase-binding sites at each end arranged in different configurations. In an initial stage, a multimeric MuA complex is assembled on these sites using domain I$\beta\gamma$ (Figure 3). Three of the six synapsed sites (including the two terminal sites) together with a MuA tetramer compose the transpososome core (see 94). MuA bound to the remaining three sites can be removed without compromising activity. Assembly is facilitated by transitory interaction with the enhancer (internal activation sequence) by way of domain Iα of MuA (Figure 3). The enhancer DNA sequence, located > 1 kilobase from the left end of the phage genome, binds MuA and presents it to the nascent complex (left end–enhancer–right end complex). Formation of this complex (LER) requires divalent metal ions and two small host proteins, IHF and HU, in addition to a supercoiled donor DNA molecule carrying both ends in their natural configuration. IHF acts indirectly at the level of the enhancer by influencing protein binding at this site, whereas HU binds at the Mu left end and directly facilitates transpososome formation on the Mu ends. Assembly can be arrested in vitro at a stage called the stable synaptic complex (SSC or type 0 complex) by use of Ca^{2+} instead of Mg^{2+} in the reaction (110). The transition from LER to SSC is blocked either by mutation of the terminal base pair or by mutation in domain IIIα of MuA (see "DNA Recognition" above). This implies a conformational change involving domain IIIα in which the Mu ends are engaged for catalysis. In the SSC, the two Mu ends are held together by a tetramer of MuA, no strand cleavage has yet occurred, and the Mu-host junction appears unwound or otherwise distorted (95, 139, 170).

For IS*10*, the short terminal IRs, which are not strictly equivalent, are defined as outside and inside ends (OE and IE) with respect to their relationship within the parental transposon Tn*10*. They are distinguished by different dispositions of *dam* methylation sites and the presence of an IHF-binding site proximal to OE. A precleavage synaptic complex has been identified using short linear OE carrying fragments including the IHF site (135). Formation of this complex (paired ends complex), composed of two ends bridged by Tpase, is dependent on IHF but not on divalent metal ions. It is thought that this is a transitory complex because titration of IHF from this complex changes its conformation and renders it competent for subsequent target capture and strand transfer (83). It is worthwhile noting that in addition to IS*10*, several elements also carry IHF-binding sites within or proximal to their terminal IRs. In the case of Tn*1000*, IHF binding to an IR proximal site has been shown to enhance Tpase binding (174).

The types of nucleoprotein complexes formed with other elements during synapsis are not well characterized. However, for retroviruses, analysis of the preintegration complex of murine leukemia virus has implicated both IN and host proteins in a higher-order structure that also includes an extensive region of the retroviral LTRs (172). For HIV-1, the host HMG I(Y) protein has been identified in such structures and appears to be required for integration (53). These host factors may play a similar architectural role to IHF and HU in IS*10* and Mu transposition.

Target Capture and the Catalytic Steps

The strand cleavage and transfer steps must be orchestrated in a precise manner. For Mu, progression through these steps is well characterized. Following the formation of the SSC, strand cleavage occurs at both ends in the presence of the appropriate metal ion to generate a cleaved donor complex (CDC). This may be facilitated by local unwinding of the DNA at the Mu termini, which occurs in the SSC (170), probably driven by the supercoiling in the donor DNA molecule (169). In addition to the terminal binding sites, MuA was shown to protect ∼10 bp of neighboring host DNA (113). This complex gives rise to a strand transfer complex (STC) in the presence of target DNA. At this stage, the 3′ transposon ends have been inserted into the target DNA, and MuA protection is extended by 20 bp into the target (113). The MuA IIβ domain is implicated in the transition between the cleaved donor complex and strand transfer complex because mutation of exposed lysine residues (K506, 515, 529, and 530) to alanine in this domain (see "Catalytic Domain" above) blocks this step (84).

Capture of a target molecule can occur at any step in the assembly process (116) and is stimulated by a second Mu protein, MuB. MuB binds target DNA in an ATP-dependent manner and delivers it to the transposome by interaction with domain IIIβ of MuA (Figure 3). ATP hydrolysis is not required for this step. However, interaction with MuA stimulates hydrolysis, which releases MuB from the target DNA (1, 178). These properties give rise to a phenomenon known as target immunity, which prevents insertion into a target molecule already carrying a Mu copy. Here, binding of MuA to Mu ends in the target stimulates the MuB ATPase activity and purges MuB from the immune target.

Strand cleavage and transfer by MuA is also directly stimulated by MuB (12, 107, 149). The in vitro reaction results in concerted integration of both ends. Recent complementation studies using a supercoiled substrate and a mixture of wild-type MuA and a catalytically inactive mutant have suggested not only that the same catalytic core is used for both cleavage and strand transfer but that both reactions are carried out by the MuA monomer located on the terminal binding site of the opposite end (118). In other words, cleavage and strand transfer occurs in *trans* (3, 138).

Several additional observations reinforce the notion of an intimately interwound Mu transpososome architecture. Complementation data suggest that a competent active site is assembled from region IIα of one MuA monomer and region IIIα of a second (179). In addition to its nonspecific DNA binding, domain IIIα exhibits weak endonuclease activity, which suggests that, in binding, it may render the appropriate phophodiester bond more susceptible to core-catalyzed cleavage (177). It has also been established that domain IIβ is not part of the same complementation group as IIα (119) but like IIIα, is supplied by a different Tpase subunit (84). Moreover, the nucleoprotein complexes become increasingly resistant to denaturation as they progress through the cycle. Finally, a mutation of the terminal nucleotides at one end abolishes cleavage of both ends (149). The interwound transpososome organization thus imposes concerted reactions at both extremities, an important constraint if the transposon is to avoid nonproductive events involving a single end.

After strand transfer, MuA is liberated from the complex and exchanged for the replication machinery necessary to complete formation of the cointegrate transposition product. This process involves several host proteins, namely the chaperone ClpX and components of the replication apparatus (PriA, PriB, DnaT, DnaBC) and occurs in a highly ordered way (72, 85, 99).

Transposition of IS10, like that of Mu, proceeds through a series of well-defined steps, with a progressive increase in stability. In the presence of Mg^{2+} (or Mn^{2+} but not Ca^{2+}), the precleavage synaptic complex (PEC) assembled in vitro undergoes double-strand cleavage to generate complexes with single- (SEB) or double- (DEB) end breaks. Kinetic data are consistent with the notion that the DEB is generated from the SEB. Cleavage of each strand at the ends occurs in a sequential manner, with cleavage of the transferred strand taking place before that of the complementary strand (17). This allows the liberated $3'$OH of the first strand to cleave the second strand (see "Mechanism" above). DEB complexes are competent for strand transfer which occurs in a coordinated way (135). The DEB complexes are equivalent to a characteristic early intermediate in IS10 transposition, the excised transposon fragment or ETF, observed both in vivo and in vitro, in which the entire element is excised from the donor backbone and in which the two transposon ends are held together by a Tpase bridge (29, 63). Cleavage requires both ends but does not necessarily occur simultaneously at each end (64). Moreover, a single catalytic site appears to be repeatedly used for the consecutive hydrolysis and transesterification reactions. In contrast to MuA, a Tpase dimer may catalyze the entire transposition reaction (both $5'$ and $3'$ cleavages and strand transfer), each acting on a distinct end without conspicuous domain sharing between monomers (18). This suggests that the transpososome architecture in the case of IS10 is less constrained than that of Mu. This notion is supported by the observation that, in contrast to Mu, a mutation of the terminal nucleotides at one end does not significantly affect cleavage of the wild-type end (64). For IS10, the absence of a mechanism to ensure concerted cleavage at both ends, as in the case of Mu, is compensated by the late entry of the target into the synaptic complex subsequent to IR cleavage (64). Stable synaptic complexes (SSCs) including target DNA have been detected only with the DEB complex, indicating that prior cleavage at both ends is a requirement for target capture (136).

Although neither the stoichiometric composition nor the temporal evolution of the retroviral synaptic complex is known in detail, some information is available concerning its activity and structure. IN activity requires a physical DNA end. Extending the ends of the viral DNA substrate severely reduces the processing reaction, whereas mismatches in the terminal region can increase activity (141, 166). These observations suggest that unwinding or DNA distortion assists IN activity. Other data clearly indicate that the complex is multimeric and that monomers collaborate with each other within the assembled structure. This is implied from complementation assays of mutant retroviral IN derivatives. Although the C-terminal domain seems to be able to act together with an active site on the same (in *cis*) or another IN monomer in the complex (in *trans*), the N-terminal domain is required to act in *trans*, and part of the C-terminal domain is required in *cis* (45, 46, 161). Moreover, a combination of cross-linking and complementation experiments has suggested that the active site of an IN monomer may act in *trans* on substrates bound to a partner monomer in the complex, further substantiating this view. Finally, like Mu, mutation at one end abolishes cleavage of both ends, reinforcing the idea of close collaboration between monomers (86, 115).

Modeling the known structures of the separated domains with suitable DNA molecules has led to the proposal that the active complex of IN may have an octameric architecture, with a tetramer acting on each LTR of the viral DNA (67). In this model, two of the four core domains assembled on each end are directly involved in catalysis and recognize the reactive terminal CA3'. The remaining two intervene indirectly by interacting with target DNA (68).

For other systems, knowledge of the steps involved in progression through the transposition cycle is sketchy. For Tn7, the nature of the synaptic and catalytic complexes has yet to be determined. However, like Mu, target capture for Tn7 is relatively elaborate. Tn7 transposition occurs in two alternative conservative modes: site-specific integration at a single chromosomal site, *att*Tn7, and random insertion. The site-specific integration pathway has been extensively studied and involves four Tn7-encoded proteins (34). In addition to the strand-specific Tpases, TnsA and B, these include TnsD, which binds directly to the target DNA site, and TnsC, which is recruited to the target site by TnsD. Interestingly, *Escherichia coli* ribosomal protein L29 and acyl carrier protein (ACP) have also been implicated in Tn7 transposition both in vivo and in vitro, and L29 stimulates binding of TnsD to its target attTn7 site (142). TnsC appears to assume a similar function to that of the MuB protein in both target capture and immunity, and, like MuB, it exhibits Tpase- (TnsB) dependent ATPase activity (147). However, unlike bacteriophage Mu, transposition in vitro into attTn7 not only necessitates the presence of the target DNA before donor strand cleavage but also requires the complete set of transposition proteins. This ensures tight control in the commitment of the transpososome to successful transposition (34).

IS*911* transposition probably involves two sets of cleavage and strand transfer reactions: one that generates the single-strand transfer product and the other that leads to integration of the transposon circles. This implies the formation of two distinct kinds of synaptic complex. Current evidence from both in vivo and in vitro studies indicates that the Tpase, OrfAB, alone is sufficient for the first set of reactions that are essential for subsequent circle formation (124, 126, 153). The first synaptic complex that comes into play presumably involves only OrfAB and must show an asymmetric functional organization because only one IS end undergoes cleavage, whereas the other functions as a target (see "Transposition Cycle" above). After evolution of this complex for accomplishing strand cleavage and transfer, it must presumably be disassembled to allow second-strand processing and to generate the transposon circle. The single-strand transfer product forms a branched structure resembling that formed in the Mu strand transfer product, and this step, like the transition of the Mu (STC) to the replication complex, may occur in a highly ordered way.

A second type of synaptic complex must then be assembled on the circle junction. For the second set of reactions, a second IS*911*-encoded protein, OrfA, is required (153, 154). The IS*911* Tpase, OrfAB, like that of other members of the IS*3* family, is a fusion protein produced by translational frame shifting between two consecutive reading frames (see 106). OrfA is the product of the first reading frame and therefore carries the N-terminal domain of OrfAB, including the leucine zipper motif (Figure 3). This motif is required for the formation of OrfAB and OrfA homomultimers and for generating heteromultimers between the two proteins essential for circle integration (65). Although circle integration requires both proteins, OrfAB alone is sufficient for cleavage of both ends in the circle junction, suggesting that OrfA is involved in target capture.

CONCLUSIONS AND PERSPECTIVES

Over the past several years, it has become clear that many transposable elements have adopted an identical chemistry for their displacement. This is reflected in similarities in the primary sequence of their Tpases and INs. The accumulating data concerning the structure of the catalytic domains of these enzymes demonstrate an even higher degree of similarity in their topological organization. As yet, the number of known structures is quite limited, especially in view of the range and diversity of these elements. It will, therefore, be of considerable interest to determine and compare the structures of Tpases from the major groups of transposons.

Of the structures so far determined, not a single example of a full-length protein has been obtained. In the most complete case, Inh of IS*50* (37a), 55 N-terminal amino acids that determine sequence-specific binding to the transposon ends are not present. For IN and MuA, the structures of all the individual domains are available, although the manner in which these are connected is a matter of conjecture. A major challenge will therefore be to determine how the different domains of these enzymes are articulated. Moreover, because transposition invariably necessitates the assembly of higher-order nucleoprotein complexes, it will be of importance to ascertain the way in which the individual components are integrated into such complexes and to define both protein-protein and protein-DNA interactions involved.

Transposition is a dynamic process. Lessons from phage Mu and Tn*10* have shown that transposition proceeds through a series of well-defined steps. Not only do these need to be defined for the other important transposition systems, but the structural details of these transitions need to be documented, as is currently being undertaken in the case of certain site-specific recombinases (59). In regard to the chemistry of the reactions, although a two-metal ion mechanism is currently favored, this has yet to receive direct experimental proof. In particular, knowledge of the transition states would obviously provide invaluable information.

Finally, although not addressed in this review and often overlooked in a purely in vitro approach, the way in which transposition activity is controlled both by the element itself and by the host represents an important aspect of transposon biology.

ACKNOWLEDGMENTS

We thank the members of the Mobile Genetic Elements Group (R Alazard, C Turlan, C Normand, G Duval-Valentin, and P Rousseau), A Diaz, A Lopez and J-F Tocanne for discussions and AJ Carpousis, A Diaz, O Fayet, and C Normand for reading the manuscript. Fred Dyda kindly supplied Figure 5*B*. This work was supported by grants from the Centre National pour la Recherche Scientifique (CNRS, France), Région Midi-Pyrenées, and l'Association pour la Recherche sur le Cancer (ARC).

Visit the Annual Reviews home page at http://www.AnnualReviews.org

LITERATURE CITED

1. Adzuma K, Mizuuchi K. 1991. Steady-state kinetic analysis of ATP hydrolysis by the B protein of bacteriophage mu. Involvement of protein oligomerization in the ATPase cycle. *J. Biol. Chem.* 266:6159–67

2. Agrawal A, Eastman QM, Schatz DG. 1998. Transposition mediated by RAG1 and RAG2 and its implications for the evolution of the immune system. *Nature.* 394:744–51

3. Aldaz H, Schuster E, Baker TA. 1996. The interwoven architecture of the Mu transposase couples DNA synapsis to catalysis. *Cell* 85:257–69

4. Andrake MD, Skalka AM. 1995. Multimerization determinants reside in both the catalytic core and C terminus of avian sarcoma virus integrase. *J. Biol. Chem.* 270:29299–306

5. Andrake MD, Skalka AM. 1996. Retroviral integrase, putting the pieces together. *J. Biol. Chem.* 271:19633–36

6. Arciszewska LK, McKown RL, Craig NL. 1991. Purification of TnsB, a transposition protein that binds to the ends of Tn7. *J. Biol. Chem.* 266:21736–44

7. Asante-Appiah E, Skalka AM. 1997. A metal-induced conformational change and

activation of HIV-1 integrase. *J. Biol. Chem.* 272:16196–205

8. Atkinson PW, Warren WD, O'Brochta DA. 1993. The hobo transposable element of Drosophila can be cross-mobilized in houseflies and excises like the Ac element of maize. *Proc. Natl. Acad. Sci. USA* 90:9693–97

9. Bainton R, Gamas P, Craig NL. 1991. Tn7 transposition in vitro proceeds through an excised transposon intermediate generated by staggered breaks in DNA. *Cell* 65:805–16

10. Baker TA, Luo L. 1994. Identification of residues in the Mu transposase essential for catalysis. *Proc. Natl. Acad. Sci. USA* 91:6654–58

11. Baker TA, Mizuuchi K. 1992. DNA-promoted assembly of the active tetramer of the Mu transposase. *Genes Dev.* 6:2221–32

12. Baker TA, Mizuuchi M, Mizuuchi K. 1991. MuB protein allosterically activates strand transfer by the transposase of phage Mu. *Cell* 65:1003–13

13. Beall EL, Rio DC. 1997. Drosophila P-element transposase is a novel site-specific endonuclease. *Genes Dev.* 11:2137–51

14. Beall EL, Rio DC. 1998. Transposase makes critical contacts with, and is stimulated by, single-stranded DNA at the P element termini in vitro. *EMBO J.* 17:2122–36

15. Becker HA, Kunze R. 1997. Maize activator transposase has a bipartite DNA binding domain that recognizes subterminal sequences and the terminal inverted repeats. *Mol. Gen. Genet.* 254:219–30

16. Berg DE, Howe MM. 1989. *Mobile DNA.* Washington, DC: Am. Soc. Microbiol.

17. Bolland S, Kleckner N. 1995. The two single-strand cleavages at each end of Tn10 occur in a specific order during transposition. *Proc. Natl. Acad. Sci. USA* 92:7814–18

18. Bolland S, Kleckner N. 1996. The three chemical steps of Tn10/IS10 transposition

involve repeated utilization of a single active site. *Cell* 84:223–33

19. Braam LA, Reznikoff WS. 1998. Functional characterization of the Tn5 transposase by limited proteolysis. *J. Biol. Chem.* 273:10908–13

20. Brown PO. 1997. *Integration, Retroviruses,* ed. JM Coffin, SH Hughes, HE Varmus, pp. 161–203. Plainview, NY: Cold Spring Harbor Lab. Press

21. Bujacz G, Alexandratos J, Qing ZL, Clement-Mella C, Wlodawer A. 1996. The catalytic domain of human immunodeficiency virus integrase: ordered active site in the F185H mutant. *FEBS Lett.* 398:175–78

22. Bujacz G, Alexandratos J, Wlodawer A, Merkel G, Andrake M, et al. 1997. Binding of different divalent cations to the active site of avian sarcoma virus integrase and their effects on enzymatic activity. *J. Biol. Chem.* 272:18161–68

23. Bujacz G, Jaskolski M, Alexandratos J, Wlodawer A, Merkel G, et al. 1995. High-resolution structure of the catalytic domain of avian sarcoma virus integrase. *J. Mol. Biol.* 253:333–46

24. Bujacz G, Jaskolski M, Alexandratos J, Wlodawer A, Merkel G, et al. 1996. The catalytic domain of avian sarcoma virus integrase: conformation of the active-site residues in the presence of divalent cations. *Structure* 4:89–96

25. Bushman FD, Engelman A, Palmer I, Wingfield P, Craigie R. 1993. Domains of the integrase protein of human immunodeficiency virus type 1 responsible for polynucleotidyl transfer and zinc binding. *Proc. Natl. Acad. Sci. USA* 90:3428–32

26. Cai M, Zheng R, Caffrey M, Craigie R, Clore GM, Gronenborn AM. 1997. Solution structure of the N-terminal zinc binding domain of HIV-1 integrase. *Nat. Struct. Biol.* 4:567–77

27. Chaconas G, Lavoie BD, Watson MA. 1996. DNA transposition: jumping gene

machine, some assembly required. *Curr. Biol.* 6:817–20

28. Chalmers R, Guhathakurta A, Benjamin H, Kleckner N. 1998. IHF modulation of Tn10 transposition: sensory transduction of supercoiling status via a proposed protein/DNA molecular spring. *Cell* 93:897–908

29. Chalmers RM, Kleckner N. 1994. Tn10/IS10 transposase purification, activation, and in vitro reaction. *J. Biol. Chem.* 269:8029–35

30. Chow SA, Vincent KA, Ellison V, Brown PO. 1992. Reversal of integration and DNA splicing mediated by integrase of human immunodeficiency virus. *Science* 255:723–26

31. Clubb RT, Schumacher S, Mizuuchi K, Gronenborn AM, Clore GM. 1997. Solution structure of the Iγ subdomain of the Mu end DNA-binding domain of phage Mu transposase. *J. Mol. Biol.* 273:19–25

32. Coen ES, Robbins TP, Almeida J, Hudson A, Carpenter R. 1989. Consequences and mechanisms of transposition in Antirrhinum majus. In *Mobile DNA*, ed. D Berg, M Howe, pp. 413–36. Washington, DC: Am. Soc. Microbiol.

33. Colot V, Haedens V, Rossignol JL. 1998. Extensive, nonrandom diversity of excision footprints generated by Ds-like transposon Ascot-1 suggests new parallels with V(D)J recombination. *Mol. Cell Biol.* 18:4337–46

34. Craig NL. 1996. Transposon Tn7. *Curr. Top. Microbiol. Immunol.* 204:27–48

35. Craigie R, Mizuuchi K. 1987. Transposition of Mu DNA: joining of Mu to target DNA can be uncoupled from cleavage at the ends of Mu. *Cell* 51:493–501

36. Craigie R, Mizuuchi M, Mizuuchi K. 1984. Site-specific recognition of the bacteriophage Mu ends by the Mu A protein. *Cell* 39:387–94

37. Dahm SC, Uhlenbeck OC. 1991. Role of divalent metal ions in the hammerhead RNA cleavage reaction. *Biochemistry* 30:9464–69

37a. Davies DR, Braam LM, Reznikoff WS, Rayment I. 1999. Three dimensional structure of a Tu5-related protein determined to 2.9 Å resolution. *J. Biol. Chem.* 274:11904–13

38. Doak TG, Doerder FP, Jahn CL, Herrick G. 1994. A proposed superfamily of transposase genes: transposon-like elements in ciliated protozoa and a common "D35E" motif. *Proc. Natl. Acad. Sci. USA* 91:942–46

39. Drelich M, Haenggi M, Mous J. 1993. Conserved residues Pro-109 and Asp-116 are required for interaction of the human immunodeficiency virus type 1 integrase protein with its viral DNA substrate. *J. Virol.* 67:5041–44

40. Drelich M, Wilhelm R, Mous J. 1992. Identification of amino acid residues critical for endonuclease and integration activities of HIV-1 IN protein in vitro. *Virology* 188:459–68

41. Dyda F, Hickman AB, Jenkins TM, Engelman A, Craigie R, Davies DR. 1994. Crystal structure of the catalytic domain of HIV-1 integrase: similarity to other polynucleotidyl transferases. *Science* 266:1981–86

42. Eijkelenboom AP, Lutzke RA, Boelens R, Plasterk RH, Kaptein R, Hard K. 1995. The DNA-binding domain of HIV-1 integrase has an SH3-like fold. *Nat. Struct. Biol.* 2:807–10

43. Eijkelenboom AP, van den Ent FM, Vos A, Doreleijers JF, Hard K, et al. 1997. The solution structure of the amino-terminal HHCC domain of HIV-2 integrase: a three-helix bundle stabilized by zinc. *Curr. Biol.* 7:739–46

44. Ellison V, Brown PO. 1994. A stable complex between integrase and viral DNA ends mediates human immunodeficiency virus integration in vitro. *Proc. Natl. Acad. Sci. USA* 91:7316–20

45. Ellison V, Gerton J, Vincent KA, Brown PO. 1995. An essential interaction between distinct domains of HIV-1 integrase me-

diates assembly of the active multimer. *J. Biol. Chem.* 270:3320–26

46. Engelman A, Bushman FD, Craigie R. 1993. Identification of discrete functional domains of HIV-1 integrase and their organization within an active multimeric complex. *EMBO J.* 12:3269–75

47. Engelman A, Craigie R. 1992. Identification of conserved amino acid residues critical for human immunodeficiency virus type 1 integrase function in vitro. *J. Virol.* 66:6361–69

48. Engelman A, Hickman AB, Craigie R. 1994. The core and carboxyl-terminal domains of the integrase protein of human immunodeficiency virus type 1 each contribute to nonspecific DNA binding. *J. Virol.* 68:5911–17

49. Engelman A, Mizuuchi K, Craigie R. 1991. HIV-1 DNA integration: mechanism of viral DNA cleavage and DNA strand transfer. *Cell* 67:1211–21

50. Engels WR. 1996. P elements in Drosophila. *Curr. Top. Microbiol. Immunol.* 204:103–23

51. Engels WR, Johnson-Schlitz DM, Eggleston WB, Sved J. 1990. High-frequency P element loss in Drosophila is homolog dependent. *Cell* 62:515–25

52. Esposito D, Craigie R. 1998. Sequence specificity of viral end DNA binding by HIV-1 integrase reveals critical regions for protein-DNA interaction. *EMBO J.* 17:5832–43

53. Farnet CM, Bushman FD. 1997. HIV-1 cDNA integration: requirement of HMG I(Y) protein for function of preintegration complexes in vitro. *Cell* 88:483–92

54. Fayet O, Ramond P, Polard P, Prere MF, Chandler M. 1990. Functional similarities between retroviruses and the IS3 family of bacterial insertion sequences? *Mol. Microbiol.* 4:1771–77

55. Fletcher TM, Soares MA, McPhearson S, Hui H, Wiskerchen M, et al. 1997. Complementation of integrase function in HIV-1 virions. *EMBO J.* 16:5123–38

56. Gerton JL, Ohgi S, Olsen M, DeRisi J, Brown PO. 1998. Effects of mutations in residues near the active site of human immunodeficiency virus type 1 integrase on specific enzyme-substrate interactions. *J. Virol.* 72:5046–55

57. Gierl A. 1996. The En/Spm transposable element of maize. *Curr. Top. Microbiol. Immunol.* 204:145–59

58. Goldgur Y, Dyda F, Hickman AB, Jenkins TM, Craigie R, Davies DR. 1998. Three new structures of the core domain of HIV-1 integrase: an active site that binds magnesium. *Proc. Natl. Acad. Sci. USA* 95:9150–54

59. Gopaul DN, Guo F, Van DG. 1998. Structure of the Holliday junction intermediate in Cre-loxP site-specific recombination. *EMBO J.* 17:4175–87

60. Goryshin IY, Reznikoff WS. 1998. Tn5 in vitro transposition. *J. Biol. Chem.* 273:7367–74

61. Grindley ND, Leschziner AE. 1995. DNA transposition: from a black box to a color monitor. *Cell* 83:1063–66

62. Hagemann AT, Craig NL. 1993. Tn7 transposition creates a hotspot for homologous recombination at the transposon donor site. *Genetics* 133:9–16

63. Haniford DB, Benjamin HW, Kleckner N. 1991. Kinetic and structural analysis of a cleaved donor intermediate and a strand transfer intermediate in Tn10 transposition. *Cell* 64:171–79

64. Haniford D, Kleckner N. 1994. Tn10 transposition in vivo: temporal separation of cleavages at the two transposon ends and roles of terminal basepairs subsequent to interaction of ends. *EMBO J.* 13:3401–11

65. Haren L. 1998. *Relations structure/fonction des facteurs protéiques exprimés par l'elément transposable bacterien IS911*. PhD thesis. Univ. Paul Sabatier, Toulouse, France

66. Haren L, Polard P, Ton-Hoang B, Chandler M. 1998. Multiple oligomerisation domains in the IS911 transposase: a leucine

zipper motif is essential for activity. *J. Mol. Biol.* 283:29–41

67. Heuer TS, Brown PO. 1997. Mapping features of HIV-1 integrase near selected sites on viral and target DNA molecules in an active enzyme-DNA complex by photo-cross-linking. *Biochemistry* 36:10655–65

68. Heuer TS, Brown PO. 1998. Photo-cross-linking studies suggest a model for the architecture of an active human immunodeficiency virus type 1 integrase-DNA complex. *Biochemistry* 37:6667–78

69. Hiom K, Melek M, Gellert M. 1998. DNA transposition by the RAG1 and RAG2 proteins: a possible source of oncogenic translocations. *Cell* 94:463–70

70. Jenkins TM, Engelman A, Ghirlando R, Craigie R. 1996. A soluble active mutant of HIV-1 integrase: involvement of both the core and carboxyl-terminal domains in multimerization. *J. Biol. Chem.* 271:7712–18

71. Jenkins TM, Esposito D, Engelman A, Craigie R. 1997. Critical contacts between HIV-1 integrase and viral DNA identified by structure-based analysis and photo-crosslinking. *EMBO J.* 16:6849–59

72. Jones JM, Nakai H. 1997. The phiX174-type primosome promotes replisome assembly at the site of recombination in bacteriophage Mu transposition. *EMBO J.* 16:6886–95

73. Junop MS, Haniford DB. 1996. Multiple roles for divalent metal ions in DNA transposition: distinct stages of Tn10 transposition have different Mg2+ requirements. *EMBO J.* 15:2547–55

74. Junop MS, Haniford DB. 1997. Factors responsible for target site selection in Tn10 transposition: a role for the DDE motif in target DNA capture. *EMBO J.* 16:2646–55

75. Kalpana GV, Goff SP. 1993. Genetic analysis of homomeric interactions of human immunodeficiency virus type 1 integrase using the yeast two-hybrid system. *Proc. Natl. Acad. Sci. USA* 90:10593–97

76. Katz RA, Merkel G, Kulkosky J, Leis J, Skalka AM. 1990. The avian retroviral IN protein is both necessary and sufficient for integrative recombination in vitro. *Cell* 63:87–95

77. Kaufman PD, Rio DC. 1992. P element transposition in vitro proceeds by a cut-and-paste mechanism and uses GTP as a cofactor. *Cell* 69:27–39

78. Kennedy AK, Guhathakurta A, Kleckner N, Haniford DB. 1998. Tn10 transposition via a DNA hairpin intermediate. *Cell* 95:125–34

79. Kennedy AK, Haniford DB. 1996. Isolation and characterization of IS10 transposase separation of function mutants: identification of amino acid residues in transposase that are important for active site function and the stability of transposition intermediates. *J. Mol. Biol.* 256:533–47

80. Khan E, Mack JP, Katz RA, Kulkosky J, Skalka AM. 1991. Retroviral integrase domains: DNA binding and the recognition of LTR sequences. *Nucleic Acids Res.* 19:851–60 (Erratum. 1991. *Nucleic Acids Res.* 19:1358)

81. Kim K, Harshey RM. 1995. Mutational analysis of the att DNA-binding domain of phage Mu transposase. *Nucleic Acids Res.* 23:3937–43

82. Kim K, Namgoong SY, Jayaram M, Harshey RM. 1995. Step-arrest mutants of phage Mu transposase. Implications in DNA-protein assembly, Mu end cleavage, and strand transfer. *J. Biol. Chem.* 270:1472–79

83. Kleckner N, Chalmers RM, Kwon D, Sakai J, Bolland S. 1996. Tn10 and IS10 transposition and chromosome rearrangements: mechanisms and regulation in vivo and in vitro. *Curr. Top. Microbiol. Immunol.* 204:49–82

84. Krementsova E, Giffin MJ, Pincus D, Baker TA. 1998. Mutational analysis of the Mu transposase. Contributions of two distinct regions of domain II to recombination. *J. Biol. Chem.* 273:31358–65

85. Kruklitis R, Welty DJ, Nakai H. 1996. ClpX protein of *Escherichia coli* activates bacteriophage Mu transposase in the strand transfer complex for initiation of Mu DNA synthesis. *EMBO J.* 15:935–44

86. Kukolj G, Skalka AM. 1995. Enhanced and coordinated processing of synapsed viral DNA ends by retroviral integrases in vitro. *Genes Dev.* 9:2556–67

87. Kulkosky J, Jones KS, Katz RA, Mack JP, Skalka AM. 1992. Residues critical for retroviral integrative recombination in a region that is highly conserved among retroviral/retrotransposon integrases and bacterial insertion sequence transposases. *Mol. Cell Biol.* 12:2331–38

88. Kulkosky J, Katz RA, Merkel G, Skalka AM. 1995. Activities and substrate specificity of the evolutionarily conserved central domain of retroviral integrase. *Virology* 206:448–56

89. Kunze R. 1996. The maize transposable element activator (Ac). *Curr. Top. Microbiol. Immunol.* 204:161–94

90. Kunze R, Behrens U, Courage-Franzkowiak U, Feldmar S, Kuhn S, Lutticke R. 1993. Dominant transposition-deficient mutants of maize activator (Ac) transposase. *Proc. Natl. Acad. Sci. USA* 90:7094–98

91. Kwon D, Chalmers RM, Kleckner N. 1995. Structural domains of IS10 transposase and reconstitution of transposition activity from proteolytic fragments lacking an interdomain linker. *Proc. Natl. Acad. Sci. USA* 92:8234–38

92. LaFemina RL, Callahan PL, Cordingley MG. 1991. Substrate specificity of recombinant human immunodeficiency virus integrase protein. *J. Virol.* 65:5624–30

93. Lampe DJ, Churchill ME, Robertson HM. 1997. A purified mariner transposase is sufficient to mediate transposition in vitro (erratum). *EMBO J.* 16:4153

94. Lavoie BD, Chaconas G. 1996. Transposition of phage Mu DNA. *Curr. Top. Microbiol. Immunol.* 204:83–102

95. Lavoie BD, Chan BS, Allison RG, Chaconas G. 1991. Structural aspects of a higher order nucleoprotein complex: induction of an altered DNA structure at the Mu-host junction of the Mu type 1 transpososome. *EMBO J.* 10:3051–59

96. Lee CC, Beall EL, Rio DC. 1998. DNA binding by the KP repressor protein inhibits P-element transposase activity in vitro. *EMBO J.* 17:4166–74

97. Lee SP, Xiao J, Knutson JR, Lewis MS, Han MK. 1997. Zn2+ promotes the self-association of human immunodeficiency virus type-1 integrase in vitro. *Biochemistry* 36:173–80

98. Leschziner AE, Griffin TI, Grindley NF. 1998. Tn552 transposase catalyzes concerted strand transfer in vitro. *Proc. Natl. Acad. Sci. USA* 95:7345–50

99. Levchenko I, Luo L, Baker TA. 1995. Disassembly of the Mu transposase tetramer by the ClpX chaperone. *Genes Dev.* 9:2399–408

100. Lodi PJ, Ernst JA, Kuszewski J, Hickman AB, Engelman A, et al. 1995. Solution structure of the DNA binding domain of HIV-1 integrase. *Biochemistry* 34:9826–33

101. Lohe AR, Sullivan DT, Hartl DL. 1996. Subunit interactions in the mariner transposase. *Genetics* 144:1087–95

102. Lutzke RP, Plasterk RA. 1998. Structure-based mutational analysis of the C-terminal DNA-binding domain of human immunodeficiency virus type 1 integrase: critical residues for protein oligomerization and DNA binding. *J. Virol.* 72:4841–48

103. Lutzke RA, Vink C, Plasterk RH. 1994. Characterization of the minimal DNA-binding domain of the HIV integrase protein. *Nucleic Acids Res.* 22:4125–31

104. Machida C, Machida Y. 1989. Regulation of IS1 transposition by the insA gene product. *J. Mol. Biol.* 208:567–74

105. Maekawa T, Amemura-Maekawa J, Ohtsubo E. 1993. DNA binding domains in Tn3 transposase. *Mol. Gen. Genet.* 236:267–74

106. Mahillon J, Chandler M. 1998. Insertion sequences. *Microbiol. Mol. Biol. Rev.* 62:725–74

107. Maxwell A, Craigie R, Mizuuchi K. 1987. B protein of bacteriophage mu is an ATPase that preferentially stimulates intermolecular DNA strand transfer. *Proc. Natl. Acad. Sci. USA* 84:699–703

108. May EW, Craig NL. 1996. Switching from cut-and-paste to replicative Tn7 transposition. *Science* 272:401–4

109. Mizuuchi K. 1992. Polynucleotidyl transfer reactions in transpositional DNA recombination. *J. Biol. Chem.* 267:21273–76

110. Mizuuchi K. 1992. Transpositional recombination: mechanistic insights from studies of Mu and other elements. *Annu. Rev. Biochem.* 61:1011–51

111. Mizuuchi K. 1997. Polynucleotidyl transfer reactions in site-specific DNA recombination. *Genes Cells* 2:1–12

112. Mizuuchi K, Adzuma K. 1991. Inversion of the phosphate chirality at the target site of Mu DNA strand transfer: evidence for a one-step transesterification mechanism. *Cell* 66:129–140

113. Mizuuchi M, Baker TA, Mizuuchi K. 1991. DNase protection analysis of the stable synaptic complexes involved in Mu transposition. *Proc. Natl. Acad. Sci. USA* 88:9031–35

114. Mizuuchi M, Baker TA, Mizuuchi K. 1992. Assembly of the active form of the transposase-Mu DNA complex: a critical control point in Mu transposition. *Cell* 70:303–11

115. Murphy JE, Goff SP. 1992. A mutation at one end of Moloney murine leukemia virus DNA blocks cleavage of both ends by the viral integrase in vivo. *J. Virol.* 66:5092–95

116. Naigamwalla DZ, Chaconas G. 1997. A new set of Mu DNA transposition intermediates: alternate pathways of target capture preceding strand transfer. *EMBO J.* 16:5227–34

117. Nakayama C, Teplow DB, Harshey RM. 1987. Structural domains in phage Mu transposase: identification of the site-specific DNA-binding domain. *Proc. Natl. Acad. Sci. USA* 84:1809–13

118. Namgoong SY, Harshey RM. 1998. The same two monomers within a MuA tetramer provide the DDE domains for the strand cleavage and strand transfer steps of transposition. *EMBO J.* 17:3775–85

119. Namgoong SY, Kim K, Saxena P, Yang JY, Jayaram M, et al. 1998. Mutational analysis of domain IIβ of bacteriophage Mu transposase: domains IIα and IIβ belongs to different catalytic complementation groups. *J. Mol. Biol.* 275:221–32

120. Olasz F, Stalder R, Arber W. 1993. Formation of the tandem repeat (IS30)2 and its role in IS30-mediated transpositional DNA rearrangements. *Mol. Gen. Genet.* 239:177–87

121. Piccirilli JA, Vyle JS, Caruthers MH, Cech TR. 1993. Metal ion catalysis in the Tetrahymena ribozyme reaction. *Nature* 361:85–88

122. Pietrokovski S, Henikoff S. 1997. A helix-turn-helix DNA-binding motif predicted for transposases of DNA transposons. *Mol. Gen. Genet.* 254:689–95

123. Plasterk RH. 1996. The Tc1/mariner transposon family. *Curr. Top. Microbiol. Immunol.* 204:125–43

124. Polard P, Chandler M. 1995. An in vivo transposase-catalyzed single-stranded DNA circularization reaction. *Genes Dev.* 9:2846–58

125. Polard P, Chandler M. 1995. Bacterial transposases and retroviral integrases. *Mol. Microbiol.* 15:13–23

126. Polard P, Ton-Hoang B, Haren L, Betermier M, Walczak R, Chandler M. 1996. IS911-mediated transpositional recombination in vitro. *J. Mol. Biol.* 264:68–81

127. Reimmann C, Haas D. 1987. Mode of replicon fusion mediated by the dupli-

cated insertion sequence IS21 in *Escherichia coli. Genetics* 115:619–25

128. Reznikoff WS. 1993. The Tn5 transposon. *Annu. Rev. Microbiol.* 47:945–63

129. Rezsohazy R, Hallet B, Delcour J, Mahillon J. 1993. The IS4 family of insertion sequences: evidence for a conserved transposase motif. *Mol. Microbiol.* 9:1283–95

130. Rice P, Craigie R, Davies DR. 1996. Retroviral integrases and their cousins. *Curr. Opin. Struct. Biol.* 6:76–83

131. Rice P, Mizuuchi K. 1995. Structure of the bacteriophage Mu transposase core: a common structural motif for DNA transposition and retroviral integration. *Cell* 82:209–20

132. Rowland SJ, Dyke KG. 1990. Tn552, a novel transposable element from Staphylococcus aureus. *Mol. Microbiol.* 4:961–75

133. Rowland SJ, Sherratt DJ, Stark WM, Boocock MR. 1995. Tn552 transposase purification and in vitro activities. *EMBO J.* 14:196–205

134. Saedler H, Gierl A. 1996. *Transposable Elements.* Berlin: Springer-Verlag

135. Sakai J, Chalmers RM, Kleckner N. 1995. Identification and characterization of a pre-cleavage synaptic complex that is an early intermediate in Tn10 transposition. *EMBO J.* 14:4374–83

136. Sakai J, Kleckner N. 1997. The Tn10 synaptic complex can capture a target DNA only after transposon excision. *Cell* 89:205–14

137. Sarnovsky RJ, May EW, Craig NL. 1996. The Tn7 transposase is a heteromeric complex in which DNA breakage and joining activities are distributed between different gene products. *EMBO J.* 15:6348–61

138. Savilahti H, Mizuuchi K. 1996. Mu transpositional recombination: donor DNA cleavage and strand transfer in trans by the Mu transposase. *Cell* 85:271–80

139. Savilahti H, Rice PA, Mizuuchi K. 1995. The phage Mu transpososome core: DNA requirements for assembly and function. *EMBO J.* 14:4893–903

140. Schumacher S, Clubb RT, Cai M, Mizuuchi K, Clore GM, Gronenborn AM. 1997. Solution structure of the Mu end DNA-binding $I\beta$ subdomain of phage Mu transposase: modular DNA recognition by two tethered domains. *EMBO J.* 16:7532–41

141. Scottoline BP, Chow S, Ellison V, Brown PO. 1997. Disruption of the terminal base pairs of retroviral DNA during integration. *Genes Dev.* 11:371–82

142. Sharpe PL, Craig NL. 1998. Host proteins can stimulate Tn7 transposition: a novel role for the ribosomal protein L29 and the acyl carrier protein. *EMBO J.* 17:5822–31

143. Sherman PA, Fyfe JA. 1990. Human immunodeficiency virus integration protein expressed in *Escherichia coli* possesses selective DNA cleaving activity. *Proc. Natl. Acad. Sci. USA* 87:5119–23

144. Sherratt D. 1989. Tn3 and related transposable elements: site-specific recombination and transposition. In *Mobile DNA*, ed. D. Berg, M. Howe, pp. 163–84. Washington, DC: Am. Soc. Microbiol.

145. Shibagaki Y, Chow SA. 1997. Central core domain of retroviral integrase is responsible for target site selection. *J. Biol. Chem.* 272:8361–69

146. Stalder R, Caspers P, Olasz F, Arber W. 1990. The N-terminal domain of the insertion sequence 30 transposase interacts specifically with the terminal inverted repeats of the element. *J. Biol. Chem.* 265:3757–62

147. Stellwagen AE, Craig NL. 1997. Avoiding self: two Tn7-encoded proteins mediate target immunity in Tn7 transposition. *EMBO J.* 16:6823–34

148. Surette MG, Buch SJ, Chaconas G. 1987. Transpososomes: stable protein-DNA complexes involved in the in vitro transposition of bacteriophage Mu DNA. *Cell* 49:253–62

149. Surette MG, Harkness T, Chaconas G. 1991. Stimulation of the Mu A protein-mediated strand cleavage reaction by the Mu B protein, and the requirement of DNA nicking for stable type 1 transpososome formation. In vitro transposition characteristics of mini-Mu plasmids carrying terminal base pair mutations. *J. Biol. Chem.* 266:3118–24

150. Szeverenyi I, Bodoky T, Olasz F. 1996. Isolation, characterization and transposition of an (IS2)2 intermediate. *Mol. Gen. Genet.* 251:281–89

151. Tavakoli NP, DeVost J, Derbyshire KM. 1997. Defining functional regions of the IS903 transposase. *J. Mol. Biol.* 274:491–504

152. Ton-Hoang B. 1998. *Etude du mécanisme de transposition de la séquence d'insertion bactérienne IS911: role d'une forme circulaire dans sa mobilité.* PhD thesis. Univ. Paul Sabatier, Toulouse, France

153. Ton-Hoang B, Betermier M, Polard P, Chandler M. 1997. Assembly of a strong promoter following IS911 circularization and the role of circles in transposition. *EMBO J.* 16:3357–71

154. Ton-Hoang B, Polard P, Chandler M. 1998. Efficient transposition of IS911 circles in vitro. *EMBO J.* 17:1169–81

155. Uchiyama Y, Iwai S, Ueno Y, Ikehara M, Ohtsuka E. 1994. Role of the Mg2+ ion in the *Escherichia coli* ribonuclease HI reaction. *J. Biochem. (Tokyo).* 116:1322–29

156. van den Ent FM, Vos A, Plasterk RH. 1998. Mutational scan of the human immunodeficiency virus type 2 integrase protein. *J. Virol.* 72:3916–24

157. van Gent DC, Elgersma Y, Bolk MW, Vink C, Plasterk RH. 1991. DNA binding properties of the integrase proteins of human immunodeficiency viruses types 1 and 2. *Nucleic Acids Res.* 19:3821–27

158. van Gent DC, Groeneger AA, Plasterk RH. 1992. Mutational analysis of the integrase protein of human immunodefi-

ciency virus type 2. *Proc. Natl. Acad. Sci. USA* 89:9598–602

159. van Gent DC, Mizuuchi K, Gellert M. 1996. Similarities between initiation of V(D)J recombination and retroviral integration. *Science* 271:1592–94

160. van Gent DC, Oude Groeneger AA, Plasterk RH. 1993. Identification of amino acids in HIV-2 integrase involved in site-specific hydrolysis and alcoholysis of viral DNA termini. *Nucleic Acids Res.* 21:3373–77

161. van Gent DC, Vink C, Groeneger AA, Plasterk RH. 1993. Complementation between HIV integrase proteins mutated in different domains. *EMBO J.* 12:3261–67

162. van Luenen HG, Colloms SD, Plasterk RH. 1994. The mechanism of transposition of Tc3 in *C. elegans. Cell* 7984:293–301

163. van Pouderoyen G, Ketting RF, Perrakis A, Plasterk RHA, Sixma TK. 1997. Crystal structure of the specific DNA-binding domain of Tc3 transposase of C. elegans in complex with transposon DNA. *EMBO J.* 16:6044–54

164. Vink C, Lutzke RA, Plasterk RH. 1994. Formation of a stable complex between the human immunodeficiency virus integrase protein and viral DNA. *Nucleic Acids Res.* 22:4103–10

165. Vink C, Oude Groeneger AM, Plasterk RH. 1993. Identification of the catalytic and DNA-binding region of the human immunodeficiency virus type I integrase protein. *Nucleic Acids Res.* 21:1419–25

166. Vink C, van Gent DC, Elgersma Y, Plasterk RH. 1991. Human immunodeficiency virus integrase protein requires a subterminal position of its viral DNA recognition sequence for efficient cleavage. *J. Virol.* 65:4636–44

167. Vos JC, Plasterk RH. 1994. Tc1 transposase of Caenorhabditis elegans is an endonuclease with a bipartite DNA binding domain. *EMBO J.* 13:6125–32

168. Vos JC, van Luenen HG, Plasterk RH.

1993. Characterization of the Caenorhabditis elegans Tc1 transposase in vivo and in vitro. *Genes Dev.* 7:1244–53

169. Wang Z, Harshey RM. 1994. Crucial role for DNA supercoiling in Mu transposition: a kinetic study. *Proc. Natl. Acad. Sci. USA* 91:699–703

170. Wang Z, Namgoong SY, Zhang X, Harshey RM. 1996. Kinetic and structural probing of the precleavage synaptic complex (type 0) formed during phage Mu transposition. Action of metal ions and reagents specific to single-stranded DNA. *J. Biol. Chem.* 271:9619–26

171. Watson MA, Chaconas G. 1996. Three-site synapsis during Mu DNA transposition: a critical intermediate preceding engagement of the active site. *Cell* 85:435–45

172. Wei SQ, Mizuuchi K, Craigie R. 1997. A large nucleoprotein assembly at the ends of the viral DNA mediates retroviral DNA integration. *EMBO J.* 16:7511–20

173. Weinreich MD, Mahnke-Braam L, Reznikoff WS. 1994. A functional analysis of the Tn5 transposase. Identification of domains required for DNA binding and multimerization. *J. Mol. Biol.* 241:166–77

174. Wiater LA, Grindley ND. 1991. Gamma delta transposase. Purification and analysis of its interaction with a transposon end. *J. Biol. Chem.* 266:1841–49

175. Wiegand TW, Reznikoff WS. 1994. Interaction of Tn5 transposase with the transposon termini. *J. Mol. Biol.* 235:486–95

176. Wolfe AL, Felock PJ, Hastings JC, Blau CU, Hazuda DJ. 1996. The role of manganese in promoting multimerization and assembly of human immunodeficiency virus type 1 integrase as a catalytically active complex on immobilized long terminal repeat substrates. *J. Virol.* 70:1424–32

177. Wu Z, Chaconas G. 1995. A novel DNA binding and nuclease activity in domain III of Mu transposase: evidence for a catalytic region involved in donor cleavage. *EMBO J.* 14:3835–43

178. Yamauchi M, Baker TA. 1998. An ATP-ADP switch in MuB controls progression of the Mu transposition pathway. *EMBO J.* 17:5509–18

179. Yang JY, Kim K, Jayaram M, Harshey RM. 1995. A domain sharing model for active site assembly within the Mu A tetramer during transposition: the enhancer may specify domain contributions. *EMBO J.* 14:2374–84

180. Yang W, Steitz TA. 1995. Recombining the structures of HIV integrase, RuvC and RNase H. *Structure* 3:131–34

181. Yin JC, Krebs MP, Reznikoff WS. 1988. Effect of dam methylation on Tn5 transposition. *J. Mol. Biol.* 199:35–45

182. Yurieva O, Nikiforov V. 1996. Catalytic center quest: comparison of transposases belonging to the Tn3 family reveals an invariant triad of acidic amino acid residues. *Biochem. Mol. Biol. Int.* 38:15–20

183. Zerbib D, Jakowec M, Prentki P, Galas DJ, Chandler M. 1987. Expression of proteins essential for IS1 transposition: specific binding of InsA to the ends of IS1. *EMBO J.* 6:3163–69

184. Zheng R, Jenkins TM, Craigie R. 1996. Zinc folds the N-terminal domain of HIV-1 integrase, promotes multimerization, and enhances catalytic activity. *Proc. Natl. Acad. Sci. USA* 93:13659–64

Annu. Rev. Microbiol. 1999. 53:283–314

TRANSMISSIBLE SPONGIFORM ENCEPHALOPATHIES IN HUMANS

Ermias D. Belay

Division of Viral and Rickettsial Diseases, National Center for Infectious Diseases,
Centers for Disease Control and Prevention, Atlanta, Georgia 30333;
e-mail: ebb8@cdc.gov

Key Words   Creutzfeldt-Jakob disease, fatal familial insomnia,
Gerstmann-Sträussler-Scheinker syndrome, new variant CJD, prion diseases

■ **Abstract**   Creutzfeldt-Jakob disease (CJD), the first transmissible spongiform encephalopathy (TSE) to be described in humans, occurs in a sporadic, familial, or iatrogenic form. Other TSEs in humans, shown to be associated with specific prion protein gene mutations, have been reported in different parts of the world. These TSEs compose a heterogeneous group of familial diseases that traditionally have been classified as familial CJD, Gerstmann-Sträussler-Scheinker syndrome, or fatal familial insomnia. In 1996, a newly recognized variant form of CJD among young patients (median age, 28 years) with unusual clinical features and a unique neuropathologic profile was reported in the United Kingdom. In the absence of known CJD risk factors or prion protein gene abnormalities, the UK government concluded that the clustering of these cases may represent transmission to humans of the agent causing bovine spongiform encephalopathy. Additional epidemiologic and recent laboratory data strongly support the UK government's conclusion.

CONTENTS

0066-4227/99/1001-0283$08.00

INTRODUCTION

Transmissible spongiform encephalopathies (TSEs) are a group of rapidly progressive, invariably fatal, neurodegenerative diseases that affect both humans and animals. Most TSEs are characterized by a long incubation period and a neuropathologic feature of multifocal spongiform changes, astrogliosis, neuronal loss, and absence of inflammatory reaction. TSEs in humans include Creutzfeldt-Jakob disease (CJD), kuru, Gerstmann-Sträussler-Scheinker syndrome (GSS), fatal familial insomnia (FFI), and new variant CJD (nvCJD). Kuru has been described only in the Fore population of New Guinea. For many years after its first recognition in 1957, kuru was the most common cause of death among women in the affected population, but it is disappearing because of the cessation of ritualistic cannibalism that had facilitated disease transmission (52, 54, 79, 94). TSEs described in animals include scrapie in sheep and goats, transmissible mink encephalopathy, chronic wasting disease in deer and elk, bovine spongiform encephalopathy (BSE, commonly known as "mad-cow" disease), exotic ungulate spongiform encephalopathy, and feline spongiform encephalopathy in cats, albino tigers, pumas, and cheetahs. The reported ungulate and feline spongiform encephalopathies appear to represent transmission of the BSE agent to these animals.

After descriptions of the transmissibility of scrapie, several theories were suggested to explain the causative agent of TSEs. For years, many researchers proposed that TSEs were caused by "slow viruses"; however, no viruses could be isolated. In the mid-1960s, the possibility that the agent causing scrapie was devoid of nucleic acids or could simply be a protein was suggested (2, 122). Griffith proposed three possible mechanisms by which such an infective protein lacking nucleic acids could support its own replication (65). Subsequent studies demonstrated that the scrapie agent has many chemical properties in common with protein molecule, supporting the protein hypothesis (127, 131). In 1982, Prusiner introduced the term prion to describe the small proteinaceous infectious particle that he showed was an essential component of the scrapie agent (127). Based on what has now become the prion theory, TSEs result from accumulation, in the neurons, of an abnormal isoform of a cellular glycoprotein known as the prion protein (142). The formation of the abnormal isoform is believed to be triggered by mutations in the prion protein gene, a chance error during prion protein gene expression, or transmission of the pathogenic prions. Once the abnormal isoform is formed or acquired, it is believed to promote the conversion of the cellular prion protein in neighboring neurons through an autocatalytic process (132).

Recently, TSEs attracted considerable public attention because of a 1996 report in the United Kingdom that BSE may have spread to humans and caused a newly recognized variant form of CJD (153). BSE was first recognized in 1986 in Great Britain where a total of 173,952 cases (111) were confirmed during 1988–1998 in over 34,500 herds.

This review article describes the clinical and epidemiologic characteristics of TSEs in humans and summarizes the available scientific evidence for a causal link between nvCJD and BSE.

THE ETIOLOGIC AGENT OF TRANSMISSIBLE SPONGIFORM ENCEPHALOPATHIES

Characteristics of the Agent

Many studies have been done to identify the agent causing TSEs. The success in experimentally transmitting scrapie to rodents facilitated etiologic studies focused on characterizing the agent that causes scrapie. The potential transmissibility of the scrapie agent, its proliferation in the animal host, and the retention of scrapie infectivity after filtration led to the suggestion that scrapie might be caused by "slow viruses" or viroids (52, 128). However, efforts to identify viruses or scrapie-specific nucleic acids by using conventional laboratory techniques were unsuccessful (1, 109, 127). Studies have demonstrated the extreme resistance of scrapie infectivity to radiation, nucleases, and other reagents damaging to nucleic acids (7, 127). These findings, coupled with those of other studies that showed diminution of scrapie infectivity with procedures that modified proteins and linkage of inherited TSEs with point genetic mutations, suggested that the TSE agent is neither a virus nor a viroid (128).

The Prion Theory

Because of the unusual properties of the agent that causes TSEs, the term prion (**pro**teinaceous **in**fectious particles) was introduced to differentiate it from conventional infectious agents (127). Several studies have identified a protease-resistant polypeptide in subcellular fractions of hamster brain enriched with scrapie infectivity (10, 51, 104, 129). Absence of the protease-resistant polypeptide in the brains of healthy control animals indicated that this protein is required for scrapie infectivity (10). The molecular characteristics of this protein were shown to be similar to those of a cellular glycoprotein that is most commonly found in neurons (117, 120) but also in other cells of mammals and birds. The pathogenic form of the protein, designated PrP-res or simply prion, is primarily distinguished from its cellular isoform, designated PrP-C, in its three-dimensional structure (75, 76, 119). Spectroscopic measurements of PrP-C from purified fractions of hamster brain demonstrated that 42% of PrP-C is composed of α-helix and is almost devoid of β-sheet (3%) (119). In contrast, PrP-res purified from hamster brain infected with the scrapie agent is composed of 43% β-sheet and 30% α-helix (119). PrP-C is found attached on the surface of neurons with a glycoprotein molecule (138). It is encoded by a group of genes located on the short arm of chromosome 20, and its function is unknown. PrP-res, however, is found in the cytoplasm of affected cells (142) and is resistant to heat, radiation, proteolytic enzymes, and conventional disinfectants such as alcohol, formalin, and phenol (6, 7, 127).

The fundamental event in the occurrence of TSEs seems to be a conformational change in PrP-C (25). Animal studies have suggested that conformational change results from biochemical interaction between PrP-C and PrP-res (132, 145). Although the nature of this interaction is poorly understood, it has been postulated that infecting PrP-res may serve as a template in directing the formation of new PrP-res (145). Ablation of the prion protein gene in mice renders them resistant to infection with the scrapie agent, indicating that the presence of PrP-C is essential for propagation of the infectious agent (24, 130). Intracerebral grafting of PrP-C-expressing neuroectodermal tissue in mice devoid of PrP-C allows the replication of inoculated PrP-res (12).

The prion theory clearly challenges our biological understanding of infectious diseases. Some researchers regard the mere thought that a protein devoid of nucleic acids could dictate its own replication as heretical. Others point out that the existence of different prion "strains" in scrapie and other TSEs indicates the presence of a tightly bound, hitherto unrecognized RNA or DNA molecule in the prion particle (151). Suggestions have also been made that the biological properties of prion "strains" may be encrypted within the conformation of PrP-res (128, 145). However, additional studies are required to substantiate these claims or explain the strain diversity of prions, including the variable phenotypical disease expression of some TSEs within the same species.

CREUTZFELDT-JAKOB DISEASE

History

Patients with rapidly progressive neurodegenerative illnesses were first reported by the German neurologists Creutzfeldt and Jakob in the early 1920s (134). The first patient described by Creutzfeldt was a 22-year-old woman with progressive dementia, tremors, spasticity, ataxia, and possibly myoclonus. Subsequently, Jakob reported five patients, the first three of whom were described as having neuropathologic features similar to Creutzfeldt's case, with a diffuse non-inflammatory disease process and foci of tissue destruction (134). Retrospective review of brain sections of four of Jakob's patients indicated that two of them had the typical neuropathologic features of CJD; neuropathologic review of Creutzfeldt's case was inconclusive (85, 134).

Incidence

CJD is the most common form of TSEs in humans and occurs worldwide, with an estimated incidence of one case/1 million population per year. CJD is reported in almost equal ratios between the sexes, although older males ($\geq$60 years of age) appear to have a higher incidence of disease (69). Brown and co-workers have reported a peak age of onset between 55 and 75 years (mean: 61.5 years) (15). Analysis of the multiple cause-of-death data compiled by the National Center for Health Statistics, Centers for Disease Control and Prevention (CDC), from 1979 through 1994, showed an average annual CJD death rate of 0.95 deaths/1 million population and a median age at the time of death of 68 years in the United States (69); CJD death rates remained stable during this 16-year period (Table 1). The age-adjusted CJD death rate for whites was significantly higher than that for blacks (69).

Clinical Features

CJD has been recognized to occur sporadically, or through iatrogenic transmission, or as a familial form. Affected patients usually present with a rapidly progressive dementia, visual abnormalities, or cerebellar dysfunction, including muscle incoordination and gait and speech abnormalities. During the course of the disease, most patients develop pyramidal and extrapyramidal dysfunction with abnormal reflexes, spasticity, tremors, and rigidity; some patients may also show behavioral changes with agitation, depression, or confusion. These symptoms often deteriorate very rapidly, and patients develop a state of akinetic mutism during the terminal

TABLE 1 Creutzfeldt-Jakob disease (CJD) deaths and death rates[a] by age group and 4-year periods in the United States, 1979–1994

Age groups (years)	1979–1982		1983–1986		1987–1990		1991–1994	
	Number of deaths	Death rate	Number of deaths	Death rate	Number of deaths	Death rate	Number of deaths	Death rate
0–44	22	0.04	10	0.02	21	0.03	27	0.04
45–59	160	1.17	170	1.26	148	1.06	174	1.13
$\geq$60	584	4.04	703	4.52	796	4.84	827	4.83

[a]Death rates are expressed per million persons.

stages of the illness. Myoclonus, the most constant physical sign, is present in nearly 90% of CJD patients (85). CJD is invariably fatal, with a median illness duration of 4 months (mean: 7.6 months); death occurs within 12 months of illness onset in ~85–90% of patients (15). Although a nonspecific, diffusely abnormal electroencephalogram (EEG) tracing is seen in all patients, serial EEG recordings will demonstrate the typical diagnostic pattern in ~75–85% of patients toward the latter part of the illness (15). The diagnostic EEG tracing shows one-cycle to two-cycles per second triphasic sharp-wave discharges (13, 15), which in conjunction with the clinical picture is considered to be diagnostic of CJD. Recently, a new immunoassay has been developed to detect the presence in the cerebrospinal fluid of 14-3-3 protein, which appears to be a marker for CJD. The 14-3-3 protein is a highly conserved protein found in insects, plants, and mammals. In humans and other mammals, 14-3-3 is a normal neuronal protein consisting of several isoforms. Antibodies against the 14-3-3 protein do not cross-react with PrP-res, confirming that these two proteins are different. In patients with dementia, the sensitivity of the 14-3-3 immunoassay in detecting CJD patients was reported to be 96%; the specificity varied from 96% to 99% (74). However, confirmatory diagnosis of CJD requires demonstration of the typical neuropathology or the presence of PrP-res in brain tissue obtained at biopsy or autopsy. The typical neuropathology consists of a microscopic picture of spongiform changes, gliosis, and neuronal loss in the absence of inflammatory reaction (6, 87). The presence of amyloid plaques can be demonstrated in ~5% of CJD patients (39). The presence of PrP-res in biopsy or autopsy brain samples can be demonstrated by immunodiagnostic tests, such as immunohistochemical staining, histoblot, or Western blot techniques (87).

SPORADIC CREUTZFELDT-JAKOB DISEASE

Possible Causes

The sporadic form of CJD accounts for ~85% of all CJD cases. Brown and co-workers reported a mean age at onset of 60 years after their analysis of 232 experimentally transmitted cases of sporadic CJD; the mean duration of illness was 8 months (Table 2) (17). Although the exact cause is unknown, two hypotheses have been suggested to explain the occurrence of sporadic CJD (128). The first one is the possibility of an age-related somatic mutation of the prion protein gene that might result in the formation of PrP-res. Such mutations can randomly occur in the population at a rate of nearly one per million, which roughly corresponds to the incidence of sporadic CJD. The second suggested explanation is the spontaneous conversion of PrP-C into PrP-res in a single neuron or a group of neurons, possibly after a chance error during prion protein gene expression. The PrP-res is then believed to initiate a chain reaction resulting in the spread of disease to other susceptible neurons. The surprisingly stable and uniform incidence of nonfamilial forms of CJD in many different countries and the absence of recognizable transmission patterns to account for a substantial proportion of cases have been strong arguments favoring the spontaneous occurrence of sporadic CJD.

Search for Risk Factors

Consumption of meat and other organs, including animal brain, liver, and kidney, and exposure to blood and blood products have been examined as possible sources of infection in sporadic CJD patients. Several case-control studies done to examine these and other possible modes of transmission have not provided conclusive results (9, 37, 83). A re-analysis of pooled data from previous three case-control studies did not show a statistically significant association of CJD with employment as a health professional, including medical doctors, nurses, dentists, laboratory workers, and ambulance personnel (152). The analysis also found no association of CJD with head trauma, blood transfusion, history of surgery, or consumption of organ meat, including brain, liver, and kidney. The largest case-control study to date conducted in six European countries by the European Union Collaborative Study Group of CJD showed no significant association of CJD with a history of blood transfusion, consumption of beef, veal, lamb, or pork, or occupational exposure to animals or animal products, including butchers and slaughterhouse and farm workers. None of the health professions evaluated in the study were associated with a significant risk of CJD, including physicians, neuropathologists, nurses, laboratory technicians, and dentists. Although raw meat and brain consumption were shown to be significantly associated with an increased risk of CJD, after a conditional regression analysis, brain consumption that would be expected to pose a greater risk was no longer significantly associated (150).

Blood and Blood Products

Experimental transmission of CJD to laboratory animals intracerebrally inoculated with blood obtained from CJD patients indicated the possible presence of the CJD agent in human blood in low concentrations (100, 143). However, review of the experimental studies has raised questions about the validity of these transmissions. In contrast to studies of human blood, three studies have clearly confirmed the infectivity of blood derived from experimentally infected guinea pigs or mice after intracerebral inoculation of healthy animals (89, 99, 144). Recently, the infectivity in rodent models of fractionated blood, including buffy coat, plasma, and plasma fractions I, II, and III derived from experimentally infected animals, has been reported (137). Whether the results of these animal studies provide any conjectural knowledge on blood-borne transmission of the CJD agent in humans is unknown. Several epidemiologic studies indicate that the risk, if any, of CJD transmission through blood or blood products in the human population must be low (137). No convincing evidence exists for any transmission of CJD to a human recipient of blood or blood products (69, 137).

In 1995, a long-term follow-up German study indicated no evidence of CJD transmission either to 27 patients who definitely or to 8 patients who probably received a blood unit from a donor who died of CJD (67). The donor was a 62-year-old man who died from CJD in 1991, and donated 55 units of blood between 1971 and 1991. At least seven of the patients who definitely received a blood unit

TABLE 2 Clinical and neuropathologic features of transmissible spongiform encephalopathies (TSE) in humans

TSE in humans	Mean age at onset (years)	Mean duration of illness (months)	Distinctive clinical features	Presence of typical EEG	Neuropathology
Sporadic CJD (17)	60	8[a]	Dementia, myoclonus, cerebellar dysfunction	75–85% of patients	Spongiform changes, gliosis, neuronal loss; amyloid plaques in 5% of patients
Familial CJD codon 178 (21)	46	23	Resembles sporadic CJD	Extremely rare	Resembles sporadic CJD
codon 183 (115)	45	50	Personality change, dementia, Parkinsonism	Rare	Spongiform change, neuronal loss, mild gliosis mainly in frontal and temporal lobes
codon 200 (20, 108)	55	8	Resembles sporadic CJD	64–74% of patients	Resembles sporadic CJD
Iatrogenic CJD hGH-associated (126)	30[b]	—	Onset with gait abnormalities and ataxia	5% of patients	Resembles sporadic CJD; but more frequent amyloid plaques
dura-associated	39[c]	10[d]	Resembles sporadic CJD	Resembles sporadic CJD	Resembles sporadic CJD
nvCJD (153, 160, 161)	29	16	Onset with psychiatric symptoms, parasthesia or dysesthesia; delayed development of neurologic signs	None	Numerous amyloid plaques surrounded by vacuoles ("florid" plaques); spongiform changes most evident in basal ganglia and thalamus

FFI (98)	49	13	Sleep disturbances and autonomic dysfunction	Rare	Marked neuronal loss and mild gliosis predominantly in the thalamus; rare spongiform changes or plaques
GSS					
codon 102 (18, 53)	48	60	Predominantly have gait abnormalities and ataxia	Extremely rare	Numerous amyloid plaques usually multicentric; spongiform changes severe to absent; gliosis and neuronal loss
codon 105 (80, 81, 156)	44	106	Spastic paraparesis and dementia	None	Numerous amyloid plaques; variable neurofibrillary tangles; absent spongiform changes; gliosis and neuronal loss
codon 117 (116, 149)	38[e]	41	Dementia, pyramidal and extrapyramidal signs; Parkinsonism (in the US family)	Extremely rare	Numerous multicentric or unicentric plaques; occasional neurofibrillary tangles; variable spongiform changes, gliosis and neuronal loss
codon 198 (47, 56, 71)	52	72	Ataxia, Parkinsonism, and dementia	Extremely rare	Numerous amyloid plaques and neurofibrillary tangles; spongiform changes, gliosis, neuronal loss

[a]Brown and coworkers have reported a median illness duration of 4 months after their analysis of 232 neuropathologically verified CJD cases (15).

[b]Based on the age at onset of 12 US patients; younger ages at onset have been reported in nine UK patients (mean age: 22 years) and 24 French patients (mean age: 19 years) (126).

[c]Based on the age at onset of 64 dura mater graft-associated CJD cases, 43 in Japan (32) and 21 in other countries (see Table 3).

[d]Based on 24 dura mater graft-associated CJD cases with known illness duration (see Table 3).

[e]A mean age at onset of 41 years was reported for a French-Alsatian family with GSS associated with a prion protein gene mutation at codon 117 (148).

from this donor survived 10 years or longer; one of these patients survived for at least 20 years. Analysis of CDC's national multiple cause-of-death data also showed that none of the 4,164 patients with CJD who died from 1979 through 1996 were reported to have hemophilia A, hemophilia B, thalassemia, or sickle cell disease, diseases in which patients have increased exposure to blood products (69, 137). If CJD were transmitted through blood products, these groups of patients would be expected to have been at increased risk. In 1995, CDC alerted over 120 US hemophilia treatment centers about the importance of CJD surveillance among hemophilia patients. This alert and subsequent surveillance of deaths of hemophilia patients, including neuropathologic examination of brain tissues from at least 30 hemophilia patients who died of non-CJD conditions, did not detect any evidence of CJD among the hemophilia population (137). Also in 1995, CDC and the American Red Cross initiated a long-term follow-up study of recipients of blood components derived from donors who subsequently develop CJD. As of June 1998, none of the 196 persons who received blood components derived from 15 donors with CJD were reported to have died of CJD. Forty-two of these recipients were reported to have lived for ≥5 years after receipt of a blood component (137), including four recipients who had survived for 13, 14, 16, and 25 years. In addition, several case-control studies have indicated that the proportion of CJD patients with a history of blood transfusion was not significantly different from that of controls (46, 66, 150, 152), indicating that blood transfusion is not a major risk factor for CJD.

Novel Classification

A recent analysis of the clinicopathologic profile, the prion protein gene polymorphism at codon 129, and the molecular characteristics of PrP-res in several hundred sporadic CJD patients indicated that the haplotype at codon 129 and the size of the PrP-res fragment correlate with distinct clinical and pathologic phenotypes. Based on these correlations, Parchi & co-workers proposed a novel classification scheme for sporadic CJD composed of six groups (121). The majority of sporadic CJD patients were classified in the first group: Met/Met or Met/Val at codon 129 and PrP-res fragment type 1; clinically, these patients had prominent dementia and myoclonus, frequent diagnostic EEG, a mean illness duration of 4.4 months, and a pathologic picture that mainly affected the cerebral cortex, deep nuclei, and the cerebellum. In contrast, sporadic CJD patients classified in the third group, for example, Met/Val at codon 129 and PrP-res fragment type 2, mostly had ataxia and dementia, no diagnostic EEG, and a pathologic picture with widespread distribution and the presence of PrP-res–positive amyloid plaques. One of the groups with the least number of patients was a group with val/val at codon 129 and PrP-res fragment type 1. Sporadic CJD patients classified in this group had an illness characterized by prominent dementia, longer duration, absence of the diagnostic EEG picture, and young age at onset (121).

FAMILIAL CREUTZFELDT-JAKOB DISEASE

The familial form of CJD accounts for 5–15% of CJD patients. Because of its autosomal dominant inheritance pattern, there is commonly a family history of CJD in these patients. Familial CJD is most frequently associated with mutations at codon 200 and less frequently with mutations at codon 178, 208, or 210 of the prion protein gene (19, 50, 53, 55, 60, 61, 125). A CJD patient with a double mutation at codons 180 and 232 was reported in Japan (68). Familial CJD with codon 200 mutation has been reported in geographical clusters among Libyan Jews in Israel (62, 108), Spanish families in rural Chile (53, 61), and in central Slovakia (61, 63). Isolated familial cases have been reported in Canada, France, Japan, the United Kingdom, and the United States (53, 61). In families with codon 200 mutations, ~56% of the carriers develop CJD (61). This form of familial CJD occasionally exhibits "generational skip" and occurs in a seemingly sporadic pattern. The age at onset (mean, 55 years), the frequent occurrence of the diagnostic EEG, and the duration of illness (mean, 8 months) in familial CJD patients with codon 200 mutation closely resemble those of sporadic CJD patients (Table 2) (20, 108).

Familial CJD with codon 178 mutation has been reported in families originating from England, Finland, France, Hungary, and the Netherlands (60). A pedigree analysis of the original Finish family indicated that codon 178 mutation could have a disease penetration rate of ~100% (59). This form of familial CJD occurs when the mutant allele coding for asparagine at codon 178 also codes for valine at codon 129 (53, 64). Compared with sporadic CJD patients, familial CJD patients with codon 178 mutation tend to have illness onset at an earlier age (mean, 46 years), longer duration of illness (mean, 23 months), and a nondiagnostic EEG tracing (Table 2) (21).

Recently, a novel prion protein gene mutation at codon 183 with atypical clinicopathologic features was reported in a Brazilian family (115). Seventeen family members over three generations may have been affected, including six patients whose clinical data were consistent with the disease and three patients confirmed by pathologic examination of the brain. Most patients presented with personality changes followed by progressive dementia and a Parkinsonian syndrome. The mean age at onset of the nine patients was 44.8 years, and the mean duration of illness was 4.2 years. In two of the three patients in whom brain tissues were examined, spongiform changes, neuronal loss, and mild gliosis were predominantly seen in the frontal and temporal lobes.

IATROGENIC CREUTZFELDT-JAKOB DISEASE

The transmissible nature of CJD was first described in 1968, after intracerebral inoculation of a brain biopsy tissue from a CJD patient into a chimpanzee (57). However, a possible person-to-person transmission of the CJD agent was not reported until 1974, when Duffy & co-workers described a 55-year-old patient who developed CJD 18 months after receiving a corneal transplant obtained from a donor with CJD (44). Subsequently, other modes of iatrogenic transmission of the CJD agent were reported, including the use of contaminated EEG depth electrodes (8), neurosurgical instruments (154), cadaveric pituitary-derived gonadotropin (34) and human growth hormone (hGH) (26), and dura mater grafts (27, 146).

Human Growth Hormone-Associated Creutzfeldt-Jakob Disease

Recognition in the United States of the first three hGH-associated CJD cases (26, 58, 82, 147) raised concern in many other countries because of a potentially large number of patients who may have received contaminated hGH. All three US patients had growth failure secondary to growth hormone deficiency and received hGH treatment between 1963 and 1980. Investigation of these and additional CJD cases who received hGH through the National Hormone and Pituitary Program did not identify a single lot of product common to all patients (26, 48). Although the production of discrete lots of hGH involved processing of a batch of ~5000–20,000 pituitary glands, combining fractions rich in hGH derived from several batches was a common practice (48). Fradkin & co-workers estimated that, between 1963 and 1985, at least 140 infected pituitary glands may have been processed and randomly distributed among many hGH lots (48). In 1985, the use of hGH was discontinued in the United States, and a follow-up study of ~8000 patients who received hGH through the National Hormone and Pituitary Program between 1963 and 1985 was initiated (110); the study cohort of hGH recipients included 6,282 patients for whom the hGH treatment was confirmed. As of April 1999, 20 of the ~8000 hGH recipients had died of CJD (LB Schonberger, personal communication). Worldwide, >84 hGH-associated CJD cases have been reported, including cases who received imported US products in Brazil and New Zealand, and cases who received locally produced hGH in Australia, France, and the United Kingdom (14, 77). Although hGH has now been replaced by a hormone synthesized with recombinant technology, additional hGH-associated CJD cases are expected because of the long incubation period.

Dura Mater Graft–Associated Creutzfeldt-Jakob Disease

In February 1987, the first recognized dura mater graft-associated CJD case was reported in a 28-year-old woman who developed CJD 19 months after a neurosurgical procedure involving implantation of Lyodura, a brand of dura mater graft produced by B Braun Melsungen AG of Germany (27, 146). A telephone survey conducted by CDC investigators indicated that the procedures used by US processors of dura mater grafts were different from those of B Braun Melsungen AG (28); the US processors avoided batch processing of dura mater grafts obtained from different donors and maintained records that allowed identification and tracing of donors of each graft. Because of these differences between the processing of

Lyodura and of other similar products, CDC cautioned that the use of Lyodura might carry an increased risk of CJD transmission (28). In June 1987, representatives of B Braun Melsungen AG reported that, as of May 1, 1987, their procedures for collecting and processing dura were revised to reduce the risk of CJD transmission (29, 78).

Subsequent case reports of Lyodura-associated CJD patients in Germany (91), Italy (103), Japan (112, 158), New Zealand (29), Spain (30, 101), the United Kingdom (155), and the United States (90) suggested that Lyodura processed before May 1987 was in fact associated with a substantial risk of CJD transmission. The epidemiologic characteristics of dura mater graft-associated CJD patients reported in journals published in English and unpublished US cases investigated by CDC are summarized in Table 3; most of the patients were reported to have received Lyodura processed before 1987. A 1996 nationwide CJD survey in Japan identified 43 dura mater graft-associated CJD patients with illness onset from September 1985 to May 1996; at least 41 of these patients received Lyodura processed before May 1987 (32). The latency period between receipt of a dura mater graft and onset of CJD in the 43 patients ranged from 16 to 193 months (mean, 89 months); the mean age at onset of the patients was 53 years (32). Worldwide, as of September 1998, at least 64 dura mater graft-associated CJD cases have been reported; 57 of these patients were reported to have received Lyodura. Four patients most likely received a non-Lyodura graft, and the type of dura in three patients was unknown (Table 3). In addition, as of June 1998, ~21 unpublished Lyodura-associated CJD cases have been identified, including cases in Argentina, Australia, Canada, Japan, and the United Kingdom (159). The mean age at onset of the 64 dura mater graft-associated CJD cases was 38.5 years; the mean incubation period was 5.8 years.

Stringent donor selection procedures, including exclusion of donors at high risk for CJD and avoidance of batch processing of dura mater grafts, are essential in ensuring the safety of dura mater grafts. Neuropathologic screening of donors to detect asymptomatic TSEs and institution of a sodium hydroxide inactivation step would further minimize the risk of CJD transmission. Because complete inactivation of the CJD agent in an intact tissue such as dura mater may not be achieved, treatment with sodium hydroxide should not be regarded as a substitute for careful clinical and neuropathologic screening of donors. Even the most stringent donor screening and dura mater processing may not totally eliminate the potential for an infectious graft. Surgeons should be aware of this possibly inherent risk of CJD transmission by dura mater grafts and may want to consider the alternative use of autologous fascia lata, temporalis fascia, or synthetic substitutes (32, 78).

Route of Infection in Iatrogenic-Creutzfeldt-Jakob Disease

The length of incubation period and the clinical presentation in iatrogenic CJD patients seem to vary depending on the portal of entry. In hGH-associated CJD cases where infection is through the peripheral route, the mean incubation period

TABLE 3   Dura mater graft-associated Creutzfeldt-Jakob disease (CJD) cases with known epidemiologic characteristics[a]

Country	Age at onset (years)	Sex	Time of dura implantation	Time of CJD onset	Latency period (months)	Illness duration (months)	Pathologic confirmation	Dura mater graft used
Australia (139)	26	M	1982	1994	144	9	No[b]	Lyodura
	61	M	1982	1987	58	—	Yes	Lyodura
France (3, 40, 84)	25	M	Jul. 17, 1986	Dec. 1993	89	8	Yes	Lyodura[c]
	52	F	1984	Mar. 1995	132	5	Yes	Unknown
	59	M	Jan. 16, 1989	Jul. 1992	42	5	Yes	Unknown[c]
Germany (91)	26	F	1984	—	84	2	No[b]	Lyodura
Italy (103, 124)	27	M	May 1985	Jan. 1989	44	—	Yes	Lyodura
	32	F	Nov. 1981	—	111	8	Yes	Non-Lyodura
Japan (32, 112, 137a, 141, 157, 158)	25	F	Mar. 15, 1984	Dec. 1986	33	18	Yes	Lyodura
	35	F	Feb. 20, 1985	Sep. 1989	56	27	Yes	Lyodura
	47	F	Sep. 1985	Nov. 1994	110	17	Yes	Lyodura
	52	M	Dec. 24, 1984	Mar. 1993	99	10	Yes	Lyodura
	65	F	Sep. & Oct. 1991	Feb. 1994	28	20	No[b]	Non-Lyodura[d]
	68	M	Dec. 1985	May 1996	125	8	Yes	Lyodura
	68	F	Sep. 1986	1996	120	—	Yes	Lyodura
New Zealand (29)	25	M	1985	May 1988	31	2	Yes	Lyodura

Spain (30, 33a, 101)	18	M	Apr. 18, 1983	1991	79	3	Yes	Lyodura
	19	M	Nov. 22, 1983	1985	16	21	Yes	Lyodura
	34	M	Dec. 5, 1983	1992	105	—	Yes	Lyodura
	41	F	May 1989	Oct. 1992	41	4	Yes	Lyodura
	57	F	Jun. 24, 1984	1987	43	25	Yes	Lyodura
United Kingdom (45, 155)	30	M	Oct. 1985	Aug. 1989	46	4	Yes	Lyodura
	44	F	1983	1991	96	4	Yes	Lyodura
	51	F	1969	1978	84	6	No[b]	Unknown
United States (5, 27, 90, 146)	28	F	Apr. 1985	Nov. 1986	19	4	Yes	Lyodura
	28	F	Sep. 1985	1990	60	18	Yes	Lyodura
	39	F	Jul. 13, 1992	Jun. 1998	72	3	Yes	Non-Lyodura
	72	M	Nov. 21, 1990	May 1995	54	2	No[b]	Non-Lyodura

[a] Additional dura mater graft-associated CJD cases not included in this table have been reported (see text for details).

[b] The CJD diagnosis in these patients was based on clinical data and a CSF test positive for protein 130/131 (in the Australian patient) or the presence of the diagnostic EEG tracing (in the other four patients).

[c] The Lyodura in the 25-year-old patient was used for embolization of the external carotid artery for a nasopharyngeal angiofibroma (3) and the unknown dura mater graft in the 59-year-old for embolization of the fourth and fifth left posterior intercostal arteries (40).

[d] Although the brand of dura mater graft used in this patient could not be determined, the use of Lyodura processed before May 1987 was reported to be unlikely (32).

has been estimated to be as long as 12 years or longer (14). These patients commonly present with gait abnormalities and ataxia (22, 48). Dementia, a common presenting symptom in sporadic CJD patients, is a late manifestation and is mild (49). Because of the resemblance of this clinical presentation with kuru, it was suggested that the peripheral route of infection may be responsible for a predominantly cerebellar dysfunction at the time of clinical presentation. In contrast, in dura mater graft-associated CJD cases, where the infective tissue is directly placed in the cranium, the mean incubation period was shown to be around 6 years. Although patients with dura mater graft-associated CJD tend to be younger than sporadic CJD patients (mean age, 38.5 and 60.0 years, respectively), their symptomatology is almost identical (45).

POLYMORPHISM AT CODON 129

The human prion protein gene has been shown to exhibit polymorphism for methionine or valine at codon 129. A number of studies have indicated that this polymorphism may play a significant role in determining host susceptibility and the phenotypical disease expression of familial, iatrogenic, or sporadic CJD. Three different studies involving analysis of the prion protein gene of 41 hGH and gonadotropin-associated CJD cases showed that 38 (92.7%) of the cases were homozygous for either methionine (21 cases) or valine (17 cases), and 3 (7.3%) cases were heterozygous for methionine and valine (Met/Val) (16, 35, 41). In addition, Brown and co-workers reported that 12 dura mater graft-associated CJD cases tested for codon 129 polymorphism were found to be homozygous for methionine (16). In comparison, among the 261 healthy controls used in the same studies, 126 (48.3%) were shown to be homozygous for either methionine (97 controls) or valine (29 controls), and 135 (51.7%) were heterozygous with Met/Val. Previously, Palmer & co-workers had shown that, among the 22 sporadic CJD cases they tested, 21 (95.5%) cases were homozygous for either methionine (16 patients) or valine (5 patients), and 1 case (4.5%) was heterozygous with Met/Val (118). Hence, persons who are homozygous for either methionine or valine at codon 129 seem to be predisposed to developing sporadic or iatrogenic CJD. In contrast, heterozygosity with Met/Val at codon 129 seems to be protective against both sporadic and iatrogenic CJD.

In addition, the methionine and valine polymorphism at codon 129 when combined with a mutation at codon 178 has been shown to influence the phenotypical disease expression of inherited TSEs in humans (64). When the mutant allele of the prion protein gene coding for asparagine at codon 178 also codes for valine at codon 129, it produces a clinical picture consistent with CJD rather than FFI. In contrast, when the mutant allele coding for asparagine at codon 178 also codes for methionine at codon 129, it produces a clinical picture consistent with FFI rather than CJD.

A study in Britain suggests that polymorphism at codon 129 may influence the pattern of PrP-res deposition and astrocytosis in sporadic CJD patients (97). Patients with valine homozygosity appeared to be associated with greater PrP-res deposition and astrocytosis in deep gray matter areas. In contrast, patients with methionine homozygosity showed greater PrP-res deposition and astrocytosis in cortical areas. Patients with Met/Val heterozygosity had a mixed deep gray matter and cortical brain pathology. The distribution of spongiform change was not associated with any particular polymorphism at codon 129.

GERSTMANN-STRÄUSSLER-SCHEINKER SYNDROME

GSS, a familial disease with autosomal dominant inheritance, was first described in 1936 by Gerstmann, Sträussler, & Scheinker (134). Since then, the name GSS has been used to describe a heterogeneous group of neurodegenerative disorders with a familial origin, infrequent myoclonus and diagnostic EEG, and a neuropathologic feature of numerous amyloid plaques. GSS is considered a variant of the familial form of CJD, but it is primarily associated with mutations at codon 102 and less frequently with mutations at codon 105, 117, 145, 198, or 217 of the prion protein gene (43, 53). It occurs at an estimated annual incidence of 5 cases/100 million population. Neurologic signs and symptoms that are commonly reported in GSS patients include cerebellar ataxia, gait abnormalities, dementia, dysarthria, ocular dysmetria, and hyporeflexia or areflexia in the lower extremities (85). However, the different prion protein gene mutations in GSS patients are associated with a widely variable clinical presentation, age at onset, and duration of illness (Table 2). The relatively frequent prion protein gene mutations associated with GSS and their clinicopathologic features are briefly described here.

Codon 102

The prion protein gene mutation at codon 102, the most frequently seen mutation in GSS, is associated with a predominantly cerebellar dysfunction at the time of clinical presentation, early age at onset (mean, 48 years), and a prolonged duration of illness (mean, 5 years) (18, 53); patients commonly present with slowly progressive gait abnormalities and ataxia. This form of GSS has been reported in families from Austria, Britain, Canada, France, Germany, Israel, Italy, Japan, and the United States (18, 53, 55, 70, 88). Pedigree and genetic analyses of a 39-year-old woman with GSS indicated that the original Austrian family reported by Gerstmann and co-workers carries a prion protein mutation at codon 102 (86).

Codon 105

A prion protein gene mutation at codon 105 with an unusual clinical picture consisting of spastic paraparesis, cerebellar dysfunction, and dementia has been reported in three Japanese families (80, 81, 156). The age at onset of the patients ranged from 38 to 48 years (mean, 44 years), with a median illness duration of 7 years (mean, 8.8 years). The neuropathology of these patients showed the presence of numerous amyloid plaques and variable degrees of neurofibrillary tangles in the absence of spongiform changes (80, 81).

Codon 117

GSS patients with a prion protein gene mutation at codon 117 predominantly present with dementia associated with extrapyramidal and pyramidal signs (72). Although some patients may develop ataxia, it is relatively a minor feature. This form of GSS has been reported in a large US family of German descent (73, 116) and a French-Alsatian family (43, 149). A 22-year follow-up of the US family indicated that at least 24 family members over five generations might have been affected with a similar disease (116). The neuropathologic profile of the proband of this family was consistent with GSS. The mean age at onset of 20 affected family members was 38 years (range, 22–58 years); the mean duration of illness was 3.4 years (range, 1.5–7 years) (116). Hsiao & Prusiner have described this form of GSS as "dementing GSS" and the codon 102 variant as "ataxic GSS" (72). A second US family with GSS associated with a prion protein gene mutation at codon 117 with early onset gait dysfunction, prominent ataxia, and pseudobulbar and Parkinsonian signs has been reported (102).

Codon 198

The largest and probably most studied family with GSS is the Indiana kindred. A pedigree analysis of this kindred with 1230 members identified 67 affected patients over six generations (47). Most of these patients had symptoms of cerebellar ataxia, dementia, and Parkinsonian features. The mean age at onset of 16 affected members of the kindred was 52 years (range, 34–71 years) (47). Several affected members had neuropathologic findings consistent with GSS (4, 42, 56). Subsequent studies demonstrated the presence of a mutation at codon 198 of the prion protein gene in at least 11 well-documented affected members of the Indiana kindred (42, 71).

FATAL FAMILIAL INSOMNIA

Patients with severe dementia and bilateral symmetrical degeneration of the thalamus have been reported since 1939 (140). Whether these earlier cases shared a similar neuropathology with cases later described as FFI is less clear. The name FFI was first used in 1986 by Lugaresi and co-workers to describe a 52-year-old man who presented with progressive insomnia and autonomic dysfunction, followed by dysarthria, tremor, and myoclonus; the patient's two sisters and many other relatives over three generations had died of a similar disease (96). Neuropathologic examination of the patient and one of his sisters showed neuronal degeneration and reactive astrocytosis confined to the anterior and dorsomedial nuclei of the thalamus; no spongiform changes or inflammatory infiltrates were noted (96). Subsequent study of other affected family members of these siblings, with a total of 22 probable and 7 neuropathologically confirmed cases over five generations, showed a pattern consistent with autosomal dominant inheritance (98); the presence of a mutation at codon 178 of the prion protein gene was later demonstrated in some of these patients (107). Before Lugaresi and co-workers' description of FFI, a US kindred with an illness that had some features consistent with FFI was reported. At least two of the seven affected members had a neuropathologic picture that included marked gliosis and neuronal loss in the dorsomedial and midline thalamic nuclei and a mutation at codon 178 of the prion protein gene (95, 123).

To date, 24 kindreds with FFI associated with a mutation at codon 178 of the prion protein gene were reported worldwide, including cases in Australia, Austria, Britain, France, Germany, Italy, Japan, and the United States (11, 54a, 105, 106, 114, 123, 133, 135). The mean age at onset of affected family members of nine of the FFI kindreds was 49.3 years (range, 25–72 years); the mean duration of illness was 12.5 months (range, 5–25 months). The predominant feature in most FFI patients is involvement of the thalamus associated with severe sleep disturbances, often with intractable insomnia, and autonomic dysfunction (98). The sleep disorder is characterized by loss of the slow-wave and rapid-eye-movement phases of the sleep cycle (98, 107). The autonomic dysfunction commonly seen in FFI patients includes hyperhidrosis, hyperthermia, tachycardia, and hypertension (98, 107). Cerebellar dysfunction, enacted dream states, myoclonus, and pyramidal signs are also reported in affected patients. A loss of circadian rhythm in the production of growth hormone, prolactin, and melatonin can occur. The diagnostic EEG seen in a majority of CJD patients is rarely seen in FFI patients. The neuropathology shows marked neuronal loss in the thalamus and mild gliosis; spongiform changes or plaques are rarely demonstrated (98). Minimal to mild gliosis has been demonstrated in both the cerebrum and cerebellum. The absence of spongiform changes and the unusual clinical presentation had raised questions about whether FFI should be classified as one of the TSEs. Demonstration of the presence of PrP-res in the brain of affected patients has since resolved this dilemma (107). The proteinase K–treated PrP-res fragment associated with FFI is distinguished by a different pattern of glycosylation and size (113). Although rare sporadic cases are reported, FFI is primarily associated with an allele that has a specific prion protein gene mutation at codon 178 in combination with methionine at the polymorphic codon 129 (64, 113). The polymorphism at codon 129 of the non-mutant allele in FFI patients influences the duration of the disease. The illness duration has been reported to be significantly shorter in FFI patients with methionine homozygosity at codon 129 (mean 12 ± 4 months) compared with patients with methionine and valine heterozygosity (mean 21 ± 15 months). The age at onset was not significantly different in the two groups of patients (54a).

NEW VARIANT CREUTZFELDT-JAKOB DISEASE

Clinicopathologic Features

New variant CJD (nvCJD) was first reported on March 20, 1996, when the UK government's expert advisory committee announced its conclusion that the agent causing BSE might have spread to humans, based on recognition of 10 persons with a newly recognized variant form of CJD during February 1994 to October 1995. In April 1996, Will & co-workers described the detailed clinical and neuropathologic features of these 10 patients, who ranged in age from 16 to 39 years (median age, 28 years) (153). Their proposal that nvCJD was a new disease entity was based on a cluster of these young patients with unusual clinical features in the absence of known recognizable CJD risk factors or prion protein gene abnormalities and the unique but uniform neuropathologic profile observed in all the patients. The unusual clinical features of the 10 nvCJD patients included prominent behavioral changes at the time of clinical presentation, with subsequent onset of neurologic abnormalities, including ataxia within weeks or months, dementia and myoclonus late in the illness, an illness duration of at least 6 months, and nondiagnostic EEG changes (31, 153). The unique neuropathologic findings included spongiform changes most evident in the basal ganglia and thalamus with sparse distribution throughout the cerebral cortex, and widespread kuru-type amyloid plaques in both the cerebellum and cerebrum. The morphology of the plaques, with a pale periphery and surrounding spongiform lesions, was different from that generally seen in patients with classic CJD but resembles the "florid" plaques described in scrapie. Immunohistochemical analysis showed accumulation of PrP-res in high densities in both the cerebellum and cerebrum. Intensified CJD surveillance in other European countries did not identify cases with similar clinicopathologic profile, except one case in France (33).

As of February 28, 1999, a total of 38 confirmed cases and 2 probable cases of nvCJD had been reported in the United Kingdom (Figure 1). The age at onset of the nvCJD patients ranged from 16 to 52 years (median age, 28 years) (161). A report of the first 21 nvCJD patients indicated that they all had early onset behavioral change symptoms and a nondiagnostic or normal EEG tracing (161). The median

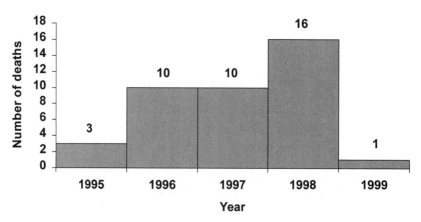

Figure 1 New variant Creutzfeldt-Jakob disease deaths in the United Kingdom, January 1995 to February 1999. Data from Department of Health, United Kingdom, Monthly Creutzfeldt-Jakob disease figures, April 12, 1999, London.

illness duration of these patients was 14 months. An analysis of the first 14 nvCJD patients revealed that nine of them presented with behavioral change symptoms, such as agitation, aggression, anxiety, apathy, depression, emotional lability, insomnia, poor concentration, paranoid delusion, recklessness, or withdrawal; 5 of these patients had a combination of at least two of the behavioral change symptoms (160). Four of the 14 nvCJD patients presented with dysesthesia or parasthesia, and one patient presented with forgetfulness. All the other patients who did not present with behavioral changes developed the behavioral change symptoms early in the illness. None of the 14 patients developed overt neurologic signs within the 4-month period after illness onset. The time interval from illness onset to development of overt neurologic signs ranged from 4- to 24.5 months (median, 6.3 months) (161). The first neurologic sign was usually ataxia with rapid progression of the illness, including development of global cognitive impairment, involuntary movements, incontinence of urine, progressive immobility, unresponsiveness, and mutism. Almost at the same time as the appearance of clear neurologic signs, 12 of the 14 nvCJD patients had delusions, such as a belief that snipers were in the kitchen, patient's baby had recently died, or microscopic people were inside a patient's body. These delusional beliefs were usually fleeting, lasting for hours or days. Magnetic resonant imaging results of some of the nvCJD patients showed typical abnormalities, including high signals in the posterior thalamic areas on T2-weighted images. Interestingly, protein 14-3-3 in the cerebrospinal fluid was detected in only 2 of the 5 patients tested (161). Detection of this protein has been reported to be highly specific and sensitive for the diagnosis of classic CJD (40). In one study, the 14-3-3 protein was detected in the cerebrospinal fluid of all 10 BSE-infected cattle tested and in only 1 of 6 healthy controls (93). All nvCJD patients so far tested for codon 129 polymorphism were homozygous for methionine (161).

Very strong epidemiologic and laboratory evidence is accumulating for a causal association between BSE and nvCJD. The epidemiologic evidence is mainly based on the geographic clustering of nvCJD patients in the United Kingdom where the overwhelming majority of BSE cases were reported. The laboratory evidence was provided by experimental data from (*a*) a French study that involved inoculation of macaques with brain tissues obtained from BSE-infected cattle (92); (*b*) a Western blot analysis of infecting prions obtained from BSE-infected animals and nvCJD patients (36); and (*c*) an ongoing animal model study involving panels of inbred mice that is being conducted in Scotland (23).

Geographic Clustering

The occurrence of unexplained spongiform encephalopathy in at least 13 UK decedents <30 years of age between 1995 and 1997 is extraordinary. In the United States, for a population ~4 times greater than that of the United Kingdom, from 1990 through 1994, mortality data analysis identified one CJD death in a person <30 years of age, a patient who received hGH injections (31). The absence of similar cases in other BSE-free countries despite intensive surveillance efforts supports the link between the nvCJD outbreak and the BSE outbreak in the United Kingdom. The interval between the initial widespread exposure to potentially contaminated food (1984–1989) and illness onset of the initial nvCJD patients (1994–1996) is consistent with known incubation periods for CJD.

Experimental Study Using Macaques

In June 1996, French researchers reported an experimental study of three cynomologous macaques (two adults and one neonate) that were intracerebrally inoculated with brain homogenates obtained from BSE-infected cattle (92). The adult macaques developed a neurologic disease 150 weeks postinoculation that was characterized by behavioral changes, including depression, edginess, and voracious appetite, followed by ataxia, gait abnormalities, tremors, and late onset myoclonus. Neuropathologic examination revealed numerous "florid" plaques in all the three macaques with plaque morphology and distribution indistinguishable from that seen in nvCJD patients. Two other macaques inoculated with the sporadic CJD agent showed a different neuropathologic picture with numerous lesions in the cortex and no plaques. Because cynomologous macaques represent the nearest available experimental model to humans, the findings of this study provided significant laboratory evidence for a possible link between nvCJD and BSE.

Western Blot Analysis

In October 1996, Collinge and co-workers reported that PrP-res obtained from sporadic, iatrogenic, and nvCJD patients can be distinguished by different banding patterns on Western blot analysis (36). Types-1 and -2 Western blot banding patterns were seen in PrP-res obtained from sporadic CJD patients, a predominantly type-3 pattern in iatrogenic CJD patients, and a type-4 pattern in all 10 nvCJD patients tested. Western blot analysis of PrP-res obtained from cattle and domestic cats naturally infected with BSE and mice and a macaque with experimental BSE showed the type 4 pattern, indicating that PrP-res obtained from BSE-infected cattle has similar molecular properties as PrP-res obtained from the nvCJD patients. The type-4 pattern was distinguishable from the type-3 pattern mainly in the band intensity or glycoform ratio.

Experimental Study Using Inbred Mice

The strongest and most compelling evidence for a causal association between BSE and nvCJD was provided by an ongoing study that is being conducted in Scotland by Bruce and co-workers (23). In this elaborate study, the authors reported that a panel of inbred mice inoculated with the agent causing BSE or nvCJD exhibited indistinguishable patterns of incubation periods and "lesion profile." In previous studies, three panels of inbred mice (RIII, C57BL, and VM) and a panel of cross-bred mice (C57BL × VM) were challenged with brain homogenates obtained from eight cattle with BSE and three domestic cats and two exotic ruminants with TSEs.

A uniform pattern of incubation periods was observed in these mice; the shortest incubation period was for RIII mice (range of mean incubation period, 302–335 days), followed sequentially by C57BL, VM, and C57BL × VM mice (23). A similar lesion profile, a vacuolation score on a scale of 0 to 5 in nine areas of the brain, was also observed in the mice. Subsequently, the authors challenged similar panels of mice with brain homogenates obtained from three nvCJD and six sporadic CJD patients. All the surviving RIII mice injected with nvCJD brain homogenates developed neurologic disease with a strikingly uniform pattern of incubation periods (range, 288–351 days) and lesion profile as the RIII mice experimentally infected with BSE of cattle and TSEs of cats and exotic ruminants. In contrast, all the mice that were inoculated with brain homogenates obtained from the six sporadic CJD patients did not show any signs of neurologic disease after 600–800 days (23). However, a large majority of the mice that survived beyond 500 days had neuropathologic evidence of disease transmission; the lesion profile of two of these RIII mice was distinct from that seen in RIII mice infected with the agents causing BSE and nvCJD. A longer period of observation is required to determine whether the non-RIII mice inoculated with the nvCJD agent will develop neurologic disease within the predicted pattern of incubation periods. If the nvCJD agent continues to behave in a similar fashion as the BSE agent, the next panel of mice that would develop neurologic disease should be the C57BL strain, followed sequentially by the VM and C57BL × VM strains. Some of the C57BL are reported to have already shown signs of neurologic disease (23).

CREUTZFELDT-JAKOB DISEASE SURVEILLANCE IN THE UNITED STATES

CDC conducts CJD surveillance through periodic review of the national multiple cause-of-death data compiled by the National Center for Health Statistics. Mortality data analysis is an efficient way of conducting CJD surveillance because CJD is invariably fatal, >85% of patients die within a year of onset, and its diagnosis is better ascertained at the time of death. Studies have shown that death certificate reviews identify ≥80% of CJD deaths in the United States (31, 38).

After the occurrence of nvCJD was announced in the United Kingdom, CDC conducted an active CJD surveillance in its four established Emerging Infections Program sites and the Atlanta Metropolitan Active Surveillance Project in Georgia (31, 136). This active CJD surveillance involved review of available death certificate data during 1991–1995 and contact by phone, mail, or fax of neurologists, neuropathologists, and pathologists to identify patients who died from CJD during 1991–1995. Between 90–100% of these specialists in the surveillance areas were contacted. In addition, clinical and neuropathologic records for each CJD patient aged <55 years were sought for review. The annual number of CJD deaths was stable in the five areas during 1991–1995, and the average annual CJD death rate was 1.2/1 million population. No nvCJD cases were identified.

In addition to regular epidemiologic reviews of the National Center for Health Statistics multiple cause-of-death data, since 1996 CDC has augmented surveillance for the nvCJD by conducting follow-up investigation of CJD decedents <55 years of age. This surveillance was initiated with the support of the Council of State and Territorial Epidemiologists. Since April 1996, review of clinical records of 76 of 86 CJD patients who died during 1994–1996 has been completed. Neuropathology records were reviewed for 34 of these patients (ED Belay, unpublished data). Also, in collaboration with CDC, the American Association of Neuropathologists helped establish a National Prion Disease Pathology Surveillance Center at Case Western Reserve University, Cleveland, Ohio and alerted its members in 1996 about the nvCJD neuropathology and requested reports of any such cases, regardless of the clinical diagnosis or age of the patient. A similar notification and request was also sent to US members of the United States and Canadian Academy of Pathologists. These surveillance efforts have not detected evidence of the occurrence of nvCJD in the United States.

ACKNOWLEDGMENTS

I am very grateful for the support I received from Dr. Lawrence B. Schonberger during the preparation of this manuscript and for his critical review of the draft. I also thank Robert C. Holman for statistical analysis of the data used in Table 1, Abtin Shahriari for his assistance in the preparation of Table 2, and John P. O'Connor for editorial review.

Visit the Annual Reviews home page at http://www.AnnualReviews.org

LITERATURE CITED

1. Aiken JM, Marsh RF. 1990. The search for scrapie agent nucleic acid. *Microbiol. Rev.* 54:242–46

2. Alper T, Haig DA, Clarke MC. 1966. The exceptionally small size of the scrapie agent. *Biochem. Biophys. Res. Commun.* 22:278–84

3. Antoine JC, Michel D, Bertholon P, Mosnier JF, Laplanche JL, et al. 1997. Creutzfeldt-Jakob disease after extracranial dura mater embolization for a nasopharyngeal angiofibroma. *Neurology* 48:1451–53

4. Azzarelli B, Muller J, Ghetti B, Dyken M, Conneally PM. 1985. Cerebellar plaques in familial Alzheimer's disease (Gerstmann-Sträussler-Scheinker variant?). *Acta Neuropathol.* 65:235–46

5. Belay ED, Dobbins JG, Malecki J, Buck BE, Bell M, et al. 1998. *Creutzfeldt-Jakob disease (CJD) in a recipient of a U.S.-processed dura mater graft: cause or coincidence?* Presented at Joint Inst. for Food Safety and Appl. Nutr. Workshop on TSE Risks in Relation to Source Materials, Processing, and End-product Use, June 8–9, 1998, College Park, MD

6. Bell JE, Ironside JW. 1993. Neuropathology of spongiform encephalopathies in humans. *Br. Med. Bull.* 49:738–77

7. Bellinger Kawahara, Diener TO, McKinley MP, Groth DF, Smith DR, Prusiner SB. 1987. Purified scrapie prions resist inactivation by procedures that hydrolyze, modify, or shear nucleic acids. *Virology* 160:271–74

8. Bernoulli C, Siegfried J, Baumgartner G, Regli F, Rabinowicz T, et al. 1977. Danger of accidental person-to-person transmission of Creutzfeldt-Jakob disease by surgery. *Lancet* 1:478–79

9. Bobowick AR, Brody JA, Matthews MR, Roos R, Gajdusek DC. 1973. Creutzfeldt-Jakob disease: a case-control study. *Am. J. Epidemiol.* 98:381–94

10. Bolton DC, McKinley MP, Prusiner SB. 1982. Identification of a protein that purifies with the scrapie prion. *Science* 218:1309–11

11. Bosque PJ, Vnencak Jones CL, Johnson MD, Whitlock JA, McLean MJ. 1992. A PrP gene codon 178 base substitution and a 24-bp interstitial deletion in familial Creutzfeldt-Jakob disease. *Neurology* 42:1864–70

12. Bandner S, Raeber A, Sailer A, Blättler T, Fischer M, et al. 1996. Normal host prion protein (PrP^C) is required for scrapie spread within the central nervous system. *Proc. Natl. Acad. Sci. USA* 93:13148–51

13. Brown P. 1993. EEG Findings in Creutzfeldt-Jakob disease. *JAMA* 269:3168

14. Brown P. 1996. Environmental causes of human spongiform encephalopathy. In *Methods in Molecular Medicine: Prion Diseases,* eds. HF Baker, RM Ridley, pp. 139–54. Totowa, NJ: Humana

15. Brown P, Cathal F, Castaigne P, Gajdusek DC. 1986. Creutzfeldt-Jakob disease: clinical analysis of a consecutive series of 230 neuropathologically verified cases. *Ann. Neurol.* 20:597–602

16. Brown P, Cervenáková L, Goldfarb G, McCombie WR, Rubenstein R, et al. 1994. Iatrogenic Creutzfeldt-Jakob disease: an example of the interplay between ancient genes and modern medicine. *Neurology* 44:291–93

17. Brown P, Gibbs J, Rodgers-Johnson P, Asher DM, Sulima M, et al. 1994. Human spongiform encephalopathy: the National Institutes of Health series of 300 cases of experimentally transmitted disease. *Ann. Neurol.* 35:513–29

18. Brown P, Goldfarb LG, Brown WT, Goldgaber D, Ruberstein R, et al. 1991. Clinical and molecular genetic study of a large German kindred with Gerstmann-Sträussler-Scheinker syndrome. *Neurology* 41:375–79

19. Brown P, Goldfarb LG, Gajdusek DC. 1991. The new biology of spongiform encephalopathy: infectious amyloidoses with a genetic twist. *Lancet* 337:1019–22

20. Brown P, Goldfarb LG, Gibbs CJ, Gajdusek DC. 1991. The phenotypic expression of different mutations in transmissible familial Creutzfeldt-Jakob disease. *Eur. J. Epidemiol.* 7:469–76

21. Brown P, Goldfarb LG, Kovanen J, Haltia M, Cathala F, et al. 1992. Phenotypic characteristics of familial Creutzfeldt-Jakob disease associated with the codon 178[Asn] PRNP mutation. *Ann. Neurol.* 31:282–85

22. Brown P, Preece MA, Will G. 1992. "Friendly fire" in medicine: hormones, homografts, and Creutzfeldt-Jakob disease. *Lancet* 340:24–27

23. Bruce ME, Will RG, Ironside JW, McConnell I, Drummond D, et al. 1997. Transmission to mice indicate that 'new variant' CJD is caused by the BSE agent. *Nature* 389:498–501

24. Büeler H, Aguzzi A, Sailer A, Greiner RA, Augenried P, et al. 1993. Mice devoid of PrP are resistant to scrapie. *Cell* 73:1339–47

25. Caughey B, Raymon G. 1991. The scrapie-associated form of PrP is made from a cell surface precursor that is both protease- and phosphlipase-sensitive. *J. Biol. Chem.* 266:18217–23

26. Centers for Disease Control. 1985. Fatal degenerative neurologic disease in patients who received pituitary-derived human growth hormone. *Morbid. Mortal. Wkly. Rep.* 34:359–60, 365–66

27. Centers for Disease Control. 1987. Rapidly progressive dementia in a patient who

received a cadaveric dura mater graft. *Morbid. Mortal. Wkly. Rep.* 36:49–50, 55

28. Centers for Disease Control. 1987. Update: Creutzfeldt-Jakob disease in a patient receiving a cadaveric dura mater graft. *Morbid. Mortal. Wkly. Rep.* 36:324–25

29. Centers for Disease Control. 1989. Update: Creutzfeldt-Jakob disease in a second patient who received a cadaveric dura mater graft. *Morbid. Mortal. Wkly. Rep.* 38:37–38, 43

30. Centers for Disease Control and Prevention. 1993. Creutzfeldt-Jakob disease in patients who received a cadaveric dura mater graft-Spain, 1985–1992. *Morbid. Mortal. Wkly. Rep.* 42:560–63

31. Centers for Disease Control and Prevention. 1996. Surveillance for Creutzfeldt-Jakob disease-United States. *Morbid. Mortal. Wkly. Rep.* 45:665–68

32. Centers for Disease Control and Prevention. 1997. Creutzfeldt-Jakob disease associated with cadaveric dura mater grafts-Japan, January 1979–May 1996. *Morbid. Mortal. Wkly. Rep.* 46:1066–69

33. Chazot E, Broussolle E, Lapras CI, Blattler T, Aguzzi A, Kopp N. 1996. New variant of Creutzfeldt-Jakob disease in a 26-year-old French man. *Lancet* 347:1181

33a. Clavel M, Clavel P. 1996. Creutzfeldt-Jakob disease transmitted by dura mater graft. *Eur. Neurol.* 36:239–40

34. Cochius JI, Mack K, Burns RJ, Alderman CP, Blumbergs PC. 1990. Creutzfeldt-Jakob disease in a recipient of human pituitary-derived gonadotropin. *Aust. N.Z. J. Med.* 20:592–93

35. Collinge J, Palmer MS, Dryden AJ. 1991. Genetic predisposition to iatrogenic Creutzfeldt-Jakob disease. *Lancet* 337:1441–42

36. Collinge J, Sidle KCL, Heads J, Ironside J, Hill AF. 1996. Molecular analysis of prion strain variation and the aetiology of 'new variant' CJD. *Nature* 383:685–90

37. Davanipour Z, Alter M, Sobel E, Asher D, Gajdusek DC. 1985. A case-control study of Creutzfeldt-Jakob disease, dietary risk factors. *Am. J. Epidemiol.* 122:443–52

38. Davanipour Z, Smoak C, Bohr T, Sobel E, Liwnicz B, Chang S. 1995. Death certificates: an efficient source for ascertainment of Creutzfeldt-Jakob disease cases. *Neuroepidemiology* 14:1–6

39. DeArmond SJ, Prusiner SB. 1995. Etiology and pathogenesis of prion diseases. *Am. J. Pathol.* 146:785–811

40. Defebvre L, Destée A, Caron J, Ruchoux MM, Wurtz A, Remy J. 1997. Creutzfeldt-Jakob disease after an embolization of intercostal arteries with cadaveric dura mater suggesting a systemic transmission of the prion agent. *Neurology* 48:1470–71

41. Deslys J-P, Marce D, Dormont D. 1994. Similar genetic susceptibility in iatrogenic and sporadic Creutzfeldt-Jakob disease. *J. Gen. Virol.* 75:23–27

42. Dlouhy SR, Hsiao K, Farlow MR, Foroud T, Conneally PM, et al. 1992. Linkage of the Indiana kindred of Gerstmann-Sträussler-Scheinker disease to the prion protein gene. *Nat. Gen.* 1:64–67

43. Doh-ura K, Tateishi J, Sasaki H, Kitamoto T, Sakaki Y. 1989. PRO-LEU change at position 102 of prion protein is the most common but not the sole mutation related to Gerstmann-Sträussler-Scheinker syndrome. *Biochem. Biophys. Res. Commun.* 163:974–79

44. Duffy P, Wolf J, Collins G, DeVoe AG, Streeten B, Cowen D. 1974. Possible person-to-person transmission of Creutzfeldt-Jakob disease. *N. Engl. J. Med.* 290:692–93

45. Esmonde T, Lueck CJ, Symon L, Duchen LW, Will RG. 1993. Creutzfeldt-Jakob disease and lyophilised dura mater grafts: report of two cases. *J. Neurol. Neurosurg. Psychiatry* 56:999–1000

46. Esmonde TFG, Will RG, Slattery JM, Knight R, Harries-Jones R, et al. 1993. Creutzfeldt-Jakob disease and blood transfusion. *Lancet* 341:205–7

47. Farlow MR, Yee RD, Dlouhy SR,

Conneally PM, Azzarelli B, Ghetti B. 1989. Gerstmann-Sträussler-Scheinker disease. I. Extending the clinical spectrum. *Neurology* 39:1446–52

48. Fradkin JE, Schonberger LB, Mills JL, Gunn WJ, Piper JM, et al. 1991. Creutzfeldt-Jakob disease in pituitary growth hormone recipients in the United States. *JAMA* 265:880–84

49. Frasier SD, Foley TP Jr. 1994. Clinical review of 58 Creutzfeldt-Jakob disease in recipients of pituitary hormones. *J. Clin. Endocrinol. Metab.* 78:1277–79

50. Furukawa H, Kitamoto T, Hashiguchi H, Tateishi J. 1996. A Japanese case of Creutzfeldt-Jakob disease with a point mutation in the prion protein gene at codon 210. *J. Neurol. Sci.* 141:120–22

51. Gabizon R, McKinley MP, Groth DF, Prusiner SB. 1988. Immunoaffinity purification and neutralization of scrapie prion infectivity. *Proc. Natl. Acad. Sci. USA* 85:6617–21

52. Gajdusek DC. 1977. Unconventional viruses and the origin and disappearance of kuru. *Science* 197:943–60

53. Gajdusek DC. 1996. Infectious amyloides: subacute spongiform encephalopathies as transmissible cerebral amyloidoses. In *Fields Virology Third Edition,* ed. BN Fields, DM Knipe, PM Howley, pp. 2851–900. Philadelphia: Lippincott-Raven

54. Gajdusek DC, Zigas V. 1957. Degenerative disease of the central nervous system in New Guinea. The endemic occurrence of "kuru" in the native population. *N. Engl. J. Med.* 257:974–78

54a. Gambetti P, Lugaresi E. 1998. Conclusions of the symposium *Brain Pathol.* 8:571–75

55. Ghetti B, Piccardo P, Frangione B, Bugiani O, Giaccone G, et al. 1996. Prion protein amyloidosis. *Brain Pathol.* 6:127–45

56. Ghetti B, Tagliavini F, Masters CL, Beyreuther K, Giaccone G, et al. 1989.

Gerstmann-Sträussler-Scheinker disease. II. Neurofibrillary tangles and plaques with PrP-amyloid coexist in an affected family. *Neurology* 39:1453–61

57. Gibbs CJ Jr, Gajdusek DC, Asher DM, Alpers MP, Beck E, et al. 1968. Creutzfeldt-Jakob disease (spongiform encephalopathy): transmission to the chimpanzee. *Science* 161:388–89

58. Gibbs CJ Joy A, Heffner R, Franko M, Miyazaki M, Jr, et al. 1985. Clinical and pathological features and laboratory confirmation of Creutzfeldt-Jakob disease in a recipient of pituitary-derived human growth hormone. *N. Engl. J. Med.* 313:734–38

59. Goldfarb LG, Brown P, Cervenáková L, Gajdusek DC. 1994. Molecular genetic studies of Creutzfeldt-Jakob disease. *Mol. Neurobiol.* 8:89–97

60. Goldfarb LG, Brown P, Haltia M, Cathala F, McCombie WR, et al. 1992. Creutzfeldt-Jakob disease cosegregates with the codon 178^{Asn} PRNP mutation in families of European origin. *Ann. Neurol.* 31:274–81

61. Goldfarb LG, Brown P, Mitrovà E, Cervenáková L, Goldin L, et al. 1991. Creutzfeldt-Jakob disease associated with the PRNP codon 200 mutation: an analysis of 45 families. *Eur. J. Epidemiol.* 7:477–86

62. Goldfarb LG, Korczyn AD, Brown P, Chapman J, Gajdusek DC. 1990. Mutation in codon 200 of scrapie amyloid precursor gene linked to Creutzfeldt-Jakob disease in Sephardic Jews of Libyan and non-Libyan origin. *Lancet* 336:637–38

63. Goldfarb LG, Mitrovà E, Brown P, Toh BH, Gajdusek DC. 1990. Mutation in codon 200 of scrapie amyloid precursor gene in two clusters of Creutzfeldt-Jakob disease in Slovakia. *Lancet* 336:514–15

64. Goldfarb LG, Peterson RB, Tabaton M, Brown P, LeBlanc AC, et al. 1992. Fatal familial insomnia and familial Creutzfeldt-Jakob disease: disease

phenotype determined by a DNA polymorphism. *Science* 258:806–8

65. Griffith JS. 1967. Self-replication and scrapie. *Nature* 215:1043–44

66. Harries-Jones R, Knight R, Will RG, Cousens S, Smith PG, Matthews WB. 1988. Creutzfeldt-Jakob disease in England and Wales, 1980–1984: a case-control study of potential risk factors. *J. Neurol. Neurosurg. Psychiatry* 51:1113–19

67. Heye N, Hensen S, Muller N. 1994. Creutzfeldt-Jakob disease and blood transfusion. *Lancet* 343:298–99

68. Hitoshi S, Nagura H, Yamagouchi H, Kitamoto T. 1993. Double mutations at codon 180 and codon 232 of the PrP gene in an apparently sporadic case of Creutzfeldt-Jakob disease. *J. Neurol. Sci.* 93:208–12

69. Holman RC, Khan AS, Belay ED, Schonberger LB. 1996. Creutzfeldt-Jakob disease in the United States, 1979–1994: Using national mortality data to assess the possible occurrence of variant cases. *Emerg. Infect. Dis.* 2:333–37

70. Hsiao K, Baker HF, Crow TJ, Poulter N, Owen F, et al. 1989. Linkage of a prion protein missense variant to Gerstmann-Sträussler-Scheinker syndrome. *Nature* 338:342–45

71. Hsiao K, Dlouhy SR, Farlow MR, Cass C, Da Costa M, et al. 1992. Mutant prion proteins in Gerstmann-Sträussler-Scheinker disease with neurofibrillary tangles. *Nat. Genet.* 1:68–71

72. Hsiao K, Prusiner SB. 1990. Inherited human prion diseases. *Neurology* 40:1820–27

73. Hsiao KK, Cass C, Schellenberg GD, Bird T, Devine-Gage E, et al. 1991. Prion protein variant in a family with the telencephalic form of Gerstmann-Sträussler-Scheinker syndrome. *Neurology* 41:681–84

74. Hsich G, Kenney K, Gibbs CJ, Lee KH, Harrington MG. 1996. The 14-3-3 brain protein in cerebrospinal fluid as a marker for transmissible spongiform encephalopathies. *N. Engl. J. Med.* 335:924–30

75. Huang Z, Gabriel JM, Baldwin MA, Fletterick RJ, Prusiner SB, Cohen FE. 1994. Proposed three-dimensional structure for the cellular prion protein. *Proc. Natl. Acad. Sci. USA* 91:7139–43

76. Huang Z, Prusiner SB, Cohen FE. 1995. Scrapie prions: a three-dimensional model of an infectious fragment. *Fold. Des.* 1:13–19

77. Huillard d'Aignaux J, Alpérovitch A, Maccario J. 1998. A statistical model to identify the contaminated lots implicated in iatrogenic transmission of Creutzfeldt-Jakob disease among French human growth hormone recipients. *Am. J. Epidemiol.* 147:597–604

78. Janssen RS, Schonberger LB. 1991. Creutzfeldt-Jakob disease from allogenic dura: a review of risks and safety. *J. Oral Maxillofac. Surg.* 49:274–75

79. King HOM. 1975. Kuru, epidemiological developments. *Lancet* 2:761–63

80. Kitamoto T, Amano N, Terao Y, Nakazato Y, Isshiki T, et al. 1993. A new inherited prion disease (PrP-P105L Mutation) showing spastic paraparesis. *Ann. Neurol.* 34:808–13

81. Kitamoto T, Ohta M, Doh-ura K, Hitoshi S, Terao Y, Tateishi J. 1993. Novel missense variants of prion protein in Creutzfeldt-Jakob disease or Gerstmann-Sträussler-Scheinker syndrome. *Biochem. Biophys. Res. Comm.* 191:709–14

82. Koch TK, Berg BO, DeArmond SJ, Gravina RF. 1985. Creutzfeldt-Jakob disease in a young adult with idiopathic hypopituitarism. Possible relation to the administration of cadaveric human growth hormone. *N. Engl. J. Med.* 313:731–33

83. Kondo K, Kuroiwa Y. 1982. A case control study of Creutzfeldt-Jakob disease: association with physical injuries. *Ann. Neurol.* 11:377–81

84. Kopp N, Streichenberger N, Deslys JP, Laplanche JL, Chazot G. 1996. Creutzfeldt-

Jakob disease in a 52-year-old woman with florid plaques. *Lancet* 348:1239–40

85. Kretzschmar HA. 1993. Human prion diseases (spongiform encephalopathies). *Arch. Virol.* 7:261–93 (Suppl.)

86. Kretzschmar HA, Honold G, Seitelberger F, Feucht M, Wessely P, et al. 1991. Prion protein mutation in family first reported by Gerstmann, Sträussler, and Scheinker. *Lancet* 337:1160

87. Kretzschmar HA, Ironside JW, DeArmond SJ, Tateishi J. 1996. Diagnostic criteria for sporadic Creutzfeldt-Jakob disease. *Arch. Neurol.* 53:913–20

88. Kretzschmar HA, Kufer P, Riethmüller G, DeArmond SJ, Prusiner SB, Schiffer D. 1992. Prion protein mutation at codon 102 in an Italian family with Gerstmann-Sträussler-Scheinker syndrome. *Neurology* 42:809–10

89. Kuroda Y, Gibbs CJ, Amyx HL, Gajdusek DC. 1983. Creutzfeldt-Jakob disease in mice: persistent viremia and preferential replication of virus in low-density lymphocytes. *Infect. Immun.* 41:154–61

90. Lane KL, Brown P, Howell DN, Crain BJ, Hulette CM, et al. 1994. Creutzfeldt-Jakob disease in a pregnant woman with an implanted dura mater graft. *Neurosurgery* 34:737–40

91. Lang CJG, Schüler P, Engelhardt A, Spring A, Brown P. 1995. Probable Creutzfeldt-Jakob disease after a cadaveric dural graft. *Eur. J. Epidemiol.* 11:79–81

92. Lasmézas CI, Deslys JP, Demaimay R, Adjou KT, Lamoury F, et al. 1996. BSE transmission to macaques. *Nature* 381:743–44

93. Lee KH, Harrington MG. 1997. 14-3-3 and BSE. *Vet. Rec.* 140:206–7

94. Liberski PP, Gajdusek DC. 1997. Kuru: forty years later, a historical note. *Brain Pathol.* 7:555–60

95. Little BW, Brown PW, Rodgers-Johnson P, Perl DP, Gajdusek DC. 1986. Familial myoclonic dementia masquerading as Creutzfeldt-Jakob disease. *Ann. Neurol.* 20:231–39

96. Lugaresi E, Medori R, Montagna P, Baruzzi A, Cortelli P, et al. 1986. Fatal familial insomnia and dysautonomia with selective degeneration of the thalamic nuclei. *N. Engl. J. Med.* 315:997–1003

97. MacDonald ST, Sutherland K, Ironside JW. 1996. Prion protein genotype and pathological phenotype studies in sporadic Creutzfeldt-Jakob disease. *Neuropathol. Appl. Neurobiol.* 22:285–92

98. Manetto V, Medori R, Cortelli P, Montagna P, Tinuper P, et al. 1992. Fatal familial insomnia: clinical and pathologic study of five new cases. *Neurology* 42:312–19

99. Manuelidis EE, Gorgacz EJ, Manuelidis L. 1978. Viremia in experimental Creutzfeldt-Jakob disease. *Science* 200:1069–71

100. Manuelidis EE, Kim JH, Mericangas JR, Manuelidis L. 1985. Transmission to animals of Creutzfeldt-Jakob disease from human blood. *Lancet* 2:896–97

101. Martinez-Lage F, Sola J, Poza M, Esteban JA. 1993. Pediatric Creutzfeldt-Jakob disease: probable transmission by a dural graft. *Child's Nerv. Syst.* 9:239–42

102. Mastrianni JA, Curtis MT, Oberholtzer JC, Da Costa MM, DeArmond S, et al. 1995. Prion disease (PrP-A117V) presenting with ataxia instead of dementia. *Neurology* 45:2042–50

103. Masullo C, Pocchiari M, Macchi G, Alema G, Piazza G, Panzera MA. 1989. Transmission of Creutzfeldt-Jakob disease by dural cadaveric graft. *J. Neurosurg.* 71:954–55

104. McKinley MP, Bolton DC, Prusiner SB. 1983. A protease-resistant protein is a structural component of the scrapie prion. *Cell* 35:57–62

105. McLean CA, Storey E, Gardner RJM, Tannenberg AEG, Cervenáková L, Brown P. 1997. The D178N (cis-129M) "fatal familial insomnia" mutation associated with diverse clinicopathologic phenotypes in an Australian kindred. *Neurology* 49:552–58

106. Medori R, Montagna P, Tritschler HJ, LeBlanc A, Cortelli P, et al. 1992. Fatal familial insomnia: a second kindred with mutation of prion protein gene at codon 178. *Neurology* 42:669–70

107. Medori R, Tritschler HJ, LeBlanc LeBlanc A, Villare F, et al. 1992. Fatal familial insomnia, a prion disease with a mutation at codon 178 of the prion protein gene. *N. Engl. J. Med.* 326:444–49

108. Meiner Z, Gabizon R, Prusiner SB. 1997. Familial Creutzfeldt-Jakob disease, codon 200 prion disease in Libyan Jews. *Medicine* 76:227–37

109. Meyer N, Rosenbaum V, Schmidt B, Gills K, Mirenda C, et al. 1991. Search for a putative scrapie genome in purified prion fractions reveals a paucity of nucleic acids. *J. Gen. Virol.* 72:37–50

110. Mills JL, Fradkin J, Schonberger L, Gunn W, Thomson RA, et al. 1990. Status report on the US human growth hormone recipient follow-up study. *Horm. Res.* 33:116–20

111. Ministry of Agriculture, Fisheries and Food, United Kingdom. 1998. *BSE Enforcement Bull.* 24, 2 pp.

112. Miyashita K, Inuzuka T, Kondo H, Saito Y, Fujuta N, et al. 1991. Creutzfeldt-Jakob disease in a patient with a cadaveric dural graft. *Neurology* 41:940–41

113. Monari L, Chen SG, Brown P, Parchi P, Peterson RB, et al. 1994. Fatal familial insomnia and familial Creutzfeldt-Jakob disease: different prion proteins determined by a DNA polymorphism. *Proc. Natl. Acad. Sci. USA* 91:2839–42

114. Nagayama M, Shinohara Y, Furukawa H, Kitamoto T. 1996. Fatal familial insomnia with a mutation at codon 178 of the prion protein gene: first report from Japan. *Neurology* 47:1313–16

115. Nitrini R, Rosemberg S, Passos-Bueno MR, Tiexeira da Silva LS, Iughetti P, et al. 1997. Familial spongiform encephalopathy associated with a novel prion protein gene mutation. *Ann. Neurol.* 42:138–46

116. Nochlin D, Sumi SM, Bird TD, Snow AD, Leventhal CM, et al. 1989. Familial dementia with PrP-positive amyloid plaques: a variant of Gerstmann-Sträussler-Scheinker syndrome. *Neurology* 39:910–18

117. Oesch B, Westaway D, Wälchli M, McKinley MP, Kent SBH, et al. 1985. A cellular gene encodes scrapie PrP 27-30 protein. *Cell* 40:735–46

118. Palmer MS, Dryden AJ, Hughes JT, Collinge J. 1991. Homozygous prion protein genotype predisposes to sporadic Creutzfeldt-Jakob disease. *Nature* 351:340–42

119. Pan KM, Baldwin M, Nguyen J, Gasset M, Serban A, et al. 1993. Conversion of α-helices into β-sheets features in the formation of the scrapie prion proteins. *Proc. Natl. Acad. Sci. USA* 90:10962–66

120. Pan KM, Stahl N, Prusiner SB. 1992. Purification and properties of the cellular prion protein from Syrian hamster brain. *Protein Sci.* 1:1343–52

121. Parchi P, Giese A, Capellari S, Brown P, Schulz-Schaeffer WJ, et al. 1998. The molecular and clinico-pathologic spectrum of phenotypes of sporadic Creutzfeldt-Jakob disease (sCJD). *Neurology* 50:S44.001 (Abstr.)

122. Pattison IH, Jones KM. 1967. The possible nature of the transmissible agent of scrapie. *Vet. Rec.* 80:2–9

123. Petterson RB, Tabaton M, Berg L, Schrank B, Torack RM, et al. 1992. Analysis of the prion protein gene in thalamic dementia. *Neurology* 42:1859–63

124. Pocchiari M, Masullo C, Salvatore M, Genuardi M, Galgani S. 1992. Creutzfeldt-Jakob disease after non-commercial dura mater graft. *Lancet* 340:614–15

125. Pocchiari M, Salvatore M, Cutruzzolá F, Genuardi M, Travaglini AC, et al. 1993. A new point mutation of the prion protein gene in Creutzfeldt-Jakob disease. *Ann. Neurol.* 34:802–7

126. Preece M. 1993. Human pituitary growth hormone and Creutzfeldt-Jakob disease. *Horm. Res.* 39:95–98

127. Prusiner SB. 1982. Novel proteinaceous infectious particles cause scrapie. *Science* 216:136–44

128. Prusiner SB. 1996. Prions. In *Fields Virology Third Edition,* ed. BN Fields, DM Knipe, PM Howley, pp. 2901–50. Philadelphia: Lippincott-Raven

129. Prusiner SB, Bolton DC, Groth DF, Bowman KA, Cochran P, McKinley MP. 1982. Further purification and characterization of scrapie prions. *Biochemistry* 21:6942–50

130. Prusiner SB, Groth D, Serban A, Koehler R, Foster D, et al. 1993. Ablation of the prion protein (PrP) gene in mice prevents scrapie and facilitates production of anti-PrP antibodies. *Proc. Natl. Acad. Sci. USA* 90:10608–12

131. Prusiner SB, McKinley MP, Groth DF, Bowman KA, Mock NI, et al. 1981. Scrapie agent contains a hydrophobic protein. *Proc. Natl. Acad. Sci. USA* 78:6675–79

132. Prusiner SB, Scott M, Foster D, Pan PK, Groth D, et al. 1990. Transgenic studies implicate interactions between homologous PrP isoforms in scrapie prion replication. *Cell* 63:673–86

133. Reder AT, Mednick AS, Brown P, Spire JP, Cauter EV, et al. 1995. Clinical and genetic studies of fatal familial insomnia. *Neurology* 45:1068–75

134. Richardson EP, Masters CL. 1995. The nosology of Creutzfeldt-Jakob disease and conditions related to the accumulation of PrP^{CJD} in the nervous system. *Brain Pathol.* 5:33–41

135. Rossi G, Macchi G, Porro M, Giaccone G, Bugiani M, et al. 1998. Fatal familial insomnia, genetic, neuropathologic, and biochemical study of a patient from a new Italian kindred. *Neurology* 50:688–92

136. Schonberger LB. 1998. New variant Creutzfeldt-Jakob disease and bovine spongiform encephalopathy. *Infect. Dis. Clin. North Am.* 12:111–21

137. Schonberger LB, Belay ED. 1998. *Summary of the clinicopathological features of transmissible spongiform encephalopathies (TSEs) in humans and Creutzfeldt-Jakob disease and blood safety data.* Handout presented to Advisory Comm. on Blood Safety and Avail., U.S. Dep. of Health and Human Serv., Aug. 27–28, 1998, Washington, DC

137a. Shimizu S, Hoshi K, Muramoto T, Homma M, Ironside JW, et al. 1999. Creutzfeldt-Jakob disease with florid-type plaques after cadaveric dura mater grafting. *Arch. Neurol.* 56:357–62

138. Shyng SL, Heuser JE, Harris DA. 1994. A glycolipid-anchored prion protein is endocytosed via clathrin-coated pits. *J. Cell Biol.* 125:1239–50

139. Simpson DA, Masters CL, Ohirich G, Purdie G, Stuart G, Tannenberg AEG. 1996. Iatrogenic Creutzfeldt-Jakob disease and its neurosurgical implications. *J. Clin. Neurosci.* 3:118–23

140. Stern K. 1939. Severe dementia associated with bilateral symmetrical degeneration of the thalamus. *Brain* 62:157–71

141. Takashima S, Tateishi J, Taguchi Y, Inoue H. 1997. Creutzfeldt-Jakob disease with florid plaques after cadaveric dural graft in a Japanese woman. *Lancet* 350:865–66

142. Taraboulos A, Serban D, Prusiner SB. 1990. Scrapie prion proteins accumulate in the cytoplasm of persistently infected cultured cells. *J. Cell. Biol.* 110:2117–32

143. Tateishi J. 1985. Transmission of Creutzfeldt-Jakob disease from human blood and urine into mice. *Lancet* 2:1074

144. Tateishi J, Sato Y, Koga M, Doi H, Ohta M. 1980. Experimental transmission of human subacute spongiform encephalopathy to small rodents. *Acta Neuropathol. (Berl.)* 51:127–34

145. Telling GC, Parchi P, DeArmond SJ,

Cortelli P, Montagna P, et al. 1996. Evidence for the conformation of the pathologic isoform of the prion protein enciphering and propagating prion diversity. *Science* 274:2079–82

146. Thadani V, Penar PL, Partington J, Kalb R, Janssen R, et al. 1988. Creutzfeldt-Jakob disease probably acquired from cadaveric dura mater graft. *J. Neurosurg.* 69:766–69

147. Tintner R, Brown P, Hedley-Whyte ET, Rappaport EB, Piccardo CP, Gajdusek DC. 1986. Neuropathologic verification of Creutzfeldt-Jakob disease in the exhumed American recipient of human pituitary growth hormone: epidemiologic and pathogenetic implications. *Neurology* 36:932–36

148. Tranchant C, Doh-ura K, Warter JM, Steinmetz G, Chevalier Y, et al. 1992. Gerstmann-Sträussler-Scheinker disease in an Alsatian family-clinical and genetic studies. *J. Neurol. Neurosurg. Psychiatry* 55:185–87

149. Tranchant C, Sergeant N, Wattez A, Mohr M, Warner JM. 1997. Neurofibrillary tangles in Gerstmann-Sträussler-Scheinker syndrome with the A117V prion gene mutation. *J. Neurol. Neurosurg. Psychiatry* 63:240–46

150. Van Duijn M, Delasnerie-Lauprêtre N, Masullo C, Zerr I, de Silva R, et al. 1998. Case-control study of risk factors of Creutzfeldt-Jakob disease in Europe during 1993–95. *Lancet* 351:1081–85

151. Weissmann C. 1991. A "unified theory" of prion propagation. *Nature* 352:679–83

152. Wientjens DPWM, Davanipour Z, Hofman A, Kondo K, Matthews WB, et al. 1996. Risk factors for Creutzfeldt-Jakob disease: a reanalysis of case-control studies. *Neurology* 46:1287–91

153. Will RG, Ironside JW, Zeidler, Cousens SN, Estibeiro K, et al. 1996. A new variant of Creutzfeldt-Jakob disease in the UK. *Lancet* 347:921–25

154. Will RG, Matthews WB. 1982. Evidence for case-to-case transmission of Creutzfeldt-Jakob disease. *J. Neurol. Neurosurg. Psychiatry* 45:235–38

155. Willison HJ, McLaughlin JE. 1991. Creutzfeldt-Jakob disease following cadaveric dura mater graft. *J. Neurol. Neurosurg. Psychiatry* 54:940

156. Yamada M, Itoh Y, Fujigasaki H, Naruse S, Kaneko K, et al. 1993. A missense mutation at codon 105 with codon 129 polymorphism of the prion protein gene in a new variant of Gerstmann-Sträussler-Scheinker disease. *Neurology* 43:2723–24

157. Yamada M, Itoh Y, Suematsu N, Matsushita M, Otomo E. 1997. Panencephalopathic type of Creutzfeldt-Jakob disease associated with cadaveric dura mater graft. *J. Neurol. Neurosurg. Psychiatry* 63:524–27

158. Yamada S, Aiba T, Endo Y, Hara M, Kitamoto T, Tateishi J. 1994. Creutzfeldt-Jakob disease transmitted by a cadaveric dura mater graft. *Neurosurgery* 34:740–44

159. Zeidler M. 1998. *Human dura mater, a WHO report.* Presented at Joint Inst. for Food Safety and Appl. Nutr. Workshop on TSE Risks in Relation to Source Materials, Processing, and End-product use, June 8–9, 1998, College Park, MD

160. Zeidler M, Johnstone EC, Bamber RWK, Dickens CM, Fisher CJ, et al. 1997. New variant Creutzfeldt-Jakob disease: psychiatric features. *Lancet* 350:908–10

161. Zeidler M, Stewart GE, Barraclough CR, Bateman DE, Bates D, et al. 1997. New variant Creutzfeldt-Jakob disease: neurological features and diagnostic tests. *Lancet* 350:903–7

Annu. Rev. Microbiol. 1999. 53:315–51

BACTERIAL BIOCATALYSTS: Molecular Biology, Three-Dimensional Structures, and Biotechnological Applications of Lipases

K-E. Jaeger

Lehrstuhl Biologie der Mikroorganismen, Ruhr-Universität, D-44780 Bochum, Germany; e-mail: karl-erich.jaeger@ruhr-uni-bochum.de

B. W. Dijkstra

Laboratory of Biophysical Chemistry, Rijksuniversiteit Groningen, NL-9747 AG Groningen, The Netherlands; e-mail: bauke@chem.rug.nl

M. T. Reetz

Max-Planck-Institut für Kohlenforschung, D-45470 Mülheim an der Ruhr, Germany; e-mail: reetz@mpi-muelheim.mpg.de

Key Words   secretion, directed evolution, enantioselectivity

■ **Abstract**   Bacteria produce and secrete lipases, which can catalyze both the hydrolysis and the synthesis of long-chain acylglycerols. These reactions usually proceed with high regioselectivity and enantioselectivity, and, therefore, lipases have become very important stereoselective biocatalysts used in organic chemistry. High-level production of these biocatalysts requires the understanding of the mechanisms underlying gene expression, folding, and secretion. Transcription of lipase genes may be regulated by quorum sensing and two-component systems; secretion can proceed either via the Sec-dependent general secretory pathway or via ABC transporters. In addition, some lipases need folding catalysts such as the lipase-specific foldases and disulfide-bond–forming proteins to achieve a secretion-competent conformation. Three-dimensional structures of bacterial lipases were solved to understand the catalytic mechanism of lipase reactions. Structural characteristics include an α/β hydrolase fold, a catalytic triad consisting of a nucleophilic serine located in a highly conserved Gly-X-Ser-X-Gly pentapeptide, and an aspartate or glutamate residue that is hydrogen bonded to a histidine. Four substrate binding pockets were identified for triglycerides: an oxyanion hole and three pockets accommodating the fatty acids bound at positions sn-1, sn-2, and sn-3. The differences in size and the hydrophilicity/hydrophobicity of these pockets determine the enantiopreference of a lipase. The understanding of structure-function relationships will enable researchers to tailor new lipases for biotechnological applications. At the same time, directed evolution in combination with appropriate screening

0066-4227/99/1001-0315$08.00

systems will be used extensively as a novel approach to develop lipases with high stability and enantioselectivity.

CONTENTS

The authors dedicate this article to Professor Dr. Ulrich K. Winkler on the occasion of his 70th birthday. Uli Winkler was among the first to isolate bacterial lipases from *Serratia marcescens* and *Pseudomonas aeruginosa* and to study their physiology and biochemical properties.

INTRODUCTION

Nearly 100 years ago, a landmark report was published by C Eijkmann (34) describing the following simple experiment: beef tallow spread on the ground of a glass plate was overlaid with agar that was inoculated with different bacteria. After 3–4 days of incubation, Ca, Na, and NH_4 soaps had formed. Eijkmann concluded that lipases had been produced and secreted by the bacteria, among them *Bacillus pyocyaneus* (today named *Pseudomonas aeruginosa*), *Staphylococcus pyogenes aureus* (*S. aureus*), *B. prodigiosus* (today *Serratia marcescens*), and *B. fluorescens* (today *P. fluorescens*). *B. coli communis* (*Escherichia coli*) was found to be lipase negative (34). Only a few lipase-producing bacteria were further characterized (95), but research was intensified when it became generally accepted that lipases remain enzymatically active in organic solvents (166), making them ideal tools for the organic chemist. The aim of this review is not to discuss every lipase described in the literature but rather to present recent information on selected and novel lipases. It covers the period from 1994, when a comprehensive review on bacterial lipases appeared (58), until 1998. During this time a number of novel lipases were cloned and characterized, considerable progress was made in understanding the regulation of lipase gene expression, and detailed knowledge became available concerning folding and secretion. Moreover, three-dimensional structures of lipases were solved and used to explain functional characteristics. These developments, as well as other aspects such as the important role of lipases in biotechnological applications, are discussed, with special emphasis on enantioselective biotransformations. The enormous interest in lipases is reflected by a rapidly growing number of excellent review articles and monographs covering molecular biology, biochemical characterization, three-dimensional structures, and biotechnological applications of lipases from prokaryotic and eukaryotic origins (4, 46, 58, 59, 66, 102, 121, 122, 125, 165).

DEFINITION OF A LIPASE

What exactly is a lipase? At present, there is no satisfying answer to this simple question. Lipolytic reactions occur at the lipid-water interface where lipolytic substrates usually form an equilibrium between monomeric, micellar, and emulsified states. Until recently, two criteria have been used to classify a lipolytic enzyme as a "true" lipase (EC 3.1.1.3): (*a*) It should be activated by the presence of an interface, that is, its activity should sharply increase as soon as the triglyceride substrate forms an emulsion. This phenomenon was termed "interfacial activation" (124). (*b*) It should contain a "lid" (see below), which is a surface loop of the protein covering the active site of the enzyme and moving away on contact with the interface (17, 24, 158). However, these obviously suggestive criteria proved to be unsuitable for classification, mainly because a number of exceptions were described of enzymes having a lid but not exhibiting interfacial activation (159). Therefore, lipases are simply defined as carboxylesterases catalyzing the hydrolysis (and synthesis) of long-chain acylglycerols (37). There is no strict definition available for the term "long-chain," but glycerolesters with an acyl chain length of ≥ 10 carbon atoms can be regarded as lipase substrates, with trioleoylglycerol being the standard substrate. Hydrolysis of glycerolesters with an acyl chain length of < 10 carbon atoms with tributyrylglycerol (tributyrin) as the standard substrate usually indicates the presence of an esterase (62). It should be emphasized that most lipases are perfectly capable of hydrolyzing these esterase substrates.

SCREENING FOR LIPASE ACTIVITY

Hydrolysis

Microbiologists generally want to use a simple and reliable plate assay allowing the identification of lipase-producing bacteria. The most widely used substrates are tributyrin and triolein, which are emulsified mechanically in various growth media and poured into a petri dish. Lipase production is indicated by the formation of clear halos around the colonies grown on tributyrin-containing agar plates (6) and orange-red fluorescence visible on irradiation with a conventional UV hand lamp at $\lambda = 350$ nm on triolein plates, which additionally contain rhodamine B (74). Lipase activity in bacterial culture supernatants is determined by hydrolysis of p-nitrophenylesters of fatty acids with various chain lengths ($\geq$C-10) and spectrophotometric detection of p-nitrophenol at 410 nm. However, care must be taken to interpret the results because these fatty acid monoester substrates are also hydrolyzed by esterases. This problem can be overcome by using the triglyceride derivative 1,2-O-dilauryl-rac-glycero-3-glutaric acid resorufin ester (available from Boehringer Mannheim Roche GmbH, Germany), yielding resorufin, which can be determined spectrophotometrically at 572 nm or fluorometrically at 583 nm. A number of novel fluorogenic alkyldiacylglycerols were synthesized and used for analysis of both lipase activity and stereoselectivity (167). A more laborious but reliable method for identifying a "true" lipase is the determination of fatty acids liberated from a triglyceride, usually trioleoylglycerol, by titration (62). Automated equipment allows the parallel assay of a large number of samples. Determination of kinetics of lipolysis requires a tight control of the interfacial quality achieved by using the monolayer technique: A lipid film is spread at the air/water interface in a so-called "zero-order" trough consisting of a substrate reservoir and a reaction compartment. Lipase-catalyzed hydrolysis of the lipid monolayer results in changes of the surface pressure, which can be readjusted automatically by a computer-controlled barostat (111). It should be emphasized that this technique requires expensive equipment and experienced personnel.

Synthesis

Biotechnological applications of lipases prompt a demand for techniques to determine their activity and, if relevant, stereoselectivity. A standard reaction is the lipase-catalyzed esterification of an alcohol with a carboxylic acid, e.g. the formation of octyl laurate from lauric acid and n-octanol reacted in an organic solvent (114). The initial rate of ester formation can be determined by gas chromatography. No single method is available to determine the enantioselectivity of a lipase-catalyzed organic reaction. Generally, the enantioselectivity of product formation is determined either by gas chromatography or high performance liquid chromatography (HPLC), with chirally modified columns. Basically, two types of enantioselective lipase-catalyzed reactions are possible: (*a*) desymmetrization of prochiral substrates in hydrolysis or acylation reactions and (*b*) kinetic resolution of racemic mixtures, hydrolysis or acylation again being the two options. Recently, a number of screening methods for lipases have been reviewed, including those which allow the conventional determination of enantioselectivity and regioselectivity (1). However, without modification they are not suitable for high throughput screening.

CLASSIFICATION OF BACTERIAL LIPASES

A search of available data banks (GenBank, Swiss Protein Sequence Database, The Protein Information Resource, The Protein Research Foundation Database, and The Brookhaven Protein Databank) revealed 217 entries of lipolytic enzymes from bacteria (search completed in November 1998). Many of the entries turned out to be redundant, and, finally, 47 different lipases were identified and grouped into six families based on amino acid sequence homology (Table 1) (JL Arpigny & K-E Jaeger, submitted for publication).

Family I comprises a total of 22 members subgrouped into 6 subfamilies. Subfamilies I.1 and I.2 extend the previously described *Pseudomonas* group I and II lipases encoded in an operon together with their cognate intramolecular chaperones, which have been designated Lif (lipase-specific foldases) (58). These lipases are secreted via the type II pathway, whereas those belonging to subfamily I.3 use the type I secretion pathway (see below). The *Bacillus* lipases grouped in subfamily I.4 are the smallest lipases known, with a molecular mass of 19.6 kDa. They seem to be well suited for biotechnological applications as is *B. thermocatenulatus* lipase belonging to subfamily I.5 (126). Lipases from *Staphylococcus hyicus* and *S. aureus* belong to the best characterized lipases originating from gram-positive bacteria (140). Family II has been described before as a novel family of lipolytic enzymes with unknown function (157). We have recently added to this family as a novel member an esterase located in the outer membrane of *P. aeruginosa* (S Wilhelm, J Tommassen, K-E Jaeger, submitted for publication). This enzyme is an autotransporter belonging to a previously identified family of channel-forming bacterial virulence factors (89). Determination of the three-dimensional structure for the *Streptomyces scabies* enzyme indicated that the tertiary fold of esterases belonging to this family may be substantially different from the α/β hydrolase fold found in most lipases. The members of family III contain extracellular lipases from *Streptomyces* spp. and the psychrophilic strain of *Moraxella* sp. Members of family IV belong to the group of the cold-adapted lipases exhibiting similarity to mammalian hormone-sensitive lipases (80). They contain the active-site serine residue in a consensus pentapeptide GDSAG, which is located close to the N terminus of the protein, and, in addition, they have another strictly conserved HGGG motif

TABLE 1   Families of lipolytic enzymes

Family	Sub-family	Enzyme-producing species	Accession no.	Similarity[b] (%)	Properties
I	1	Pseudomonas aeruginosa[a]	D50587	100	True lipases
		Pseudomonas fluorescens C9	AF031226	95	
		Vibrio cholerae	X16945	57	
		Acinetobacter calcoaceticus	X80800	43	
		Pseudomonas fragi	X14033	40	
		Pseudomonas wisconsinensis	U88907	39	
		Proteus vulgaris	U33845	38	
	2	Burkholderia glumae[a]	X70354	35	
		Chromobacterium viscosum[a]	Q05489	35	
		Burkholderia cepacia[a]	M58494	33	
		Pseudomonas luteola	AF050153	33	
	3	Pseudomonas fluorescens SIKW1	D11455	14	
		Serratia marcescens	D13253	15	
	4	Bacillus subtilis	M74010	16	
		Bacillus pumilus	A34992	13	
	5	Bacillus stearothermophilus	U78785	15	
		Bacillus thermocatenulatus	X95309	14	
		Staphylococcus hyicus	X02844	15	Phospholipase
		Staphylococcus aureus	M12715	14	
		Staphylococcus epidermidis	AF090142	13	
	6	Propionibacterium acnes	X99255	14	
		Streptomyces cinnamoneus	U80063	14	
II		Pseudomonas aeruginosa	AF005091	100	o.m.-bound esterase
		Aeromonas hydrophila	P10480	31	Acyltransferase
		Salmonella typhimurium	AF047014	17	o.m.-bound esterase
		Photorhabdus luminescens	P40601	17	
		Streptomyces scabies[a]	M57297	15	
III		Streptomyces exfoliatus[a]	M86351	100	Extracellular lipase
		Streptomyces albus	U03114	82	
		Moraxella sp.	X53053	33	Lipase 1

(continued)

Family	Sub-family	Enzyme-producing species	Accession no.	Similarity[b] (%)	Properties
IV		*Moraxella sp.*	X53868	100	Lipase 2
		Archaeoglobus fulgidus	AE000985	28	Carboxylesterase
		Alicyclobacillus acidocaldarius	X62835	25	
		Pseudomonas sp. *B11-1*	AF034088	24	
		Alcaligenes eutrophus	L36817	24	
		Escherichia coli	AE000153	20	Esterase
V		*Moraxella* sp.	X53869	100	Lipase 3
		Psychrobacter immobilis	X67712	88	
		Pseudomonas oleovorans	M58445	34	PHA-depolymerase
		Haemophilus influenzae	U32704	34	Putative esterase
		Sulfolobus acidocaldarius	AF071233	25	Esterase
		Acetobacter pasteurianus	AB013096	15	Esterase
VI		*Pseudomonas fluorescens*[a]	S79600	100	Esterases
		Synechocystis sp.	D90904	24	
		Spirulina platensis	S70419	22	
		Rickettsia prowazekii	Y11778	16	
		Chlamydia trachomatis	AE001287	15	

[a]Lipolytic enzymes with known 3-D structure.
[b]Similarities of amino acid sequences were determined with the program Megalign (DNAStar), with the first member of each family arbitrarily set at 100%.

of unknown function located immediately upstream of the active site consensus motif. It is interesting that families IV and V contain lipases belonging to either psychrophilic (*Moraxella* sp., *Pseudomonas* sp. B11-1, and *P. immobilis*) or thermophilic (*Archaeoglobus fulgidus* and *Sulfolobus acidocaldarius*) bacteria. The members of family V show structural similarities to dehalogenases, haloperoxidases, and epoxide hydrolases presumably exhibiting the α/β hydrolase fold as the characteristic tertiary structure. Family VI lists esterases that have partly been identified from genome sequences (*Synechocystis* sp., *Chlamydia trachomatis*). These esterases are small proteins presumably located in the bacterial cytoplasm with similarity to mammalian lysophospholipases.

Determination and comparison of specific activities and substrate specificities of different lipases are absolutely required to investigate the physiological function of lipases as well as to judge their usefulness for biotechnological applications.

However, serious doubts remain as to the comparability of results published by different laboratories. As outlined above, no standard substrate or assay system is available to determine specific lipase activities or to distinguish lipases from esterases. For this reason a recent study that compares the enzyme activities of 11 bacterial lipases obtained from both gram-positive and gram-negative bacteria is very important (141). All tested lipases showed high activity toward the short-chain triglyceride tributyrylglycerol and the long-chain trioctanoylglycerol, with lipases from *P. aeruginosa* and *Proteus vulgaris* being the most active. Phospholipids like 1,2-dioctanoyl-*sn*-glycero-3-phosphocholine were either poor substrates (for lipases from *S. aureus*, *S. epidermidis*, *S. warneri*, *Acinetobacter calcoaceticus*, *Arthrobacter* sp., *P. aeruginosa*, *Burkholderia cepacia*, and *P. vulgaris*) or not hydrolyzed at all (by lipases from *B. thermocatenulatus* and *S. marcescens*). The only exception was *S. hyicus* lipase, which readily hydrolyzed phospholipids with high specific activity (141). In summary, it is obvious that many lipase genes are being found by accident and the corresponding proteins remain largely uncharacterized. Novel lipase-producing microorganisms have been described (10, 61, 149), but the cognate lipase genes and proteins still need to be isolated.

REGULATION OF LIPASE GENE EXPRESSION

Until recently, a significant number of prokaryotic lipase genes have been cloned, but the molecular mechanisms regulating their expression remain largely unknown. In *S. aureus* a global regulatory network was identified that coordinately controls the transcription of genes involved in pathogenesis. Two interacting genetic loci termed *agr* and *sar* were identified that activate the transcription of genes encoding extracellular proteins when the bacteria enter the stationary growth phase (50). Lipase gene expression seems to be regulated by the *agr* locus encoding a cell density-dependent regulatory system consisting of a two-component system with *agrC* as the signal transducer and *agrA* as the response regulator. A peptide pheromone encoded by *agrD* serves as the autoinducer. Because a mutation in *agrC* originally termed *xpr* was found to repress the extracellular lipase activity (142), it can be concluded that lipase expression in *S. aureus* is regulated by a quorum sensing system mediated through the two-component regulator *agrC/A* (92). Recently, the alternative sigma factor σ^B was shown to repress lipase synthesis; however, it remained unclear whether lipase gene expression or lipase protein secretion was affected (76).

A variety of *Streptomyces* strains produce extracellular lipase, a process that is dependent on the growth phase. Transcription of the gene *lipA* of *S. exfoliatus* M11 was dependent on the presence of the downstream gene *lipR* encoding a transcriptional activator that belongs to the LuxR family of bacterial regulators (132). This lipase-specific transcriptional regulator switches on lipase synthesis at the onset of the stationary growth phase, probably enabling *Streptomyces* to make use of triacylglycerol storage compounds (101, 132).

The insect pathogen *Xenorhabdus nematophilus* that exists in the intestine of the parasitic nematode *Steinernema carpocapsae* produces a lipase. Its biosynthesis is stimulated by N-β-hydroxybutanoyl homoserine lactone (HBHL), known as the autoinducer of the luminescent system of *Vibrio harveyi* (30). This finding suggests a quorum sensing type of lipase gene regulation that would involve, apart from the signal molecule HBHL, at least two genes encoding an autoinducer synthetase and a transcriptional activator, respectively.

A. calcoaceticus, a gram-negative soil bacterium, produces a number of lipolytic enzymes, among them an extracellular lipase encoded by *lipA* (72). The effects of various physiological factors on lipase gene expression were studied by using transcriptional *lipA::lacZ* fusions integrated into the *A. calcoaceticus* chromosome. Fatty acids produced by hydrolysis of the lipase substrate triolein strongly repressed expression of *lipA* (71). These findings were explained by proposing the existence of an as yet unidentified regulatory protein that is believed to repress lipase transcription on binding of a fatty acid (71).

In *P. aeruginosa*, transcription of the lipase operon *lipA/H* from promotor P1 requires the presence of the alternative sigma factor σ^{54} (60). Recently, primer extension analysis confirmed this result and also revealed the presence of a second promotor P2 located $\sim$300 base pairs (bp) upstream of P1 (H Duefel, P Braun, W Quax, K-E Jaeger, manuscript in preparation). Because σ^{54}-dependent promotors require the presence of a cognate transcriptional activator (138), the existence of a lipase regulator *lipR* was proposed (60). Extracellular lipase is secreted when *P. aeruginosa* enters the stationary growth phase (145). This suggests a cell density-dependent regulation of lipase gene expression involving one of the quorum sensing systems identified in *P. aeruginosa* (150). Studies with *lipA::lacZ* fusions revealed that lipase gene expression is controlled via the *rhlR/I* (also named *vsmR/I*) system, including the autoinducer N-butyryl-homoserine lactone (BHL) (127), which controls the synthesis of *P. aeruginosa* rhamnolipid (97, 98) as well as several extracellular enzymes (15, 81). Characterization of a Tn*5* mutant exhibiting reduced lipase activity recently led to the identification of *lipR* encoding a transcriptional activator, which may be part of a two-component system. A second gene located upstream of *lipR*, which we tentatively name *lipQ*, encoding a putative signal transducing protein, was identified by searching the *Pseudomonas* genome project (www.pseudomonas.com/) (H Duefel, P Braun W Quax, K-E Jaeger, manuscript in preparation). In addition, the global regulator gene *gacA* was identified, which acts as part of another two-component system (119). Overproduction of GacA markedly increased extracellular lipase activity, presumably via activation of the *rhlR/I* system. Figure 1 summarizes the current status of the regulatory network controlling expression of the lipase operon *lipA/H* in *P. aeruginosa*.

In conclusion, a cell density-dependent regulation of extracellular lipase synthesis emerges as a general regulatory mechanism, especially in pathogenic bacteria with lipases qualifying as virulence factors. Additional activators or repressors of transcription responding to a variety of environmental signals seem to exist, making lipase gene expression a highly regulated system.

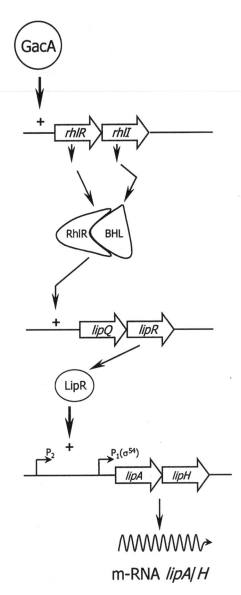

Figure 1   Model for the regulatory network controlling expression of the lipase operon *lipA/H* in *Pseudomonas aeruginosa* (see text for details).   Stimulation of gene transcription is indicated by +.

SECRETION AND FOLDING

Lipases are extracellular enzymes, which must, therefore, be translocated through the bacterial membranes to reach their final destination. At present, three main secretion pathways have been identified (11, 107, 123), and it appears that lipases can use at least two of them. More recently, it became obvious that correct folding of lipases is needed to ensure proper secretion. Periplasmic folding catalysts include intramolecular Lif chaperones (58) and, at least in *P. aeruginosa*, accessory proteins involved in disulfide bond formation (A Urban, M Leipelt, T Eggert, K-E Jaeger, submitted for publication).

ABC Exporters

Lipases from *P. fluorescens* (32) and *S. marcescens* (3, 84) lack a typical N-terminal signal peptide, and they are secreted by an ABC exporter (also named type I secretion pathway) consisting of three different proteins (11). In *S. marcescens* the inner membrane protein LipB containing an ATP-binding cassette (ABC protein) confers the substrate specificity to the system. Additional components include LipC as a membrane fusion protein (MFP), which can be associated with both the inner and the outer membrane, and LipD as an outer membrane protein (OMP). Both lipase and metalloprotease are secreted through this ABC transporter (2); however, the cell-surface layer protein SlaA seems to be its natural substrate (64).

Secretion Across the Inner Membrane

Many lipases of both gram-positive and gram-negative bacteria possess an N-terminal signal sequence mediating their secretion through the inner membrane by means of the Sec translocase. In *E. coli* this is a multisubunit protein complex, consisting of the soluble dimeric SecA and a membrane-embedded complex formed by SecY, E, D, G, and F (31). A similar Sec-translocase exists in *Bacillus* species (139), which is presumably also involved in secretion of lipases.

Secretion Across the Outer Membrane

After being secreted through the inner membrane of gram-negative bacteria, lipases fold in the periplasm into an enzymatically active conformation. Subsequently, they are transported through the outer membrane by means of a complex machinery called the secreton, which consists of ≤ 14 different proteins forming the type II or general secretion pathway (107). In *P. aeruginosa*, lipase is secreted through such a secreton (38, 60), here encoded by 12 *xcp* genes organized in two divergently transcribed operons (38, 153). The Xcp proteins are located both in the inner and the outer membrane, with XcpQ forming a multimeric pore with a diameter of 95 Å (12). Recently, a similar secreton involved in secretion of *P. alcaligenes* lipase was identified (43). When additional copies of the secreton were expressed from a cosmid introduced into *P. alcaligenes*, the production of extracellular lipase increased significantly (43). This effect, which has also been observed for *P. aeruginosa* lipase, suggests that the number of secreton complexes present in wild-type strains is probably low, thereby posing restrictions on the production of extracellular enzymes. Although similar multicomponent secretons have been identified in lipase-producing *A. hydrophila*, *X. campestris*, and *V. cholerae* (9, 104), their involvement in lipase secretion still has to be demonstrated.

Folding Catalysts

Specific intramolecular chaperones designated Lif proteins are required for folding of some lipases in the periplasm (58). These foldase proteins are usually encoded in an operon with their cognate lipases and have been identified in *P. aeruginosa* (18, 56, 57, 103, 164), *P. wisconsinensis* (M Hazbon, H Duefel, P Cornelis, K-E Jaeger, manuscript in preparation), *B. cepacia* (63), *B. glumae* (40, 41), *A. calcoaceticus* (73), and *V. cholerae* (99). The best studied Lif proteins are those from *P. aeruginosa* and *B. glumae*, which contain a hydrophobic N-terminal segment anchoring them to the inner membrane (40, 127). Truncated Lif proteins lacking this membrane anchor segment are still able to catalyze folding of lipases (133; F Rosenau, M Eller, K-E Jaeger, unpublished data), indicating that the membrane anchor domain may function to prevent secretion of Lif proteins. Lipase-Lif complexes have been immunoprecipitated with either antilipase or anti-Lif antisera (54, 55). The lipase:Lif ratio is as yet unknown; chemical cross-linking suggested a 1:1 complex, with Ca^{2+} ions needed for complex formation (134), whereas a ratio of 1:4 was suggested from in vitro refolding experiments (55). The interaction of lipase with Lif is specific, as shown by the inability of Lifs to activate heterologous lipases. The only known exception is the lipase from *P. alcaligenes*, which can partly be activated when coexpressed in *E. coli* together with *P. aeruginosa* Lif (35). Domain swapping revealed that an 138-amino-acid C-terminal domain of *B. glumae* Lif determines its substrate specificity (35). In an elegant study, *P. aeruginosa* Lif was randomly mutated by error-prone polymerase chain reaction (PCR) and coexpressed with the lipase gene in *E. coli*. Analysis of nonfunctional Lif mutant proteins allowed identification of residues Tyr-99 and Arg-115 as being essential for the interaction with lipase (135). Interestingly, a number of strictly conserved amino acid residues were identified in various Lif proteins (58), one of which is Tyr-99 in *P. aeruginosa* Lif (135).

Lipases produced by the gram-positive bacteria *S. aureus*, *S. epidermidis*, and *S. hyicus* are synthesized as pre-proenzymes, with an N-terminal pro-region of about 260 amino acids acting as a folding catalyst (45). The pro-region not only facilitates translocation of these lipases through the cytoplasmic membrane but also protects them from proteolytic attack. For *S. hyicus* lipase, removal of the pro peptide by an extracellular metalloprotease generates the 46-kDa mature lipase (7).

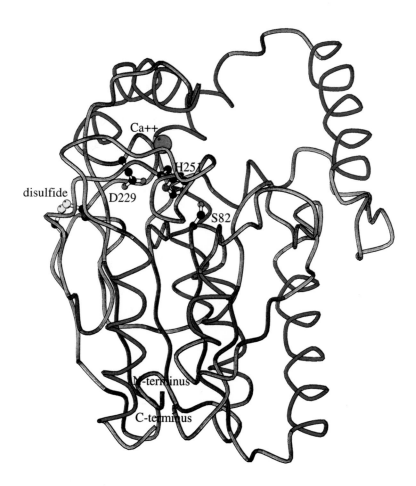

Figure 4   X-ray structure of *Pseudomonas aeruginosa* lipase. Indicated are the N- and the C- terminus, and the catalytic triad formed by the nucleophile (S82), the acid (D229) and the histidine residue (H251). In addition, the bound Ca^{++}-ion and the disulfide bond formed between residues C183 and C235 are shown.

This pro-region is also capable of binding to heterologous proteins and protecting them from proteolytic degradation (86). When the pre-pro part of *S. hyicus* lipase was fused to *E. coli* OmpA, efficient secretion was observed not only in *S. hyicus* but also in *B. subtilis* (93).

Intramolecular chaperones have been studied extensively in various proteases, where they are found as short N-terminally located pro-peptides (137), which functionally resemble the pro-peptides found in lipases from gram-positive bacteria. They can be distinguished from Lif proteins, which are (*a*) not covalently attached to the N terminus of the protein and (*b*) anchored in the cytoplasmic membrane, thereby rendering them reusable. The main function of these chaperone proteins is to help the lipases in overcoming a kinetic barrier along their folding pathway. In addition, they may also function as temporary competitive inhibitors of their cognate enzymes.

Exported proteins including lipases often contain disulfide bonds, which are formed in the periplasm. In *E. coli* a complex system consisting of Dsb proteins mediates disulfide bond formation (108). DsbA is a thiol:disulfide oxidoreductase that oxidizes the Cys-SH residues to form a disulfide bond; DsbC is a thiol:disulfide isomerase that isomerizes wrongly oxidized disulfide bonds (94). The involvement of DsbA and DsbC in folding and secretion of lipase has been studied in *P. aeruginosa* (A Urban, M Leipelt, T Eggert, K-E Jaeger, submitted for publication). This lipase contains a single disulfide bond. In a *dsbA* mutant of *P. aeruginosa*, extracellular lipase production was reduced to 10% of the wild-type level as demonstrated by enzyme activity assays and Western blotting. In a *dsbA dsbC* double mutant, a residual lipase activity of only 1% of the wild-type activity was detected. Interestingly, a *P. aeruginosa dsbC* mutant exhibited a twofold increase in extracellular lipase activity (A Urban, M Leipelt, T Eggert, K-E Jaeger, submitted for publication). *A. hydrophila* lipase is secreted via the type II secretion pathway and also contains one disulfide bond (16). Exchange of both cysteine residues to serine did not reduce the enzymatic activity or prevent secretion; however, it rendered this lipase more sensitive to urea denaturation and proteolytic degradation (16). Similar results were obtained with *P. aeruginosa* lipase (K Liebeton, A Zacharias, K-E Jaeger, submitted for publication). Both Cys-183Ser and Cys-235Ser mutants still exhibited residual lipase activity, as did a Cys-183Ser/Cys-235Ser double mutant. Refolding to enzymatic activity of mutant lipases expressed in *E. coli* was not significantly impaired, indicating an important role of the disulfide bond for lipase stabilization in a secretion-competent rather than an enzymatically active form.

Figure 2 summarizes present knowledge on the mechanism of lipase secretion via the type II secretion pathway in *P. aeruginosa*. Important questions remain to be answered: (*a*) Does a specific secretion signal exist in lipases? (*b*) Which factors determine the specificity of the interaction between lipase and its cognate Lif protein? (*c*) Is lipase secreted through the Xcp complex separately or as a

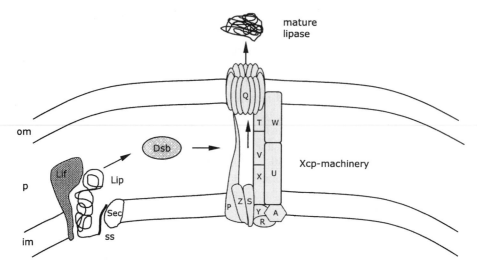

Figure 2 Model for the *Pseudomonas aeruginosa* lipase secretion pathway. Lipase (Lip) is secreted as prelipase including a signal sequence (ss) through the bacterial inner membrane (im) by a Sec-dependent mechanism. Interaction with the lipase-specific foldase Lif and Dsb proteins assists folding and formation of disulfide bonds in the periplasm (p). The final secretion through the outer membrane (om) is mediated by the Xcp machinery.

complex with Lif? (*d*) What is the mechanism leading to the close association with lipopolysaccharide observed for lipases isolated from bacterial culture supernatants (145)?

THREE-DIMENSIONAL STRUCTURES OF BACTERIAL LIPASES

Knowledge of their three-dimensional structures and the factors that determine their regiospecificity and enantiospecificity are traditionally essential to tailor lipases for specific applications. Human pancreatic lipase (163) and the lipase from the fungus *Rhizomucor miehei* (14, 28) were the first X-ray structures of lipases elucidated. Various other lipase structures of fungal origin followed, from *Geotrichum candidum* (130), *Fusarium solani* (91), *Candida rugosa* (48), *Candida antarctica* (155), *Humicola lanuginosa* (26), and *Rhizopus delemar* (26). In contrast, X-ray structures of bacterial lipases came only slowly. The first one, from *B. glumae*, appeared in 1993 (96). Several years later, it was followed by the lipases from *Chromobacterium viscosum* (77) [a lipase that appeared to be identical to the *B. glumae* lipase (151)] and *B. cepacia* (69, 79, 129). Recently, the X-ray structures of a lipase from *Streptomyces exfoliatus* (162), *Streptomyces scabies* (161) and an esterase from *Pseudomonas fluorescens* were published. The latter enzyme can hydrolyze triacetin and tributyrin but cannot hydrolyze triglycerides of longer-chain fatty acids (68).

The Fold of Lipases

When the first lipase structures became known, it already appeared that they had very similar folds despite a lack of sequence similarity (22, 27, 143). A more extensive comparison with X-ray structures of other enzymes as diverse as haloalkane dehalogenase (39), acetylcholinesterase (147), dienelactone hydrolase (106), and serine carboxypeptidase (85) revealed that these enzymes all share the same folding pattern. Because they all catalyze a hydrolysis reaction, the common folding pattern was named the α/β hydrolase fold (100).

The canonical α/β hydrolase fold consists of a central, mostly parallel β sheet of eight strands with the second strand antiparallel (Figure 3). The parallel strands $\beta3$ to $\beta8$ are connected by α helices, which pack on either side of the central β sheet. The β sheet has a left-handed superhelical twist such that the surface of the sheet covers about half a cylinder and the first and last strands cross each other at an angle of $\sim90°$. The curvature of the β sheet may differ significantly among the various enzymes, and also, the spatial positions of topologically equivalent α helices may vary considerably. Excursions of the peptide chain at the C-terminal ends of strands

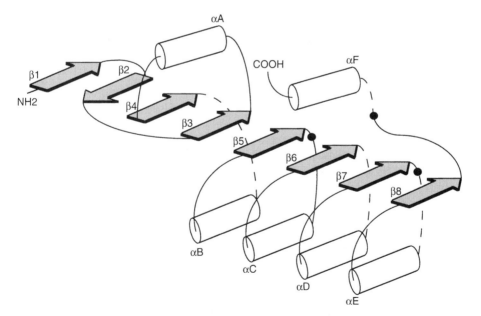

Figure 3 Canonical fold of α/β hydrolases (100). α Helices are indicated by *cylinders*, and β strands are indicated by *shaded arrows*. The topological position of the active-site residues is shown by a *solid circle*; the nucleophile is the residue after β strand 5, the Asp/Glu residue is after β strand 7, and the histidine residue is in the loop between $\beta8$ and αF.

in the C-terminal half of the β sheet form the binding subdomains of the α/β hydrolase fold proteins. They differ substantially in length and architecture, in agreement with the large substrate diversity of these enzymes.

A detailed overview of the α/β hydrolase fold in lipases has been given by Schrag & Cygler (128). The bacterial lipase structures known so far obey the α/β hydrolase fold with some variations. The lipases from *B. glumae*, *B. cepacia*, and *C. viscosum* have six parallel β strands in the central β sheet of the α/β hydrolase fold [β3 to β8 (Figure 3)] (69, 77, 96). The *P. fluorescens* carboxylesterase contains seven strands, corresponding to strands β2 to β8 of the canonical fold (68), and the *S. exfoliatus* lipase has the full canonical α/β hydrolase fold with one extra antiparallel β strand added as strand 9. However, greater variation may be expected, for instance, for the smaller lipases from *Bacillus* (110) and the thioesterase from *E. coli* (87) and the higher-molecular-mass lipases from *Staphylococcus* species (109).

Recently, the X-ray structure of *P. aeruginosa* lipase was determined (DA Lang, K-E Jaeger, BW Dijkstra, in preparation). In contrast to the *B. glumae* and *B. cepacia* lipases, which belong to family I.2 lipases (58), the *P. aeruginosa* is a family I.1 lipase. Lipases of families I.1 and I.2 show ~60% amino acid sequence homology to each other [44% identity (148)]. Family I.1 lipases are somewhat smaller (about 285 amino acid residues instead of 320). *Pseudomonas* lipases are widely used in industry, especially for the production of chiral chemicals, which serve as basic building blocks in the synthesis of pharmaceuticals, pesticides, and insecticides (152). These enzymes show distinct differences in regioselectivity and enantioselectivity (120), despite a high amino acid sequence homology. For instance, Rogalska et al (120) found that the lipase from *C. viscosum* [which is identical to the *B. glumae* lipase (151)] reacts unspecifically [*sn*-3 (*R*)/*sn*-1 (*S*)] with trioctanoin, whereas the lipases from *B. cepacia* and *P. aeruginosa* (58) are absolutely specific for *sn*-1 fatty acid chains of natural substrates. The X-ray structure of *P. aeruginosa* shows the conserved α/β hydrolase folding pattern. Compared with the family I.2 lipases, the C-terminal antiparallel β sheet [residues 214 to 228 (96)] is missing, revealing more compact packing of the molecule. The active site is quite likely solvent accessible (Figure 4; see color insert).

The Catalytic Residues of Lipases

The active site of the α/β hydrolase fold enzymes consists of three catalytic residues: a nucleophilic residue (serine, cysteine, or aspartate), a catalytic acid residue (aspartate or glutamate), and a histidine residue, always in this order in the amino acid sequence (100). This order is different from that observed in any of the other proteins that contain catalytic triads. In lipases the nucleophile has so far always been found to be a serine residue, whereas the catalytic acid is either an aspartate or a glutamate residue.

The nucleophilic serine residue is located in a highly conserved Gly-X-Ser-X-Gly pentapeptide (100), which forms a sharp, γ-like turn between β5 of the canonical α/β hydrolase central β sheet and the following α helix. The γ turn is characterized by energetically unfavorable main chain ϕ and φ torsion angles of the nucleophile. $\beta5$, the γ turn and the following α helix form the most conserved structural feature of the α/β hydrolase fold. Because close contacts exist between the residues, two positions before and two positions behind the nucleophile, one and usually both residues at these positions are glycines. They are occasionally substituted for other small residues such as alanine, valine, serine, or threonine (23, 39, 82, 156). The strand-nucleophile-helix arrangement has been named the "nucleophile elbow." It positions the nucleophilic residue free of the active site surface and allows easy access on one side by the active site histidine residue and on the other by the substrate. The sharp turn also optimally positions the nucleophile at the N-terminal end of the following helix, thereby helping to stabilize the tetrahedral intermediate and the ionized form of the nucleophile.

In the prototypic α/β hydrolase fold, the catalytic acid (Asp or Glu) occurs in a reverse turn after strand 7 of the central β sheet (100). It is hydrogen-bonded to the active site histidine. However, the topological position of the catalytic acid seems to be variable: in pancreatic lipase it is positioned after strand 6 (163). This observation prompted Schrag et al (131) to redesign the active site of *G. candidum* lipase by shifting the position of the catalytic acid from strand 7 to strand 6. The double mutant retained $\sim$10% of the wild-type activity, confirming the variability of the position of the catalytic acid. Similarly, the functionality of the catalytic acid in dehalogenase (Asp-260 positioned after strand 7) could be rescued by an Asn-148Asp mutation, a residue which is located after strand 6 (75). The X-ray structure of *B. glumae* lipase revealed Asp-263 as the catalytic acid residue that is hydrogen bonded to the catalytic histidine (96). Its position is not at the end of $\beta7$, but there are several secondary structure elements lying between the end of $\beta7$ and Asp-263. Site-directed mutagenesis experiments showed, however, only a modest reduction in activity of an Asp-263Ala mutant (42). Because the carboxyl group of Glu-287, which is located after strand 8, is close to the catalytic histidine, it is conceivable that Glu-287 takes over the role of Asp-263, either directly or via a water molecule (96).

The third catalytic residue in lipases is the catalytic histidine. This residue is located in a loop after β strand 8 of the α/β hydrolase fold. The length and conformation of this loop are variable (100).

The Catalytic Mechanism

Lipases are hydrolases acting on the carboxyl ester bonds present in acylglycerols to liberate fatty acids and glycerol. As detailed above, their active site consists of a Ser-His-Asp/Glu catalytic triad. This catalytic triad is similar to that observed in serine proteases, and therefore catalysis by lipases is thought to proceed along a similar path as in serine proteases. Hydrolysis of the substrate takes place in two steps (Figure 5).

It starts with an attack by the oxygen atom of the hydroxyl group of the nucleophilic serine residue on the activated carbonyl carbon of the susceptible lipid ester

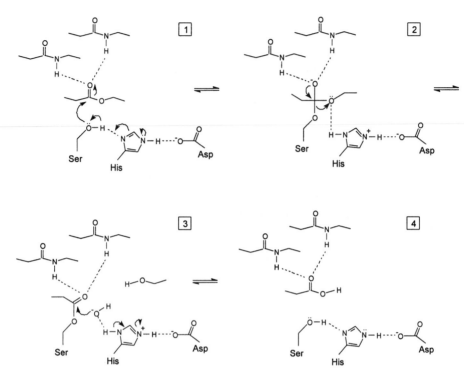

Figure 5 Reaction mechanism of lipases. [1] Binding of lipid, activation of nucleophilic serine residue by neighboring histidine and nucleophilic attack of the substrate's carbonyl carbon atom by Ser O⁻. [2] Transient tetrahedral intermediate, with O⁻ stabilized by interactions with two peptide NH groups. The histidine donates a proton to the leaving alcohol component of the substrate. [3] The covalent intermediate ("acyl enzyme"), in which the acid component of the substrate is esterified to the enzyme's serine residue. The incoming water molecule is activated by the neighboring histidine residue, and the resulting hydroxyl ion performs a nucleophilic attack on the carbonyl carbon atom of the covalent intermediate. [4] The histidine residue donates a proton to the oxygen atom of the active serine residue, the ester bond between serine and acyl component is broken, and the acyl product is released.

bond (Figure 5). A transient tetrahedral intermediate is formed, which is characterized by a negative charge on the carbonyl oxygen atom of the scissile ester bond and four atoms bonded to the carbonyl carbon atom arranged as a tetrahedron (Figure 5). The intermediate is stabilized by the helix macrodipole of helix C (see Figure 3), and hydrogen bonds between the negatively charged carbonyl oxygen atom (the "oxyanion") and at least two main-chain NH groups (the "oxyanion hole"). One of the NH groups is from the residue just behind the nucleophilic serine; the other one is from the residue at the end of strand β3 (65, 79, 129). The nucleophilicity of the attacking serine is enhanced by the catalytic histidine, to which a proton from the serine hydroxyl group is transferred. This proton transfer is facilitated by the presence of the catalytic acid, which precisely orients the imidazole ring of the histidine and partly neutralizes the charge that develops on it. Subsequently, the proton is donated to the ester oxygen of the susceptible bond, which thus is cleaved. At this stage the acid component of the substrate is esterified to the nucleophilic serine (the "covalent intermediate"), whereas the alcohol component diffuses away (Figure 5). The next stage is the deacylation step, in which a water molecule hydrolyzes the covalent intermediate. The active-site histidine activates this water molecule by drawing a proton from it. The resulting OH^- ion attacks the carbonyl carbon atom of the acyl group covalently attached to the serine (Figure 5). Again, a transient negatively charged tetrahedral intermediate is formed, which is stabilized by interactions with the oxyanion hole. The histidine donates a proton to the oxygen atom of the active serine residue, which then releases the acyl component. After diffusion of the acyl product the enzyme is ready for another round of catalysis (Figure 5).

Evidence for this mechanism has come from various studies, particularly inhibitor binding to lipases and their structural analysis (25, 33, 47, 79, 88). In addition, a crystallographic analysis of the reaction catalyzed by haloalkane dehalogenase, another α/β hydrolase enzyme, provided definitive evidence for the occurrence of a covalent intermediate (160).

Interfacial Activation

Lipolytic enzymes are characterized by their drastically increased activity when acting at the lipid-water interface of micellar or emulsified substrates (124), a phenomenon called interfacial activation. This increase in enzymatic activity is triggered by structural rearrangements of the lipase active-site region, as witnessed from crystal structures of lipases complexed by small transition state analogs (17, 24, 158). In the absence of lipid-water interfaces, the active site is covered by a so-called "lid." However, in the presence of hydrophobic substances, the lid is opened, making the catalytic residues accessible to substrate and exposing a large hydrophobic surface. This hydrophobic surface is presumed to interact with the lipid interface. The lid may consist of a single helix (17, 24), or two helices (69, 129), or a loop region (49). However, not all lipases show this interfacial activation. Notable exceptions are the 19-kDa lipase from *B. subtilis* (83), cutinase (91), and guinea pig pancreatic lipase (53). These lipases lack a lid that covers the active site in the absence of lipid-water interfaces.

Substrate Binding

Extensive research has been carried out to identify the binding regions of the acyl and alcohol portions of the substrate in the various lipases and to rationalize the observed enantioselectivity. The X-ray structures of *R. miehei* lipase complexed with a C6 phosphonate inhibitor (25), of *C. rugosa* lipase with a long sulfonyl chain (49), of the human pancreatic lipase/colipase complex covalently inhibited by the two enantiomers of a C11 alkyl chain phosphonate (33), and of porcine pancreatic lipase covalently inhibited by ethylene glycol monooctylether (51) represent important steps in mimicking the natural tetrahedral intermediates. However, none of those compounds resembled a true triglyceride. A first structural view of lipase stereoselectivity toward secondary alcohols was obtained by Cygler et al (21), who succeeded in complexing (R)- and (S)-methyl ester hexylphosphonate transition state analog to C. rugosa lipase. In the fast-reacting, (R)-enantiomer, a hydrogen bond is present between the alcohol oxygen of the substrate and the NE2 atom of the active site histidine. This hydrogen bond is absent in the slow-reacting, (S)-enantiomer, and therefore it was suggested that this hydrogen bond is responsible for the stereospecificity of the enzyme. Uppenberg et al (156) obtained similar results with C. antarctica lipase B complexed with a long-chain polyoxyethylene detergent. However, molecular dynamics calculations by Uppenberg et al suggest that the hydrogen bond is not only present in the fast-reacting (R)-enantiomer but also in the slow reacting (S)-enantiomer. Therefore, they concluded that the enzyme's enantioselectivity cannot simply be explained by the presence or absence of this hydrogen bond but that other factors such as the size of the alcohol-binding pocket may play a role as well. Longhi et al (88) prepared a complex of cutinase with an enantiopure triglyceride analog with three C4 alkyl chains, R_C-(R_P,S_P)-1,2-dibutylcarbamoylglycero-3-O-p-nitrophenyl butylphosphonate (90). Although this inhibitor was expected to reveal the stereospecific substrate interactions with the protein, it unfortunately was bound in the active site in an exposed position, with the alkyl chains not interacting with any amino acid residues of the enzyme.

A breakthrough came with the work of Lang et al (78, 79), who determined the X-ray structure of B. cepacia lipase in complex with an analog of medium alkyl chain length, R_C-(R_P,S_P)-1,2-dioctylcarbamoyl-glycero-3-O-p-nitrophenyl octylphosphonate (TC8). The enzyme is in the open conformation with the lid displaced to allow access of the substrate analog. Density for the complete inhibitor is visible, allowing an unambiguous definition of the substrate-binding mode. Four binding pockets were detected: an oxyanion hole and three pockets that accommodate the sn-1, sn-2, and sn-3 fatty acid chains (Figure 6). The boomerang-shaped active site (69, 129) is divided into a large hydrophobic groove, in which the sn-3 acyl chain snugly fits, and a part that embeds the inhibitor's alcohol moiety. The alcohol-binding pocket can be subdivided into a mixed hydrophilic/hydrophobic cleft for the sn-2 moiety of the substrate and a smaller hydrophobic groove for the sn-1 chain. Van der Waals interactions are the main forces that keep the radyl groups of the triglyceride analog in position. In addition, a hydrogen bond between the ester oxygen atom of the sn-2 chain and the NE2 atom of the active site histidine contributes to fix the position of the inhibitor. This hydrogen bond is equivalent to the one observed by Cygler et al (21) in C. rugosa lipase.

The bound lipid analog assumes the bent-tuning-fork conformation preferred by lipids at an interface (105). Because the sn-2 pocket provides the most intimate interactions with the substrate, this pocket is presumably the one that predominantly determines the enzyme's stereopreferences.

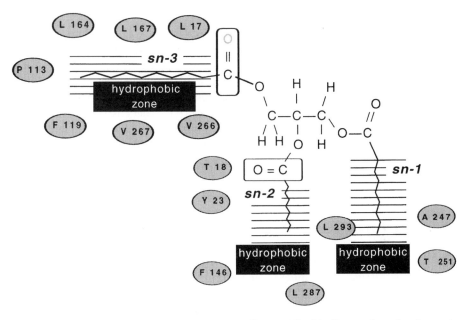

Figure 6   Active site of *Burkholderia cepacia* lipase. The binding pockets for the *sn*-1, *sn*-2, and *sn*-3 moieties of the lipid substrate are indicated. Also indicated are the residues lining these binding pockets.

Structural Determinants of Enantiomeric Selectivity

B. cepacia lipase is one of the most widely used enzymes for the enantiomeric resolution of esters of secondary alcohols (152). It has a preference for R_C over S_C compounds (79), but both enantiomers can be converted, demonstrating that the less preferred enantiomer can also be productively bound in the active site. Modeling of an S_C-TC8 compound in the active site of *B. cepacia* lipase revealed that the acyl chain bound to the primary hydroxyl group of the glycerol moiety clashes with the hydrophobic side chains of Leu-287 and Ile-290 (78, 79). To prevent this unfavorable interaction, either the amino acid side chains have to be moved out of the way or the substrate should undergo a conformational change, for instance, via a rotation about a single C-C bond. Hirose et al (52) have carried out site-directed mutagenesis experiments to probe the importance of various amino acid residues for the stereoselectivity of *B. cepacia* lipase. They succeeded in changing the enzyme's enantioselectivity from an R_C to S_C specificity by introducing a combination of three mutations, Val-266Leu, Leu-287Ile, and Phe-221Leu. Val-266 is located at the entrance of the acyl pocket (*sn*-3 pocket), whereas Leu-287 is at the beginning of the *sn*-2 pocket. Phe-221 is at the surface of the enzyme, ~20 Å away from the inhibitor. Whereas the Leu-287Ile and Val-266Leu substitutions can be envisaged to affect the size and width of the *sn*-2 and *sn*-3 pockets, respectively, Phe-221 seems too far away to influence directly the enzyme's stereospecificity. Nevertheless, the Phe-221Leu mutation on its own was reported to decrease slightly the enzyme's enantioselectivity (52). Clearly, further research is required to elucidate the role of Phe-221 as an enantioselectivity-determining factor.

In the past, it has been found that *B. cepacia* lipase shows the largest enantioselectivity if one of the substituents differs significantly in size from the other (67). This observation can now be rationalized; the enzyme contains a large hydrophobic groove in which the *sn*-3 acyl chain fits, a mixed hydrophilic/hydrophobic cleft for the *sn*-2 moiety of the substrate, and a smaller hydrophobic groove for the *sn*-1 chain. The differences in size and the hydrophilicity/hydrophobicity of the various pockets determine the enzyme's enantiopreferences and regiopreferences.

BIOTECHNOLOGICAL APPLICATIONS OF LIPASES

The commercial use of lipases is a billion-dollar business that comprises a wide variety of different applications. In the area of detergents, about 1000 tons of lipases are sold every year (44). Lipases also play an important role in the production of food ingredients (59). An example is the *Rhizomucor miehei* lipase-catalyzed transesterification reaction replacing palmitic acid by stearic acid to provide the stearic-oleic-stearic triglyceride with the desired melting point for use in chocolate (cocoa butter substitute) (19). Other applications of increasing interest include use of lipases in removing the pitch from pulp in the paper industry (36), in flavor development for dairy products and beverages, and in synthetic organic chemistry (13, 29, 35a, 59, 125, 144, 152).

Lipases as Catalysts in Synthetic Organic Chemistry

The number of reports concerning the use of lipases as catalysts in synthetic organic chemistry is increasing considerably (13, 29, 35, 125, 144). In addition to regioselective hydrolysis, acylation, or transesterification, which includes protective and deprotective methodologies, an incredibly wide variety of enantioselective processes have been reported. As outlined above, the two types of enantioselective organic transformations catalyzed by lipases are reactions of prochiral substrates and the kinetic resolution of racemates. Originally, prochiral or chiral alcohols or carboxylic acid esters served as the two main classes of compounds. However, during the last dozen years the scope has been extended to include cyanohydrins, chlorohydrins, diols, α- and β-hydroxy acids, amines, diamines, and amino alcohols. In principle, the most important classes of organic compounds can thus be produced enantioselectively by using lipase-catalysis. Either an aqueous medium is chosen (hydrolysis reactions), or organic solvents (70) (acylation or transesterification) are used. Typical catalysts for these synthetic organic reactions are bacterial lipases from *P. aeruginosa*, *P. fluorescens*, and other *Pseudomonas* species, *B. cepacia*, *C. viscosum*, *B. subtilis*, *Achromobacter* sp., *Alcaligenes* sp., and

S. marcescens. In view of the large number of publications from academic institutions concerning enantioselective reactions, it may be surprising that only a few cases of industrial processes are known. An example of considerable economic importance concerns the production of the calcium antagonist Diltiazem, a major pharmaceutical used in the treatment of high blood pressure. The key step is the kinetic resolution of a racemic mixture of a chiral epoxy ester based on the preferential hydrolysis of one enantiomeric form (125, 136) (Figure 7). Specifically, the bacterial lipase from *S. marcescens* catalyzes the hydrolysis of the (2*S*,3*R*)-configurated methyl *p*-methoxyphenylglycidate, which is easily separated from the desired (2*R*,3*S*) ester. At maximum theoretical conversion (~50%), the enantiomeric excess (ee) of (2*R*,3*S*)-methyl-*p*-methoxyphenylglycidate is >98%. Since 1993, Tanabe manufactured more than 50 tons of this chiral building block per year, which is then converted into Diltiazem.

Another notable example of the industrial application of a lipase in enantioselective organic chemistry concerns the BASF AG process for preparing optically

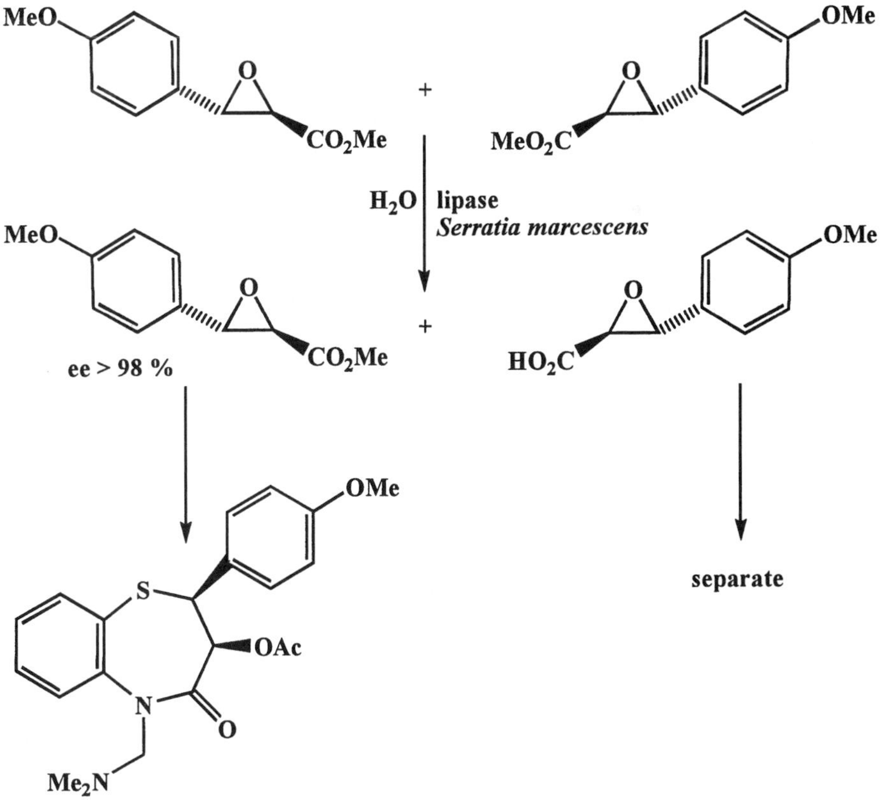

Figure 7 Kinetic resolution of methyl-*p*-methoxyphenylglycidate by *Serratia marcescens* lipase as a key step in the synthesis of Diltiazem.

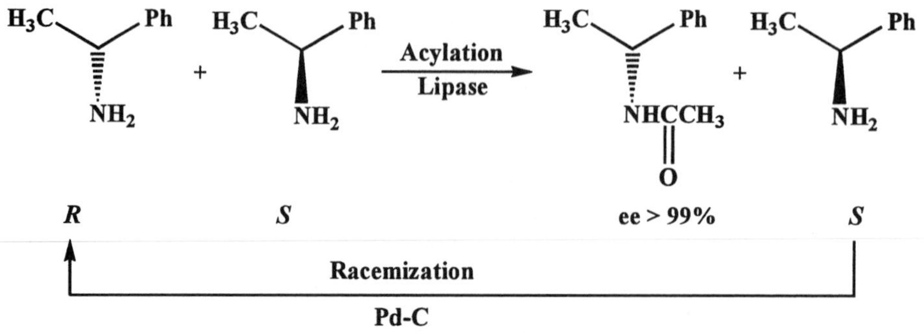

Figure 8 Dynamic kinetic resolution of racemic phenylethyl amine based on enantioselective lipase-catalyzed acylation and racemization by palladium on charcoal (Pd-C).

active amines (8). The lipase from *Burkholderia plantarii* catalyzes the enantioselective acylation of a fairly wide variety of racemic amines (ee > 95%). Perhaps one disadvantage of kinetic resolution is the fact that only 50% of the total material is used. However, methods are starting to be developed that allow for >50% conversion. The underlying principle is "dynamic kinetic resolution" (13, 144, 146). Accordingly, the system not only contains the lipase as an enantioselective biocatalyst but also a second catalyst that causes the rapid racemization of the substrate. Because the product is not racemized under the reaction conditions, 100% conversion to a single enantiomeric product is theoretically possible. This principle opens up completely new perspectives for biocatalysis and indeed for synthetic organic chemistry. In practice, this is difficult to achieve. One stringent requirement is the compatibility of the enzyme with the second catalyst. Nevertheless, several interesting publications illustrating the principle have appeared recently (13, 144, 146). An example is the dynamic resolution of racemic phenylethyl amine based on *N*-acylation catalyzed by the lipase and racemization catalyzed by palladium on charcoal (Pd-C) (116) (Figure 8). Essentially complete enantioselectivity was observed at a conversion of 70%, the side product (30%) being acetophenone.

Immobilization Techniques

It has been claimed that the success of a lipase-catalyzed enantioselective preparation of a certain pharmaceutical depends on immobilization and recyclization of the biocatalyst (20). It is likely that this statement is general. Indeed, in the above mentioned preparation of the chiral intermediate used in the industrial synthesis of Diltiazem, the lipase from *S. marcescens* was supported in a spongy matrix, which was used in a two-phase membrane bioreactor. Other forms of immobilization of lipases have also been described (114). A recent example of a highly efficient technique is based on encapsulation of lipases in hydrophobic sol-gel materials (35a, 112). Accordingly, mixtures of $Si(OCH_3)_4$ and liphophilic alkyl-derivatives $RSi(OCH_3)_3$ are hydrolyzed in the presence of a lipase. The alkyl-modified silica gel, which is produced in the sol-gel process, grows around the enzyme, leading to encapsulation. Such materials display dramatically increased enzyme stability and activity. In acylation and transesterification reactions, lipase activities of 500–10,000% relative to the use of a traditional nonimmobilized enzyme powders in organic solvents are typical. In addition to separation and recyclization by simple filtration, magnetic separation is also possible, provided that nanoparticles of iron oxide are included in the encapsulation (118). Although the reasons for the increased enzyme activities have not been unambiguously elucidated, it is clear that the sol-gel materials need to contain hydrophobic alkyl groups. In the absence of such hydrophobic groups in the matrix as a consequence of using $Si(OCH_3)_4$ as the sol-gel precursor, <5% activity is observed. Thus, a type of interfacial activation may be simulated in the alkyl-modified microenvironment of the lipase, although this needs to be proven.

Directed Evolution of Enantioselective Lipases

Although many different compounds are amenable to lipase-catalyzed enantioselective synthesis, limitations obviously exist owing to substrate specificity. If a given substrate shows an unacceptable level of enantioselectivity, it may help to vary the conditions of the reaction (e.g. solvent, temperature). Nevertheless, such empirical attempts cannot be viewed as a general method.

A recent novel approach towards increasing the enantioselectivity of a lipase-catalyzed organic transformation is based on directed evolution. This technique was used previously to develop enzymes with improved stability and activity (5, 115, 154). Mutations are introduced into genes encoding proteins of interest as catalysts by error-prone PCR or recombinative methods such as DNA shuffling. After expression of these mutated genes in suitable microbial hosts, the production of functional biocatalysts results. Selection or simple screening procedures are then used to identify the "best" mutant enzyme in a large library of potential candidates, and the procedure is repeated n times (n is the number of mutational steps) until the desired catalytic features have been attained. In principle, it should be possible to apply the relevant techniques of molecular biology to create superior biocatalysts having increased degrees of enantioselectivity. However, enantioselectivity is not a simple parameter to deal with. Indeed, several problems need to be addressed, including the all important issue of developing rapid assay systems capable of identifying large numbers of enantioselective biocatalysts.

In a classic paper concerning the creation of enantioselective enzymes by directed evolution, the lipase-catalyzed enantioselective hydrolysis of racemic p-nitrophenyl-2-methyldecanoate was chosen as the test reaction (117) [Figure 9a]. The bacterial lipase from *P. aeruginosa* PA01, an enzyme composed of 285 amino acids, was used as the catalyst, the wild-type showing an enantioselectivity of only 2% ee in favor of the (*S*)-configured 2-methyldecanoic acid. This means that the lipase has essentially no preference for either of the enantiomeric forms of the racemic ester.

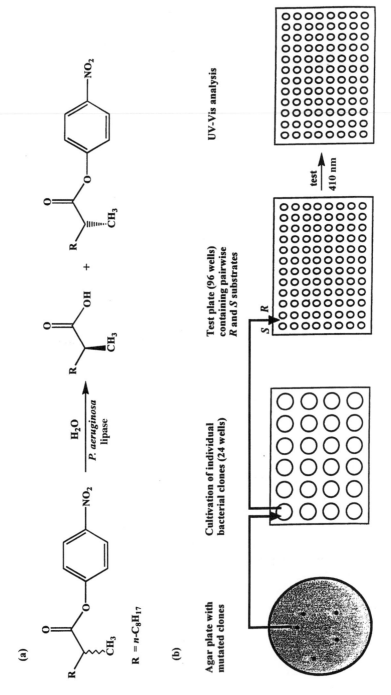

Figure 9 Creation of an enantioselective lipase by directed evolution. (*a*) Kinetic resolution of *p*-nitrophenyl-2-methyldecanoate used as the test reaction. (*b*) Principle of the screening system.

The minimal structural change that mutagenesis could bring about would be the substitution of a single amino acid somewhere along the 285-member amino acid chain. Because the optimal position and type of amino acid substitution cannot be predicted, random mutagenesis was chosen in hope of obtaining a library of mutant lipases in which at least a few members show enhanced enantioselectivity in the test reaction (117). The size of such a library can be calculated to be 5415. Experimentally, the strategy we choose was a low mutagenesis frequency, expecting that a substantial increase in enantioselectivity would result in each generation. By using the error-prone PCR (ep PCR), the lipase gene consisting of 933 bp was subjected to random mutagenesis. The mutated genes were ligated into a suitable expression vector, amplified in *E. coli*, and expressed in *P. aeruginosa*. In the first round of mutagenesis, ~1000 mutants were isolated.

The difficult problem of rapidly screening these for enantioselectivity in the test reaction was solved in the following manner (117). The 96 wells of commercially available microtiter plates were loaded pairwise, with the enantiomerically pure (*R*)- and (*S*)-substrate dissolved separately in dimethylformamide together with the culture supernatants of the respective lipase mutants in Tris/HCl buffer. This means that ≤48 mutants/microtiter plate could be screened by measuring the absorption of the *p*-nitrophenolate anion at 410 nm as a function of time (8–10 min). Obviously, if the slopes of the recorded lines of a given pair of reactions are identical, then the enzyme does not distinguish between the two enantiomeric forms of the substrate. Conversely, if the slopes differ considerably, then a certain degree of enantioselectivity must be operating. Figure 9*b* summarizes the principle of the screening system.

Of the 1000 mutants of the first generation tested, about 12 showed enhanced enantioselectivity. The exact values were determined by hydrolyzing the racemic ester in the presence of the corresponding mutant lipases and analyzing the reaction products by gas chromatography on chirally modified capillary columns. The most selective mutant showed an ee value of 31% in favor of the (*S*)-acid. Subsequent mutagenesis experiments gave rise to improved mutants in the second generation (57% ee), third generation (75% ee), and fourth generation (81% ee). The latter corresponds to an enantiomeric product ratio of better than 90:10 and an E value of 11.3 (117). A further increase in enantioselectivity can be expected by applying the following strategies: (*a*) more extensive screening of larger mutant libraries; (*b*) further generations of mutants; (*c*) application of recombinative methods like DNA shuffling; and (*d*) application of cassette mutagenesis. Indeed, initial experiments along these lines, especially the combination of mutagenesis types, have led to the creation of mutant lipases showing ee values of >90%. The results clearly show that directed evolution is a rational and viable way to obtain engineered lipases showing dramatically enhanced enantioselectivity in a given reaction. One of the prime virtues is the fact that no knowledge of the three-dimensional structure of the enzyme nor of the reaction mechanism is necessary. In contrast to site-specific mutagenesis, there is also no need to develop intuition or theoretical models as to the position or type of amino acid exchange. Thus, in a certain sense the method based on directed evolution is strictly rational. Work is in progress to see how general this novel approach to enantioselectivity is. This includes the development of new screening systems for the high throughput evaluation of enantioselectivity. Such research is necessary because the system described here applies only to chiral esters. One approach is based on black body radiation using an appropriate IR camera. Indeed, enantioselectivity of the lipase-catalyzed acylation of a chiral alcohol has been detected by this technique (113). Nevertheless, other approaches to assay enantioselectivity need to be explored, including the possible use of phage display in selection processes.

CONCLUSIONS AND FUTURE DIRECTIONS

The world of bacterial lipases is rapidly expanding. An impressive number of lipase genes have been identified, and many lipase proteins were biochemically characterized. Because lipase reactions take place at an interface, the quality of which greatly influences the obtained results, it is still difficult to compare directly specific enzyme activities as well as regioselectivities or stereoselectivities determined in different laboratories. Although lipases obtained from selected *Pseudomonas*, *Staphylococcus*, and *Bacillus* species can already be produced at an industrial scale, there is still a lack of knowledge concerning regulation and, in particular, mechanisms governing folding and secretion of lipases. The existing three-dimensional structures of lipases allow the identification of domains and amino acid residues involved in substrate binding, catalysis, and enantiopreference, thereby enabling researchers to tailor lipases for selected applications by using site-directed mutagenesis. Because this approach is limited to only a few specific cases, the creation of lipases with novel properties by directed evolution constitutes a more general approach. Indeed, it is likely that this method will become increasingly important. Undoubtedly, there is a steadily increasing demand to identify, characterize, and produce lipases for a variety of biotechnological applications, with special emphasis on enantioselective biotransformations. Therefore, as a first step, standard assay systems should be developed, allowing one to test hydrolysis and synthesis reactions catalyzed by a given lipase. Furthermore, at least one system that allows for heterologous expression and secretion of different lipases needs to be developed. Finally, a data bank should be built comprising sequences of available genes and proteins, a description of their biochemical properties, including specific activities and stereoselectivities, and options available to express and produce these lipases.

ACKNOWLEDGMENTS

Lipase research in the lab of KEJ is continuously supported by the European Commission in the framework of the programs BRIDGE (BIOT-CT-91-0272) and BIOTECH (BIOT-CT91-272 and BIO4-CT98-0249). The authors thank Susanne

Tetling for preparation of Figures 1 and 2, Dietmar Lang for providing Figure 4, and Uli Winkler for critical reading of the manuscript.

Visit the Annual Reviews home page at http://www.AnnualReviews.org

LITERATURE CITED

1. Ader U, Andersch P, Berger M, Goergens U, Haase B, et al. 1997. Screening techniques for lipase catalyst selection. See Ref. 122, pp. 351–86

2. Akatsuka H, Binet R, Kawai E, Wandersman C, Omori K. 1997. Lipase secretion by bacterial hybrid ATP-binding cassette exporters: molecular recognition of the *lipBCD*, PrtDEF, and HasDEF exporters. *J. Bacteriol.* 179:4754–60

3. Akatsuka H, Kawai E, Omori K, Shibatani T. 1995. The three genes *lipB*, *lipC*, and *lipD* involved in the extracellular secretion of the *Serratia marcescens* lipase which lacks an N-terminal signal peptide. *J. Bacteriol.* 177:6381–89

4. Alberghina L, Lotti M. 1998. Lipases and lipids: structure, specifity and applications in biocatalysis. *Chem. Phys. Lipids* 93:1–216 (Suppl.)

5. Arnold FH. 1998. Design by directed evolution. *Acc. Chem. Res.* 31:125–31

6. Atlas RM, ed. 1996. *Handbook of Microbiological Media*. Boca Raton, FL: CRC. 1440 pp. 2nd ed.

7. Ayora S, Lindgren PE, Götz F. 1994. Biochemical properties of a novel metalloprotease from *Staphylococcus hyicus* subsp. *hyicus* involved in extracellular lipase processing. *J. Bacteriol.* 176:3218–23

8. Balkenhohl F, Ditrich K, Hauer B, Ladner W. 1997. Optically active amines via lipase-catalyzed methoxyacetylation. *J. Prakt. Chem. Chem. Ztg.* 339:381–84

9. Barras F, van Gijsegem F, Chatterjee AK. 1994. Extracellular enzymes and pathogenesis of soft-rot *Erwinia*. *Annu. Rev. Phytopathol.* 32:201–34

10. Berger JL, Lee BH, Lacroix C. 1995. Identification of new enzyme activities of several strains of *Thermus* species. *Appl. Microbiol. Biotechnol.* 44:81–87

11. Binet R, Létoffé S, Ghigo JM, Delepelaire P, Wandersman C. 1997. Protein secretion by gram-negative bacterial ABC exporters—a review. *Gene* 192:7–11

12. Bitter W, Koster M, Latijnhouwers M, de Cock H, Tommassen J. 1998. Formation of oligomeric rings by XcpQ and PilQ, which are involved in protein transport across the outer membrane of *Pseudomonas aeruginosa*. *Mol. Microbiol.* 27:209–19

13. Boland W, Frößl C, Lorenz M. 1991. Esterolytic and lipolytic enzymes in organic synthesis. *Synthesis* 1049–72

14. Brady L, Brzozowski AM, Derewenda ZS, Dodson E, Dodson G, et al. 1990. A serine protease triad forms the catalytic center of a triacylglycerol lipase. *Nature* 343:767–70

15. Brint JM, Ohman DE. 1995. Synthesis of multiple exoproducts in *Pseudomonas aeruginosa* is under the control of RhIR-RhIL, another set of regulators in strain PAO1 with homology to the autoinducer-responsive LuxR-Luxl family. *J. Bacteriol.* 177:7155–63

16. Brumlik MJ, van der Goot GF, Wong KR, Buckley JT. 1997. The disulfide bond in the *Aeromonas hydrophila* lipase/acyltransferase stabilizes the structure but is not required for secretion or activity. *J. Bacteriol.* 179:7155–63

17. Brzozowski AM, Derewenda U, Derewenda ZS, Dodson GG, Lawson DM, et al. 1991. A model for interfacial activation in lipases from the structure of a fungal lipase-inhibitor complex. *Nature* 351:491–94

18. Chihara-Siomi M, Yoshikawa K, Oshima-Hirayama N, Yamamoto K, Sogabe Y,

et al. 1992. Purification, molecular cloning, and expression of lipase from *Pseudomonas aeruginosa*. *Arch. Biochem. Biophys.* 296:505–13

19. Coleman MH, Macrae AE. 1980. *U.S. Patent No. 1 577 933*

20. Cowan D. 1996. Industrial enzyme technology. *Trends Biotechnol.* 14:177–78

21. Cygler M, Grochulski P, Kazlauskas RJ, Schrag JD, Bouthillier F, et al. 1994. A structural basis for the chiral preferences of lipases. *J. Am. Chem. Soc.* 116:3180–86

22. Cygler M, Schrag JD, Ergan F. 1992. Advances in structural understanding of lipases. *Biotech. Gen. Eng. Rev.* 10:143–84

23. Dartois V, Baulard A, Schanck K, Colson C. 1992. Cloning, nucleotide sequence and expression in *Escherichia coli* of a lipase gene from *Bacillus subtilis* 168. *Biochim. Biophys. Acta* 1131:253–60

24. Derewenda U, Brzozowski AM, Lawson DM, Derewenda ZS. 1992. Catalysis at the interface: the anatomy of a conformational change in a triglyceride lipase. *Biochemistry* 31:1532–41

25. Derewenda U, Swenson L, Green R, Wei Y, Dodson GG, et al. 1994. An unusual buried polar cluster in a family of fungal lipases. *Nat. Struct. Biol.* 1:36–47

26. Derewenda U, Swenson L, Wei YY, Green R, Kobos PM, et al. 1994. Conformational lability of lipases observed in the absence of an oil-water interface—crystallographic studies of enzymes from the fungi *Humicola lanuginosa* and *Rhizopus delemar*. *J. Lipid Res.* 35:524–34

27. Derewenda ZS, Derewenda U. 1991. Relationships among serine hydrolases: evidence for a common structural motif in triacylglycerol lipases and esterases. *Biochem. Cell. Biol.* 69:842–51

28. Derewenda ZS, Derewenda U, Dodson GG. 1992. The crystal and molecular structure of the *Rhizomucor miehei* triglyceride lipase at 1.9 Å resolution. *J. Mol. Biol.* 227:818–39

29. Drauz K, Waldmann H, eds. 1995. *Enzyme Catalysis in Organic Synthesis: A Comprehensive Handbook*, Vols. 1, 2. New York: VCH Publ. 504 pp. 545 pp.

30. Dunphy G, Miyamoto C, Meighen E. 1997. A homoserine lactone autoinducer regulates virulence of an insect-pathogenic bacterium, *Xenorhabdus nematophilus* (*Enterobacteriaceae*). *J. Bacteriol.* 179:5288–91

31. Duong F, Eichler J, Price A, Leonard RM, Wickner W. 1997. Biogenesis of the gram-negative bacterial envelope. *Cell* 91:567–73

32. Duong F, Soscia A, Lazdunski A, Murgier M. 1994. The *Pseudomonas fluorescens* lipase has a C-terminal secretion signal and is secreted by a three-component bacterial ABC-exporter system. *Mol. Microbiol.* 11:1117–26

33. Egloff M-P, Marguet F, Buono G, Verger R, Cambillau C, et al. 1995. The 2.46 Å resolution structure of the pancreatic lipase-colipase complex inhibited by a C11 alkyl phosphonate. *Biochemistry* 34:2751–62

34. Eijkmann C. 1901. Über Enzyme bei Bakterien und Schimmelpilzen. *Centralbl. Bakteriol.* 29:841–48

35. El Khattabi M, Ockhuijsen C, Bitter W, Jaeger K-E, Tommassen J. 1999. Specificity of the lipase-specific foldases of *Pseudomonas* spp. *Mol. Gen. Genet.* In press

35a. Faber K, ed. 1997. *Biotransformation in Organic Chemistry*. Berlin: Springer-Verlag. 406 pp. 3rd ed.

36. Farrell RL, Hata K, Wall MB. 1997. Solving pitch problems in pulp and paper processes by the use of enzymes or fungi. *Adv. Biochem. Eng. Biotechnol.* 57:197–212

37. Ferrato F, Carriere F, Sarda L, Verger R. 1997. A critical reevaluation of the phenomenon of interfacial activation. See Ref. 122, pp. 327–47

38. Filloux A, Michel G, Bally M. 1998. GSP-dependent protein secretion in gram-negative bacteria: the Xcp-system of *Pseu-*

domonas aeruginosa. FEMS Microbiol. Rev. 22:177–98

39. Franken SM, Rozeboom HJ, Kalk KH, Dijkstra BW. 1991. Crystal structure of haloalkane dehalogenase: an enzyme to detoxify halogenated alkanes. *EMBO J.* 10:1297–302

40. Frenken LGJ, Bos JW, Visser C, Müller W, Tommassen J, Verrips CT. 1993. An accessory gene, *lipB*, required for the production of active *Pseudomonas glumae* lipase. *Mol. Microbiol.* 9:579–89

41. Frenken LGJ, de Groot A, Tommassen J, Verrips CT. 1993. Role of *lipB* gene product in the folding of the secreted lipase of *Pseudomonas glumae. Mol. Microbiol.* 9:591–99

42. Frenken LGJ, Egmond MR, Batenburg AM, Bos JW, Visser C, Verrips CT. 1992. Cloning of the *Pseudomonas glumae* lipase gene and determination of the active site residues. *Appl. Environ. Microbiol.* 58:3787–91

43. Gerritse G, Ure R, Bizoullier F, Quax W. 1998. The phenotype enhancement method identifies the Xcp outer membrane secretion machinery from *Pseudomonas alcaligenes* as a bottleneck for lipase production. *J. Biotechnol.* 64:23–28

44. Godfrey T, West S, eds. 1996. *Industrial Enzymology. The Application of Enzymes in Industry.* New York: Stockton. 512 pp. 2nd ed.

45. Götz F, Verheij HM, Rosenstein R. 1998. Staphylococcal lipases: molecular characterization, secretion, and processing. *Chem. Phys. Lipids* 93:15–20

46. Gosh PK, Saxena RK, Gupta R, Yadav RP, Davidson S. 1996. Microbial lipases: production and applications. *Sci. Prog.* 79: 119–57

47. Grochulski P, Bouthillier F, Kazlauskas RJ, Serreqi AN, Schrag JD, et al. 1994. Analogs of reaction intermediates identify a unique substrate binding site in *Candida rugosa* lipase. *Biochemistry* 33:3494–500

48. Grochulski P, Li Y, Schrag JD, Bouthill-

ier F, Smith P, et al. 1993. Insights into interfacial activation from an open structure of *Candida rugosa* lipase. *J. Biol. Chem.* 268:12843–47

49. Grochulski P, Li Y, Schrag JD, Cygler M. 1994. Two conformational states of *Candida rugosa* lipase. *Protein Sci.* 3:82–91

50. Heinrichs JH, Bayer MG, Cheung AL. 1996. Characterization of the *sar* locus and its interaction with *agr* in *Staphylococcus aureus. J. Bacteriol.* 178:418–23

51. Hermoso J, Pignol D, Kerfelec B, Crenon I, Chapus C, et al. 1996. Lipase activation by nonionic detergents. *J. Biol. Chem.* 271:18007–16

52. Hirose Y, Kariya K, Nakanishi Y, Kurono Y, Achiwa K. 1995. Inversion of enantioselectivity in hydrolysis of 1,4-dihydropyridines by point mutation of lipase PS. *Tetrahedron Lett.* 36:1063–66

53. Hjort A, Carrière F, Cudrey C, Wöldike H, Boel E, et al. 1993. A structural domain (the lid) found in pancreatic lipases is absent in the guinea pig (phospho)lipase. *Biochemistry* 32:4702–7

54. Hobson AH, Buckley CM, Jørgensen ST, Diderichsen B, McConnell DJ. 1995. Interaction of *Pseudomonas cepacia* DSM3959 lipase with its chaperone, LimA. *J. Biochem. (Tokyo)* 118:575–81

55. Ihara F, Okamoto I, Akao K, Nihira T, Yamada Y. 1995. Lipase modulator protein (LimL) of *Pseudomonas* sp. Strain 109. *J. Bacteriol.* 177:1254–58

56. Ihara F, Okamoto I, Nihira T, Yamada Y. 1992. Requirement in *trans* of the downstream *limL* gene for activation of lactonizing lipase from *Pseudomonas* sp. 109. *J. Ferment. Bioeng.* 73:337–42

57. Iizumi T, Nakamura K, Shimada Y, Sugihara A, Tominaga Y, Fugase T. 1991. Cloning, nucleotide sequencing, and expression in *Escherichia coli* of a lipase and its activator genes from *Pseudomonas* sp. KWI-56. *Agric. Biol. Chem.* 55:2349–57

58. Jaeger K-E, Ransac S, Dijkstra BW, Colson C, van Heuvel M, Misset O. 1994. Bacte-

rial lipases. *FEMS Microbiol. Rev.* 15:29–63

59. Jaeger K-E, Reetz MT. 1998. Microbial lipases form versatile tools for biotechnology. *Trends Biotechnol.* 16:396–403

60. Jaeger K-E, Schneidinger B, Liebeton K, Haas D, Reetz MT, et al. 1996. Lipase of *Pseudomonas aeruginosa*: molecular biology and biotechnological application. In *Molecular Biology of Pseudomonads*, ed. T Nakazawa, K Furukawa, D Haas, S Silver, pp. 319–330. Washington, DC: Am. Soc. Microbiol. 526 pp.

61. Jarvis GM, Thiele JH. 1997. Qualitative Rhodamine B assay which uses tallow as a substrate for lipolytic obligately anaerobic bacteria. *J. Microbiol. Methods* 29:41–47

62. Jensen RG. 1983. Detection and determination of lipase (acylglycerol hydrolase) activity from various sources. *Lipids* 18:650–57

63. Jørgensen S, Skov KW, Diderichsen B. 1991. Cloning, sequence, and expression of a lipase gene from *Pseudomonas cepacia*: lipase production in heterologous hosts requires two *Pseudomonas* genes. *J. Bacteriol.* 173:559–67

64. Kawai E, Akatsuka H, Idei A, Shibatani T, Omori K. 1998. *Serratia marcescens* S-layer protein is secreted extracellularly via ATP-binding cassette exporter, the Lip system. *Mol. Microbiol.* 27:941–52

65. Kazlauskas RJ. 1994. Elucidating structure-mechanism relationships in lipases: prospects for predicting and engineering catalytic properties. *Trends Biotechnol.* 12:464–72

66. Kazlauskas RJ, Bornscheuer UT. 1998. Biotransformations with lipases. In *Biotechnology*, Vol. 8a, *Biotransformations I*, ed. DR Kelly, pp. 37–191. Weinheim: Wiley-VCH/New York: VCH Publ. 607 pp.

67. Kazlauskas RJ, Weissfloch ANE, Rappaport AT, Cuccia LA. 1991. A rule to predict which enantiomer of a secondary alcohol reacts faster in reactions catalysed by cholesterol esterase, lipase from *Pseudomonas cepacia*, and lipase from *Candida rugosa*. *J. Org. Chem.* 56:2656–65

68. Kim KK, Song HK, Shin DH, Hwang KY, Choe S, et al. 1997. Crystal structure of carboxylesterase from *Pseudomonas fluorescens*, an α/β hydrolase with broad substrate specificity. *Structure* 5:1571–84

69. Kim KK, Song HK, Shin DH, Hwang KY, Suh SW. 1997. The crystal structure of a triacylglycerol lipase from *Pseudomonas cepacia* reveals a highly open conformation in the absence of a bound inhibitor. *Structure* 5:173–85

70. Klibanov AM. 1990. Asymmetric transformations catalyzed by enzymes in organic solvents. *Acc. Chem. Res.* 23:114–20

71. Kok RG, Nudel CB, Gonzalez RH, Nutgeren-Roodzant I, Hellingwerf KJ. 1996. Physiological factors affecting production of extracellular lipase (LipA) in *Acinetobacter calcoaceticus* BD413: fatty acid repression of *lipA* expression and degradation of LipA. *J. Bacteriol.* 177:6025–35

72. Kok RG, van Thor JJ, Nutgeren-Roodzant I, Brouwer MBW, Egmond MR, et al. 1995. Characterization of the extracellular lipase, LipA, of *Acinetobacter calcoaceticus* BD413 and sequence analysis of the cloned structural gene. *Mol. Microbiol.* 15:803–18

73. Kok RG, van Thor JJ, Nutgeren-Roodzant I, Vosman B, Hellingwerf KJ. 1995. Characterization of lipase-deficient mutants of *Acinetobacter calcoaceticus* BD413: identification of a periplasmic lipase chaperone essential for the production of extracellular lipase. *J. Bacteriol.* 177:3295–307

74. Kouker G, Jaeger K-E. 1987. Specific and sensitive plate assay for bacterial lipases. *Appl. Environ. Microbiol.* 53:211–13

75. Krooshof GH, Kwant EM, Damborsky J, Koca J, Janssen DB. 1997. Repositioning the catalytic triad aspartic acid of haloalkane dehalogenase: effects on stability, kinetics, and structure. *Biochemistry* 36:9571–80

76. Kullik I, Giachino P, Fuchs T. 1998. Deletion of the alternative sigma factor σ^B in *Staphylococcus aureus* reveals its function as a global regulator of virulence genes. *J. Bacteriol.* 180:4814–20

77. Lang D, Hofmann B, Haalck L, Hecht H-J, Spener F, et al. 1996. Crystal structure of a bacterial lipase from *Chromobacterium viscosum* ATCC 6918 refined at 1.6 Å resolution. *J. Mol. Biol.* 259:704–17

78. Lang DA, Dijkstra BW. 1998. Structural investigations of the regio- and enantioselectivity of lipases. *Chem. Phys. Lipids* 93:115–22

79. Lang DA, Mannesse MLM, de Haas GH, Verheij HM, Dijkstra BW. 1998. Structural basis of the chiral selectivity of *Pseudomonas cepacia* lipase. *Eur. J. Biochem.* 254:333–40

80. Langin D, Laurell H, Holst LS, Belfrage P, Holm C. 1993. Gene organisation and primary structure of human hormone-sensitive lipase: possible significance of a sequence homology with a lipase from *Moraxella* TA 144, an Antarctic bacterium. *Proc. Natl. Acad. Sci. USA* 90:4897–901

81. Latifi A, Winson MK, Foglino M, Bycroft BW, Stewart GSAB, et al. 1995. Multiple homologues of LuxR and LuxI control expression of virulence determinants and secondary metabolites through quorum sensing in *Pseudomonas aeruginosa* PAO1. *Mol. Microbiol.* 17:333–43

82. Lawson DM, Derewenda U, Serre L, Ferri S, Szittner R, et al. 1994. Structure of a myristoyl-ACP-specific thioesterase from *Vibrio harveyi. Biochemistry* 33:9382–88

83. Lesuisse E, Schanck K, Colson C. 1993. Purification and preliminary characterization of the extracellular lipase of *Bacillus subtilis* 168, an extremely basic pH-tolerant enzyme. *Eur. J. Biochem.* 216:155–60

84. Li X, Tetling S, Winkler UK, Jaeger K-E, Benedik MJ. 1995. Gene cloning, sequence analysis, purification and secretion by *Escherichia coli* of an extracellular lipase from *Serratia marcescens. Appl. Environ. Microbiol.* 61:2674–80

85. Liao DI, Breddam K, Sweet RM, Bullock T, Remington SJ. 1992. Refined atomic model of wheat serine carboxypeptidase-II at 2.2-Ångstrom resolution. *Biochemistry* 31:9796–812

86. Liebl W, Götz F. 1986. Lipase directed export of *Escherichia coli* β-lactamase in *Staphylococcus carnosus. Mol. Gen. Genet.* 204:166–73

87. Lin TH, Chen CP, Huang RF, Lee YL, Shaw JF, et al. 1998. Multinuclear NMR resonance assignments and the secondary structure of *Escherichia coli* thioesterase/protease I: a member of a new subclass of lipolytic enzymes. *J. Biomol. NMR* 11:363–80

88. Longhi S, Mannesse M, Verheij HM, de Haas GH, Egmond M, et al. 1997. Crystal structure of cutinase covalently inhibited by a triglyceride analogue. *Protein Sci.* 6:275–86

89. Loveless BJ, Saier JMH. 1997. A novel family of channel-forming, autotransporting, bacterial virulence factors. *Mol. Membr. Biol.* 14:113–23

90. Mannesse MLM, Boots JWP, Dijkman R, Slotboom AJ, van der Hijden HTWV, et al. 1995. Phosphonate analogues of triacylglycerols are potent inhibitors of lipase. *Biochim. Biophys. Acta* 1259:56–64

91. Martinez C, de Geus P, Lauwereys M, Matthyssens M, Cambillau C. 1992. *Fusarium solani* cutinase is a lipolytic enzyme with a catalytic serine accessible to solvent. *Nature* 356:615–18

92. McNamara PJ, Iandolo JJ. 1998. Genetic instability of the global regulator *agr* explains the phenotype of the *xpr* mutation in *Staphylococcus aureus* KSI9051. *J. Bacteriol.* 180:2609–15

93. Meens J, Herbort M, Klein M, Freudl R. 1997. Use of the pre-pro part of *Staphylococcus hyicus* lipase as a carrier for secretion of *Escherichia coli* outer membrane protein A (OmpA) prevents

proteolytic degradation of OmpA by cell-associated protease(s) in two different Gram-positive bacteria. *Appl. Environ. Microbiol.* 63:2814–20

94. Missiakis D, Raina S. 1997. Protein folding in the bacterial periplasm. *J. Bacteriol.* 179:2465–71

95. Nashif SA, Nelson FE. 1953. The extracellular lipases of some Gram-negative non-sporeforming rod-shaped bacteria. *J. Dairy Sci.* 36:698–706

96. Noble MEM, Cleasby A, Johnson LN, Egmond MR, Frenken LGJ. 1993. The crystal structure of triacylglycerol lipase from *Pseudomonas glumae* reveals a partially redundant catalytic aspartate. *FEBS Lett.* 331:123–28

97. Ochsner UA, Koch AK, Fiechter A, Reiser J. 1994. Isolation and characterization of a regulatory gene affecting rhamnolipid biosurfactant synthesis in *Pseudomonas aeruginosa. J. Bacteriol.* 176:2044–54

98. Ochsner UA, Reiser J. 1995. Autoinducer-mediated regulation of rhamnolipid biosurfactant synthesis in *Pseudomonas aeruginosa. Proc. Natl. Acad. Sci. USA* 92:6424–28

99. Ogierman MA, Fallarino A, Riess T, Willimas SG, Attridge SR, Manning P. 1997. Characterization of the *Vibrio cholerae* El Tor lipase operon *lipAB* and a protease gene downstream of the *hyl* region. *Microbiology* 140:931–43

100. Ollis DL, Shea E, Cygler M, Dijkstra B, Frolow F, et al. 1992. The α/β hydrolase fold. *Protein Eng.* 5:197–211

101. Olukoshi ER, Packter NM. 1994. Importance of stored triacylglycerols in *Streptomyces*: possible carbon source for antibiotics. *Microbiology* 140:931–43

102. Ortaggi G, Jaeger K-E. 1997. Microbial lipases in the biocatalysis. *J. Mol. Catal. B: Enzym.* 3:1–212

103. Oshima-Hirayama N, Yoshikawa K, Nishioka T, Oda J. 1993. Lipase from *Pseudomonas aeruginosa*: production in

Escherichia coli and activation in vitro with a protein from the downstream gene. *Eur. J. Biochem.* 215:239–46

104. Overbye LJ, Sandkvist M, Bagdasarian M. 1993. Genes required for extracellular secretion of enterotoxin are clustered in *Vibrio cholerae. Gene* 132:101–6

105. Pascher I. 1996. The different conformations of the glycerol region of crystalline acylglycerols. *Curr. Opin. Struct. Biol.* 6: 439–48

106. Pathak D, Ollis D. 1990. Refined structure of dienelactone hydrolase at 1.8 Å. *J. Mol. Biol.* 214:497–525

107. Pugsley AP. 1993. The complete secretory pathway in gram-negative bacteria. *Microbiol. Rev.* 57:50–108

108. Raina S, Missiakis D. 1997. Making and breaking disulfide bonds. *Annu. Rev. Microbiol.* 51:179–202

109. Ransac S, Blaauw M, Dijkstra BW, Slotboom AJ, Boots J-WP, et al. 1995. Crystallization and preliminary X-ray analysis of a lipase from *Staphylococcus hyicus. J. Struct. Biol.* 114:153–55

110. Ransac S, Blaauw M, Lesuisse E, Schanck K, Colson C, et al. 1994. Crystallization and preliminary X-ray analysis of a lipase from *Bacillus subtilis. J. Mol. Biol.* 238:857–59

111. Ransac S, Ivanova M, Verger R, Panaiotov I. 1997. Monolayer techniques for studying lipase kinetics. See Ref.122, pp. 263–92

112. Reetz MT. 1997. Entrapment of biocatalysts in hydrophobic sol-gel materials for use in organic chemistry. *Adv. Mater.* 9:943–54

113. Reetz MT, Becker MH, Kühling KM, Holzwarth A. 1998. Time-resolved IR-thermographic detection and screening of enantioselectivity in catalytic reactions. *Angew. Chem. Int. Ed. Engl.* 37:2647–50

114. Reetz MT, Jaeger K-E. 1998. Overexpression, immobilization and biotechnological application of *Pseudomonas* lipases. *Chem. Phys. Lipids* 93:3–14

115. Reetz MT, Jaeger K-E. 1999. Superior biocatalysts by directed evolution. *Top. Curr. Chem.* 200:31–57

116. Reetz MT, Schimossek K. 1996. Lipase-catalyzed dynamic kinetic resolution of chiral amines: use of palladium as the racemization catalyst. *Chimia* 50:668–69

117. Reetz MT, Zonta A, Schimossek K, Liebeton K, Jaeger K-E. 1997. Creation of enantioselective biocatalysts for organic chemistry by in vitro evolution. *Angew. Chem., Int. Ed. Engl.* 36:2830–32

118. Reetz MT, Zonta A, Vijayakrishnan V, Schimossek K. 1998. Entrapment of lipases in hydrophobic magnetite-containing sol-gel materials: magnetic separation of heterogeneous biocatalysts. *J. Mol. Catal. A: Chem.* 134:251–58

119. Reimmann C, Beyeler M, Latifi A, Winteler H, Foglino M, et al. 1997. The global activator GacA of *Pseudomonas aeruginosa* PAO positively controls the production of the autoinducer N-butyryl-homoserine lactone and the formation of the virulence factors pyocyanin, cyanide, and lipase. *Mol. Microbiol.* 24:309–19

120. Rogalska E, Cudrey C, Ferrato F, Verger R. 1993. Stereoselective hydrolysis of triglycerides by animal and microbial lipases. *Chirality* 5:24–30

121. Rubin B, Dennis EA, eds. 1997. *Lipases, Part A: Biotechnology. Methods Enzymol.*, Vol. 284. San Diego: Academic. 408 pp.

122. Rubin B, Dennis EA, eds. 1997. *Lipases, Part B: Enzyme Characterization and Utilization. Methods Enzymol.*, Vol. 286. San Diego: Academic. 563 pp.

123. Salmond GPC, Reeves PJ. 1993. Membrane traffic wardens and protein secretion in gram-negative bacteria. *Trends Biochem. Sci.* 18:7–12

124. Sarda L, Desnuelle P. 1958. Action de la lipase pancréatique sur les esters en émulsion. *Biochim. Biophys. Acta* 30:513–21

125. Schmid RD, Verger R. 1998. Lipases: Interfacial enzymes with attractive applications. *Angew. Chem. Int. Ed. Engl.* 37:1608–33

126. Schmidt-Dannert C, Rua ML, Wahl S, Schmid RD. 1997. *Bacillus thermocatenulatus* lipase: a thermoalkalophilic lipase with interesting properties. *Biochem. Soc. Trans.* 25:178–82

127. Schneidinger B. 1997. *Überexpression und transkriptionelle Regulation des Lipaseoperons von Pseudomonas aeruginosa und funktionelle Charakterisierung der Lipase-spezifischen Foldase Lip H.* PhD thesis. Ruhr-Univ. Bochum. 111 pp.

128. Schrag JD, Cygler M. 1997. Lipases and α/β hydrolase fold. *Methods Enzymol.* 284:85–107

129. Schrag JD, Li Y, Cygler M, Lang D, Burgdorf T, et al. 1997. The open conformation of a *Pseudomonas* lipase. *Structure* 5:187–202

130. Schrag JD, Li Y, Wu S, Cygler M. 1991. Ser-His-Glu triad forms the catalytic site of the lipase from *Geotrichum candidum. Nature* 351:761–64

131. Schrag JD, Vernet T, Laramee L, Thomas DY, Recktenwald A, et al. 1995. Redesigning the active site of *Geotrichum candidum* lipase. *Protein Eng.* 8:835–42

132. Servin-Gonzales L, Castro C, Perez C, Rubio M, Valdez V. 1997. *bldA*-dependent expression of the *Streptomyces exfoliatus* M11 lipase gene (*lipA*) is mediated by the product of a contiguous gene, *lipR*, encoding a putative transcriptional activator. *J. Bacteriol.* 179:7816–26

133. Shibata H, Kato H, Oda J. 1998. Molecular properties and activity of amino-terminal truncated forms of lipase activator protein. *Biosci. Biotechnol. Biochem.* 62:354–57

134. Shibata H, Kato H, Oda J. 1998. Calcium ion-dependent reactivation of a *Pseudomonas* lipase by its specific modulating protein, LipB. *J. Biochem.* 123:136–41

135. Shibata H, Kato H, Oda J. 1998. Random

mutagenesis on the *Pseudomonas* activator protein, LipB: exploring amino acid residues required for its function. *Protein Eng.* 11:467–72

136. Shibatani T, Nakamichi K, Matsumae H. 1990. *European Patent Application EP 362556*

137. Shinde U, Liu J, Inouye M. 1997. Protein folding mediated by intramolecular chaperones. In *Molecular Chaperones in the Life Cycle of Proteins*, ed. AL Fink, Y Goto, pp. 467–90. New York: Marcel Dekker. 626 pp.

138. Shingler. 1996. Signal sensing by σ^{54}-dependent regulators: depression as a control mechanism. *Mol. Microbiol.* 19: 409–16

139. Simonen M, Palva J. 1993. Protein secretion in *Bacillus* species. *Microbiol. Rev.* 57:109–37

140. Simons J-WFA, Boots J-WP, Kats MP, Slotboom AJ, Egmond MR, Verheij HM. 1997. Dissecting the catalytic mechanism of staphylococcal lipases: chain length selectivity, interfacial activation and cofactor dependence. *Biochemistry* 36:14539–50

141. Simons J-WFA, van Kampen MD, Riel S, Götz F, Egmond MR, Verheij HM. 1998. Cloning, purification and characerization of the lipase *Staphylococcus aureus*. *Eur. J. Biochem.* 253:675–83

142. Smeltzer MS, Gill SG, Iandolo JJ. 1992. Localization of a chromosomal mutation affecting expression of extracellular lipase in *Staphylococcus aureus*. *J. Bacteriol.* 174:4000–6

143. Smith LC, Faustinella F, Chan L. 1992. Lipases: three-dimensional structure and mechanism of action. *Curr. Opin. Struct. Biol.* 2:490–96

144. Stecher H, Faber K. 1997. Biocatalytic deracemization techniques. Dynamic resolutions and stereoinversions. *Synthesis* [vol.]:1–16

145. Stuer W, Jaeger K-E, Winkler UK. 1986. Purification of extracellular lipase from *Pseudomonas aeruginosa. J. Bacteriol.* 168:1070–74

146. Stürmer R. 1997. Enzymes and transition metal complexes in tandem—a new concept for dynamic kinetic resolution. *Angew. Chem. Int. Ed. Engl.* 36:1173–74

147. Sussman JL, Harel M, Frolow F, Oefner C, Goldman A, et al. 1988. Atomic structure of acetylcholinesterase from *Torpedo californica*: a prototypic acetylcholine-binding protein. *Science* 253:872–79

148. Svendsen A, Borch K, Barfoed M, Nielsen TB, Gormsen E, et al. 1995. Biochemical properties of cloned lipases from the *Pseudomonas* family. *Biochim. Biophys. Acta* 1259:9–17

149. Svetlitshnyi V, Rainey F, Wiegel J. 1996. *Thermosynthropha lipolytica* gen. nov., a lipolytic, anaerobic, alkalitolerant, thermophilic bacterium utilizing short- and long chain fatty acids in syntrophic coculture with a methanogenic archaeum. *Int. J. Syst. Bacteriol.* 46:1131–37

150. Swift S, Throup JP, Williams P, Salmond GPC, Stewart GSAB. 1996. Quorum sensing: a population-density component in the determination of bacterial phenotype. *Trends Biochem. Sci.* 21:214–19

151. Taipa MA, Liebeton K, Costa JV, Cabral JMS, Jaeger KE. 1995. Lipase from *Chromobacterium viscosum*: biochemical characterization indicating homology to the lipase from *Pseudomonas glumae*. *Biochim. Biophys. Acta* 1256:396–402

152. Theil F. 1995. Lipase-supported synthesis of biologically active compounds. *Chem. Rev.* 95:2203–27

153. Tommassen J, Filloux A, Bally M, Murgier M, Lazdunski A. 1992. Protein secretion in *Pseudomonas aeruginosa. FEMS Microbiol. Rev.* 103:73–90

154. Trower MK, ed. 1996. In vitro mutagenesis protocols. *Methods Mol. Biol.* 57:390 pp.

155. Uppenberg J, Hansen MT, Patkar S, Jones TA. 1994. The sequence, crystal structure determination and refinement of two

crystal forms of lipase B from *Candida antarctica*. *Structure* 2:293–308

156. Uppenberg J, Ohrner N, Norin M, Hult K, Kleywegt GJ, et al. 1995. Crystallographic and molecular-modeling studies of lipase B from *Candida antarctica* reveal a stereospecificity pocket for secondary alcohols. *Biochemistry* 34:16838–51

157. Upton C, Buckley JT. 1995. A new family of lipolytic enzymes. *Trends Biochem. Sci.* 20:178–79

158. van Tilbeurgh H, Egloff M-P, Martinez C, Rugani N, Verger R, et al. 1993. Interfacial activation of the lipase-procolipase complex by mixed micelles revealed by X-ray crystallography. *Nature* 362:814–20

159. Verger R. 1997. 'Interfacial activation' of lipases: facts and artifacts. *Trends Biotechnol.* 15:32–38

160. Verschueren KHG, Seljée F, Rozeboom HJ, Kalk KH, Dijkstra BW. 1993. Crystallographic analysis of the catalytic mechanism of haloalkane dehalogenase. *Nature* 363:693–98

161. Wei Y, Schottel JL, Derewenda U, Swenson L, Patkar S, Derewenda ZS. 1995. A novel variant of the catalytic triad in the *Streptomyces scabies* esterase. *Nat. Struct. Biol.* 2:218–23

162. Wei Y, Swenson L, Castro C, Derewenda U, Minor W, et al. 1998. Structure of a microbial homologue of mammalian platelet-activating factor acetylhydrolases: *Streptomyces exfoliatus* lipase at 1.9 angstrom resolution. *Structure* 6:511–19

163. Winkler FK, D'Arcy A, Hunziker W. 1990. Structure of human pancreatic lipase. *Nature* 343:771–74

164. Wohlfarth S, Hoesche C, Strunk C, Winkler UK. 1992. Molecular genetics of the extracellular lipase of *Pseudomonas aeruginosa* PAO1. *J. Gen. Microbiol.* 138:1325–35

165. Woolley P, Petersen SB, eds. 1994. *Lipases: Their Structure, Biochemistry and Application.* London: Cambridge Univ. Press. 363 pp.

166. Zaks A, Klibanov AM. 1984. Enzymatic catalysis in organic media at 100°C. *Science* 224:1249–51

167. Zandonella G, Haalck L, Spener F, Paltauf F, Hermetter A. 1997. New fluorescent glycerolipids for a dual wavelength assay of lipase activity and stereoselectivity. *J. Mol. Catal. B: Enzym.* 3:127–30

Annu. Rev. Microbiol. 1999. 53:353–87

CONTRIBUTIONS OF GENOME SEQUENCING TO UNDERSTANDING THE BIOLOGY OF *HELICOBACTER PYLORI*

Zhongming Ge

Division of Comparative Medicine, Massachusetts Institute of Technology, Cambridge, Massachusetts 02139

Diane E. Taylor

Department of Medical Microbiology and Immunology, University of Alberta, Edmonton, Alberta, Canada T6G 2H7: e-mail: diane.taylor@ualberta.ca

Key Words genomic organization, virulence determinants, shuttle vectors, bacterial evolution

■ **Abstract** About half of the world's population carries *Helicobacter pylori*, a gram-negative, spiral bacterium that colonizes the human stomach. The link between *H. pylori* and, ulceration as well as its association with the development of both gastric cancer and mucosa-associated lymphoid tissue lymphoma in humans is a serious public health concern. The publication of the genome sequences of two stains of *H. pylori* gives rise to direct evidence on the genetic diversity reported previously with respect to gene organization and nucleotide variability from strain to strain. The genome size of *H. pylori* strain 26695 is 1,6697,867 bp and is 1,643,831 bp for strain J99. Approximately 89% of the predicted open reading frames are common to both of the strains, confirming *H. pylori* as a single species. A region containing ~45% of *H. pylori* strain-specific open reading frames, termed the plasticity zone, is present on the chromosomes, verifying that some strain variability exists. Frequent alteration of nucleotides in the third position of the triplet codons and various copies of insertion elements on the individual chromosomes appear to contribute to distinct polymorphic fingerprints among strains analyzed by restriction fragment length polymorphisms, random amplified polymorphic DNA method, and repetitive element–polymerase chain reaction. Disordered chromosomal locations of some genes seen by pulsed-field gel electrophoresis are likely caused by rearrangement or inversion of certain segments in the genomes. Cloning and functional characterization of the genes involved in acidic survival, vacuolating toxin, *cag*-pathogenicity island, motility, attachment to epithelial cells, natural transformation, and the biosynthesis of lipopolysaccharides have considerably increased our understanding of the molecular genetic basis for the pathogenesis of *H. pylori*. The homopolymeric nucleotide tracts and dinucleotide repeats, which potentially regulate the on- and off-status of the target genes by the strand-slipped mispairing mechanism, are often found in the genes encoding the outer-membrane proteins, in enzymes for lipopolysaccharide synthesis, and within DNA modification/restriction systems. Therefore, these genes may be involved in the *H. pylori*–host interaction.

CONTENTS

INTRODUCTION

Helicobacter pylori is one of the most common bacterial pathogens of humans, infecting approximately half of the population worldwide (25). This microorganism has been recognized as the cause of chronic superficial gastritis, chronic active gastritis, and peptic ulcers (116), and it is associated with the development of gastric cancer and mucosa-associated lymphoid tissue lymphoma (47, 126). Studies on the molecular genetics of *H. pylori* have revealed that there is extensive sequence diversity at the levels of genome and individual genes and that such diversity may contribute to the pathogenicity of *H. pylori*. The high prevalence, medical importance, and unique genetic phenomena of this bacterial pathogen have stimulated interest in this field. A recent survey revealed that ∼7.5% of all publications in microbiological research between 1991 and 1997 were related to the genus *Helicobacter* and the annual publication rate of papers on *H. pylori* has tripled over that period (50). It may not be surprising that the complete genome sequences of two *H. pylori* strains, 26695 (158) and J99 (3), have been determined after Warren & Marshall (173) first described this organism 15 years ago. The genome sequence of strain 26695 is available on the world-wide web at http://www.tigr.org and that of strain J99 was determined by Genome Therapeutic Corp. (licensed to Astra Research Center Boston, Cambridge, MA) and has been released at http://www.astra-boston.com/hpylori or http://www.genomecorp.com/hpylori.

   H. pylori colonizes the human gastric mucosa and can persist in the stomach of patients with or without the clinical symptoms for a lifetime. How does this microorganism survive in such a unique niche (an acidic environment), escape elimination by the host immune response, and provoke a continuum of human diseases? Studies on the molecular genetics of *H. pylori* in the past several years have yielded a number of exciting findings that relate to these questions. The publication of the genome sequences of two *H. pylori* strains has enabled the construction of the entire gene organization of this pathogen and provided a genetic basis for understanding the biological processes, pathogenicity, and evolutionary history of this organism. Because of space limitations, the reader is referred to several excellent reviews for early work on the genetics and virulence factors of *H. pylori* (26, 94, 151) and to individual papers for detailed descriptions of the genome features of two *H. pylori* strains: for strain 26695 refer to Tomb et al (158) and for strain J99 refer to work by Alm et al (3). This review focuses on the contributions of these published *H. pylori* genome sequences, as well as the contributions of recent findings from genetic studies, to the current understanding of genomic organization in general, to evolutionary biology, and particularly to determinants of pathogenicity.

GENOMIC ORGANIZATION

Physical-Genetic Mapping of *H. pylori* Genomes

In the early 1990s, pulsed-field gel electrophoresis in combination with restriction digestion became a powerful tool to reveal structures of microbial genomes in terms of size and localization of known genes on the chromosome. Taylor et al (152) first published the genome map (1.7 Mbp) of a clinical *H. pylori* isolate UA802 with two restriction enzymes, *Not*I and *Nru*I. Based on data from 11 strains, the genome size of *H. pylori* strains was shown to range from 1.6 to 1.73 Mbp with an average of 1.67 Mbp. In addition, a high-resolution physical-genetic map of NCTC11638 was constructed using an orderly array of cosmid clones (18). Later, physical-genetic maps of the additional three strains, UA861, NCTC 11637, and NCTC11639, were made, and 17 known *H. pylori* genes were positioned on the chromosomes of these respective strains (83). Figure 1 presents an up-to-date comparison of the genetic maps of six *H. pylori* strains, including five previously determined (83) and two recently sequenced (3, 158). Comparison of the strain maps shows quite a degree of diversity in gene order, although certain genes (*vacA*, *katA*, *pfr*, and *hpa*) are usually found in the same 25% of the genomes (83). In general J99 and 26695 appear to be more closely related to one another than the other strains examined, based on the 21 different markers (Figure 1). However, only a small fraction of genes have been compared with this method, which is much less precise than the total genome sequence comparison performed for strains 26695 and J99 (see below).

Genomic Sequences of Two *H. pylori* Strains—Overview

H. pylori strains 26695 and J99 are both *cagA*⁺ *vacA*⁺ and were isolated from two geographically separate regions of the United States (3). The complete genome

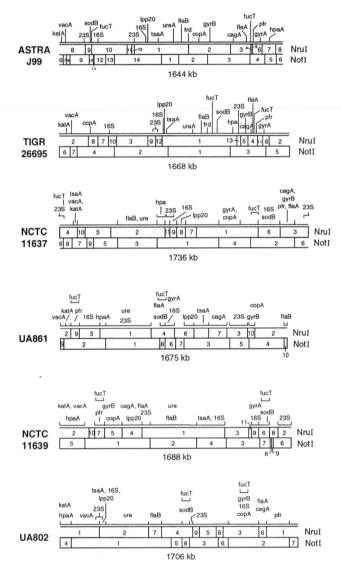

Figure 1 Genomic maps of *Helicobacter pylori* strains J99, 26695, NCTC11637, NCTC11638, UA802, and UA 861. The *katA* gene was used as a reference point for the alignment of all six maps. The positions of *Nru*I and *Not*I fragments and gene locations of strains J99 and 26695 were determined by their sequence data (3, 158). In the maps of NCTC11637, NCTC11639, UA802, and UA 861, only the approximate locations of genes are shown. The functions of these genes are shown in Table 2.

TABLE 1   General comparison of the genomes of *H. pylori* strains J99 and 26655

Genome features*	Strain J99	Strain 26695
Size	1,643,831 bp	1,667,867 bp
(G+C)%	39%	39%
Predicted ORFs	1495	1590
With predicted function	875	895
With unknown function	275	290
H. pylori-specific	345	367
ORFs common to the two strains	1406	1406
Strain-specific	89	117

*Data were summarized from Refs. 3 and 158.

sequences of these two unrelated strains allow direct comparison of the general features of *H. pylori* for the first time. Table 1 outlines the main features with respect to *H. pylori* genome comparisons. The genome sizes of both strains are consistent with the size range of 1.6–1.73 megabases based on the genomes of various *H. pylori* isolates determined by pulsed-field electrophoresis (152); the genome of *H. pylori* 26695 is ~24 kb longer than that of strain J99. These two genomes share an identical GC content (39%). The genomes of 26695 and J99 contain 1590 and 1495 predicted open reading frames (ORFs), respectively (3, 158). Approximately two-thirds of these genomes display significant sequence similarity to known genes with assigned functions in public databases, one-fifth are without known functions, and one-fourth are *H. pylori* specific (3). There are 1398 ORFs common to both the strains; ~50% of them exhibit >96% amino acid (aa) sequence identity. J99 has 89 strain-specific ORFs, and 26695 has 117, 46–48% of which are located in the *H. pylori* strain-specific DNA region, termed the "plasticity zone." Such a plasticity zone is present as a contiguous sequence in strain J99, whereas in strain 26695 it is split into two segments by an insertion of an ~600-kb sequence containing the heptamer repeat, 5S rRNA gene, and IS*605* and IS*606* elements, which suggests that the IS-mediated rearrangement occurred in this region during the evolution of this strain (3).

Unique Features of the Genomic Organization

One intriguing feature of the genomic organization is that ~1% of the genome of *H. pylori* encodes a family of 32 outer-membrane proteins that are well conserved between these two strains (70, 158). Some members of this family, called porins, may play an important role in antibiotic susceptibility of *H. pylori* (45). BabA encoded by *babA2* in *H. pylori* strain CCUG17875, a member of the outer-membrane protein family, has been reported to serve in an adhesin-mediating attachment process of *H. pylori* to Lewis[b] (Le[b]) blood group antigen

(81). Therefore, these outer-membrane proteins appear to be involved in the *H. pylori*-host interaction and could thereby contribute to the pathogenesis of *H. pylori* (70, 80).

The second feature is that the genome contains >20 homologs associated with DNA restriction and modification systems identified in other bacteria, including type I, type II, and type III systems (3, 158). Minor differences in these systems exist between the two strains (3). For example, *H. pylori* strain J99 appears to have two unique DNA restriction/modification systems (3). The role of these enzymes is unclear at present. It has been suggested that the enzymes are involved in the breakdown of intracellular and/or intercellular DNA or that they are necessary for stimulating the formation of recombinants by DNA fragmentation (14).

Third, the plasticity zone, a region containing 46–48% of strain-specific sequences, was identified in the genomes of *H. pylori* (3). The plasticity zone is a contiguous region in J99 but is separated into two segments by insertion of an ~600-kb sequence in strain 26695 (3). Many of the annotated genes in the plasticity zone appear not to be involved in pathogenesis but rather are homologous to restriction/modification enzymes (3). The role of this unique region in the pathophysiology of *H. pylori* infection needs to be further determined.

The last feature is that the homopolymeric tracts and dinucleotide repeats are frequently present in the *H. pylori* genes encoding cell-surface-associated proteins or enzymes involved in the biosynthesis of lipopolysaccharide (LPS) and/or associated with DNA restriction/modification systems. Twenty-six genes in the genomes of both strains 26695 and J99 have been shown to contain these tracts or repeats (3, 131, 158). These sequences potentially regulate the on/off switch of the respective genes by a strand-slipped mispairing mechanism and thereby could contribute to phase variation (4, 14, 131, 158). As seen in other bacteria, variation of expression of cell-surface-associated proteins by *H. pylori*, such as a flagellar protein adhesins, and ABC transporter, would be expected to affect motility, colonization, and pathogen-host interaction (131). However, it is unclear whether these predicted features are functional in all of the genes of *H. pylori*. It has been demonstrated that expression of Le antigens Le^x and Le^y by *H. pylori* is phase variable (4). This phenomenon could be explained by the on/off switch of both α1,2- and α1,3-fucosyltransferase genes (170; Z Ge, Z Shen, Q Jiang, DE Taylor, & JG Fox, submitted for publication) (see below).

Genomic Diversity in *H. pylori*—an Overstated Fact

Early studies of *H. pylori* genomes have revealed that extensive nucleotide sequence variation is present among the genomes of different *H. pylori* strains, referred to as macrodiversity. Several lines of evidence support this notion. (*a*) As mentioned above, no overall gene order could be identified from strain to strain (83). (*b*) *H. pylori* isolates from different patients display distinct restriction fragment length polymorphisms fingerprintings. These fingerprints are generated by conventional gel electrophoresis of *H. pylori* DNA digested with frequent cutting enzymes (95, 102, 127) and/or by pulsed-field gel electrophoresis with the digestion of infrequent cutting enzymes such as *Not*I and *Nru*I (152). (*c*) Methods based on the divergence of target genomic sequences, including randomly amplified polymorphic DNA-polymerase chain reaction (PCR) (1), oligofingerprinting (104), and repetitive-element PCR (66), are capable of discriminating among clinical isolates. These results demonstrated that significant nucleotide sequence diversity is present in the genomes from strain to strain.

The availability of the genome sequences of two *H. pylori* strains permits direct comparison of the genome organization of the different strains of this microorganism.

Approximately 85% of genes (1267 out of 1495 in strain J99) have the same neighboring genes in both genomes, and only 1.8% of these genes (27 out of 1495) have the same neighboring genes on one side and lack a common gene on the other side, caused by a difference in the physical structure of the two genomes (3). In addition, nine conserved gene clusters each containing >50 genes are present in the two genomes, representing 46% of the genes common to the two strains. Recently, other work also suggest that the order of specific genes could be conserved among strains. The orders of 40 genes in the *cag* region (a putative pathogenicity island, namely *cag*-PAI) appear to be conserved (2, 3, 20, 158). In addition, a gene cluster containing at least four genes arranged in the order of *ftsH-pss-copAP* is conserved among 14 of 15 strains that were analyzed by PCR (Z Ge & DE Taylor, unpublished data). In this gene cluster, *ftsH* encodes a metalloprotease involved in various cellular activities (58); the *pss* encodes a phosphatidylserine synthase involved in the biosynthesis of phosphatidylethanolamine accounting for 70–80% of the total phospholipids in the membrane of *Escherichia coli* (59); *copA* encodes a P-type ATPase associated with copper export in *H. pylori* (54, 57). Knockout of these three genes with a chloramphenicol acetyltransferase cassette (172) suggests that *ftsH* and *pss*, but not *copA*, are essential for cell viability. Recently, Beier et al (11) proposed that these genes and additional upstream genes, namely *cheY* coding for the flagella motor protein and *hsm* coding for a heat shock protein (Hsp) with putative methyltransferase activity, may constitute an operonic unit that is regulated in response to environmental stress such as temperature and copper upshifts. Therefore, some gene blocks or operons could be conserved by selective pressure over the course of bacterial evolution.

How can the discrepancy between the genomic diversity seen by various nucleotide sequence-based approaches and the relative conservation of the genomic organization of these two strains be explained? Based on the finding that the sequence differences between two strains are primarily located at the third position of the triplet codons, Alm et al (3) suggest that such differences contribute to genomic variations shown by restriction fragment length polymorphisms, repetitive-element PCR, and (random amplified polymorphic DNA) PCR (1, 95, 102, 127, 152). However, these differences could not be translated into protein sequences given that 310 proteins common to the two strains display >98% sequence identity, whereas only 8 genes share ≥98% nucleotide sequence identity (3). In addition, the gene order of the homologous regions of the two chromosomes could be aligned by artificial inversion and/or transposition of 10 segments of the J99 genomic sequences ranging from 1 to 83 kb (3). These data suggest that *H. pylori* has undergone a low level of evolutionary divergence (3). The disorganized gene order among strains could be caused by inversion and/or transposition of some chromosomal fragments mediated by various IS elements. The endpoints of these presumably rearranged regions in strains J99 and 26695 are mostly located in intergenic regions and are often associated with IS elements, repeated sequences, or restriction/modification genes on one or multiple chromosomes (3). On the other hand, the strain-specific regions of these two *H. pylori* chromosomes are of a low G+C content and also are associated with IS*605* elements in strain 26695, which suggests that these regions, like *cag*-PAI, could be acquired by the individual strains during horizontal DNA transfer (2, 3, 20, 158).

FUNCTIONALLY CHARACTERIZED GENES

In addition to the completion of the genome sequences of two *H. pylori* strains with the annotated genes, in the past decade a number of the *H. pylori* genes have been cloned and functionally characterized. Although difficulties were encountered in the early 1990s (151), now with improved molecular biological techniques >60 genes with assigned functions have been reported. Table 2 contains the majority of these genes, as of October 1998. Nevertheless, those genes published in non-English journals and/or not available in public databases such as the National Library of Medicine's PubMed are not included. The criteria for selection of the genes include (*a*) the availability of the complete nucleotide sequence of the gene and (*b*) confirmation of its function by experimental evidence. In the following section, we review the exciting findings obtained from these studies with emphasis on current knowledge of the genes implicated in *H. pylori* pathogenesis.

Survival in an Acidic Environment

H. pylori colonizes the surface of gastric epithelium including the crypts in the stomach. For colonization, this pathogen must safely penetrate the mucus layer through an extremely acidic environment—the stomach (pH < 2). One of the mechanisms evolved by *H. pylori* over its evolution to overcome this barrier is to produce a highly effective urease. The essential nature of this enzyme for survival of *H. pylori* in the stomach was demonstrated using animal models. Eaton et al (39) and Eaton & Krakowka (40) reported that the isogenic urease-negative mutants of *H. pylori* generated by inactivation of *ureG* with a kanamycin-resistance (Km^r) cassette were unable to colonize gnotobiotic piglets. Similarly, the *ureB*-disrupted mutant failed to colonize the mouse stomach (159). It has been proposed that a protective role of urease is to generate ammonia by hydrolyzing urea from gastric juice. Ammonia can neutralize gastric acid and thereby allow *H. pylori* to safely traverse the mucus layer to the epithelium surface (108). In addition, the

TABLE 2   The *Helicobacter pylori* genes have been characterized

Genes	Functions	References
abcABCD	ABC transporter involved in urease activity	72
amiE	Ammonia production	143
babA/B	An adhesin mediating the binding of *H. pylori* to gastric epithelium	81
cagA	Associated with virulence	24, 160
cagI	The first half of the *cag*-PAI, IL-8 induction	21
cagII	The second half of the *cag*-PAI, IL-8 induction	2
cdrA	Cell division	150
chey	Essential for cell motility	11
*coA-TA**	CoA-transferase subunit A	23
*coA-TB**	CoA-transferase subunit B	23
copAP	Copper-transport	54, 57
dapE	N-succinyl-L-diaminopimelic acid desuccinylase	88
flaA/B	Flagellin A/B	99, 147
flbA	Regulating of expression of FlaA/B	134
flgE	Flagellum-anchoring protein	119
fliI	Involved in Flagellar export	82
frdCAB	Energy-generating machinery under anaerobic condition	55
ftsH	Metalloprotease, essential for cell viability	58
fucT	α1,3-Fucosyltransferase involved in the biosynthesis of Le antigens Le^x/Le^y	53, 105
fucT2	α1,2-Fucosyltransferase involved in the biosynthesis of Le^y	170
fur	A repressor in the transcription of iron-regulated promoters	12
galE	UDP-galactose-4-epimerase	92
gyrA	DNA replication, ciprofloxacin-resistance	113
hpaA	N-acetylneuraminyllactose-binding hemagglutinin; a flagellar sheath protein	42, 84, 118
hpn	Ni^{2+}-binding	64
hsp60	Heat-shock protein	101
hspA	Chaperinin protein; Ni^{2+}-transport	86, 149
hspB	Chaperonin protein	86
katA	Catalase	119
nixA	Ni^{2+}-transport	109

(*continued*)

TABLE 2 *(continued)*

Genes	Functions	References
oorDABC operon	2-Oxoglutarate:acceptor oxidoreductase	79
orf2comB123	Involved in DNA transformation by natural competence	75
pfr	Nonhaem-iron ferritin involved in metal resistance	13
picB	Induction of the IL-8 production	162
porCDAB operon	Pyruvate:flavodoxin oxidoreductase	79
pss	Phosphatidylserine synthase	59
recA	Homologous recombination	133, 156
rexA	An oxygen-insensitive NADPH nitroreductase	68
ribBA	Iron acquisition	175
sod	Superoxide dismutase	145
topA	Topoisomerase I	146
ureA/B	Structural genes for urease	22, 93
ureC	Phosphoglucosamine mutase	37, 93
ureD	Unknown	93
ureEFGHI	Essential for functional urease	34, 143
uvrB	DNA damage repair	157
vacA	Vacuolating cytotoxin	31, 128, 132, 155
23S/5S rRNA genes	Protein synthesis/macrolide antibiotics resistance	153
wbcJ	Involvement in the synthesis of LPS	106

ammonia could also serve as a nitrogen source for incorporation into amino acids, given that the urease-deficient mutant did not colonize the neutral pH stomach of gnotobiotic piglets (40). Recently, the *glnA* gene encoding glutamine synthetase, which catalyzes the formation of glutamine by assimilating ammonia into glutamate, has been functionally characterized (52). Disruption of the *glnA* appears to lead to loss of viability of *H. pylori*, further supporting the notion that ammonia assimilation may also be critical for successful colonization of *H. pylori* in the gastric mucosa (40, 52). Furthermore, urease has been shown to play an important role in chemotactic motility (114). In contrast to the parental *H. pylori* strain, the *ureB*-disrupted mutant with normal flagella did not swarm on motility agar and markedly decreased its chemotactic motility towards urea, sodium bicarbonate, and potassium bicarbonate in a viscous environment (3% polyvinylpyrrolidone) mimicking the gastric mucous layer (114). Subsequently, the treatment of *H. pylori* cells with urease inhibitors gave rise to a significant decrease (with 1 μM fluorofamide) or the complete abolition (with 10 μM acetohydroxamic acid) in the chemotactic response to 10 mM sodium bicarbonate (114). Although such an inhibitory effect by these urease inhibitors could be caused by their interaction with some other components involved in this process, these results indicate that urease is essential for chemotactic motility.

At least seven gene products of *H. pylori* are required for a catalytically active urease. The genes *ureA* and *ureB* encode two structural subunits (22, 93), which constitute a multimeric apoenzyme of 550 kDa (76). The five other genes (*ureE, ureF, ureG, ureH,* and *ureI*) were previously reported to encode accessory proteins serving to incorporate nickel ions into the apoenzyme (34). A recent study has demonstrated, however, that the product of *ureI* is not necessary for the synthesis of the active enzyme but appears to be essential for in vivo survival of *H. pylori* strain SS1 in the mouse model (144). The urease can be localized to the cytoplasm, periplasm, and cellular surface of *H. pylori* (16, 38, 71, 129). Both the internal and the cellular surface-associated ureases are essential for the survival of *H. pylori* in an acid environment (90, 135). How the urease is translocated on the cellular surface appears to be controversial. One proposed mechanism is that such a process is achieved by "altruistic autolysis" (90), in which the urease is released from some *H. pylori* cells and then absorbed onto the surface of those intact cells. In contrast, Vanet & Labigne (165) suggested that this process is caused by specific secretion.

Because the presence of nickel ions is essential for activity of urease, the intracellular nickel concentration in *H. pylori* is well regulated. Several genes involved in nickel ion transport and binding, including *nixA*, an ABC transporter, a P-type ATPase, *hspA*, and *hpn*, have been identified (108). NixA, composed of 308 aa, is a membrane-bound, high-affinity nickel transport protein (109). Site-directed mutations of the conserved Asp-Glu-His residues in the transmembrane domains of NixA markedly reduce or abolish Ni^{2+} import, indicating that these residues play a crucial role in the NixA functioning (49). An ABC transporter operon likely consisting of four genes (*abcABCD*) has been demonstrated to play an important role in activity of urease (72). Insertional mutagenesis with a Km^r cassette in *nixA*, *abcC*, or *abcD* significantly reduces urease activity (42–72%), whereas abolition of urease activity was achieved by double mutations in both *nixA* and *abcC* (10, 72). Such an effect is caused by the loss of availability of nickel ions because the synthesis of urease in the mutants was not affected. Similarly, knockout of a membrane-bound P-type ATPase abolished the urease activity (107, 108). Another *H. pylori* gene, *hspA*, appears to be involved in Ni^{2+} transport (149). Its product HspA is a GroES-like protein containing two domains, the N-terminal domain A homologous to other GroES bacterial proteins, and a unique C-terminal domain B (149). Domain B can bind to several divalent cations with a high specific affinity for nickel ions (86). The gene *hpn* encodes a histine-rich, divalent ion-binding protein with a molecular mass of 7.1 kDa. The precise function of this gene is not clear at present because its disruption did not alter the urease activity (64).

In addition to urease, several other genes have been implicated in acid tolerance of *H. pylori*. Transcription (sixfold at pH 2.5) and expression (at pH 2.0) of *hsp70* encoding a DnaK-related protein, as well as the production of a GroEL-like Hsp60, were significantly upregulated with acid-shock (78). Enhanced expression of both Hsp70 and Hsp60, two surface-exposed proteins, is correlated with the switch of receptor-binding specificity from gangliotetraosylceramide, gangliotriaosylceramide, and phosphatidylethanolamine in vitro at neutral pH to sulfoglycolipids (77). Thus, it has been proposed that these proteins would protect *H. pylori* from stomach peristalsis and long exposure to an acid environment by facilitating the association of this organism with stomach mucus (78). In addition, it was reported that the region containing two genes, *recA* coding for a recombinase and *eno* coding for an enolase, were also involved in resistance of *H. pylori* to low pH (156). Inactivation of *recA* with a Km^r cassette renders the mutant 10-fold more sensitive to acid killing at pH 4.0 than the wild-type strain. This effect appears to be unrelated to the presence of urea, which suggests that the mechanism for the *recA* protective role, probably by repairing DNA damage, is distinct from that mediated by urease. A recent study suggested that the LPSs also have a role in the acid resistance of *H. pylori* (106). The gene presumably involved in the conversion of GDP-mannose to GDP-fucose, namely *wbcJ*, is transcriptionally upregulated by a low pH (pH 4.0). Inactivation of this gene with a Km^r cassette led to loss of O-antigen LPS expression in the mutants and also the absence of Le^x and Le^y antigens on the surface of the mutant cells. The isogenic *wbcJ* null mutants did not survive at pH 3.5. It was speculated that the role of *wbcJ* in acid tolerance is to protect *H. pylori* during severe acid stress (106).

Vacuolating Cytotoxin

Vacuolating cytotoxin (VacA) encoded by the *H. pylori* gene *vacA* is considered one of the major virulence factors in pathogenesis (31, 128, 132, 155). Newly synthesized VacA is 139–140 kDa and contains three domains: a signal sequence, a secreted cytotoxin (~90 kDa), and a cell-associated domain (~47 kDa) (26). Secreted VacA molecules are cleaved into an N-terminal fragment of ~37 kDa and a C-terminal fragment of ~58 kDa, which are associated by noncovalent interactions (155). It has been proposed that VacA is an AB-type toxin, in which the subunit B (the C-terminal 58-kDa fragment) is responsible for cell surface binding and the membrane translocation of the enzymatically active subunit A (the N-terminal 37-kDa fragment) into the cytosol (100). Purified 95-kDa VacA forms homooligomeric complexes with a flowerlike appearance, which was imaged by deep-etch electron microscopy (30, 100) and more recently in combination with three-dimensional reconstruction (electron tomography) (96). Two models have been proposed to explain this VacA oligomer structure. In the first model, this oligomer contains either six or seven copies of the 95-kDa polypeptide (19, 96, 100). In contrast, 12 or 14 95-kDa polypeptides were considered to be assembled into two interlocked six-membered arrays in the second model (30).

The mechanism for VacA to induce vacuolation in the epithelial cells is not fully understood yet. The short exposure of VacA to acid as low as pH 1.5 activates its cytotoxicity, presumably by conformational change (36). The VacA oligomers are disassociated by acidification (pH 3.0) into monomers, which was suggested to be an important feature of its activation (30). Vacuoles in target cells can be induced by either internalized VacA (51) or cytosolic VacA introduced by both microinjection and expression of the recombinant plasmid carrying the *vacA* gene (35), indicating that this toxin is active in the cytoplasm. Both the 37- and the 58-kDa subunits of VacA had the ability to insert in the lipid bilayer when they were incubated with liposomes containing either phosphatidylchlorine or asolectin (110). VacA also binds preferentially to the negatively charged asolectin liposome similar to the external layer of the plasma membrane (110), which suggests that VacA is translocated through the plasma membrane. Host factors are also required for vacuolating toxic activity of VacA. The presence of both the vacuolar-type ATPase and Rab7, which is a small GTPase protein normally associated with late endosomal compartments (123), is essential for such vacuolating activity (124). The vacuoles induced by VacA also contain lysosomal membrane proteins such as Lgp110 (111). This finding indicates that the VacA induces the formation of a mixed endolysomal compartment (111). A membrane-bound 140-kDa protein present in human gastric cancer cell lines (AZ-521 and AGS) has been identified as a potential receptor for binding of VacA to target cells because this protein was coimmunoprecipitated with VacA by anti-VacA antibodies (177). In addition, Seto et al (138) reported that the formation of larger vacuoles induced by VacA was inhibited by an antibody against the epidermal growth factor receptor; that the 37- and 58-kDa fragments of VacA are associated with the epidermal growth factor receptor was demonstrated by immunoprecipitation. Thus, VacA may be internalized into cells by endocytosis mediated by the epidermal growth factor receptor (138). Taken together, current evidence suggests that under acidic conditions the oligomers of VacA are dissociated into monomers that bind to unknown host factors on the cell surface (probably a 140-kDa protein or the epidermal growth factor receptor) (30, 35). Subsequently, the bound VacA is internalized, probably by endocytosis, and translocated to the cytosol where VacA, with the aid of host factors such as V-type ATPase and Rab7, could induce the endosome-lysosome fusion (30, 51, 111).

H. pylori isolates can be grouped into Tox[+] and Tox[−] strains, respectively. Tox[+] strains have the ability to induce cytoplasmic vacuolation in HeLa cells in vitro (9). Approximately 50% of *H. pylori* isolates are Tox[+] (98), although all *H. pylori* isolates contain the gene encoding VacA (*vacA*) (26). Tox[+] strains as well as their sonicated supernatants induce severe gastric epithelial damage in the mouse, whereas similar damage is not induced by Tox[−] strains, *vacA*-disrupted mutants of Tox[+] strains, or their supernatants (63, 103, 155). The difference in cytotoxicity is related to the unique genotypes of different strains and may be ascribed to their genetic variation. The *vacA* genes from Tox[+] and Tox[−] strains display much higher sequence identity at both the nucleotide and amino acid levels between members of the same type than between different types (9), particularly in the ∼250-aa region of the C-terminal 85-kDa fragment (9). Five characteristic motifs, including s1a, s1b, s2 (derived from the signal sequence region), m1, and m2 (derived from the middle region of *vacA*), have been identified in the *vacA* sequences (9). Most strains containing the s1a/m1 *vacA* type give rise to high vacuolating activity in in vitro assay. In contrast, the s1/m2 and s2/m2 strains have little or no in vitro cytotoxic activity. The concentration of VacA probably also contributes to this difference, given that Tox$^+$ strains contain much more VacA in the cells than Tox$^-$ strains (27, 28). The lower concentration of VacA in the cytoplasm and supernatants could be caused by weaker transcription of the Tox$^-$ *vacA* than of Tox$^+$ strains (48). The weak transcription of *vacA* on Tox$^-$ strains also appears to be independent of promoter strength, which suggests that the other factors are involved in this effect (48). Like Tox$^+$ VacA, Tox$^-$ cytotoxin is produced, secreted, and assembled into oligomeric complexes and disassociated by acidification (30, 122). Tox$^-$ VacA expressed by a transfected plasmid carrying the *vacA* gene is also toxic to HeLa cells, which suggests that lack of its cytotoxic activity is caused by the inability of this toxin to reach its cytosolic target site, caused by its failure to be internalized and translocated (122). This notion is supported by the evidence demonstrating that the s1/m2 toxin from *H. pylori* strain 95-54 induces vacuolation for RK-13 cells as well as adherent human cells derived from human gastric biopsies but not for HeLa cells (122). It was suggested that this effect is caused by the presence of a specific receptor(s) for the s1/m2 toxin on the surface of RK-13 cells; this toxin is not present in HeLa cells (122). Pagliaccia et al (122) pointed out that the m2 form of VacA may play a more pathogenic role in *H. pylori* infection than that previously recognized and that the conclusion regarding toxicity of the m2 VacA based on the HeLa cell assay needs to be reconsidered.

cag-PAI

A cytotoxin-associated gene A (*cagA*) encodes a 120- to 128-kDa protein (CagA) that is surface exposed and immunodominant (24, 160). Approximately 45% of *H. pylori* strains produce both CagA and VacA (24, 29, 32, 160). Expression of CagA by *H. pylori* strains is highly associated with peptic ulceration (24, 29, 32, 160). Based on the status of *cagA$^+$ vacA$^+$*, *H. pylori* clinical isolates were divided into two types, type I strains with the phenotype of CagA$^+$ VacA$^+$ and type II strains without this phenotype (176). Type I strains tend to be associated with more severe disease than type II strains. However, in some cases the expression of CagA is not necessarily associated with the clinical symptoms (43, 69, 125). In addition, inactivation of *cagA* did not alter the vacuolating cytotoxin activity within the mutants (161) and did not have an important role in colonization, induction of gastric epithelial lesion, stimulation of inflammation in the infected mice, or production of interleukin-8 by epithelial cells (20, 63, 161). The exact pathophysiological role of *cagA* and its gene product is still unclear. It has been suggested that the presence of CagA is only indicative of increased virulence of *H. pylori* strains (33, 140). On the other hand, Karita et al (89) reported that expression of CagA was maximal after brief acid shock (5 min). Upregulation of CagA by pH 6.0

appears to decrease survival ability of *H. pylori* in response to pH 3.0, because pH 3.0 is more effective on killing the CagA$^+$ cells than its isogenic *cagA$^-$* derivative (87). In addition, administration of the recombinant CagA with the mutated heat-labile enterotoxin of *E. coli* confers effective protection against reinfection by *H. pylori* (63). Furthermore, variations in the *cagA* sequence are present among some *H. pylori* strains (44, 164, 178), and the types of the *cagA* may be associated with the severity of atrophy (178). Thus, further studies are required to determine the precise role of this peculiar protein in *H. pylori* infection.

In 1996, Censini et al (20) identified a 40-kb region containing *cagA* on the chromosome of type I strains as a pathogenicity island, *cag*-PAI. The *cag*-PAI in *H. pylori* strain CCUG 17874 can be divided into two regions, namely *cagI* (containing *cagA*) and *cagII*, by insertion of an intervening sequence flanking IS*605* or by IS*605* (20). The *cagI* region encodes 16 predicted ORFs, including genes *cagA* (24, 160) and *picB* (162). The *cagII* region, which has been recently characterized by Akopyants et al (2), contains 15 predicted ORFs. The *cag*-PAI is present in the genomes of strains 26695 (158) and J99 (3). The contiguous or IS*605*-divided *cag*-PAIs are flanked by a 31-bp direct repeat (2, 3, 20, 158). The gene organizations of these *cag*-PAIs are essentially conserved among strains (2, 3). The proteins encoded by the genes residing within the *cag*-PAI contain the motifs found in various bacterial proteins, including translocases, sensors, permeases, and proteins involved in pilus or flagellum assembly (2, 20). Most notably, several ORFs in the *cag*-PAI encode proteins with a significant similarity to proteins involved in the export of the transfer DNA in *Agrobacterium tumefaciens*, pertussis toxin in *Bordetella pertussis*, and conjugative plasmid DNA (2, 3, 20, 158). Contribution of the *cag*-PAI in the virulence of *H. pylori* is indicated by the fact that many genes in the *cag*-PAI are involved in induction of interleukin-8 production by gastric epithelial cells, a proinflammatory cytokine (2, 20, 137, 162), and tyrosine phosphorylation (136, 137). The *cag*-PAI likely activates NF-κB, a transcription activator for the production of cytokines, and thereby stimulates the production of interleukin-8 (65, 139). Two protypes of *H. pylori* strains have been proposed: in addition to the absence of IS*605* from the chromosomes of both types, type I containing a contiguous *cag*-PAI with enhanced virulence and type II lacking this region with reduced virulence (20, 25). Various intermediate forms of the *cag*-PAI are identified within *H. pylori* strains (20). It is possible that disassociation of *cagA$^+$* strains with severe diseases in some cases is caused by the presence of a partial or dysfunctional *cag*-PAI. It has been hypothesized that *H. pylori* acquired the intact *cag*-PAI initially by horizontal transfer and over its evolution various intermediate types have been created by rearrangement of the chromosome mediated by the direct repeats and IS*605* (20, 25).

Genes Involved in the Biosynthesis of Lewis Antigens

Lewis carbohydrate antigens, including Le^a, Le^b, Le^x, and Le^y, are either mono- or difucosylated glycoconjugates present on the surface of eukaryotic cells. Most

H. pylori strains express Le^x or Le^y or both, although occasional isolates may also produce Le^a. Le^b seen in some strains may be the result of nonspecificity with monoclonal antibodies (112). These structures are part of the bacterial LPS (5–8, 112, 141, 142, 174). Levels of expression of these Le antigens vary among strains (5, 141, 174), and phase variation is involved in the biosynthesis of Le^x and Le^y (4). Because these bacterial Le antigens are similar to those found on the surface of the gastric epithelia, this molecular mimicry could help *H. pylori* escape elimination by the host immune response and also could cause autoimmune response (141), which thereby contributes to the pathogenicity of *H. pylori*.

The biosynthetic pathway of Le^x is identical to that found in humans, in which β1,4-galactosyltransferase catalyzes the addition of galactose to *N*-acetylglucos-amine, followed by transferring fucose from GDP-fucose to Galβ1-4GlcNAc (LacNAc) by α1,3-fucosyltranferase (21). An intact α1,3-fucosyltransferase gene (*fucT*) from *H. pylori* strain NCTC11639 has been cloned and functionally expressed in *E. coli* cells (53). A similar gene from strain NCTC11637, which appears to be in an off state because of the deletion of a single cytosine in the poly(C) tract proximal to the 5′-end of *fucT* (see below), has also been cloned and characterized by Martin et al (105). Two copies of *fucT* are present within the genomes of both strains 26695 and J99 (158). The sequences of *HpfucTs* from strains AH244, Sydney SS1, and UA802 have been determined and compared with known *fucTs* (Z Ge, Z Shen, Q Jiang, DE Taylor, JG Fox, submitted for publication). All the *fucTs* contain the 5′-terminal polymeric tract of cytosine residues that likely contribute to the off- or onstatus of the individual genes by the strand-slipped mispairing mechanism (14, 131, 158). The deduced amino acid sequences of HpFucTs are variable in length because of the difference of the numbers of the heptad repeat (Z Ge et al, submitted for publication). These tandem heptamers reside within the C-terminal ∼100-aa region of Hp-FucT, which is absent from its eukaryotic counterparts (53, 105). A typical heptad sequence is DDLRNY, with minor substitutions among several strains except for that in UA802 Fuc-T (Figure 2).

Leucine residues present within these heptad repeats constitutes a leucine-zipper domain resembling those found in various eukaryotic transcription activators. Because this domain mediates dimerization of these eukaryotic proteins, we speculate that the leucine zipper domain of HpFuc-T could have a similar function. Variation of this leucine zipper domain in amino acid sequence and length could also contribute to the function of this enzyme (53). Additional experiments are required to address this hypothesis.

Although α1,3-FucT is necessary for Le^x synthesis, both α1,2- and α1,3-FucTs are involved in Le^y synthesis. In the complete genome sequence of strain 26695, no putative α1,2-FucT was annotated (158). Berg et al (14) suggested that two ORFs, HP0093 and HP0094, constitute the intact α1,2-*fucT* gene, which is also present in the genome of strain J99. More recently, the essential role of this gene in the synthesis of Le^y has been demonstrated by Wang et al (170) who reported that disruption of this gene with a chloramphenicol acetyltransferase cassette abolished

```
                                                                    fucT

11639FucT  (348)  NPFIFCRDLNEPLVTI  DDLRVNY (DDLRVNY)9    ERLLSKATPLLELSQNT   on
26695FucT  (349)  NPFIFYRDLHEPLISI  DDLRVNY (DDLRVNY)9    DRLLQNASPLLELSQNT   on
11637FucT  (348)  NPFIFYRDLNEPLVSI  DNLRINY  DNLRINY      ERLLQNASPLLELSQNT   off
   AHFucT  (347)  NPSTLYRDLHDPLVSI  DDLRVNY (DDLRVNY)14   DRLLQNASPLLELSQNT   off
  SydFucT  (345)  NPFIFYRDLNEPLVAI  DDLRVNY (DDLRINY)2    ERLLQNASPLLELSQNT   off
26695FucT1 (350)  TSFMWEYDLHKPLVSI  DDLRVNY  DDLRVNY      DRLLQNASPLLELSQNT   on
  802FucT  (347)  NPFVFYRDLNEPLVSI  DDLRADY (NNLRADY)4    DRLLQNRSPLLELSQNT   on

      Jun         NRIAASKCRKRKLERI  ARLEEKV  KTLKAQN      SELASTANMLTEQVAQL
      Fos         NKMAAAKCRNRRRELT  DTLQAET  DQLEDKK      SALQTEIANLLKEKEKL
                        *              *                        *    *
```

Figure 2 Amino acid sequence comparison of the FucTs region containing the heptad repeats (letters in boldface type). *Numbers in parentheses* correspond to the positions of the first amino acid residues within the respective intact FucTs. Small numbers denote the copies of this heptad. The conserved leucine residues among the *H. pylori* FucTs, as well as Fos and Jun, two protooncogene products (62, 163), are indicated by *asterisks*. *On* and *Off* represent that the intact or truncated FucTs are encoded by the respective *fucTs*. *H. pylori* strains: 11639, NCTC11639 (53); 26695, strain 26695 (158); 11637, NCTC11637 (105); AH, AH244 (Z Ge et al, submitted for publication); Syd, Sydney SS1 (Z Ge et al, submitted for publication); and 802, UA802 (Z Ge et al, submitted for publication).

expression of Le^y in strains UA802 and 26695. Three unique nucleotide sequences were identified in the middle region of the α1,2-*fucT*: a poly(C) tract similar to that present in the 5'-end region of α1,3-*fucT*, a TAA repeat sequence, and a characteristic structure for generating a translational −1 frameshift similar to that used by *E. coli dnaX* encoding the ι and γ subunit of DNA polymerase III (Figure 3) (14, 131, 158, 170). A transitional frameshift cassette, including a heptamer sequence (AAAAAAG) acting as frameshift cassette, a putative Shine-Delgarno sequence, and a stem-loop structure both for stimulating such a frameshift, exist within α1,2-*fucT* of certain *H. pylori* strains (170) (Figure 3). Therefore, both frameshift mutation by the addition or deletion of one or more nucleotides in the repeat sequences and translational frame shifting may be involved in the on- and offswitch of this gene and thereby may contribute to phase variation of Le^x/Le^y expression in *H. pylori* (4, 131, 170). Based on sequence analyses of the α1,2-*fucT*s from three additional strains, a model was proposed to explain the Le$^{x/y}$ phenotypes of the respective strains (170) (Table 3).

In this model, the synthesis of Le^y requires the sequential addition of fucose onto the precursor molecule (LacNAc) at α1,3- and α1,2-linkage positions catalyzed by corresponding FucTs. Turning off either *fucT* gene will give rise to an Le^y-negative phenotype, as seen in strains UA1174 and UA1207. If both genes are on or partially on (symbolized as [On] in Table 3), the levels of expression of Le^x and Le^y will depend on the relative concentration (activities) of the two enzymes. The α1,3-FucT level in UA802 was proposed to be low so that the majority of the Le^x would be converted to Le^y by a relatively higher level of its α1,2-FucT. In strain 26695, the α1,2-*fucT* gene, which is partially on, confers a relatively lower level of α1,2-FucT and is responsible for the Le^{x+}/Le^{y+} phenotype.

The gene *wbcJ*, presumably responsible for the conversion of GDP-mannose to GDP-fucose, has been characterized (106). The deduced WbcJ is a 319-aa protein with demonstrated sequence similarity to several bacterial proteins involved in the biosynthesis of GDP-L-fucose (106). The *wbcJ*-defective mutants of *H. pylori* lose their ability to express both O-antigen and Le^x/Le^y. This is not surprising because GDP-fucose is a substrate of α1,2- and 1,3-FucTs for synthesis of Le^x/Le^y. It is interesting that the transcription of *wbcJ* is acid inducible and the *wbcJ*-disrupted

TABLE 3 A model of molecular basis of variable expression of Le^y

H. pylori strains	Lewis phenotype		α-1,2-*fucT* genotype			
	Lewis X	Lewis Y	ORF	Translation frameshift	On/Off	Proposed α(1,3) *fucT*
UA802	−	+	Intact	−	On	[On]
26695	+	+	Frameshifted	+	[On]	On
UA1174	+	−	Frameshifted	−	Off	On
UA1207	−	−	Intact	−	On	Off

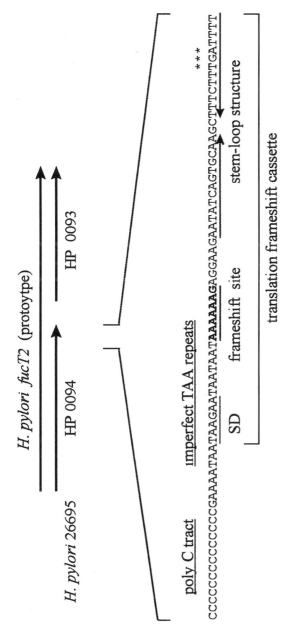

poly C tract   imperfect TAA repeats

CCCCCCCCCCGAAAATAATAAGAATAATAATAA**AAAAAA**GAGGAAGAATATCAGTGCA*A*GCTTTCTTTGATTTT

            SD   frameshift site   stem-loop structure

translation frameshift cassette

Figure 3   Frameshifting elements within the *H. pylori* 26695 *fucT2* gene. The central region of the *fucT2* gene contains a hypermutable sequence [a poly(C) tract and TAA repeats] that could frequently shift into and out of the coding frame of this gene by a strand-slippage mispairing mechanism. Because of the variation of the repeat number relative to that of the prototype gene, the reading frame of 26695 *fucT2* encounters a TGA stop codon (for HP0094) marked with asterisks, which leads to a truncated FucT2 (gene-off). However, the existence of a translation frameshift cassette immediately behind the hypermutable region provides another mechanism by which the translation of the gene could be shifted (at a frequency of ~50%) back to the prototype reading frame. Thus, the HP0094 stop codon could be bypassed and the full-length protein could be produced (gene-on).

mutants became susceptible to a low pH (3.5), which suggests that the LPS plays an important role in acid tolerance of *H. pylori* (106).

In addition to its function in the synthesis of Le^y (type 2 structure), the α1,2-FucT may also be involved in the synthesis of Le^b by the addition of fucose to type 1 precursor (Le^a) through a 1 → 2 linkage (73). This function would be confirmed by showing that this enzyme is capable of using type 1 receptor. Unfortunately, determination of α1,2-FucT activity with both type 1 and type 2 receptors by biochemical assays and cell lysates prepared from *H. pylori* and *E. coli* presumably containing the overexpressed α1,2-FucT seems to be very difficult. This could be caused by several factors, including assay conditions that are not suitable for this bacterial enzyme, its low concentration in cells, transient expression, and high instability. In addition, α1,4-FucT activity is required for the synthesis of Le^a structure present on the surface (112). So far no gene or enzyme of α1,4-*fucT* has been identified. There is no apparent candidate gene coding for this enzyme in the genome sequences of strains 26695 and J99. Many questions regarding the biosynthesis of Le antigens in *H. pylori* remain to be resolved in the future. Current evidence indicates that *H. pylori* develops efficient mechanisms such as two copies of α1,3-*fucT* and multiple regulation of α1,2-FucT activity to generate mosaic Le antigens on its surface and phase variation. Such a system could offer *H. pylori* the flexibility in response to microenvironmental change over time in the same individual or to adapt to the stomach in a new host.

Motility and Adhesion

H. pylori contains five or six monopolar flagella consisting of two structural subunits, a major 53-kDa FlaA and a 54-kDa FlaB (94, 99). The genes encoding these two flagellins are unlinked on the chromosome and transcriptionally controlled under different promoters (σ^{28} for *flaA* and σ^{54} for *flaB*) (94). The flagella are covered by a sheath containing a 29-kDa protein encoded by *hpaA* that was previously reported to code for a putative *N*-acetylneuraminyllactose-binding protein (42, 84, 120). The *H. pylori flaE* encodes a 87-kDa flagellar hook protein (121). This hook protein is required for assembly of FlaA and FlaB and motility of *H. pylori* because inactivation of *flaE* renders the mutants nonmotile, and, therefore, FlaA and FlaB stay in the cytoplasm (121). Transcription of *flaA* and *flaB* is regulated by *flaA* coding for a 732-aa polypeptide similar to members of the LcrD/FlbF family involved in flagellar biosynthesis and secretion of virulence proteins (134). The gene *cheY* has been shown to encode the flagellar motor switch protein CheY (11). This gene is located in a putative stress-responsive operon containing several genes including a homolog to L11 methyltransferase, a metalloprotease (FtsH), phosphatidylserine synthase (Pss), and copper transportation proteins (CopA and CopP) (11, 54, 57–59). The isogenic *cheY*$^-$ mutant is nonmotile, indicating that this gene product plays a crucial role in bacterial motility (11). The essential role of both *flaA* and *flaB* in the establishment of persistent colonization of *H. pylori* was demonstrated by Eaton et al (41) in the gnotobiotic piglet model. The *H. pylori* mutants containing the inactivation of either *flaB* or *flaA* or both by insertional mutagenesis with both Km^r and chloramphenicol acetyltransferase cassettes could only colonize for 2–4 d but not for 10 d.

For establishment of colonization, *H. pylori* needs to adhere to the surface of the mucous epithelial cells and the mucus layer lining the gastric epithelium (15). *H. pylori* binds to the fucosylated blood group antigen Le^b present on human gastric epithelial cells in situ (17). An adhesin, namely BabA, has been reported to mediate this specific binding between *H. pylori* and the host Le^b antigen (81). BabA, which is encoded by *babA2* in *H. pylori* strain CCUG17875, is a membrane-bound protein composed of 721 aa residues. A second gene, designated *babA1*, is essentially identical with *babA2* except for lack of an insertion of 10 bp present in *babA2* (81). The absence from *babA1* of this 10-bp insertion leads to the truncated form of BabA, which loses the Le^b antigen-binding activity. BabA belongs to a family of ~30 well-conserved outer-membrane proteins sharing extensive sequence homology in the N- and C-terminal domains (3, 81, 158). It is unclear whether BabA is solely an adhesin mediating attachment of *H. pylori* to the gastric epithelial cells. Given the fact that ~34% of isolates did not have this Le^b antigen-binding activity in the study of Ilver et al (81), other adhesins may be required for adherence of *H. pylori* to the stomach. It has been reported that a 16-kDa surface protein of *H. pylori*, which is activated by neutrophils, binds to sulfated carbohydrates on salivary mucin, which suggests that this protein is also an adhesin for *H. pylori* (115). It seems likely that *H. pylori* express multiple adhesins, which could allow them to adapt to different living environments.

DNA Transfer and Shuttle Vectors

Many strains of *H. pylori* are able to take up *H. pylori* DNA by natural competence in vitro, during the process of a natural transformation (117, 171). Transformation frequencies of chromosomal DNA between strains vary depending on the selective markers used: 4×10^{-4} per viable cell for Rif^R and Str^R, and 3.0×10^{-5} per viable cell for Mtr^R (171). The process of natural transformation remains poorly understood although more information is becoming available. An operonic unit consisting of four genes, designated *orf2*, *comB1*, *comB2*, and *comB3*, has been identified by using *blaM* transposon shuttle mutagenesis (75, 118). The *orf2* encodes a small polypeptide of 37-aa residues, whereas the genes *comB1*, *comB2*, and *comB3* encode 29-, 38-, and 42-kDa proteins, respectively. ComB3 shares significant sequence similarity to several proteins involved in conjugative or Ti plasmid DNA transfer and pertussis toxin secretion in *Pseudomonas aeruginosa*, *A. tumefaciens*, and *B. pertussis* (75). The *comB2* and *comB3* genes display sequence homology to HP0528 and HP0527, respectively, which are a part of the *cagII*-PAI of *H. pylori* strain 26695. Knockout of the individual genes in this operon leads to transformation efficiency reduced by 90–100%, demonstrating that these genes play a key role in natural transformation of *H. pylori*. DNA transfer between cells of *H. pylori* is also likely conducted by a conjugation-like mechanism described by Kuipers et al (91). After mating between donor and recipient *H. pylori* cells, accessible DNase treatment reduced but did not abolish the DNA transfer, and such DNA transfer occurred in a bidirectional manner (91). The authors suggested that a DNase-resistant conjugationlike mechanism be involved in DNA transfer between *H. pylori* cells. Thus this phenomenon may contribute to genetic variations among *H. pylori* strains (91). After the foreign *H. pylori* DNA is introduced into a recipient cell, *recA* is responsible for homologous recombination of this DNA into the target chromosomal DNA in *H. pylori* (133, 156). Additional studies are required to document the molecular basis of the DNase-resistant conjugationlike process in *H. pylori*.

A shuttle vector, which can replicate in both *E. coli* and *H. pylori*, is an essential tool for functional studies of *H. pylori* genes by complementation experiments. The shuttle vector carrying the mutated *H. pylori* genes created in *E. coli* can be introduced into recipient *H. pylori* by the procedures of natural transformation and electroporation, as described previously (61), or by mobilization using a broad-host-range plasmid (74). As the genome sequences of two *H. pylori* strains with the annotated genes are available, such a vector will accelerate our ability to determine precise functions of individual genes. Two groups have recently constructed such shuttle vectors, which contain two replication origins: one derived from the *H. pylori* cryptic plasmid and another derived from ColE1 (74, 97). The vector pHP489K with a Km^r marker constructed by Lee et al (97) has been used to restore urease activity in the $ureAB^-$ mutants. Transformation efficiency of this vector by electroporation is variable from strain to strain, with an average of $\sim 4 \times 10^3$ colony formation units per μg of DNA. In *H. pylori*, the copy number of this vector alone was estimated to be eight to nine copies per cell, and, with the insertion of *ureAB*, it fell to three copies per cell (97). Vectors, pHel2 (5.0 kb) containing a chloramphenicol acetyltransferase gene and pHel3 (5.6 kb) with a Km^r gene, have been developed and shown to be stable in both *E. coli* and *H. pylori* (74). There are several notable features with respect to transformation efficiency of these two shuttle vectors by natural competence. (*a*) The incoming plasmid isolated from *H. pylori* gave rise to a higher transformation efficiency ($\sim 3.5 \times 10^{-5}$ per viable cell) than that from *E. coli* DH5α ($\sim 3.5 \times 10^{-7}$ per viable cell). (*b*) In some strains unidirectional transformation preference occurs: pHel2 from strain P12 transformed P1 with high frequency (7.0×10^{-5}), whereas this vector from strain P1 transformed P12 at a frequency of 1.9×10^{-8}, 50-fold lower than that with pHel2 from itself. (*c*) By transconjugation using a broad-host-range plasmid for mobilization, pHPel2 from an *E. coli* DH5α donor was established in the *H. pylori* P79 recipient at the same frequency as pHPel2 from the *H. pylori* P1 donor. These results further support our previous hypothesis that there are different restriction-modification systems among *H. pylori* strains (151, 171).

It is noteworthy that the application of these vectors in complementation experiments has certain limitations. They may work only in certain strains because of the barrier of different restriction-modification systems. Also they may be unsuitable for expressing the genes not containing their own promoters because there are no efficient, universal, or controllable promoters in these vectors. Nevertheless, the availability of these shuttle vectors provides the basis for future experimental improvement.

EVOLUTION

H. pylori is a gram-negative bacterium and belongs to the Proteobacteria. However, many predicted proteins of *H. pylori* share significant sequence similarity to Archaebacteria, gram-positive bacteria, and eukaryotes (158). An example of such proteins is a phosphatidylserine synthase that displays significant sequence homology with its counterparts from *Methanococcus jannaschii* (34% identity and 54% similarity), as well as from *Bacillus subtilis* and *Saccharomyces cerevisiae*, but none with its counterparts from *E. coli* and *Haemophilus influenzae* (59). In addition, the properties of this protein, including its membrane-bound nature and Mn^{2+}-dependent catalytic activity, are identical with those in *B. subtilis* (59). Two possible explanations for such observations were suggested by Tomb et al (158). (*a*) These discordant genes may be introduced by lateral gene transfer over the evolutionary history of *H. pylori*. (*b*) *H. pylori* may retain more ancient forms of enzymes than other species in the division of Proteobacteria because of its early divergence from their common ancestor. It is possible that both models may contribute to the formation of the modern genomes of *H. pylori* strains. Some chromosomal segments, such as the *cag-PAI* and the plasticity zone, contain a relatively low G+C content, harbor IS elements, and are flanked by two direct repeats. These structural features are indicative of the occurrence of lateral gene transfer (the first model) (3, 20, 25). On the other hand, the presence of enzymes representing ancient forms but lacking these features, like Pss, suggests that *H. pylori* inherited these enzymes from its ancestor by the early divergence model.

The conservation of most genes at levels of amino acid and nucleotide sequences between two completely sequenced strains confirms *H. pylori* as a single species. Random order of the genetic markers among different strains, which was determined by pulsed-field gel electrophoresis (83), can be explained by the alignment of the corresponding regions between two sequenced genomes of *H. pylori* by rearrangement of 10 segments of the J99 genome (3). However, macrodiversity in *H. pylori* is evidently more significant than in any other bacteria, including *Neisseria gonorrhoeae* and *Neisseria meningitidis*, which contain considerable microdiversity, but have a fairly conserved gene order on their chromosome between strains (83). Such genomic macrodiversity among strains may be ascribed to rearrangement and/or transposition of the chromosomes mediated by transposons (3, 14, 20). Indeed, two IS elements, IS*605* and IS*606*, containing genes (*tnpA* and *tnpB*) encoding the putative transposases, have already been found on the chromosomes of some *H. pylori* strains. Various copies of these IS elements, as a complete or partial form, exist within different strains (3, 20). This hypothesis is strengthened by the presence of many intermediate forms of *cag*-PAI among strains (25).

In addition, the microenvironment may be another driving force for contributing to the macrodiversity of the *H. pylori* genomes. Huynen et al (80) compared complete genome sequences among *H. pylori*, *E. coli*, and *H. influenzae*, and found that 63% (123 genes) of *H. pylori*-specific gene products are potentially involved in the *H. pylori*–host interaction. Similarly, it was suggested that the strain-specific genes in *H. pylori* also play an important role in the survival of individual strains in the specific host (3). Therefore, acquisition of the species- or strain-specific DNA blocks could contribute to the genomic diversity.

There is extensive sequence diversity within individual loci such as *vacA*, *flaA*, and *flaB*, and urease genes among *H. pylori* strains (60). Recombinational genomes of *H. pylori* strains, known as panmictic structures (80, 130, 148), were demonstrated by multilocus enzyme electrophoresis (67), PCR-restriction fragment length polymorphism comparison of six genetic markers (including *flaA*, *flaB*, *vacA*, *cagA*, IS*605*, and ribosomal DNA), and nucleotide sequence comparison of three gene fragments (*flaA*, *flaB*, and *vacA*) (148). It was suggested that such genomic structures in *H. pylori* were generated mainly by horizontal gene DNA transfer with subsequent recombination between strains, because this recombination in *H. pylori* is likely much more frequent than mutations (148). This notion is supported by the fact that most *H. pylori* strains are naturally transformable and that multiple *H. pylori* strains in some cases can be present in the single host (46, 85, 154).

CONCLUDING REMARKS AND PERSPECTIVES

The availability of the complete genome sequences of two *H. pylori* strains will have profound impact on all the aspects of research from the basic biology of *H. pylori* and its identification of virulence factors to the development of new therapeutic strategies for eradication of this organism. Information obtained from direct comparison of these two genomes is exceptionally useful. For example, such a comparison could shed light on both the basic machinery for maintaining the life cycle of *H. pylori* and the genetic basis for the lifestyle (toxigenic and nontoxigenic) of individual *H. pylori* strains. Also these comparisons will help scientists to identify new drug targets for eradication of this pathogen and discover candidate proteins more suitable for the development of vaccines. Furthermore, comparative analyses among the genome sequences of *H. pylori* strains and among the genomes of *H. pylori* and other microorganisms such as *H. influenzae*, *B. subtilis*, *M. jannaschii*, and *E. coli* will provide insight into the evolutionary history of this organism.

The publication of the genome sequences of *H. pylori* does not provide the answers to the biological and pathogenic mysteries of *H. pylori*; rather it opens an avenue for answering relevant questions. Approximately one-third of the predicted ORFs need to be functionally characterized (3). As Berg et al (14) pointed out, the functions of many genes are annotated based solely on sequence homology with known genes from other organisms and therefore may not represent their true functions in vivo. In addition, many genes may have evolved a new function in addition to that deduced from their counterparts in other organisms. For example, α1,3-FucT contains a C-terminal ~100-aa region that is absent from its counterparts in eukaryotes (53). The presence of the leucine zipper domain within this region suggests that this protein could function as a transcriptional activator (Z Ge et al, submitted for publication). Comparison of the genome sequences between virulent and less virulent strains and between *H. pylori* and other gastric *Helicobacter* spp., such as *Helicobacter nemestrineae*, *Helicobacter felis*, and *Helicobacter mustelae*, will define the factors essential for survival in the gastric environment and for host specificity and virulence.

The publication of the genome sequence of *H. pylori* brings the research of *H. pylori* into a new era. Our research strategies will be enhanced from simply cloning genes of interest to functional characterization by means of molecular biological techniques combined with relevant biochemical approaches. With the improvement of animal models and shuttle vectors, the information derived from these genome sequences will accelerate our understanding of the biology of *H. pylori* and the development of novel and efficient therapeutic strategies for controlling this pathogen.

ACKNOWLEDGMENTS

We thank R Alm, C-C Chien, Q Jiang, D Rasko, M Rooker, TJ Trust, & G Wang for their help during the preparation of this review. This work was supported by grants from the National Cancer Institute of Canada, with funds from the Terry Fox Run and the Canadian Bacterial Diseases Network (to DET). DET is a recipient of a Medical Scientist Award from the Alberta Heritage Foundation for Medical Research.

Visit the Annual Reviews home page at http://www.AnnualReviews.org

LITERATURE CITED

1. Akopyanz N, Bukanov NO, Westblom TU, Kresovich S, Berg DE. 1992. DNA diversity among clinical isolates of *Helicobacter pylori* detected by PCR-based RAPD fingerprinting. *Nucleic Acids Res.* 20:5137–42

2. Akopyants NS, Clifton SW, Kersulyte D, Crabtree JE, Youree BE, et al 1998. Analyses of the *cag* pathogenicity island of *Helicobacter pylori*. *Mol. Microbiol.* 28:37–53

3. Alm RA, Ling LL, Moir DT, King BL, Brown ED, et al. Genomic sequence comparison of two unrelated isolates of the human gastric pathogen *Helicobacter pylori*. *Nature* 397:176–80

4. Appelmelk BJ, Shiberu B, Trinks C, Tapsi N, Zheng PY, et al. 1998. Phase variation in *Helicobacter pylori* lipopolysaccharide. *Infect. Immun.* 66:70–76

5. Appelmelk BJ, Simoons-Smit I, Negrini R, Moran AP, Aspinall GO, et al. 1996. Potential role of molecular mimicry between *Helicobacter pylori* lipopolysaccharide and host Lewis blood group antigens in autoimmunity. *Infect. Immun.* 64:2031–40

6. Aspinall GO, Monteiro MA. 1996. Lipopolysaccharides of *Helicobacter pylori* strains P466 and MO19: structures of the O antigen and core oligosaccharide regions. *Biochemistry* 35:2498–504

7. Aspinall GO, Monteiro MA, Pang H, Walsh EJ, Moran AP. 1994. O antigen chains in the lipopolysaccharide of the *Helicobacter pylori* NCTC11637. *Carbohydr. Lett.* 1:151–56

8. Aspinall GO, Monteiro MA, Pang H, Walsh EJ, Moran AP. 1996. Lipopolysaccharide of the *Helicobacter pylori* type strain NCTC 11637 (ATCC 43504): structure of the O antigen chain and core oligosaccharide regions. *Biochemistry* 35:2489–97

9. Atherton JC, Cao P, Peek RM Jr, Tummuru MK, Blaser MJ, et al. 1995. Mosaicism in vacuolating cytotoxin alleles of *Helicobacter pylori*: association of specific *vacA* types with cytotoxin production and peptic ulceration. *J. Biol. Chem.* 270:17771–77

10. Bauerfeind P, Garner RM, Mobley HLT. 1996. Allelic exchange mutagenesis of *nixA* in *Helicobacter pylori* results in reduced nickel transport and urease activity. *Infect. Immun.* 64:2877–80

11. Beier D, Spohn G, Rappuoli R, Scarlato V. 1997. Identification and characterization of an operon of *Helicobacter pylori* that is involved in motility and stress adaptation. *J. Bacteriol.* 179:4679–83

12. Bereswill S, Lichte F, Vey T, Fassbinder F, Kist M. 1998. Cloning and characterization of the *fur* gene from *Helicobacter pylori*. *FEMS Microbiol. Lett.* 159:193–200

13. Bereswill S, Waidner U, Odenbreit S, Lichte F, Fassbinder F, et al. 1998. Structural, functional and mutational analysis of the *pfr* gene encoding a ferritin from *Helicobacter pylori*. *Microbiology* 144:2505–16

14. Berg DE, Hoffman PS, Appelmelk BJ, Kusters JG. 1997. The *Helicobacter pylori* genome sequence: genetic factors for long life in the gastric mucosa. *Trends Microbiol.* 5:468–74

15. Blaser MJ. 1993. *Helicobacter pylori*: microbiology of a "slow" bacterial infection. *Trends Microbiol.* 1:255–60

16. Bode G, Malfertheiner P, Lehnhardt G, Nilius M, Ditschuneit H. 1993. Ultrastructural localization of urease of *Helicobacter pylori*. *Med. Microbiol. Immunol.* 182:233–42

17. Borén T, Falk P, Roth KA, Larson G, Normark S. 1993. Attachment of *Helicobacter pylori* to human gastric epithelium mediated by blood group antigens. *Science* 262:1892–95

18. Bukanov NO, Berg DE 1994. Ordered cosmid library and high-resolution physical-genetic map of *Helicobacter pylori* strain NCTC11638. *Mol. Microbiol.* 11:509–23

19. Burroni D, Lupetti P, Pagliaccia C, Reyrat JM, Dallai R, et al. 1998. Deletion of the major proteolytic site of the *Helicobacter pylori* cytotoxin does not influence toxin activity but favors assembly of the toxin into hexameric structures. *Infect. Immun.* 66:5547–50

20. Censini S, Lange C, Xiang Z, Crabtree JE, Borodovsky M, et al. 1996. *cag*, a pathogenicity island of *Helicobacter pylori*, encodes type I-specific and disease-associated virulence factors. *Proc. Natl. Acad. Sci. USA* 93:14648–53

21. Chan NW, Stangier K, Sherburne R, Taylor DE, Zhang Y, et al. 1995. The biosynthesis of Lewis X in *Helicobacter pylori*. *Glycobiology* 5:683–88

22. Clayton CL, Pallen MJ, Kleanthous H, Wren BW, Tabaqchali S. 1990. Nucleotide sequence of two genes from *Helicobacter pylori* encoding for urease subunits. *Nucleic Acids Res.* 18:362

23. Corthesy-Theulaz IE, Bergonzelli GE, Henry H, Bachmann D, Schorderet DF, et al. 1997. Cloning and characterization of *Helicobacter pylori* succinyl CoA:acetoacetate CoA-transferase, a novel prokaryotic member of the CoA-transferase family. *J. Biol. Chem.* 272:25659–67

24. Covacci A, Censini S, Bugnoli M, Petracca R, Burroni D, et al. 1993. Molecular characterization of the 128-kDa immunodominant antigen of *Helicobacter pylori* associated with cytotoxicity and duodenal ulcer. *Proc. Natl. Acad. Sci. USA* 90:5791–95

25. Covacci A, Falkow S, Berg DE, Rappuoli R. 1997. Did the inheritance of a pathogenicity island modify the virulence of *Helicobacter pylori*? *Trends Microbiol.* 5:205–8

26. Cover TL. 1996. The vacuolating cytotoxin of *Helicobacter pylori*. *Mol. Microbiol.* 20:241–46

27. Cover TL, Blaser MJ. 1992. Purification and characterization of the vacuolating toxin from *Helicobacter pylori*. *J. Biol. Chem.* 267:10570–75

28. Cover TL, Cao P, Lind CD, Tham KT, Blaser M. 1993. Correlation between vacuolating cytotoxin production by *Helicobacter pylori* isolates in vitro and in vivo. *Infect. Immun.* 61:5008–12

29. Cover TL, Dooley CP, Blaser MJ. 1990. Characterization of and human serologic response to proteins in *Helicobacter pylori* broth culture supernatants with vacuolizing cytotoxin activity. *Infect. Immun.* 58:603–10

30. Cover TL, Hanson PI, Heuser JE. 1997. Acid-induced dissociation of VacA, the *Helicobacter pylori* vacuolating cytotoxin, reveals its pattern of assembly. *J. Cell. Biol.* 138:759–69

31. Cover TL, Tummuru MK, Cao P, Thompson SA, Blaser MJ. 1994. Divergence of genetic sequences for the vacuolating cytotoxin among *Helicobacter pylori* strains. *J. Biol. Chem.* 269:10566–73

32. Crabtree JE, Figura N, Taylor JD, Bugnoli M, Armellini D, et al. 1992. Expression of 120 kilodalton protein and cytotoxicity in *Helicobacter pylori*. *J. Clin. Pathol.* 45:733–34

33. Crabtree JE, Xiang Z, Lindley IJ, Tompkins DS, Rappuoli R, et al. 1995. Induction of interleukin-8 secretion from gastric epithelial cells by a *cagA* negative isogenic mutant of *Helicobacter pylori*. *J. Clin. Pathol.* 48:967–69

34. Cussac V, Ferro RL, Labigne A. 1992. Expression of *Helicobacter pylori* urease genes in *Escherichia coli* grown under nitrogen-limiting conditions. *J. Bacteriol.* 174:2466–73

35. de Bernard M, Arico B, Papini E, Rizzuto R, Grandi G, et al. 1997. *Helicobacter pylori* toxin VacA induces vacuole formation by acting in the cell cytosol. *Mol. Microbiol.* 26:665–74

36. de Bernard M, Papini E, de Filippis V, Gottardi E, Telford J, et al. 1995. Low pH activates the vacuolating toxin of *Helicobacter pylori*, which becomes acid and pepsin resistant. *J. Biol. Chem.* 270:23937–40

37. de Reuse H, Labigne A, Mengin-Lecreulx D. 1997. The *Helicobacter pylori ureC* gene codes for a phosphoglucosamine mutase. *J. Bacteriol.* 179:3488–93

38. Dunn BE, Vakil NB, Schneider BG, Miller MM, Zitzer JB, et al. 1997. Localization of *Helicobacter pylori* urease and heat shock protein in human gastric biopsies. *Infect. Immun.* 65:1181–88

39. Eaton KA, Brooks CL, Morgan DR, Krakowka S. 1991. Essential role of urease in pathogenesis of gastritis induced by *Helicobacter pylori* in gnotobiotic piglets. *Infect. Immun.* 59:2470–75

40. Eaton KA, Krakowka S. 1994. Effect of gastric pH on urease-dependent colonization of gnotobiotic piglets by *Helicobacter pylori*. *Infect. Immun.* 62:3604–7

41. Eaton KA, Suerbaum S, Josenhans C, Krakowka S. 1996. Colonization of gnotobiotic piglets by *Helicobacter pylori* deficient in two flagellin genes. *Infect. Immun.* 64:2445–48

42. Evans DG, Karjalainen TK, Evans DJ, Moulds JJ, Graham DY, Lee CH. 1993. Cloning, nucleotide sequence, and expression of a gene encoding an adhesin subunit

protein of *Helicobacter pylori. J. Bacteriol.* 175:674–83

43. Evans DG, Queiroz DMM, Mendes EN, Evans DJ Jr. 1998. *Helicobacter pylori cagA* status and s and m alleles of *vacA* in isolates from individuals with a variety of *H. pylori*-associated gastric diseases. *J. Clin. Microbiol.* 36:3435–37

44. Evans DJ Jr, Queiroz DM, Mendes EN, Evans DG. 1998. Diversity in the variable region of *Helicobacter pylori cagA* gene involves more than simple repetition of a 102-nucleotide sequence. *Biochem. Biophys. Res. Commun.* 245:780–84

45. Exner MM, Doig P, Trust TJ, Hancock RE. 1995. Isolation and characterization of a family of porin proteins from *Helicobacter pylori. Infect. Immun.* 63:1567–72

46. Fantry GT, Zheng QX, Darwin PE, Rosenstein AH, James SP. 1996. Mixed infection with *cagA*-positive and *cagA*-negative strains of *Helicobacter pylori. Helicobacter* 1:98–106

47. Forman D, Newell DG, Fullerton F, Yarnell JW, Stacey AR, et al. 1991. Association between infection with *Helicobacter pylori* and risk of gastric cancer: evidence from a prospective investigation. *Br. Med. J.* 302:1302–5

48. Forsyth MH, Atherton JC, Blaser MJ, Cover TL. 1998. Heterogeneity in levels of vacuolating cytotoxin gene (*vacA*) transcription among *Helicobacter pylori* strains. *Infect. Immun.* 66:3088–94

49. Fulkerson JF Jr, Garner RM, Mobley HL. 1998. Conserved residues and motifs in the NixA protein of *Helicobacter pylori* are critical for the high affinity transport of nickel ion. *J. Biol. Chem.* 273:235–41

50. Gàlvez A, Maqueda M, Martinez-Bueno M, Valdivia E. 1998. Publication rates reveal trends in microbiological research. *ASM News* 64:269–75

51. Garner JA, Cover TL. 1996. Binding and internalization of the *Helicobacter pylori* vacuolating cytotoxin by epithelial cells. *Infect. Immun.* 64:4197–203

52. Garner RM, Fulkerson J Jr, Mobley HL. 1998. *Helicobacter pylori* glutamine synthetase lacks features associated with transcriptional and posttranslational regulation. *Infect. Immun.* 66:1839–47

53. Ge Z, Chan NW, Palcic MM, Taylor DE. 1997. Cloning and heterologous expression of an α1,3-fucosyltransferase gene from the gastric pathogen *Helicobacter pylori. J. Biol. Chem.* 272:21357–63

54. Ge Z, Hiratsuka K, Taylor DE. 1995. Nucleotide sequence and mutational analysis indicate the two *Helicobacter pylori* genes encodes a P-type ATPase and a cation-binding protein associated with copper transport. *Mol. Microbiol.* 15:97–106

55. Ge Z, Jiang Q, Kalisiak MS, Taylor DE. 1997. Cloning and functional characterization of *Helicobacter pylori* fumarate reductase operon comprising three structural genes coding for subunits C, A and B. *Gene* 204:227–34

56. Deleted in proof

57. Ge Z, Taylor DE. 1996. *Helicobacter pylori* genes *hpcopA* and *hpcopP* constitute a *cop* operon involved in copper export. *FEMS Microbiol. Lett.* 145:181–88

58. Ge Z, Taylor DE. 1996. Sequencing, expression, and genetic characterization of the *Helicobacter pylori ftsH* gene encoding a protein homologous to members of a novel putative ATPase family. *J. Bacteriol.* 178:6151–57

59. Ge Z, Taylor DE. 1997. The *Helicobacter pylori* gene encoding phosphatidylserine synthase: sequence expression, and insertional mutagenesis. *J. Bacteriol.* 179:4970–76

60. Ge Z, Taylor DE. 1998. *Helicobacter pylori*—molecular genetics and diagnostic typing. *Br. Med. Bull.* 54:31–38

61. Ge Z, Taylor DE. 1997. *H. pylori* DNA transformation by natural competence and electroporation. In *Method in Molecular Medicine*: Heliobacter pylori *protocol*, ed. CL Clayton, HLT Mobley, pp. 145–52. Totowa, NJ: Humana

62. Gentz R, Rauscher FJ III, Abate C, Curran T. 1989. Parallel association of Fos and Jun leucine zipper juxtaposes DNA binding domains. *Science* 243:1695–99

63. Ghiara P, Marchetti M, Blaser MJ, Tummuru MK, Cover TL, et al. 1995. Role of the *Helicobacter pylori* virulence factors vacuolating cytotoxin, CagA, and urease in a mouse model of disease. *Infect. Immun.* 63:4154–60

64. Gilbert JV, Ramakrishna J, Sunderman FW, Wright A, Plaut AG. 1995. Protein Hpn: cloning and characterization of a histidine-rich metal-binding polypeptide in *Helicobacter pylori* and *Helicobacter mustelae*. *Infect. Immun.* 63:1682–88

65. Glocker E, Lange C, Covacci A, Bereswill S, Kist M, Pahl HL. 1998. Proteins encoded by the *cag* pathogenicity island of *Helicobacter pylori* are required for NF-κB activation. *Infect. Immun.* 66:2346–48

66. Go MF, Chan KY, Versalovic J, Koeuth T, Graham DY, et al. 1995. Cluster analysis of *Helicobacter pylori* genomic DNA fingerprints suggests gastroduodenal disease-specific associations. *Scand. J. Gastroenterol.* 30:640–46

67. Go MF, Kapur V, Graham DY, Musser JM. 1996. Population genetic analysis *of Helicobacter pylori* by multilocus enzyme electrophoresis: extensive allelic diversity and recombinational population structure. *J. Bacteriol.* 178:3934–38

68. Goodwin A, Kersulyte D, Sisson G, Veldhuyzen van Zanten SJ, Berg DE, et al. 1998. Metronidazole resistance in *Helicobacter pylori* is due to null mutations in a gene (*rdxA*) that encodes an oxygen-insensitive NADPH nitroreductase. *Mol. Microbiol.* 28:383–93

69. Graham DY. 1997. Can therapy even be denied for *Helicobacter pylori* infection? *Gastroenterology* 113:S113–S17

70. Hancock RE, Alm R, Bina J, Trust T. 1998. *Helicobacter pylori*: a surpris-ingly conserved bacterium. *Nat. Biotechnol.* 16:216–17

71. Hawtin PR, Stacey AR, Newell DG. 1990. Investigation of the structure and localization of the urease of *Helicobacter pylori* using monoclonal antibodies. *J. Gen. Microbiol.* 136:1995–2000

72. Hendricks JK, Mobley HL. 1997. *Helicobacter pylori* ABC transporter: effect of allelic exchange mutagenesis on urease activity. *J. Bacteriol.* 179:5892–902

73. Henry S, Oriol R, Samuelsson B. 1995. Lewis histo-blood group system and associated secretary phenotypes. *Vox Sang.* 69:166–82

74. Heuermann D, Haas R. 1998. A stable shuttle vector system for efficient genetic complementation of *Helicobacter pylori* strains by transformation and conjugation. *Mol. Gen. Genet.* 257:519–28

75. Hofreuter D, Odenbreit S, Henke G, Haas R. 1998. Natural competence for DNA transformation in *Helicobacter pylori*: identification and genetic characterization of the *comB* locus. *Mol. Microbiol.* 28:1027–38

76. Hu L-T, Mobley HLT. 1990. Purification and N-terminal analysis of urease from *Helicobacter pylori*. *Infect. Immun.* 58:992–98

77. Huesca M, Borgia S, Hoffman P, Lingwood CA. 1996. Acidic pH changes receptor binding specificity of *Helicobacter pylori*: a binary adhesion model in which surface heat shock (stress) proteins mediate sulfatide recognition in gastric colonization. *Infect. Immun.* 64:2643–48

78. Huesca M, Goodwin A, Bhagwansingh A, Hoffman P, Lingwood CA. 1998. Characterization of an acidic-pH-inducible stress protein (hsp70), a putative sulfatide binding adhesin, from *Helicobacter pylori*. *Infect. Immun.* 66:4061–67

79. Hughes NJ, Clayton CL, Chalk PA, Kelly DJ. 1998. *Helicobacter pylori porCDAB* and *oorDABC* genes encode distinct pyruvate:flavodoxin and 2-oxoglut-

arate:acceptor oxidoreductases which mediate electron transport to NADP. *J. Bacteriol.* 180:1119–28

80. Huynen M, Dandekar T, Bork P. 1998. Differential genome analysis applied to the species-specific features of *Helicobacter pylori*. *FEBS Lett.* 426:1–5

81. Ilver D, Arnqvist A, Ogren J, Frick IM, Kersulyte D, et al. 1998. *Helicobacter pylori* adhesin binding fucosylated histoblood group antigens revealed by retagging. *Science* 279:373–77

82. Jenks PJ, Foynes S, Ward SJ, Constantinidou C, Penn CW, Wren BW. 1997. A flagellar-specific ATPase (FliI) is necessary for flagellar export in *Helicobacter pylori*. *FEMS Microbiol. Lett.* 152:205–11

83. Jiang Q, Hiratsuka K, Taylor DE. 1996. Variability of gene order in different *Helicobacter pylori* strains contributes to genome diversity. *Mol. Microbiol.* 20:833–42

84. Jones AC, Logan RPH, Foynes S, Cockayne A, Wren BW, Penn CW. 1997. A flagellar sheath protein of *Helicobacter pylori* is identical to HpaA, a putative *N*-acetylneuraminyllactose-binding hemagglutinin, but is not an adhesin for AGS cells. *J. Bacteriol.* 179:5643–47

85. Jorgensen M, Daskalopoulos G, Warburton V, Mitchell HM, Hazell SL. 1996. Multiple strain colonization and metronidazole resistance in *Helicobacter pylori*-infected patients: identification from sequential and multiple biopsy specimens. *J. Infect. Dis.* 174:631–35

86. Kansau I, Guillain F, Thiberge JM, Labigne A. 1996. Nickel binding and immunological properties of the C-terminal domain of the *Helicobacter pylori* GroES homologue (HspA). *Mol. Microbiol.* 22:1013–23

87. Karita M, Blaser MJ. 1998. Acid-tolerance response in *Helicobacter pylori* and differences between *cagA*+ and *cagA*− strains. *J. Infect. Dis.* 178:213–19

88. Karita M, Etterbeek ML, Forsyth MH,

Tummuru MK, Blaser MJ. 1997. Characterization of *Helicobacter pylori dapE* and construction of a conditionally lethal *dapE* mutant. *Infect. Immun.* 65:4158–64

89. Karita M, Tummuru MK, Wirth HP, Blaser MJ. 1996. Effect of growth phase and acid shock on *Helicobacter pylori cagA* expression. *Infect. Immun.* 64:4501–7

90. Krishnamurthy P, Parlow M, Zitzer JB, Vakil NB, Mobley HLT, et al. 1998. *Helicobacter pylori* containing only cytoplasmic urease is susceptible to acid. *Infect. Immun.* 66:5060–66

91. Kuipers EJ, Israel DA, Kusters JG, Blaser MJ. 1998. Evidence for a conjugation-like mechanism of DNA transfer in *Helicobacter pylori*. *J. Bacteriol.* 180:2901–5

92. Kwon DH, Woo JS, Perng CL, Go MF, Graham DY, et al. 1998. The effect of *galE* gene inactivation on lipopolysaccharide profile of *Helicobacter pylori*. *Curr. Microbiol.* 37:144–48

93. Labigne A, Cussac V, Courcoux P. 1991. Shuttle cloning and nucleotide sequences of *Helicobacter pylori* genes responsible for urease activity. *J. Bacteriol.* 173:1920–31

94. Labigne A, de Reuse H. 1996. Determinants of *Helicobacter pylori* pathogenicity. *Infect. Agents Dis.* 5:191–202

95. Langenberg W, Rauws EAJ, Wildjojokusumo A, Tytgat GNJ, Zanen HC. 1986. Identification of *Campylorbacter pyloridis* isolates by restriction endonuclease DNA analysis. *J. Clin. Microbiol.* 24:414–17

96. Lanzavecchia S, Bellon PL, Lupetti P, Dallai R, Rappuoli R, et al. 1998. Three-dimensional reconstruction of metal replicas of the *Helicobacter pylori* vacuolating cytotoxin. *J. Struct. Biol.* 121:9–18

97. Lee WK, An YS, Kim KH, Kim SH, Song JY, et al. 1997. Construction of a *Helicobacter pylori-Escherichia coli* shuttle vector for gene transfer in *Helicobacter pylori*. *Appl. Environ. Microbiol.* 63:4866–71

98. Leunk RD, Johnson PT, David BC, Kraft WG, Morgan DR. 1988. Cytotoxic ac-

tivity in broth-culture filtrates of *Campylobacter pylori. J. Med. Microbiol.* 26:93–99

99. Leying H, Suerbaum S, Geis G, Haas R. 1992. Cloning and genetic characterization of a *Helicobacter pylori* flagellin gene. *Mol. Microbiol.* 6:2863–74

100. Lupetti P, Heuser JE, Manetti R, Massari P, Lanzavecchia S, et al. 1996. Oligomeric and subunit structure of the *Helicobacter pylori* vacuolating cytotoxin. *J. Cell Biol.* 133:801–7

101. Macchia G, Massone A, Burroni D, Covacci A, Censini S, Rappuoli R. 1993. The Hsp60 protein of *Helicobacter pylori*: structure and immune response in patients with gastroduodenal diseases. *Mol. Microbiol.* 9:645–52

102. Majewski SIH, Goodwin CS. 1988. Restriction endonuclease analysis of the genome of *Campylobacter pylori* with a rapid extraction method: evidence for considerable genomic variation. *J. Infect. Dis.* 157:465–71

103. Marchetti M, Arico B, Burroni D, Figura N, Rappuoli R, et al. 1995. Development of a mouse model of *Helicobacter pylori* infection that mimics human disease. *Science* 267:1655–58

104. Marshall DG, Coleman DC, Sullivan DJ, Xia H, O'Morain CA, et al. 1996. Genomic DNA fingerprinting of clinical isolates of *Helicobacter pylori* using short oligonucleotide probes containing repetitive sequences. *J. Appl. Bacteriol.* 81:509–17

105. Martin SL, Edbrooke MR, Hodgman TC, van den Eijnden DH, Bird MI. 1997. Lewis X biosynthesis in *Helicobacter pylori*. Molecular cloning of an $\alpha(1,3)$-fucosyltransferase gene. *J. Biol. Chem.* 272:21349–56

106. McGowan CC, Necheva A, Thompson SA, Cover TL, Blaser MJ. 1998. Acid-induced expression of an LPS-associated gene in *Helicobacter pylori. Mol. Microbiol.* 30:19–31

107. Melchers K, Weitzenegger T, Buhmann A, Steinhilber W, Sachs G, et al. 1996. Cloning and membrane topolgy of a P-type ATPase from *Helicobacter pylori. J. Biol. Chem.* 271:446–57

108. Mobley HLT, Fulkerson J Jr, Hendricks JK, Garner RM. 1997. Expression of catalytically active urease by *Helicobacter pylori*. In *Pathogenesis and Host Response in Helicobacter pylori*, ed. AP Moran, CA O'Morain, pp. 58–66. Englewood, NJ: Normed Verlag

109. Mobley HLT, Garner RM, Bauerfeind P. 1995. *Helicobacter pylori* nickel-transport gene *nixA*: synthesis of catalytically active urease in *Escherichia coli* independent of growth conditions. *Mol. Microbiol.* 16:97–109

110. Molinari M, Galli C, de Bernard M, Norais N, Ruysschaert JM, et al. 1998. The acid activation of *Helicobacter pylori* toxin VacA: structural and membrane binding studies. *Biochem. Biophys. Res. Commun.* 248:334–40

111. Molinari M, Galli C, Norais N, Telford JL, Rappuoli R, et al. 1997. Vacuoles induced by *Helicobacter pylori* toxin contain both late endosomal and lysosomal markers. *J. Biol. Chem.* 272:25339–44

112. Monteiro MA, Chan KH, Rasko DA, Taylor DE, Zheng PY, et al. 1998. Simultaneous expression of type 1 and type 2 Lewis blood group antigens by *Helicobacter pylori* lipopolysaccharides: molecular mimicry between *H. pylori* lipopolysaccharides and human gastric epithelial cell surface glycoforms. *J. Biol. Chem.* 273:11533–43

113. Moore RA, Beckthold B, Wong S, Kureishi A, Bryan LE. 1995. Nucleotide sequence of the *gyrA* gene and characterization of ciprofloxacin-resistant mutants of *Helicobacter pylori. Antimicrob. Agents Chemother.* 37:457–63

114. Nakamura H, Yoshiyama H, Takeuchi H, Mizote T, Okita K, Nakazawa T. 1998. Urease plays an important role in the

chemotactic motility of *Helicobacter pylori* in a viscous environment. *Infect. Immun.* 66:4832–37

115. Namavar F, Sparrius M, Veerman EC, Appelmelk BJ, Vandenbroucke-Grauls CM. 1998. Neutrophil-activating protein mediates adhesion of *Helicobacter pylori* to sulfated carbohydrates on high-molecular-weight salivary mucin. *Infect. Immun.* 66:444–47

116. National Institutes of Health Consensus Development Panel on *Helicobacter pylori* in Peptic Ulcer Disease. 1994. *Helicobacter pylori* in peptic ulcer disease. *J. Am. Med. Assoc.* 272:65–69

117. Nodenskov-Sorensen P, Bukholm G, Bovre K. 1990. Natural competence for genetic transformation in *Campylobacter pylori. J. Infect. Dis.* 161:365–66

118. Odenbreit S, Till M, Haas R. 1996. Optimized BlaM-transposon shuttle mutagenesis of *Helicobacter pylori* allows the identification of novel genetic loci involved in bacterial virulence. *Mol. Microbiol.* 20:361–73

119. Odenbreit S, Wieland B, Haas R. 1996. Cloning and genetic characterization of *Helicobacter pylori* catalase and construction of a catalase-deficient mutant strain. *J. Bacteriol.* 178:6960–67

120. O'Toole PW, Janzon L, Doig P, Huang J, Kostrzynska M, et al. 1995. The putative neuraminyllactose-binding hemagglutinin HpaA of *Helicobacter pylori* CCUG 17874 is a lipoprotein. *J. Bacteriol.* 177:6049–57

121. O'Toole PW, Kostrzynska M, Trust TJ. 1994. Non-motile mutants of *Helicobacter pylori* and *Helicobacter mustelae* defective in flagellar hook production. *Mol. Microbiol.* 14:691–703

122. Pagliaccia C, de Bernard M, Lupetti P, Ji X, Burroni D, Cover TL. 1998. The m2 form of the *Helicobacter pylori* cytotoxin has cell type-specific vacuolating activity. *Proc. Natl. Acad Sci. USA* 95:10212–17

123. Papini E, de Bernard M, Milia E, Bugnoli M, Zerial M, et al. 1994. Cellular vacuoles induced by *Helicobacter pylori* originate from late endosomal compartments. *Proc. Natl. Acad. Sci. USA* 91:9720–24

124. Papini E, Satin B, Bucci C, de Bernard M, Telford JL, et al. 1997. The small GTP binding protein rab7 is essential for cellular vacuolation induced by *Helicobacter pylori* cytotoxin. *EMBO J.* 16:15–24

125. Park SM, Park J, Kim JG, Cho HD, Cho JH, et al. 1998. Infection with *Helicobacter pylori* expressing the *cagA* gene is not associated with an increased risk of developing peptic ulcer diseases in Korean patients. *Scand. J. Gastroenterol.* 33:923–27

126. Parsonnet J, Hansen S, Rodriguez L, Gelb AB, Warnke RA, et al. 1994. *Helicobacter pylori* infection and gastric lymphoma. *N. Engl. J. Med.* 330:1267–71

127. Paster BJ, Lee A, Fox JG, Dewhirst FE, Tordoff LA, et al. 1991. Phylogeny of *Helicobacter felis* sp. nov., *Helicobacter mustelae*, and related bacteria. *Int. J. Syst. Bacteriol.* 41:31–38

128. Phadnis SH, Ilver D, Janzon L, Normark S, Westblom TU. 1994. Pathological significance and molecular characterization of the vacuolating toxin gene of *Helicobacter pylori. Infect. Immun.* 62:1557–65

129. Phadnis SH, Parlow MH, Levy M, Ilver D, Caulkins CM, et al. 1996. Surface localization of *Helicobacter pylori* urease and a heat shock protein homolog requires bacterial autolysis. *Infect. Immun.* 64:905–12

130. Salaun L, Audibert C, Le Lay G, Burucoa C, Fauchere JL, et al. 1998. Panmictic structure of *Helicobacter pylori* demonstrated by the comparative study of six genetic markers. *FEMS Microbiol. Lett.* 161:231–39

131. Saunders NJ, Peden JF, Hood DW, Moxon ER. 1998. Simple sequence repeats in the *Helicobacter pylori* genome. *Mol. Microbiol.* 27:1091–98

132. Schmitt W, Haas R. 1994. Genetic analysis of the *Helicobacter pylori* vacuolating cytotoxin: structural similarities with the IgA protease type of exported protein. *Mol. Microbiol.* 12:307–19

133. Schmitt W, Odenbreit S, Heuermann D, Haas R. 1995. Cloning of the *Helicobacter pylori recA* gene and functional characterization of its product. *Mol. Gen. Genet.* 248:563–72

134. Schmitz A, Josenhans C, Suerbaum S. 1997. Cloning and characterization of the *Helicobacter pylori flbA* gene, which codes for a membrane protein involved in coordinated expression of flagellar genes. *J. Bacteriol.* 179:987–97

135. Scott DR, Weeks D, Hong C, Postius S, Melchers K, et al. 1998. The role of internal urease in acid resistance of *Helicobacter pylori*. *Gastroenterology* 114:58–70

136. Segal ED, Falkow S, Tompkins LS. 1996. *Helicobacter pylori* attachment to gastric cells induces cytoskeletal rearrangements and tyrosine phosphorylation of host cell proteins. *Proc. Natl. Acad. Sci. USA* 93:1259–64

137. Segal ED, Lange C, Covacci A, Tompkins LS, Falkow S. 1997. Induction of host signal transduction pathways by *Helicobacter pylori*. *Proc. Natl. Acad. Sci. USA* 94:7595–99

138. Seto K, Hayashi-Kuwabara Y, Yoneta T, Suda H, Tamaki H. 1998. Vacuolation induced by cytotxin for *Helicobacter pylori* is mediated by the EGF receptor in Hela cells. *FEBS Lett.* 431:347–50

139. Sharma SA, Tummuru MK, Blaser MJ, Kerr LD. 1998. Activation of IL-8 gene expression by *Helicobacter pylori* is regulated by transcription factor nuclear factor-κB in gastric epithelial cells. *J. Immunol.* 160:2401–7

140. Sharma SA, Tummuru MK, Miller GG, Blaser MJ. 1995. Interleukin-8 response of gastric epithelial cell lines to *Helicobacter pylori* stimulation in vitro. *Infect. Immun.* 63:1681–87

141. Sherburne R, Taylor DE. 1995. *Helicobacter pylori* expresses a complex surface carbohydrate, Lewis X. *Infect. Immun.* 63:4564–68

142. Simoons Smit IM, Appelmelk BJ, Verboom T, Negrini R, Penner JL, et al. 1996. Typing of *Helicobacter pylori* with monoclonal antibodies against Lewis antigens in lipopolysaccharide. *J. Clin. Microbiol.* 34:2196–200

143. Skouloubris S, Labigne A, de Reuse H. 1997. Identification and characterization of an aliphatic amidase in *Helicobacter pylori*. *Mol. Microbiol.* 25:989–98

144. Skouloubris S, Thiberge JM, Labigne A, de Reuse H. 1998. The *Helicobacter pylori* UreI protein is not involved in urease activity but is essential for bacterial survival in vivo. *Infect. Immun.* 66:4517–21

145. Spiegelhalder C, Gerstenecker B, Kersten A, Schiltz E, Kist M. 1993. Purification of *Helicobacter pylori* superoxide dismutase and cloning and sequencing of the gene. *Infect. Immun.* 61:5315–25

146. Suerbaum S, Brauer-Steppkes T, Labigne A, Cameron B, Drlica K. 1998. Topoisomerase I of *Helicobacter pylori*: juxtaposition with a flagellin gene (*flaB*) and functional requirement of a fourth zinc finger motif. *Gene* 210:151–61

147. Suerbaum S, Josenhans C, Labigne A. 1993. Cloning and genetic characterization of the *Helicobacter pylori* and *Helicobacter mustelae flaB* flagellin gnes and construction of *H. pylori flaA*- and *flaB*-negative mutants by electroporation-mediated allelic exchange. *J. Bacteriol.* 175:3278–88

148. Suerbaum S, Smith JM, Bapumia K, Morelli G, Smith NH, et al. 1998. Free recombination within *Helicobacter pylori*. *Proc. Natl. Acad. Sci. USA* 95:12619–24

149. Suerbaum S, Thiberge JM, Kansau I, Ferrero RL, Labigne A. 1994. *Helicobacter pylori hspA-hspB* heat-shock gene cluster: nucleotide sequence, expression,

putative function and immunogenicity. *Mol. Microbiol.* 14:959–74

150. Takeuchi H, Shirai M, Akada JK, Tsuda M, Nakazawa T. 1998. Nucleotide sequence and characterization of *cdrA*, a cell division-related gene of *Helicobacter pylori. J. Bacteriol.* 180:5263–68

151. Taylor DE. 1992. Genetics of *Campylobacter* and *Helicobacter. Annu. Rev. Microbiol.* 46:35–64

152. Taylor DE, Eaton M, Chang N, Salama S. 1992. Construction of a *Helicobacter pylori* genome map and demonstration of diversity at the genome level. *J. Bacteriol.* 174:6800–6

153. Taylor DE, Ge Z, Purych D, Lo T, Hiratsuka K. 1997. Cloning and sequence analysis of two copies of a 23S rRNA gene from *Helicobacter pylori* and association of clarithromycin resistance with 23S rRNA mutations. *Antimicrob. Agents Chemother.* 41:2621–28

154. Taylor NS, Fox JG, Akopyants NS, Berg DE, Thompson N, et al. 1995. Long-term colonization with single and multiple strains of *Helicobacter pylori* assessed by DNA fingerprinting. *J. Clin. Microbiol.* 33:918–23

155. Telford JL, Ghiara P, Dell'Orco M, Dwyer B, Comanducci M, et al. 1994. Gene structure of the *Helicobacter pylori* cytotoxin and evidence of its key role in gastric disease. *J. Exp. Med.* 179:1653–68

156. Thompson SA, Blaser MJ. 1995. Isolation of the *Helicobacter pylori recA* gene and involvement of the *recA* region in resistance to low pH. *Infect. Immun.* 63:2185–93

157. Thompson SA, Latch RL, Blaser JM. 1998. Molecular characterization of the *Helicobacter pylori uvr B* gene. *Gene* 209:113–22

158. Tomb JF, White O, Kerlavage AR, Clayton RA, Sutton GG, et al. 1997. The complete sequence of the gastric pathogen *Helicobacter pylori. Nature* 388:539–47

159. Tsuda M, Karita M, Morshed MG, Okita K, Nakazawa T. 1994. A urease-negative mutant of *Helicobacter pylori* constructed by allelic exchange mutagenesis lacks the ability to colonize the nude mouse stomach. *Infect. Immun.* 62:3586–89

160. Tummuru MKR, Cover TL, Blaser MJ. 1993. Cloning and expression of a high-molecular-mass major antigen of *Helicobacter pylori*: evidence of linkage to cytotoxin production. *Infect. Immun.* 61:1799–809

161. Tummuru MKR, Cover TL, Blaser MJ. 1994. Mutation of the cytotoxin-associated cagA gene does not affect the vacuolating cytotoxin activity of *Helicobacter pylori. Infect. Immun.* 62:2609–13

162. Tummuru MKR, Sharma SA, Blaser MJ. 1995. *Helicobacter pylori picB*, a homologue of the *Bordetella pertussis* toxin secretion protein, is required for induction of IL-8 in gastric epithelial cells. *Mol. Microbiol.* 18:867–76

163. Turner R, Tjian R. 1989. Leucine repeats and an adjacent DNA binding domain mediate the formation of functional cFos-cJun heterodimers. *Science* 243:1689–94

164. van der Ende A, Pan ZJ, Bart A, van der Hulst RW, Feller M, et al. 1998. cagA-positive *Helicobacter pylori* populations in China and the Netherlands are distinct. *Infect. Immun.* 66:1822–26

165. Vanet A, Labigne A. 1998. Evidence for specific secretion rather than autolysis in the release of some *Helicobacter pylori* proteins. *Infect. Immun.* 66:1023–27

166. Deleted in proof

167. Deleted in proof

168. Deleted in proof

169. Deleted in proof

170. Wang G, Rasko D, Sherburne R, Taylor DE. 1999. Molecular genetic basis for the variable expression of Lewis Y antigen in *Helicobacter pylori*: analysis of the $\alpha(1,2)$ fucosyltransferase gene. *Mol. Microbiol.* 31:1265–74

171. Wang Y, Roos KP, Taylor DE. 1993.

Transformation of *Helicobacter pylori* by chromosomal metronidazole resistance and by a plasmid with a selectable chloramphenicol resistance marker. *J. Gen. Microbiol.* 139:2485–93

172. Wang Y, Taylor DE. 1990. Chloramphenicol resistance in *Campylobacter coli*: nucleotide sequence, expression, and cloning vector construction. *Gene* 94:23–28

173. Warren JR, Marshall BJ. 1983. Unidentified curved bacilli on gastric epithelium in active chronic gastritis. *Lancet* i:1273–75

174. Wirth HP, Yang M, Karita M, Blaser MJ. 1996. Expression of the human cell surface glycoconjugates Lewis x and Lewis y by *Helicobacter pylori* isolates is related to *cagA* status. *Infect. Immun.* 64:4598–605

175. Worst DJ, Gerrits MM, Vandenbroucke-Grauls CMJE, Kusters JG. 1998. *Helicobacter pylori ribBA*-mediated ribo-flavin production is involved in iron acquisition. *J. Bacteriol.* 1473–79

176. Xiang Z, Censini S, Bayeli PF, Telford JL, Figura N, et al. 1995. Analysis of expression of CagA and VacA virulence factors in 43 strains of *Helicobacter pylori* reveals that clinical isolates can be divided into two major types and that CagA is not necessary for expression of the vacuolating cytotoxin. *Infect. Immun.* 63:94–98

177. Yahiro K, Niidome T, Hatakeyama T, Aoyagi H, Kurazono H, et al. 1997. *Helicobacter pylori* vacuolating cytotoxin binds to the 140-kDa protein in human gastric cancer cell lines, AZ-521 and AGS. *Biochem. Biophys. Res. Commun.* 238:629–32

178. Yamaoka Y, Kodama T, Kashima K, Graham DY, Sepulveda AR. 1998. Variants of the 3' region of the *cagA* gene in *Helicobacter pylori* isolates from patients with different *H. pylori*-associated diseases. *J. Clin. Microbiol.* 36:2258–63

Annu. Rev. Microbiol. 1999. 53:389–409

CIRCADIAN PROGRAMS IN CYANOBACTERIA: Adaptiveness and Mechanism

Carl Hirschie Johnson[1] and Susan S. Golden[2]

[1]*Department of Biology, Vanderbilt University, Nashville, Tennessee 37235;
e-mail: carl.h.johnson@vanderbilt.edu;* [2]*Department of Biology, Texas A&M University,
College Station, Texas 77843; e-mail: sgolden@bio.tamu.edu*

Key Words   clock, fitness, *kai*, competition, luciferase, *Synechococcus*

■ **Abstract**   At least one group of prokaryotes is known to have circadian regulation of cellular activities—the cyanobacteria. Their "biological clock" orchestrates cellular events to occur in an optimal temporal program, and it can keep track of circadian time even when the cells are dividing more rapidly than once per day. Growth competition experiments demonstrate that the fitness of cyanobacteria is enhanced when the circadian period matches the period of the environmental cycle. Three genes have been identified that specifically affect circadian phenotypes. These genes, *kaiA*, *kaiB*, and *kaiC*, are adjacent to each other on the chromosome, thus forming a clock gene cluster. The clock gene products appear to interact with each other and form an autoregulatory feedback loop.

CONTENTS

0066-4227/99/1001-0389$08.00

INTRODUCTION

What Is a Circadian Program?

Circadian rhythms are endogenous biological programs that time metabolic and/or behavioral events to occur at optimal phases of the daily cycle. They have three diagnostic characteristics. The first is that in constant conditions, the programs free-run with a period that is $\sim$24 h in duration. The second is that, in an appropriate environmental cycle (usually a light-dark and/or temperature cycle), the rhythm will take on the period of the environmental cycle, that is, circadian rhythms will entrain to the environmental cycle. The final characteristic is that the period of the free-running rhythm is nearly the same at different constant ambient temperatures within the physiological range; that is, circadian rhythms are temperature compensated. It is these three characteristics that define circadian rhythms, not the details of their biochemical mechanisms. Indeed, questions of considerable interest are whether circadian mechanisms have evolved more than once and, if so, whether completely different biochemical processes have been harnessed to the task in different organisms. The fascination of circadian rhythms is how a biochemical mechanism can keep time so precisely over such a long time constant ($\sim$24 h) at different ambient temperatures.

Before 1985, it was believed that circadian programs were exclusively a property of the eukaryotic domain (24). It was a reasonable assumption that rapidly growing prokaryotes would not have circadian organization, because it was thought that an endogenous timekeeper with a period close to 24 h would not be useful to organisms that divide more rapidly than once every 24 h, as do many prokaryotes. The assumption might be stated as, "Why have a timer for a cycle that is longer than your lifetime?" Although intuitive, this conclusion is flawed. It is based on the presumption that a bacterial cell is equivalent to a sexually reproducing multicellular organism, which it is not. A bacterial culture is more like a mass of protoplasm that grows larger and larger and incidentally subdivides. A mother cell does not die to make daughter cells; she *is* the daughter cells. From this perspective, it is reasonable that a 24-h temporal program could be adaptive to rapidly dividing protoplasm if the fitness of that protoplasm changes as a function of daily alterations in the environment (light intensity, temperature, etc).

Discovery of Circadian Programs in Prokaryotes

The proposal that prokaryotes might have circadian programs is not new. In the past 35 years, both *Escherichia coli* and *Klebsiella aerogenes* were proposed to exhibit circadian behavior (15, 54), but these studies were not persuasive (see below). That prokaryotic cells, either unicellular or multicellular, were too simple to express circadian behavior became a dogma, despite the fact that there were almost no published reports of rigorous tests of the proposition (24).

Studies in the late 1980s on cyanobacteria began to change this mind-set. It was not researchers who were interested in the presence of circadian timekeepers in prokaryotes who initiated this research, but those who attempted to resolve how some unicellular (or non-heterocystous, filamentous) cyanobacteria could fix nitrogen. Why was this an apparent dilemma? Photosynthesis evolves oxygen, and oxygen strongly inhibits the nitrogenase enzyme. How then can a photosynthetic unicellular organism fix nitrogen? Nitrogen fixation appeared to be doomed by the essential process that provides cellular energy. An imaginative idea to reconcile these incompatible processes surfaced in the 1980s; photosynthesis and nitrogen fixation could be separated in time—photosynthesis in the day and nitrogen fixation at night [reviewed by Golden et al (12)]. The first data suggesting that this "temporal separation" could be a metabolic program controlled by a circadian clock were from the non-heterocystous, filamentous cyanobacterium *Oscillatoria* sp. (52). Those authors found a nocturnal rhythm of nitrogenase activity that persisted in continuous light (LL). Shortly thereafter, temporal separation of photosynthesis (in the day) and nitrogenase activity (in the night) was demonstrated in the marine unicellular cyanobacteria *Synechococcus* spp. Miami BG 43511 and 43522 (39). This pattern continued in LL for at least 3 days, but these authors preferred to interpret their data in terms of regulation by the cell division cycle rather than by a circadian clock (39).

Huang and coworkers were apparently the first to clearly recognize that cyanobacteria were exhibiting circadian rhythms, and, in a series of publications beginning in 1986, they demonstrated all three salient properties in the same organism, the unicellular freshwater *Synechococcus* sp. RF-1 (6, 7, 13, 16–20). These pioneers studied the rhythm of nitrogen fixation and of amino acid uptake in this cyanobacterium, and were also the first to report the isolation of mutants affecting these processes (20). Another ground-breaking study was that of Sweeney & Borgese (55), who were the first to demonstrate temperature compensation of a daily rhythm in the marine cyanobacterium *Synechococcus* sp. WH7803.

We now know of circadian programs expressed in a number of cyanobacterial species. In addition to those already mentioned, circadian rhythms have been demonstrated in *Synechococcus* sp. strain PCC 7942 (32 and below), and the genera *Synechocystis* (2, 3), *Anabaena* (T Kondo & M Ishiura, unpublished data; 35), *Cyanothece* (47, 48), *Trichodesmium* (5, 8, 42), and possibly *Prochlorococcus* (49). What about other prokaryotes? This issue is addressed later in this review.

CIRCADIAN PROGRAMS IN CYANOBACTERIA

Nitrogen Fixation/Photosynthesis

The yin/yang rhythms of nitrogen fixation and photosynthesis fit the expectation that a major role of circadian timers is to temporally program metabolic events to occur at optimal phases of the environmental cycle. In addition to the work on *Oscillatoria* spp. and *Synechococcus* spp. Miami BG 43511/43522 and RF-1, another series of studies that supports the hypothesis of temporal separation has investigated the marine unicellular cyanobacterium *Cyanothece* sp. strain ATCC 51142 (9, 47, 48). A remarkable feature of this nitrogen-fixing cyanobacterium is the presence of large carbohydrate granules within the cells, which are easily visualized by electron microscopy. These granules accumulate progressively during daytime photosynthetic activity and dissipate during nocturnal nitrogen fixation (47).

There is, however, a counter-example to the temporal separation hypothesis among nitrogen-fixing cyanobacteria: the non-heterocystous, filamentous cyanobacterium *Trichodesmium* spp. fixes its nitrogen during the daytime, simultaneously with photosynthesis (5, 42). How do *Trichodesmium* spp. accomplish this feat? We don't know, but it underscores the fact that cyanobacteria are capable of fixing nitrogen in the presence of oxygen and therefore that other mechanisms must exist for lowering oxygen in the vicinity of nitrogenase (11). For example, *Synechococcus* sp. RF-1 increases its aerobic respiration rate whenever nitrogen fixation is under way (14). Enhanced respiration may be an additional mechanism for depleting oxygen in the vicinity of nitrogenase, even in cyanobacteria that separate photosynthesis and nitrogen fixation in time.

How does the circadian clock regulate the nitrogenase output rhythm? In *Synechococcus* sp. RF-1, *Cyanothece* spp., and *Trichodesmium* spp., there are circadian rhythms of nitrogenase messenger RNA (mRNA) abundance. As expected from the phasing of nitrogen fixation, in *Synechococcus* sp. RF-1 and *Cyanothece* spp., the peak of nitrogenase mRNA abundance is nocturnal (9, 17), whereas in *Trichodesmium* spp., it is diurnal (8). These rhythms of mRNA abundance presumably drive circadian rhythms of nitrogenase abundance and activity.

Luciferase Reporters Illuminate a Path for Genetic Analyses

Our interest in cyanobacteria as a model system for studying circadian programs stemmed from the genetic advantages that some cyanobacteria offer (12, 24, 33). The rhythms of nitrogen fixation, amino acid uptake, and carbohydrate content are reproducible, but the labor-intensive nature of the assays would dismay any but the strong-hearted from using this type of rhythm for a mutant screen [although the Huang group was undaunted and succeeded in isolating some mutants (see 20)]. To reap the benefits that a genetically tractable prokaryote would offer, we searched for a cyanobacterium that is amenable to molecular/genetic analyses and exhibits circadian rhythms of a parameter that can be assayed continuously for many cycles by an automated system.

The unicellular freshwater cyanobacterium *Synechococcus* sp. strain PCC 7942 was a good candidate for this approach. Although PCC 7942 does not fix nitrogen, it has many advantages for genetic analyses: It is transformable by circular or linear DNA, recombines at homologous sites, can receive DNA by conjugation from *E. coli* at high efficiency, can express reporter genes, and has a genome that is smaller than that of *E. coli* (12). Based on these characteristics, many genetic tools were developed (1).

We created a circadian reporter strain of *Synechococcus* sp. strain PCC 7942 by transforming it with a construct in which the *Vibrio harveyi* luciferase gene set *luxAB* is expressed under the control of the promoter for a *Synechococcus* photosystem II gene, *psbAI* (32). This reporter strain was named AMC149. The luminescence rhythm expressed by AMC149 in liquid cultures or from single colonies on agar medium is easily assayed by automated monitoring systems based on either photomultiplier tubes (1, 32) or CCD (charge coupled device) cameras (29, 33). The luminescence pattern conforms to all three salient properties of circadian rhythms: persistence in continuous conditions (LL) with a period close to 24 h, entrainability by light/dark (LD) signals, and temperature compensation (32). The luminescence rises during the day and falls during the night. We found that the luminescence rhythm is an accurate reporter of *psbAI* gene expression (36), confirming our expectation that this rhythm reflects circadian control over the promoter of the *psbAI* gene.

The Cyanobacterial Clock Is Unperturbed by Rapid Cell Division

Cultures of AMC149 that are growing with doubling times as rapidly as one division every 6–10 h continue to exhibit circadian rhythms of *psbAI* gene expression (31, 40). The amplitude of the luminescence patterns reflecting *psbAI* promoter activity is a function of the stage of the growth cycle for colonies on agar or for liquid cultures. For example, in early log phase, the luminescence patterns display a clear circadian rhythm that grows in amplitude exponentially (Figure 1). This

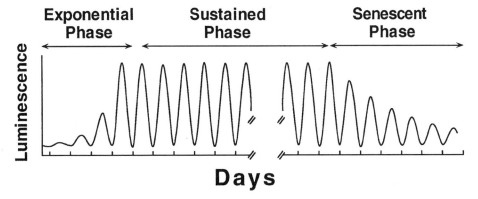

Figure 1   Rhythmic gene expression in the unicellular cyanobacterium *Synechococcus* sp. strain PCC 7942. Generalized circadian patterns of *psbAI* promoter activity monitored by luminescence of the reporter strain AMC149 are shown in three different phases of the growth cycle: exponential, sustained, and senescent.

exponential-phase pattern can be precisely modeled by assuming only two essential components: (*a*) a circadian rhythm of *psbAI* promoter activity in every cell and (*b*) an exponential increase in the number of cells in the culture (31). After the growth rate of the culture slows, the luminescence pattern stabilizes into a circadian pattern of consistent amplitude (sustained phase). It is in this growth phase that the circadian oscillator can be most easily assayed. As the culture ages, the rhythm slowly damps (senescent phase). The damping observed in the senescent phase probably results from nutrient depletion (including what is very important to the photosynthetic cyanobacteria—diminution of light intensity by high cell densities).

Not only can the cyanobacterial clock keep track of circadian time in exponentially dividing cells, it can also gate cell division. In *Synechococcus* sp. strain PCC 7942 growing in LD cycles, division occurs only during the day. In LL, however, division is blocked by the circadian clock in the early subjective night phases and is allowed in the subjective day and late subjective night (40). Therefore, not only does the clock keep track of circadian time in cells that divide two or three times a day, it controls when that division is allowed (gate open) or forbidden (gate closed). Using flow cytometry, we determined that the DNA replication rate and the growth in size of each individual cell is not rhythmic over the circadian cycle, but it is the timing of division that is controlled by the circadian program (40). In the photosynthetic marine prokaryote genus *Prochlorococcus*, there also appears to be circadian control of division in rapidly growing cells (49). Together, these data highlight the fallacy of the bias described in the second paragraph of this review; cellular events can waltz to a circadian andante even when the cells are rapidly dividing in allegro.

Global Orchestration of Gene Expression

How many genes are controlled by the circadian clock in cyanobacteria? In *Synechococcus* sp. RF-1, Huang et al (16) conducted a study of protein synthetic patterns as a function of circadian time and found >10 polypeptides that exhibit circadian rhythms of translation and are expressed in a variety of phase relationships. We therefore wondered how extensive circadian control of gene expression is in *Synechococcus* sp. strain PCC 7942 and devised a strategy to globally search for rhythmic control over promoters (38).

The bacterial luciferase gene set (*luxAB*) was inserted into the *Synechococcus* sp. strain PCC 7942 genome so as to achieve random insertions of *luxAB* throughout the chromosome (38). We screened the luminescence expression patterns from the ~800 clones whose luminescence was bright enough to be easily monitored. Unexpectedly, the luminescence expression patterns of essentially all of these 800 colonies manifested clear circadian rhythmicity. These rhythmic colonies exhibited a range of waveforms and amplitudes, and they also showed at least two predominant phase relationships. We defined Class-1 genes as those whose expression peaks at the end of the day and Class-2 genes as those peaking at the end of the

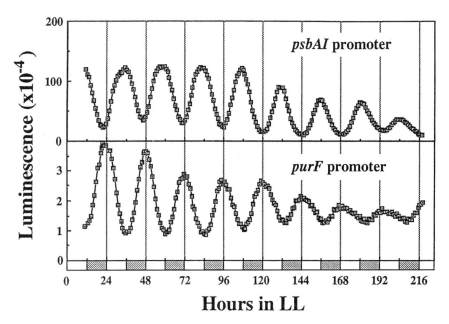

Figure 2   Luminescence patterns of the promoter activity of Class-1 and Class-2 genes as assayed with the *luxAB* reporter. The upper trace is from the Class-1 gene *psbAI*, whereas the bottom trace is from the Class-2 gene *purF*. Data were recorded in constant light (LL), but the times of expected or subjective night are shown by the gray bars along the abscissa (from 37).

night (38). This promoter-trap experiment shows that circadian programming of gene expression is pervasive in cyanobacteria.

Our original luminescent strain, which was genetically engineered with the promoter of *psbAI*, is a Class-1 gene (peak at dusk; trough at dawn), as shown in Figure 2. The transcriptional activity of the reverse-phase Class-2 genes, however, is maximal at about dawn and minimal at about dusk (Figure 2). We have identified one of the Class-2 genes as *purF*, which encodes a key regulatory enzyme in the de novo purine synthetic pathway, glutamine PRPP amidotransferase (37, 38). It is intriguing that glutamine PRPP amidotransferase is sensitive to oxygen, as is nitrogenase in *Synechococcus* sp. RF-1, and that both of the genes encoding these enzymes are expressed in the night or Class-2 phase. Therefore, even though *Synechococcus* sp. strain PCC 7942 does not fix nitrogen, we hypothesize that the Class-2 expression pattern of the *purF* gene is related to the oxygen sensitivity of glutamine PRPP amidotransferase and that it is another example of the same "temporal-separation" program exhibited by the expression of nitrogenase in some of the nitrogen-fixing cyanobacteria (25, 37).

Based on our results with the random promoter-trap experiment, we formulated a model that incorporates both nonspecific circadian control and circadian regulation by specific *trans* factors (38). In the simplest version of the model, there would be one (or a few) types of Class-1-specific *cis* elements turned on during the day by a Class-1-specific *trans* factor. A different set of Class-2-specific *cis* elements would be turned on at night by a Class-2-specific *trans* factor, and so on. Because of the large number of genes that are apparently influenced by the clock in cyanobacteria, however, it seems unlikely that each of them is controlled by a specific regulatory factor. Some global factor(s) could be involved as well. In support of our model for regulatory pathways that are specific for subsets of cyanobacterial genes, we discovered a gene whose altered expression significantly lowers the amplitude of the luminescence rhythm driven by some promoters (including that of *psbAI*), but not of luminescence rhythms driven by other promoters (such as that of *purF*). This gene encodes a sigma70-like transcription factor, *rpoD2*, and is a member of a family of sigma factor genes in *Synechococcus* sp. strain PCC 7942 (58). Apparently *rpoD2* is a component of an output pathway of the circadian oscillator that affects the rhythmic expression of a subset of clock-controlled genes in *Synechococcus* sp. strain PCC 7942.

ARE THESE CLOCK PROGRAMS ADAPTIVE?

Competition as a Test of Fitness

Although it is logical that circadian programming of the steps of gene expression, metabolic reactions, cell division, etc, could be adaptive, no rigorous test of the proposition had been performed in any organism, either eukaryotic or prokaryotic, prior to our recent work with cyanobacteria (41). We tested the value of circadian programming to reproductive fitness in cyanobacteria by using wild-type and mutant strains that exhibit different free-running periods (21, 33). The strains we used were C22a (period ~23 h), wild-type (period ~25 h), and C28a (period ~30 h). Both C22a and C28a have point mutations in the *kaiC* gene (see 21 and below). All three strains grew in pure culture at essentially the same rate in LL and in LD cycles, so there did not appear to be a significant advantage or disadvantage to having different circadian periods when the strains were grown in single-strain cultures (41).

However, when we mixed different strains together and grew them in competition with each other, a remarkable pattern emerged that depended on (*a*) the endogenous period of each strain and (*b*) the period of the LD cycle. We tested different LD regimens that had equal amounts of light and darkness but in which the frequency of the LD cycle differed: a 22-h cycle (LD 11:11), a 24-h cycle (LD 12:12), and a 30-h cycle (LD 15:15). In each case, the strain whose period most closely matched that of the LD cycle eliminated the competitor (41). Poor fitness was not necessarily associated with mutant phenotypes—in fact, the mutant strains easily defeated the wild-type if the mutant period was a better match to the LD cycle. These results were obtained for batch liquid cultures that were diluted every 8 days and for cultures maintained at constant cell density in a turbidostat (41).

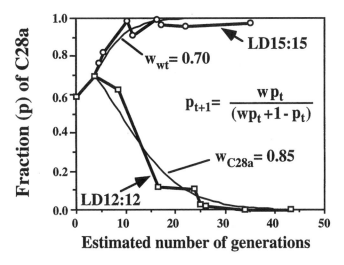

Figure 3 Kinetics of competition between wild-type (period ~25 h) and the mutant strain C28a (period ~30 h). Raw data (open-symbol points and heavy lines) of competitions in LD 12:12 vs LD 15:15 are compared with the results of modeling (thin lines) using the equation shown in the panel. (In LL, both strains do equally well.) Terms: w, relative fitness (w_{wt} for wild type, w_{C28a} for C28a); p, fraction of a given strain in the mixed population (p_t for generation t; p_{t+1} for generation t + 1). Ordinate is the percentage of colonies in the population that are C28a; abscissa is the estimated number of generations (from 41).

Estimation of the selective pressure by taking more time points suggested that the selective coefficient was surprisingly strong. As shown in Figure 3, a simple model that assumes (a) the selective pressure to be constant and (b) a lag in the initiation of selection suggested to a first approximation that the relative fitness of the less successful strain could be as low as 0.7– 0.8 (w_{wt} ~ 0.7 in LD15:15, w_{C28a} ~ 0.85 in LD12:12). Because the growth rates of the strains in pure culture are not different by a factor of 20–30%, the modeling result depicted in Figure 3 indicates that we are observing a case of soft selection in which the poorer fitness of inferior genotypes is most obvious under competition. We cannot rule out that small, presently unmeasurable differences exist between the growth rates of these strains in pure culture, but soft selection seems to be the predominant mechanism responsible for the strong selection under competition. The results are unlikely to be caused by an unrelated mutation that is deleterious for growth because (a) mutant strains can outgrow wild types in the appropriate combination of biological and environmental periods and (b) a genetic test confirmed that only the differences between the *kaiC* alleles of these strains are responsible for the competitive advantage/disadvantage (41).

How do the victorious strains win? Other than demonstrating that soft selection is operating, we do not yet know the mechanism of the selection. We do know that the phasing of *psbAI* gene expression is disrupted within strains in non-optimal LD cycles (41). This is consistent with the idea that the circadian program orders cellular processes to optimally match environmental cycles; when this order is disturbed, fitness is reduced. We conclude that the circadian pacemaker in cyanobacteria confers a significant competitive advantage when the period of the clock matches that of the environmental cycle, thus achieving optimal phasing of cellular events. This is the first rigorous demonstration in any organism of an advantage conferred by a circadian system to fitness.

ROLES OF LIGHT AND DARK

One of the big three properties of circadian clocks is their ability to be entrained to the environmental cycle. Entrainment means that the period of the biological clock becomes equal to that of the environmental cycle. Light and dark signals are usually considered to be the primary signals that set the phase of circadian clocks. In some organisms, the photopigments involved in circadian entrainment are known, such as phytochromes, rhodopsins, or cryptochromes (10, 23, 43, 51, 53, 56). In many cases, however, the identity of the relevant circadian photopigments remains a mystery (23, 43). The photopigments mediating the entrainment of cyanobacterial clocks fall into the latter group.

In *Synechococcus* sp. strain PCC 7942, we have attempted to glean clues as to the identity of clock photopigments by action spectroscopy. Our preliminary data suggest that blue and red light are most effective in setting the phase of the cyanobacterial clock, whereas green and far-red light are ineffective (T Kondo & C Johnson, unpublished observations). The phasing effect of red light was not reversed by far-red light, nor was the effect of blue light reversed by red light [as reported for light-regulated gene expression by Tsinoremas et al (57)]. The spectrum does not coincide with that expected for photosynthesis in cyanobacteria, nor does it fit with the behavior of a classically acting phytochrome. In *Synechococcus* sp. RF-1, red light of 680 nm was also found to be an effective phasing agent, and far-red light of 730 nm did not reverse red-light phasing (7). These data suggest that there are specific, unknown pigments that are the eyes of the clock of *Synechococcus* spp. strains PCC 7942 and RF-1.

Because almost all tests for circadian rhythmicity in cyanobacteria have used LL as the constant condition, it is appropriate to ask whether this circadian clock requires the presence of light to run. Most cyanobacteria are obligate photoautotrophs, including *Synechococcus* sp. strain PCC 7942. In constant darkness (DD), all luciferase reporter strains derived from PCC 7942 show rapidly damped oscillations in DD. Is this because the clock has stopped, or is it possible that the central timekeeper is still running and the outputs are turned off as an energy-saving device? One line of inquiry to answer this question in *Synechococcus* sp. strain PCC 7942 used light pulses that can reset the phase of the clock as probes of the pacemaker's phase in DD. The data suggested that the clock continued to run in this photoautotroph, even in DD (30).

Is the presence of light a necessity for the unimpeded precession of the clock, or is it merely that a certain metabolic rate must be maintained for the clock to express itself? This question has been addressed in cyanobacterial species that can grow heterotrophically. After a period of adaptation, *Synechocystis* sp. strain PCC 6803 and *Cyanothece* spp. can grow heterotrophically: on glucose for *Synechocystis* and on glycerol for *Cyanothece*. Using the *dnaK::luxAB* reporter strain of *Synechocystis* sp. growing on glucose, Aoki & coworkers were able to show that the luminescence rhythm persisted for many cycles in DD (3). Similarly, the rhythms of nitrogenase activity and carbohydrate content persisted robustly for >4 days in *Cyanothece* spp. grown on glycerol (48). Therefore, in these species of cyanobacteria it is clear that the clockwork can run in DD if metabolism is maintained and that the clock is not directly light dependent. It seems likely that, in the absence of other energy sources, DD is perceived as a signal to shut down unnecessary processes. But the central clock of photoautotrophic cyanobacteria appears to have favored status and continues to run in DD unlinked to its energy-guzzling outputs—rather like an automobile engine idling in neutral gear.

GENETIC DISSECTION OF THIS "CLOCKWORK GREEN"

The *kai* Clock Gene Cluster

Using the *PpsbAI::luxAB* reporter strain (AMC149) described above, we screened >500,000 clones of *Synechococcus* sp. strain PCC 7942 that had been treated with the mutagen ethylmethanesulfonate. Over 100 mutants exhibiting various circadian phenotypes, including arhythmia, altered waveforms, and atypical periods (ranging between 14- and 60 h) were isolated (33). Most of these mutants grow apparently as well as the wild type and exhibit no other obvious phenotype besides circadian anomalies.

Efficient rescue of mutant phenotypes is possible in *Synechococcus* spp. by the introduction of libraries of wild-type *Synechococcus* DNA, and we succeeded in rescuing >30 mutants of various phenotypes. DNA fragments from several rescued mutants complemented other mutant phenotypes, including short-period, long-period, and arhythmic phenotypes. These rescue experiments allowed us to pinpoint a cluster of three adjacent genes, named *kaiA, kaiB,* and *kaiC* (21; kai means "rotation" or "cycle" in Japanese). All of the mutants that have been complemented so far can be rescued by a plasmid carrying the entire *kaiABC* cluster, and 19 mutations were mapped by DNA sequencing to the three *kai* genes. All are missense mutations resulting from single nucleotide exchanges. Most of the mutations are recessive (rescue by wild-type DNA is complete), but a few are semidominant, such as that of the 60-h-period mutant C60a. Each of the three genes has at least two clock mutations mapped to it, and the largest gene, *kaiC*, has many mutations that include all the possible clock phenotypes: short period, long period, low amplitude, and arhythmia. No significant similarity was found among the *kai* genes and any other previously reported genes in prokaryotes or eukaryotes, except that there is a possible homolog of the *kaiC* gene among unidentified open reading frames in the genomic sequences of archaebacteria (see below). Moreover, there are two P-loop motifs in the *kaiC* gene. This motif, [G or A]XXXXGK[T or S], is a GTP/ATP nucleotide-binding region (46). The *kaiABC* cluster appears to be a clock-specific region of the chromosome in cyanobacteria, because deletion of the entire cluster or of any one of the *kai* genes separately does not affect viability (in single-strain cultures), but does cause arhythmicity (21).

Figure 4 summarizes much of our knowledge about the relationships among these components (21). Promoter activities were found in the upstream regions of both the *kaiA* and *kaiB* genes. The *kaiA* promoter gives rise to a monocistronic

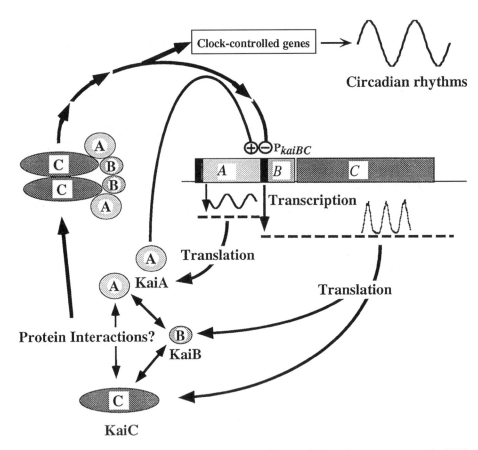

Figure 4   Feedback model for the circadian oscillator of *Synechococcus* sp. strain PCC 7942. Feedback of KaiA (positive) and a possible KaiABC complex (negative) on the *kaiBC* promoter is depicted. The *kaiA* gene is transcribed as a single transcript, whereas the *kaiB* and *kaiC* genes are cotranscribed. Both the *kaiA* and *kaiBC* promoters are rhythmically active (figure modified from 21). Because the KaiA, KaiB, and KaiC proteins interact, a multimeric complex is shown (for convenience, homodimers are shown interacting to form a larger complex).

kaiA mRNA, whereas the *kaiB* promoter produces a dicistronic *kaiBC* mRNA. Both *kaiA* and *kaiBC* transcripts are rhythmically abundant. Inactivation of any single *kai* gene abolished these rhythms and lowered *kaiBC* promoter activity. Continuous overexpression of *kaiC* repressed the *kaiBC* promoter (negative feedback), whereas *kaiA* overexpression enhanced it (positive feedback). Overexpression of the *kaiC* gene for a few hours reset the phase of the rhythms. Consequently, the level of KaiC expression is directly linked to the phase of the oscillation.

One clue to the mechanism of this clockwork is that the Kai proteins appear to interact. Yeast two-hybrid and in vitro binding assays indicate that the KaiA, KaiB, and KaiC proteins interact both homotypically and heterotypically (22). One long period mutant exhibits an altered heterotypic interaction between KaiA and KaiB; this result suggests that inter-Kai contact is important to the clock mechanism (22). Using a novel method based on resonance energy transfer to assay protein-protein interactions, we have confirmed that KaiB polypeptides interact (59). Taken together, these results suggest that there is negative feedback control of *kaiC* expression by the KaiC protein to generate a circadian oscillation in cyanobacteria involving protein-protein interactions and that KaiA sustains the oscillation by enhancing *kaiC* expression.

The model is still very preliminary, however. It is likely that this feedback model—which borrows many of its features from clock models of eukaryotes—is oversimplified. For example, the model implies that the Kai protein levels are rhythmic and that changes in Kai protein levels within the physiological range will elicit phase resetting. Mutations of the genes encoding Kai proteins that affect protein-protein interactions should alter circadian properties. Does KaiC bind ATP and/or GTP, and, if so, does modification of the P-loop motif disrupt normal time-keeping? Are there other molecular players, such as Kai-interacting transcriptional factors? These and other predictions demand extensive testing.

The *kai* genes apparently encode products that are crucial for the activity of a clockwork in cyanobacteria. But is this the only clock mechanism in cyanobacterial cells? This question cannot be dismissed lightly, because multiple clock mechanisms have been discovered in the unicellular eukaryotic alga *Gonyaulax* spp. (44). So far, no concrete evidence indicates the presence of more than one clock mechanism in *Synechococcus* spp. For example, most mutations that affect the period of one reporter strain similarly affect the period of other reporter strains; for example, the periods of the *PpsbAI::luxAB, PpurF::luxAB,* and *PkaiBC::luxAB* reporter strains respond similarly to mutations in *kaiABC* (21; C Johnson & S Golden, unpublished data). Nevertheless, we must remain alert to the possibility that new data may indicate that even in simple prokaryotic unicellular organisms, multiple circadian oscillators could coexist.

Other Clock Genes?

One of the questions posed in the previous section was whether there are more genes involved in the circadian mechanism of *Synechococcus* sp. strain PCC 7942. Although we have achieved saturation mutagenesis with ethylmethanesulfonate, we do not have enough pieces to fit the puzzle together yet. For example, if feedback regulation of the *kai* cluster is crucial, then what are the transcription factors? None of the Kai proteins has a DNA-binding motif. It is possible that a Kai protein complex interacts with a general transcription factor. We might not have discovered this factor in our screens to date if it is an essential transcription factor that confers a lethal phenotype when mutated. To identify other components in the circadian clockwork, imaginative new screens (and possibly new mutagens) need to be tested. For example, transposon mutagenesis has identified a histidine protein kinase gene whose inactivation causes a short-period phenotype (O Schmitz & SS Golden, unpublished observations). Another tactic is to search for Kai-interacting proteins by a yeast two-hybrid (or other) screen.

Yet another strategy to uncover new genes that encode components of the clockwork is to isolate extragenic suppressors. One suppressor that emerged from the initial attempts to rescue clock mutations is the period-extender gene, *pex* (34). When the copy number of the *pex* gene is increased, *pex* lengthens the period of wild-type and other strains by 2 h. Disruption of *pex* shortens the period by 1 h, and overexpression lengthens the period by up to 3 h. No meaningful homologs to *pex* were detected in DNA or protein databases (34). Screens that have been specifically designed to find other extragenic suppressors (e.g. by mutation of clock mutants to find altered phenotypes) have not yielded non-*kai* candidates, but the hunt continues.

EVOLUTIONARY ASPECTS

Evolution of the *kaiC* Gene

We have found strong candidates for homologs to the *kaiC* gene in the genomic sequences of the archaea *Methanobacterium thermoautotrophicum* (50), *Methanococcus jannaschii* (4), *Pyrococcus horikoshii* (27), and *Archaeoglobus fulgidus* (28). As shown in Figure 5 in which the archaebacterial sequences are aligned roughly end to end with those of the *kaiC* gene from the cyanobacteria *Synechococcus* sp. (21) and *Synechocystis* sp. (26), the amino acid sequence identities between the cyanobacteria and these archaebacteria range between 25.7% and 34.2% over the entire length of the proteins. More important, there is conservation of the P-loop motif GXXXXGK[T or S], which is a GTP/ATP nucleotide-binding region (46) and is indicated on Figure 5 by the double underlines. This motif appears to be highly conserved among the *kaiC* genes of the cyanobacterial species and the putative *kaiC* homologs of the archaea. The *kaiC* homolog of *Methanococcus* sp. is shorter than the others, and the P-loop motif appears only once in the *Methanococcus* sequence, whereas it appears twice in the sequences of all the other genes.

Prompted by considerations relating to the endosymbiotic theory of the origin of eukaryotic organelles, we searched the databases for possible *kai* homologs in eukaryotes. At present, no encouraging candidates have appeared. In particular, there are no apparent homologs in the chloroplast genomes of tobacco or other higher plants, nor is there any significant hybridization between *kai* DNA and the chloroplast genome of the eukaryotic alga *Chlamydomonas* (Y Xu, S Surzycki & CH Johnson, unpublished observations). Therefore, there are no data that yet encourage the view that the clockwork that evolved in cyanobacteria was transferred to eukaryotes by endosymbiosis or another mechanism of gene transfer.

Are There Circadian Timepieces in Other Prokaryotes?

As discussed in the first section of this review, it used to be thought that prokaryotic organization and lifestyle were incompatible with circadian clocks. As addressed herein, we now realize that circadian programming is an integral part of cyanobacterial organization and that this temporal program has adaptive value. But what about non-photosynthetic prokaryotes? This is a challenging question, with momentous implications for understanding the early evolution of circadian rhythmicity. The previous suggestion that *E. coli* might have a circadian clock (15) was based on an old study (45) in which *E. coli* cells were grown in rich nutrient medium in an apparatus whose temperature control was almost certainly poor. Therefore, the possible daily trends noted by Halberg & Conner (15) are likely to be merely a result of a diurnal cycle of temperature, to which the growth rate of *E. coli* would be exquisitely sensitive. (Incidentally, there are no kai homologs in the genome of *E. coli*.) A later study that was more careful to control temperature purported to discover circadian rhythms of growth rate in *Klebsiella* spp. (54), but a circadian trend is not obvious from the raw data and requires extensive statistical analyses to uncover. Moreover, the only circadian property addressed in that study was an ~24-h oscillation; the other salient properties—temperature compensation and entrainment—were neglected. At this writing, there is no persuasive evidence that circadian clocks reside in prokaryotes other than cyanobacteria.

Nonetheless, we think that there is a good chance that circadian clocks will be found in other eubacteria for which a daily timekeeper will enhance fitness. The keys are to find the proper conditions and to choose an appropriate parameter to measure. What about the third great domain of biology, the *Archaea*? Despite the fact that the ecological niche occupied by some archaebacteria, e.g. the halobacteria, is not very different from that occupied by cyanobacteria [i.e. both live in an aqueous habitat and derive their energy from the daily sunlight cycle (in the case of halobacteria, under anaerobic conditions)], there is no direct evidence that archaebacteria have invented an endogenous daily clock. In fact, it is not clear that anyone has yet looked seriously for circadian clocks in archaebacteria. The existence of putative *kaiC* homologs in *Archaea* encourages a concerted hunt for clocks, but we must also caution that the *Archaea* in which a *kaiC* homolog has been found are methanogens and/or extremophiles that inhabit environments for which an endogenous 24-h clock would not appear to be adaptive (e.g. deep-sea methane vents, hot springs, etc). Discovering circadian clocks in archaebacteria could help us to understand much about the evolution of circadian rhythms.

```
                        10         20         30         40         50         60         70         80         90
Synechococcus     ---------- MTSAEMTSPN N-NSEHQAIA KMRTMIBGFD D-ISHGGLPI GRSTLVSGTS GTGKTLFSIQ FLYNGIIEFD EPGVFVTFEE
Synechocystis     ---------- MNLPIVNERN RPDVPRKGVQ KIRTVIBGFD E-ITHGGLPI GRTTLVSGTS GTGKTLLAVQ FLYQGIHHFD YPGLFITFEE
Archaeoglobus     ---------- --------ME AVKTGIBGFD D--IFGGFYR GQIILLAGNP GSGKTTFCAK FLYBGARRFG ENGLYISIGE
Pyrococcus        MLLIVGTPFN HIYDLSKDLK YPTKTKLRGK MTKFGIEYFD KYILRGGFPN GSMILIVGEP GTGKTILSAT YLYNGAKKFG EKGMYISLAE
Methanobacterium  ---------- ---MDFENIE KAPTGIKGLD M-ITEGGFPR GRNTLIYGGP GTGKTFIAME EPGVMVSFDE
Methanococcus     ---------- --------MK RVKTGIPGMD E-ILHGGIPE RNVVLLSGGP GTGKSIFCQQ FLYKGVVDYN EPSILVALEE
                                                   *** * ***    **  * *****                   *

                       100        110        120        130        140        150        160        170        180
Synechococcus     TPQDIIKNAR SFGWDLAKLV DEGKLFILDA SP-------D PEGQEVVGGF DLSALIERIN YAIQKYRARR VSIDSVTSVF QQYDASSVVR
Synechocystis     SPSDIIENAY SFGWDLQQLI DDGKLFILDA SP-------D PEGQEVVGTF DLSALIERIQ YAVRKYKAKL VSIDSVTAVF QQYDAASVVR
Archaeoglobus     SKEEFYEYMK KLGMDFEELE KTGSFKYVEM LAP------- -----TSED ALMQLSRELT KNALELKATR IVIDSISPLL SMNP-ETARA
Pyrococcus        TKWEFYKGMK QLGMDFEELE NNGLFKFIDL VT-------- -----VSKD AVEKEIELLM SEISSFHPKR IVIDSISVFT QLLGAEKTRI
Methanobacterium  AMENLIENFR SSDRLLERLI DEGLLFIEDA SRG-----MD PD----AGSY SLEALKLRLE DAIRRTGARR VVLDKVDNLF DGTERQGMIG
Methanococcus     HPVQIRENMR QFGWDIRKLE EEGKFAIIDA FTYGIGSAAK REKYVVNDPN DERELIDVLK TAINDIGAKR IGIDSVTTLY INKP-MLAR
                            *  *     *     ** *       ** *            ** *            *****             **

                       190        200        210        220        230        240        250        260        270
Synechococcus     RELFRLVARL KQIGATTVMT TERIEEYGPI ARVGVEEFVS DNVVLRNVL EGE-RRRRTL EILKLRGTSH MKGEYPFTIT D-HGINIFPL
Synechocystis     REIFRLVARL KQLQVTSIMT TERVEEYGPI ARFGVEEFVS DNVVLRNVL EGE-RRRRTV EILKLRGTTH MKGEYPFTIT H-DGINIFPL
Archaeoglobus     ILHNALKTIS RELKSVVLMT EEMPIGETRI G-QGIEEFVV DGVIVLRLEV PEAGAPVRTM SVLKLRGKPL DRAVVNFEIG PPSGVRVLMH
Pyrococcus        FLHTILGRFI KANDATALLI AEKPIGSKTI G-YGVEEFVV DGVIILKYKS FGE-VTRRVM EIPKMRRRKI ERAEYEYVIT K-NGIEFLTI
Methanobacterium  FELRELIRWL NGMGVTSVFT SG---DYGGS PTHRLEEYIS DCVIHLHTF EGQ-VGTRHL RIVKYRGSGH GLNRYPFIIT R-RGASIFPI
Methanococcus     RTVFLLKRVI SGLGCTAIFT SQISVGERGF GGPGVEHAVD GIIRLDIGH DGE-LK-RSL IVWKMRGTSH SLKRHPFDIT N-GEITVYPD
                      *   ***   **     *   **   *        **    **        *
```

Figure 5 Deduced amino acid sequences of the products of *kaiC* gene from *Synechococcus* sp. strain PCC 7942 and its putative homologs from the cyanobacterium *Synechocystis* sp. strain PCC 6803 and the archaea *Archaeoglobus fulgidus*, *Pyrococcus horikoshii*, *Methanobacterium thermoautotrophicum*, and *Methanococcus jannaschii* (data from References 4, 21, 26–28, 50). An asterisk underlies residues that are identical in the majority of these proteins. Putative ATP/GTP-binding site motifs (P-loop, GXXXXGKT/S) are indicated by a double underline.

```
                       280        290        300        310        320        330        340        350        360
Synechococcus      GAMRLTQR-S SNVRVSSGVV RLDEMCGGGF FKDSIILATG ATGTGKTLLV SRFVENACAN KERAILFAYE ESRAQLLRNA YSWGMDFEEM
Synechocystis      GAMRLTQR-S SNARISSGVQ TLDEMCGGGF FKDSIILATG ATGTGKTLLV SKFLQEGCRQ RERAILFAYE ESRAQLSRNA SSWGIDFFEM
Archaeoglobus      GIEELESNID FNNKIATGID GFDELLGGGI IRGTATAFVG PSGGGKTVLM LSAAANVAIN GENVTYISFE EPRQQIEETL KFLG--YGEV
Pyrococcus         PELKISEKDY TSEKITTGIE RLDEMLGGGI YKGSSVLIVG MTGTGKTTFS LHFAIANALQ GRKVAYITFE EPIDQIIRSA KNYGMPIEEV
Methanobacterium   TSISLDYS-V SRELVSTGIP TLDEMLGGGV YRGSAVLVSG TTGAGKTSLL SKFAYESCRR GERCLFFSNE EPADQIVRNM ESIGIKLGEF
Methanococcus      KVLKLR----
                      *           * *   **** ***  * * *   ** *** *    *          *          *           **

                       370        380        390        400        410        420        430        440        450
Synechococcus      ERQNLLKIVC AYPESAGLED HLQIIKSEIN DFKPARIAID SLSALARG-- ---VSNNAFR QFVIGVTGYA KQEEITGLFT NTSDQFMGAH
Synechocystis      EHKGLLKLLC TYPESAGLED HLQMIKSEIS EFKPSRIAID SLSALARG-- ---VTNNAFR QFVIGVTGYA KQEEITGFFT NTTDQFMGAH
Archaeoglobus      EG---LEILS LNPRMISLRA LYDILSKTVL DHR-TMLFID GLNAIRRE-- ----FGEAFH RVRDVVFQM  KKNGITVVIS LIGGTIKET-
Pyrococcus         LG-RNLKIFA WVPESKTPVH TYIKIKELIE EFKPEALIID SLTALREH-- ---LDEKELT KMLRYLQLLT KEKRITTYFT FTEEAKFDV-
Methanobacterium   LG-DKLLIHS DRPTSLGLEA HLTVMQDLIM DFNPDSVLVD PVTGLAGAGG SPETRNENAK HLFIRLTDFL KGRGITSIFS YLIASPFTA-
Methanococcus      ----------
                      * *        * * **      * * **   *  *  * *  *      *                * **          * **       *

                       460        470        480        490        500        510        520        530        540
Synechococcus      SITDSHISTI TDTIILLQYV EIRGEMSRAI NVFKMRGSWH DKAIREFMIS DKGPDIKDSF RNFERIISGS PTRITVDEKS ELSRIVRGVQ
Synechocystis      SITESHISTI TDTILMLQYV EIRGEMSRAI NVFKMRGSWH DRGIREYSIS HDGPDIRDSF RNYERIISGS PTRISVDEKS ELSRIVRGVK
Archaeoglobus      ---LLSTI   VDNVVELRVV EKDGELRREI AVRKARMSRA SNEVKRLVFD -GKPAVR--- ---------- ---------- ----------
Pyrococcus         -VPFTGASTM VDVIILLKYE IENEKIEKRL AIIKARGSNH SRKIHKYEIT NRGVEIYE-- ---------- ----------W- ----------
Methanobacterium   TTTELKLSSL IDTWIVLESI RANGEYRRSL RILKSRGMNH SSSVAEVRFT DRGILIKG-- -------GS- ---------- ----------
Methanococcus      ----------
                      **          *           * *     ****       *               *                     **    * **

                       550
Synechococcus      EKGPESZ
Synechocystis      DKTAE--
Archaeoglobus      ------
Pyrococcus         ------
Methanobacterium   ------
Methanococcus      ------
```

Figure 5 *(continued)*

CONCLUDING REMARKS

Just as fruitfly pupae metamorphose and eclose as winged adults, so has the circadian clock field undergone a dramatic transformation in the past 10 years. Much of that transformation is caused by the rapid progress on identifying clock components in *Drosophila melanogaster*, *Neurospora crassa*, plants, and mammals. More subtle transformations, but no less profound, include the dislodging of several entrenched dogmas, including the idée fixe that prokaryotes were too simple to have complex timepieces and that rapidly growing cells cannot keep track of circadian time. Studies of cyanobacteria have single-handedly disposed of these dogmas and have also provided the first rigorous demonstration of the fitness value of circadian programming. We expect that, with the shattering of the prokaryotic barrier, new insights into the distribution, mechanism, adaptiveness, and evolution of circadian clocks will be forthcoming.

ACKNOWLEDGMENTS

We thank the members of our laboratories who allowed us to refer to their results before publication and to our long-term collaborators, Dr. Takao Kondo and Dr. Masahiro Ishiura, for a delightful and productive collaboration. Our research has been supported by grants from the NIMH (MH43836 and MH01179 to CHJ) and NSF (MCB-9219880 and MCB 9513367 to CHJ; MCB-9311352 and MCB-9513367 to SSG). NSF and JSPS supported a joint USA-Japan Co-operative Program grant for travel among our labs and those of Drs. Kondo and Ishiura (NSF INT9218744 and JSPS BSAR382). The Human Frontier Science Program funds collaborative research of our labs with those of Drs. Kondo and Ishiura (grant #RG-385/96).

Visit the Annual Reviews home page at http://www.AnnualReviews.org

LITERATURE CITED

1. Andersson CR, Tsinoremas NF, Shelton J, Lebedeva NV, Yarrow J, et al. 1999. Application of bioluminescence to the study of circadian rhythms in cyanobacteria. *Methods Enzymol.* In press
2. Aoki S, Kondo T, Ishiura M. 1995. Circadian expression of the *dnaK* gene in the cyanobacterium *Synechocystis* sp. strain PCC 6803. *J. Bacteriol.* 177:5606–11
3. Aoki S, Kondo T, Wada H, Ishiura M. 1997. Circadian rhythm of the cyanobacterium *Synechocystis* sp. strain PCC 6803 in the dark. *J. Bacteriol.* 179:5751–55
4. Bult CJ, White O, Olsen GJ, Zhou L, Fleischmann RD, et al. 1996. Complete genome sequence of the methanogenic archaeon, *Methanococcus jannaschii*. *Science* 273: 1058–73
5. Capone DG, Zehr JP, Paerl HW, Bergman B, Carpenter EJ. 1997. *Trichodesmium*, a globally significant marine cyanobacterium. *Science* 276:1221–29
6. Chen T-H, Chen T-L, Hung L-M, Huang T-C. 1991. Circadian rhythm in amino acid uptake by *Synechococcus* RF-1. *Plant Physiol.* 97:55–59

7. Chen T-H, Pen S-Y, Huang T-C. 1993. Induction of nitrogen-fixing rhythm *Synechococcus* RF-1 by light signals. *Plant Sci.* 92:179–82

8. Chen Y-B, Dominic B, Mellon MT, Zehr JP. 1998. Circadian rhythm of nitrogenase gene expression in the diazotrophic filamentous nonheterocystous cyanobacterium *Trichodesmium* sp. strain IMS 101. *J. Bacteriol.* 180:3598–605

9. Colon-Lopez MS, Sherman DM, Sherman LA. 1997. Transcriptional and translational regulation of nitrogenase in light-dark- and continuous-light-grown cultures of the unicellular cyanobacterium *Cyanothece* sp. strain ATCC 51142. *J. Bacteriol.* 179:4319–27

10. Emery P, So WV, Kaneko M, Hall JC, Rosbash M. 1998. CRY, a *Drosophila* clock and light-regulated cryptochrome, is a major contributor to circadian rhythm resetting and photosensitivity. *Cell* 95:669–79

11. Gallon JR. 1992. Reconciling the incompatible: N_2 fixation and O_2. *New Phytol.* 122:571–609

12. Golden SS, Ishiura M, Johnson CH, Kondo T. 1997. Cyanobacterial circadian rhythms. *Annu. Rev. Plant Physiol. Plant Mol. Biol.* 48:327–54

13. Grobbelaar N, Huang T-C, Lin H-Y, Chow T-J. 1986. Dinitrogen-fixing endogenous rhythm in *Synechococcus* RF-1. *FEMS Microbiol. Lett.* 37:173–77

14. Grobbelaar N, Lin H-Y, Huang T-C. 1987. Induction of a nitrogenase activity rhythm in *Synechococcus* and the protection of its nitrogenase against photosynthetic oxygen. *Curr. Microbiol.* 15:29–33

15. Halberg F, Conner RL. 1961. Circadian organization and microbiology: variance spectra and a periodogram on behavior of *Escherichia coli* growing in fluid culture. *Proc. Minn. Acad. Sci.* 29:227–39

16. Huang T-C, Chen H-M, Pen S-Y, Chen T-H. 1994. Biological clock in the prokaryote *Synechococcus* RF-1. *Planta* 193:131–36

17. Huang T-C, Chow T-J. 1990. Characterization of the rhythmic nitrogen-fixing activity of *Synechococcus* sp. RF-1 at the transcriptional level. *Curr. Microbiol.* 20:23–26

18. Huang T-C, Grobbelaar N. 1995. The circadian clock in the prokaryote *Synechococcus* RF-1. *Microbiology* 141:535–40

19. Huang T-C, Tu J, Chow T-J, Chen T-H. 1990. Circadian rhythm of the prokaryote *Synechococcus* sp. RF-1. *Plant Physiol.* 92:531–33

20. Huang T-C, Wang S-T, Grobbelaar N. 1993. Circadian rhythm mutants of the prokaryotic *Synechococcus* RF-1. *Curr. Microbiol.* 27:249–54

21. Ishiura M, Kutsuna S, Aoki S, Iwasaki H, Andersson CR, et al. 1998. Expression of a gene cluster *kaiABC* as a circadian feedback process in cyanobacteria. *Science* 281:1519–23

22. Iwasaki H, Taniguchi Y, Kondo T, Ishiura M. 1999. Physical interactions among circadian clock proteins, KaiA, KaiB and KaiC, in cyanobacteria. *EMBO J.* 18:1137–45

23. Johnson CH. 1995. Photobiology of circadian rhythms. In *CRC Handbook of Organic Photochemistry and Photobiology*, ed. WM Horspool, P-S Song, pp. 1602–10. Boca Raton, FL: CRC

24. Johnson CH, Golden SS, Ishiura M, Kondo T. 1996. Circadian clocks in prokaryotes. *Mol. Microbiol.* 21:5–11

25. Johnson CH, Golden SS, Kondo T. 1998. Adaptive significance of circadian programs in cyanobacteria. *Trends Microbiol.* 6:407–10

26. Kaneko T, Sato S, Kotani H, Tanaka A, Asamizu E, et al. 1996. Sequence analysis of the genome of the unicellular cyanobacterium *Synechocystis* sp. strain PCC6803. II. Sequence determination of the entire genome and assignment of potential protein-coding regions. *DNA Res.* 3:109–36

27. Kawarabayasi Y, Sawada M, Horikawa H, Haikawa Y, Hino Y, et al. 1998. Complete

sequence and gene organization of the genome of a hyper-thermophilic archaebacterium, *Pyrococcus horikoshii* OT3. *DNA Res.* 5:147–55

28. Klenk HP, Clayton RA, Tomb J-F, White O, Nelson KE, et al. 1997. The complete genome sequence of the hyperthermophilic, sulphate-reducing archaeon *Archaeoglobus fulgidus. Nature* 390:364–70

29. Kondo T, Ishiura M. 1994. Circadian rhythms of cyanobacteria: monitoring the biological clocks of individual colonies by bioluminescence. *J. Bacteriol.* 176:1881–85

30. Kondo T, Ishiura M, Golden SS, Johnson CH. 1994. Circadian rhythms of cyanobacteria expressed from a luciferase reporter gene. In *Evolution of Circadian Clock*, ed. T Hiroshige, K Honma, pp. 59–73. Sapporo, Jpn.: Hokkaido Univ. Press

31. Kondo T, Mori T, Lebedeva NV, Aoki S, Ishiura M, Golden SS. 1997. Circadian rhythms in rapidly dividing cyanobacteria. *Science* 275:224–27

32. Kondo T, Strayer CA, Kulkarni RD, Taylor W, Ishiura M, et al. 1993. Circadian rhythms in prokaryotes: luciferase as a reporter of circadian gene expression in cyanobacteria. *Proc. Natl. Acad. Sci. USA* 90:5672–76

33. Kondo T, Tsinoremas NF, Golden SS, Johnson CH, Kutsuna S, Ishiura M. 1994. Circadian clock mutants of cyanobacteria. *Science* 266:1233–36

34. Kutsuna S, Kondo T, Aoki S, Ishiura M. 1998. A period-extender gene, *pex*, that extends the period of the circadian clock in the cyanobacterium *Synechococcus* sp. strain PCC 7942. *J. Bacteriol.* 180:2167–74

35. Lee DY, Rhee GY. 1999. Circadian rhythm in growth and death of the cyanobacterium *Anabaena flos aquae. J. Phycol.* In press

36. Liu Y, Golden SS, Kondo T, Ishiura M, Johnson CH. 1995. Bacterial luciferase as a reporter of circadian gene expression in cyanobacteria. *J. Bacteriol.* 177:2080–86

37. Liu Y, Tsinoremas NF, Golden SS, Kondo T, Johnson CH. 1996. Circadian expression of genes involved in the purine biosynthetic pathway of the cyanobacterium *Synechococcus* sp. strain PCC 7942. *Mol. Microbiol.* 20:1071–81

38. Liu Y, Tsinoremas NF, Johnson CH, Lebedeva NV, Golden SS, et al. 1995. Circadian orchestration of gene expression in cyanobacteria. *Genes Dev.* 9:1469–78

39. Mitsui A, Kumazawa S, Takahashi A, Ikemoto H, Arai T. 1986. Strategy by which nitrogen-fixing unicellular cyanobacteria grow photoautotrophically. *Nature* 323:720–22

40. Mori T, Binder B, Johnson CH. 1996. Circadian gating of cell division in cyanobacteria growing with average doubling times of less than 24 hours. *Proc. Natl. Acad. Sci. USA* 93:10183–88

41. Ouyang Y, Andersson CR, Kondo T, Golden SS, Johnson CH. 1998. Resonating circadian clocks enhance fitness in cyanobacteria. *Proc. Natl. Acad. Sci. USA* 95:8660–64

42. Roenneberg T, Carpenter EJ. 1993. Daily rhythm of O_2-evolution in the cyanobacterium *Trichodesmium thiebautii* under natural and constant conditions. *Mar. Biol.* 117:693–97

43. Roenneberg T, Foster RG. 1997. Twilight times: light and the circadian system. *Photochem. Photobiol.* 66:549–61

44. Roenneberg T, Morse D. 1993. Two circadian oscillators in one cell. *Nature* 362:362–64

45. Rogers LA, Greenbank GR. 1930. The intermittent growth of bacterial cultures. *J. Bacteriol.* 19:181–90

46. Saraste M, Sibbald PR, Wittinghofer A. 1990. The P-loop—a common motif in ATP- and GTP-binding proteins. *Trends Biochem. Sci.* 15:430–34

47. Schneegurt MA, Sherman DM, Nayar S, Sherman LA. 1994. Oscillating behavior

of carbohydrate granule formation and dinitrogen fixation in the cyanobacterium *Cyanothece* sp. strain ATCC 51142. *J. Bacteriol.* 176:1586–97

48. Schneegurt MA, Sherman DM, Sherman LA. 1997. Growth, physiology, and ultrastructure of a diazotrophic cyanobacterium, *Cyanothece* sp. strain ATCC 51142, in mixotrophic and chemoheterotrophic cultures. *J. Phycol.* 33:632–42

49. Shalapyonok A, Olson RJ, Shalapyonok LS. 1998. Ultradian growth in *Prochlorococcus* spp. *Appl. Environ. Microbiol.* 64:1066–69

50. Smith DR, et al. 1997. Complete genome sequence of *Methanobacterium thermoautotrophicum* ΔH: functional analysis and comparative genomics. *J. Bacteriol.* 179:7135–55

51. Somers DE, Devlin PF, Kay SA. 1998. Phytochromes and cryptochromes in the entrainment of the Arabidopsis circadian clock. *Science* 282:488–90

52. Stal LJ, Krumbein WE. 1985. Nitrogenase activity in the non-heterocystous cyanobacterium *Oscillatoria* sp. grown under alternating light-dark cycles. *Arch. Microbiol.* 143:67–71

53. Stanewsky R, Kaneko M, Emery P, Beretta B, Wager-Smith K, et al. 1998. The *cry*[b]

mutation identifies cryptochrome as a circadian photoreceptor in *Drosophila*. *Cell* 95:681–92

54. Sturtevant RP. 1973. Circadian variability in *Klebsiella* demonstrated by cosinor analysis. *Int. J. Chronobiol.* 1:141–46

55. Sweeney BM, Borgese MB. 1989. A circadian rhythm in cell division in a prokaryote, the cyanobacterium *Synechococcus* WH7803. *J. Phycol.* 25:183–86

56. Thresher RJ, Vitaterna MH, Miyamoto Y, Kazantsev A, Hsu DS, et al. 1998. Role of mouse cryptochrome blue-light photoreceptor in circadian photoresponses. *Science* 282:490–94

57. Tsinoremas NF, Schaefer MR, Golden SS. 1994. Blue and red light reversibly control *psbAI* expression in the cyanobacterium *Synechococcus* sp. strain PCC 7942. *J. Biol. Chem.* 269:16143–47

58. Tsinoremas NF, Ishiura M, Kondo T, Andersson CR, Tanaka K, et al. 1996. A sigma factor that modifies the circadian expression of a subset of genes in cyanobacteria. *EMBO J.* 15:2488–95

59. Xu Y, Piston D, Johnson CH. 1999. A bioluminescence resonance energy transfer (BRET) system-application to interacting circadian clock proteins. *Proc. Natl. Acad. Sci. USA* 96:15–16

Annu. Rev. Microbiol. 1999. 53:411–46

CONSTRUCTING POLYKETIDES: From Collie to Combinatorial Biosynthesis

Ronald Bentley

Department of Biological Sciences, University of Pittsburgh, Pittsburgh, Pennsylvania 15260; e-mail: rbentley@pitt.edu

J. W. Bennett

Department of Cell and Molecular Biology, Tulane University, New Orleans, Louisiana 70118; e-mail: Jbennett@mailhost.tcs.tulane.edu

Key Words   genetic engineering, hybrid polyketides, polyacetate hypothesis, polyketide synthases, secondary metabolites

■ **Abstract**   In a new golden age, polyketides are investigated and manipulated with the tools of molecular biology and genetics; hybrid polyketides can be produced. Pharmaceutical companies hope to find new and useful polyketide products, including antibiotics, anthelminthics, and immunosuppressants. This review describes the past developments (largely chemical) on which the present investigations are based, attempts to make sense of the expanding scope of polyketides, looks at the shifting research focus around polyketides, presents a working definition in biosynthetic terms, and takes note of recent work in combinatorial biosynthesis. Also discussed is the failure of the classical enzymological approach to polyketide biosynthesis.

CONTENTS

0066-4227/99/1001-0411$08.00

INTRODUCTION

Broadly defined, "natural product" means any organic material of biological (as opposed to synthetic) origin; in practice, the term is usually restricted to secondary metabolites. The latter term was originally used by plant physiologists to classify botanicals (e.g. dyes, fragrances, and medicinals) with no obvious function in the plants that produced them. It now encompasses a heterogeneous group of compounds, usually of low relative molecular mass, and made mostly but not exclusively by organisms without a nervous system (i.e. bacteria, fungi, and plants). The notion of secondary metabolism was embraced by microbiologists in the 1960s (30), with attention focusing on antibiotics and other bioactive microbial products (12, 31, 37, 46, 89, 108, 124, 141). Although largely interchangeable, to this day the rubric "natural product" has a more chemical connotation than "secondary metabolite."

Polyketides are secondary metabolites par excellence. The terms were coined at approximately the same time, "secondary metabolite" in 1891 and "polyketide" in 1907. Some of the earliest research in biosynthetic theory and organic synthesis involved secondary metabolites in general or polyketides in particular. Both terms fell into disuse but almost half a century later were revived with more extensive meanings. The etymology, connotation, and circumscription of secondary metabolite, as well as problems with its usage have been considered (10). This review addresses similar issues for polyketide, and takes a broad historical approach to the polyketide saga.

What is a polyketide? The original meaning was a compound with polyketomethylene groups, $(CH_2\text{–}CO)_n$; such materials were said to contain "multiple keten groups" (41, 42). Also included were compounds derived from polyketomethylene structures, for example, by addition or loss of water or by decarboxylation. Loss of water typically produced cyclization of the carbon skeleton and therefore made possible further extensive modification. Eventually, it was learned that the central carbon skeleton of polyketides was formed by iterative decarboxylative condensations of malonic acid (or substituted malonic acid) thioesters, yielding polyacetate, polypropionate, or polybutyrate chains (32). Consequently, the contemporary polyketide concept is based in biosynthesis: Polyketides are united by the way in which they are made. These common underlying biosynthetic processes have so much potential variation that strikingly different chemical end products are produced. A new definition of polyketide will be proposed later.

From ancient times, some polyketides were known indirectly by their biological activities. The purgative materials in cascara, rhubarb, and senna are usually polyketide-derived anthracenes. Traditional antispasmodics, long used in the Middle East to treat angina, contain significant amounts of the polyketide khellin (5,8-dimethoxy-2-methyl-6,7-furanochromone). Perhaps the most famous example of a polyketide in history was the use of coniine-containing hemlock to execute Socrates (BCE 399)—an instance of lethal ingestion rather than the lethal injection that is used today.

The pharmaceutical potential of polyketides has been increasingly appreciated and exploited in modern times as summarized by Cane (35): "The polyketides comprise a significant fraction, not only of the total number of microbial metabolites which have been identified with physiological activities, but of the much smaller number which have found the greatest commercial application...." These latter include antibacterials (e.g. erythromycin, tetracycline, and tylosin), immunosuppressants [e.g. mycophenolic acid, rapamycin and tacrolimus (FK 506)], anticancer agents (e.g. daunomycin), antifungal agents (e.g. amphotericin and griseofulvin), cholesterol-lowering agents (e.g. lovastatin), and veterinary products (e.g. avermectin and monensin). Because the global market for microbially derived pharmaceuticals was $28 billion in 1996 (144), the hope of commercial profit has lured the pharmaceutical industry into funding much research—both basic and applied.

Academic and industrial laboratories have mined two rich lodes of discovery in polyketide biosynthesis. During the first "gold rush," roughly 1945–1970, the isotope tracer technique uncovered basic biosynthetic mechanisms and established a relationship between polyketide and fatty acid biosynthesis. In view of the role of acetate, the term polyacetate was often invoked to describe all compounds with a common biosynthetic origin from acetyl-coenzyme A (CoA); polyacetate and polyketide tended to be used as synonyms. With this usage, fatty acids and prostaglandins (which are derived from fatty acids) were both classified as polyketides. Although this capacious definition was often used (108–111, 133), nearly all contemporary usage classifies fatty acids as "primary metabolites" (i.e. metabolites common to all or most species and vital for growth and/or function). We use the narrower definition, limiting discussion to secondary metabolites and excluding acetate-derived materials in primary metabolism. We also do not use acetogenin (adjective, "acetogenous"), which is another term proposed for polyacetate products (119). Meant to imply "genesis from acetate," it is doubly misleading. "Genin" is a class name for the noncarbohydrate residue of steroid glycosides (e.g. digitogenin from digitonin); many polyketides have their origins as polypropionates, polybutyrates, and the like. In his inimitable fashion John Bu'Lock dubbed acetogenin a "cacologism" (32); unfortunately, acetogenin has been revived as a term for certain plant polyketides (67).

During the last 20 years, the second lode of polyketide research has been mined with the powerful tools of genetics and molecular biology. Polyketide biochemistry had long lagged behind the structural chemistry, and it was organic chemists who had conducted much of the "biosynthetic" research. Although a very few poorly characterized enzyme preparations were obtained, only with genetic engineering did it become possible to isolate, clone, sequence, and express the enzymes of polyketide synthesis. As will be described, gene clusters direct the synthesis of both simple and complex polyketides. Functional analysis of natural pathways is now possible, and gene manipulations which "mix-and-match" polyketide synthase (PKS) genes from different species lead to novel pathways. In Leonard Katz' felicitous phrase, we now can do "chemistry by genetics" (quoted in 118).

This paper, a meta-review, attempts to introduce microbiologists and molecular biologists to the history of polyketide research, and to make sense of the expanding scope and shifting research focus about polyketides. The polyketide and natural products literature is enormous, precluding an exhaustive summary. In addition to books cited earlier, the older biosynthetic work is well covered (68, 70, 95, 154). Two general sources for the recent molecular research are the second edition of the text edited by Strohl (143) and the thematic issue of *Chemical Reviews* dealing with polyketide and polypeptide biosynthesis (36). We have downplayed the biological activities of polyketides and the work on microbial product discovery (104, 107, 113) and have focused on the chemical and molecular biological approaches to polyketide biosynthesis, and on etymological considerations. The molecular biology of polyketide biosynthesis was reviewed by Katz & Donadio (83) and by Hutchinson & Fujii (78) in previous volumes of the *Annual Review of Microbiology*. The ongoing journals, *Journal of Antibiotics*, *Journal of Natural Products*, and *Natural Products Reports*, are wonderful resources for polyketide research.

THE EARLY YEARS

John Norman Collie (1859–1942), a truly fascinating character from the late Victorian era, distinguished himself in two worlds. As a mountaineer and an explorer, he made many first ascents and discovered the great Columbia ice field in Canada (102, 151). Two mountains carry his name: Sgurr Thormaid (Norman's Peak) on the Isle of Skye and Mount Collie in Yoho National Park, Canada. His chemistry career was also remarkable (16). He explored the chemistry of dehydroacetic acid, which he believed to be the lactone of tetraacetic acid, $CH_3-CO-(CH_2-CO)_3-OH$. Although this structure was incorrect, he gained an appreciation of the chemical reactivity of compounds containing several $-CH_2-CO-$ groups. A few of his reaction products were identical with or similar to some natural products. He proposed a biosynthetic theory based on this reactivity and coined the terms ketide and polyketide. His invention of these terms has strange features.

Collie's 1907 lecture, "Derivatives of the Multiple Ketene Group," was transcribed as follows (41): "The group $-CH_2-CO-$ (which the author proposes to call the '*ketide*' group) can be made to yield by means of the simplest reactions a very large number of interesting compounds, all belonging to classes which are largely represented amongst the compounds obtained from plants." Polyketides were then defined as $H[CH_2-CO]_xOH$. After the lecture, MO Foster asked for a more precise definition of polyketides, because Collie had also used that term to cover various "lactones, pyrone derivatives, benzenoid compounds, and substituted pyridines." Foster noted that polyketide "naturally invited comparison with the words polymethylene and polypeptide, and from that standpoint a polyketide would be a substance built up of the complex $[-CH_2-CO-]_x$ with the necessary terminal groups or atoms." There is no record of a reply by Collie. In a subsequent detailed paper (42), Collie's first paragraph is nearly the same as the one just quoted, but ketide group is replaced with keten group and in the title ketene became keten. Neither ketide nor polyketide is used in this full paper, which—confusingly—is often cited as the first mention of the polyketide hypothesis. Collie did not explain the change from ketide to keten between his oral and written renditions.

The terms ketene (a class of compounds) and keten (an individual compound) had only just been coined by Staudinger (136)—"I name the class of substances thus discovered, ketene, and of course the simplest representative is $CH_2{=}CO$, keten; therefore, the material I isolated, $(C_6H_5)_2C{=}CO$, is diphenylketen." Collie was certainly familiar with this work from the literature; moreover, his colleagues Alfred Stewart and Norman Wilsmore had provided the first evidence for keten itself (167), and Wilsmore (166) lectured on ketene in the same 1907 meeting of the Chemical Society where Collie introduced ketide and polyketide.

Collie apparently never used the terms ketide and polyketide in any of his many subsequent *Journal of the Chemical Society* papers. However, for the fourth edition (1920) of Stewart's textbook, *Recent Advances in Organic Chemistry*. Collie contributed material that was identified in separate paragraphs of that and subsequent editions with a dagger (†). Some of this material did use the ketide/polyketide terminology. Thus, in a section of the fourth edition headed "The Polyketides," Stewart described diacetylacetone as a tetraketide (139); unfortunately, this material is a triketide. In the seventh edition, Collie's material described triacetic acid as a triketide (140). Much later, Turner proposed "triketide, tetraketide, pentaketide, etc, to denote compounds derived from three, four, five, etc, 'C_2-units'" commenting that "introduction of new terminology should not be undertaken lightly" (156). Turner was apparently unaware that Stewart and Collie had already used similar terms.

Collie's experimental work is confusing and neither easy to follow nor to describe. In 1893, he and Myers (44) isolated orcinol and possibly orsellinic acid on treatment of dehydroacetic acid with strong alkali; diacetylacetone, a triketide, was an intermediate (16). Diacetylacetone was also converted to dimethylpyrone; furthermore, two molecules of diacetylacetone condensed to acetyldihydroxydimethylnaphthalene (40, 43). In describing the diacetylacetone → naphthalene condensation, Collie laid the foundation for his polyketide biosynthetic hypothesis by writing as follows: "*Polymerization* and *condensation* are probably the two chief types of change which are instrumental in forming many of the multitudinous natural compounds, condensation being usually the outcome of the union of carbon atoms in consequence of the elimination of water" (40). Aldehydes and ketones were identified as most readily dehydrated. Collie further noted that "It is well known how easily the acetyl group present in aldehyde and acetone condenses with itself. . . and numberless instances of a similar nature might be given. . . ." Thus, several years before his polyketide hypothesis, he emphasized the reactivity of acetyl groups. He stated "that when it is possible to show that by merely using the simplest kind of change, namely *condensation*, acetyl groups will produce pyridine, benzene, and naphthalene derivatives, we have imitated nature in a remarkable manner" (40). At a much later date such syntheses would have been described as "biomimetic."

Although not explicitly stated, the hypothesis can be summarized as follows: Ketomethylene compounds tend to undergo inter- and intramolecular condensations by elimination of water. Products from such condensations can undergo further modifications, for example, gain or loss of water, decarboxylation, or loss of keten groups. Such processes produce pyrones as well as benzene and naphthalene structures. With ammonia, some ketomethylene-derived compounds produce ring structures containing nitrogen atoms. In vivo, polyketides were thought to derive from carbohydrates by reactions involving loss or gain of water. Glucose [$HOCH_2$–$(CHOH)_4$–CHO] might undergo loss of three H_2O, leading to a keten structure [CH_2:$C(OH)$–CH:$C(OH)$–CH:C:O]; rehydration with one H_2O would lead to the triketide, triacetic acid [$H(CH_2CO)_3OH$]. Although much of Collie's work concerned the triketide diacetylacetone (obtained from dehydroacetic acid), he did consider the possible cyclization of a tetraketide to an anethol analog and of a heptaketide to an anthracene compound (Figure 1).

Collie's hypothesis came as the 19th century was turning to the 20th—a time when strong winds of change were sweeping over the scientific landscape and the long-standing vitalist concept was being overthrown. At that time there was a substantial gap between the pure chemists and medical chemists, as evidenced by the aphorism, "Tierchemie ist Schmierchemie" (animal chemistry is grease chemistry) (61). In 1907, no proteins had been crystallized, and their chemical nature was obscure. However, Emil Fischer had synthesized an octadecapeptide (3 leucines, 15 glycines) and followed with his trailblazing work on carbohydrates. The first steps to nucleic acid structure were taken by Phoebus Levine & Walter Jacobs in 1908–1909 with their work on inosinic acid (for historical review, see Reference 61). Clearly, the vital biological role of macromolecules was beginning to be appreciated. One historian terms the period 1897–1912 as "The New Biochemistry" (85).

Despite these stirrings, there was in 1907 no evidence as to how macromolecules and natural products were formed, and the concept of enzyme catalysis was very unclear. Considerations based on chemical composition and chemical structures led to some fruitful speculation, however, and the possibility that small molecules might polymerize to larger ones began to emerge. One example was Wallach's 1887 theory that many natural products were derived by polymerization of isoprene (127)—a proposal eventually validated as both the empirical and biosynthetic isoprene rules. The other contribution from the turn of the century was of course Collie's polyketide proposal. Not only did it confirm the importance of polymerization in natural product biosynthesis but it also called for a role for simple, well-understood chemical reactions such as dehydration, decarboxylation, condensation, and oxidation/reduction. These groundbreaking ideas remained fallow for many years. Robert Robinson (121) later returned to the biosynthetic role of polymerization by noting that "so far as we can see now the chemists of the future must concentrate *on the study of polymerization*" (emphasis added). That

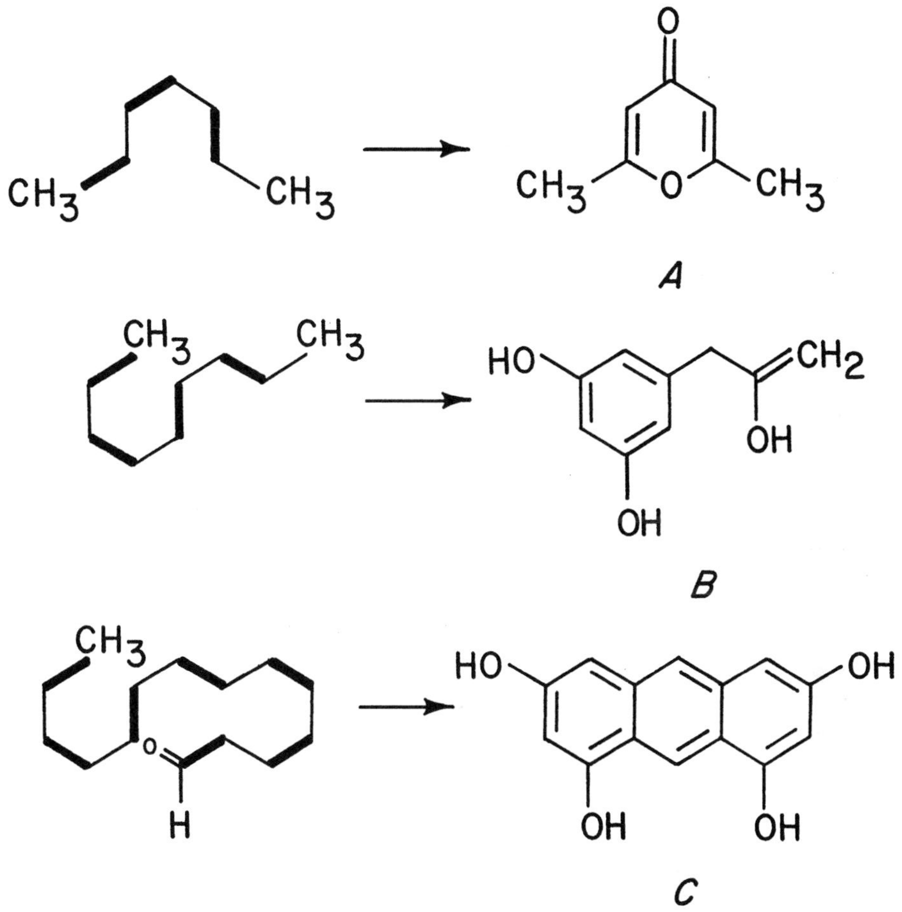

Figure 1 Selected examples of Collie's polyketide cyclizations. The bold lines indicate –CH_2–CO– groups or, if terminal, CH_3–CO– groups. The actual formation of dimethylpyrone (A) from diacetylacetone (a triketide) is shown on the top line. Suggested cyclizations of a tetraketide to a benzene structure (B) and of a heptaketide to an anthracene (C) are shown on the middle and bottom lines, respectively. The structure (B) was described by Collie as an anethole analog, but the chemical resemblance is remote.

small units polymerize into more complex structures (e.g. polyketide and fatty acid formation), and on a larger scale into macromolecules (e.g. nucleic acids, polysaccharides, and proteins), is now a dominant theme in biochemistry and molecular biology. With the benefit of hindsight, it is clear that this unifying paradigm originated in the study of natural products.

    In the first part of the twentieth century there were two major figures in the study of fungal metabolism—Harold Raistrick and Jackson Foster; however, only

Raistrick gave any serious attention to Collie's ideas on the role of acetyl groups and polyketides (117). The only other individual to take note of Collie before 1953 was the distinguished organic chemist Robert Robinson (121, 122). By that time the role of acetic acid in both primary and secondary metabolism was a "hot topic" and acetyl-coenzyme A (CoA) had been identified as the biologically active form of acetate. It was in that environment that Arthur J. Birch & F. W. Donovan (23) proposed a "polyacetate hypothesis." Birch, an Australian-born organic chemist, worked at Oxford (with Robert Robinson) and at Cambridge (with Alexander R. Todd). At Cambridge the "natural-products atmosphere" interested him in natural product structure determination and the new topic of biosynthesis. He decided to focus on biosynthesis on his return to Australia (22).

Birch realized that campnospermol, a compound that is present in Tigaso oil from *Campnosperma brevipetiolata*, contains an obvious oleic acid side chain (18 carbons) that is linked to a phenolic unit by $-CH_2-$ (22). Because fatty acids were well known to be derived from acetate, he envisioned a continuation of the head-to-tail condensation leading to the formation of the aromatic ring. Birch later noted that the campnospermol structure led him to "...draw on paper the simplest possible β-polyketo acid chain that could then be schematically cyclized according to respectable organic mechanisms" (22). Polyketones formed by the head-to-tail linkage of acetate units were proposed to cyclize by an aldol reaction ($-CH_2-$ + $-CO-$) or by acylation (CH_2- + $-CO-OH$) to phenols (23). Tetraacetic acid could form orsellinic acid or 6-methylsalicylic acid (6-MSA) via an aldol reaction, or the phloroglucinol xanthoxylin by an acyl reaction (Figure 2). This mechanism for orsellinic acid formation is basically the same as that given by Stewart & Graham (139, 140) with Collie's input (see earlier). The modification of carboxylic acids was also considered; thus, decarboxylation of orsellinic acid would produce orcinol, and 6-MSA would lead to *m*-cresol. Compounds with fused-ring systems could be derived from polyacetate compounds as already suggested by Collie (Figure 1). Neither the 1953 Birch & Donovan paper (23) nor those verifying the theory by use of radioactive acetate (25, 26) use the word "polyketide."

Birch & Donovan (23) appreciated that their ideas, although stemming from consideration of an obscure Australian botanical, had far reaching potential—from the use of biosynthetic relationships as a guide in structure determination to the realization that acetate units could be added to units derived by other metabolic pathways. Birch wrote "Within a few minutes these ideas reduced to order what had previously seemed an unrelated jumble of structures of many natural products. It was an immensely satisfying and emotional moment" (22).

After this epiphany (probably mid 1951), Birch experienced disappointments. Although some ideas were described in unpublished lectures (Birmingham, Liverpool), the Birch & Donovan hypothesis was not published until 1953 (23); it was previously "rejected by some of the best journals" (21). Also, in late 1951, while in Cambridge, Birch learned of Collie's priority through Percy Maitland (21). Initially, Birch described Collie's 1907 paper as "remarkable" and cited Collie's cyclizations of β-polyketones to aromatic compounds (23). Later he wrote that "Collie's ideas could, in fact, have saved a great deal of work since 1907 if

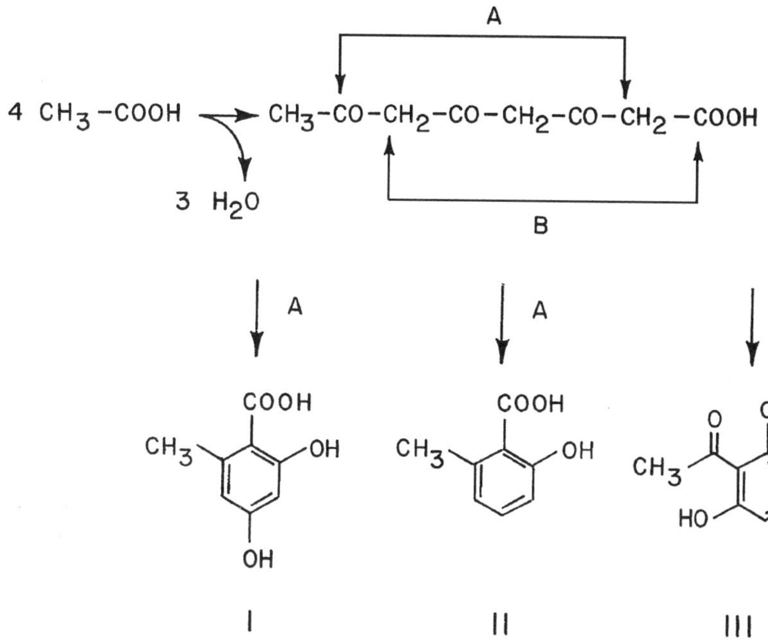

Figure 2 The Birch & Donovan polyacetate hypothesis (22, 23). Aldol type cyclization ($-CH_2-$ + $-CO-$) shown as A leads to formation from tetraacetic acid of orsellinic acid, I, or 6-methylsalicylic acid, II. An acylation process ($-CH_2-$ + $-CO-OH$), shown as B, leads to phloroacetophenone, III, and after methylation of two hydroxyls, to xanthoxylin. Note the further possibility to produce orcinol from I and *m*-cresol from II by simple decarboxylation.

anyone had had the faith to apply them" (21) and in an interview confessed: "I have even wondered, for example, whether my polyketide ideas may have unconsciously expanded Collie's ideas that I had read in passing many years previously, with no interest at the time. But I have no conscious recollection of ever having heard of him or of his work when I evolved my ideas" (22).

Birch was less generous to the biochemical tradition. By 1945, a reasonable mechanism for fatty acid synthesis was available (120), and the head-to-tail nomenclature for acetate condensations had been widely used (86, 150). The Birch & Donovan proposal, therefore, had evolved at a time of intense interest by biochemists in the overall biochemical role of acetate. Birch stated that he was partly prompted to the polyacetate hypothesis by personal awareness of the work of John Cornforth and George Popják on fatty acid and cholesterol biosynthesis. Although noting that fatty acids and steroids derived from acetate units (23), Birch & Donovan cited only a single paper (87) concerned with the biosynthesis of cholesterol from acetate. This passing reference to personal communications and the citing of only a single paper dealing with the biochemical role of acetate seems ungenerous. A 1948 review was available that had noted the use of acetate in many biosynthetic reactions (13).

In fact, Birch, a self-described "egotistical scientist" (22) had an ongoing feud with biochemists. He complained that he was "never regarded as one of the club by 'real biochemists'" and opined that "biochemists do not appear to read chemical papers" (22). Eventually, he retracted his initially generous appraisal of Collie, a bona fide "real chemist," writing that he "perhaps overreacted initially to give Collie credit" and claiming that Collie had postulated polyketides but not acetate units (22). This claim is erroneous and prejudiced in Birch's own favor. Although Collie's ideas were unfocused, he had clearly conceived the build-up of the series, acetic acid, acetoacetic acid, triacetic acid, and tetraacetic acid, and had noted "how easily the acetyl group condenses with itself" (40). Moreover, his use of "acetyl groups" is closer to acetyl-CoA than is Birch's use of "acetate units." Ultimately, it was Birch's mentor, Robinson, who had a broader and more generous vision. In 1955, he proposed the biosynthesis of certain branch-chain fatty acids from both propionate and acetate (123). Robinson based his suggestion on "Rittenberg's theory for the *n*-fatty acids," which, he said, had raised Collie's polyacetate hypothesis "to a much higher plane" (123).

THE FIRST LODE: Isotopes and Antibiotics

Tracer Studies

The decade of the 1950s was the heyday for isotopic tracer analysis and intermediary metabolism; all the major pathways by which small precursors were incorporated into large macromolecules were elucidated. Fatty acid biosynthesis, which shares so many mechanistic similarities to polyketide synthesis, is a case in point; it was shown to occur by sequential condensations of activated three carbon malonyl units with concomitant decarboxylation. By 1955, the isotope tracer technique was also well established for the study of fungal metabolites. Use of labeled acetate was known for citrate formation (*Aspergillus niger*, 1950) (60), for indirect formation of precursors for isoleucine and valine (*Neurospora crassa*, 1951) (150), for itaconic acid production (*Aspergillus terreus*, 1953) (51, 52) and for kojic acid formation (*Aspergillus flavus-oryzae*, 1953) (6). These molds used acetate itself, obviating the need to prepare labeled acetyl-CoA, and were easy to manipulate, growing rapidly in controlled laboratory culture. Labeled acetates gave good incorporations of radioactivity, and since mass yield of the fungal secondary metabolites being studied was usually substantial, adequate quantities of labeled product were available for chemical degradations. Such degradations were used to good effect for itaconic acid (51, 52) and for kojic acid (4, 5). Birch & Donovan ignored all of this work when they selected a fungal metabolite, 6-MSA, for the experimental verification of the proposed head-to-tail linkage of acetate units for biosynthesis of depsides and depsidones (25, 26). Similar to orsellinic acid from lichens, 6-MSA was originally isolated from *Penicillium griseo-fulvum* in 1931 (2). In a single experiment, this organism was fed sodium [1-^{14}C]acetate. Although the percent conversion of ^{14}C from acetate to 6-MSA was not given, a rough calculation suggests that it was only 0.08%. Nevertheless, the chemical degradations were consistent with the presence of four ^{14}C atoms in 6-MSA and with the expected alternating distribution pattern. These and many other tracer investigations verified the polyacetate mechanism for the biosynthesis of some secondary metabolites. The early fungal work using isotopic tracer analysis through 1962 has been reviewed (15) and a recent historical perspective is also available (17).

In 1959, Lynen, a leading figure in fatty acid biosynthesis, extended the concept of an acetate plus polymalonate condensation to the plant polyketide eleutherinol (90). Not realizing the need for a "starter" unit, he derived the 15 carbon atoms of eleutherinol from 8 malonyl-CoA units. Experimental verification of the use of malonate in polyketide biosynthesis was rapidly demonstrated for several metabolites from filamentous fungi: penicillic acid (18), orsellinic acid (105), 6-MSA (33), and stipitatonic acid (14). In addition, a requirement for malonyl-CoA in the action of 6-MSA synthase (6-MSAS) was established (91). Generalizations about basic polyketide construction were extended with the realization that in many simple cases, the chain is constructed from a single starter unit, $R'-CH_2-CO-S-X$, with elongation by $R'-CH(COOH)-CO-S-X$ extender units; $X-SH$ is normally either CoA or an acyl carrier protein (ACP). In the extension process, CO_2 is lost. Thus, for the basic acetate/malonate condensation the reaction is as follows ($R'' = COOH$):

$$CH_3-CO-SX + n - 1\ CH_2R''-CO-SX \rightarrow H-(CH_2-CO)_n-SX$$

"Starter"   "Extender"

$$+ n - 1\ CO_2 + n - 1\ X-SH$$

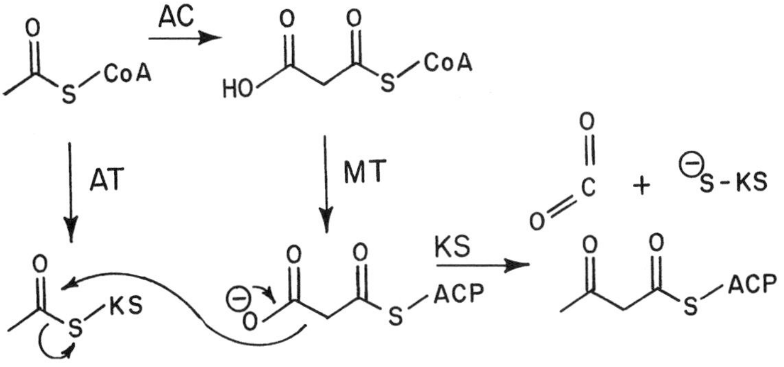

Figure 3 Initial reactions in fatty acid and polyketide biosynthesis. In this and Figures 4 and 6 the chemical convention is used that a bond drawn as a line, dashed line or wedge, and without attached substituent, indicates a methyl group. Acetyl-coenzyme A is converted to malonyl-coenzyme A by acetyl-coenzyme A carboxylase (AC). Acetyltransferase (AT) transfers an acetyl group to the SH group of the ketosynthase (KS). Malonyl transferase (MT) transfers a malonyl group from malonyl-CoA to acyl carrier protein (ACP). The acetyl-KS and malonyl-ACP yield acetoacetyl-ACP under the influence of KS; CO_2 and ^-S-KS are eliminated in this decarboxylative condensation.

In mechanistic terms this process is the attack of a nucleophile, $^-CH_2$–CO–SX, formed from the malonate component. The basic chemistry for polyketide or fatty acid biosynthesis is shown in Figures 3 and 4.

Although polyketides are readily recognized by experienced scientists, a simple, concise definition is elusive. The understanding of the construction process provides a possible definition (or more accurately, description) in biosynthetic terms rather than those of structural chemistry. The following is proposed as a formal, albeit lengthy, definition of polyketide.

"Polyketides are secondary metabolites with carbon chains formed biosynthetically by extension of a "starter" unit with –CHR–CO– "extender" units where R = H or an alkyl group; the alkyl group is frequently either CH_3 or C_2H_5. The starter unit can range from a simple structure, e.g. acetyl-CoA or propionyl-CoA, to a more complex material, e.g. C_6–C_3 and C_7 units derived from the shikimate pathway or C_7-N units probably derived from an aminoshikimate pathway. The extender units usually originate from a malonic acid or substituted malonic acid thioester formed with the SH group of acyl carrier protein (ACP); the ACP esters in turn derive from CoA esters. Insertion of the extender units into the chain requires a condensation reaction involving decarboxylation of the malonate structure. Some plant polyketides are formed by pathways not requiring the use of ACP. Polyketide chains may contain –CHR–CO– units in which R = H at some positions and R = an alkyl group at others.

The initially formed chains can retain all of the introduced CO groups, thus having the general structure R′–(CHR–CO)$_n$–S–X, where R′ derives from a starter; alternatively, one or more of the CO groups can be transformed enzymatically.

→

Figure 4   Biosynthesis of fatty acids and polyketides. Enzyme activities are abbreviated as follows: KS, ketosynthase; KR, ketoreductase; DH, dehydratase; ER, enoylreductase; ACP, acyl carrier protein; *A*, the acetyl derivative of the KS enzyme; and *B*, malonyl-ACP. The first three lines of structures show the sequential reactions used for the biosynthesis of fatty acids and an initial polyketide (polyketomethylene); note that for the former the enzymes used are KS + KR + DH + ER while for the latter only the KS enzyme activity is used. The lower three lines show how various combinations of these enzyme activities can be used to provide a wide variety of structures. They represent the synthesis of an arbitrarily chosen, partially reduced hexaketide, which has also retained three oxygen functions in addition to the terminal –CO–S–. This material is shown in the box and below:

$$CH_3–CO–CH_2–CO–CH_2–CH(OH)–CH_2–CH{=}CH–CH_2–CH_2–CO–S–X$$

Five cycles using various combinations of the KS enzyme activity and the other activities are required. In sequence these cycles are as follows: 1, KS; 2, KS; 3, KS + KR; 4, KS + KR + DH; 5, KS + KR + DH + ER. For convenience, the tetraketide derivative, $CH_3–CO–CH_2–CO–CH_2–CO–CH_2–CO–S–ACP$ is abbreviated to R–CO–S–ACP as indicated by the equal sign.

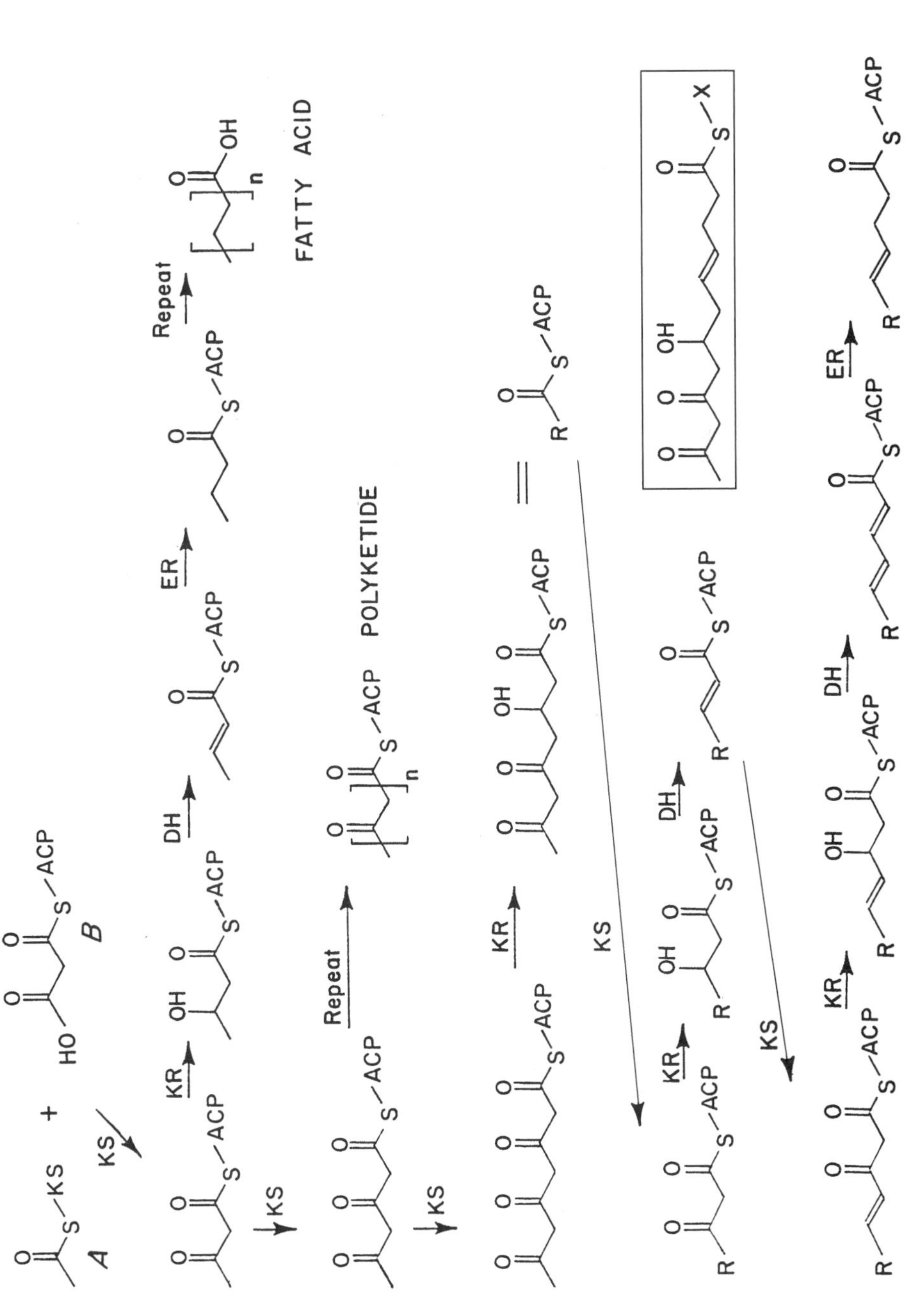

FATTY ACID

POLYKETIDE

Possible enzyme activities include: ketoreductases (KRs), –CO– → –CHOH–; dehydratases (DHs), –CHOH–CH$_2$– → –CH=CH–; and enoylreductases (ERs), –CH=CH– → –CH$_2$–CH$_2$–. In many cases, chains undergo cyclization. The enzymes building the initial chain or backbone are termed polyketide synthases PKSs. Further post-PKS "tailoring" processes are very common; they include hydration, dehydration, oxidation, reduction, decarboxylation, methylation, glycosidation, addition of isopentenyl groups, and expansion or contraction of ring systems."

Antibiotics

Concurrent with the experimental validation of the polyacetate/polyketide hypothesis, the antibiotic search was in its heyday; major classes of polyketide drugs were isolated and brought to market. Almost all of these compounds were produced by actinomycetes, largely of the genus *Streptomyces*, and included the tetracyclines and anthracyclines, the macrolides (e.g. carbomycin A, erythromycin, spiramycins, and tylosin), the polyenes (e.g. amphotericin B, candicidin B, fungimycin, nystatin, and trichomycin A), and the ansamycins (e.g. rifamycin). Pharmaceutical companies devoted considerable effort not only to screening of natural products and microbiological scale up, but also to "hard core" natural products chemistry (19, 53, 97, 124, 159). It would be difficult to exaggerate the importance of commercial considerations in the support of polyketide research.

The erythromycins are a good example. These broad spectrum antibiotics produced by *Saccharopolyspora erythraea* (formerly *Streptomyces erythreus*) inhibit bacterial protein synthesis by binding to the 50S ribosomal subunit. Structural elucidation revealed a large lactone ring; for example, the aglycone ring of erythromycin (erythronolide) has 13 carbon atoms. Woodward coined the term macrolide (in his original paper in German, Makrolide) for such compounds (169). In many cases, the macrolide carbon chain carried methyl groups at every other carbon atom; there were also attached hydroxyl groups some of which were glycosylated. Without crediting Robinson's (123) previously described suggestion, Woodward (168) proposed that the recurring pattern resulted from a use of propionate units. A similar suggestion of a role for "propionic acid or its biogenetic equivalent" was made by Gerzon et al (65) who cited both Robinson (123) and Woodward (168). Birch et al (24) suggested that a macrolide of this type was in effect a polyacetate structure modified by methylation.

The methylation hypothesis was soon discredited and propionate was established as an erythromycin precursor by the usual isotope tracer studies (48, 66, 157, 158). Moreover, Lynen (90) suggested the polymerization of the carboxylation product of propionyl-CoA, methylmalonyl-CoA, by analogy with the acetate/malonate situation. Again not realizing the role of a starter unit, he derived erythronolide A from seven units of methylmalonyl-CoA; methylmalonate was soon confirmed as an intermediate (47). The use of propionate/methylmalonate or butyrate/ethylmalonate has the result that with each extender unit, a center of chirality is introduced. Thus, deoxyerythronolide B can be considered formally to be derived from the heptaketide, $H(CH[CH_3]-CO)_7-S-CoA$. It contains 10 chiral centers, 6 of which are attributable to the $-CH[CH_3]-$ units.

Sometimes the polyketide biosynthetic process begins with a starter unit without a structural relationship to the extending unit; e.g. C_6-C_3 and C_7 units from the shikimate pathway, C_7-N units probably from the aminoshikimate pathway, amino acids, pyruvate, succinate, and isobutyrate. Rarely, a propionate starter unit is extended by malonate units (e.g. in the biosynthesis of 6-ethylsalicylic acid, homoalternariol, homoorsellinic acid, and pyrromycinone). Some structures, particularly bacterial polyketides, contain blocks of more than one aliphatic acid type (e.g. the polyether antibiotic, monensin A) has five acetate, seven propionate, and one butyrate unit.

Although the classical tracer techniques (i.e. isotope incorporation, chemical degradation) were highly productive, they were also time consuming. With newer technologies, secondary metabolites labeled with heavy isotopes (e.g. ^2H) could be introduced directly into a mass spectrometer, and rapid scanning of the entire range of fragmentation ions was possible. Thus, the tedious chemical degradations could be avoided (148). Such techniques began to be used from 1966 onwards. In addition, nuclear magnetic resonance methods for stable isotopes (particularly ^{13}C) and even some radioactive ones (^3H) were developed and again chemical degradations were not needed (148). Recent results with such methods have been reviewed (134).

In summary, the isotope tracer work showed that polyketides are constructed through repetitive decarboxylative head-to-tail condensations of activated carboxylic acids. Frequently, the resultant chains of 6–50 carbon atoms are cyclized, and almost invariably, they are subsequently modified by addition or subtraction of a variety of chemical moieties. Despite the common mode of construction, the exceedingly different end-products may be almost unrecognizable as related polyketides.

"CLASSICAL" ENZYMOLOGY

Historical Appraisal

For decades, the study of polyketide formation was almost entirely dependent on speculation and data from tracer incorporation studies. The penchant to propose whole biosynthetic pathways this way has been referred to as "incontinent biogenetic speculation" (88). Considering the large number of polyketides investigated, and the extensive resources of pharmaceutical companies, there was for many years a surprising dearth of evidence concerning either pathway intermediates or the necessary enzymes. In principle, the classical techniques for enzymological investigations should have been applicable to polyketide synthesizing enzymes. They are the availability of a suitable assay, the preparation of active cell-free extracts, and the subsequent purification of the enzyme culminating (hopefully) in crystallization and in vitro demonstration of the catalytic activity. These methods were crucial for the successful characterization of the enzymes of intermediary (primary) metabolism; fatty acid biosynthesis is a case in point. Of ubiquitous occurrence, and distinguished by essentially uniform reaction mechanisms and product patterns, fatty acid metabolism became the paradigm for showing that anabolic and catabolic modes varied in living systems. Fatty acid synthases (FASs) were purified from several biological models. It is important that, although the reaction path for fatty acid biosynthesis was basically identical in all organisms studied (Figures 3 and 4), the enzyme architecture was remarkably different. Based largely on work in *Escherichia coli*, bacteria were found to use a FAS in which each step is catalyzed by a structurally distinct and functionally different polypeptide (type II FAS). The constituent enzymes dissociate when the cell is disrupted, facilitating the study of the individual enzymes that work together in vivo in a "multienzyme complex." In contrast, based largely on work in rat liver, yeast, and a few other nonbacterial systems, most eukaryotes possess a type I FAS in which the different enzymes are associated with several multifunctional polypeptide chains. For example, the yeast system consists of a trifunctional subunit and a pentafunctional β subunit, the various constituent functions being combined on one polypeptide chain, yielding a multifunctional enzyme. The types I and II paradigm for fatty acid biosynthetic enzymes is standard fare in biochemistry textbooks, most of which ignore polyketides entirely.

In describing the parallel enzymology of polyketide synthesis, two classes of enzyme activity are distinguished: (*a*) PKSs form an initial polyketide product from the basic units such as acetyl-CoA/malonyl-CoA, propionyl-CoA/methylmalonyl-CoA, and the like (Figures 3 and 4). (*b*) Enzymes modifying a preformed polyketide (e.g. by decarboxylation, methylation, ring expansion and contraction) are termed the post-PKS auxiliary, decorating, or tailoring enzymes. Before the application of recombinant DNA technology, relatively few types of either class were isolated, and only in the form of crude extracts. As recently as 1986, Haslam wrote that "...the enzymology [of secondary metabolism] has remained almost virgin territory" (69).

Now, nearing the end of the twentieth century, as DNA sequence and derived polypeptide data are accumulating at a logarithmic rate, it is important to remember the arduous effort required to develop modern enzymology and the paucity of information about PKSs during earlier decades. To put things in context, it can be noted that the now famous and seemingly endless series, *Methods in Enzymology*, was originally issued as a four-volume treatise beginning with Volume 1 in 1955 (45) and ending with Volume 4 in 1957 (45a). As the Latin saying has it, "Parvis . e glandibus quercus" (tall oaks from little acorns grow). Probably only one of the enzymes therein described (45)—fungal glucose oxidase—would be classified as involving secondary metabolites. By 1979, a book, *Enzymes*, contained microphotographs of "most of the crystalline enzymes for which illustrations had been published" (54). Of the 192 enzymes, only 12 were obtained from fungi and 2 from streptomycetes; only 3 of them (*o*-aminophenol oxidase, chloride peroxidase, and glucose oxidase) were concerned with secondary metabolism— and none of these concerned polyketide processes. The latest (1992) edition of *Enzyme Nomenclature* lists 3196 enzyme activities, only a handful of which refer to polyketides (161). There are no PKSs listed, perhaps because they are collections of enzyme activities; this edition does contain a very brief section discussing nomenclature for multienzymes without citing examples. However, three tailoring decarboxylases for 6-MSA (EC 4.1.1.52), orsellinic acid (EC 4.1.1.58), and stipitatonic acid (EC 4.1.1.60) are included, as well as two enzymes oxidizing the polyketide sulochrin to either (+) or (−) bisdechlorogeodin, EC 1.10.3.7 and 1.10.3.8, respectively.

Why was progress so slow? Technical difficulties in obtaining cell-free extracts from fungi are often cited as a major obstacle, but this disclaimer is too broad. Many polyketides are produced by actinomycetes and other bacteria, but few polyketide pathway enzymes were isolated from these technically amenable organisms. Moreover, numerous active enzyme systems forming primary metabolites were isolated from molds and yeasts (albeit usually from young cultures before expression of the enzymes of secondary metabolism). The ephemeral nature of the enzymes of secondary metabolism was a far more important impediment. These enzymes are usually expressed only at specific stages in the life cycle of producing species, after active growth has ceased, and often in association with morphological differentiation. Furthermore, polyketide enzyme expression is subject to influences such as nutritional status, micronutrient availability, and so forth, which vary from pathway to pathway, and from species to species. Another important limitation was the human factor. Before genetic engineering, relatively few biochemists studied polyketides.

The classical enzymologists were also hindered by the large number of chemical reactions involved in polyketide biosynthesis. Many separate chemical steps are needed to produce an initial polyketide product—for example, at least 12 for 6-MSA. In many cases, extensive further tailoring occurs. Thus, patulin formation requires 9 enzymes beyond 6-MSA. In one remarkable case, an atypical *Penicillium roqueforti* strain forms both penicillic acid and patulin (112). Thus, in addition to all of the necessary tailoring enzymes, this organism requires both orsellinic acid synthase (to form penicillic acid) and 6-MSAS (to form patulin). An extreme case occurs in aflatoxin biosynthesis (see below); after assembly of the initial polyketide, norsolorinic acid, at least 15 additional tailoring enzyme activities are required (11, 103, 155). It is difficult to believe that so profligate a use of information from DNA is without benefit to the producing organism—but it is still debated in the scientific community if materials such as aflatoxin, patulin, and penicillic acid have physiological roles in the fungi producing them.

6-Methylsalicylic Acid Synthase (6-MSAS)

The first cell-free extracts for formation of classical fungal polyketides were obtained in the following order: patulin (7), 6-MSA (91), stipitatic acid (149), alternariol (63), orsellinic acid (64), 5-methylorsellinic acid (62), and sepedonin (147). However, these were only cell-free extracts and, in fact, none of the listed activities has been obtained as a crystalline protein. Only one of them, 6-MSAS, has been studied in detail.

6-MSAS was first prepared from surface mycelium of *Penicillium urticae* (= *P. patulum*) in 1961 (91). Lynen and other colleagues returned to its study on three other occasions and obtained significant purification by conventional methods—as did other investigators. Because it was one of the few PKSs available in purified form, the gene could be cloned by screening an expression library in *E. coli* with an antibody prepared against the protein (8, 160). DNA sequencing revealed a derived protein sequence of 1774 amino acids (M_r = 190.7) similar to the type I FASs. There were four active sites: ketosynthase (KS), acetyl- (or acyl-) transferase (AT), ketoreductase (KR), and ACP; each site was colinear with the corresponding catalytic sites in rat FAS. The gene was expressed in *Streptomyces coelicolor* (9), which had become the preferred model organism for polyketide research.

6-MSAS lacked a characteristic, serine-dependent, terminal thioesterase found in other PKSs and in mammalian FASs. Such an activity would have led to inhibition by phenylmethanesulfonyl fluoride; this was not the case, and phenylmethanesulfonyl fluoride was actually used in one purification to inhibit other serine proteases. A required dehydratase activity was not identified. Dehydration probably occurs at the C_8 level and not as previously believed at the C_6 level (128); hence, the pathway is probably as follows (the last steps involve cyclization, dehydration, and 6-MSA release):

$$ES-CO-CH_2-CO-CH_2-CO-CH_3 \rightarrow$$

$$ES-CO-CH_2-CH(OH)-CH_2-CO-CH_3 \rightarrow$$

$$ES-CO-CH_2-CO-CH_2-CH(OH)-CH_2-CO-CH_3 \rightarrow ES-6-MSA.$$

MOLECULAR BIOLOGY TO THE RESCUE

Sir David Hopwood's Genetic Strategies and Generous Collaborations

Gene cloning technology facilitated an approach to the enzymology of natural products biosynthesis, previously impossible with classical biochemistry. Much of the trailblazing work by David Hopwood and his collaborators used antibiotic-producing species of the genus *Streptomyces* with an initial focus on *S. coelicolor* and its weak antibiotic, actinorhodin, a pigmented aromatic polyketide. Six classes of *act* mutants had been inferred from genetic analysis using blocked mutants and cosynthesis tests (126). The molecular analysis of the actinorhodin pathway was labor and resource intensive: Mutants were in short supply, cloning vectors had to be developed, transformation efficiencies were low, and new protocols had to be refined specifically for streptomycetes. In as much as actinorhodin (like all secondary metabolites) is not essential to the viability of the producing organism, appropriate selection strategies were also necessary.

Success was achieved with the gene for an *O*-methyltransferase, isolated by complementation using the color of undecylprodigiosin as the selectable phenotype. This gene encoding an auxiliary enzyme was the very first gene for a secondary metabolic pathway ever cloned (59). The complementation-color selection strategy was then extended to the *act* pathway. Genetic data had shown that the *act* genes were linked; molecular analysis reflected the linkage. The *act* pathway genes were cloned on a single DNA fragment (93). This was the first demonstration that polyketide pathway genes are physically "clustered" on bacterial chromosomes, confirming the earlier predictions showing that numerous genes for antibiotic biosynthesis mapped close to one another (75). Complementation was also used to isolate other polyketide pathway genes, for example those for tetracenomycin C (106) and avermectin (142).

Transformants carrying antibiotic resistance determinants were relatively easy to select and a number of genes for antibiotic resistance were cloned in the early 1980s (e.g. 20, 152, 153). The observation that resistance genes were often linked to their respective biosynthetic clusters provided the basis for a second selection strategy. The ability of antibiotic producing bacteria not to commit suicide was used to devise conditional lethal screens. In case after case, the polyketide pathway genes were clustered and resistance genes were in close proximity (34, 94, 135). Clustering ensured that once one gene of a given pathway was isolated, the entire pathway subsequently could be obtained by cloning the flanking sequences. The *act* gene cluster contained 22 structural, regulatory, resistance, and export genes (72). Moreover, *act* genes, and later other PKS genes, provided probes to search for further genes in other pathways and other species. In one early study, DNA probes for *actI* and *actIII* were hybridized to genomic DNA from 25 other streptomycetes, and Southern blot analysis revealed bands for 14 of 18 known polyketide producers (92).

The tripartite strategies developed for *S. coelicolor* and actinorhodin were rapidly extended and by 1990, either complementation, resistance gene linkage, or DNA probes had been used to clone portions of the actinorhodin, avermectin, candicidin, carbomycin, daunorubicin, erythromycin, granaticin, oxytetracycline, spiramycin, tetracenomycin, and tylosin biosynthetic clusters (19a, 76, 132, 146). The gene clusters were all located on chromosomes, despite reports suggesting that some of these pathway genes might be plasmid encoded. Regulatory genes were often associated with the gene clusters, further accelerating the pace of discovery (145). A review of this time period, emphasizing the industrial implications of cloning PKS genes, has been given by Chater (38).

In summary, natural selection had created readily exploited opportunities: the evolutionary relatedness of PKS genes underlay the success of DNA probes; gene clustering facilitated chromosome walking; linked resistance genes simplified selection by providing conditional lethal strategies. Nature was not the only benevolent factor in the remarkable development of polyketide molecular biology. David Hopwood's (from 1994, Sir David) scientific authority was matched by his ethical and collegial standards. His group developed the basic genetic systems and cloning vectors, identified the first PKS genes, published most of the ground breaking papers, and was extraordinarily generous with its strains, probes, and expertise. The rapid progress would have been impossible without Sir David's unselfish collaborations and his fostering of an open data exchange between academic and industrial laboratories.

Nearly simultaneously, the first FAS genes were also cloned and sequenced (1, 39, 71, 130, 165, 172). Because of the success of the prior enzymological work, the molecular FAS findings were almost anticlimactic; the type I and type II organization of FASs was confirmed. Moreover, as the derived protein sequences of the type II fatty acid enzymes were compared with the first PKS genes cloned from *Actinomyces* spp., a similar arrangement was found; i.e. there were separate genes for chain assembly, reduction, cyclization, etc. The prokaryotic-eukaryotic dichotomy was extended when the first eukaryotic PKS was cloned. This was the previously described 6-MSAS from *P. patulum* that had figured so prominently in the early biochemical characterization of PKSs (8, 160). This filamentous fungus was the first organism to have both its FAS and PKS characterized.

With the enzyme architecture for prokaryotic and eukaryotic PKSs following a seemingly predictable pattern, the convention of labeling PKSs as type I and type II was adopted (76). In this nomenclatural dichotomy, the basic pattern of type II PKSs consisted of a string of separate genes encoding separate enzymes with high sequence similarity to the KS, ACP, and KR proteins of fatty acid biosynthesis, together with genes for cyclases (aromatases) catalyzing ring formation from poly-β-ketone chain intermediates. These type II genes were prokaryotic, clustered, and directed the biosynthesis of aromatic polyketides. The physical organization of more than a dozen prokaryotic type II PKS gene clusters has been diagrammed (72, 78). In contrast, the first eukaryotic PKSs had catalytic domains organized serially on a single polypeptide, analogous to the multi-functional type I proteins characteristic of eukaryotic FASs. The molecular biology of FASs and PKSs from 1980 to 1990 has been compared (73, 76).

Better Chemistry Through Genetics

The Hopwood-Innes Group seeded the "big bang" of polyketide research, and as the new techniques were successfully applied to many pathways, detailed biochemical analysis brought a renaissance to natural products chemistry. A dramatic event in the early 1990s was the cloning of the erythromycin PKS by two independent groups. The Leadley group (University of Cambridge, Cambridge) used a self-resistance gene (49), whereas the Katz group (Abbott Laboratories, Chicago) used complementation of mutants inactivated in the polyketide aglycon (57). The results were electrifying. The erythromycin biosynthetic gene cluster (*eryAI*, -*II*, and -*III*) produced three enormous proteins named deoxyerythronolide B synthase (DEBS) -1, -2, and -3. Each DEBS was a multidomain polypeptide, with two functional units dubbed "modules." The modules carried catalytic activities for chain elongation and reductive modification of the resultant β-ketoacyl thioester, in serially encoded catalytic sites for KS, AT, and ACP; varying numbers of dehydratase (DH), enoylreductase (ER), and KR sites were found within each set. The number of modules directed the size of the erythromycin molecule. Because the catalytic activities were sequentially encoded on a single polypeptide this prokaryotic enzyme system was classified as a type I PKS. This designation, however, did not and does not do justice to the uniqueness of the gene-protein arrangement; the modular PKS worked in a fundamentally new way. In the erythromycin PKS, the modular enzyme and its polyketide product demonstrated a colinear relationship. Rather than using catalytic sites reiteratively, the bacterium used an assembly line approach where each round of chain extension was controlled by separate activities organized on separate modules. The protein sequence could be used to forecast which part of the system performed each step in the construction of the polyketide backbone. "An enzyme that works the way chemists do" is one apt description (118).

The remarkable built-in informational potential of synthases had been appreciated at an early date, and Hopwood introduced the term "genetic programming" to describe the way in which type II PKSs direct the primary structure of their associated polyketides by specifying the type of starter unit, the length of the carbon chain, and the cyclization pattern of the final product. With the sequencing of the erythromycin PKS, the metaphor became even more apt. "The program was hard wired to the gene sequence," Hopwood eloquently expanded, "and was expressed in the encoded protein in a series of active sites appropriately arranged in relation to each other" (72).

A chronological account of the developments that ensued after the discovery of the erythromycin modular PKS is virtually impossible. In rapid succession, modular enzyme systems were found for many other complex polyketides including avermectin, candicidin aglycon (FR-008), carbomycin (platenolide), FK 506, nemadectin, niddamycin, oleandomycin, rapamycin, rifamycin, soraphen A, spiramycin, tetranosin, and tylosin (72, 78, 79, 82). Each system provided new variations on the theme, without detracting from the central concept. The catalytic domains within each module controlled the level of oxidation. An AT domain within each module specified whether the extender unit would be malonyl-ACP, methymalonyl-ACP, or ethylmalonyl-ACP, thereby controlling the level of branching along the chain. The number and composition of modules directed formation of the various polyketide backbones; auxiliary enzymes contributed further structural diversity. Thus for rapamycin, a 31-member macrocyclic immunosuppresant from *Streptomyces hygroscopicus*, the enzyme system contained

14 modules on three huge polypeptides. Unlike erythromycin, the genetic order for specifying the three polypeptides was different from the functional order (129). Many domains contained sequences for inactive functions (3). For niddamycin, a 16-membered macrolide, the chemical structure had suggested that the polyketide backbone required an ordered condensation of acetate, propionate, butyrate, and perhaps glycolate residues (115). In the sequenced niddamycin gene cluster of *S. caelestis*, seven modules were found as expected, with AT domains specific for malonyl-CoA and methylmalonyl CoA. In both niddamycin and erythromycin, propionyl-CoA was loaded by a methylmalonate-class AT (81).

The cloning work facilitated enzymological research and in vitro reconstitutional studies have been actively pursued with the DEBS proteins of erythromycin biosynthesis being the most intensively studied. They have been purified to homogeneity, and cell-free synthesis of polyketide product has been affected by components of the native system (164) and with recombinant proteins (116, 138). A dimeric double helical structure for the DEBS proteins has been proposed and FASs may be folded similarly. In an ironic twist of scientific history, it is now "as reasonable to look to the PKS structure for guidance to the FAS structure as it is to argue in the reverse direction" (138).

In contrast to modular PKSs cloned so far, *Streptomyces venezuelae* produces macrolides with two different chain lengths: 12-membered lactone ring antibiotics, methymycin and neomethymycin, and 14-membered ring antibiotics, narbomycin and pikromycin (170, 171). These four polyketides are biosynthesized with just one PKS, encoded by a large gene cluster (*pikA*) producing five enzyme complexes called PikAI-PikAV (Figure 5). Each complex contains one or more modules (Figure 6). Employing a not completely understood mechanism, the single system terminates chain assembly at two different points to produce different size macrolactones (Figure 6). The glycosyltransferase (*desVII*) is associated with the cluster and can accept both 12- and 14-membered rings as substrates (171, 173). This high substrate flexibility facilitates glycosylation of the two lactone ring types, a property not previously observed (27). Another closely linked gene produces a different tailoring enzyme, a p450 hydroxylase, PikC, that provides yet more molecular diversity to the system by adding hydroxyl groups at two different positions in two macrolide systems (170).

As far as fungal polyketides are concerned, molecular analysis has focused mainly on the carcinogenic mycotoxins, sterigmatocystin, and aflatoxin. Sterigmatocystin, an aflatoxin precursor, is produced by at least 20 species of molds whereas aflatoxin production is limited to only three species. In each case, a minimum of 15 complex post-PKS steps are required for production of the final product (11, 155). In *Aspergillus nidulans*, the genes for the sterigmatocystin pathway enzymes are clustered and include a PKS, a FAS, five monooxygenases, four dehydrogenases, an esterase, an *O*-methyltransferase, a reductase, an oxidase, and a binding protein (29). Similar clusters for the aflatoxin pathway have been discovered in *Aspergillus flavus* and *Aspergillus parasiticus* (103). The required hexanoyl starter unit is made by a distinct and dedicated FAS (28).

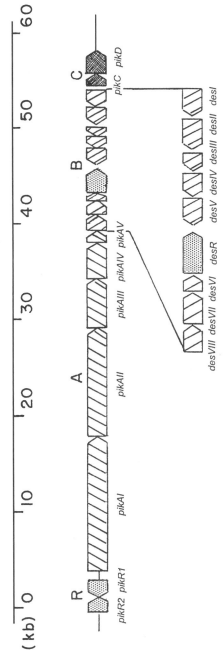

Figure 5 Organization of the PKS gene cluster (*pikA*, *pikB*, [*des*], *pikC*, and *pikR*) that encodes for methymycin, neomethymycin, narbomycin and pikromycin biosynthesis in *Streptomyces venezuelae*. Each *arrow* represents an open reading frame (ORF). The direction of transcription and relative sizes of the ORFs were deduced from nucleotide sequences (redrawn from 171).

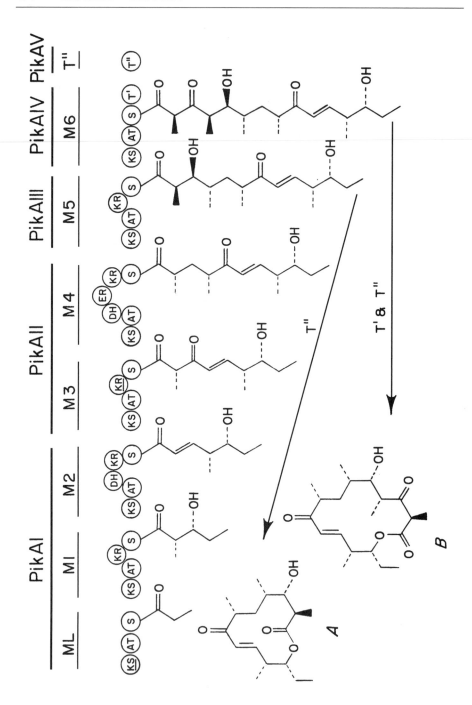

Hybrid Antibiotics and Combinatorial Biosynthesis

The application of genetics to improve the pharmaceutical properties of natural products is not new. Brute force mutagenesis and selection for high yielding strains have been the cornerstones of industrial microbiology, whereas recombinational methods such as parasexual crossing and protoplast fusion also have had some limited success. With *Streptomyces* recombinant DNA techniques, hybrid antibiotics could be created by combining individual genes, as opposed to whole genomes. Because polyketides can be built from more than one type of monomer, with many permutations of both the central PKS and the auxiliary tailoring steps, an almost limitless number of end products is theoretically possible. The set of techniques ("the genetic tool box") by which the polyketide genes are manipulated to produce libraries of polyketide products is called "combinatorial biosynthesis." Pioneers in this field include David Cane (Brown University), Richard Hutchinson (University of Wisconsin), Leonard Katz, (Khosan Biosciences), Chaitan Khosla (Stanford University), and Peter Leadlay working with Jim Staunton and coworkers at the University of Cambridge.

The first hybrid antibiotics were made by moving actinorhodin biosynthetic genes from *S. coelicolor* into other *Streptomyces* species (74, 114). The second set of hybrid antibiotics were erythromycin analogues made by shot gun cloning DNA from *Streptomyces* antibiotics into a blocked mutant of *Saccharopolyspora erythraea* unable to make erythromycin (98). These early successes involved late tailoring enzymes; that is, the accessory enzymes of post-PKS modifications.

Engineered modular PKSs have been the centerpiece of combinatorial biosynthesis with the erythromycin gene cluster becoming the model for making alterations directly on the PKS. Because genetic order is colinear with the order of biochemical reactions, directed changes (reprogramming) in the DNA produce predicted novel erythromycin derivatives. In the first demonstration of this approach, the Katz group deleted the activity of the KR domain in the fifth module of

Figure 6 Formation of macrolide antibiotics by *Streptomyces venezuelae*. (A) 10-deoxymethynolide and (B) narbonolide; further reactions (not shown) convert A to methymycin and neomethymycin, and B to narbomycin and pikromycin. The names of the genes and open reading frames derive from that of the latter antibiotic. A PKS locus, *pikA*, encodes a six module system (M, module) and a type II thioesterase, T''; ML indicates a loading module. Enzymatic domains in the PKS multifunctional protein are as follows: KS, ketosynthase (β-ketoacyl-ACP synthase); KS, a KS-like domain; AT, acyltransferase; KR, ketoreductase (β-ketoacyl-ACP reductase); KR, an inactive KR; DH, dehydratase (β-hydroxyacyl-thioester dehydratase); ER, enoyl reductase; TE', thioesterase type I; TE'', thioesterase type II. The acyl carrier protein domain is represented by S for its acyl-substituted SH group. This figure was redrawn with modifications from that of Xue et al (171). The T'' and T' & T'' indications over the arrows should be ignored to reflect the incompleteness in our understanding of the system at this stage (A Xue, personal communication).

the erythromycin PKS, yielding a 5-keto analog instead of the 5-hydroxy macrolide (56, 57). More recently in the methymycin/neomycin cluster, the deletion of the *desVI* gene resulted in the formation of new glycosylated analogs (173).

Genetic manipulation of a core set of PKSs, and the genes that control tailoring enzymes, can create novel pathways. By moving genes around, it is theoretically easy to generate an enormous variety of "unnatural natural products." Mutants can form modified full-length polyketides or make products with fewer carbon atoms (58). The starter residue has been variously manipulated including the introduction of non-native starter units (162) and replacement of the loading domains on one PKS with the comparable domains from another. Thus, the loading module of the erythromycin PKS can be exchanged with the cognate domain from the avermectin cluster (96); the latter is apparently particularly accepting of non-native starter units. Another strategy is to reposition the thioesterase (TE) domain so as to produce a shortened product (50, 82, 84). Possible manipulations include changes in the starter unit, the extender unit, ketoreduction, enoylreduction, the length of the acyl chain, and various other functionalities (82–84). In one such demonstration of "combinatorial biology," McDaniel et al (101) generated a library of 50 previously unknown macrolides by replacing the ATs and β-carbon processing domains (i.e. DH, ER, KR) of the 6-deoxyerythronolide B synthase with counterparts from the rapamycin PKS. This collection of "unnatural" natural products represents about 3% of the presently known polyketides and exceeds the total number of different macrolide ring structures yet discovered. It would have been impractical to produce them by conventional chemical methods. Alteration of the stereochemistry of chiral centers has been less successful and more research on polyketide stereochemistry is needed (137, 163).

Aromatic polyketides (e.g. actinorhodin, doxorubicin, oxytetracycline, and griseofulvin) produced by type II (iterative) PKS genes have also been manipulated. PKS programming can be studied by analyzing the polyketides produced by recombinants carrying different combinations of PKS subunits. A so-called minimal PKS has been defined (99) and a set of design rules has been promulgated for the rational design of novel molecules (72, 100). The implications for antibiotic research have been reviewed (77). Fermentation based approaches have also been used. Chemically synthesized precursors can be delivered to genetically engineered PKSs and the DEBS system (erythromycin) will incorporate a remarkable number of synthetic diketides (80).

The demonstration of combinatorial biosynthesis spurred the search for novel PKS-encoding genes, including those not producing pharmacologically useful products. The highly conserved, modular PKS genes (55, 129) can be used to screen libraries from novel species to find new PKS genes. Such an approach resulted in the isolation of a second PKS cluster from a rapamycin-producing strain of *S. hygroscopicus*, dubbed the *hyg* cluster, because the function was unknown (125). New avenues for research are opening all around. It is possible to move polyketide genes from uncultured organisms and express them in model systems or use them as raw materials for generating recombinant molecules (131). Another attractive target is the fungal symbiont of lichens. Lichens are well known both for polyketide production and notoriously slow growth. Finally, there is also the potential for expanding this approach to other natural products such as nonribosomal peptides and isoprenoids.

EPILOGUE

The study and use of polyketides have come a very long way since the beginning of the 20th century. The developments are neatly summarized by Cane's aphorism, "From Collie to *Coli*" (35) and our extension to "From Collie to Combinatorial Biosynthesis." Over almost 100 years the emphasis shifted from chemical investigations, to biochemical possibilities and now to molecular biology and genetics. The most recent phase of polyketide research has created an interface between molecular genetics and natural products chemistry that is one of the most exciting areas in contemporary science. Molecular understanding has profoundly influenced the way we think about polyketides and has created a boom in fundamental research as well as high hopes for practical rewards. We have learned that secondary metabolite biosynthesis has more in common with the pathways of primary metabolism than we suspected and we hope the new knowledge will lead to discovery of new drugs. Genomics research is also yielding unanticipated clues about hitherto unsuspected sources of polyketides. The genome of the causative agent of tuberculosis, *Mycobacterium tuberculosis*, contains many PKS genes, although no polyketides have been isolated from this species (39a). Moreover, a newly characterized polyketide macrolide toxin has been isolated from *Mycobacterium ulcerans* and implicated as a virulence factor in Buruli ulcer, a skin disease found largely in Africa and Australia (64a). The classic metabolite, 6-MSA, was earlier identified as a polyketide in *Mycobacterium phlei* (76a).

As the twenty-first century approaches, polyketides are pre-eminent among secondary metabolites, not only in their roles in medicine and biotechnology, but also in the rapid accumulation of knowledge of the ways by which their biosynthesis is effected and controlled.

As noted earlier, the generalization that natural products can be made by polymerization of simpler substances probably originated from Collie's work with polyketides. The chemistry of living systems is driven by the constraints of natural selection; as the famous statement by Theodosius Dobzhansky puts it: "Nothing in biology makes sense except in the light of evolution." Nature is opportunistic, and has had millions of years to practice combinatorial chemistry. In their delight at studying polyketide structure and now in the admiration of the power of combinatorial biosynthesis, scientists have tended to forget that evolution has driven the formation of these byzantine biosynthetic pathways. Polyketide biosynthesis is far more than the biochemical counterpart of fatty acid biosynthesis. The regular clustering of complexes of sequential metabolic enzymes reflects an evolutionary process directed by natural selection. Scientists can now mimic the mixing and matching provided by nature; but it would be hubris to envision that we can greatly improve on millions of years of evolution. As a mountain climber and explorer, Collie opened up new territories to human view. As a scientist, he opened up new vistas of intellectual excitement and of actual and possible benefits in medical practice. More than a century ago he spoke of imitating "nature in a remarkable manner" (40). Collie would be astounded, and no doubt delighted, to learn that nature can indeed be imitated and manipulated in truly remarkable ways.

ACKNOWLEDGMENTS

We thank Sir David Hopwood, Drs. Leonard Katz, David Sherman, and Yongqan (Alex) Xue for their careful reading of the manuscript, and Dr. Dick Hutchinson for providing reprints. We are grateful to Leonard Katz for helpful suggestions concerning the definition of polyketides.

Visit the Annual Reviews home page at http://www.AnnualReviews.org

LITERATURE CITED

1. Amy CM, Witkowski A, Naggert J, Williams B, Randhawa Z, Smith S. 1989. Molecular cloning and sequencing of cD-NAs encoding the entire rat fatty acid synthase. *Proc. Natl. Acad. Sci. USA* 86:3114–18

2. Anslow WK, Raistrick H. 1931. Studies in the biochemistry of micro-organisms. Part XIX. 6-Hydroxy-2-methylbenzoic acid, a product of the metabolism of glucose by *Penicillium griseo-fulvum* Dierckx. *Biochem. J.* 25:39–44

3. Aparicio JF, Molnar I, Schweke T, Konig A, Haydock SF, et al. 1996. Organization of the biosynthetic gene cluster for rapamycin in *Streptomyces hygroscopicus*: analysis of the enzymatic domains in the modular polyketide synthase. *Gene* 169:9–16

4. Arnstein HRV, Bentley R. 1950. Kojic acid biosynthesis from 1-C^{14}-glucose. *Nature* 166:948–50

5. Arnstein HRV, Bentley R. 1953. The biosynthesis of kojic acid. 1. Production from [1-^{14}C] and [3,4-^{14}C$_2$]glucose and [2-^{14}C]-1:3-dihydroxyacetone. *Biochem. J.* 54:493–508

6. Arnstein HRV, Bentley R. 1953. The biosynthesis of kojic acid. 3. The incorporation of labelled small molecules into kojic acid. *Biochem. J.* 54:517–22

7. Bassett EW, Tanenbaum SW. 1960. Acetyl-coenzyme A in patulin biosynthesis. *Biochim. Biophys. Acta* 40:535–37

8. Beck J, Ripka S, Siegner A, Schiltz E, Schweizer E. 1990. The multifunctional 6-methylsalicylic acid gene of *Penicillium patulum*. Its gene structure relative to that of other polyketide synthases. *Eur. J. Biochem.* 192:487–98

9. Bedford DJ, Schweizer E, Hopwood DA, Khosla C. 1995. Expression of a functional fungal polyketide synthase in the bacterium *Streptomyces coelicolor* A3(2). *J. Bacteriol.* 177:4544–48

10. Bennett JW, Bentley R. 1989. What's in a name? Microbial secondary metabolism. *Adv. Appl. Microbiol.* 34:1–28

11. Bennett JW, Bhatnagar D, Chang P-K. 1994. The molecular genetics of aflatoxin biosynthesis. In *The Genus Aspergillus: From Taxonomy and Genetics to Industrial Application*, ed. KA Powell, A Renwick, J. Peberdy, pp. 51–58. New York: Plenum

12. Bennett JW, Ciegler A. 1994. *Secondary*

Metabolism and Differentiation in Fungi. New York: Marcel Dekker

13. Bentley R. 1948. The function of small molecules in biosynthesis. *Ann. Rep. Chem. Soc. (London)* 45:239–67

14. Bentley R. 1961. The role of malonate in the biosynthesis of mold tropolones. *Fed. Proc.* 20:80 (Abstr.)

15. Bentley R. 1962. Biochemistry of fungi. *Annu. Rev. Biochem.* 31:589–624

16. Bentley R. 1999. John Norman Collie: chemist and mountaineer. *J. Chem. Educ.* 76:41–47

17. Bentley R. 1999. Secondary metabolite biosynthesis: the first century. *Crit. Rev. Biotechnol.* 19:1–40

18. Bentley R, Keil JG. 1961. The role of acetate and malonate in the biosynthesis of penicillic acid. *Proc. Chem. Soc.* 111–12

19. Berdy J. 1974. Recent developments of antibiotic research and classification of antibiotics according to chemical structure. *Adv. Appl. Microbiol.* 18:309–406

19a. Bibb MJ, Biró S, Motamedi H, Collins JF, Hutchinson CR. 1989. Analysis of the nucleotide sequence of the *Streptomyces glaucescens tcmI* genes provides key information about the enzymology of polyketide antibiotic biosynthesis. *EMBO J.* 8:2727–36

20. Bibb MJ, Schottel JL, Cohen SN. 1980. A DNA cloning system for interspecies gene transfer in antibiotic producing *Streptomyces. Nature* 284:526–31

21. Birch AJ. 1967. Biosynthesis of polyketides and related compounds. *Science* 156:202–6

22. Birch AJ. 1995. *To See the Obvious.* Washington, DC: Am. Chem. Soc.

23. Birch AJ, Donovan FW. 1953. Studies in relation to biosynthesis. I. Some possible routes to derivatives of orcinol and phloroglucinol. *Aust. J. Chem.* 6:360–68

24. Birch AJ, English RJ, Massy-Westropp RA, Slaytor M, Smith H. 1958. Studies in relation to biosynthesis. Part XIV. The origin of the nuclear methyl groups in my-cophenolic acid. *J. Chem. Soc.* 365–68

25. Birch AJ, Massy-Westropp RA, Moye CJ. 1955. The biosynthesis of 6-hydroxy-2-methylbenzoic acid. *Chem. Ind.* 683–84

26. Birch AJ, Massy-Westropp RA, Moye CJ. 1955. Studies in relation to biosynthesis. Part VII. 2-Hydroxy-6-methylbenzoic acid in *Penicillium griseofulvum* Dierckx. *Aust. J. Chem.* 8:539–44

27. Borman S. 1998. Polyketide system cloned, characterized. *Chem. Eng. News* 76(44):27–28

28. Brown DW, Adams TH, Keller NP. 1996. *Aspergillus* has distinct fatty acid synthases for primary and secondary metabolism. *Proc. Natl. Acad. Sci. USA* 93:14873–77

29. Brown DW, Yu J-H, Kelkar HS, Fernandes M, Nesbitt TC, et al. 1996. Twenty-five coregulated transcripts define a sterigmatocystin gene cluster in *Aspergillus nidulans. Proc. Natl. Acad. Sci. USA* 93:1418–22

30. Bu'Lock JD. 1961. Intermediary metabolism and antibiotic synthesis. *Adv. Appl. Microbiol.* 3:293–342

31. Bu'Lock JD. 1965. *The Biosynthesis of Natural Products. An Introduction to Secondary Metabolism.* New York: McGraw-Hill

32. Bu'Lock JD. 1979. Polyketide biosynthesis. In *Comprehensive Organic Chemistry*, ed. D Barton, WD Ollis, pp. 927–87. Oxford, UK: Pergamon

33. Bu'Lock JD, Smalley HM. 1961. Biosynthesis of aromatic substances from acetyl- and malonyl-CoA. *Proc. Chem. Soc.* 209–11

34. Butler MJ, Friend EJ, Hunter IS, Kaczmarek FS, Sugden DA, Warren M. 1989. Molecular cloning of resistance genes and architecture of a linked gene cluster in the biosynthesis of oxytetracycline by *Streptomyces rimosus. Mol. Gen. Genet.* 215:231–38

35. Cane DE. 1997. Introduction: polyketide and nonribosomal polypeptide biosyn-

thesis. From Collie to *Coli*. *Chem. Rev.* 97:2463–64

36. Cane DE, ed. 1997. A thematic issue: polyketide and nonribosomal polypeptide biosynthesis. *Chem. Rev.* 97:2463–705

37. Chadwick DJ, Whelan J, eds. 1992. *Secondary Metabolites: Their Function and Evolution, Ciba Found. Symp.* Vol. 171. Chichester, UK: Wiley-Intersci.

38. Chater KF. 1990 Improving prospects for yield increase by genetic engineering in antibiotic-producing *Streptomyces*. *Bio/Technology* 8:115–21

39. Chirala SS, Kuziora MA, Spector DM, Wakil SJ. 1987. Complementation of mutations and nucleotide sequence of *FAS1* gene encoding β subunit of yeast fatty acid synthase. *J. Biol. Chem.* 262:4231–40

39a. Cole ST, Brosch R, Parkhill J, Garnier T, Churcher C, et al. 1998. Deciphering the biology of *Mycobacterium tuberculosis* from the complete genome sequence. *Nature* 393:537–44

40. Collie JN. 1893. The production of naphthalene derivatives from dehydracetic acid. *J. Chem. Soc.* 63:329–37

41. Collie JN. 1907. Derivatives of the multiple ketene group. *Proc. Chem. Soc.* 23:230–31

42. Collie JN. 1907. Derivatives of the multiple keten group. *J. Chem. Soc.* 91:1806–13

43. Collie JN, Wilsmore NTM. 1896. The production of naphthalene and of isoquinoline derivatives from dehydracetic acid. *J. Chem. Soc.* 69:293–304

44. Collie N, Myers WS. 1893. The formation of orcinol and other condensation products from dehydracetic acid. *J. Chem. Soc.* 63:122–28

45. Colowick SP, Kaplan NO, eds. 1955. *Methods in Enzymology*, Vol. I. New York: Academic

45a. Colowick SP, Kaplan NO, eds. 1957. *Methods in Enzymology*, Vol. IV. New York: Academic

46. Conn EE, ed. 1981. *The Biochemistry of Plants*, Vol. 7. New York: Academic

47. Corcoran JW, Chick M. 1966. Biochemistry of the macrolide antibiotics. In *Biosynthesis of Antibiotics*, Vol. I. ed. JF Snell, pp. 159–201. New York: Academic

48. Corcoran JW, Kaneda T, Butte JC. 1960. Actinomycete antibiotics. I. The biological incorporation of propionate into the macrocyclic lactone of erythromycin. *J. Biol. Chem.* 235:PC29–PC30

49. Cortes J, Haydock SF, Roberts GA, Bevitt DJ, Leadlay PF. 1990. An unusually large multifunctional polypeptide in the erythromycin-polyketide synthase of *Saccharopolyspora erythreae*. *Nature* 346:176–78

50. Cortes J, Wiesmann KEH, Roberts GA, Brown MJB, Staunton J, Leadlay PF. 1995. Repositioning of a domain in a modular polyketide synthase to promote specific chain cleavage. *Science* 268:1487–89

51. Corzo RH. 1953. *Biogenesis of itaconic acid*. PhD thesis. Stanford Univ., Stanford, Calif. 152 pp.

52. Corzo RH, Tatum EL. 1953. Biosynthesis of itaconic acid. *Fed. Proc.* 12:470 (Abstr. 1544)

53. Demain AL, Solomon NA. 1983. *Antibiotics. Handbook of Experimental Pharmacology*. Vols. 67/1, 67/2. Berlin/New York: Springer Verlag

54. Dixon M, Webb EC. 1979. *Enzymes*, 3rd ed. New York: Academic

55. Donadio S, Katz L. 1992. Organization of the enzymatic domains in the multifunctional polyketide synthase involved in erythromycin biosynthesis in *Saccharopolyspora erythreae*. *Gene* 111:51–60

56. Donadio S, McAlpine JB, Sheldon PJ, Jackson M, Katz L. 1993. An erythromycin analog produced by reprogramming of polyketide synthesis. *Proc. Natl. Acad. Sci. USA* 90:7119–23

57. Donadio S, Staver MJ, McAlpine JB, Swanson SJ, Katz L. 1991. Modular or-

ganization of genes required for complex polyketide biosynthesis. *Science* 252:675–79

58. Donadio S, Staver MJ, McAlpine JB, Swanson SJ, Katz L. 1992. Biosynthesis of the erythromycin macrolactone and a rational approach for producing hybrid macrolides. *Gene* 115:97–103

59. Feitelson JS, Hopwood DA. 1983. Cloning of a *Streptomyces* gene for *O*-methyltransferase involved in antibiotic biosynthesis. *Mol. Gen. Genet.* 190:394–98

60. Foster JW, Carson SF. 1950. Citric acid formation by *Aspergillus niger* through condensation of 3 C_2 moieties. *J. Am. Chem. Soc.* 72:1865–66

61. Fruton JS. 1972. *Molecules and Life. Historical Essays on the Interplay of Chemistry and Biology.* New York: Wiley-Intersci.

62. Gatenbeck S, Eriksson PO, Hanson Y. 1969. Cell-free C-methylation in relation to aromatic biosynthesis. *Acta Chem. Scand.* 23:699–701

63. Gatenbeck S, Hermodsson S. 1965. Enzymic synthesis of the aromatic product, alternariol. *Acta Chem. Scand.* 19:65–71

64. Gaucher GM, Shepherd MC. 1968. Isolation of orsellinic acid synthase. *Biochem. Biophys. Res. Commun.* 32:664–71

64a. George KM, Chatterjee D, Gunawardana G, Welty D, Hayman J, et al. 1999. Mycolactone: a polyketide toxin from *Mycobacterium ulcerans* required for virulence. *Science* 283:854–57

65. Gerzon K, Flynn EH, Sigal MV, Wiley PF, Monahan R, Quarck UC. 1956. Erythromycin. VIII. Structure of dihydroerythronolide. *J. Am. Chem. Soc.* 78:6396–408

66. Grisebach H, Achenbach H, Hofheinz W. 1960. Untersuchungen zur Biogenese des Erythromycins. I. Mitt.: Der Aufbau des Lactonringes. *Z. Naturforsch. Teil B* 17:560–68

67. Gu Z-M, Zhao G-X, Oberlies NH, Zeng L,

McLaughlin JL. 1995. Annonaceous acetogenins. Potent mitochondrial inhibitors with diverse applications. *Rec. Adv. Phytochem.* 29:249–310

68. Haslam E. 1985. *Metabolites and Metabolism. A Commentary on Secondary Metabolism.* Oxford, UK: Clarendon

69. Haslam E. 1986. Secondary metabolism—fact and fiction. *Nat. Prod. Rep.* 3: 217–49

70. Herbert RB. 1981. *The Biosynthesis of Secondary Metabolites.* London: Chapman & Hall

71. Holzer KP, Liu W, Hammes GG. 1989. Molecular cloning and sequencing of chicken liver fatty acid synthase cDNA. *Proc. Natl. Acad. Sci. USA* 86:4387–91

72. Hopwood DA. 1997. Genetic contributions to understanding polyketide synthases. *Chem. Rev.* 97:2465–97

73. Hopwood DA, Khosla C. 1992. Genes for polyketide secondary metabolic pathways in microorganisms and plants. See Ref. 37, pp. 88–112

74. Hopwood DA, Malpartida F, Kieser HM, Ikeda H, Duncan J, et al. 1985. Production of "hybrid" antibiotics by genetic engineering. *Nature* 314:642–44

75. Hopwood DA, Merrick MJ. 1977. Genetics of antibiotic production. *Bacteriol. Rev.* 41:595–635

76. Hopwood DA, Sherman DH. 1990. Molecular genetics of polyketides and its comparison to fatty acid biosynthesis. *Annu. Rev. Genet.* 24:37–66

76a. Hudson AT, Campbell IM, Bentley R. 1970. Biosynthesis of 6-methylsalicylic acid by *Mycobacterium phlei*. *Biochemistry* 9:3988–92

77. Hutchinson CR. 1997. Antibiotics from genetically engineered microorganisms. See Ref. 143, pp. 683–702

78. Hutchinson CR, Fujii I. 1995. Polyketide synthase gene manipulation: a structure-function approach in engineering novel antibiotics. *Annu. Rev. Microbiol.* 49:201–38

79. Ikeda H, Omura S. 1997. Avermectin biosynthesis. *Chem. Rev.* 97:2591–609

80. Jacobsen JR, Hutchinson CR, Cane DE, Khosla C. 1997. Precursor-directed biosynthesis of erythromycin analogs by an engineered polyketide synthase. *Science* 277:367–69

81. Kakavas SJ, Katz L, Stassi D. 1997. Identification and characterization of the niddamycin polyketide synthase genes from *Streptomyces caelestis*. *J. Bacteriol.* 179:7515–22

82. Katz L. 1997. Manipulation of modular polyketide synthases. *Chem. Rev.* 97: 2557–72

83. Katz L, Donadio S. 1993. Polyketide synthesis: prospects for hybrid antibiotics. *Annu. Rev. Microbiol.* 47:875–912

84. Khosla C. 1997. Harnessing the biosynthetic potential of modular polyketide synthases. *Chem. Rev.* 97:2577–90

85. Kohler RE. 1975. The history of biochemistry: a survey. *J. Hist. Biol.* 8:275–318

86. Lipmann F. 1948–1949. Biosynthetic mechanism. *Harvey Lect. Ser.* XLIV:99–123

87. Little HW, Bloch K. 1950. Studies on the utilization of acetic acid for the biological synthesis of cholesterol. *J. Biol. Chem.* 183:33–46

88. Louden ML, Leete E. 1962. The biosynthesis of tropic acid. *J. Am. Chem. Soc.* 84:1510–11

89. Luckner M. 1984. *Secondary Metabolism in Microorganisms, Plants, and Animals.* Berlin/New York: Springer-Verlag. 3rd rev. ed.

90. Lynen F. 1959. Participation of acyl-CoA in carbon chain biosynthesis. *J. Cell. Comp. Physiol.* 54:33–49 (Suppl. 1)

91. Lynen F, Tada M. 1961. Die biochemische Grundlagen der "Polyacetat-Regel". *Angew. Chem.* 75:513–20

92. Malpartida F, Hallam SE, Kieser HM, Motamedi H, Hutchinson CR, et al. 1987. Homology between *Streptomyces* genes coding for synthesis of different polyketides used to clone antibiotic biosynthesis genes. *Nature* 325:818–21

93. Malpartida F, Hopwood DA. 1984. Molecular cloning of the whole biosynthetic pathway of a *Streptomyces* antibiotic and its expression in a heterologous host. *Nature* 309:462–64

94. Malpartida F, Hopwood DA. 1986. Physical and genetic characterization of the gene cluster for the antibiotic actinorhodin in *Streptomyces coelicolor* A3(2). *Mol. Gen. Genet.* 205:66–73

95. Mann J. 1981. *Secondary Metabolism.* Oxford, UK: Clarendon. 2nd ed.

96. Marsden AFA, Wilkinson B, Cortes J, Dunster NJ, Staunton J, Leadlay PF. 1998. Engineering broader specificity into an antibiotic-producing polyketide synthase. *Science* 279:199–202

97. Martin JF. 1977. Biosynthesis of polyene macrolide antibiotics. *Annu. Rev. Microbiol.* 31:13–38

98. McAlpine JB, Tuan JS, Brown DP, Grebner KD, Whittern DN, et al. 1987. New antibiotics from genetically engineered actinomycetes. I. 2-Norerythromycins, isolation and structural determination. *J. Antibiot.* 40:115–22

99. McDaniel R, Ebert-Khosla S, Fu H, Hopwood DA, Khosla C. 1994. Engineered biosynthesis of novel polyketides: influence of a downstream enzyme on the catalytic specificity of a minimal aromatic polyketide synthase. *Proc. Natl. Acad. Sci. USA* 91:11542–46

100. McDaniel R, Ebert-Khosla S, Hopwood DA, Khosla C. 1995. Rational design of aromatic polyketide natural products by recombinant assembly of enzymatic subunits. *Nature* 375:549–54

101. McDaniel R, Thamchaipenet A, Gustafsson C, Fu H, Betlach M, et al. 1999. Multiple genetic modifications of the erythromycin polyketide synthase to produce a library of novel "unnatural" natural products. *Proc. Natl. Acad. Sci. USA* 96:1846–51

102. Mill C. 1987. *Norman Collie. A Life in Two Worlds.* Aberdeen, Scotland: Aberdeen Univ. Press

103. Minto RE, Townsend CA. 1997. Enzymology and molecular biology of aflatoxin biosynthesis. *Chem. Rev.* 97:2537–55

104. Monaghan RL, Takacz JW. 1990. Bioactive microbial products: focus upon mechanisms of action. *Annu. Rev. Microbiol.* 44:271–301

105. Mosbach K. 1961. Die Rolle der Malonsäure in der Biosynthese der Orsellinsäure. *Naturwissenschaften* 15:525

106. Motamedi H, Hutchinson CR. 1987. Cloning and heterologous expression of a gene cluster for the biosynthesis of tetracenomycin C, the anthracycline antitumor antibiotic of *Streptomyces glaucescens. Proc. Natl. Acad. Sci. USA* 84:4445–49

107. Nisbet L. 1992. Useful functions of microbial metabolites. See Ref. 37, pp. 215–35

108. O'Hagan D. 1991. *The Polyketide Metabolites.* New York: Ellis Horwood

109. O'Hagan D. 1992. Biosynthesis of polyketide metabolites. *Nat. Prod. Rep.* 9: 447–79

110. O'Hagan D. 1993. Biosynthesis of fatty acid and polyketide metabolites. *Nat. Prod. Rep.* 10:593–624

111. O'Hagan D. 1995. Biosynthesis of fatty acid and polyketide metabolites. *Nat. Prod. Rep.* 12:1–32

112. Olivigni FJ, Bullerman LB. 1978. Production of penicillic acid and patulin by an atypical *Penicillium roqueforti* isolate. *Appl. Environ. Microbiol.* 35:435–38

113. Omura S, ed. 1992. *The Search for Bioactive Compounds from Microorganisms.* Berlin/New York: Springer-Verlag

114. Omura S, Ikeda H, Malpartida F, Kieser HM, Hopwood DA. 1986. Production of new hybrid antibiotics, mederrhodins A and B by a genetically engineered strain.

Antimicrob. Agents Chemother. 29:13–19

115. Omura S, Tsuzuki K, Nakagawa A, Lukacs G. 1983. Biosynthetic origin of carbons 3 and 4 of leucomycin aglycone. *J. Antibiot.* 36:611–13

116. Pieper R, Luo G, Cane DE, Khosla C. 1995. Cell-free synthesis of polyketides by recombinant erythromycin polyketide synthases. *Nature* 378:262–66

117. Raistrick H, Clark AB. 1919. On the mechanism of oxalic acid formation by *Aspergillus niger. Biochem. J.* 13:329–44

118. Rawls RL. 1998. Modular enzymes. *Chem. Eng. News.* 76(9):29–32. Issue of March 9

119. Richards JH, Hendrickson JB. 1964. *The Biosynthesis of Steroids, Terpenes, and Acetogenins.* New York: Benjamin

120. Rittenberg D, Bloch K. 1945. The utilization of acetic acid for the synthesis of fatty acids. *J. Biol. Chem.* 160:417–24

121. Robinson R. 1936. Synthesis in biochemistry. *J. Chem. Soc.* 1079–90

122. Robinson R. 1948. The structural relations of some plant products. *J. R. Soc. Arts* 96:795–808

123. Robinson R. 1955. *The Structural Relations of Natural Products.* Oxford, UK: Clarendon

124. Rose AH, ed. 1979. *Economic Microbiology, Vol. 3. Secondary Products of Metabolism.* New York: Academic

125. Ruan X, Stassi DL, Lax S, Katz L. 1997. A second type-I PKS gene cluster isolated from *Streptomyces hygroscopicus* ATCC 29253, a rapamycin-producing strain. *Gene* 203:1–9

126. Rudd BAM, Hopwood DA. 1979. Genetics of actinorhodin biosynthesis by *Streptomyces coelicolor* A3(2). *J. Gen. Microbiol.* 114:119–28

127. Ruzicka L. 1932. The life and work of Otto Wallach. *J. Chem. Soc.* 1582–97

128. Schorr R, Mittag M, Müller G, Schweizer E. 1994. Differential activities and in-

tramolecular location of fatty acid synthase and 6-methylsalicylic acid synthase component enzymes. *J. Plant Physiol.* 143:407–15

129. Schwecke T, Aparicio JF, Molnar I, Konig A, Khaw LE, et al. 1995. The biosynthetic gene cluster for the polyketide immunosuppressant, rapamycin. *Proc. Natl. Acad. Sci. USA* 92:7839–43

130. Schweizer M, Roberts LM, Holtke H-J, Takabayashi K, Hollerer E, et al. 1986. The pentafunctional FAS1 gene of yeast: its nucleotide sequence and order of catalytic domains. *Mol. Gen. Genet.* 203:479–86

131. Seow K-T, Meurer G, Gerlitz M, Wendt-Pienkowski E, Hutchinson CR, Davies J. 1997. A study of iterative type II polyketide synthases, using bacterial genes cloned from soil DNA: a means to access and use genes from uncultured organisms. *J. Bacteriol.* 179:7360–68

132. Sherman DH, Malpartida F, Bibb Maureen J, Kieser HM, Bibb Mervyn J, Hopwood DA. 1989. Structure and deduced function of the granaticin-producing polyketide synthase gene cluster of *Streptomyces violaceruber* TÜ22. *EMBO J.* 8:2717–25

133. Simpson TJ. 1991. The biosynthesis of polyketides. *Nat. Prod. Rep.* 8:573–602

134. Simpson TJ. 1998. Application of isotopic methods to secondary metabolic pathways. In *Biosynthesis, Polyketides and Vitamins. Top. Curr. Chem.* 195:1–48

135. Stanzak R, Matsushima P, Baltz RH, Rao RN. 1986. Cloning and expression in *Streptomyces lividans* of clustered erythromycin biosynthetic genes from *Streptomyces eyrthreus. Bio/Technology* 4:229–32

136. Staudinger H. 1905. Ketene, eine neue Körperklasse. *Ber. Deutsch. Chem. Gesellschaft* 38:1735–39

137. Staunton J. 1998. Combinatorial biosynthesis of erythromycin and complex polyketides. *Curr. Opin. Chem. Biol.* 2: 339–45

138. Staunton J, Wilkinson B. 1997. Biosynthesis of erythromycin and rapamycin. *Chem. Rev.* 97:2611–29

139. Stewart AW. 1920. *Recent Advances in Organic Chemistry*. London: Longman. 4th ed.

140. Stewart AW, Graham H. 1948. *Recent Advances in Organic Chemistry*. London: Longman. 7th ed.

141. Steyn PS, ed. 1980. *The Biosynthesis of Mycotoxins: A Study in Secondary Metabolism*. New York: Academic

142. Streicher SL, Ruby CL, Paress PS, Sweasy JB, Danis SJ, et al. 1989. Cloning the genes for avermectin biosynthesis in *Streptomyces avermitilis*. In *Genetics and Molecular Biology of Industrial Organisms*, ed. CL Hershberger, SW Queener, G Hegeman, pp. 44–52. Washington, DC: Am. Soc. Microbiol.

143. Strohl WR, ed. 1997. *Biotechnology of Antibiotics*. New York: Marcel Dekker. 2nd ed.

144. Strohl WR. 1997. Industrial antibiotics: today and the future. See Ref. 143, pp. 1–48

145. Strohl WR, Bartel PL, Li Y, Connors NC, Woodman RH. 1991. Expression of polyketide biosynthesis and regulatory genes in heterologous streptomycetes. *J. Ind. Microbiol.* 7:161–74

146. Stutzman-Engwall KJ, Hutchinson CR. 1989. Multigene families for anthracycline antibiotic production in *Streptomyces peucetius. Proc. Natl. Acad. Sci. USA* 86:3135–39

147. Takenaka S, Seto S. 1975. Biosynthesis of sepedonin by cell-free extract of *Sepedonium chrysospermum. Sci. Rep. Res. Inst. Tohoku Univ. Ser. A* 25:25–30

148. Tanabe M. 1973. Stable isotopes in biosynthetic studies. In *Biosynthesis. A Specialist Periodical Report*, Vol. 2, TA Geismann, pp. 241–299. London: The Chemical Society

149. Tanenbaum SW, Bassett EW. 1962. Cell-free biosynthesis of the tropolone ring. *Biochim. Biophys. Acta* 59:524–26

150. Tatum EL, Adelberg EA. 1951. Origin of the carbon skeletons of isoleucine and valine. *J. Biol. Chem.* 190:843–52

151. Taylor WC. 1973. *The Snows of Yesteryear.* Toronto: Holt, Rinehart & Winston

152. Thompson CJ, Skinner RH, Thompson J, Ward JM, Hopwood DA, Cundliffe E. 1982. Biochemical characterization of resistance determinants cloned from antibiotic-producing *Streptomycetes. J. Bacteriol.* 151:678–85

153. Thompson CJ, Ward JM, Hopwood DA. 1980. DNA cloning in *Streptomyces*: resistance genes from antibiotic producing species. *Nature* 286:525–27

154. Torsell KBG. 1983. *Natural Product Chemistry. A Mechanistic and Biosynthetic Approach to Secondary Metabolism.* Chichester, UK: Wiley

155. Trail F, Manhanti N, Linz J. 1995. Molecular biology of aflatoxin biosynthesis. *Microbiology* 141:755–65

156. Turner WB. 1971. *Fungal Metabolites.* London: Academic

157. Vaněk Z, Majer J, Babický A, Liebster J, Vereš K. 1958. Studies on the biosynthesis of erythromycin with the aid of substrates labeled with C^{14}. *Proc. Second Int. Conf. Peaceful Uses At. Energ.* 25:143–46

158. Vaněk Z, Půža M, Majer J, Doležilová L. 1961. Contribution to the biosynthesis of erythromycin in the presence of propionic acid-1-^{14}C. *Folia Microbiol.* 6:408–10

159. Vining LC, ed. 1983. *Biochemistry and Genetic Regulation of Commercially Important Antibiotics.* Reading, MA: Addison-Wesley

160. Wang I-K, Reeves C, Gaucher GM. 1991. Isolation and sequencing of a genomic DNA clone containing the 3′-terminus of the 6-methylsalicylic acid polyketide synthase gene of *Penicillium urticae. Can. J. Microbiol.* 37:86–95

161. Webb EC. 1992. *Enzyme Nomenclature (Recommendations of the Nomenclature Committee of the International Union of Biochemistry and Molecular Biology on the Nomenclature and Classification of Enzymes).* San Diego: Academic

162. Weissman KJ, Bycroft M, Staunton J, Leadlay PF. 1998. Origin of starter units for erythromycin biosynthesis. *Biochemistry* 37:11012–17

163. Weissman KJ, Timoney M, Bycroft M, Henefeld GP, Staunton J, Leadlay PF. 1997. The molecular basis of Celmer's rules: the stereochemistry of the condensation step in chain extension of the erythromycin polyketide synthase. *Biochemistry* 36:13848–55

164. Wiesmann KEH, Cortes J, Brown MJB, Cutter AL, Staunton J, Leadlay PF. 1995. Polyketide synthesis in vitro on a modular polyketide synthase. *Chem. Biol.* 2:583–89

165. Wiesner P, Beck K-F, Ripka S, Müller G, Lucke S, Schweizer E. 1988. Isolation and sequence analysis of the fatty acid synthetase *FAS2* gene from *Penicillium patulum. Eur. J. Biochem.* 177:69–79

166. Wilsmore NTM. 1907. Ketene. *Proc. Chem. Soc.* 23:229–30

167. Wilsmore NTM, Stewart AW. 1907. Ketene. *Nature* 75:510

168. Woodward RB. 1956. Neuere Entwicklungen in der Chemie der Naturstoffe. *Angew. Chem.* 68:13–20

169. Woodward RB. 1957. Struktur und Biogenese der Makrolide. *Angew. Chem.* 69:50–58

170. Xue Y, Wilson D, Zhao L, Liu H-w, Sherman DH. 1998. Hydroxylation of macrolactones YC-17 and narbomycin is mediated by the *pikC*-encoded cytochrome P450 in *Streptomyces venezuelae. Chem. Biol.* 5:661–67

171. Xue Y, Zhao L, Liu H-w, Sherman DH. 1998. A gene cluster for macrolide antibiotic synthesis in *Streptomyces venezuelae*: Architecture of metabolic diversity.

Proc. Natl. Acad. Sci. USA 95:12111–16

172. Yuan Z, Liu W, Hammes GG. 1988. Molecular cloning and sequencing of DNA complementary to chicken liver fatty acid synthase mRNA. *Proc. Natl. Acad. Sci. USA* 85:6328–31

173. Zhao L, Sherman DH, Liu H-w. 1998. Biosynthesis of desosamine: construction of a new methymycin/neomycin analogue by deletion of a desosamine biosynthetic gene. *J. Am. Chem. Soc.* 120:10256–57

NOTE ADDED IN PROOF

A relevant review is as follows: Khosla C, Gokhale RS, Jacobsen JR, Cane DE. 1999. Tolerance and specificity of polyketide synthases. *Annu. Rev. Biochem.* 68: 219–53. In press

Annu. Rev. Microbiol. 1999. 53:447–94

Giant Viruses Infecting Algae

James L. Van Etten[1] and Russel H. Meints[2]

[1]*Department of Plant Pathology, University of Nebraska, Lincoln, Nebraska 68583-0722; e-mail: jvanetten@unlnotes.unl.edu;* [2]*Department of Botany and Plant Pathology, Center for Gene Research and Biotechnology, Oregon State University, Corvallis, Oregon 97331-2906*

Key Words   chlorella viruses, brown algal viruses, DNA restriction/modification enzymes, virus-encoded glycosylation, lysogeny

■ **Abstract**   *Paramecium bursaria* chlorella virus (PBCV-1) is the prototype of a family of large, icosahedral, plaque-forming, double-stranded–DNA-containing viruses that replicate in certain unicellular, eukaryotic chlorella-like green algae. DNA sequence analysis of its 330, 742-bp genome leads to the prediction that this phycodnavirus has 376 protein-encoding genes and 10 transfer RNA genes. The predicted gene products of ~40% of these genes resemble proteins of known function. The chlorella viruses have other features that distinguish them from most viruses, in addition to their large genome size. These features include the following: (*a*) The viruses encode multiple DNA methyltransferases and DNA site-specific endonucleases; (*b*) PBCV-1 encodes at least part, if not the entire machinery to glycosylate its proteins; (*c*) PBCV-1 has at least two types of introns—a self-splicing intron in a transcription factor–like gene and a splicesomal processed type of intron in its DNA polymerase gene. Unlike the chlorella viruses, large double-stranded–DNA-containing viruses that infect marine, filamentous brown algae have a circular genome and a lysogenic phase in their life cycle.

CONTENTS

0066-4227/99/1001-0447$08.00   **447**

INTRODUCTION

Since the early 1970s, viruses or virus-like particles have been reported in at least 44 taxa of eukaryotic algae, which include members in 10 of the 14 classes of alga (171). However, most of these reports described isolated accounts of microscopic observations, and the virus particles were not characterized further for many reasons. This situation changed in the early 1980s with the discovery of large double-stranded–DNA (dsDNA)-containing viruses that infect and replicate in certain strains of unicellular, eukaryotic, exsymbiotic chlorella-like green algae (also referred to as zoochlorellae) and more recently with the finding of tractable virus systems in brown algae.

In 1978, Kawakami & Kawakami (60) described the appearance of large (180-nm diameter), lytic viruses in zoochlorellae after these algae were released from the protozoan *Paramecium bursaria*. No virus particles were detected in zoochlorellae growing symbiotically inside the paramecium. Independently, lytic viruses were described in zoochlorellae isolated from the green coelenterate *Hydra viridis* (93, 173) and also from *P. bursaria* (174). As in the 1978 report, viruses appeared only after the zoochlorellae were separated from their hosts. Fortunately, the zoochlorellae from *P. bursaria* can be grown free of the paramecium in culture, and ensuing experiments have revealed that these cultured algae serve as hosts for many closely related viruses. These lytic chlorella viruses can be produced in large quantities and assayed by plaque formation with standard bacteriophage techniques (159, 167). The chlorella viruses have several unique properties including the following: (*a*) They have large, 330- to 380-kb genomes (128, 195), and analyses of the recently sequenced 330,742-bp genome of the prototype chlorella virus, *Paramecium bursaria* chlorella virus 1 (PBCV-1), have revealed many interesting and unexpected putative genes. (*b*) The viruses encode multiple DNA methyltransferases and DNA site-specific (restriction) endonucleases. (*c*) PBCV-1 encodes at least part, if not all, of the intracellular machinery used to glycosylate its proteins. (*d*) PBCV-1 has at least two different types of introns, a self-splicing intron in a transcription factor (TF)-like gene and a splicesomally processed intron in its DNA polymerase gene. Of the 10 PBCV-1–encoded transfer RNA (tRNA) genes, 1 may also contain a small intron.

In addition to the chlorella viruses, a few large icosahedral, dsDNA-containing viruses that infect marine algae are under active investigation. These include viruses that infect the unicellular alga, *Micromonas pusilla* (MpV viruses) (19, 87), and viruses that infect the filamentous brown algae, *Ectocarpus* sp. (EsV viruses)

TABLE 1   Representative algal viruses

Host	Virus	Genome size (kb)	%5mC[a]	%6mA[b]
Chlorella NC64A	CA-4B	~330	0.12	ND[c]
	AL-1A	~330	0.45	ND
	PBCV-1	330	1.9	1.5
	IL-3A	~335	9.7	ND
	SC-1A	~330	1.9	7.3
	NC-1A	~335	7.1	7.3
	XZ-3A	~330	12.8	2.2
	BJ-2C	~330	12.8	11.5
	NE-8A	~330	14.3	8.1
	NYs-1	~345	47.5	11.3
	AL-2A	~330	35.8	14.6
	CA-4A	~335	39.8	19.6
	NY-2A	~380	44.9	37.0
	XZ-6E	~330	21.2	15.2
	XZ-4C	~330	46.7	20.8
	XZ-4A	~330	44.1	28.3
Chlorella Pbi	CVA-1	>300	43.1	ND
	CVB-1	>300	42.7	17.7
	CVG-1	>300	19.2	ND
	CVM-1	>300	41.9	10.1
	CVR-1	>300	14.2	ND
Ectocorpus siliculosus	EsV1	320	1	3
Feldmannia sp.	FsV-1	158	?	?[d]
	FsV-2	178	?	?
Micromonas pusilla	MpV-1	?	?	?

[a]Percentage of 5^{m}C per C plus 5mC plus deoxyuridine.

[b]Percentage of 6mA per A plus 6mA plus deoxyinosine.

[c]ND, is none detected.

[d]Question marks are unknown.

(104) and *Feldmannia* sp. (FsV viruses) (50) (Table 1). EsV viruses, which are transmitted from cell to cell in a Mendelian fashion, and FsV viruses have a lysogenic phase in their life cycles and produce virus particles only in differentiated sporangial cells of their hosts.

The chlorella viruses and the marine algal viruses are widely distributed in nature, and they have been isolated from fresh water or seawater collected throughout the world. Phylogenetic trees constructed from virus-encoded DNA polymerases indicate that all of these algal viruses probably have a common evolutionary heritage. This review focuses primarily on the chlorella and brown algal viruses. However, because the chlorella viruses and brown algal viruses have different life styles, they are described separately. Information on the early history of the algal viruses has been reviewed by Van Etten et al (171) and Fuhrman & Suttle (29). Muller et al (103) have written a recent review on certain aspects of the viruses in marine brown algae, and Suttle (153) has written a review on ecological aspects of these fascinating viruses. Finally, the first meeting ever devoted exclusively to algal viruses was held in Bergen, Norway, in June 1998.

CHLORELLA VIRUSES

Algae included in the genus *Chlorella* are among the most widely distributed and frequently encountered algae on Earth. *Chlorella* algae include small, spherical, unicellular, nonmotile, asexually reproducing green algae (164). *Chlorella* species have a rigid cell wall and typically have a single chloroplast, which sometimes contains a pyrenoid body. They have a simple developmental cycle and reproduce by mitotic division. Vegetative cells increase in size and, depending on the species and environmental conditions, divide into two, four, eight, or more progeny, which are released by rupture or enzymatic digestion of the parental walls. However, algae assigned to the genus *Chlorella* are more heterogeneous than their simple morphology suggests. This degree of heterogeneity is illustrated with two examples. (*a*) The nuclear DNAs of algae assigned to the genus have G+C contents ranging from 43% to 79% (49). However, most isolates that have been assigned to the same species contain similar G+C contents. (*b*) The cell wall polysaccharides of *Chlorella* species also vary widely (80, 158, 197), even among isolates assigned to the same species, e.g. *Chlorella vulgaris* and *Chlorella ellipsoidea* (197). Because of the heterogeneity in this genus, properties discovered in one *Chlorella* species or isolate may not apply to another species.

Most *Chlorella* species are free-living in nature. However, some forms, called zoochlorellae or *Chlorella*-like algae, live as hereditary endosymbionts within fresh water and, to less extent, marine animals (e.g. 122, 162). To our knowledge, the only hosts for the chlorella viruses are these symbiotic zoochlorellae, some of which can be cultured under laboratory conditions. These zoochlorellae have not been assigned species status.

Chlorella viruses infect either *Chlorella* strain NC64A and its equivalent strains (NC64A viruses) or *Chlorella* strain Pbi (Pbi viruses). About 50 of the NC64A viruses have been partially characterized. These viruses can be grouped into at least 16 classes identified by plaque size, antiserum sensitivity, DNA restriction patterns, sensitivity of the DNAs to restriction endonucleases, and, most importantly, the nature and abundance of methylated bases in their DNAs. One member of each of the 16 virus classes is listed in Table 1. Five Pbi viruses are also listed in Table 1.

PBCV-1 Morphology

The chlorella viruses are large icosahedra with multilaminate shells surrounding electron-dense cores. Depending on the microscopic techniques, diameters of 140–190 nm have been reported for the viruses (171). A three-dimensional reconstruction of PBCV-1 was recently created from 356 cryoelectron micrographs of PBCV-1 (Figure 1) (199). The image, which has a resolution limit of 26 angstroms, reveals several features about PBCV-1 morphology. The capsid has a distinct icosahedral shape, with a maximum diameter of 190 nm along the fivefold axes. The outer dimensions of the virion at the two- and threefold axes are both ~165 nm. The capsid consists of 1692 doughnut-like capsomers arranged in a T = 169, skew icosahedral lattice (h = 8, k = 7) (14). Twelve capsomers, each ~6.4 nm in diameter, occur as pentamers at the fivefold vertices. The remaining 1680 capsomers are trimeric structures, each about 6.7 nm in diameter and 7.5 nm high. The prominent, cylinder portion of each trimeric capsomer extends 5 nm above the surface of the capsid shell, and most appear to have axial channels. The capsomers interconnect at their bases in a contiguous shell with a thickness of 2–2.5 nm. The stoichiometry of the trimers indicates that the PBCV-1 capsid contains 5040 copies of the major capsid protein, Vp54. The difference in oligomer state and size of the two types of capsomers suggests that the pentamers are composed of one of the minor capsid proteins. In addition, unlike the trimeric capsomers, each pentamer has a cone-shaped, axial cavity at its base.

Other types of microscopy suggest that the surface of PBCV-1 has additional features beyond that of an icosahedron. For example, PBCV-1 particles prepared by the quick-freeze, deep-etch procedure have flexible, hairlike fibers that extend from at least some of the virus vertices. The tips of these appendages are swollen and may aid in attachment of the virus to the host (see Figure 3*B* and Figure 8 in reference 171). Negative staining suggests that the virions contain a distinctive 20- to 25-nm spike or tail extending from one vertex and that the DNA within the particle is retracted from this unique vertex (see Figure 3*C* of reference 171). The vertex containing this spike may be the one involved in digesting the host cell wall at the point of attachment (see below). Ultrastructural studies on a Pbi virus also suggested that one vertex is unique (7).

Physical and Chemical Properties of PBCV-1

PBCV-1 has a sedimentation coefficient of about 2300 S in sucrose density gradients (170) and an estimated molecular mass of 1×10^9 Daltons (201). From 25% to 50% of the viral particles are infectious and form plaques (170). The PBCV-1 virion contains 64% protein, 21% to 25% DNA, and 5% to 10% lipid (143). The lipid component is located inside the outer glycoprotein capsid shell and is required for virus infectivity. PBCV-1 particles contain more than 50 polypeptides, which range in size from 10 to 280 kDa (117, 143). The PBCV-1 major capsid protein weighs 53,790 Daltons (Vp54) (42; R Cerny, JL Van Etten, unpublished results)

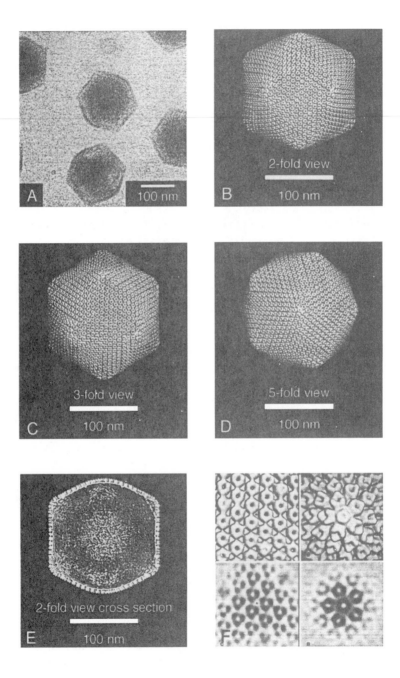

and composes about 40% of the total viral protein. Protein Vp54, along with two other PBCV-1 proteins Vp51 and Vp27.5, is myristoylated, and at least six of the viral proteins are phosphorylated (117, 118). The myristic acid is attached by amide linkages to the three PBCV-1 proteins. Surprisingly, the myristic acid of Vp54 is not attached to the N terminus but probably to an internal lysine ε-amino group (117).

At least three PBCV-1 proteins, the major capsid protein Vp54 as well as Vp280 and Vp260, are glycosylated (42, 117, 179). The glycan portion of Vp54 is on the external surface of PBCV-1 and undoubtedly contributes to the protease resistance of the virus and to its antigenicity. The glycan is probably linked to the protein by a Ser/Thr O-linkage (117, 179). The amino acid sequence of the major capsid protein predicts a molecular size of 48,116 Daltons (the N-terminal Met is removed from the protein). Therefore, the additional weight of the protein (5674 Daltons) results from glycosylation, myristoylation, and possibly other unknown post-translational modifications.

Virion proteins from NC64A virus CVK2, which is closely related to PBCV-1, were separated into 10 capsid proteins and at least 40 core proteins (196). Comparison of the N-terminal amino acids of seven of the capsid proteins with the corresponding open reading frames (ORFs) deduced from the PBCV-1 genomic sequence indicated that the N termini of six of the CVK2 proteins were proteolytically processed (146). Three of the proteins were missing the N-terminal methionine, presumably removed by methionine aminopeptidase. Two of the apparent PBCV-1 homologs were missing a hydrophobic 25- to 30-amino-acid signal peptide-like region at their amino ends. Each of the predicted proteolytic cleavage sites for both of these proteins contains a lysine with an acidic amino acid on its N-terminal side. Seven of the insoluble core proteins had strong DNA-binding activity (193).

PBCV-1 Genome

The PBCV-1 genome is a linear, 330-kb, nonpermuted dsDNA with covalently closed hairpin termini (35, 128). Some of the chlorella viruses have genomes as large as 380 kb (128, 195). The termini of the PBCV-1 genome consist of

←—————————————————————————————————

Figure 1   *A*. Micrograph of vitrified chlorella virus PBCV-1. This image shows a well-defined outer capsid and a nonuniformly distributed interior mass. *B–D*. Shaded-surface view of a three-dimensional reconstruction viewed along (*B*) a twofold axis, (*C*) a threefold axis, and (*D*) a fivefold axis. *E*. Reconstruction of a cross section of the virus from a twofold view. *F*. Close-up views along threefold (*left*) and fivefold (*right*) axes of the PBCV-1 density map, represented as *shaded-surfaces* (*top*) or as density projections (*bottom*). The trimeric nature of the hexavalent capsomers is clearly evident. Projection images include planar slabs of density, 3.3 nm (*left*) and 4.4 nm (*right*) thick, and include the most radially extended features in each view.

35-nucleotide-long, incompletely base-paired, covalently closed hairpin loops that exist in one of two forms; the two forms are complementary when the 35 nucleotide sequences are inverted (flip-flop) (208). An identical 2221-bp inverted repeat is adjacent to each hairpin end (152). The remainder of the PBCV-1 genome contains primarily single-copy DNA (35).

The sequences of the inverted-repeat regions differ among NC64A viruses. For example, 8 of 37 viruses did not hybridize with the PBCV-1 repeat element, and the remaining 29 viruses hybridized to various degrees (152). Furthermore, the genome of NC64A virus CVK1 has inverted terminal repeats of about 1 kb; the sequence of this repeat has no obvious resemblance to the PBCV-1 repeat region (194).

The PBCV-1 genome (not counting the 35-nucleotide-long hairpin ends) was recently sequenced (330,742 bp) and analyzed (68, 77, 78, 82, 83). The PBCV-1 genome contains 701 potential protein-coding regions or ORFs, defined as continuous stretches of DNA that translate into a polypeptide initiated by an ATG translation start codon and extend for 65 or more codons. The ORFs were numbered consecutively starting with ORF a1L at the extreme left terminus of the genome and ending with ORF a692R. The 701 ORFs are higher than the last numbered ORF a692R because 10 small ORFs were inadvertently overlooked during the initial analysis. In addition, correction of a DNA-sequencing error revealed that two adjacent ORFs are fused (L Sun & JL Van Etten, unpublished results). The letters R and L following the ORF number indicate that the transcript is either in a left-to-right or right-to-left orientation, respectively.

The 701 ORFs were divided into 376, mostly nonoverlapping ORFs (major ORFs, labeled "A"), which are believed to encode proteins, and 325 minor ORFs (labeled "a") which may or may not encode proteins. When ORFs overlapped extensively, the larger ORF was classified as a major ORF and the smaller ORF(s) was classified as a minor ORF. The 50 nucleotides preceding the start codon of most (293 of 376) of the major ORFs have A+T concentrations >70%. In contrast, the 50 bases preceding the start codon of most (289 of 325) of the minor ORFs have A+T concentrations <70%. Previous studies indicate that promoter regions of protein encoding ORFs are usually A+T rich (135).

The 376 major ORFs are evenly distributed along the genome and, with one exception, there is little intergenic space between them. The exception is a 1788-bp sequence near the middle of the genome. This region, which has numerous stop codons in all reading frames, encodes 10 tRNA genes. The middle 900 bp of this intergenic region also resemble a "CpG island" (2).

The PBCV-1 genome shows interspersed leftward and rightward transcription throughout the genome. Overall, 183 major PBCV-1 ORFs are transcribed in the left-to-right orientation and 193 in the right-to-left orientation. The first four PBCV-1 ORFs, a1L, A2L, A3R, and a4l, reside in the inverted terminal repeat region of the PBCV-1 genome and hence are duplicated at the right end of the genome (83, 152).

One unusual feature of chlorella virus DNAs is that they contain relatively high levels of methylated bases (175). In fact, chlorella viruses can be distinguished from one another by the site specificity and abundance of DNA methylation (Table 1) (see reviews in 108, 109, 171). DNA from each chlorella virus contains 5-methylcytosine (5mC) in amounts ranging from 0.1% to 47.5% of the total cytosine. Many viruses also contain N^6-methyladenine (6mA) in amounts ranging from 1.45% to 37% of the total adenine (Table 1). Regardless of the level of methylation, all of the NC64A viruses have a G+C content of ~40%, and the Pbi viruses have a G+C content of ~46% (125).

PBCV-1 Life Cycle

PBCV-1 attaches rapidly and irreversibly to the external surface of *Chlorella* NC64A cell walls but not to other *Chlorella* strains (91). The virus attaches to the cell wall via one of its hexagonal vertices, digests the wall at the attachment point, and releases viral DNA into the host leaving an empty capsid on the cell surface. Because the virus also attaches to and digests cell wall fragments, which have been boiled or extracted by harsh procedures, the wall-degrading enzyme(s) must be packaged in the virus particles (90, 126). Release of the virus DNA must require a host function because attachment to wall fragments and digestion of the wall at the point of attachment does not release viral DNA.

Stereo views reveal that PBCV-1 attaches to the wall by the hairlike fibers that originate from at least some virus vertices. These micrographs suggest that the tips of the hairlike fibers are responsible for the initial recognition and attachment of the virus to the host receptor (see Figure 8 of reference 171). Preliminary results indicate that the host receptor for PBCV-1 is the sugar portion of a lipopolysaccharide-like component (TS Hagge & RL Pardy, unpublished results). The discovery that *Chlorella* NC64A cell walls have a lipopolysaccharide component (129) was unexpected because lipopolysaccharides had previously been found only in the walls of gram-negative bacteria.

Two observations indicate that the infecting PBCV-1 DNA and probably DNA-associated proteins are targeted to the nucleus, where early virus transcription occurs. First, PBCV-1 does not encode a recognizable RNA polymerase gene(s), nor were we able to detect RNA polymerase activity in isolated virions (J Rohozinski & JL Van Etten, unpublished results). Second, a small intron with the splice site sequences characteristic of a nuclear-spliced mRNA is present in the PBCV-1 DNA polymerase gene (41). Presumably this intron would need to be excised in the infected cell nucleus. The signal(s) involved in targeting the infecting virus DNA to the nucleus is unknown. However, the process is rapid because early virus mRNAs are synthesized within 5–10 min postinfection (p.i.). Virus replication does not require a functional host nucleus because PBCV-1 replicates, albeit poorly and with a small burst size, in UV-irradiated cells (168). PBCV-1 proteins are synthesized on cytoplasmic ribosomes and not on organelle ribosomes because cycloheximide, but not chloramphenicol, inhibits virus replication (170).

Virus DNA replication begins ~1 h p.i. and is followed by transcription of late virus genes (134, 166). Virus DNA synthesis presumably takes place in the nucleus

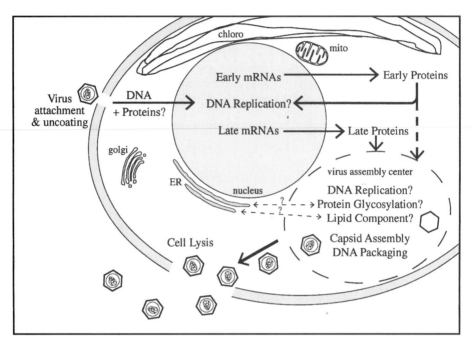

Figure 2 Proposed life cycle for chlorella virus PBCV-1. The virus uncoats at the surface of the alga, and the DNA, possibly with associated proteins, goes to the nucleus where early gene transcription begins within 5–10 min after infection. The early mRNAs are transported to the cytoplasm for translation, and the early proteins presumably return to the nucleus to initiate DNA replication beginning 60–90 min after infection, followed by late gene transcription. Late mRNAs are transported to the cytoplasm for translation, and many of these late proteins are targeted to the cytoplasmically located virus assembly centers, where virus capsids are formed. However, some of the virus capsid proteins are glycosylated, myristoylated, and phosphorylated at some intracellular location before assembly into capsids. Virus DNA also has to move from the nucleus to the virus assembly centers for additional replication and packaging into virions. Infectious PBCV-1 particles appear in the cytoplasm 50–60 min before release from the cell, which occurs by localized lysis of the cell wall.

along with late transcription. An intracellular traffic pattern for PBCV-1 replication is proposed in Figure 2. Host RNA and protein syntheses are inhibited at least 75% after PBCV-1 infection. Chloroplast rRNAs, but not cytoplasmic rRNAs, are degraded beginning ~30 min p.i. Host nuclear and chloroplast DNAs are degraded beginning 1 h p.i., even though the total DNA in the cell increases 4- to 10-fold by 4 h p.i. (88, 166). Thus PBCV-1 DNA synthesis requires a large increase in deoxyribonucleoside triphosphates (dNTP) after infection. Degraded host DNA may supply some of these intermediates; however, virus replication requires 4- to 10-fold the dNTPs that recycling can provide. As noted below, PBCV-1 encodes many putative enzymes involved in dNTP metabolism.

PBCV-1 infection rapidly inhibits host CO_2 fixation (170) and photosynthesis (136). PBCV-1 does not require host photosynthesis for replication because the virus replicates in dark-grown algae as well as light-grown cells treated with the photosynthetic inhibitor DCMU (170). However, the virus burst size is reduced $\sim$50% under these conditions. Like most bacterial viruses (48), PBCV-1 replicates most efficiently in actively growing host cells and poorly in stationary-phase cells (170)

The host cells change cytologically beginning $\sim$1 to 2 h p.i.; the nucleus, mitochondria, and Golgi apparatus become appressed to the chloroplast, leaving one or more finely granulated electron-translucent areas in the cytoplasm (92). Because PBCV-1 virions ultimately assemble in these translucent areas, the areas have been designated virus assembly centers. Capsids usually form at the periphery of the virus assembly centers, occasionally producing a rosette pattern. Complete virus capsids appear to assemble before DNA packaging. Presumably DNA is packaged through an opening in the capsid. By 4 to 5 h p.i., the cells contain many filled virus particles, which are distributed throughout the cytoplasm. Progeny PBCV-1 are first released $\sim$4 to 5 h p.i., and the majority of infectious virus particles are released by 8 h p.i. PBCV-1 release involves localized lysis of the cell wall, presumably by a late viral gene product(s). Mechanical disruption of the cells releases infectious virions 30–50 min before spontaneous lysis (170). Consequently, infectious PBCV-1 is assembled inside the host, and the virus does not acquire its lipid components and glycoproteins by budding through an exterior host membrane. The typical PBCV-1 burst size is 200–350 plaque-forming units (PFU) (170).

Because PBCV-1 is assembled in specific regions in the cytoplasm, the cytoskeleton may participate in either targeting the virus proteins to the virus assembly centers or assembling virus proteins into capsids. However, several cytoskeleton-disrupting agents, which inhibit either tubulin or actin functions, had no effect on the formation of PBCV-1 virus assembly centers or PBCV-1 replication even at concentrations several fold higher than those that inhibited host chlorella growth (110). As noted above, some PBCV-1–encoded proteins are extensively post-translationally modified, for example, the N-terminal methionine or signal peptides are removed, and proteins are phosphorylated, myristoylated, and/or glycosylated. The intracellular site(s) where these events take place is unknown. One difference between PBCV-1 and other glycoprotein-containing viruses is that PBCV-1 encodes at least part, if not all, of the machinery required to glycosylate its three structural glycoproteins (see below).

PBCV-1 Transcription

PBCV-1 transcription can be divided into early and late stages. The junction between these stages is 60 to 90 min p.i., which coincides with the initiation of virus DNA synthesis (134). It is interesting that the incorporation of [^{3}H]adenine into polyadenylate-containing RNA also changes abruptly at about 1 h p.i. Early virus transcripts are probably polyadenylated, but late transcripts are virtually devoid of polyadenylate segments (176).

Only two studies (134, 135) have examined PBCV-1 RNA synthesis. Consequently only general statements can be made about virus transcription. (*a*) Viral infection rapidly inhibits host RNA synthesis. (*b*) PBCV-1 transcription is programmed, and early transcripts appear within 5–10 min p.i. (*c*) Late viral transcription begins ~60 to 90 min p.i., after viral DNA synthesis begins. (*d*) Some early transcripts continue to be synthesized after virus DNA synthesis begins. (*e*) Early and late virus genes are interspersed throughout the PBCV-1 genome. (*f*) The sum of the sizes of the mRNAs hybridizing to viral DNA probes is often 40% to 60% larger than the probe, suggesting that PBCV-1 either contains overlapping genes, transcribes both strands of DNA, or extensively processes RNA after transcription or that some transcripts are polycistronic. Transcription studies with individual PBCV-1 genes indicate that some genes produce the expected size mRNA as either an early or late gene, whereas other genes produce complex RNA patterns, even with single-stranded–DNA (ssDNA) probes. (*g*) Transcriptional mapping of seven PBCV-1 genes indicates that the start sites are within 150 nucleotides upstream from the translational start codon, producing mRNAs with 5′ untranslated regions of as little as 14 nucleotides up to 149 nucleotides; transcripts also extend beyond the translational stop codon (42, 43, 135). To date, no early or late PBCV-1 promoters have been identified, although Schuster et al (135) noted that the regions upstream from the translation start codons for five major ORFs are A+T rich. As noted above, the content of the 50 bases preceding most of the major PBCV-1 ORFs is at least 70% A+T. (*h*) Codon usage by PBCV-1 is biased to codons ending in XXA/U (63%) over those ending in XXC/G (37%) (135). This bias is expected, because PBCV-1 DNA is 40% G+C. In contrast, the codon usage of the host tubulin gene has a strong bias to codons ending in XXC/G (67%) (RH Meints, MV Graves, unpublished results). This result is also consistent with the 67% G+C content of the host nuclear DNA (175). Therefore, the fact that PBCV-1 encodes 10 tRNA genes is not surprising.

Similarities of PBCV-1 Open Reading Frames to Proteins in the Databases

Computer analyses of the 376 PBCV-1 major ORFs indicate that ~40% resemble proteins in databases, including many interesting and unexpected enzymes. Table 2 lists 77 genes that match database ORFs of known function and, in a few cases, indicates whether the PBCV-1 ORF is transcribed early (E) or late (L). The predicted enzyme activities of some PBCV-1–encoded proteins have been confirmed by complementation of *Escherichia coli* mutants and/or bioassay of recombinant proteins (indicated with an asterisk in Table 2). Additional PBCV-1 ORFs have significant homology to ORFs of unknown function.

Some PBCV-1–encoded proteins resemble those of bacteria and phages, such as DNA restriction endonucleases and methyltransferases. However, other virus-encoded proteins resemble those of eukaryotic organisms and their viruses, such as translation elongation factor-3 (EF-3), RNA guanyltransferase, and proliferating cell nuclear antigen (PCNA). The PBCV-1 genome thus is a mosaic of prokaryotic and eukaryotic genes, suggesting considerable gene exchange in nature during the evolution of these viruses. A brief description of some of the PBCV-1 encoded proteins follows.

DNA Replication- and Repair-Associated Proteins   PBCV-1 encodes at least 11 ORFs that are involved in either replicating or interacting with DNA, including a B-family (α-like) DNA polymerase (41), an ATP DNA ligase (53), a type-II DNA topoisomerase, three helicases, two ORFs that resemble PCNA proteins, a cyclin A/cdk-associated protein, and a pyrimidine dimer-specific glycosylase endonuclease (PDG) (30).

The gene encoding the PBCV-1 DNA polymerase contains a 101-nucleotide intron with 5'-AG/GUGAG and 3'-GCAG/U splice site sequences, as well as a possible branch point UCAC sequence (41). These sequences are hallmarks of spliceosomal spliced mRNA introns (44). A similar 86-bp intron is also present in the DNA polymerase gene of another NC64A virus NY-2A (41). Experiments in progress indicate that the DNA polymerase genes from 38 other NC64A viruses also have an intron at the same position (Y Zhang, DB Burbank, JL Van Etten, unpublished results). Interestingly, the DNA polymerase gene from five viruses that infect *Chlorella* Pbi lack the intron, as do the DNA polymerase genes from the brown algal virus FsV (75a) and several micromonas viruses and chrysochromulina viruses (16).

ATP-dependent DNA ligases, which are encoded by several viruses—including vaccinia virus, African swine fever virus (ASFV), and bacteriophages T3, T4, and T7—join two DNA fragments by catalyzing the formation of internucleotide phosphodiester bonds. They range in size from 268 amino acids in *Haemophilus influenzae* (17) to 1070 amino acids in *Xenopus laevis* (76). The 298-amino-acid-residue PBCV-1 enzyme, which is the second smallest DNA ligase described to date, has been expressed in *E. coli*, and the recombinant enzyme has been extensively characterized (53, 147–149).

The 1061-amino-acid-residue PBCV-1 type-II DNA topoisomerase has >45% amino acid identity to type-II topoisomerases from several eukaryotic organisms. Type-II topoisomerases are ATP dependent and act by passing a dsDNA segment through a transient double-strand break (127). ASFV is the only other virus infecting a eukaryotic cell that encodes a type-II DNA topoisomerase (31). However, the PBCV-1 and ASFV enzymes have only 23% amino acid identity. The PBCV-1–encoded enzyme is essential for virus replication because type-II topoisomerase inhibitors, such as amsacrine and etoposide, but not type-I topoisomerase inhibitors, inhibit plaque formation (M Nelson, Y Zhang, T Morehead, JL Van Etten, unpublished results). The PBCV-1 DNA topoisomerase II is $\geq$130 amino acids smaller than the other type-II topoisomerases currently in the databases.

PBCV-1 encodes two ORFs that resemble PCNA proteins. PCNA is one of several proteins involved in forming a processive DNA replication complex (64, 150), and it is also involved in DNA repair (4). The two PBCV-1 ORFs resemble PCNA

TABLE 2   Putative open reading frames encoded by chlorella virus PBCV-1[a]

DNA replication & repair			Phosphylation/dephosphorylation		
(E) A185R	DNA polymerase		(L) A248R*	Phosphorylase B kinase
(E) A544R*	DNA ligase		A277L	Ser/Thr protein kinase
(E) A583L*	DNA topoisomerase II		A278L	Ser/Thr protein kinase
A193L	PCNA		A282L	Ser/Thr protein kinase
A574L	PCNA		A289L	Ser/Thr protein kinase
A153R	Helicase		A305L	Tyr phosphatase
A241R	Helicase		A614L	Protein kinase
A548L	Helicase		A617R	Tyr-protein kinase
(E) A50L*	T4 endonuclease V		Cell Wall Degrading	
A39L	CyclinA/cdk-associated protein		(E) A181/182R*	Chitinase
A638R	Endonuclease		(L) A260R*	Endochitinase
Nucleotide metabolism			(L) A292L*	Chitosanase
(E) A169R*	Aspartate transcarbamylase		A94L	β-1,3 Glucanase
A476R	Ribo. reductase (small subunit)		DNA restriction/modification	
A629R	Ribo. reductase (large subunit)		A251R*	Adenine DNA methylase (M.CviAII)
A427L	Thioredoxin		A252R*	Restriction endonuclease (R.CviAII)
A438L	Glutaredoxin		(E) A517L*	Cytosine DNA methylase (M.CviAIII)
A551L	dUTP pyrophosphatase		(L) A530R*	Cytosine DNA methylase (M.CviAIV)
A596R	dCMP deaminase		(E) A581R*	Adenine DNA methylase (M.CviAI)
A200R	Cytosine deaminase		(E) A579L*	Restriction endonuclease (R.CviAI)
A416R	dG/dA kinase		A683L	Cytosine DNA methylase (M.CviAV)
A363R	Phosphohydrolase		Sugar and lipid manipulation	
A392R	ATPase		(L) A64R	Galactosyl transferase
A554L	ATPase		(E) A98R*	Hyaluronan synthase
A674R	*Dictyostelium* Thy protein		(E) A100R*	Glucosamine synthase
Transcription			A114R	Fucosyltransferase
A107L	RNA transcription factor TFIIB		(E) A118R	GDP-D-mannose dehydratase
A125L	RNA transcription factor TFIIS		A222R	Cellulose synthase
A166R	Exonuclease		A295L	Fucose synthase
A422R	Endonuclease		(E) A473L	Cellulose synthase
(E) A103R*	RNA guanyltransferase		(E) A609L*	UDP-glucose dehydrogenase
A464R	RNase III		A49L	Glycerophosphoryl diesterase
Protein synthesis, modification & degradation			A53R	2-Hydroxyacid dehydrogenase
(E,L) A666L	Translation elongation factor-3		A271L	Lysophospholipase
A44L	26S protease subunit			
A85R	Prolyl 4-hydroxylase alpha-subunit			
A105L	Ubiquitin C-terminal hydrolase			
A448L	Protein disulfide isomerase			
A623L	10 tRNAs Ubiquitin-like fusion protein			

(Continued)

TABLE 2 (*Continued*)

Miscellaneous		Pro-rich proteins
(E,L) A207R[*]	Ornithine decarboxylase	A35L
(L) A237R[*]	Homospermidine synthase	A41R
A78R	β-Alanine synthase	A67R
A245R	Cu/Zn-superoxide dismutase	A145R
A284L[*]	Amidase	A189R
A445L	Ubiquinous biosynthesis protein	A246R
A465R	Yeast ERVI protein	A316R
A598L	Histidine decarboxylase	A333L
A250R	K$^+$ Ion channel protein	A384L
A625R	Transposase	A405R
Contain aukyrin-like repeats		A428L
A5R		A488R
A7L		A500L
A8L		A678R
A682R		Gene families
A330R		16 with 2 members
A607R		8 with 3 members
A672R		3 with 6 members
Virion structural proteins		2 with 8 members
A34R		
A122R		
A168R		
A203R		
(L) A430L		
A523R		
(L) A622L		

[a]E and L refer to early and late genes, respectively. An asterisk means that the gene encodes a functional enzyme as determined either by complementation or by enzyme activity of a recombinant protein.

proteins from other organisms more than each other (26% amino acid identity), indicating that PBCV-1 probably acquired the two genes independently.

Perhaps one of the more interesting PBCV-1 genes is one that resembles the bacteriophage T4 *denV* gene (30). The *denV* gene encodes a well-characterized, pyrimidine dimer-specific glycosylase, endonuclease enzyme called endonuclease V, that initiates repair of UV-induced thymidine dimers in DNA (79, 200). Although T4 endonuclease V-like enzyme activity has been discovered in several microorganisms, these enzymes differ significantly from T4 endonuclease V in size and amino acid sequence. The discovery of the PBCV-1–encoded T4 endonuclease V homolog (41% amino acid identity) led to structural and functional comparisons between the two enzymes. The PBCV-1 enzyme cleaves both *cis-syn* and *trans-syn*-II cyclobutane pyrimidine dimers (89), whereas the T4 enzyme cleaves only the *cis-syn* cyclobutane pyrimidine isomer. The PBCV-1 enzyme is also more processive than the T4 enzyme (89).

Transcription-Associated Proteins   PBCV-1 does not encode a recognizable RNA polymerase or an RNA polymerase component, but it does encode TFIIB- and TFIIS-like elements, together with three ORFs that resemble endo- or exonucleases, as well as RNase III. The lack of a virus-encoded RNA polymerase supports the notion that the infecting virus DNA is targeted to the nucleus and that transcription is initiated by a host RNA polymerase, possibly in conjunction with one or more virus-packaged transcription factors.

PBCV-1 also encodes an mRNA-capping enzyme that catalyzes the transfer of GMP from GTP to the 5′ diphosphate end of RNA (52). In its size, amino acid sequence, and biochemical properties, the PBCV-1 enzyme is more closely related to the yeast RNA guanylyltransferases than to the multifunctional capping enzymes encoded by other large DNA viruses such as the poxviruses and ASFV. To cap its transcripts, PBCV-1 must either encode additional 5′-processing activities or else rely on the host alga to provide these functions. However, the 5′-end structure of PBCV-1 mRNAs is unknown. The PBCV-1-capping enzyme has been crystallized (46, 47).

Protein Synthesis, Modification, and Degradation   The chlorella viruses are the first viruses to encode a translation elongation factor protein. An 896-amino-acid-residue ORF from PBCV-1 has 95% amino acid identity with an EF-3–like protein from NC64A virus CVK2 (192) and 39% to 47% amino acid identity to EF-3 from fungi (8). Fungi require EF-3 for growth, and the protein stimulates EF-1α-dependent binding of aminoacyl-tRNA to the ribosome. Like fungal EF-3 proteins, the CVK2 and PBCV-1 enzymes have an ABC transporter family signature and two ATP/GTP binding-site motifs. The CVK2-encoded EF-3 gene is expressed at both early and late stages of viral infection but not when viral DNA replication begins (192). Possibly the virus-encoded EF-3 alters the host protein synthetic machinery so that viral mRNAs are preferentially translated after virus infection.

PBCV-1 also encodes five other ORFs that resemble proteins involved in either post-translational modification [excluding phosphorylation and glycosylation (see below)], prolyl 4-hydroxylase α-subunit and protein disulfide isomerase or protein-degrading events, a ubiquitin carboxy-terminal hydrolase, a ubiquitin-like fusion protein, and a 26S protease subunit.

Finally, PBCV-1 encodes 10 tRNAs: 3 for Lys, 2 for Asn, and 1 each for Leu, Ile, Tyr, Arg, and Val. The Tyr-tRNA gene is predicted to have a 13-nucleotide intron. None of these tRNAs have a CCA sequence encoded at the 3′ end of the acceptor stem of the tRNA. Typically these 3 nucleotides are added separately to tRNAs. Some NC64A viruses encode up to 14 tRNAs (T Yamada, K Nishida, M Fujie, S Usami, unpublished results). The tRNA genes are cotranscribed at both early and late stages of virus replication as a large precursor RNA and processed via various intermediates. Codon usage analyses of several virus-encoded proteins indicate a strong correlation between the abundance of virus-encoded tRNAs and the virus protein codons.

Nucleotide Metabolism-Associated Proteins   PBCV-1 encodes at least 13 enzymes involved in nucleotide metabolism. These enzymes are important because the DNA concentration in a virus-infected cell increases rapidly after infection (166). Consequently, large quantities of dNTPs need to be synthesized de novo during this time period. The PBCV-1–encoded enzymes include aspartate transcarbamylase, both subunits of ribonucleotide reductase, thioredoxin, glutaredoxin, dUTP pyrophosphatase (dUTPase), deoxycytidylate (dCMP) deaminase, cytosine deaminase, deoxyguanosine/adenosine kinase, nucleoside triphosphate phosphohydrolase, and two ATPases.

PBCV-1 is the first virus known to encode aspartate transcarbamylase, the key regulatory enzyme in the de novo biosynthetic pathway of pyrimidines. The PBCV-1 gene has been expressed in *E. coli*, and the recombinant protein has the expected enzyme activity (71). The gene is expressed early in PBCV-1 replication.

PBCV-1, like many large DNA-containing viruses such as the poxviruses, ASFV, herpes virus, and lymphocystis disease virus, encodes both ribonucleotide reductase subunits. Both subunits are also encoded by the brown algal virus FsV (AM Lee & RH Meints, unpublished results). Because inhibitors of ribonucleotide reductase activity, such as hydroxyurea and *cis*-platinum (M Nelson, N Padhye, JL Van Etten, unpublished results), reduce PBCV-1 plaque formation, the enzyme is important for virus replication.

PBCV-1 encodes two putative enzymes, dUTPase and dCMP deaminase, that produce dUMP, the substrate for thymidylate synthetase. Interestingly, PBCV-1 does not encode a thymidylate synthetase. However, PBCV-1 ORF A674R is about the same size and has 29% amino acid identity to the Thy1 protein from *Dictyostelium discoideum*. Thy1 complements the thymidine growth requirement of a *Dictyostelium* mutant (26). A674R also has 51% amino acid identity to an ORF from a *Synechocystis* sp. (57) and about 25% amino acid identity to ORFs from *Brevibacterium lactofermentum* and *Corynebacterium glutamicum* (115). It seems likely that all of these putative proteins participate in a thymidine synthesis pathway that lacks a traditional thymidylate synthetase.

Phosphorylation/Dephosphorylation-Associated Proteins   Seven PBCV-1 ORFs have similarities to protein kinases, and one ORF resembles a phosphatase. At least five of these ORFs are predicted to be Ser/Thr protein kinases. Interestingly, five of these ORFs are tightly clustered on the PBCV-1 genome (82). The gene for a 35-kDa Ser/Thr protein kinase (ORF A248R) was expressed in *E. coli* and shown to have both autophosphorylation and exogenous phosphorylation activities (118). This gene was expressed late in the PBCV-1 life cycle. This protein kinase was not packaged in nascent virions, although disrupted virions had protein kinase activity. Three protein kinase activities have been detected in virions of the NC64A virus CVK2 (193).

As compared with other viruses, this large number of virus-encoded proteins involved in phosphorylation/dephosphorylation is unusual. Because protein phosphorylation is a common mechanism for regulating cellular processes, we assume that the PBCV-1–encoded proteins are involved in one or more signal transduction pathways important for virus replication.

Sugar- and Lipid-Manipulating Proteins   A major surprise was the discovery that 12 PBCV-1 ORFs have high identity to enzymes involved in manipulating sugars, synthesizing polysaccharides, or metabolizing lipids. Three enzymes, glutamine:fructose-6-phosphate amidotransferase (GFAT), UDP-glucose dehydrogenase (UDP-GlcDH), and hyaluronan synthase, are involved in the synthesis of hyaluronan, a linear polysaccharide composed of alternating β1,4-glucuronic acid and β1,3-*N*-acetylglucosamine groups. The three PBCV-1 genes have been expressed in *E. coli*, and each recombinant protein has the expected enzyme activity (23, 70). The three genes are transcribed early in PBCV-1 infection, and hyaluronan accumulates on the external surface of the host *Chlorella* cells (41a). This was unexpected because, heretofore, hyaluronan had been found only in the extracellular matrix of vertebrates (73) and the extracellular capsules of a few bacterial pathogens (54, 131). It is interesting that the predicted amino acid sequences of the PBCV-1–encoded GFAT and UDP-GluDH enzymes resemble bacterial enzymes, whereas the amino acid sequence of the PBCV-1–encoded hyaluronan synthase most resembles the vertebrate enzyme (70). These observations suggest that the GFAT and UDP-GlcDH genes were acquired by the viruses separately from the hyaluronan synthase gene.

Two of the putative PBCV-1–encoded enzymes, GDP-D-mannose dehydratase and fucose synthase, compose a pathway that converts GDP-D-mannose to GDP-L-fucose (1). These genes have been expressed in *E. coli*, and recombinant proteins are currently being assayed for enzyme activity.

Three of the PBCV-1 ORFs resemble lipid-metabolizing enzymes, a glycerophosphoryl diesterase, a 2-hydroxyacid dehydrogenase, and a lysophospholipase. PBCV-1 deletion mutants, which lack the neighboring glycerophosphoryl diesterase and the 2-hydroxyacid dehydrogenase genes, grow in the laboratory (69).

Cell Wall–Degrading Enzymes   PBCV-1 encodes five ORFs potentially involved in degrading *Chlorella* cell walls, including two chitinases, a chitosanase, a β-1,3-glucanase, and an amidase. We initially reported that PBCV-1 encoded three chitinase genes (82). However, re-examination of the PBCV-1 sequence revealed that two adjacent chitinase ORFs are components of a single ORF (L Sun, JL Van Etten, unpublished results). All five genes have been expressed in *E. coli*, and the recombinant proteins have the expected enzyme activities (L Sun, Y Ye, JL Van Etten, unpublished results).

The chitosanase gene, a late gene, has been characterized from another NC64A virus, CVK2 (196). The CVK2 gene encodes two proteins, a 37-kDa protein of the expected size from the DNA gene sequence and a 65-kDa protein. The larger protein is packaged in nascent virions, whereas the smaller protein is found only in the infected cells. Interestingly, PBCV-1 encodes the same chitosanase gene flanked by two ORFs that also exists in virus CVK2. However, the CVK2 genome has an extra ORF inserted immediately downstream of its chitosanase gene, which contributes to the bigger 65-kDa protein, presumably as a read-through product. It will be interesting to determine the size of the PBCV-1 chitosanase protein.

The presence of chitinase and chitosanase genes in PBCV-1 and CVK2 was unexpected because chitin, a linear homopolymer of β-1,4-linked N-acetylglucosamine residues that is a normal component of fungal cell walls and the exoskeleton of insects and crustaceans (37), rarely occurs in algal walls (51). However, Kapaun & Reisser (58) reported a chitin-like glycan in the cell wall of *Chlorella* Pbi. We assume that these enzymes degrade the host cell wall during virus infection and/or virus release.

Restriction-Modification (R-M) Enzymes   As discussed below, the chlorella viruses encode multiple R-M genes. PBCV-1 encodes two 6mA DNA methyltransferases (MTases), each with a corresponding restriction endonuclease and three 5mC DNA MTases.

Miscellaneous Proteins   PBCV-1 is predicted to encode several additional enzymes, including the polyamine-metabolizing enzymes ornithine decarboxylase (ODC) and homospermidine synthase. ODC catalyzes decarboxylation of ornithine to putrescine, which is a key regulatory step in polyamine biosynthesis (22). ODC, which turns over rapidly, is regulated by an antizyme-protein:protein interaction in mammalian cells (18). The predicted size of the PBCV-1–encoded ODC is smaller (372 codons) than other ODCs (ca. 460 codons), and the smaller size can be attributed to four deletions. The PBCV-1 *odc* gene has been expressed in *E. coli*, and the recombinant protein is enzymatically active (T Morehead and JL Van Etten, unpublished results).

The PBCV-1–encoded homospermidine synthase (*hss*) gene has also been expressed in *E. coli*, and the recombinant protein synthesizes homospermidine from two molecules of putrescine (A Kaiser, M Vollmert, D Tholl, KW Nickerson, MV Graves, J Gurnon, JL Van Etten, unpublished results). The *hss* mRNA is expressed late during PBCV-1 infection, and PBCV-1 virions accumulate homospermidine. However, homospermidine, a rare polyamine, is not the major polyamine in the virions. The significance of a PBCV-1–encoded pathway that synthesizes homospermidine is unknown.

PBCV-1 also encodes an ORF with 50% amino acid identity to Cu/Zn-superoxide dismutases from a variety of aerobic organisms. Superoxide dismutases convert superoxide radical anions into molecular oxygen and hydrogen peroxide (6). Presumably, the PBCV-1–encoded enzyme reduces superoxide accumulation in sunlight. Two other large dsDNA-containing viruses, *Autographa californica* nuclear polyhedrosis virus (161) and vaccinia virus (10), also encode Cu/Zn-superoxide dismutases.

PBCV-1 ORF A250R encodes a 94-amino-acid-residue protein that resembles a small K^+ ion channel protein. This protein is apparently important for PBCV-1 growth, because the two classical K^+ ion channel protein inhibitors, amantadine and Ba^{2+} ions, inhibit PBCV-1 plaque formation (M Nelson, Y Zhang, unpublished results).

Gene Families or Gene Duplications   A total of 29 of the PBCV-1 ORFs resemble 1 or more other PBCV-1 ORFs, suggesting that they might be either gene families or gene duplications. A total of 16 families have two members, 8 families have three members, 3 families have six members, and 2 families have eight members. One family with six members contains ORFs with multiple ankyrin-like repeats, that is, G-TPLH-AA–GH—(V/A)–LL–GA–(N/A)-like sequences (114). Five members in another family resemble the major capsid protein of PBCV-1. One intriguing possibility is that, in its natural environments, PBCV-1 has an additional host(s) and that these capsid-like genes are important for infection of this other host(s).

PBCV-1 Deletion Mutants

Not all of the PBCV-1 genes are required for virus replication in the laboratory. For example, four spontaneously derived, antigenic variants of PBCV-1 that contain 27- to 37-kb deletions at the left end of the 330-kb genome were isolated (69). Two of these mutants had deletions that began at map positions 4.9 kb or 16 kb and ended at position 42 kb. In total, the two deleted regions, which probably resulted from recombination, encoded 28 putative protein-encoding ORFs. The other two mutants, which probably arose from nonhomologous recombination, lacked the entire left-terminal 37-kb of the PBCV-1 genome, including the 2.2-kb inverted-repeat region. The deleted left terminus was replaced by the transposition of an inverted 7.7- or 18.5-kb copy from the right end of the PBCV-1 genome. The deleted regions encoded 26 single-copy ORFs, of which 23 were common to those deleted in the first two mutant viruses. Taken together, the results indicate that 40 kb of single-copy DNA encoding 31 ORFs at the left end of the genome, or 12% of the PBCV-1 genome, are unnecessary for PBCV-1 replication. However, replication of these mutants is attenuated because their burst sizes are about half those of the parent virus. Apparently the inverted terminal repeats in PBCV-1 can vary in size, and virus DNA packaging tolerates large changes in genome size. Similar results were obtained with NC64A virus CVK1 (145, 194). CVK1-infected cells exposed to UV radiation produced 30- to 45-kb deletion mutants. Like PBCV-1, these deletions occurred in the left terminal portion of the virus genome.

The sizes and locations of the deletions and deletion/transpositions found in the chlorella viruses resemble those of deletion mutants in the poxviruses (for a review, see 163) and ASFV (e.g. 9, 11). Like PBCV-1, poxviruses and ASFV have inverted terminal repeats and covalently closed hairpin ends. Models proposed to explain the generation of deletions and deletion/transpositions in the poxvirus genomes (140, 163) may be relevant to the chlorella viruses.

Viruses Encode DNA Methyltransferases and DNA Site-Specific Endonucleases

The chlorella virus genomes contain different levels of 5mC and 6mA (Table 1). Therefore, it is not surprising that the viruses encode multiple 5mC and 6mA DNA methyltransferases (MTases). However, it was unexpected that the viruses also encode DNA site-specific (restriction) endonucleases. Thus the virus-infected chlorella are a source of DNA site-specific endonucleases and the first source from a nonprokaryotic system (for reviews, see 108, 109).

Some of the chlorella virus-encoded endonucleases have recognition and cleavage specificities identical to bacterial type II restriction endonucleases [e.g. R.*Cvi*AI (/GATC) (185) and R.*Cvi*BI (G/ANTC) (187)], others are heteroschizomers of bacterial endonucleases [e.g. R.*Cvi*AII (C/ATG) (205) and R.*Cvi*QI (G/TAC) (189)], and still others have novel recognition sites [e.g. R.*Cvi*JI (RG/CY) (186) and R.CviRI (TG/CA) (56)]. Statistically, R.*Cvi*JI is the first site-specific endonuclease to recognize a three-base sequence, and, under certain conditions, called R.*Cvi*JI*, the enzyme also cleaves RGCR and YGCY sequences (156, 186). Because R.*Cvi*JI and R.*Cvi*JI* cleave DNA so frequently, the enzyme has novel uses. For example, partial digestion of DNA with R.*Cvi*JI* has been used to produce DNA "shotgun" libraries (27, 34).

Two of the site-specific endonucleases [NYs1-nickase (/CC) (188) and NY2A-nickase (R/AG) (206)] cleave only one strand of dsDNA; that is, they are base-specific nicking enzymes. Like bacterial restriction endonucleases, all of the viral endonucleases are inhibited by either 5mC or 6mA in their recognition site.

The characteristics of the chlorella virus R-M systems can be summarized as follows: (*a*) Some viruses, such as NY-2A, encode as many as 10 different DNA MTases and at least two site-specific DNA endonucleases (206). The 380-kb NY-2A genome is predicted to contain 434 protein-encoding sequences. At a minimum, the NY-2A genome contains 12 R-M genes, or 1 of every 36 NY-2A genes is an R-M gene, making the NY-2A genome one of the most concentrated sources of R-M genes known. Even PBCV-1, with a low level of methylated bases (1.9% 5mC and 1.5% 6mA), has seven genes involved in methylation and cleavage: two 6mA MTases with corresponding site-specific endonucleases, two functional 5mC MTases, and one nonfunctional 5mC MTase. (*b*) Viruses with high levels of methylated bases encode some MTases that recognize short (2- to 3-bp) target sites, including M.*Cvi*PI from virus NYs-1, which recognizes GC sequences (191). (*c*) Not all MTases have a companion site-specific endonuclease. We estimate that 25–30% of the MTases have a cognate endonuclease. (*d*) Under laboratory conditions, at least some R-M genes can be deleted without affecting viral replication (13, 107). (*e*) Viral DNAs also contain nonfunctional 5mC MTase genes. For example, PBCV-1 encodes a nonfunctional 5mC MTase that differs by six amino acids from a functional 5mC MTase M.*Cvi*JI present in virus IL-3A. Changing one of the six amino acids restored enzyme activity (207).

About 15 chlorella virus Mtases, as well as three companion site-specific endonucleases, have been cloned and sequenced. Bacterial type II R-M genes (182) have many similarities to the chlorella virus R-M genes, including the following: (*a*) Bacterial R-M genes are always located near one another, although the spacing and relative orientation of the two genes can vary. The genes for the three sequenced virus R-M systems are also adjacent to one another. In the PBCV-1 *Cvi*AI system, the divergent M.*Cvi*AI and R.*Cvi*AI genes are separated by 82 bases (78), whereas, in the *Cvi*AII system, the TAA termination codon of the M.*Cvi*AII gene overlaps the ATG translation start site of the R.*Cvi*AII endonuclease gene (205). The R.*Cvi*JI endonuclease gene from virus IL-3A begins 18 bases after the M.*Cvi*JI gene stop codon (142, 157). (*b*) The virus 6mA- and 5mC-MTases contain the same highly conserved amino acid motifs found in bacterial 6mA (63, 84) and 5mC (67, 116) MTases. (*c*) Some of the virus 6mA MTases have significant similarity to bacterial DNA MTases. For example, M.*Cvi*AI has up to 39% amino acid identity to several bacterial 6mA MTases that also methylate GATC sequences, including M.*Pgi*I, M.*Dpn*II, and M.*Lla*II. M.*Cvi*AII (C^mATG) has 36% amino acid identity to M.*Nla*III (C^mATG) and type-IIS M.*Fok*I (GG^mATG and C^mATCC) (205). These similarities are significant because some bacterial isomethylator pairs that share the same recognition sequence have <20% amino acid identity (182).

The similarity between bacterial and viral genes suggests that bacteria and chlorella viruses have exchanged genes in the past. Consistent with this, most of the virus MTases cloned to date use their own promoters when expressed in *E. coli*. In addition, the five cloned NY-2A virus MTases have a sequence that resembles the bacterial -35 (TTGACA) promoter sequence (only one base different in each case). However, the similarity and the 16–19 nucleotide spacing to a bacterial consensus -10-like (TATAAT) sequence are poor for these five genes (206).

Although the evidence suggests that bacterial and chlorella virus MTases may have common evolutionary origins, the amino acid similarity data and the promoter sequence data suggest that the MTase genes have a long association with the chlorella viruses. Comparison of the G+C content of the five NY-2A MTase genes with total NY-2A DNA also supports this hypothesis. If the G+C contents of these five genes differed significantly from total NY-2A DNA, then the MTase genes were probably acquired relatively recently (3). In contrast, if the G+C contents are similar to the total genomic DNA, the genes have probably existed in the genome for a long time. The *m.Cviqi, m.Cviqiii, m.Cviqv, m.Cviqvi,* and *m.Cviqvii* genes have 38%, 37%, 42%, 38%, and 41% G+C contents, respectively, which are close to the 41.5% G+C content of the entire NY-2A genome (206).

Most, but not all, of the chlorella virus MTases that methylate the same sequence have ≥75% amino acid identity. For example, the 6mA C^mATG MTases, M.*Cvi*AII, M.*Cvi*SII, and M.*Cvi*QVII, from viruses PBCV-1, SC-1A, and NY-2A, respectively, have 80–94% amino acid identity in pairwise comparisons (119, 206). However, exceptions occur. For example, M.*Cvi*SI and M.*Cvi*QV, both of which methylate TGC^{m}A sequences, have only 32% amino acid identity (119, 206). Assuming that chlorella virus MTase genes were originally acquired from bacteria, it seems likely that these latter two genes were acquired independently.

The biological function(s) of the virus-encoded DNA site-specific endonucleases and MTases is unknown. Bacterial R-M systems confer resistance to foreign DNAs and DNA viruses. In fact, the term "restriction" refers to the role of these endonucleases and MTases in excluding foreign DNA. Bacterial MTases prevent self-digestion of host DNA. Two functions have been considered for the chlorella virus R-M enzymes. (*a*) Chlorella virus endonucleases help degrade host DNA, thus providing deoxynucleotides for recycling into virus DNA. Methylation of nascent virus DNAs by the cognate MTases protects the DNA from self-digestion. (*b*) Chlorella virus endonucleases prevent infection of a cell by a second virus.

Three observations are consistent with the first hypothesis. (*a*) Chlorella nuclear and chloroplast DNAs, but not virus DNA, are digested by the virus-encoded site-specific endonuclease(s) in vitro. (*b*) In vivo degradation of host nuclear and chloroplast DNAs coincides with the appearance of DNA site-specific endonuclease activity (185). (*c*) Initiation of virus DNA synthesis in vivo coincides with the appearance of DNA MTase activity (190).

The isolation of three independently derived deletion mutants of virus IL-3A, which had lost their MTase (M.*Cvi*JI) and site-specific endonuclease (R.*Cvi*JI) activities, allowed us to test the host DNA degradation hypothesis directly (13). If R.*Cvi*JI activity was essential for host DNA degradation, nuclear and/or chloroplast DNA should be preserved or at least degraded more slowly in cells infected with the three mutants than in cells infected with wild-type IL-3A. However, both nuclear and chloroplast DNA levels decreased at nearly identical rates after infection with each of the viruses (13). Therefore, we concluded that R.*Cvi*JI activity was not essential for host DNA degradation. However, this finding does not exclude participation of the enzyme in the degradation process.

To determine whether the endonuclease(s) are involved in excluding infection of a cell by a second virus, *Chlorella* cells were inoculated with pairs of viruses and plaques arising from infective centers were distinguished by immunoblotting (15). These experiments revealed that chlorella viruses, like certain bacteriophage (e.g. 24), exclude one another. However, this exclusion was independent of the known site-specific endonuclease activities (15).

Therefore, the biological functions of the DNA MTases and DNA site-specific endonucleases are unknown. However, if a virus contains a functional 5mC MTase gene, then its expression is apparently required for virus growth, because there is a direct correlation between increasing 5mC concentrations in the virus genome and sensitivity of virus replication to 5-azacytidine (13). It should also be noted that many viral DNAs are more heavily methylated than is necessary for protection from their site-specific endonucleases. The reason for this apparent excess methylation is unknown.

Glycosylation of PBCV-1 Structural Proteins Is Unusual

Structural proteins of many viruses, such as herpesviruses, poxviruses, paramyxoviruses, and orthomyxoviruses, as well as PBCV-1, are glycosylated. Viral glycoproteins are involved in various functions, including virus assembly, release of newly formed virus particles, and attachment to susceptible hosts. Viral proteins are generally glycosylated by host-encoded glycosyltransferases located in the rough endoplasmic reticulum and Golgi apparatus (25, 130). The glycoproteins are then transported to a host membrane and the nascent viruses acquire these glycoprotein(s) by budding through the membrane, usually as they are released from the cell. Consequently, the glycan portion of virus glycoproteins is generally specific to the host (e.g. 25, 120, 130).

PBCV-1, however, encodes at least part, if not all, of the machinery required to glycosylate its structural proteins. The initial observation that led to this conclusion arose from antibody studies. Polyclonal antiserum prepared against intact PBCV-1 virions completely inhibits PBCV-1 plaque formation by agglutinating the particles. However, spontaneously derived antiserum-resistant, plaque-forming mutants of PBCV-1 occur at a frequency of about 10^{-6} (179). These antiserum-resistant mutants fall into four serologically distinct groups (117, 179). Polyclonal antisera prepared against each of these four antigenic variants reacted exclusively with the immunizing virus. The major capsid protein Vp54 as well as the two other viral glycoproteins Vp280 and Vp260 or their equivalents from the mutants migrated in a distinctive, but predictable fashion on sodium dodecyl sulfate-polyacrylamide gel electrophoresis. Western blot analyses of the mutant capsid proteins, before and after deglycosylation, revealed that the antigenic differences reflected differences in the carbohydrate moiety(ies) of the major capsid protein and the two minor glycoproteins. The ratio of 6 and/or 7 sugars (fucose, galactose, glucose, xylose, mannose, and arabinose and/or rhamnose) associated with the major capsid protein of PBCV-1 and the mutants also varied in a predictable manner related to their serology and the migration of the major capsid protein (179).

Two other observations are relevant. First, unlike many viruses that acquire their glycoprotein(s) by budding through a membrane, intact infectious PBCV-1 particles accumulate inside the cell 30–40 min before virus release (170). Second, even though the PBCV-1 major capsid protein contains ∼10% carbohydrate, the protein appears homogeneous on sodium dodecyl sulfate-polyacrylamide gel electrophoresis, and it has a single mass peak at 53,790 (R Cerny, JL Van Etten, unpublished results). This is unusual because glycoproteins are typically microheterogeneous.

Because all of the PBCV-1 antigenic variants were grown in the same host and the predicted amino acid sequences of the major capsid protein from PBCV-1 and two of the antigenic variants were identical (179), glycosylation of chlorella virus proteins differs from other viruses. A simple explanation is that the antigenic variants arose from mutations in a common PBCV-1–encoded pathway involved in glycosylation, presumably a family of glycosyltransferases. Obviously, identification of the PBCV-1 glycosylation genes is a high priority. Unfortunately, methods for tagging specific PBCV-1 genes are currently unavailable. As an alternative strategy, we isolated additional PBCV-1 antigenic variants, anticipating that some of these mutants might contain easily detected deletions in common DNA restriction fragments, thus locating some of the glycosylation genes.

This strategy resulted in the isolation of four antigenic variants of chlorella virus PBCV-1, which contained 27- to 37-kb deletions in the left end of the 330-kb genome (these deletion mutants were described above). Analysis of these mutants suggested that one of the five putative genes located between map positions 37 and 42.2 is involved in glycosylation and that another gene should be in the first 37 kb of DNA. It is interesting that the 638-codon ORF A64R at about map position 35 has some similarity with portions of a galactosyltransferase from yeasts (MV Graves, C Berndt, JL Van Etten, unpublished results). Furthermore, ORF A64R contains short motifs present in glycosyltransferases from other organisms (202), indicating that the *a64r* gene product might be responsible for converting one virus phenotype to another. We will test this prediction once techniques for doing gene disruption and gene replacement experiments are available.

NATURAL HISTORY OF THE CHLORELLA VIRUSES

Several unexpected discoveries about the diversity of the chlorella viruses and their genes are mentioned in this section. The natural history of these viruses is poorly understood, and major advances are needed in our understanding of the biology of these viruses before we can appreciate the ecology and explain the evolutionary origin of these viruses.

The two known hosts for the chlorella viruses, *Chlorella* NC64A and *Chlorella* Pbi, normally exist as hereditary endosymbionts in green isolates of the protozoan *P. bursaria*. However, the zoochlorella are resistant to virus infection when they are in the symbiotic relationship and are infected only when they are separated from the ciliate (124). In the symbiotic unit, algae are enclosed individually in perialgal vacuoles surrounded by a host-derived membrane (122). The initial establishment of a successful symbiotic relationship and the long-term maintenance of symbiosis require that the algae resist digestion by the paramecium. Reassociation studies with different *Chlorella* species and alga-free *P. bursaria* indicate that only the original symbiotic algae readily reestablish symbiosis with the ciliate (122). Other chlorellae are digested. Although there have been numerous studies and considerable speculation on the factor(s) that allow the *P. bursaria* to distinguish chlorellae that are suitable for symbiosis formation from those that are not, the specific recognition factor(s) is unknown.

Chlorella NC64A and *Chlorella* Pbi were originally isolated from American and European strains, respectively, of *P. bursaria*. Viruses that infect *Chlorella* NC64A neither infect nor attach to *Chlorella* Pbi, and Pbi viruses neither infect nor attach to *Chlorella* NC64A (124). Because the viruses can distinguish the two

Chlorella isolates, we hypothesized that the receptor for the viruses might also serve as the recognition factor for the paramecium. This hypothesis was incorrect, however, because both *Chlorella* NC64A and *Chlorella* Pbi reestablish stable symbiotic relationships with either American or European isolates of *P. bursaria* (124).

NC64A viruses have been isolated from fresh water collected in the United States (133, 169, 177), China (204), Japan, Brazil (195), Australia (JL Van Etten, J Rohozinski, unpublished results), Argentina, and Israel (Y Zhang, M Nelson, JL Van Etten, unpublished resullts). Pbi viruses were initially found in fresh water collected in Europe (123, 125) and more recently in water collected in Australia, Canada, and either the northern United States (Minnesota, Wisconsin, and Montana) or in higher altitudes in the western United States (M Nelson, Y Zhang, JL Van Etten, unpublished results). Therefore, the initial assumption that Pbi viruses were limited to Europe and the NC64A viruses were present only in the Americas and Eastern Asia is incorrect. In fact, two water samples collected in Australia contained both NC64A and Pbi viruses (JL Van Etten, J Rohozinski, unpublished results). The most important factors influencing the distribution of NC64A and Pbi viruses are probably latitude and altitude.

Typically, the chlorella virus titer in nature is 1–100 PFU/ml, but titers as high as 40,000 PFU/ml have been obtained. The natural concentration of chlorella viruses is not static, but fluctuates with the seasons; the highest titers have typically been found in late spring (177, 195). However, a water sample was recently collected through the ice on a pond in Ohio, with a titer of >1000 PFU/ml (M Nelson, unpublished results).

It is not known whether the zoochlorellae can exist outside the paramecium in natural environments. Consequently, we do not even know whether these zoochlorellae are the normal virus hosts in native waters; it is possible that NC64A and Pbi viruses replicate in another host(s). We anticipated that some of the PBCV-1–encoded proteins might help identify another host(s), if it exits. As noted above, PBCV-1 encodes three enzymes that are involved in the synthesis of hyaluronan, a glycosaminoglycan. Furthermore, hyaluronan appears as hairlike material on the external surface of the infected algae (MV Graves, J Heuser, DE Burbank, P DeAngelis, JL Van Etten, published results, 41a), suggesting that the polysaccharide plays an important role in virus survival. We have considered two biological explanations for the extracellular hyaluronan: (*a*) The polysaccharide prevents uptake of virus-infected zoochlorellae by *P. bursaria*. Presumably, such infected algae would lyse inside the paramecium, and the released virions would be digested by the protozoan, which would be detrimental to virus survival. (*b*) The viruses have another host that acquires the virus because it is attracted to or binds the hyaluronan on the virus-infected algae. A complicating factor in understanding the biological importance of the hyaluronan is the recent discovery that not all infectious NC64A viruses encode hyaluronan synthetase (MV Graves, DE Burbank, P DeAngelis, JL Van Etten, published results, 41a). However, all chlorella viruses encode the GFAT and UDP-GlcDH enzymes that synthesize hyaluronan precursors.

The high titers of the chlorella viruses in some indigenous waters are surprising given that the viruses are constantly exposed to solar radiation. This radiation should damage viral DNAs, inactivating the virus. For example, inactivation of bacteriophages and cyanophages occurs at rates of 0.4–0.8% h^{-1} in full sunlight (154, 155). The chlorella viruses have apparently adapted to solar radiation by having access to two independent DNA repair systems (30). (*a*) PBCV-1 encodes a pyrimidine dimer-specific glycosylase (PDG) that initiates UV-induced thymidine dimer repair. This DNA repair system functions in both the light and the dark. (*b*) PBCV-1 also uses the host photolyase to repair UV-induced thymidine dimers. Thus, PBCV-1 can replicate whenever suitable hosts are encountered, both in the day and at night. We expected the PDG enzyme to be packaged in the virion and accompany the virus DNA into the host, where it could initiate DNA repair. However, attempts to detect the PDG protein in PBCV-1 virions were unsuccessful (30). The *pdg* gene is expressed early after virus infection.

PBCV-1 also encodes two other enzymes that may aid survival in direct sunlight. (*a*) The virus encodes a putative Cu/Zn superoxide dismutase that should protect DNA from reactive oxygen species. (*b*) A recent report (45) indicates that the polyamine spermine functions as a free-radical scavenger, thereby reducing DNA damage. Possibly this is why PBCV-1 encodes two polyamine biosynthetic enzymes, as well as packages the polyamines in the virions.

The PBCV-1–encoded PDG repair enzyme is also interesting for another reason. As mentioned above, the enzyme cleaves both *cis-syn* and *trans-syn*-II cyclobutane pyrimidine dimers, whereas its bacteriophage T4 endonuclease V homolog cleaves only the *cis-syn* cyclobutane pyrimidine isomer (89). These differences prompted the search for *pdg* genes from 41 other NC64A viruses, which were isolated from diverse geographic regions. All the chlorella viruses contained the *pdg* gene (30; L Sun, SR Lloyd, JL Van Etten, unpublished results). Unexpectedly, however, *pdg* genes from 15 of the 41 viruses contained a 98-nucleotide intron that is 100% conserved among the viruses, and another 4 viruses contained an 81-nucleotide intron in the same position that is nearly 100% identical (one intron differed by one nucleotide). Surprisingly, the *pdg* gene protein-coding regions (exons) were less conserved than their corresponding introns. The introns in the *pdg* gene have 5'-AG/GTATGT and 3'-TTGCAG/AA splice site sequences, which are characteristic of nucleus-located, spliceosomally processed pre-mRNA introns.

There is no obvious geographic correlation between the *pdg* intron-containing and intron-lacking viruses. Of the 98-nucleotide intron-containing viruses, 13 were collected throughout the United States in 1983–1985. Viruses isolated from Australia in 1995 and Argentina in 1996 contained an identical intron. Fortuitously, some water samples had both 98-nucleotide intron-containing and intron-lacking viruses. For example, water samples collected in Massachusetts, North Carolina, Alabama, Illinois, and California had both intron-containing and intron-lacking viruses. The 100% identity of the 98-nucleotide intron sequence in 15 viruses and the near 100% identity of an 81-nucleotide intron sequence in another 4 viruses imply that either the intron was acquired recently or there is strong selective pressure to maintain the DNA sequence of the intron once it is in the *pdg* gene. However, the abilities of intron-containing and intron-lacking viruses to repair UV-damaged DNA in the dark are indistinguishable (L Sun, J Gurnon, SR Lloyd, JL Van Etten, unpublished results). These findings contradict the widely accepted dogma that intron sequences are more variable than exon sequences.

Something similar to the sporadic occurrence of a highly conserved intron in the *pdg* gene is also observed with a self-splicing intron. The PBCV-1 TFIIS-like gene contains a 400-nucleotide self-splicing group IB intron (77). Group IB introns were originally discovered in two related NC64A viruses, CVU1 and CVB11 (198). The intron in CVU1 has 98% nucleotide identity and is in the exact same position in a TFIIS-like gene as the PBCV-1 intron, whereas the CVB11 intron has 80% nucleotide identity and is in an unidentified open reading frame. Three other NC64A viruses isolated in Japan at the same time as CVU1 and CVB11 lacked the intron. It is fascinating that two NC64A viruses, PBCV-1 isolated in the United States in 1981 and CVU1 isolated in Japan in about 1990, have a nearly identical intron located at the same place in the same gene, whereas this same intron is either located in a different gene (virus CVB11) or absent in three other Japanese chlorella viruses (198).

Recently, several hundred additional NC64A viruses, isolated in Japan, have been screened for the self-splicing intron; the intron is present in ~8% of these isolates (111). The intron was inserted in the TFIIS-like gene in ~60% of these viruses, and ~40% had the intron in the same unidentified reading frame gene as CVB11. In a few of the viruses, the major capsid protein gene contained the intron, and a couple of viruses had two copies of the intron. Nucleotide sequence analysis of the introns and the flanking regions indicated that the intron sequences are apparently under strong constraint by the exons, i.e. introns in the same gene had >99% sequence identity, whereas introns in different genes were only 72–78% identical.

The various levels of methylated bases observed in the chlorella virus genomes are another example of the natural diversity of these viruses that is difficult to explain with our current understanding of their natural history. As noted above the methylated base levels of virus genomes range from 0.1% 5mC and undetected levels of 6mA to 45% 5mC and 37% 6mA. The concentrations of 5mC and 6mA are a criterion for grouping the viruses into different classes (Table 1). Like the situation with the introns, there is no obvious geographic correlation for methylation levels in virus DNAs. Typically viruses isolated from one water sample have similar levels of methylated bases. However, exceptions are common. For example, four of six plaques originally picked from a water sample collected in New York fell into different classes. The level of methylation ranged from virus NY-2C, which contained 0.4% 5mC and no detectable 6mA, to NY-2A, which contained 45% 5mC and 37% 6mA.

Presumably the virus-encoded DNA methylation and site-specific endonuclease enzymes confer an advantage to the viruses in their native environment. We have conducted one simple experiment to determine whether the virus-encoded enzymes confer an advantage in the laboratory. Cultures of *Chlorella* NC64A were inoculated with various ratios [multiplicity of infection (MOI) of 0.1–1, 1–1, and 1–0.1] of virus IL-3A and an IL-3A mutant deficient in its R-M system, R.*Cvi*JI and M.*Cvi*JI (one-step growth curves for the two viruses are identical). Fresh cultures of the algae were inoculated with portions of the lysates and, after five serial transfers, the lysates were plaqued. In most of these experiments 80–95% of the plaques were wild-type virus (C Weinfeldt, JL Van Etten, unpublished results). Thus M.*Cvi*JI and R.*Cvi*JI appear to confer an advantage to the virus under these conditions.

Although the PBCV-1 genome is the only chlorella virus to be completely sequenced, a detailed physical map of virus CVK2 is available (K Nishida, Y Kimura, T Kawasaki, M Fujie, T Yamada, published results, 110a), and portions of several other NC64A viruses have been sequenced. Comparisons among these viruses have led to the unexpected finding that "anonymous" genes are often either inserted or deleted between otherwise colinear genes. This phenomenon, which has been observed by several laboratories, is illustrated with two examples. (*a*) As described above the chitosanase gene, which is a late gene, has been characterized from NC64A virus CVK2 (196). PBCV-1 encodes the same chitosanase gene, which is flanked by two ORFs that also exist in CVK2. However, the CVK2 genome has an extra ORF inserted immediately downstream of its chitosanase gene, which is absent in PBCV-1. (*b*) Likewise, an extra ORF is inserted between two otherwise colinear genes in NC64A virus SC-1A, which is absent in PBCV-1 (118). We do not have an explanation for the apparent appearance and/or disappearance of these viral genes, but this finding indicates that the total number of genes encoded by the chlorella virus group exceeds that of any one isolate. In many ways, this phenomenon resembles the presence or absence of the introns in certain virus genes.

CHARACTERISTICS OF MARINE ALGAL VIRUSES

Host Algae

There are predicted to be >100,000 species of marine algae of which only 30,000 have been identified. Photosynthetic plankton alone produce 40% of the photosynthate on the planet. It is estimated that these organisms produce 10^{12} tons of cell wall material yearly, of which 10^{11} tons are cellulose (38).

Despite the importance of marine algae, pathogens, including viruses and virus-like particles, are only beginning to be studied. A slowly growing literature indicates that eukaryotic marine algal viruses are large dsDNA viruses, similar to the chlorella viruses. Laboratory infection of marine algae by viruses has been demonstrated only in single-celled *Micromonas pusilla* (19, 20) and in filamentous *Sorocarpus* sp. and *Ectocarpus* sp. (104, 112). Consequently, not much is known about the infection process. Infection occurs in *Ectocarpus siliculosus* only when the host cell wall is mechanically damaged or lacking (104; RH Meints, unpublished results). Only viruses of the brown alga genera *Ectocarpus* and *Feldmannia* have been studied experimentally. These viruses infect uniseriate, branched, filamentous forms in the order *Ectocarpales* (5, 98). The biology of these virus-algal systems is interesting because the viral genomes are inherited in a Mendelian fashion and virus particles appear only in the reproductive organs of the algae. In general, virus infections do not affect vegetative growth of the brown algae.

Ectocarpus Virus   Members of the genus *Ectocarpus* inhabit temperate seacoasts throughout the world. The most studied member, *E. siliculosus*, has a complex sexual life cycle consisting of alternating generations. In the most parsimonious version of its life cycle, the haploid growth phase begins with the formation of male and female filamentous gametophytes from single-celled meiospores. The gametophytes develop distinctive, multicellular structures called gametangia, which produce motile isogametes, discharging them into the seawater. After fusion of male and female isogametes into diploid zygotes, the zygotes grow into diploid filamentous sporophytes that develop sporangia of two types, plurilocular and unilocular. Each compartment of the plurilocular sporangia can produce a diploid spore that, upon germination, produces another sporophyte. The nucleus in the unilocular sporangia undergoes meiosis, followed by multiple rounds of mitosis, yielding 100 or more haploid meiospores per unicellular sporangium. The motile meiospores are discharged into the seawater, and each one develops into either a male or female gametophyte. The textbook version of this life cycle is probably rare in nature, with more complex paths being common. As an example, unfertilized gametes can also develop into haploid parthenosporophytes that can propagate independently, and polyploid forms are common (97).

In 1990, Müller & colleagues (104) reported that large (130- to 140-nm diameter), densely packed, polyhedral virus particles (EsV) appeared in the gametangia and both unilocular and plurilocular sporangia of an isolate of *E. siliculosus* but not in vegetative cells. Because the virus particles displaced the normal reproductive organs, the plants were sterile and could propagate only vegetatively. Occasionally, the virulence of the virus infection was reduced, and these plants produced functional meiospores and gametes. Genetic evidence suggested that the EsV genome was maintained in these meiospores and gametes and was transmitted to daughter generations. Fusion of a gamete from an infected plant with a gamete from an uninfected plant resulted in a sporophyte that formed normal sporangia-containing meiospores. These individual meiospores gave rise to healthy and infected gametophytes in a 1:1 ratio, strongly suggesting that the virus symptoms and sex alleles segregated independently (99). This same experiment was repeated recently with EsV-specific DNA probes, and the same results were obtained, that is, only 50% of the progeny gametophytes contained viral DNA (12). These two experiments suggest that one copy of the EsV genome is associated with host chromosomes. Furthermore, it means that EsV and its host can coexist in a nonlethal and pandemic manner because meiosis in heterozygotes eliminates virus from 50% of the progeny.

The appearance of virus particles and their release from reproductive organs can be induced by increasing the temperature from 12°C to 18–20°C, transferring the alga to fresh culture medium, or reducing salinity of the culture medium (99, 102). The viruses do not appear to be released from reproductive cells by virus-induced lysis. However, both sporangia and gametangia are designed to open and release their contents, typically zoospores; thus these cells might be considered ideal sites for virus replication. Good quantities of EsV can be produced in culture, 1.7×10^{13} to 3.2×10^{13} virus particles g^{-1} fresh weight of infected algae (99). The intracellular site of virus replication is unknown, but presumably replication begins in the nucleus. Capsid assembly and DNA packaging occur in the cytoplasm, but only after degeneration of the nuclear membrane (104). The viral assembly process in a related alga, *Hincksis hincksia*, has recently been described in a series of elegant electron and fluorescence micrographs (184).

Purified EsV can infect motile gametes or meiospores of *E. siliculosus* that lack a cell wall, but not cell-wall containing vegetative cells (104). When gametes or meiospores swim into a cloud of freshly released virus particles, flagellar activity stops, and the paralyzed cells settle to the bottom. These infected cells develop into normal-appearing filamentous plants except that the reproductive organs are filled with virus particles rather than spores. The EsV host range is not limited to *E. siliculosus*, however, because the virus can infect motile gametes of at least two other filamentous brown algae. Infection of zoospores of a *Kuckuckia* sp., which is closely related to *Ectocarpus* sp., produces infected plants that resemble those of EsV-infected *Ectocarpus* sp., that is, virus particles form only in reproductive organs (100). The virus particles grown in *Kuckuckia* sp. can infect healthy *Ectocarpus* zoospores; thus EsV maintains infectivity to *Ectocarpus* sp. after replication in another host.

EsV can also infect zoids (zoospores) of the filamentous brown alga *Feldmannia simplex*. [Other unrelated *Feldmannia* species have their own viruses (see below).] EsV infection of *Feldmannia simplex* resulted in deformed plants, including cells containing abnormal mitochondria and chloroplasts, multiple nuclei, and multiple vacuoles. Intact EsV particles are absent in these infected plants (105). The symptoms produced by EsV infection of *F. simplex*, which differ from those produced by genuine Feldmannia viruses (see below), persisted as the filaments continued to grow. Occasionally, a few filaments produced normal-appearing multicellular zoidangia, and the resultant zoids developed into healthy-appearing plants. Surprisingly, however, these apparently healthy *F. simplex* plants contained EsV DNA, even after 2 years of growth (101). The ability of EsV to infect several algae means it could serve as a vector for nonsexual gene transfer among organisms, as suggested by Reanney (121). There are many examples of gene transduction in bacteria, but there is much less information on viruses as potential vectors for DNA transfer among eukaryotes.

Many *E. siliculosus* isolates and related *Ectocarpus* sp. collected from various ocean coasts throughout the world contain virus particles similar to EsV (106, 113, 137; RH Meints, unpublished results). Thus these viruses constitute an inherent element in *Ectocarpus* sp. and other brown algal populations, with implications for their genetics and evolution.

The characterization of the EsV virus particles is just beginning. EsV particles have an outer shell with two layers surrounding an electron-dense core; the thin outer layer surrounds a broader, more diffuse inner layer (61). EsV particles contain at least 16 proteins, of which at least 3 are glycoproteins. The gene for one of the EsV glycoproteins has been sequenced; the gene is predicted to encode a protein of 72 kDa, which is larger than the 60-kDa capsid protein. This observation suggests proteolytic processing (61).

The EsV genome is a large, 320-kb, circular dsDNA molecule with a 50% G+C content (72). The dsDNA is interrupted by multiple single-stranded regions of various lengths, most of which are randomly distributed over the entire length of the DNA (62). The single-stranded regions, which may result from packaging incompletely replicated DNA, make the DNA very fragile. Like the chlorella virus DNAs, the EsV DNA has methylated bases, 1% of the cytosines are 5mC, and 3% of the adenines are 6mA (72).

Feldmannia Virus   One isolate of the genus *Feldmannia* for which uninfected algae have not been found has been designated *Feldmannia* sp. In 1992 Henry & Meints (50) described a virus, FsV, in an isolate of this marine brown alga that had originally been isolated from the New Zealand coast in 1984. The alga was maintained in the laboratory for >7 years before virus studies were initiated. This 120- to 150-nm–diameter, icosahedral dsDNA virus was present exclusively in meiotic unilocular sporangium cells. Each single-celled reproductive sporangium contains 1×10^6–5×10^6 virions. Virions have not been detected in vegetative cells of either the sporophyte or gametophyte and, in contrast to *Ectocarpus* sp., are not produced in the plurilocular sporangia of the gametophyte or sporophyte. FsV particles appear to be released by rupture of the sporangium wall, probably by the normal mechanism for spore release. Like *Ectocarpus* sp., one would expect diploid sporophytic plants to develop both plurilocular (mitotic) sporangium-producing diploid spores and unilocular (meiotic) sporangium-producing haploid spores, whereas haploid gametophytes produce only plurilocular gametangia whose spores are the gametes for the sexual cycle.

In the FsV-infected plants, this life cycle is altered. The biologically important characteristic of the *Feldmannia* system is that only single-celled sporangia from virus-infected sporophyte plants fail to develop normally, and meiospores are not produced. Instead each unilocular sporangium is filled with virus particles, rendering these organisms asexual. Viruses are not produced in the gametophyte generation but appear to be latent in this generation; in contrast, virions are produced in the sporophyte by the apparent excision of the viral genome from the host genome by an as-yet-unknown mechanism, followed by viral genome replication. Once viral production begins, death of the cell is a certainty.

Analysis of marine algal viruses at the molecular level is preliminary, and a summary of the information on FsVs follows.

1. Like EsV particles, purified FsV particles have a capsid with two layers surrounding an electron-dense core; the thin outer layer surrounds a broader, more diffuse inner layer (50).

2. Clamped homogeneous electric field electrophoresis of DNA from purified FsV reveals two genome size classes (158 and 178 kb). These genomes are variably abundant in preparations from cultures grown at 18°C, whereas the smaller genome is virtually absent in cultures grown at 9°C (50).

3. Restriction maps of both FsV genome size classes and a subvariant of each map as circles (55). Mapping and hybridization studies indicate that genomes of the two size classes are similar and that the major difference in the two size classes results from the number of copies of a recurring 173-bp repeat element (74).

4. Characterization of a 4.6-kb region that was common to all genome size class variants revealed that three ORFs hybridized to Northern blots of RNA isolated from virus-producing sporophytes (55). ORF 1 encodes a polypeptide of 75.4 kDa and contains regions with significant homologies to a "RING" zinc finger motif and a nucleotide-binding site (65). RING zinc finger motifs are also found in other viruses, for example, poxviruses, where they are critical for virulence (138), and *Ectromelia* viruses, where they are involved in DNA replication (139). Because of its Ring signature, we speculate that this FsV protein participates in protein-protein interactions.

5. Transcriptional studies with four FsV genome-spanning cosmid probes on northern blots revealed 6 major and 18 minor transcripts (Y Jia, RH Meints, unpublished results). One ORF has significant similarity to the major structural protein of chlorella virus PBCV-1 (42), the major structural protein of several iridescent viruses (28, 85, 132, 151), and the capsid protein of ASFV (81). This homology suggests that FsV is closely related to PBCV-1 and at least distantly related to these invertebrate and vertebrate viruses.

6. The FsV DNA polymerase gene was cloned and sequenced. The ORF encodes 986 amino acids and contains all of the conserved 3′ to 5′ exonuclease domains and catalytic domains found in B-family (α-like) DNA polymerases. The codons for the FsV DNA polymerase have some bias towards G/C in the third position. At the 3′ end of the gene, there is a TTTTTNT sequence motif; this motif serves as a transcription termination signal for vaccinia virus early genes (203).

7. A lambda library prepared from a *Feldmannia* haploid gametophyte that contained no visible virus particles carries 1–5 copies of the viral genome presumably integrated into the host chromosomes. A second family of 173-bp repeats, related to but different from those mentioned above, was inserted into a viral ORF that contains all 12 catalytic motifs conserved in

most Ser/Thr protein kinases and a potential autophosphorylation site (75a). Perhaps this kinase functions in diverting spore formation to virus production.

8. Despite more that 100 sequential serial passages of our culture lines, we have not isolated a gametophyte free of the viral genome. This observation, along with inheritance evidence for EsV-1 (99), encouraged us to search for the sites of FsV integration into the host genome. We predicted that the point of diversity between the known sequence of the integrated virus and unknown host sequence would be the integration junction of host and virus DNAs. Such junctions were found, and DNA sequencing revealed that the same site is used for integration of both viral size classes, but that the host chromosomal integration sites varied (RG Ivey, RH Meints, unpublished results). Furthermore, recombination appears to occur without rearrangement, analogous to *Int-att, CRE-lox*, and yeast 2μm *FLP-frt* (21).

Together the data suggest an integration/excision mechanism that uses an integrase/recombinase and conservative site-specific recombination. Unlike in previously described systems, we expect blunt-end cutting and ligation or a single-base-pair overlap with specificity residing in flanking sequences.

EVOLUTIONARY RELATIONSHIP OF THE ALGAL VIRUSES TO EACH OTHER AND TO OTHER VIRUSES

The chlorella viruses, the brown algal viruses, and the micromonas viruses have several common properties. (*a*) They are large icosahedral particles (130–190 nm in diameter). (*b*) They have large dsDNA genomes (160–380 kb). (*c*) The genomes of many contain methylated bases. (*d*) Capsid assembly and DNA packaging occur in the cytoplasm. (*e*) The virus-host systems are global in nature. The chlorella viruses have been found in fresh water from at least five continents, and the viruses infecting the two groups of marine algae have been found in tropical and subtropical seawater.

However, differences exist between the chlorella viruses and the brown algal viruses as well. (*a*) The chlorella viruses enter their single-celled host by digesting its cell wall at the point of attachment, whereas brown algal viruses have no obvious lytic activity and infect only algal cells in a stage of host development during which these cells lack a cell wall. (*b*) The 320- to 380-kb chlorella virus genomes are linear, nonpermuted molecules with cross-linked hairpin ends. In contrast, brown algal viruses have circular dsDNA genomes. (*c*) Most importantly, the brown algal viruses have a lysogenic phase, whereas there is no evidence that the chlorella viruses have integrated genomes. We do, however, occasionally observe a carrier state or pseudolysogeny state with the chlorella viruses in the laboratory (Y Xia, JL Van Etten, unpublished results).

Despite the differences between the chlorella viruses and brown algal viruses, phylogenetic trees with the DNA polymerase gene (16, 75) indicate that the algal viruses, including viruses that infect *Micromonas pusilla* and *Chrysochromulina* spp., are more closely related to each other than to other dsDNA viruses and that they form a distinct clade, suggesting a common, albeit ancient, ancestor (Figure 3). However, the algal viruses fall into several clades, which correlate with their algal hosts. The chlorella viruses have family status with the name *Phycodnaviridae* (165). It has been proposed that the brown algal viruses and the micromonas viruses be included in the *Phycodnaviridae* family and that the groups be assigned separate genera. Based on DNA polymerase gene sequences, herpesviruses are the most closely related out group (16). However, phylogenetic analyses of other genes such as the major capsid protein genes suggest that the *Phycodnaviridae* may be most closely related to the iridoviruses (RH Meints, unpublished results).

Algal viruses share several properties with certain other large dsDNA-containing viruses. Iridoviruses, which commonly infect insects and some animals, have the following properties in common with the chlorella viruses: aquatic habitats, icosahedral morphology, a large dsDNA genome, and an internal lipid component that makes up 5–10% of the viral weight (39). Frog virus 3 (FV3), the most widely studied iridovirus, is one of the few viruses that infects a eukaryotic organism and contains methylated bases; 20% of the cytosines in the FV3 genome are methylated (181). FV3 also encodes a 5mC DNA methyltransferase (59, 180). Furthermore, phylogenetic trees constructed with the major capsid proteins from several iridoviruses indicate that the iridoviruses are distantly related to the algal viruses (85, 86, 160). However, there are many differences between FV3 and PBCV-1 (172). Most importantly, the structures of their DNAs differ. FV3 DNA is linear and circularly permuted (40), whereas PBCV-1 DNA is linear and nonpermuted, with covalently closed hairpin ends, and brown algal viruses have a circular genome.

The PBCV-1 genome shares two characteristics with the vaccinia virus (a poxvirus) genome, even though the virions are morphologically distinct. Like PBCV-1, vaccinia virus DNA contains covalently closed hairpin termini and inverted terminal repetition (32, 33). However, the terminal repetitive region of vaccinia virus DNA is larger (∼10 kb) than that of PBCV-1 DNA (2.2 kb), even though the PBCV-1 genome is about twice the size of the vaccinia virus genome. The vaccinia virus genome has many identical direct tandem repeats of 50–125 bases in its termini (183). Although PBCV-1 termini contain several direct repeats, with the exception of 118- and 39-base repeats, they are smaller and less extensive (152).

ASFV (presently unclassified) is the only other large polyhedral virus containing a dsDNA genome with hairpin termini and inverted terminal repetition (9, 36, 144). However, the chlorella viruses differ from ASFV in several aspects: (*a*) PBCV-1 DNA is about twice the size of ASFV DNA. (*b*) PBCV-1 DNA contains methylated bases, whereas AFSV does not. (*c*) ASFV lacks glycoproteins. (*d*) ASFV infection occurs via endocytosis, whereas the chlorella viruses uncoat at the cell surface. Comparisons of ASFV-encoded and PBCV-1–encoded ORFs indicate low similarity between the two viruses. For example, even though these two are the only viruses infecting a eukaryotic host that encode a type-II DNA topoisomerase, there is only 23% amino acid identity between the two enzymes.

The infection process of the chlorella viruses differs from that of all DNA viruses known to infect eukaryotes, but resembles that of bacterial viruses in that uncoating occurs at the cell wall. In terms of structure, attachment, and penetration, the chlorella viruses resemble bacteriophages of the *Tectiviridae* family (94). However, tectiviruses are much smaller (~65 nm in diameter) and contain only ~40 kb of DNA.

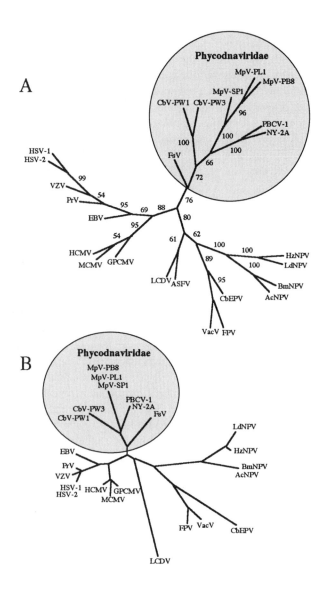

Finally, it should be noted that a forthcoming review by Villarreal (178) suggests that DNA viruses have played a significant role in the evolution of their hosts by contributing virus genes to the host. In his review, he specifically focuses on the origin of genes encoding enzymes involved in DNA replication, and he suggests that the algal virus DNA polymerase might be related to the original progenitor of eukaryotic delta DNA polymerases. This conclusion supports the notion that the algal viruses have a long evolutionary history.

CONCLUDING REMARKS

Studies on the algal viruses have revealed many unexpected properties, and undoubtedly many more of these viruses exist in nature and await discovery. Several practical benefits have also resulted from studies on these viruses: (*a*) The chlorella viruses are a new source of DNA restriction endonucleases, some of which recognize novel cleavage sites. For example, the restriction endonuclease *Cvi*JI has been used to diagnose a common mutation that causes the mitochondrial inherited genetic disease myoclonic epilepsy and ragged-red fiber disease (141). Another enzyme *Cvi*RI has been used to diagnose a common mutation producing a hereditary defect in the enzyme thiopurine S-methyltransferase (66). People lacking this enzyme are unable to methylate aromatic and heterocylic sulfhydryl compounds, including some important thioquanine anticancer drugs. (*b*) Although not discussed in this review, the chlorella viruses are a new source of promoters for expressing genes in foreign hosts. For example, the upstream region of one of the viral adenine DNA methyltransferase genes functions extremely well in several higher plants (95) and in many bacteria (96; Y Xia, JL Van Etten, F Wagner,

---

Figure 3   *A*. Phylogenetic tree of the DNA polymerases from some large dsDNA viruses generated by using protein parsimony analysis of the 100 bootstrapped data sets. The *numbers at the branches* are the bootstrap values indicating the relative strengths of those branches. *B*. Phylogenetic tree generated by the neighbor-joining method. For constructing the distance tree in *B*, ASFV was omitted from the polymerase sequence data sets. Inclusion of ASFV created problems because it was too distantly related to other viruses. Abbreviations: FsV, *Feldmannia* sp. virus; MpV-xx, viruses that infect *Micromonas pusilla*; CbV-xx, viruses that infect *Chrysochromulina brevifilum*; PBCV-1 and NY-2A, viruses that infect *Chlorella* NC64A; HSV-1 and HSV-2, herpes simplex viruses type 1 and type 2; PrV, pseudorabies virus; VZV, varicella-zoster virus; EBV, Epstein-Barr virus; HCMV, human cytomegalovirus; GPCMV, guinea pig cytomegalovirus; MCMV, murine cytomegalovirus; VacV, vaccinia virus; CbEPV, *Choristoneura biennis* entopoxvirus; FPV, fowlpox virus; AcNPV, *Autographa californica* nuclear polyhedrosis virus; BmNPV, *Bombyx mori* nuclear polyhedrosis virus; HzNPV, *Helicoverpa zea* nuclear polyhedrosis virus; LdNPV, *Lymantria dispar* nuclear polyhedrosis virus; LCDV, lymphocystis disease virus; ASFV, African swine fever virus. (Modified with permission from *J. Phycology*—75a)

unpublished results). (*c*) The chlorella viruses encode at least two proteins that are targets for medically important drugs. Type-II DNA topoisomerases are the object of many anticancer drugs, and a K^+ ion channel defect is responsible for heart muscle arrhythmia. Because drugs that effect these two proteins inhibit chlorella virus plaque formation, the virus plaque assay has the potential to serve as a rapid method to screen for new K^+ ion channel protein inhibitors and type-II DNA topoisomerase inhibitors (M Nelson & JL Van Etten, unpublished results).

Many fundamental research questions remain to be answered about these viruses—ranging from their natural history to the regulation of gene transcription. Examples include the following: (*a*) How did these viruses acquire such a mixture of prokaryotic- and eukaryotic-like genes? (*b*) Do the viruses have another host or hosts? (*c*) How is early and late gene transcription regulated? (*d*) Why are some of the viral introns more conserved than the virus exons? (*e*) What are the mechanism and intracellular location of viral protein glycosylation? (*f*) Why do the viruses encode so many sugar-manipulating enzymes?

ACKNOWLEDGMENTS

Les Lane, Mike Nelson, Myron Brakke, and Mike Graves are acknowledged for many suggestions on early drafts of this manuscript, and Mike Graves and Jim Gurnon helped with the figures. The reconstructed micrographs used to make up Figure 1 were kindly supplied by Xiaodong Yan, Norm Olson, and Tim Baker. Takashi Yamada, Curtis Suttle, and Luis Villarreal provided copies of manuscripts before publication. Investigations were supported by NIH (JVE), NSF-EPSCoR (JVE), the University of Nebraska Biotechnology Center (JVE), Office of Naval Research (RHM), and Seq-aGrant (RHM). We personally thank all past and present undergraduate students, graduate students, postdoctoral associates, technicians, and collaborating scientists who have worked with us on these fascinating viruses.

Visit the Annual Reviews home page at http://www.AnnualReviews.org

LITERATURE CITED

1. Andrianopoulos K, Wang L, Reeves PR. 1998. Identification of the fucose synthetase gene in the colanic acid gene cluster of *Escherichia coli* K-12. *J. Bacteriol.* 180:998–1001

2. Antequera F, Bird A. 1993. CpG islands. In *DNA Methylation: Molecular Biology and Biological Significance*, ed. PJ Jost, PH Saluz, pp. 169–85. Basel: Birkhauser Verlag

3. Anton BP, Heiter DF, Benner JS, Hess EJ, Greenough L, et al. 1997. Cloning and characterization of the *Bgl*II restriction-modification system reveals a possible evolutionary footprint. *Gene* 187:19–27

4. Ayyagari R, Impellizzeri KJ, Yoder BL, Gary SL, Burgers PMJ. 1995. A mutational analysis of the yeast proliferating cell nuclear antigen indicates distinct roles in DNA replication and DNA repair. *Mol. Cell. Biol.* 15:4420–29

5. Baker JR, Evans LV. 1973. The ship fouling alga *Ectocarpus*. I. Ultrastructure and cytochemistry of plurilocular repro-

ductive stages. *Protoplasma* 77:1–13

6. Bannister JV, Bannister WH, Rotilio G. 1987. Aspects of the structure, function, and applications of superoxide dismutase. *CRC Rev. Biochem.* 22:111–80

7. Becker B, Lesemann DE, Reisser W. 1993. Ultrastructural studies on a chlorella virus from Germany. *Arch. Virol.* 130: 145–55

8. Belfield GP, Tuite MF. 1993. Translation elongation factor 3: a fungus-specific translation factor? *Mol. Microbiol.* 9:411–18

9. Blasco R, Aguero M, Almendral JM, Vinuela E. 1989. Variable and constant regions in African swine fever virus DNA. *Virology* 168:330–38

10. Blasco R, Cole NB, Moss B. 1991. Sequence analysis, expression, and deletion of a vaccinia virus gene encoding a homolog of profilin, a eukaryotic actin-binding protein. *J. Virol.* 65:4598–608

11. Blasco R, De La Vega I, Almazan F, Aguero M, Vinuela E. 1989. Genetic variation of African swine fever virus: variable regions near the ends of the viral DNA. *Virology* 173:251–57

12. Brautigam M, Klein M, Knippers R, Muller DG. 1995. Inheritance and meiotic elimination of a virus genome in the host *Ectocarpus siliculosus* (Phaeophyceae). *J. Phycol.* 31:823–27

13. Burbank DE, Shields SL, Schuster AM, Van Etten JL. 1990. 5-Azacytidine resistant mutants of chlorella virus IL-3A. *Virology* 176:311–15

14. Caspar DLD, Klug A. 1962. Physical principles in the construction of regular viruses. *Cold Spring Harbor Symp. Quant. Biol.* 27:1–24

15. Chase TE, Nelson JA, Burbank DE, Van Etten JL. 1989. Mutual exclusion occurs in a chlorella-like green alga inoculated with two viruses. *J. Gen. Virol.* 70:1829–36

16. Chen F, Suttle CA. 1996. Evolutionary relationships among large double-stranded DNA viruses that infect microalgae and other organisms as inferred from DNA polymerase genes. *Virology* 219:170–78

17. Cheng C, Shuman S. 1997. Characterization of an ATP-dependent DNA ligase encoded by *Haemophilus influenzae*. *Nucleic Acids Res.* 25:1369–13

18. Cohen SS. 1998. *A Guide to the Polyamines.* New York: Oxford Univ. Press

19. Cottrell MT, Suttle CA. 1991. Wide spread occurrence and clonal variation in viruses which cause lysis of a cosmopolitan, eukaryotic marine phytoplankter, Micromonas pusilla. *Mar. Ecol. Prog. Ser.* 78: 1–9

20. Cottrell MT, Suttle CA. 1993. Isolation of axenic strains of *Micromonas pusilla* (Prasinophyceae). *J. Phycol.* 29:385–87

21. Craig NL, Nash HA. 1983. The mechanism of phage lambda site specific recombination: site specific breakage of DNA by INT topoisomerase. *Cell* 35:795–803

22. Davis RH, Morris DR, Coffino P. 1992. Sequestered end products and enzyme regulation: the case of ornithine decarboxylase. *Microbiol. Rev.* 56:280–90

23. DeAngelis PL, Jing W, Graves MV, Burbank DE, Van Etten JL. 1997. Hyaluronan synthase of chlorella virus PBCV-1. *Science* 278:1800–3

24. Doermann AH. 1983. Introduction to the early years of bacteriophage T4, In *Bacteriophage T4,* ed. CK Mathews, EM Kutter, G Mosig, PB Berget, pp. 1–7. Washington, DC: *Am. Soc. Microbiol.*

25. Doms RW, Lamb RA, Rose JK, Helenius A. 1993. Folding and assembly of viral membrane proteins. *Virology* 193:545–62

26. Dynes JL, Firtel RA. 1989. Molecular complementation of a genetic marker in *Dictyostelium* using a genomic DNA library. *Proc. Natl. Acad. Sci. USA* 86:7966–70

27. Fitzgerald MC, Skowron P, Van Etten JL, Smith LM, Mead DA. 1992. Rapid shotgun cloning utilizing the two base recognition endonuclease CviJI. *Nucleic Acids Res.* 20:3753–62

28. Flugel RM. 1985. Lymphocystis disease

virus. *Curr. Top. Microbiol. Immunol.* 116: 133–50

29. Fuhrman JA, Suttle CA. 1993. Viruses in marine planktonic systems. *Oceanography* 6:51–63

30. Furuta M, Schrader JO, Schrader HS, Kokjohn TA, Nyaga S, et al. 1997. Chlorella virus PBCV-1 encodes a homolog of the bacteriophage T4 UV damage repair gene *DenV. Appl. Environ. Microbiol.* 63:1551–56

31. Garcia-Beato R, Freije JMP, Lopez-Otin C, Blasco R, Vinuela E, Salas ML. 1992. A gene homologous to topoisomerase II in African swine fever virus. *Virology* 188:938–47

32. Garon CF, Barbosa E, Moss B. 1978. Visualization of an inverted terminal repetition in vaccinia virus DNA. *Proc. Natl. Acad. Sci. USA* 75:4863–67

33. Geshelin P, Berns KI. 1974. Characterization and localization of the naturally occurring cross-links in vaccinia virus DNA. *J. Mol. Biol.* 88:785–96

34. Gingrich JC, Boehrer DM, Basu SB. 1996. Partial CviJI digestion as an alternative approach to generate cosmid sublibraries for large-scale sequencing projects. *BioTechniques* 21:99–104

35. Girton LE, Van Etten JL. 1987. Restriction site map of the chlorella virus PBCV-1 genome. *Plant Mol. Biol.* 9:247–57

36. Gonzalez A, Talavera A, Almendral JM, Vinuela E. 1986. Hairpin loop structure of African swine fever virus DNA. *Nucleic Acids Res.* 14:6835–44

37. Gooday GW, Humphreys AM, McIntosh WH. 1986. Role of chitinases in fungal growth. In *Chitin in Nature and Technology,* ed. R Muzzarelli, C Jeuniaux, GW Gooday, pp. 83–91. New York: Plenum

38. Goodwin TW, Mercer EI. 1983. *Introduction to Plant Biochemistry.* Oxford, UK: Pergamon. 677 pp.

39. Goohra R, Granoff A. 1979. Icosahedral cytoplasmic deoxyriboviruses. In *Comprehensive Virology,* Vol. 14, ed. H Fraenkel-Conrat pp. 347–399. New York: Plenum

40. Goorha R, Murti KG. 1982. The genome of frog virus 3, an animal DNA virus, is circularly permuted and terminally redundant. *Proc. Natl. Acad. Sci. USA* 79:248–52

41. Grabherr R, Strasser P, Van Etten JL. 1992. The DNA polymerase gene from chlorella viruses PBCV-1 and NY-2A contains an intron with nuclear splicing sequences. *Virology* 188:721–31

41a. Graves MV, Burbank DE, Roth R, Heuser J, DeAngelis PL, Van Etten JL. 1999. Hyaluronan synthesis in virus PBCV-1-infected chlorella-like green algae. *Virology* 257:15–23

42. Graves MV, Meints RH. 1992. Characterization of the major capsid protein and cloning of its gene from algal virus PBCV-1. *Virology* 188:198–207

43. Graves MV, Meints RH. 1992. Characterization of the gene encoding the most abundant in vitro translation product from virus-infected chlorella-like algae. *Gene* 113:149–55

44. Green MR. 1991. Biochemical mechanisms of constitutive and regulated premRNA splicing. *Annu. Rev. Cell Biol.* 7: 559–99

45. Ha HC, Sirisoma NS, Kuppusamy P, Zweier JL, Woster PM, Casero RA. 1998. The natural polyamine spermine functions directly as a free radical scavenger. *Proc. Natl. Acad. Sci. USA* 95:11140–45

46. Hakansson K, Doherty AJ, Shuman S, Wigley DB. 1997. X-ray crystallography reveals a large conformational change during guanyl transfer by mRNA capping enzymes. *Cell* 89:543–53

47. Hakansson K, Wigley DB. 1998. Structure of a complex between a cap analogue and mRNA guanylyl transferase demonstrates the structural chemistry of RNA capping. *Proc. Natl. Acad. Sci. USA* 95:1505–10

48. Hayes W. 1968. *The Genetics of Bacteria and Their Viruses,* 2nd ed. Oxford, UK: Blackwell Sci. 925 pp.

49. Hellmann V, Kessler E. 1974. Physiologis-

che und Biochemische Beitrage zur Taxonomie der gattung Chlorella. VIII. Die Basenzusammensetzung der DNS. *Arch. Microbiol.* 95:311–18

50. Henry EC, Meints RH. 1992. A persistent virus infection in *Feldmannia* (Phaeophyceae). *J. Phycol.* 28:517–26

51. Herth W, Mulisch M, Zugenmaier P. 1986. Comparison of chitin fibril structure and assembly in three unicellular organisms. In *Chitin in Nature and Technology,* ed. R Muzzarelli, C Jeuniaux, GW Gooday, pp. 107–20. New York: Plenum

52. Ho CK, Van Etten JL, Shuman S. 1996. Expression and characterization of an RNA capping enzyme encoded by chlorella virus PBCV-1. *J. Virol.* 70:6658–64

53. Ho CK, Van Etten JL, Shuman S. 1997. Characterization of an ATP-dependent DNA ligase encoded by chlorella virus PBCV-1. *J. Virology* 71:1931–37

54. Husmann LK, Yung DL, Hollingshead SK, Scott JR. 1997. Role of putative virulence factors of *Streptococcus pyogenes* in mouse models of long-term throat colonization and pneumonia. *Infect. Immun.* 65:1422–30

55. Ivey RG, Henry EC, Lee AM, Klepper L, Krueger SK, Meints RH. 1996. A Feldmannia algal virus has two genome size-classes. *Virology* 220:267–73

56. Jin A., Zhang Y, Xia Y, Traylor E, Nelson M, Van Etten JL. 1994. New restriction endonuclease CviRI cleaves DNA at TG/CA sequences. *Nucleic Acids Res.* 22:3928–29

57. Kaneko T, Sato S, Kotani H, Tanaka A, Asamizu E, et al. 1996. Sequence analysis of the genome of the unicellular cyanobacterium *Synechocystis* sp. strain PCC6803. II. Sequence determination of the entire genome and assignment of potential protein-coding regions. *DNA Res.* 3:109–36

58. Kapaun E, Reisser W. 1995. A chitin-like glycan in the cell wall of a *Chlorella* sp. (Chlorococcales, Chlorophyceae). *Planta* 197:577–82

59. Kaur K, Rohozinski J, Goorha R. 1995. Identification and characterization of the frog virus 3 DNA methyltransferase gene. *J. Gen. Virol.* 76:1937–43

60. Kawakami H, Kawakami N. 1978. Behavior of a virus in a symbiotic system, *Paramecium bursaria-*zoochlorella. *J. Protozool.* 25:217–25

61. Klein M, Lanka STJ, Knippers R, Muller D. 1995. Coat protein of the *Ectocarpus siliculosus* virus. *Virology* 206:520–26

62. Klein M, Lanka S, Muller D, Knippers R. 1994. Single-stranded regions in the genome of the *Ectocarpus siliculosus* virus. *Virology* 202:1076–78

63. Klimasauskas S, Timinska A, Menkevicius S, Butkiene D, Butkus V, Janulaitis A. 1989. Sequence motifs characteristic of DNA [cytosine-N4] methyltransferases: similarity to adenine and cytosine-C5 DNA methylases. *Nucleic Acids Res.* 17:9823–32

64. Krishna TSR, Kong XP, Gary S, Burgers PM, Kuriyan J. 1994. Crystal structure of the eukaryotic DNA polymerase processivity factor PCNA. *Cell* 79:1233–43

65. Krueger SK, Henry EC, Ivey RG, Meints RH. 1996. A 4.5-kbp nucleotide sequence fragment from a large dsDNA brown algal virus contains ORFs and a "RING" zinc finger motif. *Virology* 219:301–3

66. Krynetski EY, Schuetz JD, Galpin AJ, Pui CH, Relling MV, Evans WE. 1995. A single point mutation leading to loss of catalytic activity in human thiopurine S-methyltransferase. *Proc. Natl. Acad. Sci. USA* 92:949–53

67. Kumar S, Cheng X, Klimasauskas S, Mi S, Posfai J, et al. 1994. The DNA (cytosine-5) methyltransferases. *Nucleic Acids Res.* 22:1–10

68. Kutish GF, Li Y, Lu Z, Furuta M, Rock DL, Van Etten JL. 1996. Analysis of 76 kb of the chlorella virus PBCV-1 330-kb genome: map positions 182 to 258. *Virology* 223:303–17

69. Landstein D, Burbank DE, Nietfeldt JW, Van Etten JL. 1995. Large deletions in antigenic variants of the chlorella virus PBCV-1. *Virology* 214:413–20

70. Landstein D, Graves MV, Burbank DE, DeAngelis P, Van Etten JL. 1998. Chlorella virus PBCV-1 encodes functional glutamine:fructose-6-phosphate amidotransferase and UDP-glucose dehydrogenase enzymes. *Virology* 250:388–96

71. Landstein D, Mincberg M, Arad S, Tal J. 1996. An early gene of the chlorella virus PBCV-1 encodes a functional aspartate transcarbamylase. *Virology* 221:151–58

72. Lanka STJ, Klein M, Ramsperger U, Muller DG, Knippers R. 1993. Genome structure of a virus infecting the marine brown alga *Ectocarpus silisulosus*. *Virology* 193:802–11

73. Laurent TC, Fraser JRE. 1992. Hyaluronan. *FASEB J.* 6:2397–404

74. Lee AM, Ivey RG, Henry EC, Meints RH. 1995. Characterization of a repetitive DNA element in a brown algal virus. *Virology* 212:474–80

75a. Lee AM, Ivey RG, Meints RH. 1998. The DNA polymerase gene of a brown algal virus: structure and phylogeny. *J. Phycol.* 34:608–15

75b. Lee AM, Ivey RG, Meints RH. 1998. Repetitive DNA insertion in a protein kinase ORF of a latent FSV (*Feldmannia* sp. virus) genome. *Virology* 248:35–45

76. Lepetit D, Thiebaud P, Aoufouchi S, Prigent C, Guesne R, Theze N. 1996. The cloning and characterization of a cDNA encoding *Xenopus levis* DNA ligase I. *Gene* 172:273–77

77. Li Y, Lu Z, Burbank DE, Kutish GF, Rock DL, Van Etten JL. 1995. Analysis of 43 kb of the chlorella virus PBCV-1 330 kb genome: map position 45 to 88. *Virology* 212:134–50

78. Li Y, Lu Z, Sun L, Ropp S, Kutish GF, Rock DL, Van Etten JL. 1997. Analysis of 74 kb of DNA located at the right end of the 330-kb chlorella virus PBCV-1 genome. *Virology* 237:360–77

79. Lloyd RS, Linn S. 1993. Nucleases involved in DNA repair. In *Nucleases*, ed. S Linn, RS Lloyd, RJ Roberts, Vol. II, pp. 263–316. Cold Spring Harbor, NY: Cold Spring Harbor Lab. Press

80. Loos E, Meindl D. 1982. Composition of the cell wall of *Chlorella fusca*. *Planta* 156:270–73

81. Lopez-Otin C, Freije JM, Parra F, Mendez E, Vinuela E. 1990. Mapping and sequence of the gene coding for protein p72, the major capsid protein of African swine fever virus. *Virology* 175:477–84

82. Lu Z, Li Y, Que Q, Kutish GF, Rock DL, Van Etten JL. 1996. Analysis of 94 kb of the chlorella virus PBCV-1 330 kb genome: map position 88 to 182. *Virology* 216:102–23

83. Lu Z, Li Y, Zhang Y, Kutish GF, Rock DL, Van Etten JL. 1995. Analysis of 45 kb of DNA located at the left end of the chlorella virus PBCV-1 genome. *Virology* 206:339–52

84. Malone TE, Blumenthal RM, Cheng X. 1995. Structure-guided analysis reveals nine sequence motifs conserved among DNA amino-methyltransferases, and suggests a catalytic mechanism for these enzymes. *J. Mol. Biol.* 253:618–32

85. Mao J, Hedrick RP, Chinchar VG. 1997. Molecular characterization, sequence analysis, and taxonomic position of newly isolated fish iridoviruses. *Virology* 229:212–20

86. Mao J, Tham TN, Gentry GA, Aubertin A, Chinchar VG. 1996. Cloning, sequence analysis, and expression of the major capsid protein of the iridovirus frog virus 3. *Virology* 216:431–36

87. Mayer JA, Taylor FJR. 1979. A virus which lyses the marine nanoflagellate *Micromonas pusilla*. *Nature* 281:299–301

88. McCluskey K, Graves MV, Mills D, Meints RH. 1992. Replication of chlorella virus PBCV-1 and host karyotype de-

termination studies with pulsed-field gel electrophoresis. *J. Phycol.* 28:846–50

89. McCullough AK, Romberg MT, Nyaga S, Lei Y, Wood T, et al. 1998. Characterization of a novel *cis-syn* and *trans-syn* II pyrimidine dimer glycosylase/AP lyase from a eukaryotic algal virus, *Paramecium bursaria* chlorella virus-1. *J. Biol. Chem.* 273:13136–42

90. Meints RH, Burbank DE, Van Etten JL, Lamport DTA. 1988. Properties of the chlorella receptor for the virus PBCV-1. *Virology* 164:15–21

91. Meints RH, Lee K, Burbank DE, Van Etten JL. 1984. Infection of a chlorella-like alga with the virus, PBCV-1: ultrastructural studies. *Virology* 138:341–46

92. Meints RH, Lee K, Van Etten JL. 1986. Assembly site of the virus PBCV-1 in a chlorella—like green alga: ultrastructural studies. *Virology* 154:240–45

93. Meints RH, Van Etten JL, Kuczmarski D, Lee K, Ang B. 1981. Viral infection of the symbiotic chlorella-like alga present in *Hydra viridis*. *Virology* 113:698–703

94. Mindich L, Bamford D. 1988. Lipid-containing bacteriophages. In *The Bacteriophages*, Vol. 2, ed. R Calendar, pp. 475–520. New York: Plenum

95. Mitra A, Higgins DW. 1994. The chlorella virus adenine methyltransferase gene promoter is a strong promoter in plants. *Plant Mol. Biol.* 26:85–93

96. Mitra A, Higgins DW, Rohe NJ. 1994. A chlorella virus gene promoter functions as a strong promoter both in plants and bacteria. *Biochem. Biophys. Res. Commun.* 204:187–94

97. Muller DG. 1967. Generationswechsel, kernphasenwechsel und sexualitat der braunalge *Ectocarpus siliculosus* in kulturversuch. *Planta* 75:39–54

98. Müller DG. 1972. Life cycle of the brown alga *Ectocarpus fasciculatus* var. *refractus* (Kutz.) Ardis. (Phaeophyceae, Ectocarpales) in culture. *Phycologia* 11:11–13

99. Muller DG. 1991. Mendelian segregation of a virus genome during host meiosis in the marine brown alga *Ectocarpus siliculosus*. *J. Plant Physiol.* 137:739–43

100. Muller DG. 1992. Intergeneric transmission of a marine plant DNA virus. *Naturwissenschafter* 79:37–39

101. Muller DG, Brautigam M, Knippers R. 1996. Virus infection and persistence of foreign DNA in the marine brown alga *Feldmannia simplex* (Ectocarpales, Phaeophyceae). *Phycologia* 35:61–63

102. Muller DG, Frenzer K. 1993. Virus infections in three marine brown algae: *Feldmannia irregularis, F. simplex,* and *Ectocarpus siliculosus*. *Hydrobiologia* 260:37–44

103. Muller DG, Kapp M, Knippers R. 1998. Viruses in marine brown algae. *Adv. Virus Res.* 50:49–67

104. Muller DG, Kawai H, Stache B, Lanka S. 1990. A virus infection in the marine brown alga *Ectocarpus siliculosus* (Phaeophyceae). *Bot. Acta* 103:72–82

105. Muller DG, Parodi E. 1993. Transfer of a marine DNA virus from *Ectocarpus* to *Feldmannia* (Ectocarpales, Phaeophyceae): aberrant symptoms and restitution of the host. *Protoplasma* 175:121–25

106. Müller DG, Stache B. 1992. Worldwide occurrence of virus infections in filamentous marine brown algae. *Helgol. Meeresunters.* 46:1–8

107. Narva KE, Wendell DL, Skrdla MP, Van Etten JL. 1987. Molecular cloning and characterization of the gene encoding the DNA methyltransferase, M.*Cvi*BIII, from chlorella virus NC-1A. *Nucleic Acids Res.* 15:9807–23

108. Nelson M, Burbank DE, Van Etten JL. 1998. Chlorella viruses encode multiple DNA methyltransferases. *Biol. Chem.* 379:423–28

109. Nelson M, Zhang Y, Van Etten JL. 1993. DNA methyltransferases and DNA site-specific endonucleases encoded by chlorella viruses. In *DNA Methylation: Molecular Biology and Biological*

Significance, ed. PJ Jost, PH Saluz, pp. 186–211. Basel: Birkhauser Verlag

110. Nietfeldt JW, Lee K, Van Etten JL. 1992. Chlorella virus PBCV-1 replication is not affected by cytoskeletal disruptors. *Intervirology* 33:116–20

110a. Nishida K, Kimura Y, Kawasaki T, Fujie M, Yamada T. 1999. Genetic variation of chlorella viruses: variable regions localized on the CVK2 genomic DNA. *Virology* 255:376–84

111. Nishida K, Suzuki S, Kimura Y, Nomura N, Fujie M, Yamada T. 1998. Group I introns found in chlorella viruses: biological implications. *Virology* 242:319–26

112. Oliviera L, Bisalputra T. 1978. A virus infection in the brown alga Sorocarpus uvaeformis (Lyngbye) Pringsheim (Phaeophyta, Ectocarpales). *Ann. Bot.* 42:439–45

113. Parodi ER, Muller DG. 1994. Field and culture studies on virus infections in *Hincksia hincksiae* and *Ectocarpus fasciculatus* (Ectocarpales, Phaeophyceae). *Eur. J. Phycol.* 29:113–17

114. Peters LL, Lux SE. 1993. Ankyrins: structure and function in normal cells and hereditary spherocytes. *Semin. Hematol.* 30:85–118

115. Pisabarro A, Malumbres M, Mateos LM, Oguiza JA, Martin JF. 1993. A cluster of three genes (dapA, ORF2, and dapB) of *Brevibacterium lactofermentum* encodes dihydrodipicolinate synthase, dihydrodipicolinate reductase, and a third polypeptide of unknown function. *J. Bacteriol.* 175:2743–49

116. Posfai J, Bhagwat AS, Posfai G, Roberts RJ. 1989. Predictive motifs derived from cytosine methyltransferases. *Nucleic Acids Res.* 17:2421–35

117. Que Q, Li Y, Wang IN, Lane LC, Chaney WG, Van Etten JL. 1994. Protein glycosylation and myristylation in chlorella virus PBCV-1 and its antigenic variants. *Virology* 203:320–27

118. Que Q, Van Etten JL. 1995. Characterization of a protein kinase gene from two chlorella viruses. *Virus Res.* 35:291–305

119. Que Q, Zhang Y, Nelson M, Ropp S, Burbank DE, Van Etten JL. 1997. Chlorella virus SC-1A encodes at least six DNA methyltransferases. *Gene* 190:237–44

120. Rademacher TW, Parekh RB, Dwek RA. 1988. Glycobiology. *Annu. Rev. Biochem.* 57:785–838

121. Reanney DC. 1974. Viruses and evolution. *Int. Rev. Cytol.* 37:21–52

122. Reisser W, ed. 1992. *Algae and Symbioses*. Bristol, UK: Biopress. 746 pp.

123. Reisser W, Becker B, Klein T. 1986. Studies on ultrastructure and host range of a chlorella attacking virus. *Protoplasma* 135:162–65

124. Reisser W, Burbank DE, Meints RH, Becker B, Van Etten JL. 1991. Viruses distinguish symbiotic *Chlorella* spp. of *Paramecium bursaria. Endocytobiosis Cell Res.* 7:245–51

125. Reisser W, Burbank DE, Meints SM, Meints RH, Becker B, Van Etten JL. 1988. A comparison of viruses infecting two different chlorella-like green algae. *Virology* 167:143–49

126. Reisser W, Kapaun E. 1991. Entry of a chlorella-virus into its host cell. *J. Phycol.* 27:609-13

127. Roca J. 1995. The mechanisms of DNA topoisomerases. *Trends Biochem. Sci.* 20:156–60

128. Rohozinski J, Girton LE, Van Etten JL. 1989. Chlorella viruses contain linear nonpermuted double stranded DNA genomes with covalently closed hairpin ends. *Virology* 168:363–69

129. Royce CL, Pardy RL. 1996. Endotoxin-like properties of an extract from a symbiotic, eukaryotic chlorella-like green alga. *J. Endotoxin Res.* 3:437–44

130. Schlesinger MJ, Schlesinger S. 1987. Domains of virus glycoproteins. *Adv. Virus Res.* 33:1–44

131. Schmidt KH, Gunther E, Courtney HS.

1996. Expression of both M protein and hyaluronic acid capsule by group A streptococcal strains results in a high virulence for chicken embryos. *Med. Microbiol. Immunol.* 184:169–73

132. Schnitzler P, Darai G. 1993. Identification of the gene encoding the major capsid protein of fish lymphocystis disease virus. *J. Gen. Virol.* 74:2143–50

133. Schuster AM, Burbank DE, Meister B, Skrdla MP, Meints RH. 1986. et al. Characterization of viruses infecting a eukaryotic chlorella-like green alga. *Virology* 150:170–77

134. Schuster AM, Girton L, Burbank DE, Van Etten JL. 1986. Infection of a chlorella-like alga with the virus PBCV-1: transcriptional studies. *Virology* 148:181–89

135. Schuster AM, Graves M, Korth K, Ziegelbein M, Brumbaugh J, et al. 1990. Transcription and sequence studies of a 4.3-kbp fragment from a dsDNA eukaryotic algal virus. *Virology* 176:515–23

136. Seaton GR, Lee K, Rohozinski J. 1995. Photosynthetic shutdown in *Chlorella* NC64A associated with the infection cycle of *Paramecium bursaria Chlorella* virus-1. *Plant Physiol.* 108:1431–38

137. Sengco MR, Brautigam M, Kapp M, Muller DG. 1996. Detection of virus-DNA in *Ectocarpus siliculosus* and *E. fasciculatus* (Phaeophyceae) from various geographic areas. *Eur. J. Phycol.* 31:73–78

138. Senkevich TG, Koonin EV, Buller ML. 1994. A poxvirus protein with a RING zinc finger motif is of crucial importance for virulence. *Virology* 198:118–28

139. Senkevich TG, Wolffe EJ, Buller ML. 1995. Ectromelia virus RING finger protein is localized in virus factories and is required for virus replication in macrophages. *J. Virol.* 69:4103–11

140. Shchelkunov SN, Totmenin AV. 1995. Two types of deletions in orthopoxvirus genomes. *Virus Genes* 9:231–45

141. Shoffner JM, Lott MT, Lezza AMS,

Seibel P, Ballinger SW, Wallace DC. 1990. Myoclonic epilepsy and ragged-red fiber disease (MERRF) is associated with a mitochondrial DNA tRNA^{lys} mutation. *Cell* 61:931–37

142. Skowron PM, Swaminathan N, McMaster K, George D, Van Etten JL, Mead DA. 1995. Cloning and application of the two/three-base restriction endonuclease R.CviJI from IL-3A virus-infected chlorella. *Gene* 157:37–41

143. Skrdla MP, Burbank DE, Xia Y, Meints RH, Van Etten JL. 1984. Structural proteins and lipids in a virus, PBCV-1, which replicates in a chlorella-like alga. *Virology* 135:308–15

144. Sogo JM, Almendral JM, Talavera A, Vinuela E. 1984. Terminal and internal inverted repetitions in African swine fever virus DNA. *Virology* 133:271–75

145. Songsri P, Hamazaki T, Ishikawa Y, Yamada T. 1995. Large deletions in the genome of chlorella virus CVK1. *Virology* 214:405–12

146. Songsri P, Hiramatsu S, Fujie M, Yamada T. 1997. Proteolytic processing of chlorella virus CVK2 capsid proteins. *Virology* 227:252–54

147. Sriskanda V, Shuman S. 1998. Chlorella virus DNA ligase: nick recognition and mutational analysis. *Nucleic Acids Res.* 26:525–31

148. Sriskanda V, Shuman S. 1998. Specificity and fidelity of strand joining by chlorella virus DNA ligase. *Nucleic Acids Res.* 26:3536–41

149. Sriskanda V, Shuman S. 1998. Mutational analysis of chlorella virus DNA ligase: catalytic roles of domain I and motif VI. *Nucleic Acids Res.* 26:4618–25

150. Stillman B. 1994. Smart machines at the DNA replication fork. *Cell* 78:725–28

151. Stohwasser R, Raab K, Schnitzler P, Janssen W, Darai G. 1993. Identification of the gene encoding the major capsid protein of insect iridescent virus type 6 by

polymerase chain reaction. *J. Gen. Virol.* 74:873–79

152. Strasser P, Zhang Y, Rohozinski J, Van Etten JL. 1991. The termini of the chlorella virus PBCV-1 genome are identical 2.2-kbp inverted repeats. *Virology* 180:763–69

153. Suttle CA. 1999. The ecological, evolutionary and geochemical consequences of viral infection of cyanobacteria and eukaryotic algae. In *Viral Ecology*, ed. CJ Hurst, pp. xx–xx. New York: Academic. In press

154. Suttle CA, Chan AM. 1994. Dynamics and distribution of cyanophages and their effect on marine *Synechococcus* spp. *Appl. Environ. Microbiol.* 60:3167–74

155. Suttle CA, Chen F. 1992. Mechanisms and rates of decay of marine viruses in seawater. *Appl. Environ. Microbiol.* 58:3721–29

156. Swaminathan ND, George D, McMaster K, Szablewski J, Van Etten JL, Mead DA. 1994. Restriction generated oligonucleotides utilizing the two base recognition endonuclease *Cvi*JI*. *Nucleic Acids Res.* 22:1470–75

157. Swaminathan ND, Mead DA, McMaster K, George D, Van Etten JL, Skowron PM. 1996. Molecular cloning of the three base restriction endonuclease R.*Cvi*JI from eucaryotic chlorella virus IL-3A. *Nucleic Acids Res.* 24:2463–69

158. Takeda H. 1988. Classification of chlorella strains by cell wall sugar composition. *Phytochemistry* 27:3822–26

159. Tessman I. 1985. Genetic recombination of the DNA plant virus PBCV-1 in a chlorella-like alga. *Virology* 145:319–22

160. Tidona CA, Schnitzler P, Kehm R, Darai G. 1998. Is the major capsid protein of iridoviruses a suitable target for the study of viral evolution? *Virus Genes* 16:59–66

161. Tomalski MD, Eldridge R, Miller LK. 1991. A baculovirus homolog of a Cu/Zn superoxide dismutase gene. *Virology* 184:149–61

162. Trench RK. 1979. The cell biology of plant-animal symbiosis. *Annu. Rev. Plant Physiol.* 30:485–531

163. Turner PC, Moyer RW. 1990. The molecular pathogenesis of poxviruses. *Curr. Top. Microbiol. Immunol.* 163:125–151

164. van den Hoek C, Mann DG, Jahns HM. 1995. *Algae, an Introduction to Phycology*. Cambridge, UK: Cambridge Univ. Press.

165. Van Etten JL. 1995. Phycodnaviridae. In *Virus Taxonomy*, ed. FA Murphy, CM Fauguet, DHL Bishop, SA Ghabrial, AW Jarvis, et al, pp. 100–103. New York: Springer-Verlag

166. Van Etten JL, Burbank DE, Joshi J, Meints RH. 1984. DNA synthesis in a chlorella-like alga following infection with the virus PBCV-1. *Virology* 134:443–49

167. Van Etten JL, Burbank DE, Kuczmarski D, Meints RH. 1983. Virus infection of culturable chlorella-like algae and development of a plaque assay. *Science* 219:994–96

168. Van Etten JL, Burbank DE, Meints RH. 1986. Replication of the algal virus PBCV-1 in UV-irradiated chlorella. *Intervirology* 26:115–20

169. Van Etten JL, Burbank DE, Schuster AM, Meints RH. 1985. Lytic viruses infecting a chlorella like alga. *Virology* 140:135–43

170. Van Etten JL, Burbank DE, Xia Y, Meints RH. 1983. Growth cycle of a virus, PBCV-1, that infects chlorella-like algae. *Virology* 126:117–25

171. Van Etten JL, Lane LC, Meints RH. 1991. Viruses and viruslike particles of eukaryotic algae. *Microbiological Rev.* 55:586–620

172. Van Etten JL, Lane LC, Meints RH. 1991. Unicellular plants also have large dsDNA viruses. *Semin. Virol.* 2:71–77

173. Van Etten JL, Meints RH, Burbank DE, Kuczmarski D, Cuppels DA, Lane LC. 1981. Isolation and characterization of a virus from the intracellular green alga

symbiotic with *Hydra viridis*. *Virology* 113:704–11

174. Van Etten JL, Meints RH, Kuczmarski D, Burbank DE, Lee K. 1982. Viruses of symbiotic chlorella-like algae isolated from *Paramecium bursaria* and *Hydra viridis*. *Proc. Natl. Acad. Sci. USA* 79:3867–71

175. Van Etten JL, Schuster AM, Girton L, Burbank D, Swinton D, Hattman S. 1985. DNA methylation of viruses infecting a eukaryotic chlorella-like green alga. *Nucleic Acids Res.* 13:3471–78

176. Van Etten JL, Schuster AM, Meints RH. 1988. Viruses of eukaryotic chlorella-like algae. In *Viruses of Fungi and Simple Eukaryotes*, ed. Y Koltin, MJ Leibowitz, pp. 411–28, New York: Marcel Dekker

177. Van Etten JL, Van Etten CH, Johnson JK, Burbank DE. 1985. A survey for viruses from freshwater that infect a eukaryotic chlorella-like green alga. *Appl. Environ. Microbiol.* 49:1326–28

178. Villarreal LP. 1999. DNA viruses: their influence on host evolution. In *Origin and Evolution of Viruses*, ed. E Domingo, R Webster, JJ Holland, T Pickett, pp. xx–xx. London, Academic. In press

179. Wang IN, Li Y, Que Q, Bhattacharya M, Lane LC, Chaney WG, Van Etten JL. 1993. Evidence for virus-encoded glycosylation specificity. *Proc. Natl. Acad. Sci. USA* 90:3840–44

180. Willis DB, Goorha R, Granoff A. 1984. DNA methyltransferase induced by frog virus 3. *J. Virol.* 49:86–91

181. Willis DB, Granoff A. 1980. Frog virus 3 is heavily methylated at CpG sequences. *Virology* 107:250–57

182. Wilson GG, Murray NE. 1991. Restriction and modification systems. *Annu. Rev. Genet.* 25:585–627

183. Wittek R, Moss B. 1980. Tandem repeats within the inverted terminal repetition of vaccinia virus DNA. *Cell* 21:277–84

184. Wolf S, Maier I, Katsaros C, Muller DG. 1998. Virus assembly in *Hincksia*

hincksiae (Ecocarpales, Phaeophyceae). An electron and fluorescence microscopic study. *Protoplasma* 203:153–67

185. Xia Y, Burbank DE, Uher L, Rabussay D, Van Etten JL. 1986. Restriction endonuclease activity induced by PBCV-1 virus infection of a chlorella-like green alga. *Mol. Cell. Biol.* 6:1430–39

186. Xia Y, Burbank DE, Uher L, Rabussay D, Van Etten JL. 1987. IL-3A virus infection of a chlorella-like green alga induces a DNA restriction endonuclease with novel sequence specificity. *Nucleic Acids Res.* 15:6075–90

187. Xia Y, Burbank DE, Van Etten JL. 1986. Restriction endonuclease activity induced by NC-1A virus infection of a chlorella-like green alga. *Nucleic Acids Res.* 14:6017–30

188. Xia Y, Morgan R, Schildkraut I, Van Etten JL. 1988. A site-specific single strand endonuclease activity induced by NYs-1 virus infection of a chlorella-like green alga. *Nucleic Acids Res.* 16:9477–87

189. Xia Y, Narva KE, Van Etten JL. 1987. The cleavage site of the *Rsa*I isoschizomer, *Cvi*II, is G/TAC. *Nucleic Acids Res.* 15:10063

190. Xia Y, Van Etten JL. 1986. DNA methyltransferase induced by PBCV-1 virus infection of a chlorella-like green alga. *Mol. Cell. Biol.* 6:1440–45

191. Xu M, Kladde MP, Van Etten JL, Simpson RT. 1998. Cloning, characterization and expression of the gene coding for a cytosine-5-DNA methyltransferase recognizing GpC sites. *Nucleic Acids Res.* 26:3961–66

192. Yamada T, Fukuda T, Tamura K, Furukawa S, Songsri P. 1993. Expression of the gene encoding a translational elongation factor 3 homolog of chlorella virus CVK2. *Virology* 197:742–50

193. Yamada T, Furukawa S, Hamazaki T, Songsri P. 1996. Characterization of DNA-binding proteins and protein kinase

activities in chlorella virus CVK2. *Virology* 219:395–406

194. Yamada T, Higashiyama T. 1993. Characterization of the terminal inverted repeats andtheir neighboring tandem repeats in the chlorella CVK1 virus genome. *Mol. Gen. Genet.* 241:554–63

195. Yamada T, Higashiyama T, Fukuda T. 1991. Screening of natural waters for viruses which infect chlorella cells. *Appl. Environ. Microbiol.* 57:3433–37

196. Yamada T, Hiramatsu S, Songsri P, Fujie M. 1997. Alternative expression of a chitosanase gene produces two different proteins in cells infected with chlorella virus CVK2. *Virology* 230:361–68

197. Yamada T, Sakaguchi K. 1982. Comparative studies on chlorella cell walls: induction of protoplast formation. *Arch. Microbiol.* 132:1013

198. Yamada T, Tamura K, Aimi T, Songsri P. 1994. Self-splicing group I introns in eukaryotic viruses. *Nucleic Acids Res.* 22:2532–37

199. Yan X, Olson NH, Van Etten JL, Baker TS. 1998. Cryoelectron microscopy and image reconstruction of PBCV-1, an algal virus with T = 169 lattice symmetry. *Electron Microsc.* 1:775–76

200. Yasuda S, Sekiguchi M. 1970. T4 endonuclease involved in repair of DNA. *Proc. Natl. Acad. Sci. USA* 67:1839–45

201. Yonker CR, Caldwell KD, Giddings JC, Van Etten JL. 1985. Physical characterization of PBCV virus by sedimentation field flow fractionation. *J. Virol. Methods* 11:145–60

202. Yuan YP, Schultz J, Mlodzik M, Bork P. 1997. Secreted fringe-like signaling molecules may be glycosyltransferases. *Cell* 88:9–11

203. Yuen L, Moss B. 1987. Oligonucleotide sequence signaling transcriptional termination of vaccinia virus early genes. *Proc. Natl. Acad. Sci. USA* 84:6417–21

204. Zhang Y, Burbank DE, Van Etten JL. 1988. Chlorella viruses isolated in China. *Appl. Environ. Microbiol.* 54:2170–73

205. Zhang Y, Nelson M, Nietfeldt JW, Burbank DE, Van Etten JL. 1992. Characterization of chlorella virus PBCV-1*Cvi*AII restriction and modification system. *Nucleic Acids Res.* 20:5351–56

206. Zhang Y, Nelson M, Nietfeldt J, Xia Y, Burbank DE, et al. 1998. Chlorella virus NY-2A encodes at least twelve DNA endonuclease/methyltransferase genes. *Virology* 240:366–75

207. Zhang Y, Nelson M, Van Etten JL. 1992. A single amino acid change restores DNA cytosine methyltransferase activity in a cloned chlorella virus pseudogene. *Nucleic Acids Res.* 20:1637–42

208. Zhang Y, Strasser P, Grabherr R, Van Etten JL. 1994. Hairpin loop structure at the termini of the chlorella virus PBCV-1 genome. *Virology* 202:1079–82

Annu. Rev. Microbiol. 1999. 53:495–523

MECHANISMS FOR REDOX CONTROL OF GENE EXPRESSION

Carl E. Bauer

Department of Biology, Indiana University, Bloomington, Indiana 47405

Sylvie Elsen

Department of Biology, Indiana University, Bloomington, Indiana 47405

Terry H. Bird

National Cancer Institute, National Institutes of Health, Bethesda, Maryland 20892

Key Words   SoxR, FNR, FixL, NifL, OxyR, ArcB, RegB, CrtJ, NifA

■ **Abstract**   This review discusses various mechanisms that regulatory proteins use to control gene expression in response to alterations in redox. The transcription factor SoxR contains stable [2Fe-2S] centers that promote transcription activation when oxidized. FNR contains [4Fe-4S] centers that disassemble under oxidizing conditions, which affects DNA-binding activity. FixL is a histidine sensor kinase that utilizes heme as a cofactor to bind oxygen, which affects its autophosphorylation activity. NifL is a flavoprotein that contains FAD as a redox responsive cofactor. Under oxidizing conditions, NifL binds and inactivates NifA, the transcriptional activator of the nitrogen fixation genes. OxyR is a transcription factor that responds to redox by breaking or forming disulfide bonds that affect its DNA-binding activity. The ability of the histidine sensor kinase ArcB to promote phosphorylation of the response regulator ArcA is affected by multiple factors such as anaerobic metabolites and the redox state of the membrane. The global regulator of anaerobic gene expression in α-purple proteobacteria, RegB, appears to directly monitor respiratory activity of cytochrome oxidase. The aerobic repressor of photopigment synthesis, CrtJ, seems to contain a redox responsive cysteine. Finally, oxygen-sensitive rhizobial NifA proteins presumably bind a metal cofactor that senses redox. The functional variability of these regulatory proteins demonstrates that prokaryotes apply many different mechanisms to sense and respond to alterations in redox.

CONTENTS

INTRODUCTION

Cells undergo significant alterations in their physiology while growing in aerobic versus anaerobic environments. When grown aerobically, cells utilize the transcription factors OxyR and SoxR to control induction of oxygen defense proteins such as superoxide dismutase and catalase. Histidine sensor kinases such as ArcB and RegB, and the DNA-binding protein FNR, are involved in controlling numerous physiological changes that occur when cells undergo respiratory growth shifts. These global regulators control such diverse metabolic processes as respiration and photosynthesis as well as carbon and nitrogen fixation. In some species, the flavoprotein NifL functions as a redox-responding regulator that controls the activity of NifA, the activator of nitrogenase gene expression. These different "redox regulators" have different sensitivities to oxygen and/or oxygen-reactive species, thus providing differential control of target genes in response to alterations in oxygen tension (166). This review focuses on molecular mechanisms used by various redox regulators to sense the presence or absence of oxygen, oxygen-reactive species, or changes in the cellular redox state. A number of recent reviews have been published that describe details of the target genes that these various systems control. References to these complementary reviews are cited in the text.

IRON SULFUR CENTERS

SoxR-SoxS

The *Escherichia coli* SoxR/SoxS regulon provides defense against oxidative damage caused by superoxide ($O_2^{\cdot-}$). The *soxRS* locus was identified through several different genetic approaches. One approach was to isolate mutants that provide resistance to menadione, which is a redox cycling agent that generates superoxide by diverting electrons from the NADH pool to oxygen (64). Another approach involved the selection of mutants that constitutively express the *nfo* locus, which codes for endonuclease IV (21, 165). *nfo* expression was known to be induced by the superoxide-generating redox-cycling agent paraquat, so this selection yielded similar mutants as menadione selection. Subsequent analyses of these constitutive *sox* mutants (64), and of the effect of deleting the *sox* locus (74, 165), demonstrated that the *sox* locus was responsible for the induction of approximately 10 proteins that provide oxidative defense to $O_2^{\cdot-}$. In addition to endonuclease IV, the *sox* locus controls induction of Mn-superoxide dismutase, NADPH:ferredoxin oxidoreductase, and glucose-6-phosphate-dehydrogenase as well as others (reviewed in 70).

Sequence analysis indicated that the *sox* locus contains two genes, *soxS* and *soxR*, that are divergently transcribed (4, 165, 170). Divergent promoters for *soxS* and *soxR* are located in the intergenic region and share overlapping -10 promoter recognition sequences (73). Induction of oxidative defense genes by SoxR and SoxS involves a two-step transcriptional activation (125, 171). First, the redox sensor protein SoxR stimulates transcription of *soxS* under inducing conditions ($O_2^{\cdot-}$). SoxS then directly activates target genes by binding to their promoter regions (4, 45, 103, 104).

How does SoxR function as the key player responsible for sensing the presence of $O_2^{\cdot-}$? Overexpression and purification of SoxR under oxidizing conditions resulted in protein preparations with a strong red absorbance (69). Subsequent analysis indicated that SoxR exists in solution as a homodimer containing two stable [2Fe-2S] centers that are anchored to four cysteine residues near its carboxyl-terminus (19, 68, 169). Whole cell EPR measurements indicated that the iron-sulfur centers are reduced under normal physiological conditions and that they rapidly oxidize when challenged with oxidative stress conditions (34). The iron-sulfur centers become reduced again when oxidative stress conditions are removed, indicating that oxidation-reduction of these centers is reversible.

Considerable progress has occurred in determining the role of the iron-sulfur centers in controlling activity of SoxR (Figure 1*a*). In vitro transcription studies with wild-type and constitutively active mutant forms of SoxR have demonstrated that only the oxidized form of SoxR stimulates transcription of *soxS* (19, 35, 53, 72). However, DNA-binding analysis indicates that both oxidized and reduced SoxR, as well as apo-SoxR which lacks iron-sulfur centers, bind to the *soxS* promoter with comparable affinities (69). Since the presence of an iron-sulfur center does not affect SoxR binding to DNA, it must somehow affect the interaction of SoxR with RNA polymerase. One intriguing finding is that SoxR binds to the region between the σ^{70}-type -10 and -35 promoter recognition elements of the *soxS* promoter, which is very unusual for a prokaryotic activator (69). Mutational analysis further demonstrates that an unusually long 19-bp spacer sequence between the -10 to -35 motifs is critical for SoxR-dependent transcription of *soxS* (71). Additional footprint analysis using Cu-5-phenyl-1,10-phenanthroline as a probe for DNA single strands indicates that only oxidized SoxR stimulates RNA polymerase to form an open complex at the promoter (Figure 1) (68). These results indicate that the *soxS* promoter is most likely co-occupied by SoxR and

(a) **SoxR**

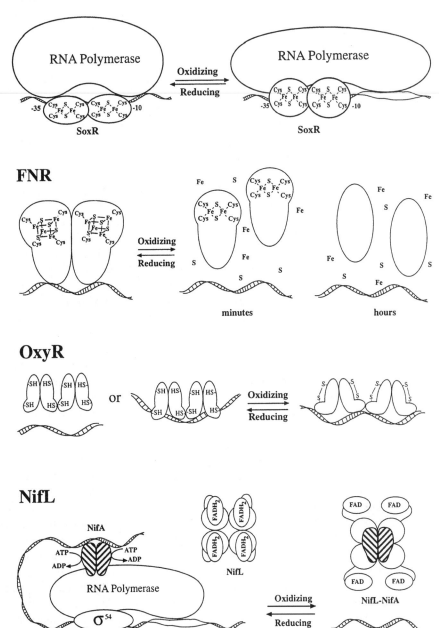

Figure 1 *(continued)*

(b) **FixL**

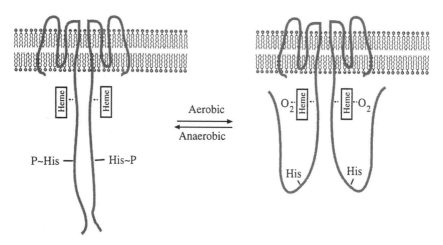

Figure 1 Models for the control of various regulators in response to redox changes. Details of individual mechanisms are given in the text.

RNA polymerase and that the oxidation/reduction state of iron sulfur centers must somehow affect the ability of SoxR to stimulate RNA polymerase to form an open complex. It remains to be determined how this occurs.

FNR

FNR is a global regulator responsible for controlling aerobic-anaerobic regulation of over 120 target genes in *E. coli* (141), details of which can be obtained in several recent reviews (65, 152, 167). FNR was initially identified by Guest and coworkers in the mid 1970s through the isolation and characterization of mutants that failed to carry out fumarate and nitrate reduction, which were subsequently designated FNR mutants (66). Since then it has been demonstrated that many genes whose expression is affected by an aerobic-to-anaerobic growth shift are controlled by FNR. Specifically, FNR is responsible for anaerobic induction of succinate dehydrogenase, fumarase, and isocitrate dehydrogenase, which, together with other enzymes in the tricarboxylic acid cycle, generate energy via oxidative phosphorylation (22, 128, 129). FNR also activates expression of anaerobic respiratory enzymes that utilize alternative terminal electron acceptors such as TMAO/DMSO, fumarate or nitrate (27, 88, 155). In addition to activating expression of many anaerobically expressed genes, FNR is also responsible for repressing several aerobic respiratory enzymes such as cytochrome oxidase (26, 50) and NADH dehydrogenase (154). Mutational analyses have defined a region in FNR that activates transcription by interacting with the α subunit of RNA polymerase (167a). Interestingly, many mutations that only affect repressing activity of FNR also map to the same region (61a).

Sequence analysis revealed striking similarities of FNR to the well-characterized catabolite activator protein (CAP) (148). One notable difference was the presence of an amino terminal extension in FNR that contains a ferredoxin-like four cysteine cluster (Cys-X_3-Cys-X_2-Cys-X_5-Cys) that has no counterpart in CAP. Mutational analysis indicated that three out of the four cysteine residues in this cluster, as well as a fourth Cys located ~100 amino acid residues away, are required for FNR activation (62, 113, 147, 153). These results led to the proposal that FNR senses redox via an iron center that is coordinated by four Cys residues (113, 153).

Several early studies indicated a role for iron for in vivo and in vitro activity of FNR (60, 61, 63, 124, 154, 164). However, initial attempts at isolating FNR resulted in protein preparations with variable and substoichiometric amounts of iron that exhibited weak DNA-binding and ability to promote transcription activation. A major advance to the biochemical analysis of FNR occurred through the isolation and characterization of FNR* mutants, by Kiley and coworkers, that constitutively activated expression of genes in the FNR regulon (8, 92, 94, 100, 101, 175). Over-expression and purification of FNR* from one of the mutant strains resulted in the first FNR preparation that contained significant amounts of iron (92). Subsequent spectral and EPR analyses of the FNR* protein indicated that it contained one stable [3Fe-4S] center per subunit (92). Because four cysteine residues are required for FNR activity, it was concluded that each FNR subunit most likely contained a [4Fe-4S] center in vivo (Figure 1). These investigators subsequently developed anaerobic chromatographic procedures that resulted in the isolation of active wild-type FNR. These FNR preparations contained one [4Fe-4S] center per subunit (93, 101). In vitro analysis of these FNR preparations indicated that the [4Fe-4S] centers are very labile and that they rapidly disassemble when exposed to oxygen (101). More recent Mössbauer spectroscopy analysis has allowed the direct in vivo observation of the oxidation/reduction state of the iron-sulfur centers in FNR (137). This study demonstrated that a growth shift from anaerobic to aerobic conditions caused the two [4Fe-4S] centers to rapidly partially disassemble (within minutes) to form two [2Fe-2S] centers (Figure 1a) (137). Prolonged incubation (hours) resulted in further disassembly of the [2Fe-2S] centers to form apoproteins that lack iron. It was also observed that shifting cells from aerobic to anaerobic conditions resulted in the rapid conversion of the [2Fe-2S] centers to [4Fe-4S] centers (137). Thus, oxidative conversion of [4Fe-4S] centers to [2Fe-2S] centers is a reversible process.

How is FNR's activity affected by conversion of [4Fe-4S] centers to [2Fe-2S] centers? Initial studies of FNR mutants indicated that DNA-binding of FNR is promoted by formation of a homodimer (Figure 1a) (8, 100, 101, 114, 175). It was also demonstrated that loss of the [4Fe-4S] centers by oxidation resulted in the conversion of FNR from a homodimer, which binds with high affinity to its target sequences, into monomers, which have a low DNA-binding affinity (92, 93). Thus, the presence of [4Fe-4S] centers is required for FNR activity.

HEME

FixL

The nitrogen fixing enzyme, nitrogenase, contains several iron sulfur centers that are oxygen labile. To keep from synthesizing an enzyme that would rapidly become inactivated by oxygen, the cells only express nitrogen fixation genes (*nif* and *fix*) under nitrogen and oxygen limiting conditions. Mechanisms used for the regulation of *nif* gene expression in response to redox vary depending on the bacterial species. The regulation can involve both transcriptional and posttranscriptional modification of *nif/fix* specific transcription factors. A detailed review of the various regulatory factors used for oxygen and nitrogen control of *nif/fix* expression that are used by different bacterial species has been written by Hans-Martin Fisher (46). A relevant point is that, in several species, oxygen regulation of *nif/fix* expression is dependent on the two-component regulatory system, FixL/FixJ (9, 29). In *Rhizobium meliloti*, phosphorylated FixJ activates expression of the transcription factors FixK and NifA, which subsequently activate *fix* and *nif* expression. A similar control occurs in *Bradyrhizobium japonicum* in which phosphorylated FixJ activates anaerobic expression of FixK2, which then activates expression of σ^{54} coded by *rpoN*. σ^{54} is then used for *nifA* expression. Consequently, FixL/FixJ plays a central role in regulating *nif/fix* gene expression in response to a reduction in oxygen tension in these species.

How does FixL respond to oxygen? As noted above, FixL and FixJ are members of a two-component signal transduction system in which FixL is a histidine sensor kinase and FixJ is a DNA-binding response regulator (29, 105, 131). In vitro transcription assays demonstrated that oxygen responsive transcription of the *R. meliloti nifA* and *fixK* genes requires only RNA polymerase, FixL and FixJ (1, 9, 32, 116, 140). This indicates that FixL is capable of directly sensing oxygen tension and that FixJ~P is capable of activating transcription. FixJ appears to be a conventional DNA-binding type response regulator that exhibits recognizable receiver and helix-turn-helix (HTH) type DNA-binding effector domains (29, 52, 90). The mechanism of FixJ~P-mediated transcription activation is not known beyond the fact that it activates RNA polymerase holoenzyme that contains σ^{70} (140).

Extensive in vitro analyses have determined that the autophosphorylation of FixL, rather than phosphotransfer to FixJ, is the step that is inhibited by oxygen (105). This was expected since autophosphorylation of FixL is the rate-limiting step in generating phosphorylated FixJ and is thus a logical point of regulation. Measurements of the rate of FixL* autophosphorylation have determined that it is 12–15 times greater under anaerobic conditions than when in the presence of oxygen (105). Kinetic studies have also demonstrated that FixL can function as a phosphatase of FixJ~P (106). Interestingly, the phosphatase activities for FixL and FixL~P are about equal. However, while FixL is unaffected by the presence of oxygen, the phosphatase activity of FixL~P is essentially absent under anaerobic conditions (106). This means that a decrease in oxygen tension would have opposite effects on the kinase and phosphatase activities of FixL such that an anaerobic environment would favor the kinase activity while simultaneously suppressing its phosphatase activity. The net result would be an accumulation of FixJ~P in the cell and activation of *fix* and *nif* gene expression.

The *R. meliloti* FixL protein contains four transmembrane domains near its aminoterminus (Figure 1*b*) (105). However, this region is apparently dispensable for oxygen-sensing since a FixL homolog containing no apparent transmembrane domain has been discovered in *B. japonicum* (5). Furthermore, amino-terminal truncated versions of *R. meliloti* FixL lacking the four transmembrane domains are capable of mediating oxygen regulation in vivo (33, 105). A significant advance to understanding the mechanism by which FixL senses oxygen tension was revealed through analysis of a truncated derivative of the *R. meliloti* protein (FixL*) that lacked all transmembrane domains. Purified preparations of the water soluble FixL* were orange-red in color, suggesting that it might be a haemoprotein (55). This was confirmed by spectral analysis which further demonstrated that the absorption peaks for FixL* shifted upon oxygenation and deoxygenation of the protein. Consequently, it was proposed that the bound heme provided a mechanism for FixL to sense oxygen and that its kinase activity could be regulated accordingly (55).

It is now known that a heme moiety is noncovalently attached to a histidine residue (His-194) in a PER-ARNT-SIM (PAS) regulatory domain of FixL that is well upstream of the site of autophosphorylation (His-285) (Figure 1*b*) (115). PAS domains have been reported in a large family of sensor proteins of all kingdoms and have been associated with light reception and regulation, regulation of circadian rhythmicity, oxygen and redox potential sensing, and protein-protein interactions (174). A deletion derivative of FixL that contains only the central heme-binding PAS domain is fully capable of binding heme. Furthermore, a fragment containing only the carboxyl-terminal kinase domain is also able to function as a kinase (116). This has led to the proposal that FixL structure is composed of modular regulatory and kinase components (116). Moreover, it was shown that the FixL kinase domain alone was capable of constitutively activating *nifA* and *fixK* transcription (116), indicating that the heme-binding regulatory domain functions to inhibit kinase activity when oxygen is present.

Several biophysical investigations have attempted to determine how the heme moiety in FixL can affect catalytic function of the kinase domain. These studies have established that the regulatory domain can exert an effect on the kinase domain that depends on the spin-state of the heme Fe atom (57). A change from a high to low spin-state of the Fe upon binding oxygen is thought to result in a slight shift in the position of the Fe atom within the porphyrin ring (57). New evidence suggests that this event is linked to the propagation of long-range changes in FixL conformation (Figure 1), mediated in part by deprotonation of one or more tyrosine residues that inactivates kinase activity (107).

Besides oxygen, the in vitro kinase activity of FixL is also influenced by other factors such as Mn ion concentration and the ATP/ADP ratio (56). Sensitivity to the ATP/ADP ratio may indicate that an adequate energy charge is required for activation of *fix* and *nif* gene expression. The effect of Mn might be important given that the intracellular Mn concentration is known to increase 1000-fold after the bacteria have colonized plant nodules (56). Superimposing the effects of energy charge and Mn ion concentration on oxygen regulation of FixL autophosphorylation could prevent inadvertent expression of nitrogen fixation genes in anaerobic environments outside of plant nodules.

FLAVIN

NifL

Instead of controlling the expression of *nifA* in response to redox (discussed above), some species have developed a posttranscriptional mechanism of controlling activity of NifA. In *Klebsiella, Azotobacter,* and *Enterobacter,* which are diazotrophic representatives of the γ-subdivision of Proteobacteria, the activity of NifA is indeed inhibited by the redox sensitive protein NifL in response to elevated levels of oxygen and fixed nitrogen in vivo (15, 39, 95, 139, 150). Details of NifL regulation have also recently been reviewed by Dixon (36).

NifL is a protein composed of two domains tethered by a Q-linker. The aminoterminal domain of NifL contains two conserved S-boxes that comprise a PAS domain. The PAS domain from NifL is most similar to those observed in the *Halobacterium halobium bat* gene product, the heme binding domain of FixL, and the regulatory flavoprotein Aer (15, 36). The NifL carboxy-terminal domain presents similarities to the transmitter domain of histidine protein kinases (15, 39, 139). Specifically, *Azotobacter vinelandii* NifL contains the five conserved blocks characteristic of histidine sensor kinases, including the conserved His residue (15, 131). Despite the similarities to sensor kinases, autophosphorylation has not been observed with purified NifL (7, 102, 168). Furthermore, mutational analyses have indicated that the conserved His residue is not required for in vivo activity of NifL, which indicates that NifL activity does not involve phosphorylation. This conclusion is also supported by sequence analysis of NifL from other species that found less conservation in the putative histidine kinase boxes. For example, *Klebsiella pneumoniae* NifL lacks three out of the five conserved kinase sequence motifs as well as the conserved His residue (15).

If NifL does not use phosphorylation to communicate with NifA, then how does it repress NifA activity? Considerable evidence suggests that NifL inactivates NifA through the formation of a complex between both proteins. Specifically, the two proteins coprecipitate with NifL or NifA specific antisera (67). Furthermore, inhibition of NifA activity also requires stoichiometric amounts of NifL (7, 59, 67, 102). This is accomplished in vivo by organizing the two genes into an operon and by using a mechanism of translational coupling that allows balanced synthesis of the two proteins (58, 59).

NifL activity is known to be affected by redox because the oxidized form of NifL inhibits NifA activity in vitro, whereas the reduced form does not (75). A major advance toward understanding the mechanism of redox sensing coincided with the discovery that NifL is a yellow flavoprotein containing FAD as a prosthetic group (Figure 1a). Purified NifL from A. vinelandii was shown to exist in solution as a tetramer containing 3.3 FAD/tetramer (75), which indicates that each subunit most likely binds one molecule of FAD. The redox potential of the flavin is -226 mV at pH8 (110) and approximately -196 mV at pH7 (36). In contrast with a relatively slow reduction rate, oxidation of NifL occurs rapidly in the presence of air in vitro, raising the possibility that NifL might be directly oxidized by intracellular oxygen. NifL can be reduced in vitro in the presence of spinach ferredoxin:NAD(P) oxidoreductase with NADH, suggesting that the in vivo electron donor might be a ferredoxin type oxidoreductase in A. vinelandii (110). Although NifL apparently doesn't contain any iron, iron has been shown to be involved in the redox response of NifL in vivo (67, 144). So an Fe-containing ferredoxin may indeed be involved in electron donation to NifL. Instead of a specific electron donor, it has also been proposed that different electron donors and/or NAD(P)H-dependent enzymes may have the capability to reduce NifL in vivo because of the relatively high redox potential of the flavin (110).

An isolated carboxyl-terminal fragment of K. pneumoniae NifL is able to inhibit NifA activity (123) indicating that this domain is responsible for causing the response. Similar deletion analysis indicates that the amino-terminal domain of NifL binds the flavin, indicating that this is the redox-sensing domain (143, 151). Presumably, alteration of the redox state of the flavin in the sensor domain results in a conformational change that ultimately affects the ability of the carboxy-terminal effector domain to interact with NifA (Figure 1a). Details of NifL interaction with NifA are still unclear, except that it appears somewhat different depending on the species. In A. vinelandii, NifL does not seem to affect binding of NifA to its target sequence (7) but does affect NifA triphosphatase activity, which is needed for the formation of open complex by σ^{54}-RNA polymerase (44). On the other hand, NifL from K. pneumoniae does affect NifA binding to the target sequence (118) but not its triphosphatase activity (13). In this case it has been suggested that NifL affects NifA binding as well as interactions with σ^{54}-RNA polymerase.

In addition to redox, NifL modulates NifA activity in response to energy and the fixed nitrogen status of the cell. Specifically, NifL of A. vinelandii is responsive to adenosine nucleotides in vitro, suggesting that NifL controls NifA activity in response to the energy status of the cell (151). The inhibitory activity of NifL is stimulated by ADP even when NifL is in the reduced form indicating that the adenosine nucleotide response overrides the redox response. The repressive activity of K. pneumoniae NifL is also affected by nitrogen status through a mechanism that involves a PII-like protein GlnK (66a). Redox sensing function of NifL appears to be independent from both the adenosine nucleotide and nitrogen-sensing functions of the protein in vitro (75, 151), whereas in vivo, NifL inhibition of NifA activity is relieved only under oxygen and nitrogen limiting conditions (114a).

DISULFIDE BOND FORMATION

OxyR

Pretreatment of exponentially grown *E. coli* and *Salmonella typhimurium* cells with low doses of H_2O_2 induces an "adaptive response" that provides resistance to normally lethal doses of H_2O_2 and other oxidants, as well as to killing by heat (2, 23, 31, 120). Adaptation to H_2O_2 induces de novo synthesis of approximately 30 proteins as evidenced by two-dimensional polypeptide protein profiles; nine of them are under control of the *oxyR* locus (23). The *oxyR* regulon is now known to include such gene products as catalase, Mn-superoxide dismutase, alkyl hydroperoxide reductase, glutathione reductase, glutaredoxin 1, Dps (DNA-binding protein from starved cells), coproporphyrinogen III oxidase, several heat-shock proteins, a sensor-regulator protein of capsular polysaccharide synthesis genes, and an arylsulfatase (2, 3, 17, 23, 120, 122, 157, 161).

The *oxyR* gene encodes a 34.4-kDa protein that is a member of the LysR family (24, 159). As is the case for most LysR members, OxyR negatively autoregulates its own expression, both in presence or absence of H_2O_2 (24, 142). OxyR also activates expression of a divergently transcribed upstream gene, *oxyS*, which encodes a small nontranslated RNA that affects expression of numerous loci (3). Purified OxyR exists in solution as a tetramer (98) and belongs to the class I activators that interact with the carboxyl-terminal region of the α-subunit of RNA polymerase to promote transcription activation. OxyR acts cooperatively to increase binding of σ^{70}—containing RNA polymerase to the *katG* and *oxyS* promoters, presumably through direct contact with the α-subunit (99, 158).

OxyR synthesis occurs even in the absence of H_2O_2 treatment, and there is no increased synthesis upon addition of H_2O_2 (156). This indicates that H_2O_2 activation of the OxyR regulon is a result of modification of preexisting OxyR (156). Direct redox sensing by OxyR was demonstrated by in vitro transcription assays, which showed that addition of 100 mM DTT prevented OxyR from activating transcription of the *katG* and *ahpC* genes, whereas removal of the DTT restored its activity (24, 156, 160). DNA footprint analyses also indicated that oxidized OxyR binds to the *katG, ahpC, gorA,* and *dps* promoters, but that reduced OxyR does not (163). Interestingly, both oxidized and reduced OxyR can bind to the *oxyR-oxyS* promoter region, but with distinct differences in their binding characteristics (163) (Figure 1a). When binding to target promoters in an oxidized state, OxyR produces a large footprint (>45 nucleotides) that is typical for members of the LysR family. Hydroxyl radical protection experiments have refined the region of DNA-binding and shown that oxidized OxyR binds to four successive major grooves on one face of the DNA helix. However, when OxyR is reduced, the footprint pattern to the *oxyR-oxyS* promoter is significantly smaller, encompassing only two major grooves that are separated by one helical turn (163). Reduced but not oxidized OxyR also promotes a significant bend in the helix axis (163). All of these studies indicate that OxyR must undergo a conformational change upon an alteration in its oxidation-reduction state.

Understanding how OxyR responds to alterations in redox has recently progressed. Several analyses determined that OxyR did not contain any metals nor other prosthetic groups that could function as a redox active center (156, 173). In vivo analysis demonstrated that mutations in four of the six conserved cysteine residues had no effect on OxyR-dependent oxidative defense (99, 173). However, a mutation of Cys-199 resulted in cells that were as hypersensitive to H_2O_2 killing as were strains lacking *oxyR*. A mutation in Cys-208 was also shown to be partially defective in oxidative resistance (99). This led to a model in which Cys-199 reacted with H_2O_2 to form a sulfenic acid derivative (Cys-SOH) that is required for DNA-binding (99). More recent MALDI-TOF mass spectrometry analyses of tryptic fragments derived from oxidized OxyR indicated that oxidation of OxyR results in the formation of an intramolecular disulfide bond between Cys-199 and Cys-208 (Figure 1*a*) (173). Presumably formation of a disulfide bond stabilizes a conformational change that affects DNA-binding. It has also been suggested that one reason why the Cys-208 mutation shows little effect on OxyR activity in vivo is that this mutation may promote intermolecular disulfide bond formation between two OxyR subunits at Cys-199. Presumably, this could cause a conformational change that partially activates DNA-binding of the Cys-208 mutant protein (30).

Zheng et al (173) also used genetic and biochemical analyses to demonstrate that reduction of oxidized OxyR is catalyzed by glutaredoxin 1, which is a thiol-disulfide oxido-reductase that utilizes reduced glutathione and NADH as reductants (76). Using defined concentrations of GSH/GSSG, the redox potential of reducing the disulfide bond in OxyR has been estimated as -185 mV, which is about 90 mV higher than the estimated -280 mV potential of the normal *E. coli* cytosol (173). This provides a basis for explaining why OxyR remains in a reduced state in *E. coli* when it has not been challenged with oxidizing agents. Recent in vivo studies by Åslund et al (6a) demonstrate that OxyR can be directly activated by hydrogen peroxide in wild-type cells. In mutants deficient in cellular disulfide-reducing systems, OxyR is oxidized even in the absence of added oxidants because of the change in cellular redox status (6a).

OxyR also regulates expression of the *grxA* gene that encodes glutaredoxin 1 (156, 157) as well as *gorA* which encodes glutathione reductase (23). So oxidized OxyR also activates expression of enzymes required for its in vivo reduction and inactivation.

PROTEINS WITH UNDEFINED REDOX SENSING MECHANISM

ArcB

The ArcB/ArcA two-component signal transduction system was discovered through a genetic search for *E. coli* mutants with elevated anaerobic expression of succinate dehydrogenase (coded by *sdh*), an enzyme of the tricarboxylic acid

(TCA) cycle (83). Subsequent investigations revealed that mutations in either *arcB* or *arcA* resulted in pleiotropic phenotypes indicative of their role as global regulators of gene expression (81, 83). The ArcB/ArcA regulon is now known to comprise more than 30 loci, including those for flavoprotein-type dehydrogenases, cytochrome oxidases, other enzymes of the TCA cycle, glyoxylate shunt, and fatty acid degradation pathways. Most of the regulation involves anaerobic repression of genes that code for enzymes involved in aerobic respiration, although ArcB/ArcA also activates some metabolism genes in a microaerobic environment. Details of ArcB/ArcA target genes have recently been reviewed by Lynch & Lin (109).

A significant overlap of genes exists within the ArcB/ArcA and FNR regulons, indicating that these two systems coordinate changes in gene expression during transitions between aerobic and anaerobic growth (22, 28, 51, 84, 86, 109 128, 129, 130). For instance, ArcA and FNR both bind to the promoter of *cydAB*, which encodes a high-affinity cytochrome oxidase that is optimally expressed when oxygen is limiting (28). It has been proposed that ArcA is responsible for activating *cydAB* transcription when oxygen levels begin to diminish (in the range of 20%–10% oxygen) but that this stimulation gives way to repression mediated by FNR as the oxygen tension is further reduced (in the range of 10%–0% oxygen) (28, 166).

ArcA is a 28-kDa response regulator with a typical amino-terminal receiver domain and a carboxy-terminal effector domain containing a HTH DNA-binding motif (83). Binding of ArcA to a number of promoters has been examined through gel-mobility shift and DNaseI footprint assays (28, 37, 108, 149). There have been indications that DNA-binding involves cooperative interactions between ArcA proteins (37, 149), and that DNA-binding is enhanced by phosphorylation as is observed for many other response regulators (108, 149).

ArcB has a mass of about 77-kDa and belongs to a small subset of histidine sensor kinases that have both transmitter and receiver domains (87). There are two putative transmembrane domains near its amino-terminus, and it is believed that only a small portion of the protein (7 amino acids) is exposed to the periplasm (87). In vitro studies employing plasma membranes enriched for full length ArcB, as well as purified truncated derivatives, have demonstrated that ArcB autophosphorylates at His-292 of the transmitter domain (80, 85). These studies also established that the phosphoryl group at His-292 can be transferred to an aspartate residue (Asp-576) in the ArcB receiver domain or to an aspartate residue (Asp-54) on ArcA (80). Early genetic analysis also indicated that phosphorylation of the aspartate residue in ArcB (Asp-576) was required to facilitate efficient transfer of a phosphate from ArcB at His-292 to ArcA at Asp-54 (80). More recent investigations into signal processing by ArcB have revealed that it is much more complex than first thought and that ArcB contains a third regulatory domain at the carboxy-terminus called the histidine phosphotransfer domain (Hpt) (79, 91). In vitro analyses have demonstrated that while His-292 is the sole site of autophosphorylation, phosphoryl groups can be transferred to His-717 in the Hpt domain (79). Georgellis et al

(54) examined phosphotransfer using several mutant forms of ArcB and concluded that the normal flux of phosphoryl groups through ArcB proceeds from His-292 to Asp-576 then to His-717. It also appears that phosphate from both His-292 and His-717 may be transferred to Asp-54 on ArcA in vivo, but that the relative contributions might be affected by certain factors present in the cytosol (discussed below). The pattern of alternating histidine and aspartate residues as donors and acceptors of phosphoryl groups resembles the multistep phosphorelay signal pathways that control spore development in *Bacillus subtilis*, virulence factors in *Bordetella pertussis*, and the prokaryotic-like osmoregulation relay found in yeast (reviewed in 6).

Despite over a decade of investigation, it is still unclear what constitutes a signal for anaerobiosis for ArcB. Current models propose that the ability of ArcB to autophosphorylate, and to participate in a phosphotransfer reaction to ArcA, is regulated by several different signals. One signal is thought to be the redox state of an electron carrier or possibly membrane proton potential (16, 82). This conclusion is based on the observation that treatment of cells with agents that artificially reduce or oxidize the aerobic respiratory chain affect expression of genes within the Arc regulon (16). The other signal is thought to be metabolites generated by anaerobic respiration. This mechanism is supported by in vitro studies that indicate that intrinsic phosphatase activity mediated by the receiver domain of ArcB is inhibited by such anaerobic metabolic products such as D-lactate, acetate, pyruvate, and NADH (80). An inhibition of phosphatase activity and concomitant increase in kinase activity would have the effect of increasing the concentration of ArcA~P in the cell leading to an alteration in gene expression (80).

Recently, Matsushika & Mizuno (112) made the observation that a mutation of His-717 in the Hpt domain of ArcB resulted in a normal in vivo response to anaerobic metabolites (fermentable sugars) but no response to alterations in redox. This raises the intriguing possibility that transfer of a phosphate to ArcA from histidine residues in the ArcB transmitter and Hpt domains involves separate regulation by metabolites and redox, respectively. In another study, Ogino et al (127) reported the characterization of the gene *sixA* (signal inhibitory factor-X) that encodes a histidyl-phosphatase with specificity for the Hpt His-717~P. Unfortunately, there was no effect of deleting *sixA* on ArcB-regulated succinate dehydrogenase expression. Thus, it remains to be seen if SixA has a significant in vivo role in the regulation of ArcB in response to redox.

RegB

When *Rhodobacter capsulatus* cells are exposed to oxygen, the synthesis of carotenoids, bacteriochlorophyll, and apoproteins of the photosystem is abruptly halted (25). Redox control of photosystem synthesis is known to occur at both transcriptional and posttranscriptional levels, with the regulation of gene expression involving several different transcription factors. A detailed description of the various transcription factors that have been identified in *R. capsulatus* is beyond the scope of this review. Consequently, this and the next sections will discuss only the redox-responding two-component regulatory system RegB/RegA and the aerobic repressor CrtJ, respectively. More in-depth discussions on the regulation of photosynthesis gene expression can be obtained from recent reviews on this topic (10, 11, 172).

Sganga & Bauer (146) first described a transcription factor that affected anaerobic induction of prokaryotic photosynthesis gene expression. This study described the isolation of *R. capsulatus* mutants that failed to induce light harvesting and reaction center apoprotein synthesis in response to anaerobiosis. The regulatory gene identified by this study coded for a response regulator termed RegA. A subsequent study by Mosley et al (121) described the identification of mutants with a similar phenotype that mapped to a different locus termed *regB*. Sequence analysis indicates that *regB* codes for a histidine sensor kinase. Highly conserved RegB and RegA homologs have now been found in a variety of additional photosynthetic α-purple Proteobacteria (*Rhodovulum sulfidophilum, Roseobacter denitrificans*, and *Rhodobacter sphaeroides*) as well as in several nonphotosynthetic α-purple species (*R. meliloti* and *B. japonicum*) (12, 42, 43, 111, 162).

In addition to controlling photosynthesis gene expression, recent genetic analyses have provided evidence that the RegB/RegA system also functions as an anaerobic redox sensor for induction of carbon and nitrogen fixation in *R. sphaeroides* (89, 138) as well as carbon fixation, nitrogen fixation and hydrogen utilization in *R. capsulatus* (RF Tabita, personal communication; S Elsen & CE Bauer, unpublished results). Regarding nitrogen fixation, RegB/RegA is needed for optimal *fixR-nifA* expression in *B. japonicum* (12) as well as for *nifA2* expression in *R. capsulatus* (S Elsen & CE Bauer, unpublished results). Thus, RegB/RegA appears to constitute a signal transduction system that functions as a global regulator of many diverse anaerobic processes in α-purple bacterial species, and, in some respects, it is similar to the Arc and FNR systems in *E. coli*.

Sequence analysis indicates that RegA homologs contain a relatively short effector domain containing a characteristic HTH DNA-binding motif that is separated from the receiver domain by a short "linker" comprised of four consecutive proline residues (12, 43, 111, 121, 162). Analysis of homologs from various species demonstrates that RegA proteins exhibit a remarkable 78.7–84.2% overall sequence identity, which is the highest level of sequence conservation observed among response regulators (111). Furthermore, there is total conservation of the putative HTH DNA-binding motif among RegA homologs (111). This indicates that there are significant constraints in the type of amino acid substitutions that are tolerated by RegA and that there must be conservation of the target DNA recognized by these various RegA homologs. To date, there is no good consensus sequence for RegA binding sites despite footprint patterns of RegA binding to a variety of promoters (40; S Du, TH Bird, S Elsen, CE Bauer, unpublished results). This suggests that DNA-binding specificity of RegA might be determined by structural features of the DNA rather than by specific nucleotide sequences. Recent footprint analyses also indicate that RegA~P binds to its target sequences 16-fold better than does unphosphorylated RegA (TH Bird, S Du, CE Bauer, unpublished results). A constitutively active mutant of RegA has also been described (RegA*) that is capable of activating transcription without need of phosphorylation (40). An unphosphorylated preparation of RegA* binds to the *puc* promoter region with approximately the same affinity as does phosphorylated wild-type RegA (TH Bird, S Du and CE Bauer, unpublished results).

Homologs of RegB exhibit sequence conservation (55.9–61.5%) that is lower than that observed with RegA homologs, but still much higher than the identity typically observed among histidine sensor kinase homologs (111). Unlike ArcB, RegB has a large (~210 amino acids) amino-terminal sensing domain that contains four or five membrane spanning regions (111, 121). This region is followed by a kinase domain containing typical H, N, G1, F, and G2 blocks of homology. In vitro experiments with truncated forms of RegB (RegB' which has one putative membrane spanning domain, or RegB" which has no apparent membrane spanning domains) have demonstrated that these proteins can autophosphorylate when mixed with $[\gamma\text{-}^{32}\text{P}]$ATP and that the phosphate moiety can be transferred to RegA (14a, 78; TH Bird, S Du, CE Bauer, unpublished results). The results of chemical treatments of RegB~P and RegA~P are consistent with the predictions that these proteins are phosphorylated on histidine and aspartate residues, respectively (78). The rate limiting step in the sensory transduction cascade appears to be autophosphorylation of RegB, which is significantly slower in vitro (50 min for half maximal phosphorylation) than phosphotransfer from RegB"~P to RegA (half maximal transfer occurs in less than one minute) (TH Bird, S Du, CE Bauer, published results, 14a). There is currently no information on dephosphorylation of RegB.

As in the case for ArcB, it is not yet clear what constitutes a signal that is perceived by RegB for anaerobiosis. Unlike FixL, there is no heme associated with the isolated truncated form of RegB (78). However, several genetic studies have implicated respiratory activity of the cbb_3-type cytochrome c oxidase as a possible signal that shuts off RegB kinase activity (20, 126). Specifically, genes encoding RegB and RegA homologs from various photosynthetic species have all been found in a three gene cluster that also contains *senC* that codes for a cbb_3-type cytochrome c oxidase assembly factor (20, 43, 111). Disruption of *senC* affects respiration as well as RegB activity in vivo (20). Null mutations in the cbb_3-type cytochrome c oxidase also result in constitutive (aerobic and anaerobic) expression of photosynthesis genes that are controlled by RegB/RegA (20, 126). The constitutive phenotype is dependent on a functional RegB or RegA, indicating that cbb_3-type cytochrome c oxidase functions upstream in the signal transduction cascade. A working model is that respiratory activity of cbb_3-type cytochrome c oxidase somehow serves as a signal to shut off RegB autophosphorylation activity.

CrtJ

CrtJ is a redox-responding protein that mediates aerobic repression of photosynthesis gene expression in *R. capulatus* and *R. sphaeroides* (18, 132, 133, 135). Specifically, CrtJ is responsible for the aerobic repression of light harvesting-II, bacteriochlorophyll and carotenoid biosynthesis genes coded by *puc*, *bch*, and *crt* genes, respectively (18, 135). More details of this regulatory circuit can also be found in a review by Bauer & Bird (11).

CrtJ is a soluble 52-kDa protein with a deduced amino acid sequence that exhibits no similarity to proteins in various database sets other than the presence of a putative HTH DNA-binding domain at its carboxy-terminus (133). Gel-mobility shift and DNase I footprint experiments have demonstrated that CrtJ binds to a conserved palindrome TGT-N_{12}-ACA located in CrtJ-regulated promoters that appear to belong to the σ^{70} class (134, 135). In general, CrtJ-regulated promoters can be divided into two types. One contains two CrtJ palindromes separated by 7 or 8 nucleotides with one palindrome spanning the -10 region and the other the -35 region (10, 134, 136). The second type has only one CrtJ palindrome located within the -10 to -35 promoter region and a second palindrome located upstream, at variable distances (76 to 240 bp) (41). For both types of promoters, DNA-binding of CrtJ involves cooperative interactions between CrtJ proteins bound to two sets of palindromes (41, 136). Presumably, cooperative CrtJ interactions that occur between distant palindromes involve the formation of a looped DNA structure (41).

In vitro experiments with purified *R. capsulatus* CrtJ have demonstrated that the protein binds to DNA in a redox-sensitive manner such that binding is significantly tighter under oxidizing conditions (20 mM ferricyanide) than under reducing conditions (10 mM sodium dithionite). If CrtJ is preincubated with DTT and then a large excess of ferricyanide is added, CrtJ recovers its high affinity DNA-binding activity indicating that reduction-oxidation of CrtJ is reversible (134). This suggests that CrtJ can directly sense redox and switch between active (oxidized) and inactive (reduced) forms in terms of its DNA-binding properties. No redox-sensing metals have been detected by EPR and atomic absorption spectrometry with purified CrtJ preparations (134). There is also no clustering of cysteine residues in CrtJ. Only two distantly separated Cys residues are conserved between *R. capsulatus* and *R. sphaeroides* CrtJ (133). Interestingly, one of the conserved Cys is located near the HTH motif. When this Cys is mutated to an aspartate, CrtJ exhibits constitutive DNA-binding activity in vivo (C Dong, CE Bauer, unpublished results). It remains to be determined if this Cys is involved in the formation of a disulfide bridge and/or could be oxidized to sulfenic acid in response to redox changes.

NifA

NifA is the transcriptional activator of nitrogen fixation (*nif/fix*) genes in many diazotrophs. NifA is a member of the σ^{54} bacterial enhancer-binding family, and it exhibits an ATPase activity needed for σ^{54}-RNA polymerase to form an open complex (reviewed in 46, 119). NifA protein from most species typically contains three domains. The amino-terminal domain is dispensable for NifA activity (14, 38, 47) and is even absent in NifA from *Rhizobium leguminosarum* biovar *trifolii* (77). In *K. pneumoniae* this domain seems to be required to prevent NifA from being inactivated by NifL under derepressing conditions (38). The central domain, which is suggested to interact with σ^{54}-RNA polymerase, possesses the conserved putative nucleotide-binding site. The carboxyl-terminal domain of NifA contains a HTH-type DNA-binding motif.

As already mentioned in this review, both the expression and activity of NifA can be subject to oxygen control. For example, *nifA* transcription is regulated by FixL/FixJ or by RegB/RegA in several rhizobial species. Furthermore, in the γ-Proteobacteria, NifA activity is modulated by interaction with the redox-regulated flavoprotein NifL. Interestingly, several diazotrophs (*Bradyrhizobium japonicum, Rhizobium meliloti,* and *R. leguminosarum* biovar *trifolii*) exhibit a second posttranscriptional mechanism of regulating NifA activity in response to oxygen (14, 49, 145). In these species, where there are no reports of NifL homologs, oxygen rapidly and irreversibly inhibits the activity of NifA (97, 117). In vivo experiments performed by Morett et al (117) demonstrated that oxygen affects not only the DNA-binding activity of NifA, but also its ability to stimulate σ^{54}-RNA polymerase to form an open complex.

Alignments between NifA homologs indicated a striking difference between oxygen-tolerant NifA proteins from *K. pneumoniae* and *A. vinelandii*, and the oxygen-sensitive proteins (46, 47, 119). These alignments showed that a conserved interdomain linker region is located between the central and carboxyl-terminal domains only in the oxygen-sensitive NifA proteins. This interdomain linker comprises part of a Cys cluster, $Cys-X_{11}-Cys-X_{19}-Cys-X_4-Cys$ (47). Deletion experiments indicate that the amino terminal region of *Herbaspirillum seropedicae* NifA appears responsible for nitrogen control of NifA activity and that the central/C-terminal domains are responsible for oxygen control of NifA activity (116a, 151a). Additional mutagenesis experiments demonstrated that these four cysteine residues are indispensable for *B. japonicum* NifA activity and that a Cys-472 mutant protein is defective in DNA binding (47, 117). The spacing between the last two conserved cysteine residues, which are located in the interlinker domain, is also critical for NifA activity (88). Fischer et al (47) have proposed that the four conserved cysteine residues could coordinate a metal cofactor that senses the redox status of the cell. Presumably a bound metal would respond to redox by triggering conformational changes in NifA structure that modulate its function. This model is supported by the observation that metal ions are required for NifA activity in vivo (47, 117, 145).

Krey et al (96) demonstrated that oxygen-regulation of NifA in *R. meliloti* might also involve an alteration of the nucleotide binding site. This is based on the observation that a mutation of a Met residue at position 217, which is located 10 residues upstream of the putative nucleotide binding site, leads to oxygen-tolerance. They propose that a change in the redox state of NifA results in a conformational change that coincides with a loss of binding or hydrolysis of the nucleotide and, hence, suppression of NifA activity.

CONCLUDING REMARKS

There currently are good working models for how several regulatory factors control gene expression in response to alterations in the redox state of the environment. FNR contains iron sulfur centers that are disrupted upon oxidation, affecting its ability to form a dimer that binds DNA. SoxR has stable iron sulfur centers that, upon changes in redox state, affect its interaction with RNA polymerase. FixL utilizes heme as a cofactor to facilitate a conformational change in response to the presence of oxygen. NifL uses a FAD as a cofactor to sense redox, which in turn affects the ability of NifL to interact with and regulate NifA activity. OxyR is in a class of its own by needing no cofactors for sensing redox. Instead, its activity is controlled through the formation and disruption of an intramolecular disulfide bond. The challenge that remains is to obtain a better understanding of the molecular mechanisms that other redox responding proteins such as ArcB, RegB, CrtJ, and NifA utilize to respond to alterations in redox. Given the diversity of molecular mechanisms that are known to date, it seems likely that additional novel regulatory mechanisms await discovery upon further study of these regulators.

ACKNOWLEDGMENTS

We thank Chen Dong and Shouying Du for comments regarding the manuscript and Robert Gunsalus, Bruce Demple, and Patricia Kiley for providing a generous supply of reprints and preprints. We also apologize to those whose work was not directly cited due to space limitations. Research in this area is supported by National Institutes of Health grants GM539040 and GM40941 to CEB.

Visit the Annual Reviews home page at http://www.AnnualReviews.org

LITERATURE CITED

1. Agron PG, Helinski DR. 1993. Oxygen regulation of *nifA* transcription in vitro. *Proc. Natl. Acad. Sci. USA* 90:3506–10
2. Altuvia S, Almirón M, Huisman G, Kolter R, Storz G. 1994. The *dps* promoter is activated by OxyR during growth and by IHF and σ^{70} in stationary phase. *Mol. Microbiol.* 13:265–72
3. Altuvia S, Weinstein-Fischer D, Zhang A, Postow L, Storz G. 1997. A small, stable RNA induced by oxidative stress: role as a pleiotropic regulator and antimutator. *Cell* 90:45–53
4. Amábile-Cuevas CF, Demple B. 1991. Molecular characterization of the *soxRS* genes of *Escherichia coli:* two genes control a superoxide stress regulon. *Nucleic Acids Res.* 19:4479–84
5. Anthamatten D, Hennecke H. 1991. The regulatory status of the *fixL-* and *fixJ*-like

genes in *Bradyrhizobium japonicum* may be different from that in *Rhizobium melitoti. Mol. Gen. Genet.* 225:38–48

6. Appleby JL, Parkinson JS, Bourret RB. 1996. Signal transduction via the multistep phosphorelay: not necessarily a road less traveled. *Cell* 86:845–48

6a. Åslund F, Zheng M, Beckwith J, Storz G. 1999. Regulation of the OxyR transcription factor by hydrogen peroxide and the cellular thiol-disulfide status *Proc. Natl. Acad. Sci. USA.* 96:6161–65

7. Austin S, Buck M, Cannon W, Eydmann T, Dixon R. 1994. Purification and in vitro activities of the native nitrogen fixation control proteins NifA and NifL. *J. Bacteriol.* 176:3460–65

8. Bates DM, Lazazzera BA, Kiley PJ. 1995. Characterization of FNR* mutant proteins indicates two distinct mechanisms for altering oxygen regulation of the *Escherichia coli* transcription factor FNR. *J. Bacteriol.* 177:3972–78

9. Batut J, Daveran-Mingot M-L, David M, Jacobs J, Garnerone AM, Kahn D. 1989. *fixK*, a gene homologous with *fnr* and *crp* from *Escherichia coli*, regulates nitrogen fixation genes both positively and negatively in *Rhizobium meliloti. EMBO J.* 8:1279–86

10. Bauer CE. 1995. Regulation of photosynthesis gene expression. In *Anoxygenic photosynthetic bacteria*, ed. RE, Blankenship, MT Madigan, CE Bauer. pp. 1221–34. The Netherlands: Kluwer

11. Bauer CE, Bird TH. 1996. Regulatory circuits controlling photosynthesis gene expression. *Cell* 85:5–8

12. Bauer E, Kaspar T, Fischer H, Hennecke H. 1998. Expression of the *fixR-nifA* operon in *Bradyrhizobium japonicum* depends on a new response regulator, RegR. *J. Bacteriol.* 180:3853–63

13. Berger DK, Narberhaus F, Kustu S. 1994. The isolated catalytic domain of NifA, a bacterial enhancer-binding protein, activates transcription in vitro: activation is inhibited by NifL. *Proc. Natl. Acad. Sci. USA* 91:103–07

14. Beynon JL, Williams MK, Cannon FC. 1988. Expression and functional analysis of the *Rhizobium meliloti nifA* gene. *EMBO J.* 7:7–14

14a. Bird TH, Du S, Bauer CE. 1999. Autophosphorylation, phosphotransfer, and DNA-binding properties of the RegB/RegA two-component regulatory system in *Rhodobacter capsulatus. J. Biol. Chem.* 274:16343–48

15. Blanco G, Drummond M, Woodley P, Kennedy C. 1993. Sequence and molecular analysis of the *nifL* gene of *Azotobacter vinelandii. Mol. Microbiol.* 9:869–79

16. Bogachev AV, Murtazina RA, Skulachev VP. 1993. Cytochrome *d* induction in *Escherichia coli* growing under unfavorable conditions. *FEBS Lett.* 336:75–78

17. Bölker M, Kahmann R. 1989. The *Escherichia coli* regulatory protein OxyR discriminates between methylated and unmethylated states of the phage Mu *mom* promoter. *EMBO J.* 8:2403–10

18. Bollivar DW, Suzuki JY, Beatty JT, Dobrowolski JM, Bauer CE. 1994. Directed mutational analysis of Bacteriochlorophyll *a* biosynthesis in *Rhodobacter capsulatus. J. Mol. Biol.* 237:622–40

19. Bradley TM, Hidalgo E, Leautaud V, Ding H, Demple B. 1997. Cysteine-to-alanine replacements in the *Escherichia coli* SoxR protein and the role of the [2Fe-2S] centers in transcriptional activation. *Nucleic Acids Res.* 25:1469–75

20. Buggy JJ, Bauer CE. 1995. Cloning and characterization of *senC*, a gene involved in both aerobic respiration and photosynthesis gene expression in *Rhodobacter capsulatus. J. Bacteriol.* 177:6958–65

21. Chan E, Weiss B. 1987. Endonuclease IV of *Escherichia coli* is induced by paraquat. *J. Bacteriol.* 84:3189–93

22. Chao G, Shen J, Tseng CP, Park SJ, Gun-

salus RP. 1997. Aerobic regulation of isocitrate dehydrogenase gene (*icd*) expression in *Escherichia coli* by the *arcA* and *fnr* gene products. *J. Bacteriol.* 179:4299–304

23. Christman MF, Morgan RW, Jacobson FS, Ames BN. 1985. Positive control of a regulon for defenses against oxidative stress and some heat-shock proteins in *Salmonella typhimurium. Cell* 41:753–62

24. Christman MF, Storz G, Ames BN. 1989. OxyR, a positive regulator of hydrogen peroxyde-inducible genes in *Escherichia coli* and *Salmonella typhimurium*, is homologous to a family of bacterial regulatory proteins. *Proc. Natl. Acad. Sci. USA* 86:3484–88

25. Cohen-Bazire G, Sistrom WR, Stanier RY. 1957. Kinetic studies of pigment synthesis by non-sulfur purple photosynthetic bacteria. *J. Cell Comp. Physiol.* 49:25–68

26. Cotter PA, Chepuri V, Gennis RB, Gunsalus RP. 1990. Cytochrome *o* (*cyoABCDE*) and *d* (*cydAB*) oxidase gene expression in *Escherichia coli* is regulated by oxygen, pH, and the *fnr* gene product. *J. Bacteriol.* 172:6333–38

27. Cotter PA, Gunsalus RP. 1989. Oxygen, nitrate and molybdenum regulation of *dmsABC* gene expression in *Escherichia coli. J. Bacteriol.* 171:3817–23

28. Cotter PA, Melville SB, Albrecht JA, Gunsalus RP. 1997. Aerobic regulation of cytochrome *d* oxidase (*cydAB*) operon expression in *Escherichia coli*: roles of Fnr and ArcA in repression and activation. *Mol. Microbiol.* 25:605–15

29. David M, Daveran ML, Batut J, Dedieu A, Domergue O, Ghai J, Hertig C, Boistard P, Kahn D. 1988. Cascade regulation of *nif* gene expression in *Rhizobium meliloti. Cell* 54:671–83

30. Demple B. 1998. A bridge to control. *Science* 279:1655–56

31. Demple B, Halbrook J. 1983. Inducible repair of oxidative DNA damage in *Escherichia coli. Nature* 304:466–68

32. de Philip P, Batut J, Boistard P. 1990.

Rhizobium meliloti FixL is an oxygen sensor and regulates *R. meliloti nifA* and *fixK* genes differently in *Escherichia coli. J. Bacteriol.* 172:4255–62

33. de Philip P, Soupene E, Batut J, Boistard P. 1992. Modular structure of the FixL protein of *Rhizobium meliloti. Mol. Gen. Genet.* 235:49–54

34. Ding H, Demple B. 1997. In vivo kinetics of a redox-regulated transcriptional switch. *Proc. Natl. Acad. Sci. USA* 94:8445–49

35. Ding H, Hidalgo E, Demple B. 1996. The redox state of the [2Fe-2S] clusters in SoxR protein regulates its activity as a transcription factor. *J. Biol. Chem.* 271:33173–75

36. Dixon R. 1998. The oxygen-responsive NifL-NifA complex: a novel two-component regulatory system controlling nitrogenase synthesis in α-Proteobacteria. *Arch. Microbiol.* 169:371–80

37. Drapal N, Sawers G. 1995. Purification of ArcA and analysis of its specific interaction with the *pfl* promoter-regulatory region. *Mol. Microbiol.* 16:597–607

38. Drummond MH, Contreras A, Michenall LA. 1990. The function of isolated domains and chimaeric proteins constructed from the transcriptional activators NifA and NtrC of *Klebsiella pneumoniae. Mol. Microbiol.* 4:29–37

39. Drummond MH, Wootton JC. 1987. Sequence of *nifL* from *Klebsiella pneumoniae*: mode of action and relationship to two families of regulatory proteins. *Mol. Microbiol.* 1:37–44

40. Du S, Bird TH, Bauer CE. 1998. DNA-binding characteristics of RegA*: a constitutively active anaerobic activator of photosynthesis gene expression in *Rhodobacter capsulatus. J. Biol. Chem.* 273:18509–13

41. Elsen S, Ponnampalam SN, Bauer CE. 1998. CrtJ bound to distant sites interacts cooperatively to aerobically repress photopigment biosynthesis and light harvesting-II gene expression in *Rhodobacter capsulatus. J. Biol. Chem.* 273:30762–69

42. Eraso JM, Kaplan S. 1994. *prrA*, a putative

response regulator involved in oxygen regulation of photosynthesis gene expression in *Rhodobacter sphaeroides*. *J. Bacteriol.* 176:32–43

43. Eraso JM, Kaplan S. 1995. Oxygen-insensitive synthesis of the photosynthetic membranes of *Rhodobacter sphaeroides*: a mutant histidine kinase. *J. Bacteriol.* 177: 2695–706

44. Eydmann T, Söderbäck E, Jones T, Hill S, Austin S, Dixon R. 1995. Transcriptional activation of the nitrogenase promoter in vitro: adenosine nucleotides are required for inhibition of NifA activity by NifL. *J. Bacteriol.* 177:1186–95

45. Fawcett W, Wolf RE. 1994. Purification of a MalE-SoxS fusion protein and identification of the control sites of *Escherichia coli* superoxide-inducible genes. *Mol. Microbiol.* 14:669–79

46. Fisher H-M. 1994. Genetic regulation of nitrogen fixation in Rhizobia. *Microbiol. Rev.* 58:352–86

47. Fisher H-M, Bruderer T, Hennecke H. 1988. Essential and non-essential domains in the *Bradyrhizobium japonicum* NifA protein; identification of indispensable cysteine residues potentially involved in redox reactivity and/or metal binding. *Nucleic Acids Res.* 16:2207–24

48. Fisher H-M, Fritsche S, Herzog B, Hennecke H. 1989. Critical spacing between two essential cysteine residues in the interdomain linker of the *Bradyrhizobium japonicum* NifA protein. *FEBS Lett.* 255: 167–71

49. Fisher H-M, Hennecke H. 1987. Direct response of *Bradyrhizobium japonicum nifA*-mediated *nif* gene regulation to cellular oxygen status. *Mol. Gen. Genet.* 209:621–26

50. Frey B, Janel G, Michelson U, Kersten H. 1989. Mutations in the *Escherichia coli fnr* and *tgt* genes: control of molybdate reductase activity and the cytochrome *d* complex by *fnr*. *J. Bacteriol.* 171:1524–30

51. Fu HA, Lin ECC. 1991. The requirement

of ArcA and Fnr for peak expression of the *cyd* operon in *Escherichia coli* under microaerobic conditions. *Mol. Gen. Genet.* 266:209–13

52. Galinier A, Garnerone A-M, Reyrat J-M, Kahn D, Batut K, Boistard P. 1994. Phosphorylation of the *Rhizobium meliloti* FixJ protein induces its binding to a compound regulatory region at the *fixK* promoter. *J. Biol. Chem.* 269:23784–89

53. Gaudu P, Weiss B. 1996. SoxR, a [2Fe-2S] transcription factor, is active only in its oxidized form. *Proc. Natl. Acad. Sci. USA* 93:10094–98

54. Georgellis D, Lynch AS, Lin ECC. 1997. In vitro phosphorylation study of the Arc two-component signal transduction system of *Escherichia coli*. *J. Bacteriol.* 179:5429–35

55. Gilles-Gonzalez MA, Ditta GS, Helinski DR. 1991. A haemoprotein with kinase activity encoded by the oxygen sensor of *Rhizobium meliloti*. *Nature* 350:170–72

56. Gilles-Gonzalez MA, Gonzales G. 1993. Regulation of the kinase activity of heme protein FixL from the two-component system FixL/FixJ of *Rhizobium meliloti*. *J. Biol. Chem.* 268:16293–97

57. Gilles-Gonzalez MA, Gonzalez G, Perutz MF, Kiger L, Marden MC, Poyart C. 1994. Heme-based sensors, exemplified by the kinase FixL, are a new class of heme protein with distinctive ligand binding and autoxidation. *Biochemistry* 33:8067–73

58. Govantes F, Andújar E, Santero E. 1998. Mechanism of translational coupling in the *nifLA* operon of *Klebsiella pneumoniae*. *EMBO J.* 17:2368–77

59. Govantes F, Molina-López JA, Santero E. 1996. Mechanism of coordinated synthesis of the antagonistic regulatory proteins NifL and NifA of *Klebsiella pneumoniae*. *J. Bacteriol.* 178:6817–23

60. Green J, Guest JR. 1993. Activation of FNR-dependent transcription by iron: an in vitro switch for FNR. *FEMS Microbiol. Lett.* 113:219–22

61. Green J, Guest JR. 1993. A role for iron in transcriptional activation by FNR. *FEBS Lett.* 329:55–58

61a. Green J, Marshall FA. 1999. Identification of a surface of FNR overlapping activating region 1 that is required for repression of gene expression. *J. Biol. Chem.* 274: 10244–48

62. Green J, Sharrocks AD, Green B, Geisow M, Guest JR. 1993. Properties of FNR proteins substituted at each of the five cysteine residues. *Mol. Microbiol.* 8:61–68

63. Green J, Trageser M, Six S, Unden G, Guest JR. 1991. Characterization of the FNR protein of *Escherichia coli*, an iron-binding transcriptional regulator. *Proc. R. Soc. Lond. B. Biol. Sci.* 244:137–44

64. Greenberg JT, Monach P, Chou JH, Josephy PD, Demple B. 1990. Positive control of a global antioxidant defense regulon activated by superoxide-generating agents in *Escherichia coli. Proc. Natl. Acad. Sci. USA* 87:6181–85

65. Guest JR, Green J, Irvine AS, Spiro S. 1996. The FNR modulon and FNR-regulated gene expression. In *Regulation of Gene Expression in Escherichia coli*, ed. ECC Lin, AS Lynch, pp. 317–42. Austin, TX: RG Landes

66. Guest JR, Lambden PR. 1976. Mutants of *Escherichia coli* unable to use fumerate as an anaerobic electron acceptor. *J. Gen. Microbiol.* 97:145–60

66a. He L, Soupene E, Ninfa A, Kustu S. 1998. Physiological role for the GlnK protein of enteric bacteria: Relief of NifL inhibition under nitrogen limiting conditions. *J. Bacteriol.* 180:6661–67

67. Henderson N, Austin S, Dixon R. 1989. Role of metal ions in negative regulation of nitrogen fixation by the *nifL* gene products from *Klebsiella pneumoniae. Mol. Gen. Genet.* 216:484–91

68. Hildalgo E, Bollinger JM, Bradley TM, Walsh CT, Demple B. 1995. Binuclear [2Fe-2S] clusters in the *Escherichia coli* SoxR protein and role of the metal centers

69. Hildalgo E, Demple B. 1994. An iron-sulfur center essential for transcriptional activation by the redox-sensing SoxR protein. *EMBO J.* 13:138–46

70. Hidalgo E, Demple B. 1996. Adaptive responses to oxidative stress: The *soxRS* and *oxyR* regulons. In *Regulation of Gene Expression in Escherichia coli*, ed. ECC Lin, A. Simon-Lynch, pp. 433–50. Austin, TX: Landes

71. Hidalgo E, Demple B. 1997. Spacing of promoter elements regulates the basal expression of the *soxS* gene and converts SoxR from a transcriptional activator into a repressor. *EMBO J.* 16:1056–65

72. Hidalgo E, Ding H, Demple B. 1997. Redox signal transduction: mutations shifting [2Fe-2S] centers of the SoxR sensor-regulator to the oxidized form. *Cell* 88: 121–29

73. Hidalgo E, Leautaud V, Demple B. 1998. The redox-regulated SoxR protein acts from a single DNA site as a repressor and an allosteric activator. *EMBO J.* 17:2629–36

74. Hidalgo E, Nunoshiba T, Demple B. 1994. Molecular genetics of the *soxRS* oxidative stress regulon of *Escherichia coli. Methods Mol. Genet.* 3:325–40

75. Hill S, Austin S, Eydmann T, Jones T, Dixon R. 1996. *Azotobacter vinelandii* NifL is a flavoprotein that modulates transcriptional activation of nitrogen-fixation genes via a redox-sensitive switch. *Proc. Natl. Acad. Sci. USA* 93:2143–48

76. Holmgren A. 1985. Thioredoxin. *Annu. Rev. Biochem.* 54:237–71

77. Iisma SE, Watson JM. 1989. The *nifA* gene product from *Rhizobium leguminosarum* biovar *trifolii* lacks the N-terminal domain found in other NifA proteins. *Mol. Microbiol.* 3:943–55

78. Inoue K, Kouadio JK, Mosley CS, Bauer CE. 1995. Isolation and in vitro phosphorylation of sensory transduction components controlling anaerobic induction

of light harvesting and reaction center gene expression in *Rhodobacter capsulatus. Biochemistry* 34:391–96

79. Ishige K, Nagasawa S, Tokishita S, Mizuno T. 1994. A novel device of bacterial signal transducers. *EMBO J.* 13:5195–202

80. Iuchi S. 1993. Phosphorylation/dephosphorylation of the receiver module at the conserved asparate residue controls transphosphorylation activity of histidine kinase in sensor protein ArcB of *Escherichia coli. J. Biol. Chem.* 268:23972–80

81. Iuchi S, Cameron DC, Lin ECC. 1989. A second global regulator gene (*arcB*) mediating repression of enzymes in aerobic pathways of *Escherichia coli. J. Bacteriol.* 171:868–73

82. Iuchi S, Chepuri V, Fu H-A, Guennis RB, Lin ECC. 1990. Requirement for terminal cytochromes for the *arc* regulatory system in *Escherichia coli*: study utilizing deletions and *lac* fusions of *cyo* and *cyd. J. Bacteriol.* 172:6020–25

83. Iuchi S, Lin ECC. 1988. *arcA* (dye), a global regulatory gene in *Escherichia coli* mediating repression of enzymes in aerobic pathways. *Proc. Natl. Acad. Sci. USA* 85:1888–92

84. Iuchi S, Lin ECC. 1991. Adaptation of *Escherichia coli* to respiratory conditions: regulation of gene expression. *Cell* 66:5–7

85. Iuchi S, Lin ECC. 1992. Purification and phosphorylation of the Arc regulatory components of *Escherichia coli. J. Bacteriol.* 174:5617–23

86. Iuchi S, Lin ECC. 1993. Adaptation of *Escherichia coli* to redox environments by gene expression. *Mol. Microbiol.* 9:1–15

87. Iuchi S, Matsuda Z, Fujiwara T, Lin ECC. 1990. The *arcB* gene of *Escherichia coli* encodes a sensor-regulator protein for anaerobic repression of the *arc* modulon. *Mol. Microbiol.* 4:715–27

88. Jones HM, Gunsalus RP. 1987. Regulation of *Escherichia coli* fumarate reductase genes (*frdABCD*) operon expression by respiratory electron acceptors and the *fnr* gene product. *J. Bacteriol.* 169:3340–49

89. Joshi HM, Tabita FR. 1996. A global two component signal transduction system that integrates the control of photosynthesis, carbon dioxide assimilation, and nitrogen fixation. *Proc. Natl. Acad. Sci. USA* 93:14515–20

90. Kahn D, Ditta G. 1991. Modular structure of FixJ: homology of the transcriptional activator domain with the -35 binding domain of sigma factors. *Mol. Microbiol.* 5:987–97

91. Kato M, Mizuno T, Shimizu T, Hakoshima T. 1997. Insights into multistep phosphorelay from the crystal structure of the C-terminal HPt domain of ArcB. *Cell* 88:717–23

92. Khoroshilova N, Beinert H, Kiley PJ. 1995. Association of a polynuclear iron sulfur center with a mutant FNR protein enhances DNA binding. *Proc. Natl. Acad. Sci. USA* 92:2499–503

93. Khoroshilova N, Popescu C, Münck E, Beinert H, Kiley PJ. 1997. Iron-sulfur cluster disassembly in the FNR protein of *Escherichia coli* by O_2: [4Fe-4S] to [2Fe-2S] conversion with loss of biological activity. *Proc. Natl. Acad. Sci. USA* 94:6087–92

94. Kiley PJ, Reznikoff WS. 1991. Fnr mutants that activate gene expression in the presence of oxygen. *J. Bacteriol.* 173:16–22

95. Kim Y-M, Ahn K-J, Beppu T, Uozumi T. 1986. Nucleotide sequence of the *nifLA* operon of *Klebsiella oxytoca* NG13 and characterization of the gene products. *Mol. Gen. Genet.* 205:253–59

96. Krey R, Pühler A, Klipp W. 1992. A defined amino acid exchange close to the putative nucleotide binding site is responsible for an oxygen-tolerant variant of the *Rhizobium meliloti* NifA protein. *Mol. Gen. Genet.* 234:433–41

97. Kullik I, Hennecke H, Fischer H-M. 1989. Inhibition of *Bradyrhizobium*

japonicum nifA-dependent *nif* gene activation by oxygen occurs at the NifA protein level and is irreversible. *Arch. Microbiol.* 151:191–97

98. Kullik I, Stevens J, Toledano MB, Storz G. 1995. Mutational analysis of the redox-sensitive transcriptional regulator OxyR: regions important for DNA binding and multimerization. *J. Bacteriol.* 177:1285–91

99. Kullik I, Toledano MB, Tartaglia LA, Storz G. 1995. Mutational analysis of the redox-sensitive transcriptional regulator OxyR: regions important for oxidation and transcriptional activation. *J. Bacteriol.* 177:1275–84

100. Lazazzera BA, Bates DM, Kiley PJ. 1993. The activity of the *Escherichia coli* transcription factor FNR is regulated by a change in oligomeric state. *Genes Dev.* 7:1993–2005

101. Lazazzera BA, Beinert H, Khoroshilova N, Kennedy MC, Kiley PJ. 1996. DNA binding and dimerization of the Fe-S containing FNR protein from *Escherichia coli* are regulated by oxygen. *J. Biol. Chem.* 271:2762–68

102. Lee HS, Narberhaus F, Kustu S. 1993. In vitro activity of NifL, a signal transduction protein for biological nitrogen fixation. *J. Bacteriol.* 175:7683–88

103. Li Z, Demple B. 1994. SoxS, an activator of superoxide stress genes in *Escherichia coli*. Purification and interaction with DNA. *J. Biol. Chem.* 269:18371–77

104. Li Z, Demple B. 1996. Sequence specificity for DNA binding by *Escherichia coli* SoxS and Rob proteins. *Mol. Microbiol.* 20:937–45

105. Lois AF, Ditta GS, Helinski DR. 1993. The oxygen sensor FixL of *Rhizobium meliloti* is a membrane protein containing four possible transmembrane segments. *J. Bacteriol.* 175:1103–09

106. Lois AF, Weinstein M, Ditta GS, Helinski DR. 1993. Autophosphorylation and phosphatase activities of the oxygen-sensing protein FixL of *Rhizobium meliloti* are coordinately regulated by oxygen. *J. Biol. Chem.* 268:4370–75

107. Lukat–Rodgers GS, Rexine JL, Rodgers KR. 1998. Heme speciation in alkaline ferric FixL and possible tyrosine involvement in the signal transduction pathway for regulation of nitrogen fixation. *Biochemistry* 37:13543–52

108. Lynch AS, Lin ECC. 1996. Transcriptional control mediated by the ArcA two-component response regulator protein of *Escherichia coli*: characterization of DNA binding at target promoters. *J. Bacteriol.* 178:6238–49

109. Lynch AS, Lin ECC. 1996. Responses to molecular oxygen. In *Escherichia coli and Salmonella: Cellular and Molecular Biology*, ed. FC Neidhardt, pp. 1526–38 Washington, DC: ASM Press

110. Macheroux P, Hill S, Austin S, Eydmann T, Jones T, et al. 1998. Electron donation to the flavoprotein NifL, a redox-sensing transcriptional regulator. *Biochem. J.* 332:413–19

111. Masuda S, Matsumoto Y, Nagashima KYP, Shimada K, Inoue K, et al. 1999. Structural and functional analysis of photosynthetic regulatory genes *regA* and *regB* from *Rhodovulum sulfidophilum, Roseobacter denitrificans* and *Rhodobacter capsulatus. J. Bacteriol.* In press

112. Matsushika A, Mizuno T. 1998. A dual-signaling mechanism mediated by the ArcB hybrid sensor kinase containing the histidine-containing phosphotransfer domain in *Escherichia coli. J. Bacteriol.* 180:3973–77

113. Melville SB, Gunsalus RP. 1990. Mutations in *fnr* that alter anaerobic regulation of electron transport-associated genes in *Escherichia coli. J. Biol. Chem.* 265:18733–36

114. Melville SB, Gunsalus RP. 1996. Isolation of an oxygen-sensitive FNR protein of *Escherichia coli*: interaction at activator and repressor sites of FNR-

controlled genes. *Proc. Natl. Acad. Sci. USA* 93:1226–31

114a. Merrick M, Hill S, Hennecke H, Hahn M, Dixon R, Kennedy C. 1982. Repressor properties of the *nifL* gene product in *Klebsiella pneumoniae. Mol. Gen. Genet.* 185:75–81

115. Monson EK, Ditta GS, Helinski DR. 1994. The oxygen sensor protein, FixL, of *Rhizobium meliloti*: role of histidine residues in heme binding, phosphorylation and signal transduction. *J. Biol. Chem.* 270:5243–50

116. Monson EK, Weinstein M, Ditta GS, Helinski DR. 1992. The FixL protein of *Rhizobium meliloti* can be separated into a heme-binding oxygen-sensing domain and a functional C-terminal kinase domain. *Proc. Natl. Acad. Sci. USA* 89:4280–84

116a. Monteiro RA, Souza EM, Funayama S, Yates MG, Pedrosa FO, Chubatsu LS. 1999. Expression and functional analysis of an N-truncated NifA protein of *Herbaspirillum seropedicae. FEBS Lett.* 447:283–86

117. Morett E, Fischer H-M, Hennecke H. 1991. Influence of oxygen on DNA binding, positive control, and stability of the *Bradyrhizobium japonicum* NifA regulatory protein. *J. Bacteriol.* 173:3478–87

118. Morett E, Kreutzer R, Cannon W, Buck M. 1990. The influence of the *Klebsiella pneumoniae* regulatory gene *nifL* upon the transcriptional activator protein NifA. *Mol. Microbiol.* 4:1253–58

119. Morett E, Segovia L. 1993. The σ^{54} bacterial enhancer-binding protein family: mechanism of action and phylogenetic relationship of their functional domains. *J. Bacteriol.* 175:6067–74

120. Morgan RW, Christman MF, Jacobson FS, Storz G, Ames BN. 1986. Hydrogen peroxide-inducible proteins in *Salmonella typhimurium* overlap with heat shock and other stress proteins.

Proc. Natl. Acad. Sci. USA 83:8059–63

121. Mosley CS, Suzuki JY, Bauer CE. 1994. Identification and molecular genetic characterization of a sensor kinase responsible for coordinately regulating light harvesting and reaction center gene expression in anaerobiosis. *J. Bacteriol.* 176:7566–73

122. Mukhopadhyay S, Shellhorn HE. 1997. Identification and characterization of hydrogen peroxide-sensitive mutants of *Escherichia coli*: genes that require OxyR for expression. *J. Bacteriol.* 179: 330–38

123. Narberhaus F, Lee H-S, Schmitz RA, He L, Kustu S. 1995. The C-terminal domain of NifL is sufficient to inhibit NifA activity. *J. Bacteriol.* 177:5078–87

124. Niehaus F, Hantke K, Unden G. 1991. Iron content and FNR-dependent gene regulation in *Escherichia coli. FEMS Microbiol. Lett.* 84:319–24

125. Nunoshiba T, Hidalgo E, Amábile-Cuevas CF, Demple B. 1992. Two-stage control of an oxidative stress regulon: the *Escherichia coli* SoxR protein triggers redox-inducible expression of the *soxS* regulatory gene. *J. Bacteriol.* 174: 6054–60

126. O'Gara JP, Eraso JM, Kaplan S. 1998. A redox-responsive pathway for aerobic regulation of photosynthesis gene expression in *Rhodobacter sphaeroides* 2.4.1. *J. Bacteriol* 180:4044–50

127. Ogino T, Matsubara M, Kato N, Nakamura Y, Mizuno T. 1998. An *Escherichia coli* protein that exhibits phosphohistidine phosphatase activity towards the HPt domain of the ArcB sensor involved in the multistep His-Asp phophorelay. *Mol. Microbiol.* 27:573–85

128. Park S-J, Chao G, Gunsalus RP. 1997. Aerobic regulation of the *sucABCD* genes of *Escherichia coli*, which encode alpha-ketogluterate dehydrogenase and succinyl coenzyme A synthetase: roles

of ArcA, Fnr, and the upstream *sdhCDAB* promoter. *J. Bacteriol.* 179:4138–42

129. Park S-J, Gunsalus RP. 1995. Oxygen, iron, carbon and superoxide control of the fumarase *fumA* and *fumC* genes of *Escherichia coli*: role of the *arcA*, *fnr*, and *soxR* gene products. *J. Bacteriol.* 177: 6255–62

130. Park S-J, Tseng C-P, Gunsalus RP. 1995. Regulation of succinate dehydrogenase (*sdhCDAB*) operon expression in *Escherichia coli* in response to carbon supply and anaerobiosis: role of ArcA and Fnr. *Mol. Microbiol.* 15:473–82

131. Parkinson JS, Kofoid EC. 1992. Communication modules in bacterial signalling proteins. *Annu. Rev. Genet.* 26:71–112

132. Penfold RJ, Pemberton JM. 1991. A gene from the photosynthetic gene cluster of *Rhodobacter sphaeroides* induces *trans*-suppression of bacteriochlorophyll and carotenoid levels in *R. sphaeroides* and *R. capsulatus*. *Curr. Microbiol.* 23:259–63

133. Penfold RJ, Pemberton JM. 1994. Sequencing, chromosomal inactivation, and functional expression in *Escherichia coli* of *ppsR*, a gene which represses carotenoid and bacteriochlorophyll synthesis in *Rhodobacter sphaeroides*. *J. Bacteriol.* 176:2869–76

134. Ponnampalam SN, Bauer CE. 1997. DNA binding characteristics of CrtJ. *J. Biol. Chem.* 272:18391–96

135. Ponnampalam SN, Buggy JJ, Bauer CE. 1995. Characterization of an aerobic repressor that coordinately regulates bacteriochlorophyll, carotenoid, and light harvesting-II expression in *Rhodobacter capsulatus*. *J. Bacteriol.* 177:2990–97

136. Ponnampalam SN, Elsen S, Bauer CE. 1998. Aerobic repression of the *Rhodobacter capsulatus bchC* promoter involves cooperative interactions between CrtJ bound to neighboring palindromes. *J. Biol. Chem.* 273:30757–61

137. Popescu C, Bates DM, Beinert H, Münck E, Kiley PJ. 1998. Mössbauer spec-

troscopy as a tool for the study of activation/inactivation of the transcription regulator FNR in whole cells of *Escherichia coli*. *Proc. Natl. Acad. Sci. USA* 95:13431–35

138. Qian Y, Tabita FR. 1996. A global signal transduction system regulates aerobic and anaerobic CO_2 fixation in *Rhodobacter sphaeroides*. *J. Bacteriol.* 178:12–18

139. Raina R, Bageshwar UK, Das HK. 1993. The *Azotobacter vinelandii nifL*-like gene: nucleotide sequence analysis and regulation of expression. *Mol. Gen. Genet.* 237:400–06

140. Reyrat JM, Blonski C, Boistard P, Batut J. 1993. Oxygen-regulated in vitro transcription of *Rhizobium meliloti nifA* and *fixK* genes. *J. Bacteriol.* 175:6867–72

141. Sawers RG, Zalelein E, Böck A. 1988. Two-dimensional gel electrophoretic analysis of *Escherichia coli* proteins: influence of various anaerobic growth conditions and the *fnr* gene product on cellular protein composition. *Arch. Microbiol.* 149:240–44

142. Schell MA. 1993. Molecular biology of the LysR family of transcriptional regulators. *Annu. Rev. Microbiol.* 47:597–626

143. Schmitz RA. 1997. NifL of *Klebsiella pneumoniae* carries an N-terminally bound FAD cofactor, which is not directly required for the inhibitory function of NifL. *FEMS Microbiol. Lett.* 157:313–18

144. Schmitz RA, He L, Kustu S. 1996. Iron is required to relieve inhibitory effects of NifL on transcriptional activation by NifA in *Klebsiella pneumoniae*. *J. Bacteriol.* 178:4679–87

145. Screen S, Watson J, Dixon R. 1994. Oxygen sensitivity and meta ion-dependent transcriptional activation by NifA protein from *Rhizobium leguminosarum* biovar *trifolii*. *Mol. Gen. Genet.* 245:313–22

146. Sganga MW, Bauer CE. 1992. Regulatory factors controlling photosynthetic reaction center and light-harvesting gene

expression in *Rhodobacter capsulatus*. *Cell* 68:945–54

147. Sharrocks AD, Green J, Guest JR. 1990. In vivo and in vitro mutants of FNR the anaerobic transcription factor of *Escherichia coli*. *FEBS Lett.* 270:119–22

148. Shaw DJ, Rice DW, Guest JR. 1983. Homology between CAP and FNR, a regulator of anaerobic respiration in *Escherichia coli*. *J. Mol. Biol.* 166:241–47

149. Shen J, Gunsalus RP. 1997. Role of multiple ArcA recognition sites in anaerobic regulation of succinate dehydrogenase (*sdhCDAB*) gene expression in *Escherichia coli*. *Mol. Microbiol.* 26: 223–36

150. Siddavattam D, Steibl H-D, Kreutzer R, Klingmüller W. 1995. Regulation of *nif* gene expression in *Enterobacter agglomerans*: nucleotide sequence of the *nifLA* operon and influence of temperature and ammonium on its transcription. *Mol. Gen. Genet.* 249:629–36

151. Söderbäck E, Reyes-Ramirez F, Eydmann T, Austin S, Hill S, Dixon R. 1998. The redox- and fixed nitrogen-responsive regulatory protein NifL from *Azotobacter vinelandii* comprises discrete flavin and nucleotide-binding domains. *Mol. Microbiol.* 28:179–92

151a. Souza EM, Pedrosa FO, Drummond M, Rigo LU, Yates MG. 1999. Control of *Herbaspirillum seropedicae* NifA activity by ammonium ions and oxygen. *J. Bacteriol.* 181:681–84

152. Spiro S. 1994. The FNR family of transcriptional regulators. *Antonie van Leeuwenhoek* 66:23–36

153. Spiro S, Guest JR. 1988. Inactivation of FNR protein of *Escherichia coli* by targeted mutagenesis in the N-terminal region. *Mol. Microbiol.* 2:701–07

154. Spiro S, Roberts RE, Guest JR. 1989. FNR-dependent repression of *ndh* gene of *Escherichia coli* and metal ion requirement for FNR-regulated gene expression. *Mol. Microbiol.* 3:601–08

155. Stewart V. 1982. Requirment of Fnr and NarL functions for nitrate reductase expression in *Escherichia coli* K-12. *J. Bacteriol.* 151:1320–25

156. Storz G, Tartaglia LA, Ames BN. 1990. Transcriptional regulator of oxidative stress-inducible genes: direct activation by oxidation. *Science* 248:189–94

157. Tao K. 1997. *oxyR*-dependent induction of *Escherichia coli grx* gene expression by peroxide stress. *J. Bacteriol.* 179:5967–70

158. Tao K, Fujita N, Ishihama A. 1993. Involvement of the RNA polymerase α subunit *C*-terminal region in cooperative interaction and transcriptional activation with OxyR protein. *Mol. Microbiol.* 7:859–64

159. Tao K, Makino K, Yonei S, Nakata A, Shinagawa H. 1989. Molecular cloning and nucleotide sequencing of *oxyR*, the positive regulatory gene of a regulon for an adaptive response to oxidative stress in *Escherichia coli*: homologies between OxyR protein and a family of bacterial activator proteins. *Mol. Gen. Genet.* 218:371–76

160. Tao K, Makino K, Yonei S, Nakata A, Shinagawa H. 1991. Purification and characterization of the *Escherichia coli* OxyR protein, the positive regulator for a hydrogen peroxide-inducible regulon. *J. Biochem.* 109:262–66

161. Tartaglia LA, Storz G, Ames BN. 1989. Identification and molecular analysis of *oxyR*-regulated promoters important for the bacterial adaptation to oxidative stress. *J. Mol. Biol.* 210:709–19

162. Tiwari RP, Reeve WG, Dilworth MJ, Glenn AR. 1996. Acid tolerance in *Rhizobium meliloti* strain WSM419 involves a two-component sensor-regulator system. *Microbiology* 142:1693–704

163. Toledano MB, Kullik I, Trinh F, Baird PT, Schneider TD, Storz G. 1994. Redox-dependent shift of OxyR-DNA contacts along an extended DNA-

binding site: a mechanism for differential promoter selection. *Cell* 78:897–909

164. Trageser M, Unden G. 1989. Role of cysteine residues and of metal ions in the regulatory functioning of FNR, the transcriptional regulator of anaerobic respiration in *Escherichia coli. Mol. Microbiol.* 3:593–99

165. Tsaneva IR, Weiss B. 1990. *soxR*, a locus governing a superoxide response regulon in *Escherichia coli* K-12. *J. Bacteriol.* 172:4197–205

166. Tseng C-P, Albrecht J, Gunsalus RP. 1996. Effect of microaerophilic cell growth conditions on expression of the aerobic (*cyoABCDE* and *cydAB*) and anaerobic (*narGHJI, frdABCD*, and *dmsABC*) respiratory pathway genes in *Escherichia coli. J. Bacteriol.* 178: 1094–98

167. Unden G, Becker S, Bongaerts J, Holighaus G, Schirawski J, Six S. 1995. O_2-sensing and O_2-dependent gene regulation in facultatively anaerobic bacteria. *Arch. Microbiol.* 164:81–90

167a. Williams SM, Savery NJ, Busby SJ, Wing HJ. 1997. Transcription activation at a class I FNR-dependent promoters: identification of the activating surface of FNR and the corresponding contact site in the C-terminal domain of the RNA polymerase α subunit. *Nucleic Acid Res.* 25:4028–34

168. Woodley P, Drummond M. 1994. Redundancy of the conserved His residue in *Azotobacter vinelandii* NifL, a histidine autokinase homologue which regulates transcription of nitrogen fixation genes. *Mol. Microbiol.* 13:619–26

169. Wu J, Dunham WR, Weiss B. 1995. Overproduction and physical characterization of SoxR, a [2Fe-2S] protein that governs an oxidative response regulon in *Escherichia coli. J. Biol. Chem.* 270:10323–27

170. Wu J, Weiss B. 1991. Two divergently transcribed genes, *soxR* and *soxS*, control a superoxide response regulon of *Escherichia coli. J. Bacteriol.* 173:2864–71

171. Wu J, Weiss B. 1992. Two stage induction of the *soxRS* (superoxide response) regulon of *Escherichia coli. J. Bacteriol.* 174:3915–20

172. Zeilstra-Ryalls J, Gomelsky M, Eraso JM, Yeliseev A, O'Gara J, Kaplan S. 1998. Control of photosystem formation in *Rhodobacter sphaeroides. J. Bacteriol.* 180:2801–09

173. Zheng M, Åslund F, Storz G. 1998. Activation of the OxyR transcription factor by reversible disulfide bond formation. *Science* 279:1718–21

174. Zhulin IB, Taylor BL, Dixon R. 1997. PAS domain S-boxes in Archae, Bacteria and sensors for oxygen and redox. *Trends Biochem. Sci.* 22:331–33

175. Ziegelhoffer EC, Kiley PJ. 1995. In vitro analysis of a constitutively active mutant form of the *Escherichia coli* global transcription factor FNR. *J. Mol. Biol.* 245:351–61

Annu. Rev. Microbiol. 1999. 53:525–49

INTERCELLULAR SIGNALING DURING FRUITING-BODY DEVELOPMENT OF *MYXOCOCCUS XANTHUS*

Lawrence J. Shimkets

Department of Microbiology, University of Georgia, Athens, Georgia 30602

Key Words   myxobacterium, fruiting body, cell-cell signaling

■ **Abstract**   The myxobacterium *Myxococcus xanthus* has a life cycle that is dominated by social behavior. During vegetative growth, cells prey on other bacteria in large groups that have been likened to wolf packs. When faced with starvation, cells form a macroscopic fruiting body containing thousands of spores. The social systems that guide fruiting body development have been examined through the isolation of conditional developmental mutants that can be stimulated to develop in the presence of wild-type cells. Extracellular complementation is due to the transfer of soluble and cell contact-dependent intercellular signals. This review describes the current state of knowledge concerning cell-cell signaling during development.

CONTENTS

INTRODUCTION

The myxobacteria display cooperative behavior and multicellular development through an intricate social network. Based on ribosomal RNA gene divergence myxobacteria predate the simple multicellular invertebrates that began to appear during the late Precambrian period ~565 million years ago (25, 107). Myxobacteria, however, never adopted an obligate multicellular existence where development is the inevitable consequence of growth. Development is induced by environmental conditions that inhibit growth and results in the production of resting cells known as myxospores. Furthermore, the fate of myxobacterial cells is not exclusively linked to the fate of the multicellular mass; individual cells are free to leave the group

0066-4227/99/1001-0525$08.00

and establish colonies of their own. It appears that the selective advantage afforded by group behavior was carefully balanced by the necessity to survive occasional bouts of a solitary existence in a world shaped by unpredictable environmental forces.

Several characteristics of multicellular organisms include self-recognition, spatial morphogenesis, use of specialized cell types, and intercellular signaling. Each of these properties is evident in the myxobacterial life cycle. Self-recognition, or the ability to distinguish sibling cells from foreign cells, is widespread in eukaryotes; it is found in vertebrates, invertebrates, protozoans, and plants (4, 12, 80). The myxobacterium *Myxococcus xanthus* lives in soil, where it preys on other bacteria for food. The protein fraction of the prey is digested with extracellular enzymes and consumed as the principle carbon, energy, and nitrogen source (21). Self recognition apparently operates at the species level. Even closely related species like *Myxococcus virescens* and *M. xanthus* engage in predatory encounters that lead to the predomination of one species (115).

Multicellular organisms construct three-dimensional structures using spatially directed cell growth, spatially directed cell movement, or both Myxobacterial cells use spatially directed cell movement to accumulate at aggregation foci (Figure 1C) where fruiting bodies will be constructed (Figure 1D). The process is similar in scope to the great animal herd migrations. Approximately 10^5 cells begin the migration to each fruiting body. Dozens of fruiting bodies may form at the same time if large territories were previously established. *M. xanthus* cells are long, thin rods that move by gliding at a rate of ~0.5 cell lengths min^{-1} (Figure 1A). The distance between fruiting bodies is usually <500 μm such that each cell traverses a linear distance of no more than 50 cell lengths to reach an aggregation center. The actual distance they travel may be much longer as their movement in one direction is punctuated by temporary reversals in direction. The signal that orchestrates directed movement into the fruiting body remains unknown. The slow rate of cell movement and the near-famine conditions that induce fruiting body formation encourage a swift conclusion.

The differentiation of specialized cell types is a device multicellular organisms use to engage in spatially discrete activities. Several types of specialized cell types have been described during *M. xanthus* development. Fruiting body development culminates with sporulation but only 10–20% of the initial cell population becomes spherical, stress resistant spores (Figure 1B). Another 10% of the starting population never enters the fruiting body and forms a differentiated population known as peripheral rods that expresses a unique suite of genes (83–85). While the precise function of these cells has not been ascertained, the territoriality of the myxobacteria suggests that peripheral rods could prevent consumption and colonization of the dormant fruiting body by other microbes. Up to 80% of the cell population dies during development (130). The extent of autolysis can be manipulated by the experimental conditions (86). Although the function of autolysis is unknown, it most likely provides carbon and energy sources for development.

Commitment to these differentiated states is controlled by a phase variation that is demarked by yellow and tan pigmentation differences. Tan cells within a

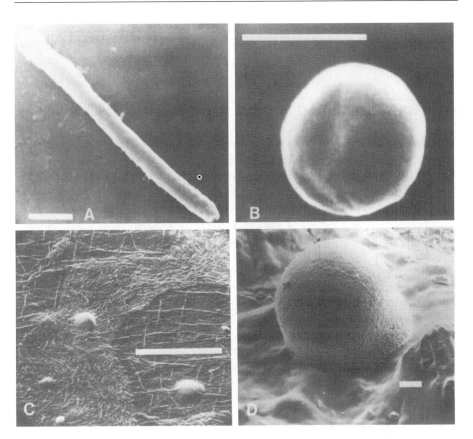

Figure 1 Scanning electron micrographs of *Myxococcus xanthus* during various stages of the life cycle. (*A*) Vegetative cell. Bar is 1.0 μm. Reprinted with permission from Shimkets & Seale (104). (*B*) Myxospore. Bar is 1.0 μm. Reprinted with permission from Shimkets & Seale (104). (*C*) Developing cells migrating toward a fruiting body. Bar is 20 μm. Reprinted with permission from Shimkets & Seale (104). (*D*) Fruiting body. Bar is 20 μm. Reprinted with permission from Brockman & Todd (10).

population of phase-variation proficient cells are preferentially recovered as spores (65). Mutants that are locked into the tan phase cannot form fruiting bodies, and, although they initiate spore formation, they are unable to develop resistance to heat and sonication. The developmental defect of a phase-locked tan mutant can be rescued by addition of phase-variation proficient cells predominated by cells in the yellow phase. In such mixtures, the tan-phase-locked mutant is preferentially represented among the population of viable spores. These results suggest that yellow cells direct fruiting body development and spore maturation through intercellular signaling.

This chapter will not attempt to be a comprehensive review of myxobacterial physiology but will focus on the developmental cell-cell signaling systems of

M. xanthus. There are a number of recent review articles and a book that describe the motility, physiology, and genetics of myxobacteria (23, 24, 41, 48, 49, 52, 91a, 95, 108–111, 112a, 128, 137).

EXTRACELLULAR COMPLEMENTATION PARADIGM

Intercellular signaling refers to the ability of one cell, or group of cells, to regulate the physiology of other cells through the use of soluble or contact-mediated signals. The first approach that was taken to study developmental cell-cell interactions involved the isolation of *M. xanthus* mutants that were unable to form spore-filled fruiting bodies alone but did sporulate when mixed with developmental mutants in a different complementation group (36, 81). Five extracellular complementation groups, A through E, exist for sporulation. Another group of mutants, group S, is defective in social motility and does not dominate the spore population when complemented extracellularly (70, 106). An example of extracellular complementation is shown in Figure 2. The *csgA* mutant (group C) and the *dsp* mutant (group S) are unable to form fruiting bodies or spores alone, but, when mixed at a ratio of 1:1, they form fruiting bodies with the normal level of spores. In this particular example, the S mutant participates in fruiting-body formation, but most of the spores are derived from the *csgA* mutant.

The remarkable efficiency with which fruiting body morphogenesis and spore differentiation occur argues that the extracellular molecule causing the complementation is not merely leaking out of the cells but has an essential extracellular role during development.

Each Complementation Group Is Blocked at a Unique Developmental Step

Mutants in each of the complementation groups are blocked early in development making it difficult to determine by morphology alone whether they arrest at precisely the same point. Kroos & Kaiser (59) used a set of reporter genes constructed by insertion of Tn*5 lac* into development-specific transcriptional units to more accurately define the time of arrest. The insertions that were chosen for this work have no detrimental effect on development. A complementation group mutation that has no effect on induction of a particular reporter is assumed to disrupt the developmental pathway after the expression of the reporter. Conversely, a mutation that blocks expression of a particular reporter is assumed to disrupt the developmental pathway before induction of the reporter. Some reporters exhibit partial expression in a particular mutant background, which suggests that the reporter promoter has multiple activation pathways.

For the sake of simplicity, 11 reporters were selected from the original set, and their expression patterns in A-, B-, C-, D-, and S-signaling mutants are shown in Figure 3. If the six complementation groups reflect six biochemical steps in the pro-

Dsp   CsgA

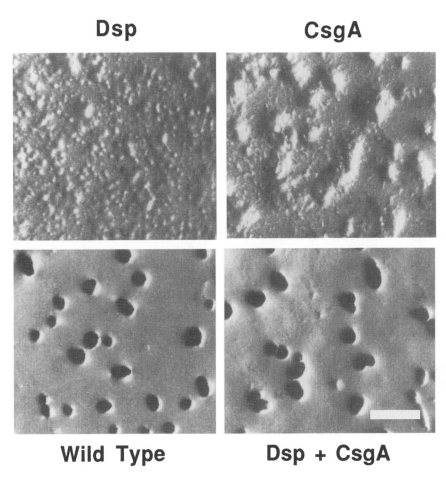

Wild Type   Dsp + CsgA

Figure 2 Extracellular complementation between *dsp* and *csgA* mutants. Cells were allowed to develop alone or were mixed at a 1:1 ratio. Reprinted with permission from Li & Shimkets (70).

duction of a single signal, one would expect the mutants in all the complementation groups to arrest at the same developmental point for lack of the essential signal. In fact, the opposite result is obtained; the expression pattern of these reporters is different in each mutant background (Figure 3). Mutants in the A and B groups arrest immediately after initiation of starvation, whereas the C, D, and S groups arrest several hours later. The effect of the E mutation was not studied in the same suite of reporters and therefore does not appear in Figure 3. However, E mutations appear to block development at a stage after the A and B groups but before the C, D, and S groups (19, 122). Because each mutant group has a different expression pattern, it appears that each group arrests at a different point in development.

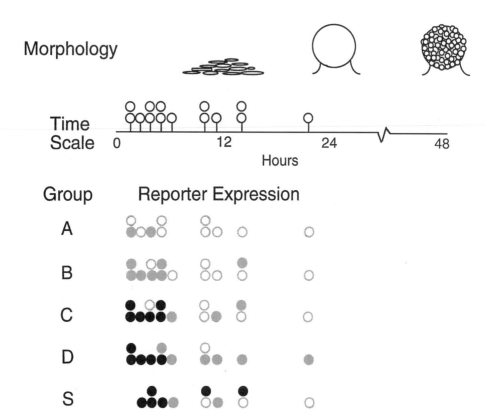

Figure 3   The effect of A, B, C, D, and S mutations on developmental gene expression. The developmental markers are derived from a set of Tn5 *lac* insertions in genes normally expressed during wild-type development. Most of the insertions do not disrupt development. The time of expression of each insertion, denoted by a *circle* above the timescale, is indicated by the position of the insertion above the timescale. The insertions depicted in this figure are ΩDK4521, 3 h; ΩDK4455, 4 h; ΩDK4531, 4 h; ΩDK4411, 5 h; ΩDK4400, 7 h; ΩDK4406, 10 h; ΩDK4506, 10 h; ΩDK4473/DK4414, 11 h; ΩDK4500/DK4403, 13 h; ΩDK4492, 13h; and ΩDK4435, 22 h. The time of expression of each insertion is the arithmetic mean of those reported elsewhere (14, 59, 70). The morphology of wild-type cells at various developmental times is depicted above the timescale. Aggregates appear 8–12 h after starvation. Fruiting bodies without spores are formed by 24 h, whereas sporulation is complete by ~48 h. The effect of A (59, 61), B (59), C (59, 70), D (14), or S mutations (70) on expression of this set of insertions is depicted by the coloration of the circles. *Black* indicates wild-type levels of expression, *gray* indicates partial expression, and *white* indicates no expression. No circle indicates that expression of that reporter has not been determined in the mutant background. Each mutant group has a unique pattern of developmental gene expression, suggesting that each mutant group arrests at a unique developmental point.

Genetic and Biochemical Analysis of the Complementation Mutants

Most of the genes responsible for the complementation mutant phenotypes have been identified and their functions will be described in the following sections (also see Table 1). It is possible to restore the wild-type developmental phenotype to the A, C, and S mutant groups by adding back particular cellular fractions during development. Biochemical complementation is exceedingly efficient and in each of the three cases restores both fruiting body morphogenesis and near-wild-type levels of spores. This constitutes a level of proof that these cellular fractions act from the outside in and thereby have an extracellular state during development.

The A Group   The A group is represented by three genetic loci, *asgA*, *asgB*, and *asgC*. The *asgA* gene encodes a protein that contains both a histidine protein kinase domain and a response regulator domain (93). The *asgB* gene encodes a putative transcription factor with a helix-turn-helix motif near the C terminus (91). The *asgC* gene encodes the major sigma factor of RNA polymerase holoenzyme, σ^{70}, also referred to as *sigA* and *rpoD* (17). The *asgC767* allele has amino acid substitution $E_{598}K$ in region 3.1. This region is suspected to be involved in (p)ppGpp-mediated gene regulation in *Escherichia coli* because suppressors of a $\Delta relA \Delta spoT$ mutation arise in an adjacent codon in *E. coli* (42).

The *asgA*, *asgB* and *asgC* mutants are rescued for fruiting-body formation and sporulation by a mixture of amino acids (62). The amino acids with the highest activities are tyrosine, proline, tryptophan, phenylalanine, leucine, and isoleucine. The amino acids are produced at an average concentration of ~ 25 μm by proteolysis of cell surface proteins. Cells secrete at least two proteases during early development and mutation of any *asg* gene reduces the production of extracellular proteases following amino acid limitation (92). The extracellular concentration of A-signal amino acids is directly proportional to the cell density (63). The properties of A-signaling were illustrated in starvation buffer, which permits the early stages of development, through the use of an A-dependent Tn5 *lac* fusion, $\Omega DK4521$. At cell densities below 2×10^8 cells ml^{-1}, expression of $\Omega DK4521$ is $\sim 25\%$ of maximum. Gene expression increases proportionally with the cell density until $\sim 10^9$ cells/ml^{-1}. If saturating A-signal is added to cells at low cell density, $\Omega DK4521$ expression increases to the maximal level. These results have led to the concept that the A-signal serves a quorum sensing function to determine whether a sufficient population of starved cells is present to make a fruiting body.

How are the A-signal amino acids sensed? The most curious feature about A-signal amino acids is the fact that one amino acid can be replaced by simply adding a higher amount of another. There is no obvious anabolic or catabolic link between the species with the highest activities. There is, however, an interesting relationship in their use in protein synthesis. To initiate protein synthesis, each amino acid is attached to the 3' end of its unique, cognate tRNA molecule. Specificity for both the amino acid and tRNA is determined by the aminoacyl tRNA

TABLE 1   Genes of *Myxococcus xanthus* extracellular complementation groups

Complementation group	Gene symbol	Function of gene product	Complementing cellular fraction
A	*asgA*	Histidine protein kinase/response regulator	Amino acid mixture
	asgB	Putative transcription factor	Amino acid mixture
	asgC (*sigA, rpoD*)	Sigma factor (σ^{70})	Amino acid mixture
B	*bsgA*	ATP-dependent protease	Unknown
C	*csgA*	Short-chain alcohol dehydrogenase	CsgA
	socA	Short-chain alcohol dehydrogenase	
D	*dsgA*	Translation initiation factor IF-3	Unknown
E	*esg*	E1α subunit branched-chain keto acid dehydrogenase	Isovaleric acid
		E1β subunit branched-chain keto acid dehydrogenase	
S	*difA*	Methyl-accepting chemotaxis protein (MCP)	Fibrils
	difB	Unknown	
	difC	CheW	
	difD	CheY-like response regulator	
	difE	CheA-like histidine protein kinase	Fibrils
	dsp	Dispersed growth, function unknown	Fibrils
	ifp-1:20	Protein associated with fibrils	
	mglA	GTP/GDP-binding protein	
	pilA (*sglB*)	Pilin, the structural subunit of pili	
	pilB	Pilus assembly	
	pilC (*sglG*)	Pilus assembly	
	pilR (*sglB*)	Response regulator; positive regulator of *pilA*	
	pilS (*sglB*)	Histidine kinase; negative regulator of *pilA*	
	pilT	Unknown function (pilus retraction?)	
	wzm (*rfbA*)	Membrane component of ABC transporter for O^- antigen biosynthesis	

(continued)

Complementation group	Gene symbol	Function of gene product	Complementing cellular fraction
	wzt (rfbB)	ATPase domain of ABC transporter for O-antigen biosynthesis	
	wbgA (rfbC)	Unknown function in O-antigen biosynthesis	
	sgl	Unsequenced social gliding genes	
	tglA	Pilus assembly	
Other *M. xanthus* genes of interest			
	fruA	Response regulator	
	frzA	CheW-like coupling protein	
	frzB	Unknown	
	frzCD	Methyl-accepting chemotaxis protein	
	frzE	Histidine protein kinase/response regulator	
	frzF	MCP methyl transferase	
	frzG	MCP methyl esterase	
	frzZ	Response regulator	
	relA	(p)ppGpp synthase	
	sasS	Histidine protein kinase	
	socD	Histidine protein kinase	
	socE	C-signaling suppressor of unknown function	
	stk	Inhibitor of cell adhesion	

synthetase given to each pair. In addition, the synthetase determines whether the amino acid is charged to the 2′- or 3′-hydroxyl of the terminal ribonucleotide. The hydroxyl group to which an amino acid is initially attached is determined by the isomeric specificity of the aminoacyl tRNA synthetase. After the initial attachment, there is a rapid and spontaneous migration of amino acids between the 2′- and 3′-hydroxyl groups with a half-life of ~0.2 ms. In *E. coli* and other organisms, tyrosine, phenylalanine, and leucine are added to the 2′-hydroxyl, whereas proline, tryptophan, and isoleucine are added to the 2′ hydroxyl ~50% of the time (27). Detection of the A-signal amino acids by *M. xanthus* might involve sampling the 2′ acyl pool of specific tRNAs.

Another approach to determine the mechanism by which A-signaling amino acids are sensed involved identification of mutations that resulted in constitutive

expression of the A-signal-dependent developmental reporter Tn5 *lac* ΩDK4521
(51). This reporter is regulated by at least two negative regulators, *sasA* and *sasS*.
The *sasA* locus encodes genes required for synthesis of the lipopolysaccharide
O-antigen (35). As described below, the *sasA* locus is essential for the S group.
The *sasS* locus encodes a histidine protein kinase (135).

The B Group   The B group is represented by a single gene, *bsgA*, which encodes
an ATP-dependent protease related to protease La (Lon) of *E. coli* (34). BsgA
is clearly intracellular; the mechanism of the extracellular complementation is
unknown (33).

The C Group   The C group contains *csgA*, which encodes a 24.6-kDa protein
with striking amino acid identity to members of the short-chain alcohol dehydrogenase family (3, 66, 67). These enzymes use NAD(H) or NADP(H) to catalyze the
interconversion of secondary alcohols and ketones or to mediate decarboxylation
(82, 90). Overproduction of SocA, another short-chain alcohol dehydrogenase,
substitutes for lack of CsgA in C signaling (67, 68). Whereas *csgA* is transcribed
in vegetative cells, there is a four- to sixfold increase in *csgA* expression during development that reaches a peak at about the time of sporulation (37). The
csgA upstream regulatory region is as large as the structural gene and appears to
be controlled by several different stimuli. Mutations in *socE* result in increased
csgA expression, whereas mutations in *bsgA* (69), *csgA* (54, 69), and *relA* (EW
Crawford Jr & LJ Shimkets, submitted for publication) diminish *csgA* expression
in a somewhat additive manner. Together a *relA* mutation and a *bsgA* mutation, or
a *relA* mutation and *csgA* mutation, abolish developmental *csgA* expression (EW
Crawford Jr & LJ Shimkets, submitted for publication).

CsgA purified from wild-type cells (55, 58) or expressed in *E. coli* as a MalE-
CsgA fusion (66) restores fruiting-body formation when added to *csgA* cells. Furthermore, anti-CsgA antibodies inhibit wild-type development (113). Together
these data provide evidence that CsgA stimulates cells from the outside. The X-ray
crystal structure is known for many members of the short-chain alcohol dehydrogenase family, especially the amino acids lining the NAD(H)- or NADP(H)-binding
pocket, because the protein crystals also contained crystallized coenzyme (18, 30–
32, 94, 120, 121, 123, 124). Short-chain alcohol dehydrogenase family members
contain a universally conserved $Sx_{12-14}YxxxK$ motif near the middle of the protein that serves a role in chemical catalysis (90). Two pieces of evidence suggest
that the dehydrogenase activity of CsgA is essential for development. First, substitution of a threonine near the N terminus that is anticipated to stabilize coenzyme
binding through hydrogen bonding (31) renders the protein developmentally inactive and unable to bind NAD^+ in vitro (66). Second, conservative substitution
of the CsgA active-site residue, S135T results in an inactive protein (66). Short-
chain alcohol dehydrogenases catalyze chemical reactions involving such a wide
variety of substrates that it is not possible to deduce the actual substrate from the

CsgA amino acid sequence alone. Presently, it is not known what the substrate is or whether the enzyme activity is extracellular or intracellular.

C-signaling is contact dependent and requires cell motility and alignment during presentation of the stimulus (56, 57). As development progresses, CsgA produces a series of developmental phenotypes that depend on the level of CsgA. At low concentrations, CsgA modulates cell reversal frequency to regulate directional movement of cells during another starvation-induced behavior known as rippling (100). High concentrations of CsgA are required for sporulation (55, 69). Perhaps the most striking role of CsgA is to arrest growth during early development (EW Crawford Jr & LJ Shimkets, submitted). Growth arrest is determined by the relative levels of CsgA and SocE, a vegetative protein of unknown function that is essential for growth. By genetically manipulating the levels of CsgA and SocE, a model emerged in which SocE works in opposition with CsgA to regulate growth arrest. Deletion of *csgA* has no adverse effects on cell growth (112) but inhibits development and expression of most developmental genes (60). Conversely, depletion of SocE inhibits cell elongation, chromosome replication, and stable RNA synthesis but does not have any adverse consequences on fruiting body development (EW Crawford Jr & LJ Shimkets, submitted for publication). During vegetative growth, SocE is more abundant than CsgA, but the situation changes during early development when *socE* transcription is curtailed and *csgA* transcription increases (EW Crawford Jr & LJ Shimkets, submitted for publication). Induction of the stringent response by amino acid deprivation inhibits *socE* transcription and induces *csgA* transcription providing coordinate but opposing control over these two transcriptional units. The choice between growth and development is normally made at a time when the stringent response is shifting the balance in favor of CsgA.

Curiously, the focal point of the contest between SocE and CsgA appears to be RelA, which has both amino acid-dependent control like in other bacteria and a novel CsgA/SocE-dependent control. The *csgA* mutants are clearly able to mount a stringent response after amino acid starvation (64; EW Crawford Jr & LJ Shimkets, submitted for publication) but are unable to participate in the CsgA/SocE-dependent pathway, which can be studied in amino acid rich growth medium (EW Crawford Jr & LJ Shimkets, submitted for publication). In response to SocE depletion, *csgA*$^+$ cells produce (p)ppGpp and undergo growth arrest and sporulation in the presence of excess nutrients. Under these same conditions, a *relA* mutation prevents (p)ppGpp accumulation, growth arrest, and sporulation, indicating that the response is dependent on RelA.

Two approaches have been taken to understand the mechanism of C-signal perception, ordering existing mutants (118) and isolating C-signal suppressors (96, 97). The results have turned up a histidine protein kinase named SocD (97), a response regulator named FruA (24a, 87, 118), an alternate short-chain alcohol dehydrogenase known as SocA (67, 68), and SocE. (EW Crawford Jr & LJ Shimkets, submitted for publication). As yet, these components have not been ordered on the developmental pathway.

The D Group   The sole member of the D group is the *dsgA* gene that encodes translation initiation factor IF3, a protein which enables the ribosome to select the initiation codon (15, 50). DsgA is present at uniform concentrations throughout growth and development, and the mechanism of extracellular complementation is unknown.

The E Group   The E group contains two genes that encode the E1α and E1β subunits of branched-chain keto acid dehydrogenase (122). This enzyme converts branched-chain keto acids derived from leucine, isoleucine, and valine to coenzyme A derivatives of the short branched-chain fatty acids isovalerate, methylbutyrate, and isobutyrate, respectively, which serve as primers for fatty acid biosynthesis. These primers are elongated through the fatty acid biosynthetic pathway to produce the three classes of branched-chain fatty acids.

Certain phosphatidylethanolamine (PE) species direct *M. xanthus* movement up steep gradients by chemotaxis (53). Directed movement is achieved by suppression of directional reversals to achieve longer runs up the gradient. Adaptation to the attractant is observed after ~1 h. *M. xanthus* cells respond to PE molecules with specific fatty acids, in particular dilauroyl PE, dioleoyl PE, and their diacyl glycerol derivatives. This work was initiated by the discovery that PE extracts of vegetative and developing cells stimulate directed movement, indicating that the *M. xanthus* cell membrane contains one or more discrete PE species that serve as autoattractants. Fatty acid profiles of vegetative and developmental PE suggest that the major PE autoattractant(s) remain to be discovered. Laurate was absent from both vegetative and developmental PE samples (<0.5%). Oleate accounted for 8.64% of the fatty acids in the vegetative PE and, based on the dose response curve with dioleoyl PE, can account for a portion of the chemotactic activity in the vegetative extract. However, oleate decreased to 1.54% in the more active developmental PE sample, suggesting the presence of an additional chemoattractant. Most of the *M. xanthus* PE species, including the species with branched-chain fatty acids, are not commercially available.

The *esg* mutants have greatly reduced levels of branched-chain fatty acids in their phospholipids and much higher levels of unsaturated fatty acids (122). Exogenous isovalerate supplied during vegetative growth restores branched-chain fatty acid biosynthesis and corrects the *esg* developmental defect, allowing directed migration and fruiting body development. It is not clear whether the developmental defect in *esg* mutants is due to a shortage of branched-chain fatty acids or an excess of unsaturated fatty acids. Future experiments investigating the biologically relevant PE autoattractants may benefit from the ability to manipulate the fatty acid composition of *esg* mutants.

The S Group   *M. xanthus* contains two motility systems known as A (adventurous) and S (social), which determine whether cells move as individuals or groups, respectively (45, 46; for review, see 41, 137). Mutations in S system genes inhibit group movement, but cells can still move in a contact-independent manner via the A system. Mutations in A system genes eliminate movement of isolated cells, but groups of cells can still move using the S system. Each system contains over a dozen loci but share only a single gene, *mglA*, which encodes a cytoplasmic GTP/GDP binding protein (39, 40). Together the A and S systems allow the cells to establish a balance between dispersal, which is useful in searching for new food sources, and contiguity, which is necessary for fruiting body development.

Mutations in S group genes disrupt the production of three types of cell surface molecules, type IV pili (47), lipopolysaccharide O-antigen (8), and fibrils (1). The S system genes also encode a sensory transduction system composed of DifA, DifC, DifD, and DifE, which have homology to the enteric chemotaxis genes including the methyl-accepting chemotaxis protein, CheW, CheY, and CheA, respectively (136). Mutations in S system genes also block fruiting-body development and sporulation (46).

Pili or fimbriae are long, thin protein appendages that assemble on only one of the cell poles (47). Mutation of the genes associated with the production of type IV pili eliminates S motility (131, 132, 134). The *pilA* gene encodes pilin, the structural subunit of pili. The *pilA* gene has a σ^{54} promoter that is positively regulated by PilR, a response regulator, and negatively regulated by PilS, a histidine protein kinase, and PilA (134). There are also a host of genes involved in pili biogenesis including *tglA*, which has homology with eukaryotic targeting proteins (98, 99). The *tglA* mutants can be transiently stimulated to become S motile by contact with *tglA*$^+$ cells (44). The *tglA* mutants make pilin but are apparently unable to assemble pili (126). Alignment of TglA-producing cells with *tglA* mutant cells results in the intercellular transfer of TglA and restoration of pili biogenesis (125).

It has been suggested that coordinated extension and retraction of pili mediates motility based on the discovery that *Pseudomonas aeruginosa* type IV pili are necessary for twitching motility (9, 129). Twitching is a form of bacterial surface translocation that consists of jerky cell movements that are independent of cell orientation unlike gliding in which cells move smoothly in the direction of their long axis. Whereas the correlation between piliation and surface translocation is clear, it would be desirable to prove that the pilus machinery provides the energy for motility through another type of experimental approach. A key protein in this regard is PilT, which is not essential for pilus assembly but is required for S motility in *M. xanthus*, and twitching motility in *P. aeruginosa* (129, 134). PilT, with its putative ATP-binding sequences, could conceivably supply energy for pilus retraction. Such a mechanism would not explain the requirement for cell proximity in *M. xanthus* to generate S motility unless the pili interact specifically with other *M. xanthus* cells.

The *sasA* locus encodes genes with identity to the ATP-binding cassette transporter family involved in lipopolysaccharide O-antigen biosynthesis (35). *M. xanthus* lipopolysaccharide contains components similar to those of other proteobacteria, including lipid A, core, and O-antigen (26). The O-antigen mutants have normal quantities of PilA and the fibril proteins recognized by monoclonal antibody mAb 2105 (8). These mutants are defective in fruiting body development and sporulation. The O-antigen becomes methylated early in development and a novel sugar, 6-O-methylgalactosamine, becomes enriched in developing cells (88, 89). The role of the O-antigen in motility and development is unknown.

Fibrils are 30–40 nm in diameter and ~1 cell length long (1). They emanate from many points on the cell surface and interconnect cells in dense assemblages. Fibrils are the principal attachment organelle and mediate cohesion between cells and adhesion to inert surfaces (1, 2). Fibrils are composed of a polysaccharide backbone containing galactose, glucosamine, glucose, rhamnose, and xylose and are decorated with at least five proteins ranging in size from 14 to 66 kDa (6, 7). Interestingly, all five proteins react with monoclonal antibody mAb 2105, although they appear to be encoded by individual genes (116). Deletion of the *ifp-1:20* gene, which encodes the 20-kDa member of this family, results in unstable fibrils. Isolated fibrils carry out ADP ribosylation of the 29-kDa fibril protein (43). Several types of S mutants fail to produce fibrils, including *dsp* and *dif* (1, 105, 106; Z Yang, X Ma, L Tong, HB Kaplan, LJ Shimkets, W Shi, submitted for publication), and many other S mutants have reduced levels of fibrils (16). Fibril synthesis is increased by contact with other cells (5) and is negatively regulated by the *stk* locus (16).

Fibrils extracted from wild-type cells (5) restore cohesion and development to fibrilless mutants such as *dsp* (13) and *dif* (Z Yang, X Ma, L Tong, HB Kaplan, LJ Shimkets, W Shi, submitted for publication). S group complementation by fibrils can also occur genetically by taking advantage of the *stk* mutation, which restores fibril synthesis and development to many S mutants (16). Fibrilless *dsp*, *dif*, and *tglA* mutants do not show excitation by dilauroyl PE (D Kearns & LJ Shimkets, submitted for publication). Excitation, the first behavior in a chemotaxis pathway, consists of reduced cell reversals in response to the attractant. Elimination of fibrils by treatment of wild-type cells with Congo red also eliminates excitation by dilauroyl PE. Restoration of fibrils to the *tglA* mutant with a *stk* mutation restores excitation. Addition of purified wild type fibrils to *dsp* mutants also restores excitation. The role of the fibrils in directed movement is unknown. One possibility is that the fibrils serve as primitive antennae to probe the surface of adjacent cells for chemical signals.

A Model For Early Development   Amino acids are the major sources of carbon, nitrogen, and energy for *M. xanthus* (21), and amino acid limitation plays a central role the initiation of development (22). The principal decision facing the cell during times of nutrient limitation is whether to divert this carbon to growth or developmental macromolecules. The choice between growth and development consists of at least three stages, starvation sensing through the stringent response, quorum sensing through A signaling, and CsgA-dependent arrest of growth (Figure 4). Each of these stages is anchored to the stringent response.

In *E. coli*, the stringent response system evaluates the availability of all 20 amino acids by sampling the amino-acylated tRNA pool during translation. Translational pausing due to limitation of amino-acylated tRNA results in production of

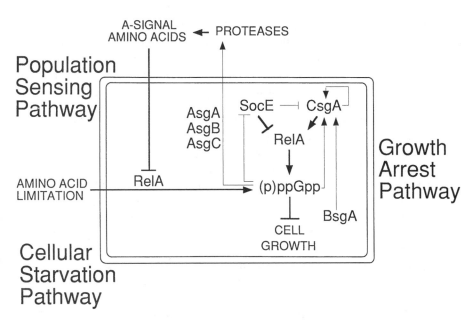

Figure 4 Control of early development. An individual *M. xanthus* cell is represented during early development to illustrate the three pathways leading to the decision to arrest growth. Amino acid limitation activates the cellular starvation pathway by stimulating the production of 3′-di-5′-(tri)diphosphate nucleotides (p)ppGpp by the synthase RelA. The major consequences of this pathway are (*a*) reduction in cell growth, (*b*) activation of the AsgA, AsgB, and AsgC proteins to allow secretion of extracellular proteases, (*c*) induction of *csgA* expression, and (*d*) inhibition of *socE* expression. The extracellular proteases hydrolyze surface associated proteins to generate a source of amino acids. A subset of these amino acids act as a quorum sensing signal to determine whether the population of starved cells is sufficiently large to form a fruiting body. The amino acids reenter the cells where they form a carbon, energy, and nitrogen source. To prevent this influx of amino acids from repressing the stringent response and development, cells regulate the activity of RelA through a second mechanism involving SocE, an inhibitor, and CsgA, an activator. The relative ratio of SocE to CsgA determines whether cells divert carbon into the growth pathway or the developmental pathway. *Bold lines* indicate regulation that is believed to be post-translational, whereas *fine lines* indicate regulation that is believed to be transcriptional. The functions of the genes given by abbreviations are given in Table 1.

3′-di-5′-(tri)diphosphate nucleotides [(p)ppGpp] by the ribosome-associated (p)ppGpp synthase RelA (11). The stringent response, in turn, leads to reduced transcription of rRNA operons and many other effects, causing a slowdown in cellular growth. Because (p)ppGpp turnover is rapid, the stringent system allows for a continuous assessment of nutritional status. Amino acid starvation also induces the stringent response in *M. xanthus* (72, 73) through a typical RelA-dependent pathway. Ectopic expression of the *E. coli relA* gene in *M. xanthus* induces a transient rise in the (p)ppGpp, which activates expression of early developmental genes (114). Inactivation of the *M. xanthus relA* homolog prevents accumulation of (p)ppGpp and development in response to nutrient deprivation (38). Starvation sensing during the stringent response sets in motion three events: (*a*) activation of the A-signaling pathway, (*b*) activation of *csgA* transcription, and (*c*) inhibition of *socE* transcription.

Because fruiting-body development requires the commitment of such a large number of cells, *M. xanthus* uses a population signaling pathway to determine whether there is a sufficient number of starved cells to complete fruiting-body development. The concept of quorum sensing, originally developed for organisms that use acylated homoserine lactones, also applies to the A-signaling system, although the A signal consists of certain extracellular amino acids. The stringent response leads to the production and export of proteases that hydrolyze cells surface proteins to generate peptides and amino acids. The extracellular concentration of each amino acid increases to 25 μM on average and serves both as a carbon source and a population of signaling molecules. Some of these amino acids serve as the A signal. When they reach a suitable concentration, expression of the A-signal-dependent genes is initiated. The integration of the cellular starvation and population starvation pathways assures that there is a sufficiently high density of starved cells to form a fruiting body (49).

The manner in which AsgA, AsgB, and AsgC are activated by the stringent response has not been examined. AsgA is a member of the two-component regulatory group although it lacks the hydrophobic, membrane-spanning domain of most sensors (71, 93). This would suggest that it senses a soluble cytoplasmic signal of nutritional stress. One possibility is that AsgA binds directly to (p)ppGpp to become activated. AsgC is the major sigma factor, and the AsgC amino acid substitution that disrupts A signaling is in a region of the protein known to be essential for responsiveness to the stringent response in *E. coli* (38). It is possible that transcription regulation by AsgC occurs by the initiating NTP concentration as was proposed for *E. coli* (28). The regulation of the protease genes has not yet been examined.

The A-signaling system also generates a carbon and energy source for development. To avoid diverting this precious resource into growth-related activities, growth is impaired by a novel regulatory mechanism established by the relative concentrations of CsgA and SocE. *socE* transcription is inactivated by the stringent response, whereas transcription of *csgA* is activated by the stringent response. When the relative balance between CsgA and SocE tips in favor of CsgA, growth arrest occurs even in the presence of excess amino acids. The mechanism of growth arrest may also involve RelA, because it is inhibited by a *relA* mutation. Growth arrest assures that the carbon generated from A-signaling is diverted into the developmental pathway. One cannot help but admire the elegant manner in which nutritional sensing and cell-cell signaling is coupled directly to carbon flow.

DEVELOPMENTAL REGULATION OF MOTILITY

The question of how cells migrate to an aggregation focus and construct a three-dimensional fruiting body is the subject of much speculation. Several recent advances have suggested that cells produce chemical signals that regulate the speed and directionality of cells. There are at least two sensory processing systems homologous to the Che chemotaxis system of *E. coli*, Frz and Dif. The sensory inputs and transduction systems are summarized in Figure 5. The Frz system contains the entire suite of Che components found in the *E. coli* chemotaxis sensory transduction system (Table 1). There are two notable differences. First, the FrzE protein is a chimeric protein containing domains from both CheA and CheY (78). FrzE is autophosphorylated in vitro, suggesting that it has the histidine protein kinase activity of CheA (79). Second, FrzCD is a soluble cytoplasmic protein unlike other members of the methylated chemotaxis protein family, which are integral membrane proteins (75). Its cytoplasmic location is curious because attractants or repellants would need to be transported into the cell to interact with it. Two

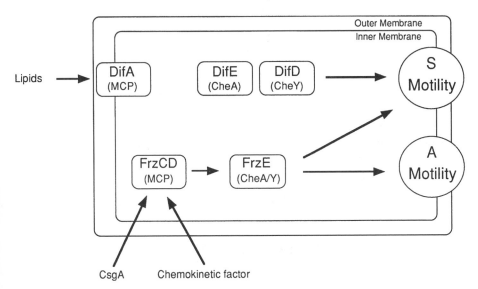

Figure 5   A model for control of motility during development. Individual cells may exhibit either S motility or A motility. There are two sensory transduction pathways with homology to the enteric Che genes. The Dif system is essential for S motility. The Frz system is not essential for either A motility or S motility but modifies the behavior of A motile and S motile cells. DifA processes information concerning the abundance of dilauroyl PE (C12:0). It is not yet known whether it stimulates DifE. The Frz system appears to process signals derived from CsgA and an unidentified chemokinetic factor. The functions of the genes given by abbreviations are listed in Table 1.

types of Frz stimuli have been reported. Methylation of FrzCD defines a discrete step in early development that is inhibited in many types of developmental mutants (29, 76, 77). FrzCD is also methylated during vegetative growth when cells undergo directed movement (101) and demethylated when cells are exposed to repellants (74, 101–103). The nature of the chemical(s) that interact with FrzCD remains unknown. The Frz system is required for response to an unidentified extracellular, chemokinetic factor that stimulates cell velocity (127) and for adaptation to PE during directed movement (53). The Frz system also responds to a CsgA-dependent stimulus (46a, 117, 118). Addition of CsgA to developing *csgA* mutants results in methylation of FrzCD within 30 min.

The Dif system is required for S motility and contains homologs of many of the Che genes (136). DifA, the methylated chemotaxis protein homolog, is an integral membrane protein with an unusually short periplasmic domain that is unlikely to bind a small molecule directly or dock with a periplasmic binding protein as has been described for *E. coli*. This situation is reminiscent of the light-sensing pathway in *Halobacterium salinarium* in which the HtrI transducer protein (with a similar small periplasmic domain) is bound to the integral membrane rhodopsin SR-I (119). Photoisomerization of retinal in SR-I causes transfer of the Schiff base proton to Asp-85 in SR-I. This signal is thought to be relayed to HtrI through electrostatic and allosteric protein-protein interactions. By analogy, it is likely that DifA interacts with a macromolecule in the membrane to initiate the signaling cascade. One possibility is that fibrils are anchored to DifA to regulate the response to dilauroyl PE because fibril mutants lack directed movement toward this compound (D Kearns & LJ Shimkets, submitted for publication). Another possibility is that dilauroyl PE interacts directly with DifA. DifA mutants still respond to dioleoyl PE suggesting the existence of another methylated chemotaxis protein.

ACKNOWLEDGMENTS

This work was supported by NSF grant MCB9601077. The author thanks Daniel B. Kearns for his insightful comments.

Visit the Annual Reviews home page at http://www.AnnualReviews.org

LITERATURE CITED

1. Arnold JW, Shimkets LJ. 1988. Cell surface properties correlated with cohesion in *Myxococcus xanthus*. *J. Bacteriol.* 170:5771–77
2. Arnold JW, Shimkets LJ. 1988. Inhibition of cell-cell interactions in *Myxococcus xanthus* by Congo red. *J. Bacteriol.* 170:5765–70
3. Baker M. 1994. *Myxococcus xanthus* C-factor, a morphogenetic paracrin signal, is similar to *Escherichia coli* 3-oxoacyl-[acyl-carrier-protein] reductase and human 17β-hydroxysteroid dehydrogenase. *Biochem. J.* 301:311–12
4. Beale G. 1990. Self and nonself recognition in the ciliate protozoan Euplotes. *Trends Genet.* 6:137–139
5. Behmlander RM, Dworkin M. 1991. Extra-

cellular fibrils and contact-mediated cell interactions in *Myxococcus xanthus*. *J. Bacteriol.* 173:7810–21

6. Behmlander RM, Dworkin M. 1994. Biochemical and structural analyses of the extracellular matrix fibrils of *Myxococcus xanthus*. *J. Bacteriol.* 176:6295–303

7. Behmlander RM, Dworkin M. 1994. Integral proteins of the extracellular matrix fibrils of *Myxococcus xanthus*. *J. Bacteriol.* 176:6304–11

8. Bowden MG, Kaplan HB. 1998. The *Myxococcus xanthus* lipopolysaccharide O-antigen is required for social motility and multicellular development. *Mol. Microbiol.* 30:275–284

9. Bradley DE. 1980. A function of *Pseudomonas aeruginosa* PAO polar pili: twitching motility. *Can. J. Microbiol.* 26: 146–54.

10. Brockman ER, Todd RL. 1974. Fruiting myxobacters as viewed with a scanning electron microscope. *Int. J. Sys. Bacteriol.* 24:118–24

11. Cashel M, Gentry DR, Hernandez VJ, Vinella D. 1996. The stringent response. In *Escherichia coli* and *Salmonella: Cellular and Molecular Biology*, ed. FC Neidhardt et al. Washington, DC: Am. Soc. Microbiol. pp. 1458–96

12. Chadwick-Furman N, Rinkevich B. 1994. A complex allorecognition system in a reef-building coral: delayed responses, reversals and nontransitive hierarchies. *Coral Reefs* 13:57–63

13. Chang B-Y, Dworkin M. 1994. Isolated fibrils rescue cohesion and development in the *dsp* mutant of *Myxococcus xanthus*. *J. Bacteriol.* 176:7190–96

14. Cheng Y, Kaiser D. 1989. *dsg*, a gene required for cell-cell interaction early in *Myxococcus* development. *J. Bacteriol.* 171:3719–26

15. Cheng Y, Kalman LV, Kaiser D. 1994. The *dsg* gene of *Myxococcus xanthus* encodes a protein similar to translation initiation factor IF3. *J. Bacteriol.* 176:1427–33

16. Dana JR, Shimkets LJ. 1993. Regulation of cohesion-dependent cell interactions in *Myxococcus xanthus*. *J. Bacteriol.* 175:3636–47

17. Davis JM, Mayor J, Plamann L. 1995. A missense mutation in *rpoD* results in an A-signalling defect in *Myxococcus xanthus*. *Mol. Microbiol.* 18:943–52

18. Dessen A, Quemard A, Blanchard JS, Jacobs WR Jr., Sacchettini JC. 1995. Crystal structure and function of the isoniazid target of *Mycobacterium tuberculosis*. *Science* 267:1638–41

19. Downard J, Ramaswamy SV, Kil K-S. 1993. Identification of *esg*, a genetic locus involved in cell-cell signaling during *Myxococcus xanthus* development. *J. Bacteriol.* 175:7762–70

20. Downard J, Toal D. 1995. Branched chain fatty acids: the case for a novel form of cell-cell signaling during *Myxococcus xanthus* development. *Mol. Microbiol.* 16:171–75

21. Dworkin M. 1962. Nutritional requirements for vegetative growth of *Myxococcus xanthus*. *J. Bacteriol.* 84:250–57

22. Dworkin M. 1963. Nutritional regulation of morphogenesis in *Myxococcus xanthus*. *J. Bacteriol.* 86:67–72

23. Dworkin M. 1996. Recent advances in the social and developmental biology of the myxobacteria. *Microbiol. Rev.* 60:70–102

24. Dworkin M, Kaiser D, eds. 1993. *Myxobacteria II*. Washington DC: Am. Soc. Microbiol.

24a. Ellehauge E, Nørregaard-Madsen M, Søgaard-Andersen L. 1998. The FruA signal transduction protein provides a checkpoint for the temporal co-ordination of intercellular signals in *Myxococc. xanthus* development. *Mol. Microbiol.* 30:807–817

25. Erwin D, Valentine J, Jablonski D. 1997. The origin of animal body plans. *Am. Sci.* 85:126–37.

26. Fink JM, Zissler JF. 1989. Characterization of lipopolysaccharide from *Myxococcus xanthus* by use of monoclonal antibodies. *J. Bacteriol.* 171:2028–32

27. Fraser TH, Julius DJ, Rich A. 1979. Determination of aminoacylation isomeric specificity using tRNA terminating in 3'-amino-3'-deoxyadenosine and 2'-amino-2'-deoxyadenosine. *Methods Enzymol.* LIX:272–82.

28. Gaal T, Bartlett MS, Ross W, Turnbough CL Jr, Gourse RL. 1997. Transcription regulation by initiating NTP concentration: rRNA synthesis in bacteria. *Science* 278:2092–97

29. Geng Y, Yang Z, Downard J, Zusman D, Shi W. 1998. Methylation of FrzCD defines a discrete step in the developmental program of *Myxococcus xanthus. J. Bacteriol.* 180:5765–68

30. Ghosh D, Pletnev VZ, Zhu DW, Wawrkak Z, Pangborn WL, et al. 1995. Structure of human estrogenic 17β-hydroxysteroid dehydrogenase at 2.20 A resolution. *Structure* 3:503–13

31. Ghosh D, Wawrzak Z, Weeks CM, Duax WL, Erman M. 1994. The refined three-dimensional structure of 3α, 20β-hydroxysteroid dehydrogenase and possible roles of the residues conserved in short chain dehydrogenases. *Structure* 2:629–40

32. Ghosh D, Weeks CM, Grochulski P, Duax WL, Erman M, et al. 1991. Three-dimensional structure of $3\alpha,20\beta$-hydroxysteroid dehydrogenase: a member of the short-chain dehydrogenase family. *Proc. Natl. Acad. Sci. USA* 88:10064–68

33. Gill RE, Bornemann MC. 1988. Identification and characterization of the *Myxococcus xanthus bsgA* gene product. *J. Bacteriol.* 170:5289–97

34. Gill RE, Karlok M, Benton D. 1993. *Myxococcus xanthus* encodes an ATP-dependent protease which is required for developmental gene transcription and intercellular signaling. *J. Bacteriol.* 175:4538–44

35. Guo D, Bowden MG, Pershad R, Kaplan HB. 1996. The *Myxococcus xanthus rfbABC* operon encodes an ATP-binding cassette transporter homolog required for O-antigen biosynthesis and multicellular development. *J. Bacteriol.* 178:1631–39

36. Hagen DC, Bretscher AP, Kaiser D. 1978. Synergism between morphogenetic mutants of *Myxococcus xanthus. Dev. Biol.* 64:284–96

37. Hagen TJ, Shimkets LJ. 1990. Nucleotide sequence and transcriptional products of the *csg* locus of *Myxococcus xanthus. J. Bacteriol.* 172:15–23

38. Harris BZ, Kaiser D, Singer M. 1998. The guanosine nucleotide (p)ppGpp initiates development and A-factor production in *Myxococcus xanthus. Genes Dev.* 12:1022–35

39. Hartzell PL. 1997. Complementation of sporulation and motility defects in a prokaryote by a eukaryotic GTPase. *Proc. Natl. Acad. Sci. USA* 94:9881–86

40. Hartzell P, Kaiser D. 1991. Function of MglA, a 22-kilodalton protein essential for gliding in *Myxococcus xanthus. J. Bacteriol.* 173:7615–24

41. Hartzell PL, Youderian P. 1995. Genetics of gliding motility and development in *Myxococcus xanthus. Arch. Microbiol.* 164:309–23

42. Hernandez J, Cashel M. 1995. Changes in conserved region 3 of *Escherichia coli* σ^{70} mediate ppGpp-dependent function *in vivo. J. Mol. Biol.* 252:536–49.

43. Hildebrandt K, Eastman D, Dworkin M. 1997. ADP-ribosylation by the extracellular fibrils of *Myxococcus xanthus. Mol. Microbiol.* 23:231–35

44. Hodgkin J, Kaiser D. 1977. Cell-to-cell stimulation of movement in nonmotile mutants of *Myxococcus. Proc. Natl. Acad. Sci. USA* 74:2938–42

45. Hodgkin J, Kaiser D. 1979. Genetics of gliding motility in *Myxococcus xanthus* (Myxobacterales): genes controlling movement of single cells. *Mol. Gen. Genet.* 171:167–76

46. Hodgkin J, Kaiser D. 1979. Genetics of gliding motility in *Myxococcus xanthus* (Myxobacterales): two gene systems con-

trol movement. *Mol. Gen. Genet.* 171: 177–91

46a. Jelsbak L, Søgaard-Andersen L. 1999. The cell surface-associated intercellular C-signal induces behavioral changes in individual *Myxococcus xanthus* cells during fruiting body morphogenesis, *Proc. Natl. Acad. Sci. USA* 96:5031–503

47. Kaiser D. 1979. Social gliding is correlated with the presence of pili in *Myxococcus xanthus. Proc. Natl. Acad. Sci. USA* 76:5952–56

48. Kaiser D. 1993. Roland Thaxter's legacy and the origins of multicellular development. *Genetics* 135:249–54

49. Kaiser D. 1996. Bacteria also vote. *Science* 272:1598–99

50. Kalman LV, Cheng YL, Kaiser D. 1994. The *Myxococcus xanthus dsg* gene product performs functions of translation initiation factor IF3 in vivo. *J. Bacteriol.* 176:1434–42

51. Kaplan H, Kuspa A, Kaiser D. 1991. Suppressors that permit A-signal-independent developmental gene expression in *Myxococcus xanthus. J. Bacteriol.* 173:1460–70

52. Kaplan HB, Plamann L. 1996. A *Myxococcus xanthus* cell density-sensing system required for multicellular development. *FEMS Microbiol. Lett.* 139:89–95

53. Kearns DB, Shimkets LJ. 1998. Chemotaxis in a gliding bacterium. *Proc. Natl. Acad. Sci. USA* 95:11957–11962.

54. Kim S, Kaiser D. 1991. C-Factor has distinct aggregation and sporulation thresholds during *Myxococcus* development. *J. Bacteriol.* 173:1722–28

55. Kim SK, Kaiser D. 1990. C-factor: a cell-cell signaling protein required for fruiting body morphogenesis of *M. xanthus. Cell* 61:19–26

56. Kim SK, Kaiser D. 1990. Cell alignment required in differentiation of *M. xanthus. Science* 249:926–28

57. Kim SK, Kaiser D. 1990. Cell motil-

ity is required for the transmission of C-factor, an intercellular signal that coordinates fruiting body morphogenesis of *Myxococcus xanthus. Genes Dev.* 4:896–905

58. Kim SK, Kaiser D. 1990. Purification and properties of *Myxococcus xanthus* C-factor, an intercellular signalling protein. *Proc. Natl. Acad. Sci. USA* 87:3635–39

59. Kroos L, Kaiser D. 1984. Construction of Tn*5 lac*, a transposon that fuses *lacZ* expression to exogenous promoters, and its introduction into *Myxococcus xanthus. Proc. Natl. Acad. Sci. USA* 81:5816–20

60. Kroos L, Kaiser D. 1987. Expression of many developmentally regulated genes in *Myxococcus* depends on a sequence of cell interactions. *Genes Dev.* 1:840–54

61. Kuspa A, Kroos L, Kaiser D. 1986. Intercellular signaling is required for developmental gene expression in *Myxococcus xanthus. Dev. Biol.* 117:267–76

62. Kuspa A, Plamann L, Kaiser D. 1992a. Identification of heat-stable A-factor from *Myxococcus xanthus. J. Bacteriol.* 174:3319–26

63. Kuspa A, Plamann L, Kaiser D. 1992b. A-Signalling and the cell density requirement for *Myxococcus xanthus* development. *J. Bacteriol.* 174:7360–69

64. LaRossa R, Kuner J, Hagan D, Manoil C, Kaiser D. 1983. Developmental cell interactions of *Myxococcus xanthus*: Analysis of mutants. *J. Bacteriol.* 153:1394–1404

65. Laue BE, Gill RE. 1995. Using a phase-locked mutant of *Myxococcus xanthus* to study the role of phase variation in development. *J. Bacteriol.* 177:4089–96

66. Lee B-U, Lee K, Mendez J, Shimkets LJ. 1995. A tactile sensory system of *Myxococcus xanthus* involves an extracellular NAD(P)$^+$-containing protein. *Genes Dev.* 9:2964–73

67. Lee K, Shimkets LJ. 1994. Cloning and characterization of the socA locus which restores development to *Myxococcus*

xanthus C-signaling mutants. *J. Bacteriol.* 176:2200–9

68. Lee K, Shimkets LJ. 1996. Suppression of a signaling defect during *Myxococcus xanthus* development. *J. Bacteriol.* 178:977–84

69. Li S, Lee B, Shimkets LJ. 1992. *csgA* expression entrains *Myxococcus xanthus* development. *Genes Dev.* 6:401–10

70. Li SF, Shimkets LJ. 1993. Effect of *dsp* mutations on the cell-to-cell transmission of CsgA in *Myxococcus xanthus*. *J. Bacteriol.* 175:3648–52

71. Li Y, Plamann L. 1996. Purification and in vitro phosphorylation of *Myxococcus xanthus* AsgA protein. *J. Bacteriol.* 178:289–92

72. Manoil C, Kaiser D. 1980. Accumulation of guanosine tetraphosphate and guanosine pentaphosphate in *Myxococcus xanthus* during starvation and myxospore formation. *J. Bacteriol.* 141:297–304

73. Manoil C, Kaiser D. 1980. Guanosine pentaphosphate and guanosine tetraphosphate accumulation and induction of *Myxococcus xanthus* fruiting body development. *J. Bacteriol.* 141:305–15

74. McBride MJ, Kohler T, Zusman DR. 1992. Methylation of FrzCD, a methyl-accepting taxis protein of *Myxococcus xanthus*, is correlated with factors affecting cell behavior. *J. Bacteriol.* 174:4246–57

75. McBride MJ, Weinberg RA, Zusman DR. 1989. "Frizzy" aggregation genes of the gliding bacterium *Myxococcus xanthus* show sequence similarities to the chemotaxis genes of enteric bacteria. *Proc. Natl. Acad. Sci. USA* 86:424–28

76. McBride MJ, Zusman DR. 1993. FrzCD, a methyl-accepting taxis protein from *Myxococcus xanthus*, shows modulated methylation during fruiting body formation. *J. Bacteriol.* 175:4936–40

77. McCleary W, McBride M, Zusman D. 1990. Developmental sensory transduction in *M. xanthus* involves methylation and demethylation of FrzCD. *J. Bacteriol.* 172:4877–87

78. McCleary W, Zusman D. 1990. FrzE of *M. xanthus* is homologous to both CheA and CheY of *Salmonella typhimurium*. *Proc. Natl. Acad. Sci. USA* 87:5898–902

79. McCleary W, Zusman D. 1990. Purification and characterization of the *M. xanthus* FrzE protein shows that it has autophosphorylation activity. *J. Bacteriol.* 172:6661–68

80. McClure BA, Haring V, Ebert PR, Anderson MA, Simpson RJ, et al. 1989. Style self-incompatibility gene products of *Nicotina alata* are ribonucleases. *Nature* 342:955–57

81. McVittie A, Messik F, Zahler SA. 1962. Developmental biology of *Myxococcus*. *J. Bacteriol.* 84:546–1

82. Neidle E, Hartnett C, Ornston LN, Bairoch A, Rekik M, Harayama S. 1992. *Cis*-diol dehydrogenases encoded by the TOL pWWO plasmid *xylL* gene and the *Acinetobacter calcoaceticus* chromosomal *benD* gene are members of the short-chain alcohol dehydrogenase superfamily. *Eur. J. Biochem.* 204:113–20

83. O'Connor K, Zusman D. 1991. Analysis of *Myxococcus xanthus* cell types by two-dimensional polyacrylamide gel electrophoresis. *J. Bacteriol.* 173:3334–41

84. O'Connor K, Zusman D. 1991. Behavior of peripheral rods and their role in the life cycle of *Myxococcus xanthus*. *J. Bacteriol.* 173:3342–55

85. O'Connor K, Zusman D. 1991. Development in *Myxococcus xanthus* involves differentiation into two cell types, peripheral rods and spores. *J. Bacteriol.* 173:3318–33

86. O'Connor KA, Zusman DR. 1988. Reexamination of the role of autolysis in the development of *Myxococcus xanthus*. *J. Bacteriol.* 170:4103–12

87. Ogawa M, Fujitani S, Mao X, Inouye S, Komano T. 1996. FruA, a putative transcription factor essential for the development of *Myxococcus xanthus*. *Mol. Microbiol.* 22:757–67

88. Panasenko SM. 1983. Protein and lipid methylation by methionine and

S-adenosylmethionine in *Myxococcus xanthus. Can. J. Microbiol.* 29:1224–28

89. Panasenko SM, Jann B, Jann K. 1989. Novel change in the carbohydrate portion of *Myxococcus xanthus* lipopolysaccharide during development. *J. Bacteriol.* 171:1835–40

90. Persson B, Krook M, Jornvall H. 1991. Characteristics of short-chain alcohol dehydrogenases and related enzymes. *Eur. J. Biochem.* 200:537–43

91. Plamann L, Davis JM, Cantwell B, Mayor J. 1994. Evidence that *asgB* encodes a DNA-binding protein essential for growth and development of *Myxococcus xanthus. J. Bacteriol.* 176:2013–20

91a. Plamann L, Kaplan HB. 1999. Cell-density signaling during early development in *Myxococcus xanthus.* In *Cell-cell signaling in bacteria*, eds. GM Dunny, SC Winans. Washington DC: American Society for Microbiology Press pp 67–82

92. Plamann L, Kuspa A, Kaiser D. 1992. Proteins that rescue A-signal-defective mutants of *Myxococcus xanthus. J. Bacteriol.* 174:3311–18

93. Plamann L, Li Y, Cantwell B, Mayor J. 1995. The *Myxococcus xanthus asgA* gene encodes a novel signal transduction protein required for multicellular development. *J. Bacteriol.* 177:2014–20

94. Rafferty JB, Simon JW, Baldock C, Artymiuk PJ, Stuitje AR, et al. 1995. Common themes in redox chemistry emerge from the X-ray structure of oilseed rape (*Brassica napus*) enoyl acyl carrier protein reductase. *Structure* 3:927–38

95. Reichenbach H, Dworkin M. 1992. The Myxobacteria. In *The Prokaryotes*, eds. HGTA Balows, M Dworkin, W Harder, K-H Schleifer. *A Handbook on the Biology of Bacteria: Ecophysiology, Isolation, Identification, Applications.* New York: Springer-Verlag. pp. 3416–87

96. Rhie HG, Shimkets LJ. 1989. Develop-

mental bypass supression of *Myxococcus xanthus csgA* mutations. *J. Bacteriol.* 171:3268–76

97. Rhie H, Shimkets L. 1991. Low-temperature induction of *M. xanthus* developmental gene expression in wild-type and *csgA* suppressor cells. *J. Bacteriol.* 173:2206–11

98. Rodriguez-Soto JP, Kaiser D. 1997. Identification and localization of the Tgl protein, which is required for *Myxococcus xanthus* social motility. *J. Bacteriol.* 179:4372–81

99. Rodriguez-Soto JP, Kaiser D. 1997. The *tgl* gene: social motility and stimulation in *Myxococcus xanthus. J. Bacteriol.* 179:4361–71

100. Sager B, Kaiser D. 1994. Intercellular C-signaling and the traveling waves of *Myxococcus. Genes Dev.* 8:2793–804

101. Shi W, Kohler T, Zusman DR. 1993. Chemotaxis plays a role in the social behavior of *Myxococcus xanthus. Mol. Microbiol.* 9:601–11

102. Shi W, Kohler T, Zusman DR. 1994. Isolation and characterization of *Myxococcus xanthus* mutants which are defective in sensing negative stimuli. *J. Bacteriol.* 176:696–701

103. Shi W, Zusman DR. 1994. Sensory adaptation during negative chemotaxis in *Myxococcus xanthus. J. Bacteriol.* 176:1517–20

104. Shimkets L, Seale TW. 1975. Fruiting-body formation and myxospore differentiation and germination in *Myxococcus xanthus* viewed by scanning electron microscopy. *J. Bacteriol.* 121:711–20

105. Shimkets LJ. 1986. Correlation of energy-dependent cell cohesion with social motility in *Myxococcus xanthus. J. Bacteriol.* 166:837–41

106. Shimkets LJ. 1986. Role of cell cohesion in *Myxococcus xanthus* fruiting body formation. *J. Bacteriol.* 166:842–48

107. Shimkets LJ. 1993. The myxobacterial genome. In *Myxobacteria II*, ed. M

Dworkin, D Kaiser. Washington DC: Am. Soc. Microbiol. pp. 85–107

108. Shimkets LJ. 1996. *Myxococcus* coadhesion and role in the life cycle. In *Bacterial Adhesion: Molecular and Ecological Diversity*, ed. M Fletcher. New York: Wiley-Liss. pp. 333–47

109. Shimkets LJ. 1997a. *Myxococcus xanthus* DK1622. In *Bacterial Genomes: Physical Structure and Analysis*, ed. FJ de Bruijn, JR Lupski, GM Weinstock. New York: Chapman & Hall. pp. 695–701

110. Shimkets, LJ. 1997b. Structure and sizes of genomes of the Archaea and Bacteria. See Shimkets 1997a, pp. 5–11

111. Shimkets LJ, Dworkin M. 1997. Myxobacterial multicellularity. In *Bacteria as Multicellular Organisms*, eds. JA Shapiro, M Dworkin. New York: Oxford Univ. Press. pp. 220–44

112. Shimkets LJ, Kaiser D. 1982. Murein components rescue developmental sporulation of *Myxococcus xanthus*. *J. Bacteriol.* 152:462–70

112a. Shimkets LJ, Kaiser D. 1999. Cell contact-dependent C-signaling in *Myxococcus xanthus*. In *cell-cell signaling in bacteria*, eds. GM Dunny, SC Winans, Washington DC: American Society for Microbiology Press. pp 83–97

113. Shimkets LJ, Rafiee H. 1990. CsgA, an extracellular protein essential for *Myxococcus xanthus* development. *J. Bacteriol.* 172:5299–306

114. Singer M, Kaiser D. 1995. Ectopic production of guanosine penta- and tetraphosphate can initiate early developmental gene expression in *Myxococcus xanthus*. *Genes Dev.* 9:1633–44

115. Smith D, Dworkin M. 1994. Territorial interactions between two *Myxococcus* species. *J. Bacteriol.* 176:1201–5

116. Smith DR, Dworkin M. 1997. A mutation that affects fibril protein, development, cohesion and gene expression in *Myxococcus xanthus*. *Microbiology* 143:3683–92

117. Sogaard-Anderson L, Kaiser D. 1996. C-factor, a cell-surface-associated intercellular signaling protein, stimulates the cytoplasmic Frz signal transduction system in *Myxococcus xanthus*. *Proc. Natl. Acad. Sci. USA* 93:2625–79

118. Sogaard-Anderson L, Slack FJ, Kimsey H, Kaiser D. 1996. Intercellular C-signaling in *Myxococcus xanthus* involves a branched signal transduction pathway. *Genes Dev.* 10:740–54

119. Spudich JL. 1994. Protein-protein interaction converts a proton pump into a sensory receptor. 1994. *Cell* 79:747–50

120. Su Y, Varughese KI, Xuong NH, Bray TL, Roche DJ, Whiteley JM. 1993. The crystallographic structure of a human dihydropteridine reductase NADH binary complex expressed in *Escherichia coli* by a cDNA constructed from it rat homologue. *J. Biol. Chem.* 268:26836–41

121. Tanaka N, Nanaka T, Nakanishi M, Deyashiki Y, Hara A, Mitsui Y. 1996. Crystal structure of the ternary complex of mouse lung carbonyl reductase at 1.8 A resolution: the structural origin of coenzyme specificity in the short-chain alcohol dehydrogenase/reductase family. *Structure* 4:33–45

122. Toal DR, Clifton SW, Roe BR, Downard J. 1995. The *esg* locus of *Myxococcus xanthus* encodes the E1α and E1ß subunits of a branched-chain keto acid dehydrogenase. *Mol. Microbiol.* 16:177–89

123. Varughese KI, Skinner MM, Whiteley JM, Matthews DA, Xuong NH. 1992. Crystal structure of rat liver dihydropteridine reductase. *Proc. Natl. Acad. Sci USA* 89:6080–84

124. Varughese KI, Xuong NH, Kiefer PM, Matthews DA, Whiteley JM. 1994. Structural and mechanistic characteristics of dihydropterine reductase: A member of the Tyr-(Xaa)$_3$-Lys-containing family of reductases and dehydrogenases. *Proc. Natl. Acad. Sci. USA* 91:5582–86

125. Wall D, Kaiser D. 1998. Alignment enhances the cell-to-cell transfer of pilus phenotype. *Proc. Nat. Acad. Sci. USA* 95:3054–58

126. Wall D, Wu SS, Kaiser D. 1998. Contact stimulation of Tgl and type IV pili in *Myxococcus xanthus. J. Bacteriol.* 180:759–61

127. Ward MJ, Mok KC, Zusman DR. 1998. *Myxococcus xanthus* displays Frz-dependent chemokinetic behavior during vegetative swarming. *J. Bacteriol.* 180:440–43

128. Ward MJ, Zusman DR. 1997. Regulation of directed motility in *Myxococcus xanthus. Mol. Microbiol.* 24:885–93

129. Whitchurch CB, Hobbs M, Livingston SP, Krishnapillai V, Mattick JS. 1991. Characterization of a *Pseudomonas aeruginosa* twitching motility gene and evidence for a specialised protein export system widespread in eubacteria. *Gene* 101:33–44

130. Wireman JW, Dworkin M. 1977. Developmentally induced autolysis during fruiting body formation by *Myxococcus xanthus. J. Bacteriol.* 129:796–802

131. Wu S, Kaiser D. 1996. Markerless deletions of *pil* genes in *Myxococcus xanthus* generated by counterselection with the *Bacillus subtilis sacB* gene. *J. Bacteriol.* 178:5817–21

132. Wu SS, Kaiser D. 1995. Genetic and functional evidence that Type IV pili are required for social gliding motility in *Myxococcus xanthus.* 18:547–58

133. Wu SS, Kaiser D. 1997. Regulation of expression of the *pilA* gene of *Myxococcus xanthus. J. Bacteriol.* 179:7748–58

134. Wu SS, Wu J, Kaiser D. 1997. The *Myxococcus xanthus pilT* locus is required for social gliding motility although pili are still produced. *Mol. Microbiol.* 23:109–21

135. Yang C, Kaplan HB. 1997. *Myxococcus xanthus sasS* encodes a sensor histidine kinase required for early developmental gene expression. *J. Bacteriol.* 179:7759–67

136. Yang Z, Geng Y, Xu D, Kaplan HB, Shi W. 1998. A new set of chemotaxis homologs is essential for *Myxococcus xanthus* social motility. *Mol. Microbiol.* In press 30:1123–1130

137. Youderian P. 1998. Bacterial motility: secretory secrets of gliding bacteria. *Curr. Biol.* 8:R408–R411

Annu. Rev. Microbiol. 1999. 53:551–75

CLOSTRIDIAL TOXINS AS THERAPEUTIC AGENTS: Benefits of Nature's Most Toxic Proteins

Eric A. Johnson

Department of Food Microbiology and Toxicology, Food Research Institute, University of Wisconsin, Madison, Wisconsin 53706; e-mail: eajohnso@facstaff.wisc.edu

Key Words *Clostridium*, neurotoxin, botulinum, pharmaceutical

■ **Abstract** Toxins are increasingly being used as valuable tools for analysis of cellular physiology, and some are used medicinally for treatment of human diseases. In particular, botulinum toxin, the most poisonous biological substance known, is used for treatment of a myriad of human neuromuscular disorders characterized by involuntary muscle contractions. Since approval of type-A botulinum toxin by the US Food and Drug Administration in December 1989 for three disorders (strabismus, blepharospasm, and hemifacial spasm), the number of indications being treated has increased greatly to include numerous focal dystonias, spasticity, tremors, cosmetic applications, migraine and tension headaches, and other maladies. Many of these diseases were previously refractory to pharmacological and surgical treatments. The remarkable therapeutic utility of botulinum toxin lies in its ability to specifically and potently inhibit involuntary muscle activity for an extended duration. The clostridia produce more protein toxins than any other bacterial genus and are a rich reservoir of toxins for research and medicinal uses. Research is underway to use clostridial toxins or toxin domains for drug delivery, prevention of food poisoning, and the treatment of cancer and other diseases. The remarkable success of botulinum toxin as a therapeutic agent has created a new field of investigation in microbiology.

CONTENTS

"In poison there is physic"
—Shakespeare, *Henry IV*   .

INTRODUCTION

Toxins and poisons that have detrimental effects on humans are produced by widely diverse animals, plants, and microorganisms (2, 69, 97, 126). A poison is defined as any substance that is harmful or lethal to a biological (living) system; a poison derived from a biological source is a toxin (18). Organisms probably synthesize toxins to acquire food, defend themselves against predation, and help infect potential hosts. Toxins likely have more subtle biological functions in the producers' physiology, because many microbial toxins evolved before the appearance of eukaryotic multicellular organisms that are the targets of many toxins. When eukaryotic organisms appeared on earth, microbial toxins offered a source of virulence factors to inflict disease and promote colonization and pathogenesis (126, 128).

Humans have long used poisons and toxins for their own benefit. Poisoned arrows and spears tipped with curare are used to capture animals for food, and poisons and toxins have been used for murderous purposes and warfare (13, 87). Among the most famous murderous and suicidal uses of toxins were the cup of hemlock extract given to Socrates (470–399 BC) and Cleopatra VII's (69–30 BC) use of an Egyptian viper to assist in her suicide. During past wars and in recent years, bioterrorism has loomed as a potentially devastating means of destroying human lives, partly owing to the high potency of certain toxins and infectious agents and the ease of preparing some of these biological agents by relatively unskilled people in primitive laboratories or factories (4, 87, 113).

Although toxins and poisons have been used for deleterious purposes, they also have a beneficial side for humankind. Toxins can be of tremendous use in preventing or relieving the symptoms of diseases and in understanding the physiology of organisms. From a clinical perspective, toxic cardiac glycosides such as digitalis from foxglove have been used for centuries as therapeutic agents to prevent heart failure (130). Curare has been used for treatment of spasmodic or convulsive disorders (24). Curare was the toxin used by Claude Bernard (1813–1878) in his pioneering studies demonstrating nerve-to-muscle signaling (7, 8). Bernard's research stimulated scientists and physicians to discover and elucidate the mechanisms of toxins and poisons and to use them for therapeutic and scientific purposes.

The discovery and characterization of protein toxins (3, 125) also contributed to the development of antitoxins and vaccines. In 1897 Ehrlich (see Reference 56) proposed that exposure to detoxified proteins that retained antigenicity could result in immunity on subsequent exposure to the active toxins. The experimental demonstration of antitoxins followed the discovery of the diphtheria and tetanus organisms (3, 33, 45, 80). Antitoxins in serum were demonstrated after administration of detoxified forms of the toxins to animals (45). Horses immunized against toxoids provided antisera for serum therapy of toxin-mediated diseases. In 1897 Ehrlich (see Reference 56) introduced the important concept of activity standardization of toxins and antitoxins to provide consistent utility and to avoid unwanted side reactions. Vaccinations have been extraordinarily successful in protecting humans against many human infectious diseases, such as tetanus, diphtheria, and whooping cough (11, 45), and microbial diseases of animals (98).

THE CLOSTRIDIA: A Rich Source of Toxins

Among the eubacteria, the clostridia produce more protein toxins than any other bacterial genus (52, 54, 78, 98, 126). Clostridial toxins are biologically active proteins that are antigenic in nature and can be neutralized with antisera raised in animals against toxoids (52). The clostridia are a diverse group of rod-shaped bacteria characterized by a gram-positive cell wall structure, formation of endospores, and a fermentative metabolism (26, 54, 67). The genus *Clostridium* presently comprises a large collection of more than 100 species, and the genus is clearly in need of taxonomic reevaluation (26, 28, 29, 118). Only ~20 clostridial species are known to cause disease in humans and animals, generally through the activity of extracellular toxins and enzymes (26, 52, 54, 80, 129). The toxigenic *Clostridium* species appear to fall within a few natural lineages based on DNA and RNA homologies, 16S rRNA sequences, and cell wall structure compared with the genus as a whole (28, 29, 54, 66, 67).

Diseases caused by pathogenic clostridia often are vivid and memorable in their presentations (17, 30, 39, 66, 78, 98, 116, 129). Clostridia that cause wound infections, particularly *Clostridium perfringens*, *C. novyi* type A, *C. bifermentans*, *C. histolyticum*, *C. sordellii*, and *C. sporogenes*, can inflict rapidly progressing lesions that result in horrendous infections such as gas gangrene (clostridial myonecrosis) (39, 78, 129). Clostridia have been associated with neutropenia, hematological malignancies, bacteremias, and promotion of or protection against colorectal carcinomas (6, 17). Clostridial infections in humans occur in hypoxic and necrotic areas of tumors and infections (75, 78, 129). Clostridia may cause severe cases of antibiotic-associated diarrhea and pseudomembraneous colitis (5, 17), sometimes requiring surgical removal of bowel sections. The primary etiologic agent is *C. difficile*, with rare cases by *C. perfringens* type C and other clostridial species.

The neuroparalytic clostridia, *C. botulinum*, *C. tetani*, and *C. argentinense*, as well as rare strains of *C. butyricum* and *C. baratii*, cause distinctive paralytic or spastic diseases (30, 39, 89, 125). Botulism is a rare and distinctive neuroparalytic disease that can weaken or paralyze every skeletal muscle in the body (27, 116). The hallmark of botulism is a descending paralysis with an extremely long duration of illness of months to years in severe cases (27, 30, 116, 125). Severe tetanus presents as violent and persistent spasms of the head, trunk, and limb muscles (11). Botulinum NT (BoNT) and tetanus NT (TeNT) are the most potent toxins known (66, 73, 84, 125). The extraordinary potency of BoNT and TeNT derives from their remarkable neurospecificity and their catalytic cleavage at exceedingly low concentrations (ca. 10^{-12}–10^{-13} M) of neuronal substrates involved in exocytosis (89, 115).

BoNTs are produced by diverse groups (species) of clostridia, whereas TeNT is produced only by the single species *C. tetani*. The diversity of clostridia that produce BoNTs indicates that toxin gene exchange occurs naturally by mobile genetic elements (29, 55, 64). It is likely that new botulinogenic clostridial strains and species will be isolated if care is taken in laboratories to culture anaerobic specimens or environmental samples and the isolates are purified and tested for toxicity. The isolation and purification of some species of clostridia are frequently difficult because clostridia tend to grow as mixed cultures with other bacterial species (65). Repeated isolation on nonselective growth media is often necessary to affirm purity of cultures (52, 54).

Several excellent reviews have thoroughly described current knowledge of the biochemistry, genetics, immunology, and pharmacology of clostridial toxins (2, 52, 53, 80, 98, 115). The bacteriological properties of the toxigenic clostridia and methods for their isolation, purification, anaerobic culture, and identification have also been described (26, 28, 54, 66, 129). This review focuses on properties of clostridial toxins related to their utility as therapeutic agents for treatment of human disease. Botulinum toxin is currently of principal interest from a therapeutic perspective, although therapeutic uses have also been proposed for TeNT and its derivative fragments, *C. botulinum* C_2 toxin, *C. difficile* toxin A, and *C. perfringens* enterotoxin. Clostridial spores and vegetative cells are also being investigated as delivery vehicles of chemotherapeutic agents for treatment of certain cancers. The greatest success so far in clostridial toxins as therapeutic agents is that with botulinum toxin type A, which is being used to treat an accelerating array of involuntary muscle disorders in humans (16, 20, 61–63, 92, 101, 109, 122). In this chapter, botulinum toxin is emphasized as a therapeutic paradigm because of the tremendous success of treatments with this toxin and its greatly increasing use, and because it is a focus of research in the author's laboratory.

SCIENTIFIC BASIS OF BOTULINUM AND TETANUS TOXINS AS THERAPEUTIC AGENTS

Neurotoxigenic clostridia were isolated from the 1890s through the 1980s (52, 54, 66, 125), and the NTs produced by these organisms have been isolated and characterized over the past 70 years (46, 73, 101). In 1928, a high-potency concentrate of BoNT/A was obtained from autolyzed *C. botulinum* by simple acid precipitation at about pH 4 of culture fluid (see Reference 101). Following this lead, "crystalline" botulinum toxin type A was purified by researchers at Fort Detrick, MD (73, 74, 101, 102), and the procedures were adapted to other botulinum serotypes (101). The research group at Fort Detrick showed that the large (900-kDa) crystalline type-A molecule could be separated into distinct neurotoxic, nontoxic, and hemagglutinating molecules when solutions of crystalline toxin were raised to pH 7.5 at the proper ionic strength (73, 101). Ultracentrifugation or chromatographic techniques were then used to isolate the type-A NT(BoNT/A) component from the crystalline complex (101, 114). Seven serotypes of BoNTs have been isolated, which are designated by the letters A through G. Each serotype of BoNT is neutralized by a specific antiserum raised against toxoids of the purified toxins, and each has a high potency of 10^7 to $\geq 10^8$ mouse LD_{50} (MLD) (103) per mg of toxin depending on the serotype (121). TeNT was originally purified in the 1940s from autolyzed culture filtrates of *C. tetani* (81). Only one serotype of TeNT has been isolated from *Clostridium* spp.

BoNTs are proteins of ~150 kDa, which naturally exist as components of toxic complexes: as the Medium (M) complex (~300 kDa) consisting of BoNT associated with a nontoxic-nonhemagglutinin protein of ~150 kDa, or as the Large (L) and Large-Large (LL) complexes (~450 and 900 kDa, respectively) in which the M complex associates with hemagglutinin protein(s) (99, 121). Some ribonucleic acid is also present in the complexes (101). The nontoxic proteins in the complexes provide stability during manipulations and during passage through the gastrointestinal tract (99, 101, 121). TeNT is produced as a single peptide of ~150 kDa, and it is not known to form complexes with nontoxic proteins. It does not appear to be stable in the intestinal tract and is not a cause of food poisoning, although *C. tetani* may still colonize and produce intestinal infections (121).

Very little is known of the environmental, nutritional, and genetic regulation of synthesis of BoNTs and TeNT, although this is an important aspect contributing to the quality of BoNTs and TeNT as pharmaceutical agents (101). Medium components and culturing conditions such as temperature affect toxin titer and toxin quality, including the type of complex formed, potency and stability of the NT, and chain fragmentation during purification and storage (101). Synthesis of BoNT/A, BoNT/E, TeNT, and proteases is regulated by nutrition, particularly by nitrogen and energy sources (101). Although suspected for some time, the presence of positive regulatory genes within the BoNT/A and TeNT gene clusters (55) has recently been confirmed (79, 80). Nutrition and regulatory genes are also involved in regulation of other clostridial toxins such as those produced by *C. difficile* A and B toxins (34, 58). An efficient conjugative gene transfer system has been adapted from *C. perfringens* for use in *C. botulinum* (19), which will enable genetic manipulation and recombinant-expression studies of BoNTs as pharmaceutical agents.

BoNTs and TeNT are produced as single-chain molecules of ~150 kDa that achieve their characteristic high toxicities of 10^7 to $\geq 10^8$ MLD/mg by post-translational proteolytic cleavage within a narrow region of the polypeptide to form a dichain molecule composed of a light-chain (L_C) (~50,000 kDa) and heavy-

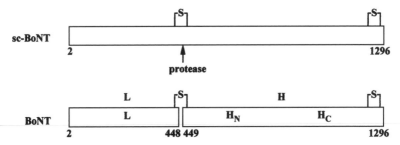

Figure 1 Designation for *Clostridium* neurotoxin nomenclature. The diagram depicts the botulinum neurotoxin (BoNT/A) molecule. The single-chain BoNT is activated by protease cleavage of a peptide bond, forming a dichain-activated molecule consisting of a heavy (H) and light (L) chain held together by a disulfide bond. The H chain can be divided into the N-terminal (H$_N$) and C-terminal (H$_C$) fragments by proteolytic cleavage. The numbers indicate the amino acids of the single-chain peptide. The botulinum serotype is indicated in the text by a letter, such as BoNT/A for type A. Drawing by M Bradshaw.

chain (H$_C$) ($\sim$100,000 kDa) linked by a disulfide bond (Figure 1) (48, 94, 99, 121). BoNTs and TeNT consist of three primary functional domains (72, 90) (Figure 1): (*a*) the catalytic domain within L$_C$, possessing endopeptidase activity; (*b*) the translocation domain located in the N-terminal region of the H$_C$; and (*c*) the receptor-binding domain located in the C-terminal region of H$_C$. The gene sequences of BoNTs and TeNTs and the corresponding amino acid sequences of all seven BoNTs and TeNT have been determined from various strains (29, 55, 94). These gene and amino acid sequences of TeNT and BoNTs have been analyzed for regions of homology (55, 94). Certain regions of the BoNTs among the various serotypes have low homology, whereas other regions have high homology, suggesting that the NTs possess expected conserved regions but also that they evolved as hybrids of various structural motifs (55, 72). TeNT and BoNTs show high homology in the regions defining the catalytic active site, the translocation domain, and the two cysteine residues forming the disulfide bond connecting the H$_C$ and L$_C$ (55, 94). The least degree of homology is in the carboxyl region of the H$_C$, which is involved in neurospecific binding. When considered with substrate specificity these differences may provide the basis for the different serotypes of BoNTs. The high homologies of amino acid sequences of NT genes isolated from diverse clostridia suggest that a single ancestral gene was dispersed by lateral transfer, and six lineages of clostridial NT evolution have been proposed (29). These observations reinforce the hypothesis that genetic transfer of the toxin genes or regions of the genes to normally benign clostridia occurs in certain environments (64, 65).

Poisoning of nerves by BoNTs and TeNT occurs by a multistep mechanism (89, 90, 115) involving (*a*) cell-binding, (*b*) internalization into endosomes probably by receptor-mediated endocytosis and/or synaptic vesicle reuptake, (*c*) membrane translocation from mature endosomes into the neuronal cytoplasm, and

(d) target recognition and catalytic cleavage of neuronal substrates. An understanding of in vivo stability and degradation of the internalized toxins contribute to their longevity of action, and understanding alteration of nerve structure and gene expression will improve this model. The postulated model for trafficking of NTs into nerves is similar to models proposed for certain other protein toxins and viruses with intracellular targets, although the mechanisms are complex and have not been fully elucidated.

Certain common biochemical and pharmacological features exist among the clostridial NTs and other protein toxins. Like many protein toxins, BoNTs and TeNT have a carbohydrate or lectin recognition domain that lies in the carboxyl-terminal half of the H_C (70, 90), and binding of NTs is initially to polysialogangliosides. It is unlikely that polysialogangliosides are the sole receptors of BoNTs and TeNT, because the binding affinities for these glycolipids ($\sim 10^{-8}$ M) are insufficiently tight to explain in vivo NT intoxication (90). In vivo poisoning by BoNTs and TeNT occurs at subpicomolar concentrations (ca. $\cdot 10^{-12}$ M). Evidence indicates that a protein receptor(s) is also involved in a multicomponent binding process (70, 90). Although many investigators have attempted to characterize the postulated high-affinity receptor complexes of BoNTs and TeNT for nearly a century (90), the in vivo complexes remain to be defined. Recent evidence suggests that the high-affinity receptor for BoNT/B is synaptogamin II embedded in a lipid environment (70), but further characterization is required using genetic and biochemical strategies to elucidate the receptor process in vivo. Like BoNTs, TeNT possesses a ganglioside recognition domain that has been elegantly mapped to specific regions of TeNT H_C by photoaffinity labeling (112) and deletion mutants (50). A protein receptor complex has not been isolated for TeNT, although isolation of such a protein could lead to the development of competitive ligands to counteract tetanus or to act as delivery vehicles for transport to the CNS. The nature of the receptor could come from studies of neurotropic viruses that use cell membrane permeases to gain access to the CNS (105, 127) and from an understanding of the ancestral genes of NT evolution.

The binding of TeNT and BoNTs to specific cell surface receptors is followed by internalization into cellular compartments by receptor-mediated endocytosis and/or synaptic vesicle reuptake (82, 90, 93). Sequestration within clathrin-coated pits does not appear to be involved in internalization of TeNT (91). BoNTs and TeNT are more rapidly internalized with electrical stimulation (35, 82), a property typical of fusion of neurotransmitter vesicles with the presynaptic plasma membrane and release of the contents (synaptic vesicle reuptake). The synaptic vesicles then acquire a new cargo of neurotransmitters and the recycling may recur in ≤ 2 min (9, 93). Synaptic-vesicle recycling has been proposed for TeNT uptake into hippocampal neurons (82). After incorporation of TeNT or BoNTs within early endosomes, acidification of the endosome environment may induce an NT structural change and formation of a membrane-spanning pore (95). This presumably allows the L_C to slip into the nerve cytosol for the mature endosome. Oligomerization of four toxin amphipathic alpha-helices is required for channel formation (95).

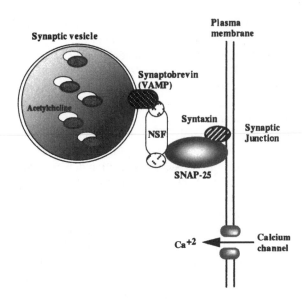

Figure 2 Simple portrait of current concept of the SNARE complex. The SNARE proteins synaptobrevin/VAMP, SNAP-25, and syntaxin form a complex with the ATPase NSF. The complexed SNARE proteins are relatively resistant to cleavage by botulinum toxin, but are susceptible when they are disassembled. The SNARE complex is involved in trafficking of the synaptic vesicle containing neurotransmitters to the presynaptic membrane, where fusion and exocytosis of the neurotransmitters occurs. See text for details. Drawing by MC Goodnough.

After internalization into the cytosol, BoNTs and TeNT specifically cleave neuronal proteins integral to vesicular trafficking and neurotransmitter release (51, 90, 131). TeNT and BoNTs comprise a unique group of zinc proteases (90) that recognize specific structural motifs of the neuronal substrates referred to as SNAREs: (*a*) vesicle-associated membrane protein (VAMP/synaptobrevin), (*b*) syntaxin, and (*c*) soluble 25-kDa *N*-ethyl-maleimide–sensitive fusion attachment protein (SNAP-25) (22, 51, 96, 117, 131). These SNAREs reside on transport vesicles or on the target membranes (Figure 2). The SNARE proteins form a multiprotein 20S complex, together with NSF that possesses ATPase activity (51, 96, 117, 131). When these proteins are assembled in the tight complex, the substrates are resistant to in vitro proteolysis by TeNT and BoNT (96), whereas the free proteins are specifically and efficiently cleaved by the NTs. The three SNAREs, VAMP, SNAP-25, and syntaxin, possess a distinct three-dimensional motif (the SNARE motif) that is required for specific proteolysis by the NTs (90). BoNTs and TeNT are unusual among zinc-dependent proteases in requiring a substrate of ≥ 16 amino acids in length (37, 106), probably because the NTs recognize the shape of the substrate and not specific peptide bonds (90).

The use of molecular biology methods combined with electrophysiology to investigate cellular secretion has led to tremendous advances in our understanding of the processes of neurotransmission and its inhibition by BoNTs and TeNT. In vivo, both fast and slow phases of membrane fusion and neurotransmitter secretion are blocked by all but one of the different BoNTs (BoNT/A) or TeNT, suggesting that the SNAREs integral for secretion must exist both in a ternary complex and free within the neuron (51, 96, 117). Although much progress has been made in understanding the mechanisms by which clostridial NTs inhibit neurotransmission, many gaps exist in our knowledge of the mechanisms of the BoNTs and TeNT, particularly in vivo within the nerve environment. The mechanism of endocytosis and trafficking of the toxins to the nerve cytosol has not been solved. BoNTs/A/B/E and TeNT undergo tyrosine phosphorylation, which affects their catalytic activity and stability (37), indicating that the neuronal environment and signaling pathways are integrated with NT activity. The 26-mer–cleavage product released by BoNT/E inhibits vesicle docking, adding further complexity to the roles of BoNTs and TeNT in neurons (37). The concentrations of SNARE proteins, whether free or in the complex, have not been quantitated. Their concentration or the presence of degradative enzymes for BoNTs and TeNT would contribute to the half-lives of the NTs. Besides being involved in exocytosis, SNAP-25 and other neuronal proteins perform several functions in the nerve, including axonal growth, neurite sprouting, maturation, synaptogenesis, and G-protein regulation of presynaptic calcium channels (59, 90, 119), and other unknown functions. Much additional research is needed to understand their mechanisms and to validate the in vitro models with in vivo systems.

CRYSTAL STRUCTURE OF BOTULINUM NEUROTOXIN TYPE A

Understanding many of the complexities of NTs in neuronal poisoning and practical strategies to improve BoNT/A as a pharmaceutical agent could be greatly assisted by elucidating the three-dimensional structures of BoNTs and TeNT. Solving the crystal structure of BoNT/A and BoNT/A complex (23, 72) proved to be an onerous endeavor. After reporting the isolation of diffracting crystals and preliminary X-ray analysis in 1991 (120), the crystal structure of the entire 1285-amino-acid dichain structure of BoNT/A was determined at 3.3 Å resolution in 1998 (72). The overall shape of BoNT/A is rectangular with dimensions of $\sim$45 Å $\times$ 105 Å $\times$ 130 Å, and the molecule shows a linear arrangement of the three functional domains with no contact between the catalytic and binding domains (Figure 3—see color insert). The three functional domains have separated and distinct structures, with the exception of an unusual loop that encircles the perimeter of the catalytic domain (72). The existence of this loop was unexpected and presents a puzzle regarding catalysis because it hides the catalytically active site. It is possible that the site opens when it contacts the substrate or when the L_C is internalized within neurons and associates with membranes, but the mechanism of catalysis will require further studies. The crystal structure revealed that the ganglioside-binding C-terminal subdomain has structural homology with proteins known to interact with sugars such as the H_C fragment of TeNT, serum amyloid P, sialidase, various lectins, and the cryia and insecticidal α-endotoxin, which bind glycoproteins and create leakage channels in membranes (72). Thermolysin and leishmanolysin had the most homology to the catalytic domain of BoNT/A. The translocation domain was distinct in structure from bacterial pore-forming toxins and showed more resemblance to coiled-coil viral proteins such as human immunodeficiency virus type 1 gp41/GCN4, influenza hemagglutinin, and the Moloney murine leukemia virus TM fragment (72). BoNT/A appears to consist of a hybrid of varied structural motifs that may have evolved by the combination of functional subunits to generate a highly toxic pathogenic molecule.

Burkhard and colleagues (23) also successfully determined the three-dimensional structure of the 900-kDa botulinum type-A complex to 15-Å resolution to understand the biological significance of the large complex. The complex is triangular, has an estimated radius of $\sim$110 Å, and possesses six distinct cylindrical lobes. The BoNT component appeared to be located in the center of the complex, where it possibly was protected from the external environment.

Although the complete three-dimensional structure of TeNT has not yet been solved at high resolution, Umland et al (123, 124) have determined the structure of the receptor-binding fragment H_C to 2.6-Å resolution. It contains two closely associated domains including a variation of the beta-trefoil motif in the C-terminal domain and a putative ganglioside-binding site.

CLINICAL DEVELOPMENT OF BOTULINUM TOXIN AS A PHARMACEUTICAL AGENT

Like many important discoveries, the development of botulinum toxin as a drug involved creative thought and experimentation, knowledge of historical toxicology, some twists of fate, more than a bit of good fortune, extreme patience and tenacity, and fruitful collaborations between physicians and research scientists (100–102, 110). The concepts of graded selective denervation by toxins and their use as therapeutic agents and as tools in human physiology actually had their beginnings in the early studies of Claude Bernard (1813–1878) with curare and other toxic substances (7, 8, 57). Bernard demonstrated that curare "kills rapidly without convulsions and at once renders the nerves inexcitable" (7, 57). Bernard showed that, although nerves become inexcitable after curarization, the muscles still react on direct stimulation (7, 8). The site of action of curare was shown to be motor nerve endings. These experiments demonstrated that curare destroyed only the action of the motor nerves without impairing that of sensory nerves or the CNS. In essence, he demonstrated selective denervation with retention of voluntary muscle activity by a naturally occurring toxin. Curare was used with limited success for treatment of tetanus in animals and humans, for epilepsy, for tremors of Parkinsonism, and as an anaesthetic (24).

The long-lasting paralytic action of type-A botulinum toxin at motor neuron synapses of striated muscle and the response of the synapses to acetylcholine

Figure 4   Edward J Schantz (1908–present). University of Wisconsin, Madison, WI.

were shown in the early 1900s (46). The paralysis or weakening by injection of botulinum toxin could be localized to individual muscle groups (e.g. the gastrocnemius leg muscle of animals), whereas other regions in the animal were unaffected. In nonfatal botulinal intoxications, poisoned animals required an extraordinarily long time for recovery. For example, the lethal dose of botulinum toxin for ~500-g guinea pigs is ~5 MLD (104) [along with humans and horses, guinea pigs are probably the most sensitive animals to botulinum toxin that we know (44, 116)]. After injection of ~1 MLD into the gastrocnemius leg muscle, 60% of the guinea pigs recovered in 7 months, and 90% recovered after 1 year (46). Administration of massive quantities of antitoxin did not reverse paralysis once it had commenced (1–2 days), but it did prevent paralysis before this latency period, depending on the serotype of botulinum toxin injected.

Purification of the 900-kDa botulinum type-A toxin complex was refined and perfected in the 1940s–1960s at Fort Detrick by Edward Schantz (Figure 4) and his colleagues (101, 102). Schantz also developed formulations for stabilization of picogram quantities of type-A complex (101, 102). The generosity of Schantz in providing purified type-A complex to qualified researchers led to significant advances in understanding various aspects of botulinum toxin.

Also in the 1940s, the action of botulinum toxin on the neuromuscular junction was clearly demonstrated in a classic investigation by using isolated nerve-muscle preparations (21). A dose response and a latency period occurred on intoxication of the phrenic nerve-hemidiaphragm of rodents, whereas single motor nerve twitches were not affected. It was further shown that (a) latency of toxin action was only slightly decreased by large doses of toxin, (b) bound toxin could not be removed from the nerve terminal by washing, (c) addition of acetylcholine to the bath restored twitch, (d) early addition of antitoxin prevented the twitch response, and (e) type-B toxin was much less toxic than type A in rat but not in guinea pig phrenic hemidiaphragm preparations (21). Thus, the basic physiology of type-A botulinum toxin was described by the early 1950s.

The concept of using botulinum toxin as a selective chemodenervating agent was probably considered for several years by neurologists (110). However, Drachman's innovative studies of the effects of botulinum toxin in chicken embryos (32) provided the conclusive experimental proof that botulinum toxin A induced selective denervation, muscle weakening, and muscle atrophy. The chick embryo model was extremely useful for investigation of botulinum paralysis, because respiratory gas exchange occurs by passive diffusion across the chorioallantoic membrane, and large quantities of botulinum toxin could be administered without detrimental effects on respiration. After injection of enormous quantities (30 μg) of botulinum toxin, the thighs and legs of the inoculated chicks were markedly shrunken, devoid of muscle mass, degenerative, and fatty (32). The cardiac muscle appeared normal, indicating that botulinum toxin did not have a generalized toxic effect on striated muscle but preferentially poisoned skeletal muscle. Drachman (32) proposed the characteristics of an "ideal (nerve) blocking agent" in terms of its (a) mode of action in blocking cholinergic transmission and known mechanism of action, (b) specificity in blocking only cholinergic transmission, (c) reversibility in not permanently impairing function or structure of nerve or muscle, (d) generality of action in blocking all motor neuron terminals of striated skeletal muscle, (e) convenience of use in requiring simple injection into desired muscle regions, (f) safety if used with appropriate precautions, and (g) apparent absence of systemic or CNS effects.

The use of botulinum toxin type A as an injectable selective muscle-weakening agent was investigated experimentally in monkeys in the late 1960s and late 1970s and in humans in the late 1970s and 1980s by Alan B Scott (Figure 5). This investigation involved a unique collaboration with Schantz beginning in the 1960s and later with Schantz and myself in the 1980s (100–102, 108). Scott's goal was to provide a pharmacological alternative to surgery in nonaccommodative strabismus by injection of extraocular muscles with neurotoxic agents. He cites (110) previous investigators who had injected ethanol or other drugs in

Figure 5   Alan B Scott, (1932–present). Smith-Kettlewell Eye Research Foundation, San Francisco, CA.

attempts to achieve this goal, but the experiments failed. Scott et al (110) found that ketamine provided surgical levels of anesthesia with preservation of an active electromyogram signal recorded by an electrode at the tip of the injection needle. Scott first tried injecting certain poisonous agents into the lateral rectus muscle of adult rhesus monkeys, including di-isopropyl-fluoro-phosphate, α-bungarotoxin, and ethanol as paralyzing agents, but these were unsatisfactory because of lack of nerve selectivity, generalized toxicity, or brevity of action (110). He was able to produce transient weakness lasting 2–8 months by injecting botulinum toxin A received from Schantz, and he demonstrated that it altered ocular alignment. After the surgery, treated monkeys awakened promptly from the light ketamine anesthesia and thrived in their cages without evidence of any systemic side effects.

Dietary habits and activity appeared normal. Two monkeys died, but these deaths were caused by viral or enteric infections in the monkey colony. Transient ptosis to the sides of the monkeys' eyes cleared within a few weeks. Three and a half months after injection, the electromyogram recorded from the injected muscle was of normal amplitude, and eye movement was also normal. In these remarkable experiments, permanent ocular alignment changes after temporary muscle paralysis were the clinical outcome without any need for incision or direct exposure of the muscle. The quantity of toxin needed to induce ocular alignment was 1000- to 100,000-fold less than the estimated LD_{50} [$\sim$0.1 μg or $\sim$28 U/ng (111); the estimated lethal dose for a 70-kg human has been estimated as 1–2 μg (101)]. Scott predicted that botulinum toxin injection was a suitable pharmacologic approach that could replace or augment existing methods of surgical correction of strabismus. Scott predicted that botulinum toxin could be used to reduce other conditions, such as lid retraction and blepharospasm, and to influence skeletal muscle groups.

In 1972, the Fort Detrick laboratories investigating biological agents were closed (102), and Schantz accepted a position at the University of Wisconsin, where he continued to provide Scott with batches of high-quality type-A botulinum toxin complex and advice on its handling and stabilization. Schantz also provided other serotypes of botulinum toxin complexes to Scott, but these were not as successful in providing a long-lasting denervation (100).

All the tools were in place for experimentation with humans. In response to a courageous request by Scott in the late 1970s, the National Institutes of Health granted him permission to inject human adult volunteers with strabismus as an alternative to surgery. In 1980, he reported the results of 67 injections of botulinum A toxin for correction of strabismus (e.g. Figure 6) (107–109). These studies established type-A botulinum toxin injections as an effective nonsurgical alternative treatment for strabismus. Botulinum toxin type A appeared to be a specific, effective, and safe chemodenervating agent for weakening of extraocular muscles. Its use became successful as an adjunct or alternative to the tedious surgical correction. After thousands of injections into patients, the US Food and Drug Administration approved one batch of toxin produced in 1979 (batch 79-11) by Schantz for use in the United States for strabismus, blepharospasm, and hemifacial spasm. By this time, botulinum toxin was being experimentally tried for many other disorders, and this single batch of toxin (79-11; $\sim$150 mg) eventually was used for >250,000 injections in humans. I joined Schantz in 1985 from Harvard Medical School and together we prepared several batches of type-A botulinum toxin that were injected with excellent results in humans.

Scott (107–109) outlined the following principles and characteristics of botulinum toxin based on his monkey and human studies: (*a*) it showed no known focal or systemic effects apart from muscle paralysis; (*b*) it apparently did not elicit antibody production in the small doses that were used; (*c*) it diffused slowly out of the injected muscle region into adjacent muscles; (*d*) the toxin acted for several weeks to months; and (*e*) the paralytic intensity was strongly correlated with the dose injected.

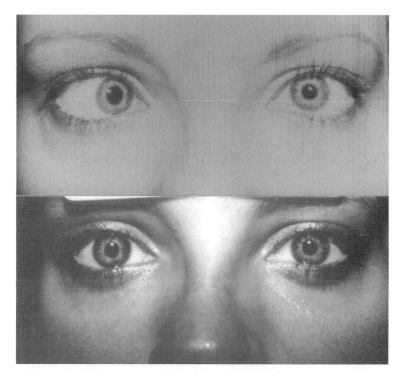

Figure 6 Representative treatment of strabismus (crossed eyes) by injection of botulinum toxin into the extraocular muscles, weakening them and allowing compensatory alignment by the agonist muscle. *Top*, Before treatment; *bottom*, after injection.

EXPANDING USE OF BOTULINUM TOXIN AS A THERAPEUTIC AGENT

As Scott predicted in 1973 (110), the use of botulinum toxin A in medicine has greatly accelerated, particularly since its limited approval in 1989 (14, 16, 20, 25, 61–63, 92, 109, 122). Botulinum toxin type A has been effective in treatment of a myriad of disorders involving involuntary muscle contractions, dystonias, and spasticity in focal or segmental muscle regions. Ironically, its primary uses are not for the approved indications. Somewhat unexpectedly, it has also been efficacious in treating pain syndromes. One of the first pain syndromes to be treated was myofascial pain (1). Tension and migraine headaches have also been treated successfully. The treatment of certain pain conditions suggests that it may affect autonomic nervous systems and the transmitters involved in pain such as substance P (60). It is also being evaluated in animal models for conditions for which muscle atrophy is desired, such as prostrate involution (31), or for convalescence in which muscle immobilization is required (20). Evidently, the numbers of disorders treated with botulinum toxin will continue to grow, particularly as we learn more of its basic effects on nerve-muscle physiology. Various secretory cells including endocrine cells and phagosomes (49) possess v-SNARE secretion systems, and BoNT or TeNT can disrupt secretion if microinjected into the cells. If BoNT or TeNTs or their component L_C could be targeted to these various cells, it may be possible to affect release of various hormones and to alter phagocytosis and other secretory processes.

PROBLEMS WITH BOTULINUM TOXIN THERAPY

The primary complications of botulinum toxin therapy have been (a) formation of antibodies and obliteration of response to type-A toxin, (b) lack of alternate botulinum serotypes with the potency and duration of action of type A, (c) diffusion of botulinum toxin to neighboring muscles with transient and sometimes debilitating ptosis, (d) lack of consistency and low specific activities of certain toxin preparations, and (e) need for repeated injection of toxin in chronic disorders. These problems and potential solutions have been debated and reviewed (14, 15, 20, 53, 122). Other serotypes of botulinum toxin than A have been tested for patients with immunity, but these other serotypes do not appear to have the potency or duration of action of type A, with the possible exception of type C (36). It is likely that technical advances in administration and improved botulinum toxin molecules and formulations can help to mitigate these problems (14, 101, 102).

USES OF OTHER *CLOSTRIDIUM* TOXINS AS PHARMACEUTICAL AGENTS

Tetanus Toxin

In experimental animal systems, dichain TeNT has been reported to enter the CNS from the circulatory system faster than any other known protein (47). Researchers have attempted to exploit this property to enable targeted delivery of proteins or small-molecular-weight ligands to the CNS. Nontoxic fragments of H_C of TeNT (and the H_C of BoNT) have been proposed as delivery vehicles for targeted transport of ligands from the circulation to peripheral or CNS neurons (10, 40, 42). The carrier determinants necessary for targeting have been localized to the 50-kDa C-terminal receptor-binding domains of TeNT H_C. The crystal structure of the receptor-binding fragment H_C of TeNT has been determined to 2.6-Å resolution (123, 124), which should be of value in designing efficient carriers. The isolated H_C retains the ability of being retrogradely transported to the CNS in a manner analogous to TeNT in animal models (10, 40, 42). Although the approach is elegant, it will probably be hindered by the presence of antibodies to H_C in humans and the need for injection into the circulatory system. The H_C of BoNT is also being investigated as a carrier molecule to peripheral nerves, which could avoid the immunity difficulty because most humans do not have antibodies to BoNT (53).

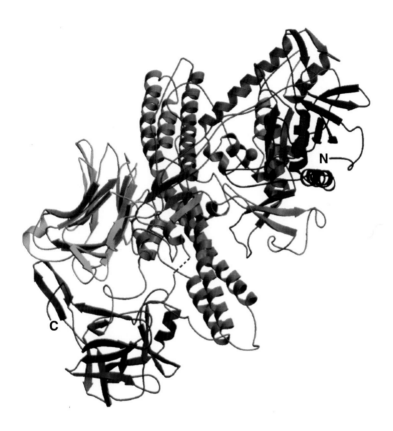

Figure 3 Backbone trace of the BoNA/A molecule. The catalytic domain is colored in blue, the translocation domain in green, the N-terminal subdomain in yellow, and the C-terminal sub-domain in red. The catalytic zinc is depicted as a ball in gray. The overall structure is 45 Å × 105 Å × 130 Å. Kindly provided by Raymond C. Stevens, University of California, Berkeley.

Injection of TeNT into the hippocampus of animals has served as a useful model of a chronic epileptic syndrome (43, 76). The syndrome is characterized by focal and generalized seizures, and the animals also experience permanent changes in performing behavioral tasks. Interestingly, the animal model mimics certain clinical psychological changes and learning disabilities associated with limbic seizures in humans. From the findings in animal models, the interesting hypothesis has been proposed that exposure to TeNT, possibly through intestinal infections, could contribute to autism in some individuals (12). This proposal highlights scattered experimental data indicating that intestinal infections by clostridia such as *C. tetani* (12) or *C. botulinum* (116) may not cause overt tetanus- or botulism-induced spastic or paralytic symptoms but could lead to chronic disease states through absorption of these extremely potent NTs.

Clostridium botulinum C_2 Toxin

During the past decade, elegant work by Aktories and his colleagues has revealed the catalytic activities of several clostridial ADP-ribosylating toxins (2). *C. botulinum* C_2 toxin and certain other clostridial toxins ADP-ribosylate actin (2). They are the most effective known agents to depolymerize actin in intact cells (2). It is interesting that *C. botulinum* C_2 toxin has been demonstrated to induce degranulation of attached mast cells. Because mast cells are responsible for release of histamine during allergic responses, it is possible that C_2 toxin or a derivative could be used to modulate allergic responses.

Clostridium difficile A Toxin

C. difficile was isolated in 1935 as a species of the normal flora of healthy infants (52). Ironically, it was not recognized as a widespread pathogen until it was culturally identified from feces of adult patients with antibiotic-induced pseudomembraneous colitis (a severe form of diarrhea and intestinal lacerations). The intestinal disease was confirmed by feeding culture supernatants in the absence of bacteria to hamsters (5). Bone and joint infections caused by *C. difficile* have also been reported (52). Although little information has been published on the organism itself (52), abundant information is available regarding its two protein toxins (2).

The two *C. difficile* toxins A and B are very large, single-chain exotoxins, having molecular masses of 270 and 308 kDa, respectively (2). Like many other clostridial toxins, *C. difficile* A and B toxins have carbohydrate-binding properties, which contribute to receptor selectivity. Toxins A and B possess glucosyltransferase activity and, in vitro, transferred one glucose moiety from UDP-glucose to Rho, Rac, and Cdc42, but other members of the Ras superfamily were not affected (2, 68).

Some evidence suggests that *C. difficile* toxin A could have pharmaceutical uses. Toxin A destroys various tumor cell lines, and it appears to inflict apoptotic-induced cell death in rat intestinal cells with a preference for cancerous cells (71). *C. difficile* toxin A has been proposed for treatment of colon and pancreatic cancers (71). *C. difficile* toxin A has also been reported to induce microvascular dysfunction (2) and to degranulate mast cells with release of histamine (2), and it potentially could be used for treatment of inflammatory disorders.

Clostridium perfringens Enterotoxin

C. perfringens is the most prolific toxin producer among the toxigenic clostridia (77), and these toxins cause a variety of human and animal diseases (52, 77). *C. perfringens* enterotoxin (CPE) is a cause of acute food poisoning, but the toxin is also of interest for biomedical purposes. It has been shown to destroy tumor cell lines and also to induce the release of certain cytokines (77). Nontoxic molecules derived from CPE have been proposed to be useful in preventing food poisoning (83) and possibly other infections. An NT (epsilon toxin) has recently been characterized from *C. perfringens* and shown to evoke excessive release of the neurotransmitter glutamate in animal models (86). This toxin could potentially be used for neurotargeting to the CNS.

TUMOR TARGETING WITH CLOSTRIDIAL SPORES AND CELLS

Species of anaerobic bacteria, particularly clostridia, are known to localize in hypoxic regions of tumors, where they cause lysis and degradation of necrotic tissue for prolonged periods (75, 85). It is likely that poor vascularization of the tumors creates favorable anaerobic conditions for growth of clostridia. Targeting of clostridial spores or cells to tumors has been considered as a unique therapy for tumor destruction (41, 85). Clostridia could also serve as carriers to bring cancer prodrugs to hypoxic regions of tumor cells.

LABORATORY SAFETY PRECAUTIONS IN WORKING WITH TOXINS

Research with and manufacturing of microbial toxins require rigorous safety precautions, standard operating procedures, and emergency protocols. Extreme care must be taken when handling toxigenic organisms and their toxins. Appropriate biosafety levels and practices are required for this purpose. Guidelines are available from the Centers for Disease Control and Prevention as well as other sources, and these must be followed to prevent laboratory accidents. The dispersal of organisms and toxins must be done prudently to avoid use of toxins for bioterrorism or other deleterious practices.

CONCLUSIONS AND PERSPECTIVES

Natural toxins affect human physiology by widely diverse mechanisms. Botulinum toxin has been used as a paradigm in this review to emphasize the potential of microbial toxins as drugs. Botulinum toxin has several desirable properties as a pharmaceutical agent, including extraordinary specificity for motor neurons innervating striated skeletal muscle, localized action, relative absence of systemic effects, remarkably high potency requiring extremely low doses for treatment, an uncommonly long duration of action of months for a single injected dose, and lack of immunogenicity when manufactured, handled, and administered properly. It also presently has some disadvantages, such as diffusion to neighboring muscles, a requirement for repeated injections, lack of consistency of biological activity, and antibody formation with improper use. It is likely that many of these disadvantages will be overcome as we better understand the physiology of botulinum toxin in relation to the human body and new-generation botulinum drugs are developed. Despite its current shortcomings, it has benefited thousands of humans with neuronal diseases and allowed them to live normal and healthy lives. The remarkably successful use of botulinum toxin as a drug and as a molecular tool for understanding cellular secretion has helped to generate much interest in microbial toxins. The clostridia, as well as many other organisms, produce a multitude of toxins that act by a myriad of mechanisms and affect tissues in various manners. As human genomics advances, it should be possible to recognize potential target sites for toxins. New toxins continue to be discovered, and in many cases the intricacies of their structure and function are only beginning to be unraveled. Continued basic and applied research in toxicology should enable us to harness many of these toxins for beneficial human purposes.

ACKNOWLEDGMENTS

Research in the author's laboratory has been supported by the National Institutes of Health, the United States Department of Agriculture, the US Department of Defense, Food Industry sponsors of the Food Research Institute—University of Wisconsin, Madison, and the College of Agriculture and Life Sciences, University of Wisconsin—Madison. The author is grateful to members of his laboratory for their research contributions, and is indebted to Edward J Schantz for his teaching and wisdom.

Visit the Annual Reviews home page at http://www.AnnualReviews.org

LITERATURE CITED

1. Acquadro MA, Borodic GE. 1994. Treatment of myofascial pain with botulinum A toxin. *Anesthesiology* 80:705–6

2. Aktories K, ed. 1997. *Bacterial Toxins. Tools in Cell Biology and Pharmacology.* London: Chapman & Hall. 308 pp.

3. Alouf JE. 1988. From diphtheritic poison (1888) to molecular toxinology (1988). *Bull. Inst. Pasteur* 86:127–44

4. Atlas RM. 1998. The medical threat of biological weapons. *Crit. Rev. Microbiol.* 24:157–68

5. Bartlett JG. 1994. *Clostridium difficile*: history of its role as an enteric pathogen and the current state of the knowledge about the organism. *Clin. Infect. Dis.* 18:S265–S272

6. Beebe JL, Koneman EW. 1995. Recovery of uncommon bacteria from blood: associ-

ation with neoplastic disease. *Clin. Microbiol. Rev.* 8:336–56

7. Bernard C. 1856. Analyse physiologique des propriétés des systèmes musculaire et verveux au moyen du curare. *CR Acad. Sci. (Paris)* 43:825–29

8. Bernard C. 1927. *An Introduction to the Study of Experimental Medicine*. (Transl. HC Green) New York: MacMillan. 226 pp.

9. Betz WJ, Wu LG. 1995. Synaptic transmission. Kinetics of synaptic-vesicle recycling. *Curr. Biol.* 5:1098–101

10. Bizzini B, Grob P, Glicksman MA, Akert K. 1980. Use of the B-II$_b$ tetanus toxin derived fragment as a specific neuropharmacological transport agent. *Brain Res.* 193:221–27

11. Bleck TP, Brauner JS. 1997. Tetanus. In *Infections of the Central Nervous System*, ed. WM Scheld, RJ Whitley, DT Durack, 34:629–53. Philadelphia/New York: Lippincott-Raven. 1064 pp. 2nd ed.

12. Bolte ER. 1998. Autism and *Clostridium tetani*. *Med. Hypoth.* 51:133–44

13. Bond RT. 1951. *Handbook for Poisoners. A Collection of Famous Poison Stories*. New York: Rinehart. 311 pp.

14. Borodic G, Johnson E, Goodnough M, Schantz E. 1996. Botulinum toxin therapy, immunologic resistance, and problems with available materials. *Neurology* 46:26–29

15. Borodic GE, Ferrante RJ, Pearce LB, Alderson K. 1994. Pharmacology and histology of the therapeutic application of botulinum toxin. In *Therapy with Botulinum Toxin*, ed. J Jankovic, M Hallett, 10:119–57. New York: Marcel-Dekker. 608 pp.

16. Borodic GE, Pearce LB, Johnson EA, Schantz EJ. 1991. Clinical and scientific aspects of botulinum A toxin. *Ophthamol. Clin. North Am.* 4:491–503

17. Borriello SP, ed. 1985. *Clostridia in Gastrointestinal Disease*. Boca Raton, FL: CRC Press. 239 pp.

18. Borzelleca JF. 1994. History of toxicology. In *Principles and Methods of Toxicology*, ed. AW Hayes, 1:1–17. New York: Raven. 1468 pp. 3rd ed.

19. Bradshaw M, Goodnough MC, Johnson EA. 1998. Conjugative transfer of the *Escherichia coli-Clostridium perfringens* shuttle vector to pJIR1457 to *Clostridium botulinum* type A strains. *Plasmid* 40:233–37

20. Brin MF. 1997. Botulinum toxin: chemistry, pharmacology, toxicity, and immunology. *Muscle Nerve* 6:S146–S168 (Suppl.)

21. Burgen ASV, Dickens F, Zatman LJ. 1949. The action of botulinum toxin on the neuromuscular junction. *J. Physiol.* 109:10–24

22. Burgoyne RD, Morgan A. 1993. Regulated exocytosis. *Biochem. J.* 293:305–16

23. Burkhard F, Chen F, Kuziemko GM, Stevens RC. 1997. Electron density map of the botulinum neurotoxin 900-kilodalton complex by electron crystallography. *J. Struct. Biol.* 120:78–84

24. Burnap TK, Little DM Jr, ed. 1968. *The Flying Death. Classic Papers and Commentary on Curare. Int. Anesthesiol. Clin.*, Vol. 6. Boston: Little, Brown

25. Carruthers A, Kiene K, Carruthers J. 1996. Botulinum A exotoxin use in clinical dermatology. *J. Am. Acad. Dermatol.* 34:788–97

26. Cato EP, George WL, Finegold SM. 1986. Genus *Clostridium* Prazmowski 1880, 23 AL. In *Bergey's Manual of Systematic Bacteriology*, ed. PHA Sneath, NS Mair, ME Sharpe, JG Holt, 1:1141–200. Baltimore, MD: Williams & Wilkins. 1599 pp.

27. Cherington M. 1998. Clinical spectrum of botulism. *Muscle Nerve* 21:701–10

28. Collins MD, East AK. 1998. Phylogeny and taxonomy of the food-borne pathogen *Clostridium botulinum* and its neurotoxins. *J. Appl. Microbiol.* 84:5–17

29. Collins MD, Lawson PA, Willems A, Cordoba JJ, Fernandez-Garayzabal, et al. 1994. The phylogeny of the genus *Clostridium*: proposal of five new genera and eleven new species combinations. *Int. J. Syst. Bacteriol.* 44:812–26

30. Dickson EC. 1919. Botulism. A clinical and experimental study. *Monogr. Rockefeller Inst. Med. Res.,* no. 8, pp. 1–117. New York: Rockfeller Inst. Med. Res.

31. Doggweiler R, Zermann D-H, Ishigooka M, Schmidt RA. 1998. Botox-induced prostratic involution. *Prostrate* 37:44–50

32. Drachman DB. 1971. Botulinum toxin as a tool for research on the nervous system. In *Neuropoisons. Their Pathophysiological Actions*, Vol. 1, *Poisons of Animal Origin*, ed. LL Simpson, pp. 325–47. New York: Plenum. 361 pp.

33. Dubos RJ, ed. 1948. *Bacterial and Mycotic Infections of Man*. Philadelphia: Lippincott. 785 pp.

34. Dupuy B, Sonenshein AL. 1998. Regulated transcription of *Clostridium difficile* toxin genes. *Mol. Microbiol.* 27:107–20

35. Eleopra R, Tugnoli V, De Grandis D. 1997. The variability in the clinical effect induced by botulinum toxin type A: the role of muscle activity in humans. *Mov. Disord.* 12:89–94

36. Eleopra R, Tugnoli V, Rosetto O, Montecucco C, et al. 1997. Botulinum neurotoxin type C: a novel effective botulinum toxin therapy in humans. *Neurosci. Lett.* 224:91–94

37. Ferrier-Montiel AV, Guitierrez LM, Apland JP, Canaves JM, Gil A, et al. 1998. The 26-mer peptide released from SNAP-25 cleavage by botulinum neurotoxin E inhibits vesicle docking. *FEBS Lett.* 435:84–88

38. Figueiredo DM, Hallewell RA, Chen LL, Fairweather NF, Savitt JM, et al. 1997. Delivery of recombinant tetanus-superoxide dismutase proteins to the central nervous system by retrograde axonal transport. *Exp. Neurol.* 145:546–54

39. Finegold SM, George WL. 1989. *Anaerobic Infections in Humans*. New York: Academic Press. 851 pp.

40. Fishman PS, Savitt JM, Farrand DA. 1990. Enhanced CNS uptake of systemically administered proteins through conjugation with tetanus C-fragment. *J. Neurol. Sci.* 98:311–25

41. Fox ME, Lemmon MJ, Mauchline ML, Davis TO, Giaccia AJ, et al. 1996. Anaerobic bacteria as a delivery system for cancer gene therapy: in vitro activation of 5-fluorocytosine by genetically engineered clostridia. *Gene Ther.* 3:173–78. Corrigendum. *Gene Ther.* 3:741

42. Francis JW, Hosler BA, Brown RH, Fishman PS. 1995. CuZn superoxide dismutase (SOD-1): tetanus toxin fragment C hybrid protein for targeted delivery of SOD-1 to neuronal cells. *J. Biol. Chem.* 270:15432–42

43. Francis PT, Lowe SL, Bowen DM, Jefferys JGR. 1990. Lack of change in neurochemical markers during the post-epileptic phase of intrahippocampal tetanus toxin syndrome in rats. *Epilepsia* 31:697–701

44. Gill DM. 1982. Bacterial toxins: a table of lethal amounts. 1992. *Microbiol. Rev.* 46:86–94

45. Grundbacher FJ. Behring's discovery of diphtheria and tetanus antitoxins. *Immunol. Today* 13:188–90

46. Guyton AC, MacDonald A. 1947. Physiology of botulinus toxin. *Arch. Neurol. Psych.* 57:578–92

47. Habermann E, Dimpfel W. 1973. Distribution of ^{125}I-tetanus toxin and ^{125}I-toxoid in rates with generalized tetanus, as influenced by antitoxin. *Naunyn Schmiedbergs Arch. Pharmacol.* 276:327–40

48. Habermann E, Dryer F. 1986. Clostridial neurotoxins: handling and action at the cellular and molecular level. *Curr. Top. Microbiol. Immunol.* 129:1–179

49. Hackam DJ, Rotstein OD, Sjolin C, Schreiber AD, Trimble WS, Grinstein S. 1998. V-SNARE-dependent secretion is required for phagocytosis. *Proc. Natl. Acad. Sci. USA* 95:11691–96

50. Halpern JL, Loftus JL. 1993. Characterization of the receptor-binding domain of tetanus toxin. *J. Biol. Chem.* 268:11188–92

51. Hanson PI, Heuser JE, Jahn R. 1997. Neurotransmitter release—four years of SNARE complexes. *Curr. Opin. Neurobiol.* 7:310–15

52. Hatheway CL. 1990. Toxigenic clostridia. *Clin. Microbiol. Rev.* 3:67–98

53. Hatheway CL, Dang C. 1994. Immunogenicity of the neurotoxins of *Clostridium botulinum*. In *Therapy with Botulinum Toxin*, ed. J Jankovic, M Hallet, 8:93–107. New York: Marcel Dekker. 608 pp.

54. Hatheway CL, Johnson EA. 1998. *Clostridium*: the spore-bearing anaerobes. In *Topley and Wilson's Microbiology and Microbial Infections*, Vol. 2, *Systematic Bacteriology*, ed. A Balows, BI Duerden, 32:731–82. London: Arnold. 1501 pp.

55. Henderson I, Davis T, Elmore M, Minton NP. 1997. The genetic basis of toxin production in *Clostridium botulinum*. See Ref. 98, pp. 261

56. Himmelweit F, eds. 1956. *Collected Papers of Paul Ehrlich*. New York: Pergamon

57. Holmstedt G, Liljestrand G, eds. 1981. *Readings in Pharmacology*, pp. 70–72. New York: Raven. 395 pp.

58. Hundsberger T, Braun V, Weidman M, Leukel P, Sauerborn M, von Eichel-Streiber C. 1997. Transcription analysis of the genes tcdA-E of the pathogenicity locus of *Clostridium difficile*. *Eur. J. Biochem.* 244:735–42

59. Igarashi M, Kozaki S, Terakawa S, Kawano K, Ide C, Komiya Y. 1996. Growth cone collapse and inhibition of neurite growth by botulinum neurotoxin C1: a t-SNARE is involved in axonal growth. *J. Cell Biol.* 134:205–15

60. Inoue M, Kobayashi M, Kozaki S, Zimmer A, Ueda H. 1998. Nociceptin/orphanin FQ-induced nociceptive responses through substance P release from peripheral nerve endings in mice. *Proc. Natl. Acad. Sci. USA* 95:10949–53

61. Jankovic J, Brin MF. 1991. Therapeutic uses of botulinum toxin. *N. Engl. J. Med.* 324:1186–94

62. Jankovic J, Brin MF. 1997. Botulinum toxin: historical perspective and potential new indications. *Muscle Nerve* 6:S129–S145 (Suppl.)

63. Jankovic J, Hallett M, ed. 1994. *Therapy with Botulinum Toxin*. New York: Marcel Dekker. 608 pp.

64. Johnson EA. 1997. Extrachromosomal virulence determinants in the clostridia. See Ref. 98, pp. 35–48

65. Johnson EA. 1999. Anaerobic fermentations. In *Manual of Industrial Microbiology and Biotechnology*, ed. AL Demain, J Davies, RM Atlas, G Cohen, CC Hershberg, et al. Washington, DC: ASM Press. In press

66. Johnson EA, Goodnough MC. 1998. Botulism. In *Topley and Wilson's Microbiology and Microbial Infections*, Vol. 3, *Bacterial Infections*, ed. L Collier, A Balows, M Sussman, 37:723–41. London: Arnold

67. Johnson JL, Francis G. 1975. Taxonomy of the clostridia: ribosomal ribonucleic acid homologies among the species. *J. Gen. Microbiol.* 88:229–44

68. Just I, Will M, Selzer J, Rax G, von Eichel-Streiber C, et al. 1995. The enterotoxin from *Clostridium difficile* (ToxA) monoglycosylates the Rho proteins. *J. Biol. Chem.* 270:13932–36

69. Klassen CD, ed. 1996. *Casarett and Doull's Toxicology. The Basic Science of Poison*. New York: McGraw-Hill. 1111 pp. 5th ed.

70. Kozaki S, Kamata Y, Watarai S, Nishiki T-I, Mochida S. 1998. Ganglioside Gt1b as a complementary receptor component for *Clostridium botulinum* neurotoxins. *Microb. Pathog.* 25:91–99

71. Kushnaryov VM, Redlich PN, Grossberg E, Sedmak JJ. 1995. Therapeutic use of *Clostridium difficile* toxin A. *U.S. Patent No. 5446672*

72. Lacy DB, Tepp W, Cohen AC, DasGupta BR, Stevens RC. 1998. Crystal structure of botulinum neurotoxin type A and implica-

tions for toxicity. *Nat. Struct. Biol.* 5:898–902

73. Lamanna C. 1959. The most poisonous poison. *Science* 130:763–72

74. Lamanna C, Eklund HW, McElroy OE. 1946. Botulinum toxin (type A); including a study of shaking with chloroform as a step in the isolation procedure. *J. Bacteriol.* 52:1–13

75. Lambin P, Theys J, Landuyt W, Rijken P, Vanderkogel E, et al. 1998. Colonisation of *Clostridium* in the body is restricted to hypoxic and necrotic areas of tumours. *Anaerobe* 4:183–88

76. Lee CL, Hrachovy RA, Smith KL, Frost JD Jr, Swann JW. 1995. Tetanus toxin-induced seizures in infant rats and their effects on hippocampal excitability in childhood. *Brain Res.* 677:97–109

77. Lindsay JA. 1996. *Clostridium perfringens* type A enterotoxin (CPE): more than just explosive diarrhea. *Crit. Rev. Microbiol.* 22:257–77

78. MacLennan JD. 1962. The histotoxic clostridial infections of man. *Bacteriol. Rev.* 26:177–276

79. Marvaud JC, Eisel U, Binz T, Niemann H, Popoff MR. 1998. *tetR* is a positive regulator of the tetanus toxin gene in *Clostridium tetani* and is homologous to *BotR*. *Infect. Immun.* 66:5698–702

80. Marvaud JC, Gibert M, Inoue K, Fujinaga Y, Oguma K, Popoff MR. 1998. *botR/A* is a positive regulator of botulinum neurotoxin and associated non-toxin protein genes in *Clostridium botulinum* A. *Mol. Microbiol.* 29:1009–18

81. Matsuda M. 1989. The structure of tetanus toxin. In *Botulinum Neurotoxin and Tetanus Toxin*, ed. LL Simpson, 4:69–92. San Diego: Academic. 422 pp.

82. Matteoli M, Verderio C, Rossetto O, Iezzi N, Coco S, Schiavo G, et al. 1996. Synaptic vesicle endocytosis mediates the entry of tetanus neurotoxin into hippocampal neurons. *Proc. Natl. Acad. Sci. USA* 93:13310–15

83. McClane BA, Hanna PC, Mietzner TA. 1997. *Clostridium perfringens* type A enterotoxin and methods of preparation and uses as a vaccine and a therapeutic agent. *U.S. Patent No. 5695956*

84. Middlebrook JL. 1989. Cell surface receptors for protein toxins. In *Botulinum Neurotoxin and Tetanus Toxin*, ed. LL Simpson, 5:95–119. San Diego: Academic. 422 pp.

85. Minton NP, Mauchline ML, Lemmon MJ, Brehm JK, Fox M, Michael NP, et al. 1995. Chemotherapeutic tumour targeting using clostridial spores. *FEMS Microbiol. Rev.* 17:357–64

86. Miyamoto O, Minami J, Toyoshima T, Nakamura T. 1998. Neurotoxicity of *Clostridium perfringens* epsilon-toxin for the rat hippocampus via the glutamergic system. *Infect. Immun.* 66:2501–8

87. Mobley JA. 1995. Biological warfare in the twentieth century: lessons from the past, challenges for the future. *Military Med.* 160:547–53

88. Montecucco C, ed. 1995. *Clostridial Neurotoxins. The Molecular Pathogenesis of Tetanus and Botulism.* Berlin: Springer-Verlag. 278 pp.

89. Montecucco C, Papini E, Schiavo G. 1991. Bacterial toxins penetrate cells via a four-step mechanism. *FEBS Lett.* 346:92–98

90. Montecucco C, Schiavo G. 1995. Structure and function of tetanus and botulinum neurotoxins. *Q. Rev. Biophys.* 28:423–72

91. Montesano R, Roth J, Robert A, Orci L. 1982. Non-coated membrane invaginations are involved in binding and internalization of cholera and tetanus toxins. *Nature* 296:651–53

92. Moore R, ed. 1995. *Handbook of Botulinum Toxin Treatment.* Oxford, U.K.: Blackwell Sci. 289 pp.

93. Mukherjee S, Ghosh RN, Maxfield FR. 1997. Endocytosis. *Physiol. Rev.* 77:759–803

94. Niemann H. 1991. Molecular biology of the clostridial neurotoxins. In *Sourcebook*

of Bacterial Toxins, ed. JE Alouf, JH Freer, 15:303–48. London: Academic. 518 pp.

95. Oblatt-Montal M, Ymazaki M, Nelson R, Montal M. 1995. Formation of ion channels in lipid bilayers by a peptide with the predicted transmembrane sequence of botulinum neurotoxin A. *Protein Sci.* 4:1490–97

96. Pellegrini LL, O'Connor V, Lottspeich F, Betz H. 1995. Clostridial neurotoxins compromise the stability of a low energy SNARE complex mediating NSF activation of synaptic vesicle fusion. *EMBO J.* 14:4705–13

97. Rappuoli R, Montecucco C. 1997. *Guidebook to Protein Toxins and Their Use in Cell Biology.* Oxford, U.K.: Oxford Univ. Press. 256 pp.

98. Rood JI, McClane BA, Songer JG, Titball RW, eds. 1997. *The Clostridia. Molecular Biology and Pathogensis.* San Diego/London: Academic. 533 pp.

99. Sakaguchi G. 1983. Clostridium botulinum toxins. *Pharmacol. Ther.* 19:165–94

100. Schantz EJ. 1994. Historical perspective. In *Therapy with Botulinum Toxin*, ed. J Jankovic, M Hallett, pp. xxiii–vi. New York: Marcel-Dekker. 608 pp.

101. Schantz EJ, Johnson EA. 1992. Properties and use of botulinum toxin and other microbial neurotoxins in medicine. *Microbiol. Rev.* 56:80–99

102. Schantz EJ, Johnson EA. 1997. Botulinum toxin: the story of its development for the treatment of human disease. *Perspect. Biol. Med.* 40:317–27

103. Schantz EJ, Kautter DA. 1978. Standardized assay for *Clostridium botulinum* toxins. *J. Assoc. Off. Anal. Chem.* 61:96–99

104. Schantz EJ, Scott AB. 1981. Use of crystalline type A botulinum toxin in medical research. In *Biomedical Aspects of Botulism*, ed. GE Lewis Jr, pp. 143–50. New York: Academic. 366 pp.

105. Scheld WM, Whitley RJ, Durack DT, eds. 1997. *Infections of the Central Nervous System.* Philadelphia: Lippincott-Raven. 1064 pp. 2nd ed.

106. Schmidt JJ, Stafford RG, Bostian KA. 1998. Type A botulinum neurotoxin proteolytic activity: development of competitive inhibitors and implications for substrate specificity at the S_1' binding subsite. *FEBS Lett.* 435:61–64

107. Scott AB. 1980. Botulinum toxin injection into extraocular muscles as an alternative to surgery. *Ophthalmol.* 87:1044–49

108. Scott AB. 1981. Botulinum toxin injection of eye muscles to correct strabismus. *Trans. Am. Ophthalmol. Soc.* 79:734–70

109. Scott AB. 1989. Clostridial toxins as therapeutic agents. In *Botulinum Neurotoxin and Tetanus Toxin*, ed. LL Simpson, 18:399–412. San Diego: Academic. 422 pp.

110. Scott AB, Rosenbaum A, Collins CC. 1973. Pharmacologic weakening of extraocular muscles. *Invest. Ophthalmol.* 12:924–27

111. Scott AB, Suzuki D. 1988. Systemic toxicity of botulinum toxin by intramuscular injection in the monkey. *Mov. Disord.* 3:333–35

112. Shapiro RE, Specht CD, Collins BE, Woods AS, Cotter RJ, Schnaar RL. 1997. Identification of a ganglioside recognition domain of tetanus toxin using a novel ganglioside photoaffinity label. *J. Biol. Chem.* 48:30380–86

113. Simon JD. 1997. Biological terrorism. Preparing to meet the threat. *JAMA* 278:428–30

114. Simpson LL. 1971. The neuroparalytic and hemagglutinating activities of botulinum toxin. In *Neuropoisons. Their Pathophysiogical Actions*, Vol. 1. *Poisons of Animal Origin*, ed. LL Simpson, 14:303–23. New York/London: Plenum. 361 pp.

115. Simpson LL, ed. 1989. *Botulinum Neurotoxin and Tetanus Toxin.* San Diego: Academic. 422 pp.

116. Smith LDS, Sugiyama H. 1988. *Botulism. The Organism, Its Toxins, the Disease.* Springfield, IL: Charles C. Thomas. 171 pp. 2nd ed.

117. Söllner T, Bennett MK, Whiteheart SW, Scheller RH, Rothman JE. 1993. A protein assembly-disassembly pathway in vitro that may correspond to sequential steps of synaptic vesicle docking, activation, and fusion. *Cell* 75:409–18

118. Stackenbrandt E, Rainey FA. 1997. Phylogenetic relationships. See Ref. 98, pp. 3–19119

119. Stanley EF, Mirotznik RR. 1997. Cleavage of syntaxin prevents G-protein regulation of presynaptic calcium channels. *Nature* 385:340–43

120. Stevens RD, Evenson ML, Tepp W, Das-Gupta BR. 1981. Crystallization and preliminary x-ray analysis of botulinum neurotoxin type A. *J. Mol. Biol.* 222:877–90

121. Sugiyama H. 1980. *Clostridium botulinum* neurotoxin. *Microbiol. Rev.* 44:419–48

122. Tsui JKC. 1996. Botulinum toxin as a therapeutic agent. *Pharmacol. Ther.* 72:13–24

123. Umland TC, Wingert LM, Swaminathan S, Furey WF, Schmidt JJ, Sax M. 1997. Structure of the receptor binding fragment Hc of tetanus neurotoxin. *Nature Struct. Biol.* 4:788–92

124. Umland TC, Wingert L, Swaminathan S, Schmidt JJ, Sax M. 1998. Crystallization and preliminary X-ray analysis of tetanus neurotoxin fragment. *Acta Crystallogr. D* 54:273–75

125. van Ermengem EP. 1897. Ueber einen neuen anaeroben Bacillus und seine Beziehungen zum Botulismus. *Z. Hyg. Infekt.krankh.* 26:1–56. (Abridged English transl.: 1979. *Rev. Infect. Dis.* 1:701–19)

126. van Heyningen WE. 1950. *Bacterial Toxins.* Oxford, U.K.: Blackwell Sci. 133 pp.

127. Vile RG, Weiss RA. 1991. Virus receptors as permeases. *Nature* 352:666–67

128. Williams PH, Clarke SC. 1998. Why do microbes have toxins? *Soc. Appl. Gen. Microbiol. Symp. Ser.* 27:1S–6S

129. Willis AT. 1969. *Clostridia of Wound Infections.* London: Butterworths. 470 pp.

130. Withering W. 1941. An account of foxglove and some of its medical uses, with practical remarks on dropsy, and other diseases. In *Classics of Cardiology*, ed. FA Willius, TE Keys. St. Louis, MO: Mosby. 858 pp.

131. Xu T, Binz T, Niemann H, Neher E. 1998. Multiple kinetic components of exocytosis distinguished by neurotoxin sensitivity. *Nature Neurosci.* 1:192–200

Annu. Rev. Microbiol. 1999. 53:577–628

VIRUSES AND APOPTOSIS

Anne Roulston,[1] Richard C. Marcellus,[1] and Philip E. Branton[1,2]

[1]*GeminX Biotechnologies Inc., Montreal, Quebec, Canada H2W 2M9;
e-mail: aroulston@geminx.com, marcellus@geminx.com, and* [2]*Department
of Biochemistry, McGill University, Montreal, Quebec, Canada H3G 1Y6;
e-mail: branton@med.mcgill.ca*

Key Words programmed cell death, viral infection, host defense, death pathways

■ **Abstract** Successful viral replication requires not only the efficient production and spread of progeny, but also evasion of host defense mechanisms that limit replication by killing infected cells. In addition to inducing immune and inflammatory responses, infection by most viruses triggers apoptosis or programmed cell death of the infected cell. This cell response often results as a compulsory or unavoidable byproduct of the action of critical viral replicative functions. In addition, some viruses seem to use apoptosis as a mechanism of cell killing and virus spread. In both cases, successful replication relies on the ability of certain viral products to block or delay apoptosis until sufficient progeny have been produced. Such proteins target a variety of strategic points in the apoptotic pathway. In this review we summarize the great amount of recent information on viruses and apoptosis and offer insights into how this knowledge may be used for future research and novel therapies.

CONTENTS

0066-4227/99/1001-0577$08.00

PROLOGUE

The objectives of all viruses are to infect target cells, replicate large numbers of progeny virions, and spread these progeny to initiate new rounds of infection. Viruses encode highly efficient proteins to optimize such replication; however, target organisms possess both systemic and cell-based defenses to limit virus infection, including immune and inflammatory processes and the execution or suicide of infected cells. In the face of such powerful host defense mechanisms, most viruses have evolved proteins that are able to inhibit or delay protective actions until sufficient viral yields have been produced. Such viral proteins, which have been created either by convergent evolution or by the capture of host sequences encoding entire proteins or individual functional domains, ablate the host response by targeting strategic points in defense pathways. Sometimes such functions are novel, but commonly they exploit mechanisms inherent in normal cellular regulation. Knowledge of these viral mechanisms provides not only means to develop new antiviral agents but also windows on critical molecular events in the cell. In addition, there is a growing awareness that viruses and viral products may be of therapeutic benefit in the treatment of human disease. This review deals with viral strategies against apoptosis or programmed cell death that are instituted by the cell host response or occur as an unavoidable consequence of the action of viral products. In addition, some viruses appear to encode products that actively induce apoptosis as part of an exit strategy to enhance virus spread. In these cases a delicate balancing act between inhibition and induction of apoptosis is performed by combinations of viral products. A number of other important recent reviews on viruses and apoptosis also provide valuable insights (1–10). See Reference 11 for in-depth analysis of the overall biology of all virus classes.

APOPTOSIS

Role of Apoptosis

Apoptosis is a genetically controlled process involved in the regulation of home-ostasis, tissue development, and the immune system by eliminating cells that are no longer useful. Apoptosis also functions by eliminating aberrant cells created by DNA damage or those infected by viral pathogens. Survival factors and internal sensors prevent normal cells from succumbing to apoptosis, but upon receipt of "instructive" signals resulting from damage or infection, death pathways are initiated.

Cell death by apoptosis is characterized by chromatin condensation, DNA fragmentation to nucleosome-sized pieces, membrane blebbing, cell shrinkage, and compartmentalization of the dead cells into membrane-enclosed vesicles or apoptotic bodies. Regrettably, space limitations permit neither a detailed discussion of apoptosis nor the appropriate citation of the hundreds of recent remarkable discoveries in this field.

Genetic Insights from Worms

Genetic analysis of the nematode *Caenorhabditis elegans* and the identification of three genes, *ced4, ced3*, and *ced9*, provided the first evidence for the existence of a carefully controlled apoptotic pathway. Ced-4 and Ced-3 promote cell death during worm development, whereas Ced-9 blocks this process. The role of the more recently identified Egl-1 is to prevent inhibition by Ced-9. Identification of mammalian homologs of these *C. elegans* genes led to insights into the biochemistry of apoptosis. *Ced-9* is related to *Bcl-2*, a gene that was known to be activated in many follicular lymphomas. The Ced-3 product is the major death-inducing factor in *C. elegans* and is a homolog of mammalian caspases, with the implication that apoptosis results from the regulated activation of specific proteolysis. The mechanism of apoptosis is remarkably conserved, although the complexity is far greater in mammals than in lower metazoans.

Caspases

Caspases are central players in apoptosis, because they catalyze many steps in the death pathway by cleavage at specific sites containing aspartic acid (12). Over a dozen caspases have been identified, some of which are important in apoptosis. They are present as inactive proenzymes that are coordinately activated by caspase-specific cleavage. Figure 1 shows that at least two general classes of apoptotic caspases exist. Upstream "initiator" caspases (caspases 2, 8, 9, and probably 10) are present in complexes with other regulatory proteins and are activated by facilitated autocatalysis in response to apoptotic signals. Downstream "effector" caspases (caspases 3, 6, and 7) are activated in a cascade through cleavage by initiator caspases. Effector caspases then cleave a number of specific substrates leading to destruction of cell-cell interactions and nuclear structure, reorganization of the cytoskeleton, inhibition of DNA synthesis, repair and splicing, degradation of DNA, and disintegration of the entire cell contents into apoptotic bodies.

Death Receptors and Upstream Regulation of Apoptosis

The best understood initiating events of apoptosis are those involving the cell surface transmembrane tumor necrosis factor (TNF) receptor (TNFR) superfamily of "death receptors" (13). The two most studied receptors are TNFR1, which is ubiquitously expressed, and Fas (CD95), which is predominantly expressed on activated T cells. Ligand binding to the extracellular domains promotes receptor trimerization. The cytoplasmic domains of these receptors contain death domains

$\longrightarrow$

Figure 1   Role of viral proteins in the induction and suppression of apoptosis. A model for apoptotic pathways has been presented. Lines with arrows denote an activating reaction; lines ending in perpendicular lines denote inhibition of the reaction; dotted lines and ? denote uncertainty of the reaction; thick lines denote physical binding between proteins. DD—death domain. DED—death effector domain. CODE: *Viral Inducers of Apoptosis.* A. Oncogene mediated up-regulation of p53 (Ad-E1A; HBV-HBx;). B. p53 stabilization mediated through the cell cycle (Ad-E1A; SV40-LT; Py-LT; HPV-E7; EBV-EBNA-3C). C. DNA damage (MVM/B19-NS1). D. Sensitivity to TNF (Ad-E1A; Py-MT; HBV-HBx; HCV-core). E. Increased FasL synthesis (HTLV-Tax; HIV-Tat). F. Increased TNFR synthesis (HHV-6/7). G. Induction of permeability transition (Ad-E4orf4). *Viral Suppressors of Apoptosis.* 1. Inhibition of interferon response (vaccinia-E3L/K3L; reovirus-σ3; influenza-NS1; HSV1-$_{\gamma 1}$34.5). 2. Inhibition of TNF membrane signals (Ad-E314.7K/RID; HCV core). 3. TNFR decoys (Cowpox Crm B/C; Shope SFV-T2; Myxoma M-T2). 4. vFLIPs (MCV-MC159/160; BHV4-BORFE2; EHV2-E8; HHV8-K13; H. HVS-orf71). 5. TNF intracellular signaling (Ad-E1B19K; EBV-LMP1; Py-ST). 6. Caspase inhibitors. 6a. (AcNPV/BmNPV p35). 6b. Serpins (Cowpox Crm A; Rabbit pox SPI1/SPI2; Vaccinia-B13R/B22R; Myxoma-Serp2). 6c. vIAPs (Baculovirus Cp-IAP/Op-IAP). 7. Bcl-2 homologues (Ad-E1B19K; ASFV-A179L/p21; HHV8-KSbcl2; HVS-ORF16; EBV-BHRF1; γHV68-M11; SV40-LT?). 8. Inhibitors of p53 (Ad-E1B55K/E4orf6; SV40-LT; HPV-E6; HBV-HBx; HCMV-IE2; HTLV1-Tax). 9. Induction of Bcl-2 expression. (HIV-Tat; EBV-EBNA4, LMP-1). 10. Enhancement of pro-survival signaling (Py-MT).

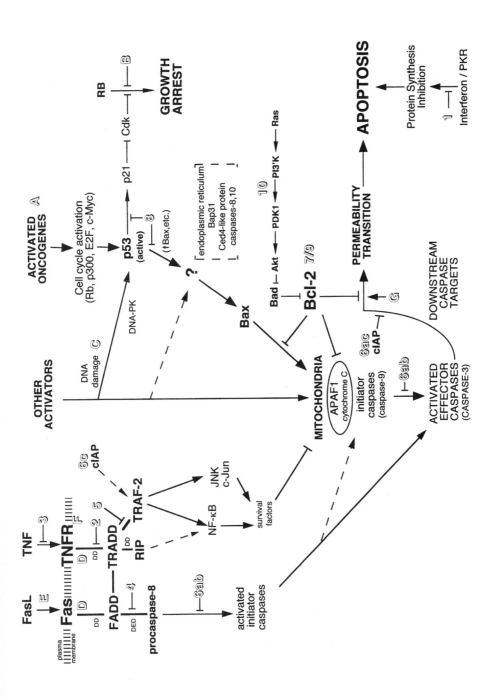

(DD) that dimerize with other DD-containing adapter proteins, such as FADD, TRADD, RIP, and RAIDD. FADD, which binds directly to Fas, also contains death effector domains (DED), which interact with DEDs in the prodomain of procaspase 8. This oligomerization results in the autoproteolytic activation of caspase 8 and subsequent activation of downstream effector caspases. TNFR1 activates caspase 8 in a similar manner, but recruits FADD indirectly through TRADD. Additionally, TNFR1 activates a survival pathway, which is initiated through TRADD binding to RIP and TRAF2. RIP and TRAF2 promote two survival pathways. One mediates activation of the kinases NIK and IKK leading to NF-κB activation, and the other involves the JNK stress kinase cascade leading to c-jun/AP-1 activation. Both NF-κB and AP-1 transactivate prosurvival genes in some cells, although, in others, activation of these pathways is proapoptotic.

p53

In addition to the death receptors, p53, normally a very short-lived protein, is a transcriptional regulator that can induce growth arrest and apoptosis in response to intracellular disruptions resulting from DNA damage, metabolite deprivation, heat shock, hypoxia and activated cellular or viral oncoproteins (14, 15). p53 causes apoptosis when it is activated and stabilized. The turn-over of p53 by ubiquitin-mediated degradation is enhanced by binding with Mdm-2; however, when Arf is expressed, it forms complexes with p53 and Mdm-2 and both stabilizes p53 and enhances its transactivation activity. Arf expression is activated by the transcription factor E2F-1, which in turn is negatively regulated by the RB (retinoblastane susceptibility gene product) tumor suppressor family. In addition to stabilization, p53 is activated by two classes of post-translational modifications. It is phosphorylated by several protein kinases, including DNA-PK, a kinase that is induced by DNA damage. It also interacts with p300, a histone acetyltransferase that seems to acetylate p53 and enhance transactivation activity. These signals activate and stabilize p53, thus altering gene expression. A major target for p53 is *p21*, which encodes an inhibitor of cyclin-dependent kinases (Cdk). Because Cdks are required to inactivate the RB family, growth arrest ensues. For DNA damage, this response likely allows cells the opportunity for DNA repair before entry into S phase. p53 also activates apoptosis through transcriptional control of Bax and likely other apoptotic inducers.

Mitochondrial Controls

Apoptotic signals converge at the mitochondria, with the release of cytochrome *c*, a critical event in the activation of caspases. In activated cells, cytochrome *c* binds to Apaf-1, the mammalian homolog of Ced-4, and through association with procaspase 9 activates the caspase cascade. This critical apoptotic signaling complex, termed the apoptosome, is regulated by Bcl-2 family members, homologs of Ced-9, which include both pro- and antiapoptotic members (reviewed in References 16, 17). The Bcl-2 subfamily, including Bcl-2, Bcl-X_L, Bcl-w, and several others, are potent suppressors of apoptosis. These proteins are characterized by up to four short sequences, termed Bcl-2 homology (BH1-4) domains, and a membrane anchor that directs them to endoplasmic reticulum (ER), mitochondrial, and nuclear membranes. BH domains function in both homomeric and heteromeric binding between family members, as well as in interactions with other regulatory proteins. There are two proapoptotic subfamilies: Bax (Bax, Bak, and Bok) and BH3 (Bik, Blk, Hrk, BNIP3, Bad, Bid, and others). Proapoptotic family members all contain BH3, although the Bax subfamily also contains BH1 and BH2. The balance between the life and death of a cell may depend on the relative levels of pro- and antiapoptotic members. The Bax and BH3 subfamilies appear to function in part by binding and inhibiting the Bcl-2 class, but both subfamilies likely have other functions relating to the stimulation of apoptosis. Some members of the Bcl-2 and Bax groups are capable of pore formation in vitro, suggesting that they may constitute components that regulate mitochondrial pores. Suppressive Bcl-2 family members block Apaf-1–mediated procaspase activation. Proapoptotic members like Bik, appear to release this inhibition.

The ER contains a second regulatory complex leading to caspase activation. A recently described Bcl-2–binding protein, Bap31, can function to regulate the activation of upstream procaspases in cooperation with an unidentified Ced-4–like protein (18, 19). Activated oncogenes including those from viruses, stimulate this control point although the precise mechanism is not understood. Once the initiator and effector caspases have been activated, they trigger a second mitochondrial response, termed the permeability transition (PT). PT involves the opening of a large pore in the mitochondria that disrupts the mitochondrial function and releases additional proapoptotic factors. This point of no return leads to the morphological changes associated with apoptotic cell death.

Survival Mechanisms

The role of the apoptotic machinery in the normal cell cycle remains an open question, but there is growing evidence that cell proliferation and cell death are linked because growth stimulation appears to prime cells for apoptosis (reviewed in Reference 20). In normal cells apoptosis is blocked through the coordinated expression of inhibitors, including survival factors. Certain survival factors, such as insulin-like growth factor-1 (IGF-1), activate the protein kinases Akt and PDK-1 via the Ras-phosphatidyl inositol 3′ kinase pathway, resulting in phosphorylation and inactivation of proapoptotic Bad. Akt also phosphorylates and blocks the activation of procaspase 9. These processes are believed to suppress apoptosis and permit normal cell cycling. In cancerous or virus-infected cells, growth stimulation is usually unaccompanied by normal survival pathways, resulting in cell death. For cancer cells to survive, they must block apoptosis, a task frequently accomplished by inactivation of p53 or overexpression of apoptotic suppressors like Bcl-2. Viruses have evolved a phenomenal array of products that yield similar effects.

THE HOST ANTIVIRAL RESPONSE

Multicellular organisms have a variety of host defenses against viral infection. Systemically, the first line of defense is generally a cell-mediated immune response. This involves the generation of cytokines such as interleukin 1 (IL-1) and TNF, which activate and recruit macrophages, NK cells, and neutrophils that then phagocytize and help to clear the infected cells. The second line of defense is mediated through the humoral immune response and involves cytotoxic T cells, antibodies generated from B cells, and numerous cytokines that regulate their response. For many primary viral infections, serious damage or death can occur before the mounting of an adequate immune response, and, thus, a third level of defense exists at the cellular level. Induction of interferon by the presence of viral nucleic acids (usually double-stranded RNA) can curtail the spread of infection by inducing an antiviral state in neighboring cells. Interferons have two modes of action that lead to the shut down of viral protein synthesis (reviewed in 21, 22). First, interferons upregulate $2'-5'$ oligoadenylate ($2'-5'$ A) synthetase, an enzyme that synthesizes $2'-5'$ A. Translation is inhibited when $2'-5'$ A–dependent RNase L is activated and degrades viral mRNA. Second, interferons also induce the synthesis of PKR, a double-strand–RNA-dependent kinase. When PKR is activated during infection, it phosphorylates the translation initiation factor eIF-2α, leading to inhibition of protein synthesis. More recently, evidence suggests that interferons may also act directly on virally infected cells by inducing apoptosis and preventing viral spread. Overexpression of RNase L in the presence of $2'-5'$ A synthetase can induce apoptosis (23–25). Similarly, overexpression of activated PKR can also cause apoptosis in healthy cells (26–30). Many viruses have evolved mechanisms to cope with cellular antiviral responses, and some have evolved to use apoptosis to their advantage.

INDUCTION OF APOPTOSIS IN VIRUS-INFECTED CELLS

Preamble

This section focuses on the induction of apoptosis by specific viral products as is summarized in Table 1. Many viral proteins perturb normal cell physiology and provide the upstream signals that trigger a death response. Apoptosis may be used by many viruses to kill cells at the end of the infectious cycle. Death by apoptosis offers several advantages for the virus. During apoptosis, the entire cellular contents, including progeny virions, are packaged into membrane-bound apoptotic bodies that are rapidly taken up by surrounding cells. This process severely limits the inflammatory response and allows the infection to spread undetected by the host organism. Virus particles enclosed within apoptotic vesicles are also protected from inactivation by host antibodies and proteases. An apoptotic mechanism of viral spread may best serve those viruses that have not evolved sophisticated

TABLE 1 Viruses and viral gene products that induce apoptosis[a]

Virus family	Virus	Protein (gene)	Mechanism	Figure code[b]	Reference
Adenoviridae	Adenovirus	E1A	Induces sensitivity to TNF	D	36–38
	"	E1A	Up-regulates p53 through stimulation of the cell cycle	A/B	31, 32
	"	E4orf4	Activates PP2A and induces apoptosis	G	45–47
	"	E3-11.6K	Unknown		52, 53
Papovaviridae	SV40	Large T antigen	Up-regulates p53 through stimulation of the cell cycle	A/B	54, 55
	Polyoma virus	Large T antigen	Up-regulates p53 through stimulation of the cell cycle	A/B	56
	"	Middle T antigen	Induces sensitivity to TNF	D	60
	Papilloma virus	E7	Up-regulates p53 through stimulation of the cell cycle	A/B	57–59
	"	E2	Suppresses E6 expression during infection		332
Parvoviridae	MVM, B19	NS1	Causes DNA strand breaks	C	66
	"	NS1	Induces cell cycle arrest in G2		67
Circoviridae	CAV	VP3 "Apoptin"	Unknown		69
Herpesviridae	HHV6/7	Unknown	Up-regulates TNFR1 in infected and bystander cells	F	79, 80
	EBV	EBNA3C	Binds Rb and promotes cell cycle progression	A/B	81

(continued)

TABLE 1 *(continued)*

Virus family	Virus	Protein (gene)	Mechanism	Figure code[b]	Reference
	"	EBNA1	When expressed in the absence of other latent viral proteins-mechanism unknown		82
Baculoviridae	AcNPV	IE-1	Unknown		84, 85
Hepadnaviridae	Hepatitis B virus	HBx	Induces sensitivity to TNF	D	87
	"	HBx	Activates MEKK which upregulates c-myc then p53	A	88
	"	"	Unknown p53-independent mechanism		90
Retroviridae	HIV-1	Tat	Upregulation of the FasL promoter	E	95
	HTLV-1	Tax			106
	HIV-1	Tat	Increases the oxidative state of infected T cells		97
	HTLV-1	Tax			107
	HIV-1	Gp160	Forms aggregates, increases intracellular Ca++		99
	HIV-1	Gp120/gp41	Crosslinks CD4 molecules on uninfected cells		100, 101
	"	Gp120/gp41	Interacts with CXCR4 chemokine receptor in neurons		102, 103
	"	Vpr	Induces cell cycle arrest in G2		105

Togaviridae	Semliki Forest virus	NS region	Unknown		113, 114, 116
	Sindbis virus	E1, E2	Upregulation of NF-κB (in neurons)		118, 120, 121, 123, 124
Flaviviridae	Hepatitis C	Core protein	Binds cytoplasmic domain of TNFR1, inhibits NFκB activation	D	132
Floating Genus: Arterivirus	PRRSV	Orf5 gene product	Unknown		133
Paramyxoviridae	Sendai virus	C protein	Unknown		140
	"	Leader region	May involve the induction of interferon		141
	NDV	Unknown	Induction of interferon and TNF		142, 143
	RSV	"	Induction of interferon and caspase-1		147, 148
Orthomyxoviridae	Influenza A/B	Unknown	Upregulation of Fas expression. May be PKR mediated		149
Reoviridae	Reovirus	σ1 (capsid)	Unknown		151, 152
	Rotavirus	NSP4	Loss of plasma membrane permeability (Necrosis?)		156
Birnaviridae	IBDV	NS and VP2	Unknown		161, 162

[a]Abbreviations: TNF, tumor necrosis factor; SV40, simian virus 40; MVM, Minute virus of mouse; CAV, Chicken anemia virus; HHV, human herpesvirus; EBV, Epstein-Barr virus; AcNPV, *Autographica californica* nuclear polyhedrosis virus; HIV, human immunodeficiency virus; HTLV, human T-cell leukemia virus; NDV, Newcastle disease virus; RSV, respiratory syncytial virus; IBDV, infectious bursal disease virus; PRRSV, Porcine reproductive and respiratory syndrome virus.

[b]See Figure 1.

antiapoptotic mechanisms or mechanisms of immune evasion. However, present evidence for apoptosis as a major mechanism of viral release is becoming convincing in only a few cases, and additional information will be required to establish the existence of this strategy with other viruses.

Adenoviridae

Adenoviruses are small DNA tumor viruses of moderate complexity that infect terminally differentiated epithelial cells. During the course of infection, a number of viral proteins stimulate apoptosis. The E1A products are essential for two early events, the stimulation of S phase to permit the use of cellular machinery for viral DNA replication and the transactivation of early viral gene expression. Induction of DNA synthesis by E1A proteins occurs through the binding and inhibition of the RB family of tumor suppressors, which suppress E2F-activated S-phase gene expression, and by complex formation with the p300/CBP family of histone acetyltransferases. Formation of these complexes causes the accumulation and activation of p53 which, if allowed to function, would cripple viral replication (31–34). p53 stabilization has been linked to the direct upregulation of *arf* expression by E2F-1 (35). Binding of p300 by E1A may affect transactivation by p53, resulting in decreased expression of the *Mdm-2* gene leading to increased p53 stability. p300 is also required for Mdm2-dependent p53 turnover. The mechanism by which E1A expression activates p53 is not known, but it is probably linked to the onset of unscheduled DNA synthesis. There is a clear parallel between the activation of cell cycle progression by E1A and the early products of other small DNA tumor viruses and the sensitization of cancer cells to apoptosis by activated oncogenes. In the absence of well-coordinated growth processes, cells are fatally primed to die unless apoptosis is blocked by downstream suppressive factors.

In addition to the induction of p53, E1A products sensitize cells to killing by TNF and FasL by an unknown mechanism requiring binding of p300/CBP (36–38). Because p300/CBP regulates gene expression, this effect may result from overexpression of factors involved in death receptor signaling or conversely from transcriptional repression of genes acting in prosurvival pathways. The larger 289R E1A protein is more proapoptotic in p53-null cells than is the 243R species that lacks the CR3 transactivation domain (39). Increased levels of both Apaf-1 and caspase 9, as well as release of cytochrome *c*, have been reported after E1A expression (40), which also activates downstream caspases (41).

In the context of a viral infection, the 289R E1A product was found to induce apoptosis by a p53-independent pathway involving the transactivation of another early viral product (42, 43) identified as E4orf4, which can independently induce apoptosis (44–47). This killing can be inhibited by Bcl-2 or Bcl-X_L, but not by zVAD-fmk, a potent pan-inhibitor of caspases, suggesting that E4orf4 may act at the terminal mitochondrial-permeability transition phase (46). The only known function of E4orf4 is to activate protein phosphatase 2A (PP2A) through binding to the Bα subunit, and an extensive genetic analysis has indicated an absolute requirement for this interaction (47; RC Marcellus, H Chan, D Paquette, GC Shore,

PE Branton, unpublished). Although E4orf4 is highly proapoptotic, it acts to limit E4 expression, and its toxic effects are probably also suppressed by adenovirus E1B-19K, a functional homolog of Bcl-2. One proposed role for E4orf4 is to kill cells at the end of the infectious cycle as a means of viral spread. Although mutants defective in E4orf4 cause early disruption of cell monolayers (49), cell death is greatly delayed relative to that of wild-type virus (45). The consequences of PP2A binding by E4orf4 are not known, but apoptosis could be induced directly by heightened dephosphorylation of critical regulators of the apoptotic pathway or indirectly by transcriptional effects caused by dephosphorylation of transcription factors.

Late in viral growth, the E3-11.6K membrane-associated glycoprotein is greatly upregulated (50). It is associated with the Golgi, endoplasmic reticulum (ER), and nuclear membranes (51) and partially colocalizes with E1B-19K (W Wold, personal communication). Viral mutants lacking E3-11.6K accumulate large amounts of virus, but are greatly delayed in cell killing (52, 53). One model is that E3-11.6K may inactivate E1B-19K (or Bcl-2), thus allowing apoptotic killing by E4orf4. Alternatively, E3-11.6 may be involved in activating other cell death pathways. Both of these proteins could ultimately be of use therapeutically to kill aberrant cells, including cancer cells.

Papovaviridae

Like adenoviruses, the small DNA tumor viruses of the *Papovaviridae* family also need to induce cellular DNA synthesis to facilitate replication, and this function accounts for much of their well-studied oncogenic potential in rodents. Members of the *Polyomavirus* genus can cause long-term persistent infections and often progressive multifocal leukoencephalopathy in their respective hosts, whereas some papillomaviruses are linked to cervical neoplasia in humans. Like adenovirus E1A proteins, one major mechanism for induction of S phase by the large-T antigens (LT) of simian virus 40 (SV40) and mouse polyomavirus of the *Polyomavirus* genus and the E7 product of human papillomavirus (HPV) of the *Papillomavirus* genus involves binding and inactivating RB family proteins. Studies in transgenic mice indicated that this complex formation is responsible for p53-dependent apoptosis by SV40 LT (54, 55, 333), polyomavirus LT (56), and HPV E7 (57, 58), and the mechanism likely resembles that observed with adenovirus E1A involving activation of Arf via E2F-1 and stabilization of p53. One report has indicated that HPV E7 causes increased apoptosis in p53-null mice, indicating that it can also induce p53-independent apoptosis (59). The polyomavirus middle-T antigen (MT) induces hypersensitivity to TNF-induced killing that is suppressed by the presence of the small-T antigen (ST) (60). The HPV E2 transcriptional regulator also elicits both p53-dependent and -independent apoptosis, the former by suppressing expression of E6, a viral protein that prevents p53-mediated effects (332). Whether induction of apoptosis by these products serves the infectious process or is simply an inescapable consequence of their expression is not known.

Parvoviridae

Parvoviruses are extremely simple DNA viruses that encode only one or two early products and limited late structural proteins. The autonomous parvoviruses replicate only in dividing host cells because, unlike papovaviruses and adenoviruses, they have no mechanism to induce S phase to allow replication of viral DNA. This feature raised hope that parvovirus infection could be used as a selective agent for killing rapidly growing cancer cells. The most commonly studied parvoviruses of this type are the minute virus of mice (MVM) and human B19 virus. B19 virus infects and causes apoptosis in erythroid progenitor cells leading to aplastic anemia, Fifth disease in children, erythema infectiosum, and arthritis in adults. The apoptosis-inducing activity of these viruses was mapped to the nonstructural proteins NS-1/NS-2 of MVM and NS-1 of B19 (61, 62). NS proteins function as transcription factors, DNA nickases, and helicases (63), and so deletion of these genes prevents viral replication. Expression of NS-1 proteins either individually or in virus infection induces apoptosis, which can be blocked by caspase inhibitors and Bcl-2 (64), and is independent of p53 (65). The mechanism of apoptosis is not known, but may result from the accumulation of DNA strand breaks (66) or by NS-1–induced cell cycle arrest in G2 (67). NS-1 killing may require specific oncogene expression (68), but further studies are required to determine its specificity and potential as an anticancer agent.

Circoviridae

An extremely interesting inducer of apoptosis has been identified in the *Circovirus* chicken anemia virus, which contains a single-stranded circular DNA genome (69). VP3 (now termed apoptin) expression causes the selective apoptotic death of cancer cells but not normal cells (70). Intracellular localization seems to play an important role in this selective killing, because deletion of the nuclear localization signal of apoptin abolishes killing. When expressed in normal cells, apoptin accumulates in the cytoplasm. In cancer cells apoptin is targeted exclusively to the nucleus where it appears to elicit its lethal effects (70). Apoptosis is p53-independent (71) and is not inhibitable by Bcl-2, Bag-1, or caspase inhibitors (72), suggesting that it may act well downstream in the apoptotic pathway. Apoptin shows little homology to other known proteins, and considerable work is underway to study its method of action and its potential as an anticancer therapeutic.

Herpesviridae

Herpesviruses are large DNA viruses that contain highly complex genomes. All herpesviruses remain latent in specific cell types, but the cell type and latency period vary considerably between viruses. Herpesviruses are divided into three subfamilies based on their host range and replication characteristics.

The *Alphaherpesvirinae* have a wide host range and short replication cycle, and these viruses efficiently destroy infected cells. This subfamily includes herpes simplex viruses (HSV), varicella-zoster virus (VZV), and bovine herpesviruses (BHV). HSV-1 induces apoptosis in peripheral blood mononuclear cells (PBMCs) (73), CD4+ T cells (74), and hepatocytes (75). In hepatocytes apoptosis occurs in the absence of de novo protein synthesis, suggesting that a viral structural protein may be involved (75). Similarly, BHV-1 infection induces apoptosis of B-lymphoma cells (76) or PBMCs (77). In both cases, viral particle attachment seems to be sufficient to induce this effect. VZV also induces apoptosis, which is not affected by Bcl-2 overexpression (78).

The *Betaherpesvirinae*, including the cytomegaloviruses and human herpesviruses, have characteristically long replication periods. These viruses are usually maintained in latent form in secretory glands and in lymphoid, kidney, and other tissues. No evidence to date implicates cytomegaloviruses as aggressive inducers of apoptosis, but infection of T cells by the human herpesvirus (HHV)-6/7 virus reportedly upregulates TNFR1, enhancing TNF-mediated cell death of both infected and bystander cells (79, 80).

Finally, the *Gammaherpesvirinae* exhibit the most limited host range because they infect primarily lymphoid cells and are associated with the induction of cellular proliferation. In humans, infection with the Epstein-Barr virus (EBV) can lead to Burkitt's lymphoma (BL) and Hodgkin's disease. Although infections are usually latent in lymphoid cells, infection of epithelial cells and fibroblasts can be lytic. Induction of apoptosis by EBV appears at least in part to result from proliferation-related apoptotic signals, perhaps involving p53. For example, the EBV nuclear antigen EBNA-3C binds RB and promotes cell cycle progression in rat embryo fibroblasts. It seems functionally similar to HPV E7 and adenovirus E1A in that, when combined with the activated *ras* oncogene, EBNA-3C makes cells highly susceptible to cell death (81). Similarly, expression of EBNA1 alone in BL cells also enhances apoptosis by an unknown mechanism (82). These data imply that these viruses may cease to suppress apoptosis when faced with suboptimal replication conditions.

Baculoviridae

To many researchers, insect baculoviruses are best known as agents for the expression of high levels of transgene products; however, they have also proven to be invaluable tools in the study of apoptosis, largely because they encode several interesting apoptotic suppressors. They also induce apoptosis, which is generally inhibitable by Bcl-2 (83). IE-1 is an immediate early protein that functions as a transcription factor involved in both early and late gene expression. IE-1–null mutants are much less cytopathic in some cells (84), and ectopic expression of IE-1 induces a characteristic apoptotic response that is suppressed by the baculovirus p35 and inhibitor of apoptosis (IAP) proteins (85). As with the DNA tumor viruses, induction of apoptosis by p35-deficient baculoviruses is linked to the stimulation of DNA replication, and there is also some evidence that late viral functions may be proapoptotic (86).

Hepadnaviridae

Hepatitis B virus (HBV) is an *Orthohephanavirus* of the *Hepadnaviridae* family of viruses, containing a circular double-stranded DNA genome. HBV causes acute and chronic infections, and a significant proportion of the latter develop into hepatic-cell carcinomas. The HBx protein is of great importance in the suppression of apoptosis; however, expression of HBx either during virus infection or alone has in some cases been found to sensitize cells to apoptosis and increase sensitivity to TNF-induced killing (87–89). This response involves activation of both c-Myc- and MEKK-dependent pathways and requires p53. In the absence of p53, HBx is growth promoting and appears to play a role in the oncogenicity of HBV. Thus, like adenovirus E1A and papovavirus LTs, stimulation of the proliferative response activates p53-dependent apoptotic pathways. However, studies in p53-null transgenic mice indicated that HBx expression can also cause apoptosis in the absence of p53 (90).

Retroviridae

The human immunodeficiency virus (HIV)-1 is the most intensely studied of the human retroviruses and infects mainly CD4+ T cells, macrophages, and some neuronal cells. Disease pathogenesis is characterized by a progressive decline in CD4+ T cells leading to a severe immunodeficiency syndrome known as AIDS. A few years ago, it was discovered that apoptosis might play an important role in the destruction of HIV-1–infected T cells (91–93). Since then, a multitude of viral proteins and mechanisms have been shown to contribute to HIV-induced apoptosis (reviewed in Reference 8).

HIV Tat seems to play an important role at many levels. Tat is a potent transactivator of the viral promoter and is secreted from infected cells. Intracellular Tat expression reportedly causes Jurkat cell apoptosis (94) and sensitizes infected T cells to CD4/TCR-mediated killing through upregulation of the FasL promoter (95). Tat also sensitizes cells to Fas-induced cell death (96) and to TNF-induced apoptosis by downregulating expression of manganese-dependent superoxide dismutase, an enzyme that reduces the intracellular oxidative state (97). The presence of extracellular Tat can also induce apoptosis in uninfected lymphocytes and potentiate TNF-induced apoptosis in neuronal cells, possibly by activation of cyclin-dependent kinases (94, 98).

At least two other HIV products enhance apoptosis. The envelope precursor protein gp160 encoded by the *env* gene is ultimately cleaved to form the gp120 and gp41 viral surface membrane proteins. gp160 forms intracellular aggregates that potentiate a rise in calcium levels (99). Calcium is believed to play a role in apoptotic signaling and may be important in communication between apoptotic machinery in the endoplasmic reticulum and mitochondria. Binding of gp120 to CD4 at the cell surface also primes T cells for apoptosis through CD4 cross-linking (100, 101) and affects neurons through interactions with the CXCR4 chemokine receptor (102, 103). Finally, Vpr is an accessory protein that enhances HIV transcription by causing cell cycle arrest in G2 (104), the cell phase in which viral transcription is most active. Vpr expression has been reported to induce G2 arrest and apoptosis in several cell types (105). It would appear that all of these proapoptotic effects enhance the host response to infection, but result in the fatal loss of both lymphocytes and neurons, contributing to HIV-associated immunodeficiency and dementia. Many of the strategies used by HIV may also be used by the related viruses HIV-2 and simian immunodeficiency virus (SIV), a simian retrovirus.

The human T-cell leukemia virus-1 (HTLV-1), which falls within a different class of retroviruses, also infects T cells, but often leads to adult T-cell leukemia. At least two mechanisms by which the HIV Tat protein induces apoptosis are also caused by the equivalent HTLV-1 transactivator, Tax. Tax expression induces apoptosis in T cells at least in part by transcriptionally upregulating the FasL promoter (106). Tax also enhances CD3/T-cell receptor-mediated killing by increasing the prooxidant state within Tax-expressing T cells (107). Induction of apoptosis is also a common feature of infection by other retroviruses of the genus *Spumavirus*, including the lymphotropic feline syncytial-forming virus (108) and simian foamy virus (SFV)-1 (109), although the mechanism of induction is not understood.

Picornaviridae

Picornaviridae are among the smallest RNA viruses known. They include poliovirus, cardioviruses, rhinoviruses (causing the common cold in humans), and enteroviruses, among others. Thus far, only three viruses of this family have been shown to induce apoptosis. Coxsackievirus B3 (a human enterovirus) causes apoptosis in HeLa cells that is inhibitable by zVAD-fmk (a pan–caspase inhibitor) (110). Theiler's murine encephalomyelitis virus (TMEV) (111) and hepatitis A virus (112) induce apoptosis that correlates with virulence and therefore may determine whether the infection is persistent or acute. In each case little is known about the mechanism or the viral products involved.

Togaviridae

Two genera of animal togaviruses exist: *Alphavirus*, including sindbis virus (SV) and Semliki Forest virus (SFV), and *Rubivirus*, including rubella virus. They are positive-sense single-stranded RNA viruses with small genomes that produce serious to lethal encephalopathies in their hosts, often involving persistent infections (reviewed in Reference 10). Both virulent and avirulent SFV strains induce typical apoptotic hallmarks in rat prostatic adenocarcinoma cells (113), and SFV also induces apoptosis in BRK cells and oligodendrocytes, although cerebral neurons appear to be killed by necrosis (114). Overexpression of Bcl-2 blocks both the apoptotic response and viral replication, promoting the formation of chronically infected cell lines and supporting the view that apoptosis may be required for completion of the infectious cycle. Infection by SFV and SV in some

Bcl-2–overexpressing lines causes caspase-3 activation and cleavage of Bcl-2, removing the BH4 domain. These effects can be blocked by the caspase inhibitor zVAD-fmk or by using a mutant form of Bcl-2 lacking the target caspase-cleavage site (115). Apoptosis appears to be induced by a nonstructural protein and is independent of p53 because apoptosis results in both SFV-infected BRK cells and the p53-null H358a human lung carcinoma cells (116). The induction of apoptosis by an early SFV protein in the absence of particle formation suggests that SFV could represent an interesting new class of reagent for anticancer therapy.

SV also induces apoptosis in most vertebrate cells apart from postmitotic neurons (117, 118). Apoptosis in infected BHK cells, N18 neuroblastomas, and AT-3 rat prostatic adenocarcinoma cells is inhibited by Bcl-2, resulting in long-term persistence (117). In animal models, SV expressing Bcl-2 as a transgene generated much lower mortality and viral replication (119). Avirulent strains of SV kill immature neurons by a Bcl-2–inhibitable process, but virulent strains are not affected by Bcl-2 even in mature neurons (118). Such neurovirulence has been mapped to a single amino acid change in the E2 glycoprotein (118) and correlates with induction of apoptosis (120). Apoptotic induction was mapped to the transmembrane sequence of the E1 and E2 proteins (121). SV-induced apoptosis activates caspases, and treatment with caspase inhibitors increases survival of infected mice without affecting viral replication severely (122). Apoptosis is also inhibited in neuronal cells by pretreatment with NF-κB antagonists (123, 124). NF-κB is induced after infection and although NF-κB stimulates prosurvival signals in most cells, it provides proapoptotic effects in neuronal cells.

Rubella induces apoptotic death of monkey VERO cells, an effect that is inhibitable by zVAD-fmk and enhanced by the kinase inhibitor staurosporine (125). Although some of the togavirus proteins involved in inducing apoptosis have been identified, molecular mechanisms have yet to be determined. Death by apoptosis appears to play a role in viral spread, and its inhibition leads to viral persistence.

Flaviviridae

The *Flaviviridae* are a family of positive-sense single-stranded RNA viruses that generally cause encephalitis, fever, hepatitis C, and other infections. The family is divided into three genera based on serological relatedness. The *Flavivirus* genus includes the Japanese encephalitis virus (JEV) and Dengue virus, which infect mostly neuronal cells. JEV triggers an apoptotic response in BHK and neuroblastoma cells requiring viral replication (126). Bcl-2 is able to block this effect in some cells (126). Dengue virus infection of mice induces apoptosis in neurons of the cortical and hippocampal regions within the CNS (127). In cultured neuroblastoma and hepatoma cells, Dengue-induced cell death may involve activation of NF-κB (128). Apoptosis has also been reported in vivo for classical swine fever virus (129) and in vitro for bovine viral diarrhea virus (130), of the genus *Pestivirus*. Finally, HCV (hepatitis C virus), a member of an unnamed genus, sensitizes several cell types to TNF- and FasL-mediated killing (131, 132). The HCV viral core protein binds to the cytoplasmic domain of TNFR1 and in some cells inhibits NF-κB activation and perhaps prosurvival pathways. However, regardless of the effects on NF-κB, this product stimulates apoptosis.

Coronaviridae

The *Coronaviridae* represent a family of positive-sense single-stranded RNA viruses that are responsible for upper respiratory infections and gastroenteritis in humans and animals. Several members of the genus *Coronaviruses* are known to activate apoptosis, including transmissible gastroenteritis coronavirus (TGEV) (134) and mouse hepatitis virus strain 3 (MHV-3) (135). In both cases, the protein and mechanism remain to be identified. However, for TGEV, inhibition of apoptosis had no effect on viral replication, suggesting that apoptosis may not play a central role in viral spread. Porcine reproductive and respiratory syndrome virus (PRRSV) is a member of a floating genus called *Arterivirus*. The *orf5* gene product p25 of PRRSV is a viral capsid protein expressed late during infection. If ectopically expressed, p25 causes a strong apoptotic response that can be overcome by Bcl-2 (133). This protein may also play an active role in cell death and virus spread.

Rhabdoviridae

Rhabdoviridae are small single-stranded RNA viruses that infect a wide range of cell types and are classified into three genera, the *Vesiculoviruses*, including vesicular stomatitis virus (VSV), the *Lyssaviruses*, otherwise known as the rabies viruses, and the *Ephermeroviruses*. Infection of mouse neuroblastoma cells with rabies virus causes upregulation of Bax and caspase 1, and it induces apoptosis (136). Rabies (137) and VSV (138) also cause apoptosis in Jurkat and HeLa cells, respectively. In both cases, glycoprotein synthesis and production of viral progeny occur simultaneously with the onset of apoptosis. Because apoptosis does not interfere with viral yield, it may serve as a mechanism of viral dissemination.

Paramyxoviridae

Paramyxoviruses are enveloped negative-sense, single-stranded RNA viruses that cause measles, mumps, respiratory, and other flu–like infections in humans. Induction of apoptosis is widespread within this virus family and in many cases may involve the interferon system. The mouse paramyxovirus-1 (Sendai) causes apoptosis by activating caspases 3 and 8 (139). It encodes a protein (C protein) that, when overexpressed alone, causes apoptosis in a variety of cultured cells (140). The first 42 nucleotides of the virus RNA leader sequence are also required for induction of apoptosis by Sendai virus (141), because deletion of this region leads to persistent infection. It is possible that this RNA sequence triggers PKR activation and apoptosis. Avian Newcastle disease virus (NDV) elicits both interferon and TNF-mediated responses (142, 143). NDV exhibits selective replication in human cancer cells, especially those expressing activated Ras, and thus may be a candidate for replicative viral anticancer therapy because it is normally non-pathogenic in humans (144, 145). Measles virus (genus *Morbillivirus*) induces an apoptotic response that involves ICAM-1, because anti-ICAM-1 antibodies reduce the effect (146). Finally, respiratory syncytial virus (RSV) infection enhances neutrophil apoptosis in vivo and upregulates interferon and caspase 1 gene expression (147, 148). More studies are required to determine whether the widespread apoptosis-inducing activity of these viruses serves as a viral exit strategy.

Orthomyxoviridae

Orthomyxoviridae are a large and varied family of RNA viruses and are the most common cause of influenza in humans and animals. Influenza A/B virus infection of HeLa cells induces apoptosis associated with increased Fas expression (149). Inhibition of PKR blocks both Fas expression and apoptosis, suggesting that this response is largely dependent on interferon-induced killing triggered by double-stranded viral RNA. Infection of MDCK cells causes apoptosis late in infection (150). Bcl-2 can inhibit apoptosis, causing a reduction in virus production and spread, suggesting that apoptosis may play a role in virus release.

Reoviridae

Reoviridae contain unique segmented RNA genomes and, although having little effect in humans, cause fever and sometimes serious disease in animals. Reoviruses induce apoptosis in vitro and in vivo, although the degree varies considerably with each strain (reviewed in Reference 5). Expression of the viral σ1 capsid protein can induce apoptosis; however, σ1 is not required for an apoptotic response, suggesting that other mechanisms, possibly involving the M2 gene product, also exist (151–153). Inhibition of reovirus strain T3D-induced apoptosis by Bcl-2 overexpression does not affect viral yield (153). Evidence also demonstrates that reoviruses can induce cell death with characteristics different from classical apoptosis. For example, reovirus strain T3A-induced apoptosis can be blocked by calpain inhibitors (154). LaCrosse virus induces cell death, which is not affected by Bcl-2, in neuronal cells (155). The NSP4 protein of rotavirus causes plasma membrane permeability loss, but this cell death appears distinct from apoptosis (156). Reovirus may also be a potential therapeutic agent for cancer because it replicates and selectively kills Ras-transformed cells (157).

Birnaviridae

Infectious bursal disease virus (IBDV) is a member of the *Avibirnavirus* genus of segmented RNA viruses. IBVD induces apoptosis after infection of cultured avian peripheral blood lymphocytes (158), bursal lymphoid cells of chicken embryos (159), and cultured chicken embryo fibroblasts or VERO cells (160). Apoptosis occurs early after infection and is independent of replication. Expression of the late structural protein VP2 results in apoptosis that is inhibitable by Bcl-2 (161). A replication-competent mutant defective in the early NS protein exhibited reduced cytopathic and apoptotic effects, suggesting that this protein is also involved directly or indirectly (162). The mechanisms of induction are not known.

Induction of apoptosis in most cases represents a major challenge to successful viral replication and is an obstacle to essentially all viruses. Clearly viruses cannot thrive unless they possess sufficient mechanisms to prevent, delay, or otherwise manipulate the apoptotic response.

SUPPRESSION OF APOPTOSIS BY VIRAL PRODUCTS

Viruses have developed a range of strategies to defend themselves against the host immune response (reviewed in 4, 163, 164) and apoptosis as summarized in Table 2. Many target interferon and inflammatory pathways, yet others seem to target specific stages of the downstream apoptotic pathway by using proteins that often mimic or counteract host cell functions.

Inhibition of the Interferon Response

Several viruses have mechanisms to protect against the apoptotic and antiviral effects of interferon. The poxvirus vaccinia encodes at least two proteins that inhibit PKR action. E3L is a double-stranded RNA-binding protein that prevents the interaction of PKR with viral RNA (28, 165–167). A similar function has been attributed to the $\sigma 3$ protein of reovirus (168). E3L can inhibit apoptosis induced by infection with an E3L mutant virus or by overexpression of activated PKR (28, 167). E3L can also inhibit apoptosis induced by RNase L and $2'$-$5'$A synthetase overexpression (25). A second vaccinia protein, K3L, is less effective than E3L in inhibiting PKR, and, because it resembles eIF-2α, it may interfere with PKR–eIF-2α interactions (166). Inhibitors from other viruses include influenza A NS1, which binds to PKR and prevents phosphorylation and inactivation of eIF-2α (169). The herpesvirus HSV-1 protein $\gamma_1 34.5$ interacts with and redirects protein phosphatase 1 to dephosphorylate eIF-2α, thus preventing the PKR-stimulated inhibition of protein synthesis and induction of apoptosis (26, 170). Additional strategies used by viruses to overcome the interferon system have recently been reviewed (4, 163, 164), and likely other strategies are still to be uncovered.

Inhibition of TNFR and Fas Responses

Viruses have evolved a myriad of proteins to inhibit various steps in the inflammatory and apoptotic response pathways mediated by TNF and FasL to protect infected cells from early death.

Plasma Membrane Signaling   Human adenoviruses use three E3 products to block the effects of TNF and FasL. Some years ago, cells infected with E3-deleted adenoviruses were noted to be much more sensitive to death by TNF and other

TABLE 2 Mechanisms viruses use to inhibit apoptosis[a]

Mechanism	Details	Viral product	Virus	Figure code[b]	Reference
Inhibition of the interferon response	Inhibition of PKR	E3L	Vaccinia virus	1	28, 165–167
	"	K3L	"	1	166
	"	σ3	Reovirus	1	168
	"	NS1	Influenza A	1	169
	Inhibits RNAse L	E3L	Vaccinia virus	1	25
	Dephosphorylates eIF-2α	γ₁34.5	HSV-1	1	26, 170
Inhibition of Fas and TNF responses	TNF receptor mimics	Crm B	Cowpox virus	3	180
	"	Crm C	Cowpox virus	3	181
	"	M-T2	myxomavirus	3	182, 184
	"	SFV-T2	Shope fibroma virus	3	183
Death signaling factors	VFLIPs	MC159/MC160	MCV	4	185–188
	"	BORFE2 (E1.1)	BHV-4	4	
	"	E8	EHV-2	4	
	"	K13	HHV-8	4	188
	"	Orf 71	HVS	4	188
	Interacts with TRAFs and TRADD	LMP-1	EBV	5	192–194
	Interacts with TNF receptor cytoplasmic domain	Core protein	HCV	2	200

	Function	Protein	Virus		References
	Interacts with FADD	E1B-19K	Adenovirus	4	190
	Blocks phospholipase A2 action	E3-14.7K	Adenovirus	2	174, 175
	Blocks phospholipase A2 action	RID	Adenovirus	2	175, 177
	Degrades Fas	RID	Adenovirus	2	179
	Blocks TNF signaling	ST	Polyomavirus	5	60
	Blocks Fas signaling	LT	SV40		201
Caspase-inhibitors	Caspase-substrate and inhibitor	p35	AcNPV BmNPV	6a	203, 204, 213, 214
	Serpin	CrmA	Cowpox virus	6b	218, 219
	"	SPI-1	Rabbitpox virus	6b	222
	"	SPI-2	Rabbitpox virus	6b	220
	"	B13R	Vaccinia virus	6b	221
	"	B22R	Vaccinia virus	6b	221
	"	Serp-2	Myxomavirus	6b	223, 224
	IAP	Cp-IAP	CpGV	6c	227
	"	Op-IAP	OpMNPV	6c	228, 235
Bcl-2 homologues	Bcl-2-like anti-apoptotic activity	E1B-19K	Adenovirus	7	251–255
	"	A179L	ASFV (BA71V)	7	258, 259
	"	p21	ASFV (Malawi)	7	260
	"	KSbcl-2	HHV-8	7	261
	"	ORF16	HVS	7	262–264
	"	BHRF1	EBV	7	265–269

(continued)

TABLE 2 *(continued)*

Mechanism	Details	Viral product	Virus	Figure code[b]	Reference
Inhibitors of p53	"	M11	γ-HV68	7	270
	"	SV40 LT?	SV40	7	271
	Binds to p53, represses transactivation and blocks apoptosis	E1B-55K	Adenovirus	8	45, 273–276, 276, 278
	Binds to p53, represses transactivation and blocks apoptosis	E4orf6	Adenovirus	8	286–288
	Degrades p53	E1B-55K/E4orf6 complex	Adenovirus	8	287–290
	Binds p53 and blocks apoptosis	LT	SV40	8	333
	Degrades p53 through ubiquitin pathway	E6	HPV	8	57, 59, 294, 295, 297
	Binds p53 and inhibits transactivation and apoptosis	HBx	Hepatitis B virus	8	300, 301
	Binds to p53 and inactivates	IE2	HCMV	8	304
	Blocks p53 mediated transcription	Tax	HTLV-1	8	309
	Sequesters p53 in cytoplasm	Unknown	HCMV	8	305
Transcriptional regulation	Blocks p53 transcription	Core protein	HCV	8	316, 317
	Up-regulates Bcl-2	EBNA-4	EBV	9	306
	Up-regulates Bcl-2 family members	LMP-1	EBV	9	196–199

Mechanism		Protein	Virus		Reference
	Up-regulates Bcl-2	Tat	HIV-1	9	314, 315
	Suppresses TcR-induced apoptosis by inhibiting NF-κB activity	Vpr	HIV-1		313
Inhibition of cell cycle progression		LRP	BHV-1		307, 308
Enhancement of survival signals	Activates PI-3K leading to Bad phosphorylation via Akt	MT	Polyomavirus	10	334, 335
Other mechanisms	Anti-oxidant	MC066L	MCV		319
Unknown mechanisms		CHOhr (CP77)	Vaccinia virus		320
		M-T4	Myxomavirus		321
		M-T5	Myxomavirus		323
		M11L	Myxomavirus		322
		N1R	Shope fibroma virus		324
		L*	TMEV		325
		Us3	HSV-1		326
		MEQ	Marek's disease virus		328
		E^{rns}	CSFV		329

[a] Abbreviations: TNF, tumor necrosis factor; BHV, bovine herpesvirus; EHV, equine herpesvirus; HHV, human herpesvirus; EBV, Epstein-Barr virus; HCV, Hepatitis C virus; AcNPV, *Autographica californica* nuclear polyhedrosis virus; BmNPV, *Bombyx mori* nuclear polyhedrosis virus; CpGV, *Cydia pomonella* granulosis virus; OpMNPV, *Orgyia pseudotsugata* polyhedrosis virus; ASFV, African swine fever virus; SV40, simian virus 40; HPV, human papillomavirus; HCMV, Human cytomegalovirus; HTLV, human T-cell leukemia virus; HIV, human immunodeficiency virus; HSV, herpes simplex virus; CSFV, Classical swine fever virus; TMEV, Theiler's murine encephalomyelitis virus; MCV, Molluscum contagiosum virus; HVS, Herpesvirus saimiri.

[b] See Figure 1.

stresses than were wild-type–infected cells (171, 172). This protection was targeted towards specific membrane-associated biochemical processes (173) and was mapped in part to the E3-14.7K protein (171), which inhibits the TNFR-stimulated release of arachidonic acid by phospholipase A_2 (PLA_2) (174, 175). Additionally, the E3-14.5K and E3-10.4K proteins form a complex termed RID (176) that blocks apoptosis by inhibiting the translocation of PLA_2 from the cytoplasm to the plasma membrane in response to TNFα (175, 177). RID also causes the rapid internalization and lysosomal degradation of Fas (178, 179).

Mimics of Death Receptors   Some viruses, especially poxviruses, can successfully evade the inflammatory response by synthesizing proteins that mimic cytokine and death receptors (reviewed in References 163, 164). Cowpox produces two secreted proteins, CrmB and CrmC, that resemble the extracellular domains of TNFRs and compete effectively for TNF ligand binding. CrmB resembles TNFR2 and is able to compete for binding of both TNFα and TNFβ (180). CrmC is expressed late in infection and binds specifically to TNFα (181). The M-T2 protein of myxomavirus also resembles the TNFR ligand-binding domain. It is secreted early during infection and binds to TNFα with an affinity similar to that of TNFR (182). A similar protein, SFV-T2 secreted by the Shope fibroma virus, also binds specifically to both TNFα and TNFβ (183). Each of these homologs may play a dual role during infection, acting as competitive anti-inflammatory molecules as well as inhibitors of TNF-induced apoptosis (182, 183). Owing to the host immune responses, the replication of viruses lacking M-T2 is severely attenuated (181). Although M-T2 is largely secreted early in infection, full-length and a truncated species of M-T2 become more efficiently retained in the cytosol at later times and are therefore unavailable for TNF binding. Intracellular M-T2 is unable to bind TNF, yet still provides full protection against apoptosis (184), suggesting that it has additional antiapoptotic functions late during infection.

Mimics of Death Signaling Factors (vFLIPS)   Both poxviruses and herpesviruses encode proteins containing high homology to DED domains responsible for linking death receptors and the FADD adapter protein to procaspase 8. These proteins are collectively termed vFLIPs and inhibit apoptosis induced by FasL and TNF. MC159 and MC160, products of molluscum contagiosum poxvirus (MCV), bind FADD and inhibit recruitment of procaspase 8 to the receptor complex. BORFE2 (E1.1) and E8 of the herpesviruses BHV-4 and EHV-2, respectively, bind to procaspase 8 and also inhibit caspase activation in response to death receptor signaling (185–188). Two other herpesviruses, Kaposi's herpesvirus (HHV-8) and herpesvirus saimiri (HVS), encode proteins K13 and orf71, respectively, which also contain DED motifs, suggesting that they may function in a similar manner (188).

Interactions with Signaling Factors   A related strategy used by some viruses involves interactions with death-signaling factors, including FADD, TRADD, and

TRAF, to block proapoptotic signaling or to promote prosurvival pathways. As discussed below, the adenovirus E1B-19K protein acts at least partially as a functional analog of Bcl-2, but, whereas Bcl-2 fails to suppress the TNFR and Fas pathways, E1B-19K is reasonably effective (189). This effect appears to be caused by an interaction with FADD that blunts signal transduction from these receptors (190). The EBV LMP-1 protein largely targets TRAF family members, and it is one of several EBV products involved in latency (191). It is an integral membrane protein containing six membrane-spanning domains, and it clusters in large aggregates in the plasma membranes during latent infection. One cytoplasmic domain of LMP-1 interacts with members of the TRAF family and promotes NF-κB prosurvival pathways (192–194). A second binds to TRADD, the adapter protein that couples RIP-TRAF pathways to the TNFR (195). LMP-1 may promote latency by increasing the expression of antiapoptotic proteins, including A20 (196), Bcl-2 (197), and Mcl-1 (198) and by downregulating synthesis of c-Myc (199).

Other Mechanisms   Other less well-characterized viral strategies to inhibit the inflammatory response exist. For example, the core protein of HCV was described above to sensitize cells to TNF-mediated apoptosis; however, a recent report argues that it also has a protective effect (200). Hepatocytes from transgenic mice expressing SV40 LT exhibit resistance to killing by anti-Fas antibody, and a protein kinase C pathway has been suggested to induce this protective effect (201). Polyoma ST lessens both MT -induced sensitivity to TNF and TNF-induced killing (60). A major function of ST is to bind and inhibit PP2A, and this effect may result from prevention of the inhibitory dephosphorylation of components of prosurvival pathways. Additionally polyoma MT directly activates pro survival signaling pathways including PI3-kinase which leads to Bad phosphorylation and inhibition of its proapoptotic function (334, 335). Finally, an unknown product of HSV-2 (but not HSV-1) inhibits surface expression of FasL by retaining it in the cytoplasm (202).

Caspase Inhibitors

A number of viral products act either directly or indirectly to inhibit caspase activity as a means of preventing apoptosis. Such inhibitors include many examples of the serpin family of protease inhibitors, homologs of the cellular IAP (cIAP) family, and the unique p35 protein.

Baculovirus p35   The p35 protein of *Autographica californica* nuclear polyhedrosis virus (AcNPV) was shown to block the apoptotic effects of AcNPV infection (203). Another baculovirus, *Bombyx mori* NPV (BmNPV), also encodes a comparable p35 product (204). Isolated p35 blocks neuronal apoptosis induced by a number of stimuli (205), interferes with normal developmental apoptosis in *Drosophila melanogaster* (206), and partially complements for *ced9* mutations in *C. elegans* (207). It is a cytosolic early protein that accumulates during infection (208). Bcl-2 is unable to prevent AcNPV-induced apoptosis by p35-defective mutants, and dominant-negative forms of p35 block suppression by wild-type p35 (209). p35 was shown to block both TNF- and FasL-induced killing, with a noted absence of caspase-induced cleavage of a prime downstream target, PARP (210, 211). p35 also has the ability to enhance cell transformation, presumably by inhibiting apoptosis (212). p35 is cleaved by Ced-3 and a number of vertebrate caspases, but instead of being released, it remains bound to the caspase in a stable inhibitory complex (213, 214). Mutation of the caspase cleavage site in p35 abolishes both p35 cleavage and caspase inhibition (215). Complex formation involves direct binding to the active site of a number of caspases, including 1, 3, 6, 7, 8, and 10 of both the initiator and effector classes (216).

Serpins   Many poxviruses encode homologs of mammalian serpins, a family of chymotrypsinlike serine proteinases (217). The best known viral serpin is CrmA of cowpox virus, which inhibits FasL- and TNF-induced apoptosis by binding to and inhibiting caspases 1 and 8 (218, 219). Other serpins include SPI-2 of rabbitpox (220) and B13R of vaccinia virus (221), which are both highly similar to CrmA and can also inhibit Fas- and TNF-induced apoptosis. Other serpins such as SPI-1 (rabbitpox) and B22R (vaccinia) share 45% amino acid identity with CrmA and SPI-2; however, the amino acids in the reactive center are different from those in CrmA/SPI-2, suggesting that they may have different specificities (221, 222). Infection of cells with a virus containing an SPI-1 mutation causes apoptosis (222), yet SPI-1 cannot protect HeLa cells from Fas- or TNF-induced apoptosis. Infection by a myxomavirus lacking Serp2 results in more rapid apoptosis and inflammatory reactions at the site of infection, suggesting that this gene may increase viral virulence by impairing host inflammatory responses and apoptosis (223, 224). In vitro, Serp2 can inhibit caspase 1, a protease involved in IL-1β processing and the inflammatory response (223). Myxomavirus and rabbitpox virus also encode Serp1 and SPI-4, respectively. These serpins, when deleted, yield attenuated strains, in vivo, mainly because the host immune response against infection proceeds unimpaired (225, 226). Interestingly, unlike p35, viral serpins exhibit specificity and are not pan-inhibitors of all caspases.

vIAPs   Both baculoviruses and certain poxviruses contain functional homologs of cIAPs. *Cydia pomonella* granulosis virus (CpGP) and *Orgyia pseudotsugata* nucleo polyhedrosis virus (OpMNPV) both encode IAP products (Cp-IAP, OpMNPV-IAP) which are able to block apoptosis induced by p35-defective mutants of (AcNPV) (227–229). Although AcNPV also encodes a homolog of IAP, it is unable to inhibit apoptosis caused by the loss of p35. The role of cIAPs in apoptosis was first revealed through their homology to baculovirus IAPs (230–232). IAPs appear to act just upstream of p35 (which can block activated caspases) by binding and inhibiting the effector procaspases and caspase 9, but not activated effector caspases (233–235). cIAPs are known to associate with TRAF family members (230), but also are effective against induction of apoptosis by non-TNFR-related processes (236–238). IAPs contain a conserved RING finger and one or more cys/his motifs, termed baculovirus IAP repeats (BIR) (227, 228). The RING finger motif and at least one BIR domain are important for antiapoptotic activity (239, 240). The ASFV poxviruses also produce proteins (ORFA224L from the Malawi strain and ORF4CL from BA71V) with single BIR motifs; however, viruses with these genes mutated appear to survive normally in infected macrophages (241, 242). The mechanism of caspase inhibition by IAPs is not currently understood.

Bcl-2 Homologs

Bcl-2 homologs containing BH-1, BH-2, and the membrane anchor domains have been identified in several viruses and seem to function as apoptotic suppressors. Early studies noted that adenovirus E1B-19K–defective mutants exhibited a cyt/deg phenotype characterized by enhanced cytopathic effect and degradation of viral and cellular DNA, identified as apoptosis (243–249). The growth-promoting effects of E1A can transform cells in the presence of E1B-19K or E1B-55K (249, 250), and this cooperation is caused by the ability of the E1B proteins (or Bcl-2) to block E1A-induced apoptosis (251). E1B-19K contains a functional BH1 domain that interacts with a series of proteins (termed NIPs) that also bind Bcl-2. E1B-19K also interacts with the proapoptotic Bcl-2 family members Bax, Bak, and Bik (251–255), but does not interact with any antiapoptotic members. This indicates that it may function like Bcl-2 to bind and inhibit proapoptotic family members. E1B-19K is localized primarily in cytoplasmic membranes and in the nuclear membrane in association with nuclear lamin A (256, 257). It is interesting that little or no E1B-19K resides in mitochondria, the proposed major site of action of Bcl-2.

Other virally encoded Bcl-2 homologs include the ASFV poxvirus A179L protein of the BA71V strain (258, 259) and p21 of the Malawi strain (260), which can block cell death in several systems. It is interesting that p21 does not contain a membrane anchor sequence. The herpesvirus product KSbcl-2 of HHV-8 can also protect cells from apoptosis, but does not dimerize with other Bcl-2 family members (261). ORF16 of HVS binds Bax and Bak (262) and can protect T cells from a variety of apoptotic stimuli, but like Bcl-2 cannot block death receptor-mediated apoptosis (263, 264). Finally, BHRF1, encoded by EBV, is a more broad-spectrum inhibitor, protecting against γ-radiation, chemotherapeutic agents, c-*myc* overexpression, and death receptor-induced apoptosis (265–269). M11 from γ-HV68 has homology to Bcl-2, but antiapoptotic activity has not yet been demonstrated (270). A region towards the carboxy terminus of SV40 LT has been found to suppress apoptosis induced by a LT amino terminal fragment, and this region shows weak homology to E1B-19K, but its mechanism of action is not known (271). This important cellular apoptotic control point is a target of many viral families.

Inhibitors of p53 and Transcriptional Regulation

In addition to interfering with existing apoptotic pathways, viruses affect apoptosis at the transcriptional level. Virally encoded transcription factors protect against cell death by directly up- or downregulating genes involved in apoptosis. For many viruses, replication depends on the induction of S phase, which often leads to increased levels of the transcription factor p53. Activated p53 leads to apoptosis and, therefore, these viruses have evolved specific means of countering p53 to maximize progeny production.

Adenoviruses   Human adenoviruses encode two products that regulate p53. E1B-55K binds p53 within its acidic transcriptional-activation domain, and mutants that fail to do so are unable to cooperate with E1A to transform cells (272–275). E1B-55K blocks p53-mediated gene expression by tethering a transcriptional repression domain to p53 (274, 276–278). The E1B-55K repression domain lies at the carboxy terminus near three phosphorylation sites that upregulate repression (277, 278; L Assefi & PE Branton, unpublished data). The mechanism of repression is not known, but does not involve histones (279). Binding of E1B-55K blocks p53-dependent apoptosis (278, 280), but actually enhances p53 DNA-binding activity, an effect that likely ensures that p53-dependent promoters are repressed below basal levels (279). The failure of E1B-55K$^-$ viral mutants to inactivate p53 forms the basis for the clinical use of such a mutant by ONYX Pharmaceuticals Inc. for the selective killing of p53-null tumors (281, 282). Selective viral replication based on p53 status is controversial (283, 284), but continuing clinical trials have been encouraging (285).

The E4orf6 protein binds to p53 in the carboxy terminal region, resulting in inhibition of p53 transactivational activity (286) and cooperates with E1A in cellular transformation (287, 288). E4orf6 also binds to E1B-55K, and this complex causes the rapid degradation of p53 by an unknown mechanism (287–290) requiring binding between all three proteins (J Roth, C König, S Wienzet, S Weigel, S Ristea & M Dobbelstein, 1998, personal communication; S Rubenwolf, M Nevels, H Wolf & T Dobner, 1998, personal communication). Finally, although E1A induces the stabilization and activation of p53, it also sequesters p300, which is required for p53-mediated transactivation of many promoters, and this interaction may contribute to the suppression of p53 function.

Simian Virus 40 and Human Papillomavirus   Like adenovirus, SV40 and HPV also need to inactivate p53, but they use different mechanisms. SV40 LT binds to the DNA-binding domain of p53, preventing p53-mediated gene activation and leading to the accumulation of high levels of nonfunctional p53 complexes (291–293). The HPV E6 protein also binds p53 (294) and targets it for degradation by the ubiquitin pathway (295), a function seen largely in the highly tumorigenic strains of HPV (296). Turnover is facilitated by concomitant binding of the cellular E6AP-100K protein (a ubiquitin ligase) linking p53 directly with the ubiquitin machinery (297). Studies in transgenic mice expressing E6 or E6 and E7 indicate that E6 blocks apoptosis (57, 59). E6 has also been reported to inhibit p53 activity independently of degradation (298), possibly via binding to a second site near the p53 carboxy terminus (299).

Hepatitis B Virus   As mentioned, the HBx protein of HBV activates p53-dependent apoptosis; however, like SV40 LT, it is also a potent inhibitor of p53. HBx binds to the carboxy terminal portion of p53 and inhibits both DNA-binding and transactivation, as evidenced by suppression of p21 Waf-1 expression (300). This binding also seems to sequester p53 in the cytoplasm, away from its DNA targets (301). Like the small DNA tumor viruses, HBV survives the effects of its proliferative functions by inhibiting p53.

Herpesviruses   Herpesvirus infection inhibits the activities of p53 at multiple levels. During human cytomegalovirus (HCMV) infection, IE1 and IE2 act synergistically as transcriptional activators of early viral gene expression, and then, in an autoregulatory loop, IE2 represses early gene transcription to allow for late gene expression. The IE1 and IE2 proteins block apoptosis induced by TNF and E1A (302). Little is known about the mechanism of IE1, but IE2 seems to target p53. IE2 causes an accumulation of p53 by transactivation of the p53 promoter (303). However, IE2 also inactivates p53 through direct binding, linking a repression domain that can partially inhibit p53-mediated gene expression (304). HCMV infection of some cells also leads to sequestration of p53 in the cytoplasm, although the mechanism of this effect is not understood (305). Other herpesviruses, like EBV, encode a protein (EBNA-4) that represses apoptosis by upregulating Bcl-2 (306). Although not directly related to transcription, the BHV-1 virus latency related protein (LRP) promotes latency in infected postmitotic neurons by blocking the cell cycle, thus avoiding induction of apoptosis (307, 308). This effect appears to result from interactions with cyclin A and cyclin E/Cdk complexes resulting in cell cycle arrest.

Retroviruses   As previously mentioned, HTLV-1 infection of T cells leads to adult T-cell leukemia after prolonged infection with the virus. The Tax protein, a major player involved in transformation, causes the stabilization of p53, but it also interferes with p53 function (309). DNA binding and intracellular localization are not affected, but in some cells Tax blocks p53-mediated transcription via a mechanism that is dependent on the ATF/CREB pathway. Tax also represses p53-mediated expression from the Bax promoter, causing decreased Bax expression and Bax-mediated apoptosis (310). Other transcription factors such as NF-κB associate with Tax and are targeted to cellular genes regulated by NF-κB, such as IL-2, which provides a potent T-cell growth factor (reviewed in 311).

Although HIV infection causes apoptosis in T cells, other cells such as myeloid cells remain chronically infected and serve as important reservoirs of HIV. Recent evidence suggests that persistent NF-κB activation may protect these cells from TNF-induced apoptosis (312). In contrast, the Vpr protein of HIV-1 causes apoptosis in resting T cells, but suppresses T-cell receptor-mediated apoptosis, and this effect has been linked to suppression of transcription factor NF-κB activity (313). HIV-1 also protects cells through Tat, which upregulates expression of Bcl-2 in lymphoid, epithelial, and neuronal cells, although the mechanism of this effect is still unclear (314, 315).

Flaviviruses   Finally, the hepatitis C virus core protein binds to and represses the p53 promoter, thus blocking p53 synthesis (316, 317). This effect may play a role in survival of hepatocarcinoma cells transformed by HCV. We expect other examples of transcriptional control to be uncovered in the near future.

Other Mechanisms

Several other viruses suppress apoptosis by a variety of unique mechanisms. Infection by the arenavirus lymphocytic choriomeningitis virus (LCMV) induces the translocation of clusters of promyelocytic leukemia protein (termed PML bodies) from the nucleus to the cytoplasm (318). PML bodies appear to be involved in transcription, and this effect makes infected cells more resistant to apoptosis induced by serum starvation. Virally induced translocation may deregulate apoptotic machinery and promote latency. The MC066L product of MCV is a selenoprotein with homology to glutathione peroxidase and acts as an antioxidant (319). MC066L has been found to protect human keratinocytes against apoptosis induced by UV irradiation and hydrogen peroxide and may play a role in survival of skin neoplasms in infected patients.

Unknown Territory

Many other virus products have been identified for which the mechanism of suppression of apoptosis is unknown. Much of this information is derived from studies showing enhanced apoptosis in viral mutants containing deletions of specific coding sequences.

Poxviruses are well represented in this category. The host range proteins of several poxviruses, including the cowpox CHOhr protein (CP77), contain ankyrin-like repeats that provide interfaces for protein-protein interaction and are found in many cytosolic and cytoskeletal proteins. Introduction of CHOhr into vaccinia virus allows replication and inhibits apoptosis in the normally nonpermissive Cto cell line (320). Myxomavirus encodes three virulence factors, M-T4 (321), M11L (322), and M-T5, which share homology with CHOhr and other host range proteins (323). In each of these cases, infection with relevant deletion mutants leads to greatly enhanced apoptosis. The RING finger-containing NIR protein of Shope fibroma virus reduces apoptosis, and it is localized to viral factories where it binds single- and double-stranded DNA (324).

Other viral families encode protective proteins as well. Theiler's murine encephalomyelitis virus (TMEV, *Picornaviridae*) encodes protein L* which promotes persistence and reduces apoptosis in infected macrophages (325). Mutants affecting the U_S3 product of HSV-1, a serine/threonine kinase active against arginine-rich sites, cause greatly enhanced apoptosis relative to wild-type HSV-1 infection (326). One or more HSV-1 proteins, possibly including structural proteins or the alpha 4 product, provide resistance to sorbitol-induced apoptosis (327). Marek's disease herpesvirus encodes MEQ, a bZIP transcription factor that has transforming activity. MEQ-transformed Rat-2 cells express high levels of Bcl-2, whereas Bax synthesis is inhibited. These cells are protected from apoptosis induced by TNF, C2-ceramide, UV irradiation, and serum withdrawal (328). Classical swine fever virus of the *Flaviviridae* family contains an envelope glycoprotein, termed E^{rns}, that is indispensable for viral attachment and entry, and it is secreted from infected cells. It is related to a family of RNases found in plants and fungi, and mutations affecting RNase activity cause enhanced apoptosis in swine kidney-infected cells (329). In contrast, a previous report suggested that this product is proapoptotic in lymphocytes (330). Finally, infection of HeLa cells by poliovirus provides protection against apoptosis induced by actinomycin D and cycloheximide (331). The viral product responsible is not known.

CONCLUDING REMARKS

The complexity and variety of viral products involved in the induction and suppression of apoptosis are remarkable. Remarkable too is the rapid progress in understanding the mechanisms involved. Hundreds of reports now appear each year related to viruses and apoptosis, a result of the growing awareness of the role of apoptosis in viral infection and of advances in understanding the molecular biology of apoptosis. Information about viral proteins has provided several critical insights into cellular apoptotic processes, and this trend will certainly continue. As viral products have evolved over countless rounds of replication under highly selective pressures, they have acquired the means to efficiently target critical steps in the apoptotic response. Agents that promote apoptosis in virus-infected cells or that block the suppression of apoptosis by viral products could be highly useful in treating viral diseases. Furthermore, these viral products or reagents that mimic their function could be invaluable for the suppression of apoptosis in diseases characterized by inappropriate cell death. Similarly, proapoptotic viral proteins could be useful for the induction of apoptosis in cancer or diseases characterized by the lack of cell death. We believe that in the next few years such therapies will begin to appear in the clinic.

Visit the Annual Reviews home page at http://www.AnnualReviews.org

LITERATURE CITED

1. Miller LK, Kaiser WJ, Seshagiri S. 1998. Baculovirus regulation of apoptosis. *Semin. Virol.* 8:445–52
2. Hardwick JM. 1998. Viral interference with apoptosis. *Semin. Cell Dev. Biol.* 9:339–49
3. Barry M, McFadden G. 1998. Apoptosis regulators from DNA viruses. *Curr. Opin. Immunol.* 10:422–30
4. Krajcsi P, Wold WS. 1998. Inhibition of tumor necrosis factor and interferon triggered responses by DNA viruses. *Semin. Cell Dev. Biol.* 9:351–58
5. Oberhaus SM, Dermody TS, Tyler KL. 1998. Apoptosis and the cytopathic effects of reovirus *Curr. Top. Microbiol. Immunol.* 233:23–49

6. McFadden G, Barry M. 1998. How poxviruses oppose apoptosis. *Semin. Virol.* 8:429–42

7. Teodoro JG, Branton PE. 1997. Regulation of apoptosis by viral gene products. *J. Virol.* 71:1739–46

8. Kaplan D, Sieg S. 1998. Role of the Fas/Fas ligand apoptotic pathway in human immunodeficiency virus type 1 disease. *J. Virol.* 72:6279–82

9. Shen Y, Shenk TE. 1995. Viruses and apoptosis. *Curr. Opin. Genet. Dev.* 5:105–11

10. Griffin DE, Hardwick JM. 1997. Regulators of apoptosis on the road to persistent alphavirus infection. *Annu. Rev. Microbiol.* 51:565–92

11. Fields BN, Knipe DM, Howley PM, eds. 1996. *Virology.* Philadelphia: Lippincott-Raven

12. Thornberry NA, Lazebnik Y. 1998. Caspases: enemies within. *Science* 281:1312–16

13. Ashkenazi A, Dixit VM. 1998. Death receptors: signaling and modulation. *Science* 281:1305–8

14. Levine AJ. 1997. p53, the cellular gatekeeper for growth and division. *Cell* 88:323–31

15. Ko LJ, Prives C. 1996. p53: puzzle and paradigm. *Genes Dev.* 10:1054–72

16. Adams JM, Cory S. 1998. The Bcl-2 protein family: arbiters of cell survival. *Science* 281:1322–26

17. Raff M. 1998. Cell suicide for beginners. *Nature* 396:119–22

18. Ng FW, Shore GC. 1998. Bcl-X_L cooperatively associates with the Bap31 complex in the endoplasmic reticulum, dependent on procaspase-8 and Ced-4 adaptor. *J. Biol. Chem.* 273:3140–43

19. Ng FW, Nguyen M, Kwan T, Branton PE, Nicholson DW, et al. 1997. p28 Bap31, a Bcl-2/Bcl-X_L- and procaspase-8-associated protein in the endoplasmic reticulum. *J. Cell Biol.* 139:327–38

20. Evan G, Littlewood T. 1998. A matter of life and cell death. *Science* 281:1317–22

21. Samuel CE, Kuhen KL, George CX, Ortega LG, Rende-Fournier R, Tanaka H. 1997. The PKR protein kinase—an interferon-inducible regulator of cell growth and differentiation. *Int. J. Hematol.* 65:227–37

22. Clemens MJ, Elia A. 1997. The double-stranded RNA-dependent protein kinase PKR: structure and function. *J. Interferon Cytokine Res.* 17:503–24

23. Diaz-Guerra M, Rivas C, Esteban M. 1997. Activation of the IFN-inducible enzyme RNase L causes apoptosis of animal cells. *Virology* 236:354–63

24. Castelli JC, Hassel BA, Wood KA, Li XL, Amemiya K, et al. 1997. A study of the interferon antiviral mechanism: apoptosis activation by the 2-5A system. *J. Exp. Med.* 186:967–72

25. Rivas C, Gil J, Melkova Z, Esteban M, Diaz-Guerra M. 1998. Vaccinia virus E3L protein is an inhibitor of the interferon (i.f.n.)-induced 2-5A synthetase enzyme. *Virology* 243:406–14

26. He B, Gross M, Roizman B. 1997. The gamma(1)34.5 protein of herpes simplex virus 1 complexes with protein phosphatase 1alpha to dephosphorylate the alpha subunit of the eukaryotic translation initiation factor 2 and preclude the shutoff of protein synthesis by double-stranded RNA-activated protein kinase. *Proc. Natl. Acad. Sci. USA* 94:843–48

27. Lee SB, Rodriguez D, Rodriguez JR, Esteban M. 1997. The apoptosis pathway triggered by the interferon-induced protein kinase PKR requires the third basic domain, initiates upstream of Bcl- 2, and involves ICE-like proteases. *Virology* 231:81–88

28. Lee SB, Esteban M. 1994. The interferon-induced double-stranded RNA-activated protein kinase induces apoptosis. *Virology* 199:491–96

29. Srivastava SP, Kumar KU, Kaufman RJ. 1998. Phosphorylation of eukaryotic translation initiation factor 2 mediates apoptosis in response to activation of the double-stranded RNA-dependent protein kinase. *J. Biol. Chem.* 273:2416–23

30. Tanaka N, Sato M, Lamphier MS, Nozawa H, Oda E, et al. 1998. Type I interferons are essential mediators of apoptotic death in virally infected cells. *Genes Cells* 3:29–37

31. Debbas M, White E. 1993. Wild-type p53 mediates apoptosis by E1A, which is inhibited by E1B. *Genes Dev.* 7:546–54

32. Lowe SW, Ruley HE. 1993. Stabilization of the p53 tumor suppressor is induced by adenovirus 5 E1A and accompanies apoptosis. *Genes Dev.* 7:535–45

33. Querido E, Teodoro JG, Branton PE. 1997. Accumulation of p53 induced by the adenovirus E1A protein requires regions involved in the stimulation of DNA synthesis. *J. Virol.* 71:3526–33

34. Chiou SK, White E. 1997. p300 binding by E1A cosegregates with p53 induction but is dispensable for apoptosis. *J. Virol.* 71:3515–25

35. de Stanchina E, McCurrach ME, Zindy F, Shieh SY, Ferbeyre G, et al. 1998. E1A signaling to p53 involves the p19(ARF) tumor suppressor. *Genes Dev.* 12:2434–42

36. Duerksen-Hughes P, Wold WS, Gooding LR. 1989. Adenovirus E1A renders infected cells sensitive to cytolysis by tumor necrosis factor. *J. Immunol.* 143:4193–200

37. Duerksen-Hughes PJ, Hermiston TW, Wold WS, Gooding LR. 1991. The amino-terminal portion of CD1 of the adenovirus E1A proteins is required to induce susceptibility to tumor necrosis factor cytolysis in adenovirus-infected mouse cells. *J. Virol.* 65:1236–44

38. Shisler J, Duerksen-Hughes P, Hermiston TM, Wold WS, Gooding LR. 1996. Induction of susceptibility to tumor necrosis factor by E1A is dependent on binding to either p300 or p105-Rb and induction of DNA synthesis. *J. Virol.* 70:68–77

39. Whalen SG, Marcellus RC, Barbeau D, Branton PE. 1996. Importance of the Ser-132 phosphorylation site in cell transformation and apoptosis induced by the adenovirus type 5 E1A protein. *J. Virol.* 70:5373–83

40. Fearnhead HO, Rodriguez J, Govek EE, Guo W, Kobayashi R, et al. 1998. Oncogene-dependent apoptosis is mediated by caspase-9. *Proc. Natl. Acad. Sci. USA* 95:13664–69

41. Boulakia CA, Chen G, Ng FW, Teodoro JG, Branton PE, et al. 1996. Bcl-2 and adenovirus E1B 19 kDA protein prevent E1A-induced processing of CPP32 and cleavage of poly(ADP-ribose) polymerase. *Oncogene* 12:529–35

42. Teodoro JG, Shore GC, Branton PE. 1995. Adenovirus E1A proteins induce apoptosis by both p53-dependent and p53-independent mechanisms. *Oncogene* 11:467–74

43. Subramanian T, Tarodi B, Chinnadurai G. 1995. p53-independent apoptotic and necrotic cell deaths induced by adenovirus infection: suppression by E1B 19K and Bcl-2 proteins. *Cell Growth Differ.* 6:131–37

44. Marcellus RC, Teodoro JG, Wu T, Brough DE, Ketner G, et al. 1996. Adenovirus type 5 early region 4 is responsible for E1A-induced p53-independent apoptosis. *J. Virol.* 70:6207–15

45. Marcellus RC, Lavoie JN, Boivin D, Shore GC, Ketner G, Branton PE. 1998. The early region 4 orf4 protein of human adenovirus type 5 induces p53- independent cell death by apoptosis. *J. Virol.* 72:7144–53

46. Lavoie JN, Nguyen M, Marcellus RC, Branton PE, Shore GC. 1998. E4orf4, a novel adenovirus death factor that induces p53-independent apoptosis by a pathway that is not inhibited by zVAD-fmk. *J. Cell Biol.* 140:637–45

47. Shtrichman R, Kleinberger T. 1998. Adenovirus type 5 E4 open reading frame 4 protein induces apoptosis in transformed cells. *J. Virol.* 72:2975–82

48. Deleted in proof

49. Muller U, Kleinberger T, Shenk T. 1992. Adenovirus E4orf4 protein reduces phos-

phorylation of c-Fos and E1A proteins while simultaneously reducing the level of AP-1. *J. Virol.* 66:5867–78

50. Tollefson AE, Scaria A, Saha SK, Wold WS. 1992. The 11,600-MW protein encoded by region E3 of adenovirus is expressed early but is greatly amplified at late stages of infection. *J. Virol.* 66:3633–42

51. Scaria A, Tollefson AE, Saha SK, Wold WS. 1992. The E3-11.6K protein of adenovirus is an Asn-glycosylated integral membrane protein that localizes to the nuclear membrane. *Virology* 191:743–53

52. Tollefson AE, Ryerse JS, Scaria A, Hermiston TW, Wold WS. 1996. The E3-11.6-kDa adenovirus death protein (ADP) is required for efficient cell death: characterization of cells infected with *adp* mutants. *Virology* 220:152–62

53. Tollefson AE, Scaria A, Hermiston TW, Ryerse JS, Wold LJ, Wold WS. 1996. The adenovirus death protein (E3-11.6K) is required at very late stages of infection for efficient cell lysis and release of adenovirus from infected cells. *J. Virol.* 70:2296–306

54. Symonds H, Krall L, Remington L, Saenz-Robles M, Lowe S, et al. 1994. p53-dependent apoptosis suppresses tumor growth and progression in vivo. *Cell* 78:703–11

55. Fromm L, Shawlot W, Gunning K, Butel JS, Overbeek PA. 1994. The retinoblastoma protein-binding region of simian virus 40 large T antigen alters cell cycle regulation in lenses of transgenic mice. *Mol. Cell. Biol.* 14:6743–54

56. Fimia GM, Gottifredi V, Bellei B, Ricciardi MR, Tafuri A, et al. 1998. The activity of differentiation factors induces apoptosis in polyomavirus large T-expressing myoblasts. *Mol. Biol. Cell* 9:1449–63

57. Pan H, Griep AE. 1994. Altered cell cycle regulation in the lens of HPV-16 E6 or E7 transgenic mice: implications for tumor suppressor gene function in development. *Genes Dev.* 8:1285–99

58. Howes KA, Ransom N, Papermaster DS, Lasudry JG, Albert DM, Windle JJ. 1994. Apoptosis or retinoblastoma: alternative fates of photoreceptors expressing the HPV-16 E7 gene in the presence or absence of p53. *Genes Dev.* 8:1300–10. Published erratum in *Genes Dev.* 8:1738

59. Pan H, Griep AE. 1995. Temporally distinct patterns of p53-dependent and p53-independent apoptosis during mouse lens development. *Genes Dev.* 9:2157–69

60. Bergqvist A, Soderbarg K, Magnusson G. 1997. Altered susceptibility to tumor necrosis factor alpha-induced apoptosis of mouse cells expressing polyomavirus middle and small T antigens. *J. Virol.* 71:276–83

61. Caillet-Fauquet P, Perros M, Brandenburger A, Spegelaere P, Rommelaere J. 1990. Programmed killing of human cells by means of an inducible clone of parvoviral genes encoding non-structural proteins. *EMBO J.* 9:2989–95

62. Ozawa K, Ayub J, Kajigaya S, Shimada T, Young N. 1988. The gene encoding the nonstructural protein of B19 (human) parvovirus may be lethal in transfected cells. *J. Virol.* 62:2884–89

63. Berns KI. 1996. In *Parvoviridae: the Viruses and Their Replication*, ed. BN Fields, DM Knipe, PM Howley, pp. 2173–97. Philadelphia: Lippincott-Raven

64. Moffatt S, Yaegashi N, Tada K, Tanaka N, Sugamura K. 1998. Human parvovirus B19 nonstructural (NS1) protein induces apoptosis in erythroid lineage cells. *J. Virol.* 72:3018–28

65. Telerman A, Tuynder M, Dupressoir T, Robaye B, Sigaux F, et al. 1993. A model for tumor suppression using H-1 parvovirus. *Proc. Natl. Acad. Sci. USA* 90:8702–6

66. Op De Beeck A, Caillet-Fauquet P. 1997. The NS1 protein of the autonomous parvovirus minute virus of mice blocks cellular DNA replication: a consequence of lesions to the chromatin? *J. Virol.* 71:5323–29

67. Op De Beeck A, Anouja F, Mousset S, Rommelaere J, Caillet-Fauquet P. 1995. The nonstructural proteins of the autonomous parvovirus minute virus of mice interfere with the cell cycle, inducing accumulation in G2. *Cell Growth Differ.* 6:781–87

68. Mousset S, Ouadrhiri Y, Caillet-Fauquet P, Rommelaere J. 1994. The cytotoxicity of the autonomous parvovirus minute virus of mice nonstructural proteins in FR3T3 rat cells depends on oncogene expression. *J. Virol.* 68:6446–53

69. Noteborn MH, Todd D, Verschueren CA, de Gauw HW, Curran WL, et al. 1994. A single chicken anemia virus protein induces apoptosis. *J. Virol.* 68:346–51

70. Danen-Van Oorschot AA, Fischer DF, Grimbergen JM, Klein B, Zhuang S, et al. 1997. Apoptin induces apoptosis in human transformed and malignant cells but not in normal cells. *Proc. Natl. Acad. Sci. USA* 94:5843–47

71. Zhuang SM, Shvarts A, van Ormondt H, Jochemsen AG, van der Eb AJ, Noteborn MH. 1995. Apoptin, a protein derived from chicken anemia virus, induces p53-independent apoptosis in human osteosarcoma cells. *Cancer Res.* 55:486–89

72. Noteborn MH, van Oorschot AA, van der Eb AJ. 1998. Chicken anemia virus: induction of apoptosis by a single protein of a single-stranded DNA virus. *Semin. Virol.* 8:497–508

73. Tropea F, Troiano L, Monti D, Lovato E, Malorni W, et al. 1995. Sendai virus and herpes virus type 1 induce apoptosis in human peripheral blood mononuclear cells. *Exp. Cell Res.* 218:63–70

74. Ito M, Watanabe M, Kamiya H, Sakurai M. 1997. Herpes simplex virus type 1 induces apoptosis in peripheral blood T lymphocytes. *J. Infect. Dis.* 175:1220–24

75. Koyama AH, Adachi A. 1997. Induction of apoptosis by herpes simplex virus type 1. *J. Gen. Virol.* 78:2909–12

76. Hanon E, Meyer G, Vanderplasschen A, Dessy-Doize C, Thiry E, Pastoret PP. 1998. Attachment but not penetration of bovine herpesvirus 1 is necessary to induce apoptosis in target cells. *J. Virol.* 72:7638–41

77. Hanon E, Vanderplasschen A, Lyaku S, Keil G, Denis M, Pastoret PP. 1996. Inactivated bovine herpesvirus 1 induces apoptotic cell death of mitogen-stimulated bovine peripheral blood mononuclear cells. *J. Virol.* 70:4116–20

78. Sadzot-Delvaux C, Thonard P, Schoonbroodt S, Piette J, Rentier B. 1995. Varicella-zoster virus induces apoptosis in cell culture. *J. Gen. Virol.* 76:2875–79

79. Secchiero P, Berneman ZN, Sun D, Nicholas J, Reitz MS, Jr. 1997. Identification of envelope glycoproteins H and B homologues of human herpesvirus 7. *Intervirology* 40:22–32

80. Inoue Y, Yasukawa M, Fujita S. 1997. Induction of T-cell apoptosis by human herpesvirus 6. *J. Virol.* 71:3751–59

81. Parker GA, Crook T, Bain M, Sara EA, Farrell PJ, Allday MJ. 1996. Epstein-Barr virus nuclear antigen (EBNA)3C is an immortalizing oncoprotein with similar properties to adenovirus E1A and papillomavirus E7. *Oncogene* 13:2541–49

82. Gregory CD, Dive C, Henderson S, Smith CA, Williams GT, et al. 1991. Activation of Epstein-Barr virus latent genes protects human B cells from death by apoptosis. *Nature* 349:612–14

83. Alnemri ES, Robertson NM, Fernandes TF, Croce CM, Litwack G. 1992. Overexpressed full-length human BCL2 extends the survival of baculovirus-infected Sf9 insect cells. *Proc. Natl. Acad. Sci. USA* 89:7295–99

84. Ribeiro BM, Hutchinson K, Miller LK. 1994. A mutant baculovirus with a temperature-sensitive IE-1 transregulatory protein. *J. Virol.* 68:1075–84

85. Prikhod'ko EA, Miller LK. 1996. Induction of apoptosis by baculovirus transactivator IE1. *J. Virol.* 70:7116–24

86. LaCount DJ, Friesen PD. 1997. Role of

early and late replication events in induction of apoptosis by baculoviruses. *J. Virol.* 71:1530–37

87. Su F, Schneider RJ. 1997. Hepatitis B virus HBx protein sensitizes cells to apoptotic killing by tumor necrosis factor alpha. *Proc. Natl. Acad. Sci. USA* 94:8744–49

88. Chirillo P, Pagano S, Natoli G, Puri PL, Burgio VL, Balsano C, Levrero M. 1997. The hepatitis B virus X gene induces p53-mediated programmed cell death. *Proc. Natl. Acad. Sci. USA* 94:8162–67

89. Kim H, Lee H, Yun Y. 1998. X-gene product of hepatitis B virus induces apoptosis in liver cells. *J. Biol. Chem.* 273:381–85

90. Terradillos O, Pollicino T, Lecoeur H, Tripodi M, Gougeon ML, et al. 1998. p53-independent apoptotic effects of the hepatitis B virus HBx protein in vivo and in vitro. *Oncogene* 17:2115–23

91. Terai C, Kornbluth RS, Pauza CD, Richman DD, Carson DA. 1991. Apoptosis as a mechanism of cell death in cultured T lymphoblasts acutely infected with HIV-1. *J. Clin. Invest.* 87:1710–15

92. Laurent-Crawford AG, Krust B, Muller S, Riviere Y, Rey-Cuille MA, et al. 1991. The cytopathic effect of HIV is associated with apoptosis. *Virology* 185:829–39

93. Ameisen JC, Capron A. 1991. Cell dysfunction and depletion in AIDS: the programmed cell death hypothesis. *Immunol. Today* 12:102–5

94. Purvis SF, Jacobberger JW, Sramkoski RM, Patki AH, Lederman MM. 1995. HIV type 1 Tat protein induces apoptosis and death in Jurkat cells. *AIDS Res. Hum. Retroviruses* 11:443–50

95. Westendorp MO, Frank R, Ochsenbauer C, Stricker K, Dhein J, et al. 1995. Sensitization of T cells to CD95-mediated apoptosis by HIV-1 Tat and gp120. *Nature* 375:497–500

96. Kolesnitchenko V, King L, Riva A, Tani Y, Korsmeyer SJ, Cohen DI. 1997. A major human immunodeficiency virus type 1-initiated killing pathway distinct from apoptosis. *J. Virol.* 71:9753–63

97. Westendorp MO, Shatrov VA, Schulze-Osthoff K, Frank R, Kraft M, et al. 1995. HIV-1 Tat potentiates TNF-induced NF-kappa B activation and cytotoxicity by altering the cellular redox state. *EMBO J.* 14:546–54

98. Li CJ, Friedman DJ, Wang C, Metelev V, Pardee AB. 1995. Induction of apoptosis in uninfected lymphocytes by HIV-1 Tat protein. *Science* 268:429–31

99. Sasaki M, Uchiyama J, Ishikawa H, Matsushita S, Kimura G, et al. 1996. Induction of apoptosis by calmodulin-dependent intracellular Ca^{2+} elevation in $CD4^+$ cells expressing gp 160 of HIV. *Virology* 224:18–24

100. Banda NK, Bernier J, Kurahara DK, Kurrle R, Haigwood N, et al. 1992. Crosslinking CD4 by human immunodeficiency virus gp120 primes T cells for activation-induced apoptosis. *J. Exp. Med.* 176:1099–106

101. Shi B, Raina J, Lorenzo A, Busciglio J, Gabuzda D. 1998. Neuronal apoptosis induced by HIV-1 Tat protein and TNF-alpha: potentiation of neurotoxicity mediated by oxidative stress and implications for HIV-1 dementia. *J. Neurovirol.* 4:281–90

102. Moutouh L, Estaquier J, Richman DD, Corbeil J. 1998. Molecular and cellular analysis of human immunodeficiency virus-induced apoptosis in lymphoblastoid T-cell-line-expressing wild-type and mutated CD4 receptors. *J. Virol.* 72:8061–72

103. Hesselgesser J, Taub D, Baskar P, Greenberg M, Hoxie J, et al. 1998. Neuronal apoptosis induced by HIV-1 gp120 and the chemokine SDF-1 alpha is mediated by the chemokine receptor CXCR4. *Curr. Biol.* 8:595–98

104. Bartz SR, Rogel ME, Emerman M 1996. Human immunodeficiency virus type 1 cell cycle control: Vpr is cytostatic and mediates G2 accumulation by a mechanism

which differs from DNA damage checkpoint control. *J. Virol.* 70:2324–31

105. Stewart SA, Poon B, Jowett JB, Chen IS. 1997. Human immunodeficiency virus type 1 Vpr induces apoptosis following cell cycle arrest. *J. Virol.* 71:5579–92

106. Chen X, Zachar V, Zdravkovic M, Guo M, Ebbesen P, Liu X. 1997. Role of the Fas/Fas ligand pathway in apoptotic cell death induced by the human T cell lymphotropic virus type I Tax transactivator. *J. Gen. Virol.* 78:3277–85

107. Los M, Khazaie K, Schulze-Osthoff K, Baeuerle PA, Schirrmacher V, Chlichlia K. 1998. Human T cell leukemia virus-I (HTLV-I) Tax-mediated apoptosis in activated T cells requires an enhanced intracellular prooxidant state. *J. Immunol.* 161:3050–55

108. Ikeda Y, Itagaki S, Tsutsui S, Inoshima Y, Fukasawa M, et al. 1997. Replication of feline syncytial virus in feline T-lymphoblastoid cells and induction of apoptosis in the cells. *Microbiol. Immunol.* 41:431–35

109. Mergia A, Blackwell J, Papadi G, Johnson C. 1997. Simian foamy virus type 1 (SFV-1) induces apoptosis. *Virus Res.* 50:129–37

110. Carthy CM, Granville DJ, Watson KA, Anderson DR, Wilson JE, et al. 1998. Caspase activation and specific cleavage of substrates after coxsackievirus B3-induced cytopathic effect in HeLa cells. *J. Virol.* 72:7669–75

111. Jelachich ML, Lipton HL. 1996. Theiler's murine encephalomyelitis virus kills restrictive but not permissive cells by apoptosis. *J. Virol.* 70:6856–61

112. Brack K, Frings W, Dotzauer A, Vallbracht A. 1998. A cytopathogenic, apoptosis-inducing variant of hepatitis A virus. *J. Virol.* 72:3370–76

113. Scallan MF, Allsopp TE, Fazakerley JK. 1997. bcl-2 acts early to restrict Semliki Forest virus replication and delays virus-induced programmed cell death. *J. Virol.* 71:1583–90

114. Glasgow GM, McGee MM, Sheahan BJ, Atkins GJ. 1997. Death mechanisms in cultured cells infected by Semliki Forest virus. *J. Gen. Virol.* 78:1559–63

115. Grandgirard D, Studer E, Monney L, Belser T, Fellay I, et al. 1998. Alphaviruses induce apoptosis in Bcl-2-overexpressing cells: evidence for a caspase-mediated, proteolytic inactivation of Bcl-2. *EMBO J.* 17:1268–78

116. Glasgow GM, McGee MM, Tarbatt CJ, Mooney DA, Sheahan BJ, Atkins GJ. 1998. The Semliki Forest virus vector induces p53-independent apoptosis. *J. Gen. Virol.* 79:2405–10

117. Levine B, Huang Q, Isaacs JT, Reed JC, Griffin DE, Hardwick JM. 1993. Conversion of lytic to persistent alphavirus infection by the *bcl-2* cellular oncogene. *Nature* 361:739–42

118. Ubol S, Tucker PC, Griffin DE, Hardwick JM. 1994. Neurovirulent strains of alphavirus induce apoptosis in *bcl-2*- expressing cells: role of a single amino acid change in the E2 glycoprotein. *Proc. Natl. Acad. Sci. USA* 91:5202–6

119. Levine B, Goldman JE, Jiang HH, Griffin DE, Hardwick J M. 1996. Bcl-2 protects mice against fatal alphavirus encephalitis. *Proc. Natl. Acad. Sci. USA* 93:4810–15

120. Lewis J, Wesselingh SL, Griffin DE, Hardwick JM. 1996. Alphavirus-induced apoptosis in mouse brains correlates with neurovirulence. *J. Virol.* 70:1828–35

121. Joe AK, Foo HH, Kleeman L, Levine B. 1998. The transmembrane domains of Sindbis virus envelope glycoproteins induce cell death. *J. Virol.* 72:3935–43

122. Nava VE, Rosen A, Veliuona MA, Clem RJ, Levine B, Hardwick JM. 1998. Sindbis virus induces apoptosis through a caspase-dependent, CrmA-sensitive pathway. *J. Virol.* 72:452–59

123. Lin KI, Lee SH, Narayanan R, Baraban JM, Hardwick JM, Ratan RR. 1995. Thiol

agents and Bcl-2 identify an alphavirus-induced apoptotic pathway that requires activation of the transcription factor NF-kappa B. *J. Cell Biol.* 131:1149–61

124. Lin KI, DiDonato JA, Hoffmann A, Hardwick JM, Ratan RR. 1998. Suppression of steady-state, but not stimulus-induced NF-κB activity inhibits alphavirus-induced apoptosis. *J. Cell Biol.* 141:1479–87

125. Pugachev KV, Frey TK. 1998. Rubella virus induces apoptosis in culture cells. *Virology* 250:359–70

126. Liao CL, Lin YL, Wang JJ, Huang YL, Yeh CT, et al. 1997. Effect of enforced expression of human *bcl-2* on Japanese encephalitis virus-induced apoptosis in cultured cells. *J. Virol.* 71:5963–71

127. Despres P, Frenkiel MP, Ceccaldi PE, Duarte Dos Santos C, Deubel V. 1998. Apoptosis in the mouse central nervous system in response to infection with mouse-neurovirulent dengue viruses. *J. Virol.* 72:823–29

128. Marianneau P, Cardona A, Edelman L, Deubel V, Despres P. 1997. Dengue virus replication in human hepatoma cells activates NF-kappaB which in turn induces apoptotic cell death. *J. Virol.* 71:3244–49

129. Summerfield A, Knotig SM, McCullough KC. 1998. Lymphocyte apoptosis during classical swine fever: implication of activation-induced cell death. *J. Virol.* 72:1853–61

130. Zhang G, Aldridge S, Clarke MC, McCauley JW. 1996. Cell death induced by cytopathic bovine viral diarrhea virus is mediated by apoptosis. *J. Gen. Virol.* 77:1677–81

131. Ruggieri A, Harada T, Matsuura Y, Miyamura T. 1997. Sensitization to Fas-mediated apoptosis by hepatitis C virus core protein. *Virology* 229:68–76

132. Zhu N, Khoshnan A, Schneider R, Matsumoto M, Dennert G, et al. 1998. Hepatitis C virus core protein binds to the cytoplasmic domain of tumor necrosis factor (TNF) receptor 1 and enhances TNF-induced apoptosis. *J. Virol.* 72:3691–97

133. Suarez P, Diaz-Guerra M, Prieto C, Esteban M, Castro JM, et al. 1996. Open reading frame 5 of porcine reproductive and respiratory syndrome virus as a cause of virus-induced apoptosis. *J. Virol.* 70:2876–82

134. Eleouet JF, Chilmonczyk S, Besnardeau L, Laude H. 1998. Transmissible gastroenteritis coronavirus induces programmed cell death in infected cells through a caspase-dependent pathway. *J. Virol.* 72:4918–24

135. Belyavsky M, Belyavskaya E, Levy GA, Leibowitz JL. 1998. Coronavirus MHV-3-induced apoptosis in macrophages. *Virology* 250:41–49

136. Ubol S, Sukwattanapan C, Utaisincharoen P. 1998. Rabies virus replication induces Bax-related, caspase dependent apoptosis in mouse neuroblastoma cells. *Virus Res.* 56:207–15

137. Thoulouze MI, Lafage M, Montano-Hirose JA, Lafon M. 1997. Rabies virus infects mouse and human lymphocytes and induces apoptosis. *J. Virol.* 71:7372–80

138. Koyama AH. 1995. Induction of apoptotic DNA fragmentation by the infection of vesicular stomatitis virus. *Virus Res.* 37:285–90

139. Bitzer M, Prinz F, Bauer M, Spiegel M, Neubert WJ, et al. 1999. Sendai virus infection induces apoptosis through activation of caspase-8 (FLICE) and caspase-3 (CPP32). *J. Virol.* 73:702–8

140. Itoh M, Hotta H, Homma M. 1998. Increased induction of apoptosis by a Sendai virus mutant is associated with attenuation of mouse pathogenicity. *J. Virol.* 72:2927–34

141. Garcin D, Taylor G, Tanebayashi K, Compans R, Kolakofsky D. 1998. The short Sendai virus leader region controls induc-

tion of programmed cell death. *Virology* 243:340–53

142. Zorn U, Dallmann I, Grosse J, Kirchner H, Poliwoda H, Atzpodien J. 1994. Induction of cytokines and cytotoxicity against tumor cells by Newcastle disease virus. *Cancer Biother.* 9:225–35

143. Lorence RM, Rood PA, Kelley KW. 1988. Newcastle disease virus as an antineoplastic agent: induction of tumor necrosis factor-alpha and augmentation of its cytotoxicity. *J. Natl. Cancer Inst.* 80:1305–12

144. Reichard KW, Lorence RM, Cascino CJ, Peeples ME, Walter RJ, et al. 1992. Newcastle disease virus selectively kills human tumor cells. *J. Surg. Res.* 52:448–53

145. Lorence RM, Katubig BB, Reichard KW, Reyes HM, Phuangsab A, et al. 1994. Complete regression of human fibrosarcoma xenografts after local Newcastle disease virus therapy. *Cancer Res.* 54:6017–21

146. Ito M, Yamamoto T, Watanabe M, Ihara T, Kamiya H, Sakurai M. 1996. Detection of measles virus-induced apoptosis of human monocytic cell line (THP-1) by DNA fragmentation ELISA. *FEMS Immunol. Med. Microbiol.* 15:115–22

147. Wang SZ, Smith PK, Lovejoy M, Bowden JJ, Alpers JH, Forsyth KD. 1998. The apoptosis of neutrophils is accelerated in respiratory syncytial virus (RSV)-induced bronchiolitis. *Clin. Exp. Immunol.* 114:49–54

148. Takeuchi R, Tsutsumi H, Osaki M, Haseyama K, Mizue N, Chiba S. 1998. Respiratory syncytial virus infection of human alveolar epithelial cells enhances interferon regulatory factor 1 and interleukin-1beta-converting enzyme gene expression but does not cause apoptosis. *J. Virol.* 72:4498–502

149. Takizawa T, Ohashi K, Nakanishi Y. 1996. Possible involvement of double-stranded RNA-activated protein kinase in cell death by influenza virus infection. *J. Virol.* 70:8128–32

150. Olsen CW, Kehren JC, Dybdahl-Sissoko NR, Hinshaw VS. 1996. *bcl-2* alters influenza virus yield, spread, and hemagglutinin glycosylation. *J. Virol.* 70:663–66

151. Tyler KL, Squier MK, Rodgers SE, Schneider BE, Oberhaus SM, et al. 1995. Differences in the capacity of reovirus strains to induce apoptosis are determined by the viral attachment protein sigma 1. *J. Virol.* 69:6972–79

152. Tyler KL, Squier MK, Brown AL, Pike B, Willis D, et al. 1996. Linkage between reovirus-induced apoptosis and inhibition of cellular DNA synthesis: role of the S1 and M2 genes. *J. Virol.* 70:7984–91

153. Rodgers SE, Barton ES, Oberhaus SM, Pike B, Gibson CA, et al. 1997. Reovirus-induced apoptosis of MDCK cells is not linked to viral yield and is blocked by Bcl-2. *J. Virol.* 71:2540–46

154. Debiasi RL, Squier MKT, Pike B, Wynes M, Dermody TS, et al. 1999. Reovirus-induced apoptosis is preceded by increased cellular calpain activity and is blocked by calpain inhibitors. *J. Virol.* 73:695–701

155. Pekosz A, Phillips J, Pleasure D, Merry D, Gonzalez-Scarano F. 1996. Induction of apoptosis by La Crosse virus infection and role of neuronal differentiation and human *bcl-2* expression in its prevention. *J. Virol.* 70:5329–35

156. Newton K, Meyer JC, Bellamy AR, Taylor JA. 1997. Rotavirus nonstructural glycoprotein NSP4 alters plasma membrane permeability in mammalian cells. *J. Virol.* 71:9458–65

157. Coffey MC, Strong JE, Forsyth PA, Lee PWK. 1998. Reovirus therapy of tumors with activated ras pathway. *Science* 282:1332–34

158. Vasconcelos AC, Lam KM. 1994. Apoptosis induced by infectious bursal disease virus. *J. Gen. Virol.* 75:1803–6

159. Vasconcelos AC, Lam KM. 1995. Apoptosis in chicken embryos induced by the

infectious bursal disease virus. *J. Comp. Pathol.* 112:327–38

160. Tham KM, Moon CD. 1996. Apoptosis in cell cultures induced by infectious bursal disease virus following in vitro infection. *Avian Dis.* 40:109–13

161. Fernandez-Arias A, Martinez S, Rodriguez JF. 1997. The major antigenic protein of infectious bursal disease virus, VP2, is an apoptotic inducer. *J. Virol.* 71:8014–18

162. Yao K, Goodwin MA, Vakharia VN. 1998. Generation of a mutant infectious bursal disease virus that does not cause bursal lesions. *J. Virol.* 72:2647–54

163. McFadden G, Lalani A, Everett H, Nash P, Xu X. 1998. Virus-encoded receptors for cytokines and chemokines. *Semin. Cell Dev. Biol.* 9:359–68

164. Barry M, McFadden G. 1998. In *Virokines and Viroceptors*, ed. DG Remick, JS Friedland, pp. 251–61. New York: Marcel Dekker

165. Chang HW, Watson JC, Jacobs BL. 1992. The E3L gene of vaccinia virus encodes an inhibitor of the interferon-induced, double-stranded RNA-dependent protein kinase. *Proc. Natl. Acad. Sci. USA* 89:4825–29

166. Davies MV, Chang HW, Jacobs BL, Kaufman RJ. 1993. The E3L and K3L vaccinia virus gene products stimulate translation through inhibition of the double-stranded RNA-dependent protein kinase by different mechanisms. *J. Virol.* 67:1688–92

167. Kibler KV, Shors T, Perkins KB, Zeman CC, Banaszak MP, et al. 1997. Double-stranded RNA is a trigger for apoptosis in vaccinia virus-infected cells. *J. Virol.* 71:1992–2003

168. Denzler KL, Jacobs BL. 1994. Site-directed mutagenic analysis of reovirus sigma 3 protein binding to dsRNA. *Virology* 204:190–99

169. Lu Y, Wambach M, Katze MG, Krug RM. 1995. Binding of the influenza virus NS1 protein to double-stranded RNA inhibits the activation of the protein kinase that phosphorylates the elF-2 translation initiation factor. *Virology* 214:222–28

170. Chou J, Roizman B. 1992. The gamma 1(34.5) gene of herpes simplex virus 1 precludes neuroblastoma cells from triggering total shutoff of protein synthesis characteristic of programed cell death in neuronal cells. *Proc. Natl. Acad. Sci. USA* 89:3266–70

171. Gooding LR, Elmore LW, Tollefson AE, Brady HA, Wold WS. 1988. A 14,700 MW protein from the E3 region of adenovirus inhibits cytolysis by tumor necrosis factor. *Cell* 53:341–46

172. Gooding LR, Sofola IO, Tollefson AE, Duerksen-Hughes P, Wold WS. 1990. The adenovirus E3-14.7K protein is a general inhibitor of tumor necrosis factor-mediated cytolysis. *J. Immunol.* 145:3080–86

173. Horton TM, Ranheim TS, Aquino L, Kusher DI, Saha SK, et al. 1991. Adenovirus E3 14.7K protein functions in the absence of other adenovirus proteins to protect transfected cells from tumor necrosis factor cytolysis. *J. Virol.* 65:2629–39

174. Zilli D, Voelkel-Johnson C, Skinner T, Laster SM. 1992. The adenovirus E3 region 14.7 kDa protein, heat and sodium arsenite inhibit the TNF-induced release of arachidonic acid. *Biochem. Biophys. Res. Commun.* 188:177–83

175. Krajcsi P, Dimitrov T, Hermiston TW, Tollefson AE, Ranheim TS, et al. 1996. The adenovirus E3-14.7K protein and the E3-10.4K/14.5K complex of proteins, which independently inhibit tumor necrosis factor (TNF)-induced apoptosis, also independently inhibit TNF-induced release of arachidonic acid. *J. Virol.* 70:4904–13

176. Gooding LR, Ranheim TS, Tollefson AE, Aquino L, Duerksen-Hughes P, et al. 1991. The 10,400- and 14,500-dalton pro-

teins encoded by region E3 of adenovirus function together to protect many but not all mouse cell lines against lysis by tumor necrosis factor. *J. Virol.* 65:4114–23

177. Dimitrov T, Krajcsi P, Hermiston TW, Tollefson AE, Hannink M, Wold WS. 1997. Adenovirus E3-10.4K/14.5K protein complex inhibits tumor necrosis factor-induced translocation of cytosolic phospholipase A$_2$ to membranes. *J. Virol.* 71:2830–37

178. Shisler J, Yang C, Walter B, Ware CF, Gooding LR. 1997. The adenovirus E3-10.4K/14.5K complex mediates loss of cell surface Fas (CD95) and resistance to Fas-induced apoptosis. *J. Virol.* 71:8299–306

179. Tollefson AE, Hermiston TW, Lichtenstein DL, Colle CF, et al. 1998. Forced degradation of Fas inhibits apoptosis in adenovirus-infected cells. *Nature* 392:726–30

180. Hu FQ, Smith CA, Pickup DJ. 1994. Cowpox virus contains two copies of an early gene encoding a soluble secreted form of the type II TNF receptor. *Virology* 204:343–56

181. Upton C, Macen JL, Schreiber M, McFadden G. 1991. Myxoma virus expresses a secreted protein with homology to the tumor necrosis factor receptor gene family that contributes to viral virulence. *Virology* 184:370–82

182. Schreiber M, Rajarathnam K, McFadden G. 1996. Myxoma virus T2 protein, a tumor necrosis factor (TNF) receptor homolog, is secreted as a monomer and dimer that each bind rabbit TNFalpha, but the dimer is a more potent TNF inhibitor. *J. Biol. Chem.* 271:13333–41

183. Smith CA, Davis T, Wignall JM, Din WS, Farrah T, et al. 1991. T2 open reading frame from the Shope fibroma virus encodes a soluble form of the TNF receptor. *Biochem. Biophys. Res. Commun.* 176:335–42

184. Schreiber M, Sedger L, McFadden G.

1997. Distinct domains of M-T2, the myxoma virus tumor necrosis factor (TNF) receptor homolog, mediate extracellular TNF binding and intracellular apoptosis inhibition. *J. Virol.* 71:2171–81

185. Bertin J, Armstrong RC, Ottilie S, Martin DA, Wang Y, et al. 1997. Death effector domain-containing herpesvirus and poxvirus proteins inhibit both Fas- and TNFR1-induced apoptosis. *Proc. Natl. Acad. Sci. USA* 94:1172–76

186. Hu S, Vincenz C, Buller M, Dixit VM. 1997. A novel family of viral death effector domain-containing molecules that inhibit both CD-95- and tumor necrosis factor receptor-1-induced apoptosis. *J. Biol. Chem.* 272:9621–24

187. Thome M, Schneider P, Hofmann K, Fickenscher H, Meinl E, et al. 1997. Viral FLICE-inhibitory proteins (FLIPs) prevent apoptosis induced by death receptors. *Nature* 386:517–21

188. Wang GH, Bertin J, Wang Y, Martin DA, Wang J, et al. 1997. Bovine herpesvirus 4 BORFE2 protein inhibits Fas- and tumor necrosis factor receptor 1-induced apoptosis and contains death effector domains shared with other gamma-2 herpesviruses. *J. Virol.* 71:8928–32

189. White E, Sabbatini P, Debbas M, Wold WS, Kusher DI, Gooding LR. 1992. The 19-kilodalton adenovirus E1B transforming protein inhibits programmed cell death and prevents cytolysis by tumor necrosis factor alpha. *Mol. Cell. Biol.* 12:2570–80

190. Perez D, White E. 1998. E1B 19K inhibits Fas-mediated apoptosis through FADD-dependent sequestration of FLICE. *J. Cell Biol.* 141:1255–66

191. Kieff E. 1996. In *Epstein-Barr Virus and Its Replication*, ed. BN Fields, pp. 2343–96. Philadelphia: Lippincott-Raven

192. Devergne O, Hatzivassiliou E, Izumi KM, Kaye KM, Kleijnen MF, et al. 1996. Association of TRAF1, TRAF2, and TRAF3 with an Epstein-Barr virus LMP1 domain

important for B-lymphocyte transformation: role in NF-κB activation. *Mol. Cell. Biol.* 16:7098–108

193. Kaye KM, Devergne O, Harada JN, Izumi KM, Yalamanchili R, et al. 1996. Tumor necrosis factor receptor associated factor 2 is a mediator of NF-kappa B activation by latent infection membrane protein 1, the Epstein-Barr virus transforming protein. *Proc. Natl. Acad. Sci. USA* 93:11085–90

194. Sandberg M, Hammerschmidt W, Sugden B. 1997. Characterization of LMP-1's association with TRAF1, TRAF2, and TRAF3. *J. Virol.* 71:4649–56

195. Izumi KM, Kaye KM, Kieff ED. 1997. The Epstein-Barr virus LMP1 amino acid sequence that engages tumor necrosis factor receptor associated factors is critical for primary B lymphocyte growth transformation. *Proc. Natl. Acad. Sci. USA* 94:1447–52

196. Laherty CD, Hu HM, Opipari AW, Wang F, Dixit VM. 1992. The Epstein-Barr virus LMP1 gene product induces A20 zinc finger protein expression by activating nuclear factor kappa B. *J. Biol. Chem.* 267:24157–60

197. Henderson S, Rowe M, Gregory C, Croom-Carter D, Wang F, et al. 1991. Induction of bcl-2 expression by Epstein-Barr virus latent membrane protein 1 protects infected B cells from programmed cell death. *Cell* 65:1107–15

198. Wang S, Rowe M, Lundgren E. 1996. Expression of the Epstein Barr virus transforming protein LMP1 causes a rapid and transient stimulation of the Bcl-2 homologue Mcl-1 levels in B-cell lines. *Cancer Res.* 56:4610–13

199. Kawanishi M. 1997. Expression of Epstein-Barr virus latent membrane protein 1 protects Jurkat T cells from apoptosis induced by serum deprivation. *Virology* 228:244–50

200. Ray RB, Meyer K, Steele R, Shrivastava A, Aggarwal BB, Ray R. 1998. Inhibition of tumor necrosis factor (TNF-alpha)-mediated apoptosis by hepatitis C virus core protein. *J. Biol. Chem.* 273:2256–59

201. Rouquet N, Allemand I, Molina T, Bennoun M, Briand P, Joulin V. 1995. Fas-dependent apoptosis is impaired by SV40 T-antigen in transgenic liver. *Oncogene* 11:1061–67

202. Sieg S, Yildirim Z, Smith D, Kayagaki N, Yagita H, et al. 1996. Herpes simplex virus type 2 inhibition of Fas ligand expression. *J. Virol.* 70:8747–51

203. Clem RJ, Fechheimer M, Miller LK. 1991. Prevention of apoptosis by a baculovirus gene during infection of insect cells. *Science* 254:1388–90

204. Kamita SG, Majima K, Maeda S. 1993. Identification and characterization of the p35 gene of *Bombyx mori* nuclear polyhedrosis virus that prevents virus-induced apoptosis. *J. Virol.* 67:455–63

205. Rabizadeh S, LaCount DJ, Friesen PD, Bredesen DE. 1993. Expression of the baculovirus p35 gene inhibits mammalian neural cell death. *J. Neurochem.* 61:2318–21

206. Hay BA, Wolff T, Rubin GM. 1994. Expression of baculovirus P35 prevents cell death in *Drosophila*. *Development* 120:2121–29

207. Sugimoto A, Friesen PD, Rothman JH. 1994. Baculovirus p35 prevents developmentally programmed cell death and rescues a ced-9 mutant in the nematode *Caenorhabditis elegans*. *EMBO J.* 13:2023–28

208. Hershberger PA, LaCount DJ, Friesen PD. 1994. The apoptotic suppressor P35 is required early during baculovirus replication and is targeted to the cytosol of infected cells. *J. Virol.* 68:3467–77

209. Cartier JL, Hershberger PA, Friesen PD. 1994. Suppression of apoptosis in insect cells stably transfected with baculovirus *p35*: dominant interference by N-terminal sequences p35(1–76). *J. Virol.* 68:7728–37

210. Beidler DR, Tewari M, Friesen PD, Poirier G, Dixit VM. 1995. The baculovirus p35 protein inhibits Fas- and tumor necrosis factor–induced apoptosis. *J. Biol. Chem.* 270:16526–28

211. Datta R, Kojima H, Banach D, Bump NJ, Talanian RV, et al. W. 1997. Activation of a CrmA-insensitive, p35-sensitive pathway in ionizing radiation-induced apoptosis. *J. Biol. Chem.* 272:1965–69

212. Resnicoff M, Valentinis B, Herbert D, Abraham D, Friesen PD, et al. 1998. The baculovirus anti-apoptotic p35 protein promotes transformation of mouse embryo fibroblasts. *J. Biol. Chem.* 273:10376–80

213. Xue D, Horvitz HR. 1995. Inhibition of the *Caenorhabditis elegans* cell-death protease CED-3 by a CED-3 cleavage site in baculovirus p35 protein. *Nature* 377:248–51

214. Bump NJ, Hackett M, Hugunin M, Seshagiri S, Brady K, et al. 1995. Inhibition of ICE family proteases by baculovirus anti-apoptotic protein p35. *Science* 269:1885–88

215. Bertin J, Mendrysa SM, LaCount DJ, Gaur S, Krebs JF, et al. 1996. Apoptotic suppression by baculovirus P35 involves cleavage by and inhibition of a virus-induced CED-3/ICE-like protease. *J. Virol.* 70:6251–59

216. Zhou Q, Krebs JF, Snipas SJ, Price A, Alnemri ES, et al. 1998. Interaction of the baculovirus anti-apoptotic protein p35 with caspases. Specificity, kinetics, and characterization of the caspase/p35 complex. *Biochemistry* 37:10757–65

217. Potempa J, Korzus E, Travis J. 1994. The serpin superfamily of proteinase inhibitors: structure, function, and regulation. *J. Biol. Chem.* 269:15957–60

218. Tewari M, Telford WG, Miller RA, Dixit VM. 1995. CrmA, a poxvirus-encoded serpin, inhibits cytotoxic T-lymphocyte-mediated apoptosis. *J. Biol. Chem.* 270:22705–8

219. Zhou Q, Snipas S, Orth K, Muzio M, Dixit VM, Salvesen GS. 1997. Target protease specificity of the viral serpin CrmA. Analysis of five caspases. *J. Biol. Chem.* 272:7797–800

220. Macen JL, Garner RS, Musy PY, Brooks MA, Turner PC, et al. 1996. Differential inhibition of the Fas- and granule-mediated cytolysis pathways by the orthopoxvirus cytokine response modifier A/SPI-2 and SPI-1 protein. *Proc. Natl. Acad. Sci. USA* 93:9108–13

221. Dobbelstein M, Shenk T. 1996. Protection against apoptosis by the vaccinia virus SPI-2 (B13R) gene product. *J. Virol.* 70:6479–85

222. Brooks MA, Ali AN, Turner PC, Moyer RW. 1995. A rabbitpox virus serpin gene controls host range by inhibiting apoptosis in restrictive cells. *J. Virol.* 69:7688–98

223. Petit PX, Susin SA, Zamzami N, Mignotte B, Kroemer G. 1996. Mitochondria and programmed cell death: back to the future. *FEBS Lett.* 396:7–13

224. Messud-Petit F, Gelfi J, Delverdier M, Amardeilh MF, Py R, et al. 1998. Serp2, an inhibitor of the interleukin-1beta-converting enzyme, is critical in the pathobiology of myxoma virus. *J. Virol.* 72:7830–39

225. Upton C, Macen JL, Wishart DS, McFadden G. 1990. Myxoma virus and malignant rabbit fibroma virus encode a serpin-like protein important for virus virulence. *Virology* 179:618–31

226. Macen JL, Upton C, Nation N, McFadden G. 1993. SERP1, a serine proteinase inhibitor encoded by myxoma virus, is a secreted glycoprotein that interferes with inflammation. *Virology* 195:348–63

227. Crook NE, Clem RJ, Miller LK. 1993. An apoptosis-inhibiting baculovirus gene with a zinc finger-like motif. *J. Virol.* 67:2168–74

228. Birnbaum MJ, Clem RJ, Miller LK. 1994. An apoptosis-inhibiting gene from a nuclear polyhedrosis virus encoding a

polypeptide with Cys/His sequence motifs. *J. Virol.* 68:2521–28

229. Clem RJ, Robson M, Miller LK. 1994. Influence of infection route on the infectivity of baculovirus mutants lacking the apoptosis-inhibiting gene p35 and the adjacent gene p94. *J. Virol.* 68:6759–62

230. Rothe M, Pan MG, Henzel WJ, Ayres TM, Goeddel DV. 1995. The TNFR2-TRAF signaling complex contains two novel proteins related to baculoviral inhibitor of apoptosis proteins. *Cell* 83:1243–52

231. Hay BA, Wassarman DA, Rubin GM. 1995. *Drosophila* homologs of baculovirus inhibitor of apoptosis proteins function to block cell death. *Cell* 83:1253–62

232. Liston P, Roy N, Tamai K, Lefebvre C, Baird S, et al. 1996. Suppression of apoptosis in mammalian cells by NAIP and a related family of IAP genes. *Nature* 379:349–53

233. Deveraux QL, Takahashi R, Salvesen GS, Reed JC. 1997. X-linked IAP is a direct inhibitor of cell-death proteases. *Nature* 388:300–4

234. Roy N, Deveraux QL, Takahashi R, Salvesen GS, Reed JC. 1997. The c-IAP-1 and c-IAP-2 proteins are direct inhibitors of specific caspases. *EMBO J.* 16:6914–25

235. Manji GA, Hozak RR, LaCount DJ, Friesen PD. 1997. Baculovirus inhibitor of apoptosis functions at or upstream of the apoptotic suppressor P35 to prevent programmed cell death. *J. Virol.* 71:4509–16

236. Orth K, Dixit VM. 1997. Bik and Bak induce apoptosis downstream of CrmA but upstream of inhibitor of apoptosis. *J. Biol. Chem.* 272:8841–44

237. Harvey AJ, Bidwai AP, Miller LK. 1997. Doom, a product of the *Drosophila mod (mdg4)* gene, induces apoptosis and binds to baculovirus inhibitor-of-apoptosis proteins. *Mol. Cell. Biol.* 17:2835–43

238. Vucic D, Kaiser WJ, Harvey AJ, Miller

LK. 1997. Inhibition of reaper-induced apoptosis by interaction with inhibitor of apoptosis proteins (IAPs). *Proc. Natl. Acad. Sci. USA* 94:10183–88

239. Clem RJ, Miller LK. 1994. Control of programmed cell death by the baculovirus genes *p35* and *iap*. *Mol. Cell. Biol.* 14:5212–22

240. Takahashi R, Deveraux Q, Tamm I, Welsh K, Assa-Munt N, et al. 1998. A single BIR domain of XIAP sufficient for inhibiting caspases. *J. Biol. Chem.* 273:7787–90

241. Chacon MR, Almazan F, Nogal ML, Vinuela E, Rodriguez JF. 1995. The African swine fever virus IAP homolog is a late structural polypeptide. *Virology* 214:670–74

242. Neilan JG, Lu Z, Kutish GF, Zsak L, Burrage TG, et al. 1997. A BIR motif containing gene of African swine fever virus, 4CL, is nonessential for growth in vitro and viral virulence. *Virology* 230:252–64

243. Ezoe H, Fatt RB, Mak S. 1981. Degradation of intracellular DNA in KB cells infected with *cyt* mutants of human adenovirus type 12. *J. Virol.* 40:20–27

244. Pilder S, Logan J, Shenk T. 1984. Deletion of the gene encoding the adenovirus 5 early region 1b 21,000-molecular-weight polypeptide leads to degradation of viral and host cell DNA. *J. Virol.* 52:664–71

245. Takemori N, Cladaras C, Bhat B, Conley AJ, Wold WS. 1984. *Cyt* gene of adenoviruses 2 and 5 is an oncogene for transforming function in early region E1B and encodes the E1B 19,000-molecular-weight polypeptide. *J. Virol.* 52:793–805

246. Subramanian T, Kuppuswamy M, Mak S, Chinnadurai G. 1984. Adenovirus *cyt*⁺ locus, which controls cell transformation and tumorigenicity, is an allele of *lp*⁺ locus, which codes for a 19-kilodalton tumor antigen. *J. Virol.* 52:336–43

247. Subramanian T, Kuppuswamy M, Gysbers J, Mak S, Chinnadurai G. 1984. 19-kDa tumor antigen coded by early region E1b of adenovirus 2 is required for effi-

cient synthesis and for protection of viral DNA. *J. Biol. Chem.* 259:11777–83

248. White E, Grodzicker T, Stillman BW. 1984. Mutations in the gene encoding the adenovirus early region 1B 19,000-molecular-weight tumor antigen cause the degradation of chromosomal DNA. *J. Virol.* 52:410–19

249. White E, Cipriani R, Sabbatini P, Denton A. 1991. Adenovirus E1B 19-kilodalton protein overcomes the cytotoxicity of E1A proteins. *J. Virol.* 65:2968–78

250. McLorie W, McGlade CJ, Takayesu D, Branton PE. 1991. Individual adenovirus E1B proteins induce transformation independently but by additive pathways. *J. Gen. Virol.* 72:1467–71

251. Rao L, Debbas M, Sabbatini P, Hockenbery D, Korsmeyer S, White E. 1992. The adenovirus E1A proteins induce apoptosis, which is inhibited by the E1B 19-kDa and Bcl-2 proteins. *Proc. Natl. Acad. Sci. USA* 89:7742–46; Published erratum appears in *Proc. Natl. Acad. Sci. USA* 89:9974.

252. Boyd JM, Malstrom S, Subramaniam T, Venkatesh LK, Schaeper U, Elangovan B, D'Sa-Eipper C, Chinnadurai G. 1994. Adenovirus E1B 19 kDa and Bcl-2 proteins interact with a common set. *Cell* 79:341–51; Published erratum appears in *Cell* 79:1120

253. Farrow SN, White JH, Martinou I, Raven T, Pun KT, et al. 1995. Cloning of a *bcl-2* homologue by interaction with adenovirus E1B 19K. *Nature* 374:731–33. Published erratum appears in *Nature* 375:431

254. Han J, Sabbatini P, White E. 1996. Induction of apoptosis by human Nbk/Bik, a BH3-containing protein that interacts with E1B 19K. *Mol. Cell. Biol.* 16:5857–64

255. Han J, Sabbatini P, Perez D, Rao L, Modha D, White E. 1996. The E1B 19K protein blocks apoptosis by interacting with and inhibiting the p53-inducible and death-promoting Bax protein. *Genes Dev.* 10:461–77

256. White E, Cipriani R. 1989. Specific disruption of intermediate filaments and the nuclear lamina by the 19-kDa product of the adenovirus E1B oncogene. *Proc. Natl. Acad. Sci. USA* 86:9886–90

257. Rao L, Modha D, White E. 1997. The E1B 19K protein associates with lamins in vivo and its proper localization is required for inhibition of apoptosis. *Oncogene* 15:1587–97

258. Revilla Y, Cebrian A, Baixeras E, Martinez C, Vinuela E, Salas ML. 1997. Inhibition of apoptosis by the African swine fever virus Bcl-2 homologue: role of the BH1 domain. *Virology* 228:400–4

259. Brun A, Rivas C, Esteban M, Escribano JM, Alonso C. 1996. African swine fever virus gene A179L, a viral homologue of *bcl-2*, protects cells from programmed cell death. *Virology* 225:227–30

260. Afonso CL, Neilan JG, Kutish GF, Rock DL. 1996. An African swine fever virus Bcl-2 homolog, 5-HL, suppresses apoptotic cell death. *J. Virol.* 70:4858–63

261. Cheng EH, Nicholas J, Bellows DS, Hayward GS, Guo HG, et al. 1997. A Bcl-2 homolog encoded by Kaposi sarcoma-associated virus, human herpesvirus 8, inhibits apoptosis but does not heterodimerize with Bax or Bak. *Proc. Natl. Acad. Sci. USA* 94:690–94

262. Nava VE, Cheng EH, Veliuona M, Zou S, Clem RJ, et al. 1997. Herpesvirus saimiri encodes a functional homolog of the human *bcl-2* oncogene. *J. Virol.* 71:4118–22

263. Derfuss T, Fickenscher H, Kraft MS, Henning G, Lengenfelder D, et al. 1998. Antiapoptotic activity of the herpesvirus saimiri-encoded Bcl-2 homolog: stabilization of mitochondria and inhibition of caspase-3-like activity. *J. Virol.* 72:5897–904

264. Kraft MS, Henning G, Fickenscher H, Lengenfelder D, Tschopp J, et al. 1998. Herpesvirus saimiri transforms human T-

cell clones to stable growth without inducing resistance to apoptosis. *J. Virol.* 72:3138–45

265. Henderson S, Huen D, Rowe M, Dawson C, Johnson G, Rickinson A. 1993. Epstein-Barr virus-coded BHRF1 protein, a viral homologue of Bcl-2, protects human B cells from programmed cell death. *Proc. Natl. Acad. Sci. USA* 90:8479–83

266. Kawanishi M. 1997. Epstein-Barr virus BHRF1 protein protects intestine 407 epithelial cells from apoptosis induced by tumor necrosis factor alpha and anti-Fas antibody. *J. Virol.* 71:3319–22

267. McCarthy NJ, Hazlewood SA, Huen DS, Rickinson AB, Williams GT. 1996. The Epstein-Barr virus gene BHRF1, a homologue of the cellular oncogene Bcl-2, inhibits apoptosis induced by gamma radiation and chemotherapeutic drugs. *Adv. Exp. Med. Biol.* 406:83–97

268. Fanidi A, Hancock DC, Littlewood TD. 1998. Suppression of c-Myc-induced apoptosis by the Epstein-Barr virus gene product BHRF1. *J. Virol.* 72:8392–95

269. Tarodi B, Subramanian T, Chinnadurai G. 1994. Epstein-Barr virus BHRF1 protein protects against cell death induced by DNA-damaging agents and heterologous viral infection. *Virology* 201:404–7

270. Virgin HWT, Latreille P, Wamsley P, Hallsworth K, Weck KE, et al. 1997. Complete sequence and genomic analysis of murine gammaherpesvirus 68. *J. Virol.* 71:5894–904

271. Conzen SD, Snay CA, Cole CN. 1997. Identification of a novel antiapoptotic functional domain in simian virus 40 large T antigen. *J. Virol.* 71:4536–43

272. Sarnow P, Ho YS, Williams J, Levine AJ. 1982. Adenovirus E1b-58kd tumor antigen and SV40 large tumor antigen are physically associated with the same 54 kd cellular protein in transformed cells. *Cell* 28:387–94

273. Kao CC, Yew PR, Berk AJ. 1990. Do-

mains required for in vitro association between the cellular p53 and the adenovirus 2 E1B 55K proteins. *Virology* 179:806–14

274. Yew PR, Berk AJ. 1992. Inhibition of p53 transactivation required for transformation by adenovirus early 1B protein. *Nature* 357:82–85

275. Lin J, Chen J, Elenbaas B, Levine AJ. 1994. Several hydrophobic amino acids in the p53 amino-terminal domain are required for transcriptional activation, binding to Mdm-2 and the adenovirus 5 E1B 55-kD protein. *Genes Dev.* 8:1235–46

276. Yew PR, Liu X, Berk AJ. 1994. Adenovirus E1B oncoprotein tethers a transcriptional repression domain to p53. *Genes Dev.* 8:190–202

277. Teodoro JG, Halliday T, Whalen SG, Takayesu D, Graham FL, Branton PE. 1994. Phosphorylation at the carboxy terminus of the 55-kilodalton adenovirus type 5 E1B protein regulates transforming activity. *J. Virol.* 68:776–86

278. Teodoro JG, Branton PE. 1997. Regulation of p53-dependent apoptosis, transcriptional repression, and cell transformation by phosphorylation of the 55-kilodalton E1B protein of human adenovirus type 5. *J. Virol.* 71:3620–27

279. Martin ME, Berk AJ. 1998. Adenovirus E1B 55K represses p53 activation in vitro. *J. Virol.* 72:3146–54

280. Marcellus RC, Teodoro JG, Charbonneau R, Shore GC, Branton PE. 1996. Expression of p53 in Saos-2 osteosarcoma cells induces apoptosis which can be inhibited by Bcl-2 or the adenovirus E1B-55 kDa protein. *Cell Growth Differ.* 7:1643–50

281. Bischoff JR, Kirn DH, Williams A, Heise C, Horn S, et al. 1996. An adenovirus mutant that replicates selectively in p53-deficient human tumor cells. *Science* 274:373–76

282. Heise C, Sampson-Johannes A, Williams A, McCormick F, Von Hoff DD, Kirn

DH. 1997. ONYX-015, an E1B gene-attenuated adenovirus, causes tumor-specific cytolysis and antitumoral efficacy that can be augmented by standard chemotherapeutic agents. *Nat. Med.* 3:639–45

283. Rothmann T, Hengstermann A, Whitaker NJ, Scheffner M, zur Hausen H. 1998. Replication of ONYX-015, a potential anticancer adenovirus, is independent of p53 status in tumor cells. *J. Virol.* 72:9470–78

284. Hall AR, Dix BR, O'Carroll SJ, Braithwaite AW. 1998. p53-dependent cell death/apoptosis is required for a productive adenovirus infection. *Nat. Med.* 4:1068–72

285. Kirn D, Hermiston T, McCormick F. 1998. ONYX-015: clinical data are encouraging. *Nat. Med.* 4:1341–42

286. Dobner T, Horikoshi N, Rubenwolf S, Shenk T. 1996. Blockage by adenovirus E4orf6 of transcriptional activation by the p53 tumor suppressor. *Science* 272:1470–73

287. Moore M, Horikoshi N, Shenk T. 1996. Oncogenic potential of the adenovirus E4orf6 protein. *Proc. Natl. Acad. Sci. USA* 93:11295–301

288. Nevels M, Rubenwolf S, Spruss T, Wolf H, Dobner T. 1997. The adenovirus E4orf6 protein can promote E1A/E1B-induced focus formation by interfering with p53 tumor suppressor function. *Proc. Natl. Acad. Sci. USA* 94:1206–11

289. Querido E, Marcellus RC, Lai A, Charbonneau R, Teodoro JG, et al. 1997. Regulation of p53 levels by the E1B 55-kilodalton protein and E4orf6 in adenovirus-infected cells. *J. Virol.* 71: 3788–98

290. Steegenga WT, Riteco N, Jochemsen AG, Fallaux FJ, Bos JL. 1998. The large E1B protein together with the E4orf6 protein target p53 for active degradation in adenovirus infected cells. *Oncogene* 16:349–57

291. Farmer G, Bargonetti J, Zhu H, Friedman P, Prywes R, Prives C. 1992. Wild-type p53 activates transcription in vitro. *Nature* 358:83–86

292. Bargonetti J, Reynisdottir I, Friedman PN, Prives C. 1992. Site-specific binding of wild-type p53 to cellular DNA is inhibited by SV40 T antigen and mutant p53. *Genes Dev.* 6:1886–98

293. Mietz JA, Unger T, Huibregtse JM, Howley PM. 1992. The transcriptional transactivation function of wild-type p53 is inhibited by SV40 large T-antigen and by HPV-16 E6 oncoprotein. *EMBO J.* 11:5013–20

294. Werness BA, Levine AJ, Howley PM. 1990. Association of human papillomavirus types 16 and 18 E6 proteins with p53. *Science* 248:76–79

295. Scheffner M, Werness BA, Huibregtse JM, Levine AJ, Howley PM. 1990. The E6 oncoprotein encoded by human papillomavirus types 16 and 18 promotes the degradation of p53. *Cell* 63:1129–36

296. Crook T, Tidy JA, Vousden KH. 1991. Degradation of p53 can be targeted by HPV E6 sequences distinct from those required for p53 binding and transactivation. *Cell* 67:547–56

297. Scheffner M, Huibregtse JM, Vierstra RD, Howley PM. 1993. The HPV-16 E6 and E6-AP complex functions as a ubiquitin-protein ligase in the ubiquitination of p53. *Cell* 75:495–505

298. Thomas M, Massimi P, Jenkins J, Banks L. 1995. HPV-18 E6 mediated inhibition of p53 DNA binding activity is independent of E6 induced degradation. *Oncogene* 10:261–68

299. Li X, Coffino P. 1996. High-risk human papillomavirus E6 protein has two distinct binding sites within p53, of which only one determines degradation. *J. Virol.* 70:4509–16

300. Wang XW, Gibson MK, Vermeulen W, Yeh H, Forrester K, et al. 1995. Abro-

gation of p53-induced apoptosis by the hepatitis B virus X gene. *Cancer Res.* 55: 6012–16

301. Elmore LW, Hancock AR, Chang SF, Wang XW, Chang S, et al. 1997. Hepatitis B virus X protein and p53 tumor suppressor interactions in the modulation of apoptosis. *Proc. Natl. Acad. Sci. USA* 94:14707–12

302. Zhu H, Shen Y, Shenk T. 1995. Human cytomegalovirus IE1 and IE2 proteins block apoptosis. *J. Virol.* 69:7960–70

303. Muganda P, Carrasco R, Qian Q. 1998. The human cytomegalovirus IE2 86 kDa protein elevates p53 levels and transactivates the p53 promoter in human fibroblasts. *Cell Mol. Biol.* 44:321–31

304. Tsai HL, Kou GH, Chen SC, Wu CW, Lin YS. 1996. Human cytomegalovirus immediate-early protein IE2 tethers a transcriptional repression domain to p53. *J. Biol. Chem.* 271:3534–40

305. Kovacs A, Weber ML, Burns LJ, Jacob HS, Vercellotti GM. 1996. Cytoplasmic sequestration of p53 in cytomegalovirus-infected human endothelial cells. *Am. J. Pathol* 149:1531–39

306. Silins SL, Sculley TB. 1995. Burkitt's lymphoma cells are resistant to programmed cell death in the presence of the Epstein-Barr virus latent antigen EBNA-4. *Int. J. Cancer* 60:65–72

307. Schang LM, Hossain A, Jones C. 1996. The latency-related gene of bovine herpesvirus 1 encodes a product which inhibits cell cycle progression. *J. Virol.* 70: 3807–14

308. Jiang Y, Hossain A, Winkler MT, Holt T, Doster A, Jones C. 1998. A protein encoded by the latency-related gene of bovine herpesvirus 1 is expressed in trigeminal ganglionic neurons of latently infected cattle and interacts with cyclin-dependent kinase 2 during productive infection. *J. Virol.* 72:8133–42

309. Mulloy JC, Kislyakova T, Cereseto A, Casareto L, LoMonico A, et al. 1998. Human T-cell lymphotropic/leukemia virus type 1 Tax abrogates p53-induced cell cycle arrest and apoptosis through its CREB/ATF functional domain. *J. Virol.* 72:8852–60

310. Brauweiler A, Garrus JE, Reed JC, Nyborg JK. 1997. Repression of *bax* gene expression by the HTLV-1 Tax protein: implications for suppression of apoptosis in virally infected cells. *Virology* 231:135–40

311. Hiscott J, Petropoulos L, Lacoste J. 1995. Molecular interactions between HTLV-1 Tax protein and the NF-κB/IκB transcription complex. *Virology* 214:3–11

312. DeLuca C, Kwon H, Pelletier N, Wainberg MA, Hiscott J. 1998. NF-κB protects HIV-1-infected myeloid cells from apoptosis. *Virology* 244:27–38

313. Ayyavoo V, Mahboubi A, Mahalingam S, Ramalingam R, Kudchodkar S, et al. 1997. HIV-1 Vpr suppresses immune activation and apoptosis through regulation of nuclear factor kappa B. *Nat. Med.* 3:1117–23

314. Zauli G, Gibellini D, Milani D, Mazzoni M, Borgatti P, et al. 1993. Human immunodeficiency virus type 1 Tat protein protects lymphoid, epithelial, and neuronal cell lines from death by apoptosis. *Cancer Res.* 53:4481–85

315. Zauli G, Gibellini D, Celeghini C, Mischiati C, Bassini A, et al. 1996. Pleiotropic effects of immobilized versus soluble recombinant HIV-1 Tat protein on CD3-mediated activation, induction of apoptosis, and HIV-1 long terminal repeat transactivation in purified CD4+ T lymphocytes. *J. Immunol.* 157:2216–24

316. Ray RB, Meyer K, Ray R. 1996. Suppression of apoptotic cell death by hepatitis C virus core protein. *Virology* 226:176–82

317. Ray RB, Steele R, Meyer K, Ray R. 1997. Transcriptional repression of p53 promoter by hepatitis C virus core protein. *J. Biol. Chem.* 272:10983–86

318. Borden KL, CampbellDwyer EJ, Salvato MS. 1997. The promyelocytic leukemia protein PML has a pro-apoptotic activity mediated through its RING domain. *FEBS Lett.* 418:30–34

319. Shisler JL, Senkevich TG, Berry MJ, Moss B. 1998. Ultraviolet-induced cell death blocked by a selenoprotein from a human dermatotropic poxvirus [see comments]. *Science* 279:102–5

320. Ink BS, Gilbert CS, Evan GI. 1995. Delay of vaccinia virus-induced apoptosis in nonpermissive Chinese hamster ovary cells by the cowpox virus CHOhr and adenovirus E1B 19K genes. *J. Virol.* 69:661–68

321. Barry M, Hnatiuk S, Mossman K, Lee SF, Boshkov L, McFadden G. 1997. The myxoma virus M-T4 gene encodes a novel RDEL-containing protein that is retained within the endoplasmic reticulum and is important for the productive infection of lymphocytes. *Virology* 239:360–77

322. Macen JL, Graham KA, Lee SF, Schreiber M, Boshkov LK, McFadden G. 1996. Expression of the myxoma virus tumor necrosis factor receptor homologue and M11L genes is required to prevent virus-induced apoptosis in infected rabbit T lymphocytes. *Virology* 218:232–37

323. Mossman K, Lee SF, Barry M, Boshkov L, McFadden G. 1996. Disruption of M-T5, a novel myxoma virus gene member of poxvirus host range superfamily, results in dramatic attenuation of myxomatosis in infected European rabbits. *J. Virol.* 70:4394–410

324. Brick DJ, Burke RD, Schiff L, Upton C. 1998. Shope fibroma virus RING finger protein N1R binds DNA and inhibits apoptosis. *Virology* 249:42–51

325. Ghadge GD, Ma L, Sato S, Kim J, Roos RP. 1998. A protein critical for a Theiler's virus-induced immune system-mediated demyelinating disease has a cell type-

specific antiapoptotic effect and a key role in virus persistence. *J. Virol.* 72:8605–12

326. Leopardi R, Van Sant C, Roizman B. 1997. The herpes simplex virus 1 protein kinase U_S3 is required for protection from apoptosis induced by the virus. *Proc. Natl. Acad. Sci. USA* 94:7891–96

327. Galvan V, Roizman B. 1998. Herpes simplex virus 1 induces and blocks apoptosis at multiple steps during infection and protects cells from exogenous inducers in a cell-type-dependent manner. *Proc. Natl. Acad. Sci. USA* 95:3931–36

328. Liu JL, Ye Y, Lee LF, Kung HJ. 1998. Transforming potential of the herpesvirus oncoprotein MEQ: morphological transformation, serum-independent growth, and inhibition of apoptosis. *J. Virol.* 72:388–95

329. Hulst MM, Panoto FE, Hoekman A, van Gennip HG, Moormann RJ. 1998. Inactivation of the RNase activity of glycoprotein E(rns) of classical swine fever virus results in a cytopathogenic virus. *J. Virol.* 72:151–57

330. Bruschke CJ, Hulst MM, Moormann RJ, van Rijn PA, van Oirschot JT. 1997. Glycoprotein E^{rns} of pestiviruses induces apoptosis in lymphocytes of several species. *J. Virol.* 71:6692–96

331. Tolskaya EA, Romanova LI, Kolesnikova MS, Ivannikova TA, Smirnova EA, et al. 1995. Apoptosis-inducing and apoptosis-preventing functions of poliovirus. *J. Virol.* 69:1181–89

332. Desaintes C, Demeret C, Goyat S, Yaniv M, Thierry F. 1997. Expression of the papillomavirus E2 protein in HeLa cells leads to apoptosis. *EMBO J.* 16:504–14

333. McCarthy SA, Symonds HS, Van Dyke T. 1994. Regulation of apoptosis in transgenic mice by simian virus 40 T antigen-mediated inactivation of p53. *Proc. Natl. Acad. Sci. USA* 91:3979–83

334. Dahl J, Jurczak A, Cheng LA, Baker

DC, Benjamin TL. 1998. Evidence of a role for phosphatidylinositol 3-kinase activation in the blocking of apoptosis by polyomavirus middle T antigen. *J. Virol.* 72:3221–26

335. Webster MA, Hutchinson JN, Rauh MJ, Muthuswamy SK, Anton M, et al. 1998. Requirement for both Shc and phosphatidylinositol 3′ kinase signaling pathways in polyomavirus middle T-mediated mammary tumorigenesis. *Mol. Cell. Biol.* 18:2344–59

Annu. Rev. Microbiol. 1999. 53:629–55

THE CYTOSKELETON OF TRYPANOSOMATID PARASITES

Keith Gull

School of Biological Sciences, University of Manchester, Manchester M13 9PT, United Kingdom; e-mail: K.Gull@man.ac.uk

Key Words   cytoskeleton, microtubule, flagellum, parasite, trypanosome, Leishmania, kinetoplast, protozoa

■ **Abstract**   Species of the trypanosomatid parasite genera *Trypanosoma* and *Leishmania* exhibit a particular range of cell shapes that are defined by their internal cytoskeletons. The cytoskeleton is characterized by a subpellicular corset of microtubules that are cross-linked to each other and to the plasma membrane. Trypanosomatid cells possess an extremely precise organization of microtubules and filaments, with some of their organelles, such as the mitochondria, kinetoplasts, basal bodies, and flagella, present as single copies in each cell. The duplication of these structures and changes in their position during life cycle differentiations provide markers and insight into events involved in determining cell form and division. We have a rapidly increasing catalog of these structures, their molecular cytology, and their ontogeny. The current sophistication of available molecular genetic techniques for use in these organisms has allowed a new functional analysis of the cytoskeleton, including functions that are intrinsic to the proliferation and pathogenicity of these parasites.

CONTENTS

0066-4227/99/1001-0629$08.00

INTRODUCTION

It is axiomatic that the cells of eukaryotic microorganisms have such fidelity of shape and form that much of their classification and identification can be based on these structural parameters. The cell shape and form of algae, filamentous fungi, and yeasts depend to various degrees on the morphogenetic processes in the surrounding cell wall. However, in many protozoa the overall form of the cell is essentially the product of the internal cytoskeleton. The ciliates have often been used as a particular model (11) for these morphogenetic processes; however, it is now recognized that the trypanosomatid flagellates are particularly interesting, experimentally tractable, and informative in this context.

The family Trypanosomatidae comprises a major group of flagellated digenetic parasites that cause diseases of humans and animals, including sleeping sickness in Africa (*Trypanosoma brucei*) and Chagas' disease in Central and South America (*Trypanosoma cruzi*). Species of *Leishmania* cause various forms of leishmaniasis in much of the tropical and subtropical world. Other genera such as the *Crithidia* are monogenetic parasites of insects. These organisms have specific cell shapes during particular parts of their life cycles. The shapes are characterized by the presence and position of the basal bodies, flagellum, and kinetoplast. A number of morphological types are recognized, the most important of which are illustrated in Figure 1. There are, however, unifying structures and morphogenetic principles that underlie all of these apparently diverse cellular forms. This review emphasizes this unity and concentrates on integrating information on a number of members of the Trypanosomatidae. Most often, however, *T. brucei* is treated as the archetypal organism, particularly when descriptions of structural organization are provided.

The shape and form of uniflagellated trypanosomatids are maintained by a subpellicular corset of microtubules, which are cross-linked to each other and to the plasma membrane. The trypanosomatid cell possesses an extremely precise organization of microtubules and filaments, with some of its organelles, such as the mitochondrion, kinetoplast, basal body, and flagellum, occurring as single copies. The duplication of these structures and changes in their position during life cycle differentiations provide markers and insight into events involved in determining cell form and division.

Examination of the cytoskeleton of a *T. brucei* procyclic cell (91) (the tsetse midgut, trypomastigote form) serves to illustrate the basic organization that characterizes those trypanosomatids that possess an attached flagellum (Figure 2). The single flagellum exits from the flagellar pocket at the posterior end of the cell and is attached to the cell body along its length by a unique cytoplasmic

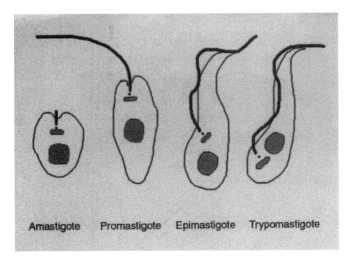

Figure 1 The major cellular forms of trypanosomatids defined by cell shape, flagellum presence (and whether the flagellum is attached or free), and then the position of the basal body, kinetoplast, and nucleus.

filament/microtubule system called the flagellum attachment zone (FAZ). The FAZ is an invariant and precisely defined structure that occurs at a unique site where, in a gap between two microtubules of the subpellicular corset, a single filament runs down the cell. At this site, the flagellar membrane and the cell body membrane are in very close contact, and the flagellum is cross-linked to this internal cytoplasmic filament. Moreover, at one side of the filament, always to its immediate left when sections are viewed from the posterior of the cell, is a unique set of four microtubules that are closely associated with the smooth endoplasmic reticulum. The flagellum contains, in addition to the axoneme, another major structure—the paraflagellar rod. Variations on this basic theme can be recognized in the cytoskeleton of other forms of trypanosomatids.

CYTOSKELETAL STRUCTURE AND MOLECULAR ORGANIZATION

The Subpellicular Corset

An extensive body of literature describes the internal subpellicular corset of microtubules in a wide range of trypanosomatids (41, 51, 104). Transverse sections can reveal >100 microtubule profiles underlying the plasma membrane and forming the subpellicular corset. The 24-nm-diameter microtubules form a helical pattern along the long axis of the cell, with a regular intermicrotubule spacing (18–22 nm), and are cross-linked to each other and to the plasma membrane. Individual

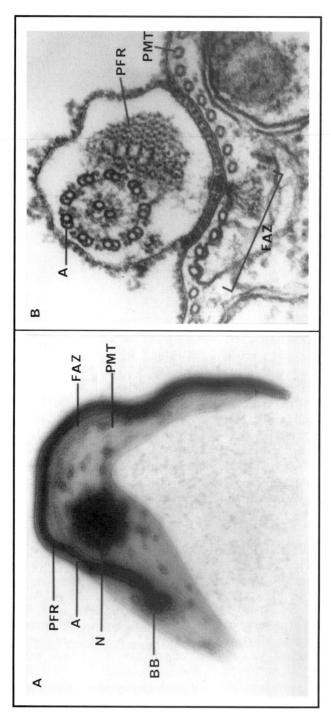

Figure 2 *A.* Negatively stained electron microscopic image of a *Trypanosoma brucei* cytoskeleton after detergent extraction. Flagellar attachment zone (*FAZ*), basal body (*BB*), paraflagellar rod (*PFR*), subpellicular microtubules (*PMT*), axoneme (*A*), nucleus (*N*). *B.* Transverse section through a *T. brucei* cell showing the flagellar attachment zone (*FAZ*). The four specialized microtubules are indicated as well as the FAZ filament and basal body (*BB*), paraflagellar rod (*PFR*), subpellicular microtubules (*PMT*), and axoneme (*A*).

microtubules within the corset of *T. brucei* procyclic cells vary in length. Examination of negatively stained cytoskeletons shows that, when a microtubule stops, those on either side of it continue and become connected partners. This construction allows for shape changes along the length of the parasite and no doubt between life cycle stages, without producing "weak spots" in the cortical array. A consequence of this form of patterning is that the number of microtubules seen in cross-sections of trypanosomes is a reflection of the diameter of the cell at that point. The cortical microtubules are very stable and remain attached to the plasma membrane during cell fractionation studies. Treatment by nonionic detergent produces a cytoskeleton, whose shape and form mimic that of the original cell (91).

Microtubules have an intrinsic polarity reflecting their constituent ($\alpha\beta$-tubulin) heterodimers. Three assays of polarity indicate that the microtubules of the *T. brucei* cortex all appear to have the same polarity—with their plus (+) ends at the posterior of the cell (78). This is the opposite orientation to that of the microtubules of the flagellar axoneme, which have their + ends (i.e. the distal end of the flagellum) at the anterior end of the cell. The subpellicular corset does not break down during cell division. Rather, new microtubules are added and intercalated into this array during the cell cycle, such that a complete microtubule corset is inherited by each daughter cell at cytokinesis in a semiconservative manner (92).

The FAZ (comprising the FAZ filament and four unique microtubules) occupies a particular point in the subpellicular corset. Because these four microtubules originate near the basal bodies, it is likely that they have the opposite polarity (+ end at the anterior end of the cell) to the main array (color Figure 3). Biochemically, these microtubules are more resistant to high-salt treatment and survive such procedures attached to the basal-body region of the flagellum (91).

The Mitotic Spindle

Certain characteristics of the mitotic spindle appear common to a number of trypanosomatids (32, 97, 103, 104). There is evidence for both kinetochore and longer "pole-to-pole" microtubules, but the microtubule-organizing centers (MTOCs) at the spindle poles are always indistinct and often appear to be located in the nucleoplasm, rather than on the inner face of the persistent nuclear envelope. However, the recent description of γ-tubulin dots in the dividing nucleus gives a first indication of discrete spindle pole MTOCs (89). The nucleolar material also persists in mitosis, becoming stretched along the spindle microtubules whose distinct structural phases can be visualized by antitubulin immunofluorescence (85).

Chromatin is indistinct, but plaque structures are consistently seen that have many of the structural and behavioral characteristics of kinetochores. Semiquantitative electron microscopic analysis of these structures, however, suggests that there is no clear relationship between their number and that of the chromosomes estimated by electrokaryotyping (97). Molecular analyses show *T. brucei* to have a set of 11 pairs of megabase chromosomes between 1 and 6 Mbp, 1–5 intermediate chromosomes (200–900 Kbp), and ~100 minichromosomes (50–150 Kbp).

The spindle, however, does not appear to possess enough microtubules for even individual microtubule/kinetochore interactions to be made. Recently, the presence and behavior of individual megabase chromosomes on the *T. brucei* spindle have been visualized by a combination of fluorescence in situ hybridization and antitubulin immunofluorescence (28). In addition, visualization of the minichromosome population showed that they are also associated with the spindle; their segregation is microtubule dependent, but the pattern is unusual. The minichromosome population is segregated with fidelity, associating first with the spindle at its center and then migrating to the extreme spindle poles. We have developed a model that could explain and accommodate this data. The lateral stacking model (36) explains the segregation of large-megabase chromosomes in a classical manner via kinetochore microtubules. The model predicts that the replicated minichromosome complexes bind preferentially at the spindle center between two antiparallel microtubules in a lateral attachment to each microtubule. This opens the possibility of several replicated minichromosomes being able to attach between antiparallel pairs of microtubules, thus allowing their segregation. After chromatid resolution, a directional (microtubule minus end-directed), poleward movement of individual minichromosomes could be accomplished by the involvement of microtubule motor proteins. The model accounts for many of the observed phenomena of chromosomes on the trypanosome spindle and may have wider implications for other spindles.

The Flagellum Attachment Zone

Structure and Biochemistry In trypanosomatids with attached flagella, there is a complex system of membrane connections, filaments, and specialized microtubules that define a FAZ along the length of the cell body.

The connection between the cell body and flagellum is seen as a series of opposed 25-nm-diameter junctional complexes on the cytoplasmic side of each membrane (91, 104). These complexes are spaced with a center-to-center periodicity of 95 nm along a line defined by the FAZ filament. When transverse sections are viewed from the posterior end of the cell, this FAZ filament has the four microtubules with attached smooth endoplasmic reticulum immediately to its left. The four microtubules originate from the basal body area and execute a half turn of the flagellar pocket to border the line of junctional complexes. The FAZ appears to be characterized by very immunogenic proteins often possessing repetitive motifs. Immunological approaches reveal a number of proteins that locate to this region of the *T. brucei* and *T. cruzi* cytoskeleton or to equivalent regions in other trypanosomatids (2, 18, 44, 51, 53, 65, 80, 110). Unfortunately, in most cases there is little information on their function within the FAZ complex. Antibodies to a 200-kDa protein have revealed the growth of the FAZ in relation to other cell cycle events in *T. brucei* (52).

Attachment and Position   The FAZ defines a particular site in the trypanosome corset. Moreover, the site is duplicated before cytokinesis and provides a positional link between the unique basal body/kinetoplast/flagellar-pocket area at the posterior of the trypanosome and the sites of cleavage initiation at the anterior (color Figure 3). This and experiments described later (5, 66, 78) have led us to argue that the FAZ acts as a critical positional and directional cellular template for cytokinesis.

Cross and colleagues have cloned the *T. brucei* gene encoding FLA1, a protein that was located along the flagellum region (69). A double-knockout deletion of the homologous gene (*GP72*) in *T. cruzi* resulted in trypanosomes that had detached flagella, impaired motility, and a reduction in the ability to infect the insect vector *Triatoma infestans* (17, 38, 68). Western blot studies showed the *T. brucei* protein to be larger than predicted (100 kDa for bloodstream forms; 80 kDa for procyclic cells). There was some evidence that this difference can be accounted for by differential glycosylation events. Moreover, gene knockout studies suggested that one allele could be deleted but not both, indicating that, in *T. brucei*, this is an essential gene (69). The distribution of this glycosylated protein along the FAZ is very reminiscent of a similarly sized glycoprotein detected by lectin-binding studies (108). Cross and colleagues (69) suggest that, given the position of the protein, its expression may be necessary for *T. brucei* cells to effect flagellum/basal body/kinetoplast segregation. Moreover, given the proposal that the FAZ may be critical for division in *T. brucei* (78), interference with such flagellum attachment morphogenesis is likely to be lethal.

The Paraflagellar Rod

Structure   In addition to the classical "9 + 2" microtubule axoneme, trypanosomatid flagella contain a large, latticelike structure, the paraflagellar rod (PFR), which runs alongside the axoneme, usually from its exit from the flagellar pocket, and is present in all life cycle stages except for the amastigote stage. The PFR is found exclusively in a restricted set of evolutionary ancient unicellular eukaryotes (kinetoplastida, euglenoids, and possibly the dinoflagellates) (3). PFR structure has been described in various organisms and is shown in Figure 2. Three distinct regions—proximal, intermediate, and distal—can be visualized relative to the axoneme. There are minor variations in this form and arrangement, but overall the PFRs of all kinetoplastids exhibit a very similar tripartite pattern of construction (1, 3, 30, 98). The PFR has a constant position relative to the flagellum axoneme, with the proximal region linked to axonemal doublets 4 through 7. This attachment is very stable, it persists on isolation of the flagellum, and it is resistant to nonionic detergent, high-salt treatment, and hypotonic shock. It is however very sensitive to mild trypsinization. In cells that possess an attached flagellum, there is a further fibrillar link between the PFR (particularly the proximal region) and the inner face of the flagellum membrane at a point precisely opposite the FAZ filament that runs up the internal face of the plasma membrane (91).

Composition   The first isolation of the PFR showed that the main components are a doublet of two proteins $\sim M_r$ 70×10^3 (83). In *T. brucei* the PFR is composed of two major (M_r 69×10^3 and 72×10^3) proteins, termed PFR-A and PFR-C, now known to be closely related in sequence. PFR-A and -C are each encoded by a similar cluster of four tandemly arranged genes (22). The 5′-untranslated region of the first gene in each cluster differs from those regions of the next three genes, which are identical to each other. Conversely, the 3′-untranslated regions of the first three genes of each cluster are identical, whereas the corresponding region of the final fourth gene is different. Hence, processing of the products of these regions produces a family of mRNAs with identical coding sequences but different 5′- and 3′-noncoding regions. This pattern of two abundant, closely related proteins is seen in *L. mexicana* and *T. cruzi*, in which the homologous genes have also been identified and sequenced (6, 31, 47, 64, 84). The homologous proteins in the major PFR doublet on gels are termed, respectively for *T. brucei*, *L. mexicana* and *T. cruzi*, slower migrating bands, PFR-C, PFR-1, and PAR-3; and faster migrating bands, PFR-A, PFR-2, and PAR-2. Monoclonal antibodies have been produced that recognize both or individual PFR proteins.

Given the complex architecture of the PFR, it is logical to assume that it is constructed from many different types of polypeptides (51). Some minor components have been identified, most usually by immunological means. The ROD-1 monoclonal antibody recognizes high-molecular-weight doublets (180×10^3 and 200×10^3) on immunoblots of *T. brucei* proteins. It seems likely that these proteins are encoded by the cDNA clone 5.20 (111), which encodes a protein with multiple 11-amino-acid imperfect repeats. Antibodies to this protein and the ROD-1 monoclonal antibody detect the outer distal face of the PFR in immunoelectron microscopy. Two other large polypeptides ($M_r >300 \times 10^3$) containing repetitive sequences have been found to locate, respectively, throughout the PFR and to the region of connections between the PFR and the axoneme (46). Two other interesting proteins of the *T. cruzi* PFR have been characterized and termed PAR-1 and PAR-4. They show very little homology to the major doublet PFR protein (31).

The major PFR proteins are highly immunogenic, and injection of purified PFR proteins has been shown to protect mice against *T. cruzi* infection. Differing cytokine levels suggest that protective immunity induced by the PFR proteins is associated with a Th-1 type response (62, 63).

Function   The role of the PFR has been the subject of much speculation, and the possibilities considered include roles in motility and in attachment to epithelia via the flagellum in the insect vector stages. The cloning of the genes encoding the major PFR proteins and the increasing sophistication of the molecular genetics in *T. brucei* and *Leishmania* spp. have now allowed direct tests of function. A role in motility of *L. mexicana* has been demonstrated by making a null mutant of PFR-2, one of the major PFR polypeptides. PFR-2 minus parasites grow and divide normally in culture and still express the PFR-1 protein (84). However, they lack most of the PFR, possessing only a residual inner-region substructure that contains small amounts of the PFR-1 protein, indicating that PFR-1 may attach to the axoneme in the absence of PFR-2. The PFR-2–null mutant displays pronounced changes in flagellar-beat waveform and forward swimming velocity. In *T. brucei*, attempts to delete both clusters of the *PFR-A* gene in the diploid genome led to the nonrecovery of progeny, which suggests that the protein has an essential function and that its absence is lethal to this organism (45). However, an antisense RNA approach has recently provided particular insight into the structure, assembly, and function of the PFR. Molecular ablation was achieved via insertion of the antisense construct at one of the two *PFR-A* loci within the diploid trypanosome genome (4). Mutant trypanosomes snl-1 showed a striking phenotype; they grew normally but sedimented to the bottom of the well and appeared paralyzed. Only a tiny amount of PFR-A mRNA and protein was present in the mutant trypanosome cell line. The snl-1 mutant cells lack the intermediate and distal regions of the PFR, and only a fraction of the smaller proximal region remains (Figure 4). However, and perhaps of some significance, the connections between this rudimentary PFR structure and the axoneme, as well as those to the FAZ, were still present. The bulk of the PFR-C protein in the mutant cell line remained soluble. Immunofluorescence studies indicated that this nonassembled PFR-C protein entered the new flagellum compartment and accumulated in a dilated flagellum tip. This "blob" of nonassembled PFR material showed cell cycle regulation. The blob was detected at the distal tip of the new flagellum and got bigger as this flagellum elongated. Division produced two daughter cells, of which only one possessed the blob. The fact that no cell entering its cell cycle possessed a blob on its flagellum means that this unassembled PFR material is resorbed at some point early in the G1 phase. There is evidence from experiments in *Chlamydomonas* spp. for the presence of transport systems that move material to and from the flagellum tip (16, 72). The behavior of PFR-C in the snl-1 mutant of *T. brucei* suggests the existence of such transport mechanisms in trypanosomatids and, further, a level of cell cycle control over their regulation.

Thus, these sets of experiments in *T. brucei* and *L. mexicana* have provided new insights into the function of the PFR-A/PFR-1 homolog in trypanosomatid parasite motility. Two different approaches, gene knockout and molecular ablation by antisense RNA, have produced intriguing phenotypes. The double-knockout experiments in *T. brucei* and *L. mexicana* suggest that, in the former, PFR-A is an essential protein (45), but it is not essential in the latter. The difficulty here lies in the problem that lack of progeny in a double-gene knockout experiment is always difficult to interpret as a definite outcome. However, if this difference between the two parasites is confirmed, then the active growth of the *snl-1* mutant would indicate that only a minuscule amount of PFR-A protein is required for survival. One could explain this by suggesting that the attached flagellum of *T. brucei* places more functional constraints on the PFR than the free flagellum of *L. mexicana*. Again, the *T. brucei* PFR-FAZ link has been implicated as part of a major structural site that is important for shape, polarity, and division in this trypanosome (78).

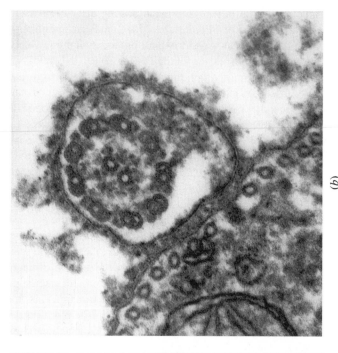

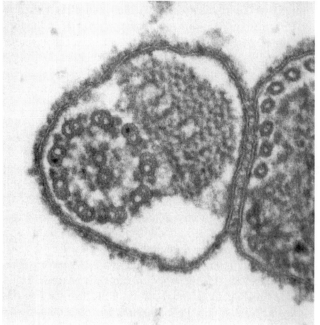

(a)

(b)

Figure 4 *a.* Electron micrograph of the paraflagellar rod of *Trypanosoma brucei* showing the three regions: proximal, intermediate, and distal. *b.* The flagellum of the *T. brucei snl-1* mutant showing the absence of normal paraflagellar rod structure.

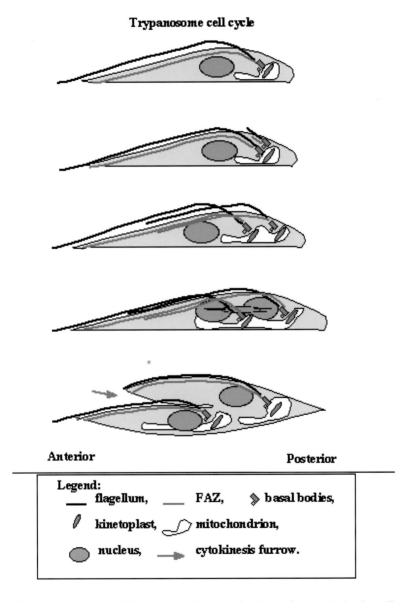

Trypanosome cell cycle

Anterior **Posterior**

Legend:
___ flagellum, ___ FAZ, ◈ basal bodies,
∅ kinetoplast, ⌒ mitochondrion,
● nucleus, → cytokinesis furrow.

Figure 4 A view of five stages of cytoskeletal morphogenesis in the cell cycle of a procyclic *T. brucei*.

Flagellum and Basal Bodies

The basal body complex is the only easily recognizable MTOC in *T. brucei*. This complex appears to be responsible both for the nucleation of the flagellar axoneme microtubules in the classical manner and for the specialized group of four microtubules that arises close to it. This concept that the basal body has additional roles as a positional MTOC to nucleate specific cytoplasmic microtubule arrays is reminiscent of the situation in many other protists (37). Interestingly, although basal body structure is well described in trypanosomatids, the concept of basal body duplication and maturation has been somewhat ignored. The existence of a second "barren" basal body is often overlooked. In their review, Vickerman & Preston drew attention to the fact that trypanosome cells sometimes exhibit four basal bodies (104). A difficulty is that the use of thin-section electron microscopy (EM) limits the ordering and interpretation of a temporal sequence. EM of negatively stained, complete cytoskeletons gave a much clearer view of the basal body cycle in procyclic *T. brucei*, and the pattern is seen to follow cell cycle-dependent replication and maturation. Cells at the beginning of the cell cycle have one basal body and one pro-basal body, and the basal body subtends the flagellum axoneme. Later in the cycle, the pro-basal body matures to a basal body, and morphogenesis of the new flagellum axoneme is initiated in addition to the formation of two pro-basal bodies, one next to each of the two mature basal bodies (91). Pro–basal body formation appears to occur almost coincident with the initiation of S phase (112), but we know little of the three-dimensional positioning phenomena that operate to ensure a correct spatial inheritance pattern with the new flagellum always at the posterior of the cell (see color Figure 3).

A lack of understanding of the basal body cycle in trypanosomes has probably led to a number of misinterpretations. For instance, reports of *T. brucei* metacyclic cells in division were based on descriptions of some cells with two basal bodies (99). However, consideration of the basal body cycle suggests that metacyclic cells, if blocked in the G1 (G0) phase of the cell cycle, will possess a basal body and a pro-basal body. Thus, observation of two basal bodies cannot be taken as evidence of division unless the section includes two axonemes subtended by these basal bodies.

Actin and Intermediate Filaments

The kinetoplastid flagellate cytoskeleton exhibits such a pronounced microtubule-based ultrastructure that the presence of classic actin microfilaments has not received the attention it deserves. Certainly there is evidence for the presence of actin genes in *T. brucei* (9) and *L. major* (21) and for their expression. However, there is very little evidence for the location of the actin protein or microfilaments. Classical actin microfilaments have not been detected by EM, but there is some evidence that genes encoding proteins associated with actin function, such as profilin (106), in other cell types are present and expressed in trypanosomatids. Hence it is likely that an important, yet still cryptic, role exists for an actomyosin microfilament system.

Similarly, there is little evidence for the presence of the intermediate filament class of proteins or structures in trypanosomatids apart from ultrastructural reports of filaments with this type of diameter, often in nonproliferative or drug-treated cells, with some evidence of immunological cross-reactivity with heterologous antibodies (71).

Kinetoplast Position and Segregation

The existence of the mitochondrial DNA in the form of the kinetoplast (90) is one of the remarkable defining features of trypanosomatids. The presence and precise positioning of the kinetoplast close to the basal body were recognized rather early as key features of trypanosomes. Even by the end of the first decade of this century, textbooks of "bacteriology" (sic) carried such statements. For example, Hewlett (43) wrote in 1911, "The nuclear apparatus is usually double, consisting of a large principal or macronucleus, and a small or micronucleus or blepharoplast: the latter is not composed of generative chromatin and is in relation to the locomotor apparatus." Moreover, the close relationship between the two structures was recognized both during division (color Figure 3) and in their repositioning between the particular life cycle stages. Biochemical and molecular characterizations of the kinetoplast subsequently reinforced the view of a possible physical attachment between the basal body and the kinetoplast. Early studies of hypotonically lysed *Leishmania* species reported the transformation of the mitochondrion into a large, swollen vesicle containing the kinetoplast. There was a clear indication of some form of attachment of the kinetoplast to the mitochondrial membrane, and, in turn, the vesicle was often attached to the basal body, even after inversion of the cell ghost (94). Robinson & Gull (77) were able to show that segregation of the replicated kinetoplast depended on microtubule-mediated separation of the replicated basal body complexes during the cell cycle. Similarly, repositioning of the kinetoplast during the bloodstream-to-procyclic form differentiation is also a microtubule-mediated process that occurs concomitant with other cell cycle events (61). The physical connection between kinetoplast and basal body complex can be most directly demonstrated by the isolation of flagella with attached kinetoplasts, using methods including even detergent-containing buffers.

The molecular identity of the high-order physical link between these two organelles remains cryptic. However, it seems most likely that it takes the form of a relationship between three specific structures: (1) a set of fibrils that lie in the lumen of the mitochondrion and connect one side of the kinetoplast to a (2) differentiated area of the mitochondrial membranes and then (3) another set of fibrils that link this differentiated membrane area to the basal body in the cytoplasm. The two fibrillar components have been difficult to visualize by thin-section EM but have been alluded to individually by authors in general descriptions of the ultrastructure of trypanosomatids. The attachment of the basal body to the mitochondrial

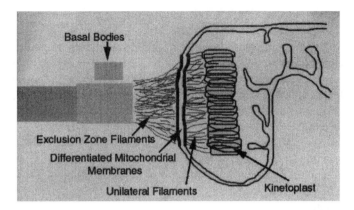

Figure 5   The tripartite attachment complex (TAC) that links the kinetoplast inside the mitochondrion to the cytoplasmic basal bodies. The three components of the TAC are the exclusion zone filaments, the differentiated regions of the mitochondrial membranes, and the unilateral filaments.

membrane has perhaps been best visualized as 11-nm fibrils by quick-freeze, deep-etch techniques (98). Recently, however, it has been possible to visualize all three components (Figure 5), and we have described the full tripartite attachment complex by using EM fixation regimes that include detergent in the initial fixative (E Ogbadoyi, DR Robinson, K Gull, unpublished observations). What are the molecular components? How is the structure replicated during the cell cycle? How is the kinetoplast attached, and how is the attachment to the tripartite attachment complex made by the new basal body in the cell cycle? We have little insight into the answers to these questions at present, but will certainly require such knowledge if we are to understand the inheritance patterns of this unique genome in both the cell cycle and life cycle. Moreover, one might speculate that the tripartite attachment complex plays a role as an organizational center around which growth, structure, and division of the mitochondrion as a whole (95) can be orchestrated.

Host-Parasite Attachment

The life cycles of many trypanosomatids include an attachment stage in their invertebrate vector or host during which they attach to tissues to complete their differentiation. Given the subpellicular corset of microtubules, it is perhaps not surprising that attachment takes place via the flagellum, which undergoes considerable modification in the process, becoming enlarged with the proliferation of filaments. There is a flattening out of the flagellum at the attachment area and formation of electron-dense attachment plaques. Both in vitro and in vivo, there often is a preferred orientation of attachment (42).

The flagella can project deeply into the microvillous zone of the epithelial lining (99), and attachment of *T. brucei* epimastigotes to the salivary gland epithelial brush border is mediated by rather elaborate branched outgrowths of the flagellar membrane. These outgrowths diminish, but the attachment plaques are maintained as the parasite differentiates to the nascent metacyclic form before the eventual release of the metacyclic parasite. Moreover, this differentiation of the epimastigotes to premetacyclic cells involves movement of the kinetoplast from a prenuclear to a postnuclear position while the parasite remains attached (102). Epimastigote forms are proliferative and therefore able to divide while attached. One might suggest that the use of the flagellum as the organ of attachment (and in particular the distal end) may be predicated by the necessity of keeping the cell body (and subpellicular microtubule cortex) free from cytoskeletal attachment structures that would compromise the division process of the cell. We are completely ignorant as to how and when the new daughter flagellum of the dividing epimastigote establishes the attachment plaques.

Detergent extraction reveals the detailed substructure of the cytoskeletal filaments and electron-dense attachment plaques (7), but the biochemical nature of the new filaments and attachment plaques is cryptic. This is an important omission in our understanding of rather critical phases in the life cycle of trypanosomatids that, in organisms such as *T. brucei*, includes the stage during which the sexual process appears to occur (35).

MICROTUBULES: Genes and Proteins

Microtubules and α- and β-Tubulin

Tubulin Genes   The genes encoding α- and β-tubulin subunits of microtubules are generally clustered in the genomes of trypanosomatids. In *T. brucei*, the tubulin genes are arranged in a cluster of tandemly repeated α/β pairs, and recent high-resolution imaging of DNA spread directly from the nucleus has shown a maximum of 19 repeats (27). A similar organization is observed in *T. cruzi*, although some pairs exist outside the main cluster. In contrast, in various *Leishmania* species, the α-tubulin and the β-tubulin genes are in separate clusters of tandemly linked repeats (19).

Post-Translational Modifications   The sequence evidence to date suggests that all of the genes in these loci encode the same α- or β-tubulin proteins within a particular organism. However, some of the earliest evidence for post-translational modifications (PTMs) of tubulin associated with cytoskeletal construction was obtained with trypanosomes. A number of PTMs of tubulin are now known and form two categories, general protein modifications (phosphorylation, acetylation, and palmytoylation) not specific to tubulin, and tubulin-specific modifications (tyrosination and formation of $\Delta 2$-tubulin, polyglutamylation, polyglycylation) (56, 58). The acetylation of α-tubulin at lysine 40 was identified initially because of the concomitant shift of the α-tubulin spot on two-dimensional gels (81, 82,

88). Immunological studies with monoclonal antibodies specific to this modified isotype showed that acetylated tubulin associated with both the subpellicular and the axonemal microtubules and that acetylation occurred concomitant with assembly of axonemal microtubules, with the modification reversible on microtubule disassembly (85, 88).

Trypanosomes also exhibit a second PTM of α-tubulin. Both α- and β-tubulins of *T. brucei* have an encoded C-terminal tyrosine. The tyrosine can be removed by a carboxypeptidase and replaced on α-tubulin by a tubulin tyrosine ligase enzyme (93). Detyrosination of α-tubulin occurs after microtubule assembly and, given the stability of both the axonemal and the subpellicular corset microtubules of trypanosomes, tyrosinated α-tubulin can act as a molecular marker for assembly of new microtubules. The use of monoclonal antibodies specific for tyrosinated α-tubulin reveals a cell cycle–related modulation of microtubule assembly, shown most clearly in some cells in mid-cycle, when the old flagellum is not detected, but the extending axoneme of the new flagellum is (93). Moreover, at the electron microscopic level, immunogold labeling of ends of individual microtubules in the subpellicular cortex showed that new microtubules invade the cytoskeletal array between old microtubules (92). In postmitotic cells, a unique form of microtubule assembly occurs, with very short microtubules being intercalated in the array. This construction pattern suggests a templated morphogenesis of the cortical microtubule array with semiconservative distribution to each of the daughter cells. Recent work suggests that there are specific spatial and temporal patterns of modified tubulins within the microtubules of the flagellum axoneme (48).

More recently, the spectrum of tubulin variants in cytoskeletons of *T. brucei* was analyzed by protein sequencing and mass spectrometry (87). β-Tubulin was shown to occur with and without its carboxy-terminal tyrosine. These tubulins were shown to be even more complex than originally envisaged because both tyrosinated and detyrosinated α- and β-tubulins are extensively glutamylated. Polyglutamylation introduces a dramatic chemical change into the tubulin molecule, namely the addition of multiple negative charges in the already very acidic carboxy terminus. The modification involves the addition of up to 20 glutamate residues to a defined glutamate near the carboxy terminus of both α- and β-tubulin (26, 107). The first glutamate of the oligoglutamyl side chain is attached to the γ-carboxyl group of a glutamate in the tubulin molecule (residues 445 of α-tubulin and 435 of β-tubulin) via its α-carboxyl group, γ/α linkage; the α-amino groups of the subsequent glutamates are thought to be joined to the α-carboxyl group of preceding glutamates, α/α linkages (74). The maximum numbers of glutamyl residues in the lateral chain are 15 and 6 for α- and β-tubulins, respectively.

Modification Functions: Ignorance and Opportunity   The current knowledge of the biochemical and cell biological aspects of this catalogue of tubulin PTMs is not matched by equivalent insight into their functions. In vitro approaches have suggested that polyglutamylation may influence processes such as the binding of kinesin or the microtubule-associated protein (MAP) tau to microtubules (55) or specific aspects of flagellum function (34). Parts of the trypanosome cytoskeleton (the axoneme and subpellicular array) conform to the general principle that stable microtubule arrays tend to be enriched in detyrosinated and acetylated tubulin. We do not know whether microtubule stability is caused by the presence of the modification or the presence of the modification is a consequence of the stability. However, even the general observation breaks down within trypanosomes when one considers the evidence (85) obtained from studies with monoclonal antibodies specific for the modified or unmodified isotypes: basal bodies (stable) appear to be tyrosinated, and spindle microtubules (dynamic) appear to be detyrosinated and acetylated! Moreover, molecular engineering of *Tetrahymena* spp. expressing α-tubulin that was unable to be acetylated at residue 40 resulted in mutant cells devoid of acetylated tubulin but expressing no novel phenotype. This is a very reasonable test of a specific function, but there remain a number of possible interpretations including a cryptic phenotype or that the main function of the tubulin acetyltransferase enzyme might be to acetylate some other protein (33). We have almost no information on the enzymes that perform these modifications. The notable exception is the mammalian cell tubulin tyrosine ligase whose properties are known in some detail and for which the gene has been cloned and sequenced (29).

The increasing sophistication of trypanosome molecular genetics means that these organisms now offer one of the best avenues for a future understanding of the tubulin PTMs. Once the enzymes responsible have been identified, then the available molecular genetic technologies of gene knockouts and antisense RNA, coupled with the well-understood cytoskeleton, will allow phenotypes of null or conditional mutations in the enzymes to be interpreted in terms of in vivo cellular function.

Evolutionary considerations are interesting, but are not particularly helpful in fitting the presence of a PTM in an organism to the particular cell biology of its microtubule cytoskeleton. Most of the modifications appear to have developed early in the evolution of eukaryotic cells and are present in many protists. Intriguing variations do occur, however, as shown by comparisons (86, 87, 105) of representatives of three ancient groups, the trichomonads (*Tritrichomonas mobilensis*), the trypanosomatids (*T. brucei*), and the diplomonads (*Giardia lamblia*). Tubulin acetylation and tubulin polyglutamylation are present in all three, whereas tyrosination has been detected only in trypanosomes and may therefore have appeared after the trichomonads had diverged from the rest of the eukaryotes. Conversely, polyglycylation has been detected in the diplomonads and not in the other two groups, suggesting that it is an ancient modification that was lost in trichomonads and trypanosomatids. Although these observations add to our views of protist evolution, they do not assist in determining the function of the modifications. This form of evolutionary analysis will be strengthened in the future as the genes encoding the modification enzymes are identified and the results of the various trypanosomatid genome projects accumulate.

MTOCs and Minor Tubulins

γ-*Tubulin*   A third member of the tubulin super family, γ-tubulin, is now accepted as a molecular indicator of MTOCs and is essential for cytoplasmic and mitotic microtubule function (24, 37). The gene encoding *T. brucei* γ-tubulin has been cloned and immunofluorescence studies showed that γ-tubulin has a restricted cellular distribution, which is modulated during the trypanosome cell cycle (89). The protein is associated with the basal bodies as might be expected in their role as MTOCs for the axoneme. When the kinetoplasts have segregated in the cell cycle, γ-tubulin staining is associated with both the old and the new sets of basal bodies. A γ-tubulin dot is present in the nucleus, and, in early mitotic cells, the dot is duplicated and forms a bar-shaped region corresponding to the developing mitotic spindle. In late mitotic cells there are two γ-tubulin dots in the elongated nucleus corresponding to the poles of the mitotic spindle.

The staining at the anterior of the cell and the punctate staining over the subpellicular array raises questions as to whether the initial nucleation of each of the individual microtubules in the subpellicular array is performed by a γ-tubulin complex. An individual microtubule might have such a complex at its minus $(-)$ end, hence explaining the punctate cell body pattern. However, there is good evidence that at least some of the microtubules in the subpellicular array exhibit two-end assembly, which does not fit easily with a maintained γ-tubulin complex capping the $-$ end. The high concentration of microtubule ends per unit of surface area at the anterior end of the trypanosome would, however, account for the discrete staining at that site. A number of models could account for morphogenesis of the subpellicular complex. They range from strict templated intercalation of new microtubules to initial nucleation via a γ-tubulin complex at the anterior end of the cell, coupled with release of the individual microtubules and their movement, growth, and invasion into the array via lateral interactions with existing microtubules. Future analysis of γ-tubulin and other MTOC mutants will be useful in exploring these phenomena.

δ-*Tubulin*   Recently, a fourth member of the tubulin superfamily (δ-tubulin) was discovered by analysis of a mutation in *Chlamydomonas* (25). The *UNI3* gene encodes a protein that shows significant homology to all three other tubulins. The *uni3-1* mutation in the δ-tubulin gene produces cells that exhibit zero, one, or two flagella instead of the consistent two of wild-type *Chlamydomonas*. Moreover, there is a lineage dependency in the numbers with a spatial pattern to the inheritance. EM reveals that the mutant possesses doublet rather than triplet microtubules at the proximal ends of its basal bodies. It will be interesting whether this and other tubulins are encoded in the genomes of the trypanosomatids, where spatial and temporal inheritance patterns of basal body/centriole replication and maturation appear to be as much part of the cell cycle as they are in algae (8, 39) and mammalian cells (54).

Microtubule-Associated Proteins

Discrete projections interconnect the microtubules of the subpellicular array and link the microtubules to the plasma membrane. MAPs associating with the subpellicular corset have often been identified by antibodies raised against complex immunogens such as the whole cytoskeleton [identification of CAP5.5 (10, 60) and the WCB210 phosphoprotein (109)] or antiserum from early immune responses of infected animals (23). This latter response is interesting in that it appears that many cytoskeletal antigens of trypanosomatids are recognized by stimulation of an existing autoimmune response in the animals. Unfortunately, this evidence (and much more that is anecdotal) emphasizes the need for caution when using polyclonal antibodies to assist in the definition of repetitive cytoskeletal proteins.

The criteria used for designation of a protein as a MAP are diverse (57) and of various levels of rigor. We have very little insight into the molecular basis for protein targeting to the subpellicular cytoskeletal corset. Some early reports of MAPs, characterized by their ability to bind to microtubules in vitro, turned out to be glycosomal enzymes (the acidic tubulin molecule binds many basic proteins in a nonspecific manner). These phenomena and other more recent data on MAPs have been reviewed (51, 76).

A family of high-molecular-weight, microtubule-associated repetitive proteins has been described in *T. brucei* (40). One member, microtubule-associated repetitive protein-1, binds to microtubules via tubulin domains other than the carboxy termini used by MAPs from mammalian brain. In vitro binding assays with recombinant protein, as well as transfection of mammalian cell lines, established that the 38-amino-acid–repeat units of the protein define a novel microtubule-binding motif. This motif is very similar in length to those of mammalian MAPs (57), but both its sequence and charge are different.

Two other MAPs that associate with the subpellicular array were again identified by immunological approaches. Protein I/6 is another repetitive MAP encoded by a single gene. It is a 33-kDa polypeptide and includes a domain of six tandemly arranged 8-amino-acid repeats (23). The protein WCB210 was shown to have a relative molecular mass of 210 kDa by Western blotting (109). In both cases immunogold-labeling studies showed the proteins to be closely associated with the cross-bridges lying between the microtubules.

In contrast, the Gb4 protein has a restricted location within the subpellicular corset of *T. brucei* at the posterior end of the microtubule array (75). Immunogold labeling is suggestive of a punctate pattern that may correspond to the ends of microtubules. Unfortunately, at present, labeling of cytoskeletons at this level of resolution is not easy, and, although it has a very clear macro localization, a function of Gb4 in capping microtubule ends remains an intriguing possibility.

Examples are emerging of the relationship between the subpellicular microtubule array and the plasma membrane in targeting membrane proteins to the cell body. The major glucose transporter of *Leishmania* spp. exists in two isoforms differing only in their cytosolic NH2-terminal domains (96). One (*iso-1*) localizes to the flagellar membrane, whereas *iso-2* localizes to the plasma membrane of the cell body and is associated with the underlying microtubule cytoskeleton. The *iso-1* form does not appear to have any association with the cytoskeleton. Deletion of the N-terminal 25 or more amino acids from *iso-1* targets the protein from the flagellum to the cell body plasma membrane and an association with the cytoskeleton. These results are suggestive of cytoskeletal binding serving as an anchor to localize the *iso-2* transporter within the cell body plasma membrane and that the flagellar targeting signal of *iso-1* diverts this transporter into the flagellar membrane and away from the pellicular microtubules. This form of molecular analysis, coupled with functional studies made amenable by new molecular genetics, is likely to uncover some novel mechanisms for targeting and trafficking of proteins that include interplay between the cytoskeleton and the overlying membrane.

FUNCTIONAL ANALYSIS OF THE CYTOSKELETON

Microtubule Inhibitors

In vitro assembly of tubulin into microtubules is the most direct assay available to study the action of microtubule inhibitors. Although tubulin is a relatively abundant protein in trypanosomatids, it is much more difficult to purify in any meaningful amounts than tubulin from mammalian brain. However, in vitro assays have been achieved by adaptation of techniques developed previously for amoebal tubulin (73, 79). Purification of *T. brucei* tubulin and its assembly in vitro into microtubules allowed a direct test of the effectiveness of inhibitors (59). Many comments in the literature (103) and anecdotal evidence suggested that trypanosomatids were resistant to the effects of well-known inhibitors such as colchicine. Indeed, colchicine and the benzimidazoles, both classical and potent inhibitors of mammalian tubulin polymerization, had only very slight effects on the polymerization of trypanosome tubulin (59). This result fits with the well-established view that protistan tubulins are remarkably resistant to the action of colchicine. The situation with the benzimidazoles is complex. The trypanosomatids are rather resistant to the action of a range of benzimidazoles, although some protozoa such as *Giardia* and *Trichomonas* spp. are susceptible to some of these inhibitors (49). Only very few studies have examined directly the susceptibility of purified protistan tubulin to inhibitors. It is known that a spectrum of susceptibility often exists to different benzimidazoles. For instance, both yeast and helminth tubulins are sensitive to the action of some benzimidazoles but resistant to others (20, 50). Attempts therefore to link tubulin gene sequences to susceptibility profiles of organisms (49) are unlikely to produce clear relationships. The recent atomic model of tubulin will, however, undoubtedly assist in studies of these problems (67).

There has been a much stronger conservation of the binding site for the vinca alkaloids vinblastine and vincristine and the macrolides maytansine, ansamitocin, and rhizoxin during evolution, and this site is undoubtedly present in trypanosome tubulin. Vinblastine and maytansine have been shown to be good inhibitors of trypanosome tubulin polymerization in vitro (59). Moreover, analysis of rhizoxin-sensitive and -resistant organisms and mutants has shown that sensitivity to this drug depends on the location of an asparagine residue at position 100 in β-tubulin (100, 101). The trypanosome β-tubulin sequence has an asparagine at this site, and studies with *T. brucei* have shown that this compound is a powerful antimicrotubule agent (78).

A number of groups have assayed a vast range of antimicrotubule inhibitors for activity against trypanosomatid flagellates and found that they show some suscep-tibility to compounds that were initially associated with herbicide activity (14). In particular, trifluralin and its analogs inhibit *Leishmania* species and *T. brucei* (13, 15). They produce bizarre cell forms but are only active at rather higher con-centrations than the macrolide series exemplified by rhizoxin. More recent work (12) has focused on the presence of an impurity, chloralin, in these preparations that is 100-fold more active than trifluralin.

Taxol, in contrast to other antimicrotubule agents, affects cytoskeletal systems by stabilizing microtubule arrays. It produces interesting effects on trypanosomes. Low concentrations of the drug allow replication of the kinetoplast, nucleus, and flagellum but have profound effects on cytokinesis (5). *T. cruzi* epimastigotes initiate cytokinesis at the anterior end but appear to block when the cleavage furrow reaches partway along the cell body. This raises the possibility of some unique site occurring at this point (the basal bodies?) which is affected by taxol-induced microtubule stabilization. Indeed this does seem to be a critical factor, because the electron micrographs in this paper (5) indicate that the kinetoplast/basal body complexes have not moved apart to the normal extent. Hence an interpretation would be that taxol may produce a hyperstabilization of the 4-microtubule/FAZ complex or other subpellicular microtubules and so inhibits complete segregation of the basal bodies. Cytokinesis then apparently fails at this point.

When mitosis is inhibited with a low concentration of the antimicrotubule in-hibitor rhizoxin, basal body, flagellum, and kinetoplast segregation can take place (78). In such drug-treated cells, cleavage still starts from the anterior end to pro-duce a nucleated cell and a flagellated cytoplast (a "zoid") from the anterior. The length of the zoid seems to be influenced by the length of the daughter flagellum. Because the FAZ filament extends from the basal body region to the anterior tip of the cell, it may be responsible for defining the required positional information at the "distant" anterior end of the cell where cytokinesis must be initiated between the tips of the old and new daughter flagella.

In this respect it is unclear whether zoid production is ever part of taxol inhibi-tion. Alternatively, there may be informative differences between the phenotypes produced by the two types of drug. Microtubule stabilization by taxol may in-hibit basal body segregation and not allow cytokinesis because it places a form of rigor on the subpellicular microtubule array. In contrast, microtubule assembly inhibition by rhizoxin may inhibit mitosis, but allow cleavage to continue, albeit in a form that produces the zoid from the posterior portion of the cell. These phenotypes are important because they indicate that trypanosomatids may exhibit rather different cell cycle check point controls to take account of their need to regulate kinetoplast replication and segregation as well as mitotic/cytokinesis events.

Molecular Genetics

In the preceding sections, I have tried to indicate at appropriate points how new developments in the molecular genetics of trypanosomatids are enabling functional analysis of the cytoskeleton. A novel approach for the analysis of function of proteins of the trypanosome cytoskeleton has recently been developed. The method, which is likely to have broad application in functional analysis of many genes in these and other organisms, involves the transfection or production in vivo of double-stranded RNA (dsRNA) corresponding to portions of the gene. In response to the introduction of such dsRNA corresponding to the 5' untranslated region of the α-tubulin gene in *T. brucei*, the corresponding α-tubulin mRNA, but not the pre-mRNA, was degraded specifically, resulting in a reduction in the translation of new α-tubulin protein. The response of the trypanosomes to this deficit was dramatic, with a phenotype involving defects in microtubule organization and failure of cytokinesis (66). This form of molecular intervention, along with antisense technologies, gene knockouts, replacements, modifications, inducible expression, and dominant negative approaches, now offers unrivaled opportunities for studying complex functions of cytoskeletal proteins in motility, morphogenesis, and pathogenicity.

CONCLUSIONS

The intrinsic order and beauty of the trypanosomatid organisms are now recognized to be produced, maintained, and modulated by the cytoskeleton. We have seen a period during which studies have concentrated on structural and positional descriptions producing in trypanosomatids, some of the best characterized protistan cells. Molecular-cytology studies have increased their pace and will benefit enormously from the various parasite genome-sequencing projects that are underway. The events orchestrated by the cytoskeleton are so intrinsic to proliferation and pathogenicity of these parasites that the next phases of functional analysis are likely to be of significance in both revealing fundamental biological principles and assisting in combating these etiological agents of disease.

ACKNOWLEDGMENTS

I would like to thank members of my group who provided illustrations. I have benefited greatly from many discussions with past and present members and record my thanks to all. Work in my laboratory is funded by grants from the Wellcome Trust and the Biotechnology and Biological Sciences Research Council.

LITERATURE CITED

1. Alshammary FJ, Shoukrey NM, Al-shewemi SE, Ibrahim EA, Alzahrani MA, Altuwaijri AS. 1995. Leishmania major, in-vitro ultrastructural-study of the paraxial rod of promastigotes. *Int. J. Parasitol.* 25:443–52

2. Baqui MMA, Takata CSA, Milder RV, Pudles J. 1996. A giant protein associated with the anterior pole of a trypanosomatid cell body skeleton. *Eur. J. Cell Biol.* 70:243–49

3. Bastin P, Matthews KR, Gull K. 1996. The paraflagellar rod of kinetoplastida: solved and unsolved questions. *Parasitol. Today* 12:302–7

4. Bastin P, Sherwin T, Gull K. 1998. Paraflagellar rod is vital for trypanosome motility. *Nature* 391:548

5. Baum SG, Wittner M, Nadler JP, Horwitz SB, Dennis JE, et al. 1981. Taxol, a microtubule stabilizing agent, blocks the replication of Trypanosoma cruzi. *Proc. Natl. Acad. Sci. USA* 78:4571–75

6. Beard CA, Saborio JL, Tewari D, Krieglstein KG, Henschen AH, et al. 1992. Evidence for 2 distinct major protein-components, PAR-1 and PAR-2, in the paraflagellar rod of Trypanosoma cruzi—complete nucleotide sequence of PAR-2. *J. Biol. Chem.* 267:21656–62

7. Beattie P, Gull K. 1997. Cytoskeletal architecture and components involved in the attachment of Trypanosoma congolense epimastigotes. *Parasitology* 115:47–55

8. Beech PL, Heimann K, Melkonian M. 1991. Development of the flagellar apparatus during the cell-cycle in unicellular algae. *Protoplasma* 164:23–37

9. Ben Amar MF, Pays A, Tebabi P, Dero B, Seebeck T, et al. 1988. Structure and transcription of the actin gene of Trypanosoma brucei. *Mol. Cell. Biol.* 8:2166–76

10. Birkett CR, Parma AE, Gerke-Bonet R, Woodward R, Gull K. 1992. Isolation of cDNA clones encoding proteins of complex structure—analysis of the Trypanosoma-brucei cytoskeleton. *Gene* 110:65–70

11. Bouck GB, Ngo H. 1996. Cortical structure and function in euglenoids with reference to trypanosomes, ciliates, and dinoflagellates. *Int. Rev. Cytol.* 169:267–318

12. Callahan HL, Kelley C, Pereira T, Grogl M. 1996. Microtubule inhibitors: structure-activity analyses suggest rational models to identify potentially active compounds. *Antimicrob. Agents Chemother.* 40:947–52

13. Chan MM, Fong D. 1994. Plant microtubule inhibitors against Trypanosomatids. *Parasitol. Today* 10:448–51

14. Chan MMY, Fong D. 1990. Inhibition of Leishmanias but not host macrophages by the antitubulin herbicide trifluralin. *Science* 249:924–26

15. Chan MMY, Triemer RE, Fong D. 1991. Effect of the anti-microtubule drug oryzalin on growth and differentiation of the parasitic protozoan Leishmania mexicana. *Differentiation* 46:15–21

16. Cole DG, Diener DR, Himelblau AL, Beech PL, Fuster JC, Rosenbaum JL. 1998. Chlamydomonas kinesin-II-dependent intraflagellar transport (IFT): IFT particles contain proteins required for ciliary assembly in Caenorhabditis elegans sensory neurons. *J. Cell Biol.* 141:993–1008

17. Cooper R, De Jesus AR, Cross GAM. 1993. Deletion of an immunodominant Trypanosoma cruzi surface glycoprotein disrupts flagellum-cell adhesion. *J. Cell Biol.* 122:149–56

18. Cotrim PC, Paranhos Baccala G, Santos MR, Mortensen C, Cano MI, et al. 1995. Organization and expression of the gene encoding an immunodominant repetitive

antigen associated to the cytoskeleton of Trypanosoma cruzi. *Mol. Biochem. Parasitol.* 71:89–98

19. Coulson RMR, Conner V, Chen JC, Ajioka JW. 1996. Differential expression of Leishmania major beta-tubulin genes during the acquisition of promastigote infectivity. *Mol. Biochem. Parasitol.* 82:227–36

20. Dawson PJ, Gutteridge WE, Gull K. 1984. A comparison of the interaction of anthelminthic benzimidazoles with tubulin isolated from mammalian tissue and the parasitic nematode Ascaridia galli. *Biochem. Pharmacol.* 33:1069–74

21. Dearruda MV, Matsudaira P. 1994. Cloning and sequencing of the Leishmania major actin-encoding gene. *Gene* 139:123–25

22. Deflorin J, Rudolf M, Seebeck T. 1994. The major components of the paraflagellar rod of Trypanosoma brucei are two similar, but distinct proteins which are encoded by two different gene loci. *J. Biol. Chem.* 269:28745–51

23. Detmer E, Hemphill A, Muller N, Seebeck T. 1997. The Trypanosoma brucei autoantigen 1/6 is an internally repetitive cytoskeletal protein. *Eur. J. Cell Biol.* 72:378–84

24. Dictenberg JB, Zimmerman W, Sparks CA, Young A, Vidair C, et al. 1998. Pericentrin and gamma-tubulin form a protein complex and are organized into a novel lattice at the centrosome. *J. Cell Biol.* 141:163–74

25. Dutcher SK, Trabuco EC. 1998. The UNI3 gene is required for assembly of basal bodies of Chlamydomonas and encodes delta-tubulin, a new member of the tubulin superfamily. *Mol. Biol. Cell* 9:1293–308

26. Eddé B, Rossier J, LeCaer J-P, Desbruyères E, Gros F, Denoulet P. 1990. Posttranslational glutamylation of alpha-tubulin. *Science* 247:83–85

27. Ersfeld K, Asbeck K, Gull K. 1998. Direct visualisation of individual gene organisation in Trypanosoma brucei by high-resolution in situ hybridisation. *Chromosoma* 107:237–40

28. Ersfeld K, Gull K. 1997. Partitioning of large and minichromosomes in Trypanosoma brucei. *Science* 276:611–14

29. Ersfeld K, Wehland J, Plessmann U, Dodemont H, Gerke V, Weber K. 1993. Characterization of the tubulin-tyrosine ligase. *J. Cell Biol.* 120:725–32

30. Farina M, Attias M, Soutopadron T, Desouza W. 1986. Further studies on the organization of the paraxial rod of Trypanosomatids. *J. Protozool.* 33:552–57

31. Fouts DL, Stryker GA, Gorski KS, Miller MJ, Nguyen TV, et al. 1998. Evidence for four distinct major protein components in the paraflagellar rod of Trypanosoma cruzi. *J. Biol. Chem.* 273:21846–55

32. Frolov AO, Karpov SA, Malysheva MN. 1996. The ultrastructure of mitosis in the free-living kinetoplastid Bodo curvifilus. *Eur. J. Protistol.* 32:498–505

33. Gaertig J, Cruz MA, Bowen J, Gu L, Pennock DG, Gorovsky MA. 1995. Acetylation of lysine 40 in alpha-tubulin is not essential in Tetrahymena thermophila. *J. Cell Biol.* 129:1301–10.

34. Gagnon C, White D, Cosson J, Huitorel P, Eddé B, et al. 1996. The polyglutamylated lateral chain of alpha-tubulin plays a key role in flagellar motility. *J. Cell Sci.* 109:1545–53.

35. Gibson W, Bailey M. 1994. Genetic exchange in Trypanosoma brucei: evidence for meiosis from analysis of a cross between drug-resistant transformants. *Mol. Biochem. Parasitol.* 64:241–52

36. Gull K, Alsford S, Ersfeld K. 1998. Segregation of minichromosomes in trypanosomes: implications for mitotic mechanisms. *Trends Microbiol.* 6:319–23

37. Hagan IM, Gull K, Glover DM. 1998. Poles apart? Spindle pole bodies and centrosomes differ in ultrastructure yet their function and regulation are conserved. In *Dynamics of Cell Division*, ed SA Endow, DM Glover, pp. 57–96. Oxford, UK: Oxford Univ. Press

38. Haynes PA, Russell DG, Cross GAM. 1996. Subcellular localization of Try-

panosoma cruzi glycoprotein Gp72. *J. Cell Sci.* 109:2979–88

39. Heimann K, Reize IB, Melkonian M. 1989. The flagellar developmental cycle in algae—flagellar transformation in Cyanophora paradoxa (Glaucocystophyceae). *Protoplasma* 148:106–10

40. Hemphill A, Affolter M, Seebeck T. 1992. A novel microtubule-binding motif identified in a high molecular weight microtubule-associated protein from Trypanosoma brucei. *J. Cell Biol.* 117:95–103

41. Hemphill A, Lawson D, Seebeck T. 1991. The cytoskeletal architecture of Trypanosoma brucei. *J. Parasitol.* 77:603–12

42. Hemphill A, Ross CA. 1995. Flagellum-mediated adhesion of Trypanosoma congolense to bovine aorta endothelial cells. *Parasitol. Res.* 81:412–20

43. Hewlett RW. 1911. *A Manual of Bacteriology: Clinical and Applied.* London: Churchill

44. Hoft DF, Donelson JE, Kirchhoff LV. 1995. Repetitive protein antigens of Trypanosoma cruzi have diverse intracellular locations. *J. Parasitol.* 81:549–54

45. HungerGlaser I, Seebeck T. 1997. Deletion of the genes for the paraflagellar rod protein PFR-A in Trypanosoma brucei is probably lethal. *Mol. Biochem. Parasitol.* 90:347–51

46. Imboden M, Muller N, Hemphill A, Mattioli R, Seebeck T. 1995. Repetitive proteins from the flagellar cytoskeleton of African trypanosomes are diagnostically useful antigens. *Parasitology* 10:249–58

47. Ismach R, Cianci CML, Caulfield JP, Langer PJ, Hein A, McMahon-Pratt D. 1989. Flagellar membrane and paraxial rod proteins of Leishmania—characterization employing monoclonal-antibodies. *J. Protozool.* 36:617–24

48. Johnson KA. 1998. The axonemal microtubules of the Chlamydomonas flagellum differ in tubulin isoform content. *J. Cell Sci.* 111:313–20

49. Katiyar SK, Gordon VR, McLaughlin GL, Edlind TD. 1994. Antiprotozoal activities of benzimidazoles and correlations with beta-tubulin sequence. *Antimicrob. Agents Chemother.* 38:2086–90

50. Kilmartin JV. 1981. Purification of yeast tubulin by self-assembly invitro. *Biochemistry* 20:3629–33

51. Kohl L, Gull K. 1998. Molecular architecture of the trypanosome cytoskeleton. *Mol. Biochem. Parasitol.* 93:1–9

52. Kohl L, Gull K. 1999. Assembly of the paraflagellar rod and flagellum attachment zone during the Trypanosoma brucei cell cycle. *J. Eukaryot. Microbiol.* 46:105–9

53. Lafaille JJ, Linss J, Krieger MA, SoutoPadron T, De Souza W, Goldenberg S. 1989. Structure and expression of two Trypanosoma cruzi genes encoding antigenic proteins bearing repetitive epitopes. *Mol. Biochem. Parasitol.* 35:127–36

54. Lange BMH, Gull K. 1996. Structure and function of the centriole in animal cells: progress and questions. *Trends Cell. Biol.* 6:348–52

55. Larcher J-C, Boucher D, Lazereg S, Gros F, Denoulet P. 1996. Interactions of kinesin motor domains with alpha- and beta-tubulin subunits at a tau-independent binding site. *J. Biol. Chem.* 271:22117–24

56. Luduena RF. 1998. Multiple forms of tubulin: different gene products and covalent modifications. *Int. Rev. Cytol.* 178:207–75

57. Maccioni RB, Cambiazo V. 1995. Role of microtubule-associated proteins in the control of microtubule assembly. *Physiol. Rev.* 75:835–64

58. MacRae TH. 1997. Tubulin post-translational modifications (enzymes and their mechanisms of action). *Eur. J. Biochem.* 244:265–78

59. Macrae TH, Gull K. 1990. Purification and assembly invitro of tubulin from Trypanosoma brucei-brucei. *Biochem. J.* 265:87–93

60. Matthews KR, Gull K. 1994. Evidence for an interplay between cell-cycle progression and the initiation of differentiation between life-cycle forms of African

trypanosomes. *J. Cell Biol.* 125:1147–56

61. Matthews KR, Sherwin T, Gull K. 1995. Mitochondrial genome repositioning during the differentiation of the African trypanosome between life cycle forms is microtubule mediated. *J. Cell. Sci.* 108:2231–9

62. Miller MJ, Wrightsman RA, Manning JE. 1996. Trypanosoma cruzi: protective immunity in mice immunized with paraflagellar rod proteins is associated with a T-helper type 1 response. *Exp. Parasitol.* 84:156–67

63. Miller MJ, Wrightsman RA, Stryker GA, Manning JE. 1997. Protection of mice against Trypanosoma cruzi by immunization with paraflagellar rod proteins requires T cell, but not B cell, function. *J. Immunol.* 158:5330–37

64. Moore LL, Santrich C, LeBowitz JH. 1996. Stage-specific expression of the Leishmania mexicana paraflagellar rod protein PFR-2. *Mol. Biochem. Parasitol.* 80:125–35

65. Muller N, Hemphill A, Imboden M, Duvallet G, Dwinger RH, Seebeck T. 1992. Identification and characterization of two repetitive non-variable antigens from African trypanosomes which are recognized early during infection. *Parasitology* 104:111–20

66. Ngo H, Tschudi C, Gull K, Ullu E. 1998. Double-stranded RNA induces mRNA degradation in Trypanosoma brucei. *Proc. Natl. Acad. Sci. USA* 95:14687–92

67. Nogales E, Wolf SG, Downing KH. 1998. Structure of the alpha beta tubulin dimer by electron crystallography. *Nature* 391:199–203

68. Nozaki T, Cross GAM. 1994. Functional complementation of glycoprotein 72 in a Trypanosoma cruzi glycoprotein 72 null mutant. *Mol. Biochem. Parasitol.* 67:91–102

69. Nozaki T, Haynes PA, Cross GA. 1996. Characterization of the Trypanosoma brucei homologue of a Trypanosoma cruzi flagellum-adhesion glycoprotein. *Mol. Biochem. Parasitol.* 82:245–55

70. Deleted in proof

71. Page AM, Lagnado JR. 1998. Novel filamentous bundles in the cytoplasm of a unicellular eukaryote, Crithidia fasciculata. *Protoplasma* 201:64–70

72. Pazour GJ, Wilkerson CG, Witman GB. 1998. A dynein light chain is essential for the retrograde particle movement of intraflagellar transport (IFT). *J. Cell Biol.* 141:979–92

73. Quinlan RA, Roobol A, Pogson CI, Gull K. 1981. A correlation between invivo and invitro effects of the microtubule inhibitors colchicine, parbendazole and nocodazole on myxamoebae of Physarum polycephalum. *J. Gen. Microbiol.* 122:1–6

74. Redeker V, LeCaer J-P, Rossier J, Promé J-C. 1991. Structure of the polyglutamyl side chain posttranslationally added to alpha-tubulin. *J. Biol. Chem.* 266:23461–66.

75. Rindisbacher L, Hemphill A, Seebeck T. 1993. A repetitive protein from Trypanosome brucei which caps the microtubules at the posterior end of the cytoskeleton. *Mol. Biochem. Parasitol.* 58:83–96

76. Robinson D, Beattie P, Sherwin T, Gull K. 1991. Microtubules, tubulin, and microtubule-associated proteins of trypanosomes. *Methods Enzymol.* 196:285–302

77. Robinson DR, Gull K. 1991. Basal body movements as a mechanism for mitochondrial genome segregation in the trypanosome cell cycle. *Nature* 352:731–33

78. Robinson DR, Sherwin T, Ploubidou A, Byard EH, Gull K. 1995. Microtubule polarity and dynamics in the control of organelle positioning, segregation, and cytokinesis in the trypanosome cell cycle. *J. Cell Biol.* 128:1163–72

79. Roobol A, Pogson CI, Gull K. 1980. In vitro assembly of microtubule proteins from yxamoebae of Physarum polycephalum. *Exp. Cell Res.* 130:203–15

80. RuizMoreno L, Bijovsky AT, Pudles J, Alves MJM, Colli W. 1995. Trypanosoma cruzi: Monoclonal antibody to cytoskeleton recognizes giant proteins of the flagellar attachment zone. *Exp. Parasitol.* 80: 605–15

81. Russell DG, Gull K. 1984. Flagellar regeneration of the trypanosome Crithidia fasciculata involves post-translational modification of cytoplasmic-alpha tubulin. *Mol. Cell. Biol.* 4:1182–85

82. Russell DG, Miller D, Gull K. 1984. Tubulin heterogeneity in the trypanosome Crithidia fasciculata. *Mol. Cell. Biol.* 4: 779–90

83. Russell DG, Newsam RJ, Palmer GCN, Gull K. 1983. Structural and biochemical characterization of the paraflagellar rod of Crithidia fasciculata. *Eur. J. Cell Biol.* 30:137–43

84. Santrich C, Moore L, Sherwin T, Bastin P, Brokaw C, et al. 1997. A motility function for the paraflagellar rod of Leishmania parasites revealed by PFR-2 gene knockouts. *Mol. Biochem. Parasitol.* 90:95–109

85. Sasse R, Gull K. 1988. Tubulin posttranslational modifications and the construction of microtubular organelles in Trypanosoma brucei. *J. Cell. Sci.* 90:577–89

86. Schneider A, Plessmann U, Felleisen R, Weber K. 1998. Posttranslational modifications of trichomonad tubulins: identification of multiple glutamylation sites. *FEBS Lett.* 429:399–402

87. Schneider A, Plessmann U, Weber K. 1997. Subpellicular and flagellar microtubules of Trypanosoma brucei are extensively glutamylated. *J. Cell Sci.* 110:431–37

88. Schneider A, Sherwin T, Sasse R, Russell DG, Gull K, Seebeck T. 1987. Subpellicular and flagellar microtubules of Trypanosoma brucei brucei contain the same alpha-tubulin isoforms. *J. Cell Biol.* 104:431–38

89. Scott V, Sherwin T, Gull K. 1997. γ-Tubulin in trypanosomes: molecular characterisation and localisation to multiple and diverse microtubule organising centres. *J. Cell Sci.* 110:157–68

90. Shapiro TA, Englund PT. 1995. The structure and replication of kinetoplast DNA. *Annu. Rev. Microbiol.* 49:117–43

91. Sherwin T, Gull K. 1989. The cell-division cycle of Trypanosoma brucei brucei—timing of event markers and cytoskeletal modulations. *Philos. Trans. R. Soc. London Ser. B.* 323:573–88

92. Sherwin T, Gull K. 1989. Visualization of detyrosination along single microtubules reveals novel mechanisms of assembly during cytoskeletal duplication in trypanosomes. *Cell* 57:211–21

93. Sherwin T, Schneider A, Sasse R, Seebeck T, Gull K. 1987. Distinct localization and cell-cycle dependence of COOH terminally tyrosinolated alpha-tubulin in the microtubules of Trypanosoma brucei brucei. *J. Cell Biol.* 104:439–46

94. Simpson L. 1972. The kinetoplast of hemoflagellates. *Int. Rev. Cytol.* 32:139–207

95. Simpson L, Kretzer F. 1997. The mitochondrion in dividing Leishmania tarentolae cells is symmetric and circular and becomes a single asymmetric tubule in non-dividing cells due to division of the kinetoplast portion. *Mol. Biochem. Parasitol.* 87:71–78

96. Snapp EL, Landfear SM. 1997. Cytoskeletal association is important for differential targeting of glucose transporter isoforms in Leishmania. *J. Cell Biol.* 139:1775–83

97. Solari AJ. 1995. Mitosis and genome partitioning in trypanosomes. *Biocell* 19:65–84

98. Soutopadron T, Desouza W, Heuser JE. 1984. Quick-freeze, deep-etch rotary replication of Trypanosoma cruzi and Herpetomonas megaseliae. *J. Cell Sci.* 69:167–78

99. Steiger RF. 1973. On the ultrastructure of Trypanosoma (Trypanozoon) brucei in the course of its life cycle and some related aspects. *Acta Trop.* 30:64–168

100. Takahashi M, Kobayashi H, Iwasaki S. 1989. Rhizoxin resistant mutants with an altered beta-tubulin gene in Aspergillus nidulans. *Mol. Gen. Genet.* 220:53–59

101. Takahashi M, Matsumoto S, Iwasaki S, Yahara I. 1990. Molecular basis for determining the sensitivity of eukaryotes to the antimitotic drug rhizoxin. *Mol. Gen. Genet.* 222:169–75

102. Tetley L, Vickerman K. 1985. Differentiation in Trypanosoma brucei: host-parasite cell junctions and their persistence during acquisition of the variable antigen coat. *J. Cell Sci.* 74:1–19

103. Vickerman K, Preston TM. 1970. Spindle microtubules in the dividing nuclei of trypanosomes. *J. Cell Sci.* 6:365–83

104. Vickerman K, Preston TM. 1976. Comparative cell biology of the kinetoplastid flagellates. In *Biology of the Kinetoplastida*, vol. 1, ed. WCA Lumsden, DA Evans, pp. 35–130. New York: Academic Press

105. Weber K, Schneider A, Westermann S, Muller N, Plessmann U. 1997. Posttranslational modifications of alpha- and beta-tubulin in Giardia lamblia, an ancient eukaryote. *FEBS Lett.* 419:87–91

106. Wilson W, Seebeck T. 1997. Identification of a profilin homologue in Trypanosoma brucei by complementation screening. *Gene* 187:201–9

107. Wolff A, deNéchaud B, Chillet D, Mazarguil H, Dsebruyères E, et al. 1992. Distribution of glutamylated alpha and beta-tubulin in mouse tissues using a specific monoclonal antibody, GT335. *Eur. J. Cell Biol.* 59:425–32

108. Woods A, Baines AJ, Gull K. 1989. Evidence for a Mr 88,000 glycoprotein with a transmembrane association to a unique flagellum attachment region in Trypanosoma brucei. *J. Cell Sci.* 93:501–8

109. Woods A, Baines AJ, Gull K. 1992. A high molecular mass phosphoprotein defined by a novel monoclonal antibody is closely associated with the intermicrotubule cross bridges in the Trypanosoma brucei cytoskeleton. *J. Cell Sci.* 103:665–75

110. Woods A, Sherwin T, Sasse R, MacRae TH, Baines AJ, Gull K. 1989. Definition of individual components within the cytoskeleton of Trypanosoma brucei by a library of monoclonal antibodies. *J. Cell Sci.* 93:491–500

111. Woodward R, Carden MJ, Gull K. 1994. Molecular characterisation of a novel, repetitive protein of the paraflagellar rod in Trypanosoma brucei. *Mol. Biochem. Parasitol.* 67:31

112. Woodward R, Gull K. 1990. Timing of nuclear and kinetoplast DNA replication and early morphological events in the cell cycle of Trypanosoma brucei. *J. Cell Sci.* 95:49–57

SUBJECT INDEX

CUMULATIVE INDEXES

CONTRIBUTING AUTHORS, VOLUMES 49–53

CHAPTER TITLES, VOLUMES 49–53

Prefatory Chapters

Animal Pathogens and Diseases

699

Applied Microbiology and Ecology

Chemotherapy and Chemotherapeutic Agents

Diversity and Systematics

Genetics and Physiology

Immunology

Morphology, Ultrastructure, and Differentiation

Organismic Microbiology

Pathogenesis and Control

Physiology, Growth, and Nutrition

Virology